Advanced Machining Processes of Metallic Materials

Advanced Machining Processes of Metallic Materials

Theory, Modelling and Applications

WIT GRZESIK

ELSEVIER

Amsterdam • Boston • Heidelberg • London • New York • Oxford
Paris • San Diego • San Francisco • Singapore • Sydney • Tokyo

Elsevier
Radarweg 29, PO Box 211, 1000 AE Amsterdam, The Netherlands
The Boulevard, Langford Lane, Kidlington, Oxford OX5 1GB, UK

First edition 2008

Library of Congress Cataloging-in-Publication Data
A catalog record for this book is available from the Library of Congress

British Library Cataloguing in Publication Data
A catalogue record for this book is available from the British Library

ISBN: 978-0-08-044534-2

For information on all Elsevier publications visit our website at books.elsevier.com

Printed and bound in the United Kingdom

Transferred to Digital Print 2010

Contents

Preface

The broad subject of manufacturing engineering and technology, including machining technology, continues to be recognized as an important and distinct area of study at universities and various technical and research institutes. After a couple of decades of neglect, this production subject has finally acquired the distinct academic stature and significance. Engineers and students have come to the conclusion that without a sound manufacturing base, no nation can hope for economic survival in an increasingly competitive international marketplace.

This book is an exploration in modern metal cutting. In addition to providing basic information on metal cutting, this book also describes the level of modern technology adopted, by varying degrees, by industry in general. Metal cutting is a dynamic technology, involving the range of disciplines of science, which must be mastered to become a practitioner of advanced machining technology. Some of these disciplines are the province of machining technologists; others concern both machine tool manufacturers and machine tool builders and users. Nonetheless, it can be helpful for all machining-related businesses to have a good grasp of the relevant issues in each area. There are eight disciplines, each of which is covered in relevant clusters of chapters.

- Materials engineering (Chapters 4, 13).
- Engineering mechanics and related disciplines (Chapters 6, 7, 8).
- Thermodynamics (Chapters 9, 12 and partially Chapter 10).
- Tribology (Chapters 11, 12 and partially Chapter 10).
- Modelling techniques (basically Chapters 5 and successively Chapters 6–14).
- Manufacturing engineering (Chapter 15 and appropriate sections involved).
- Process and motion control (Chapters 7, 18, 19 and partially Chapter 15).
- Surface engineering (Chapter 20).

In general, this book is structured into three parts: the first, including Chapters 2–14, provides fundamentals of machining, the second, including Chapters 15–19, overviews the effects of the theoretical and experimental considerations in high-level machining technology, and the third summarizes production outputs related to part quality. One appendix is provided to facilitate the comprehension of the physical phenomenon by numerous colour images.

Numerous references are provided for more detailed or more extensive information on various aspects of metal cutting and its effective applications ranging from mezo- to nano-scale. In particular, I have found the following books (in alphabetic order) to be good sources of additional information for metal cutting process and their optimal applications:

G. Boothroyd, W.A. Knight, *Fundamentals of Machining and Machine Tools*, CRC Press, Boca Raton, 2006, is an exceptional source of descriptions of various cutting-oriented phenomena and recent advances in conventional and non-conventional machining processes.

T.H.C. Childs, K. Maekawa, T. Obikawa, Y. Yamane, *Metal Machining. Theory and Applications*, Arnold, London, 2000, is a good source for reliable experimental data and modelling techniques (slip-line, FEM, AI-based) developed mainly in the UK and Japan.

M.C. Shaw, *Metal Cutting Principles*, Clarendon Press, Oxford, 1989, is a good source for scientific interpretation of physical principles of conventional machining processes based on classical mechanics, strength of materials and tribology.

D.A. Stephenson, J.S. Agapiou, *Metal Cutting Theory and Practice*, Marcel Dekker, New York, 1997, is a unique item devoted to available technology of metal cutting processes and explains how research results can be integrated in the contemporary industrial practice.

E.M. Trent, P.K. Wright, *Metal Cutting*, Butterworth-Heinemann, Boston, 2000, is a unique source for both a traditional material-based approach to the metal cutting phenomena and essential aspects of 21st century manufacturing.

According to the author's intention, this book is addressed to those studying and teaching the principles of machining processes and operations at universities, as well as providing updated theoretical and applied knowledge for those involved in the machining/manufacturing industry.

I am very grateful to all of those companies (cited by name or reference number in the figure legends and table footnotes) that granted permission for reproduction of numerous figures and tables.

I express my gratitude to my research assistant Mr K. Żak for his invaluable help in preparing illustrations and graphics. Finally and most importantly, I thank my wife, my daughters and son-in-law for their patience during the many times when my preoccupation with this book inconvenienced them.

Nomenclature

Latin Symbols

A	shape factor in Shaw's equation for heat partition
A_a	apparent area of contact between two surfaces; average value of shape factor A
A_c	cross-sectional area of the uncut chip, i.e. the cross-sectional area of the layer of material being removed by one cutting edge measured normal to the resultant cutting direction; contact area
A_m	maximum value of shape factor A
A_r	real area of contact between two surfaces
A_{sh}	area of shear plane
A_α	tool flank, i.e. the surface over which the surface produced on the workpiece passes
A_γ	tool face, i.e. the surface over which the chip flows
a	undeformed chip thickness, i.e. the thickness of the layer of material being removed by one cutting edge at the selected point measured normal to the resultant cutting force direction
a_c	chip thickness
$a_{c_{av}}$	mean undeformed chip thickness, i.e. the mean value of a_c
$a_{c_{max}}$	maximum undeformed chip thickness, i.e. the maximum value of a_c
a_e	working engagement, i.e. the instantaneous engagement of the complete tool with the workpiece, measured in the working plane P_{fe} and perpendicular to the direction of feed motion (previously known as depth of cut in a slab-milling operation)
a_f	feed engagement, i.e. the instantaneous engagement of the tool cutting edge with the workpiece, measured in the working plane P_{fe} and in the direction of feed motion (in single-point machining operations it is equal to the feed f; in multipoint tool operations, it is equal to the feed per tooth)
a_c	chip thickness (also h_{ch}), i.e. the thickness of the chip produced during machining
a_p	back engagement, i.e. the instantaneous engagement of the complete tool with the workpiece, measured perpendicular to the working plane P_{fe} (previously known as depth of cut in a single-point tool operation and width of cut in a slab-milling operation)
a_{pl}	lower limit of doc
a_{pu}	upper limit of doc
a_v	amplitude of vibration
B	groove width in a groove tool; zone where the flank is regularly worn
B_e	equivalent groove width in a groove tool
BL	length of groove backwall wear
BW	width of groove backwall wear
b	width of cut; width of the cutting edge
b_{cr}	the lowest b_{lim} obtained for the phasing most favourable for chatter generation
b_{lim}	limiting stable axial depth of cut
C	constant in upper boundary prediction for the shear angle by Oxley, constant in Shaw's equation
C_T	tool life for cutting speed of 1 m per min
C_{T1}, C_{T2}, C_{T3}	constant in general tool-life equation
C_v	cutting speed for 1 min of tool life (in m per min)

C_m	cost of machining, neglecting non-productive costs
C_{mat}	cost of material for one workpiece
C_{min}	minimum cost of production, i.e. the minimum value of C_{pr}
C_{mt}	total machining cost
C_{pr}	production cost, i.e. the average cost of producing each component on one machine tool
Cv	constant equal to the cutting speed for $T = 1\,\text{min}$
C_t	constant in the original Taylor tool-life equation
C_T	constant in the Taylor equation equal to T for $v_c = 1\,\text{m/min}$
c	rigidity constant
c_d	damping force per unit velocity, i.e. the viscous damping constant
c_p	specific heat capacity
D	tool diameter
dF	variation in the cutting force
E	Young's modulus; process activation energy
E_c	cutting energy
E_f	energy required to perform feed motion; friction energy
E_p	energy required to perform plastic deformation
E_{sh}	energy required to perform shearing
$E_{f\alpha}$	energy required to overcome friction on the flank face
$E_{f\gamma}$	energy required to overcome friction on the rake face
e	base of natural logarithm
e_c	specific cutting energy
$e_{f\gamma}$	specific friction energy related to the rake face
e_{sh}	specific cutting energy related to shearing
F	resultant cutting force
$F(t)$	periodic force
F_a	active force
F_c	cutting component of the resultant tool force F_r
$F_{c\infty}$	an asymptotic value of the cutting force F_c
F_{dyn}	force component due to chip deformation in HSC
F_f	feed force
F_m	momentum force
Fo	Fourier number
F_o	objective function
F_p	ploughing force
F_r	resultant tool force
F_{sh}	force required to shear the work material on the shear plane
F_{sN}	force perpendicular to the shear plane
F_{su}	resultant shear force in HSC
F_α	tangential force on the flank face
$F_{\alpha N}$	force perpendicular to the flank face
F_γ	frictional force on the tool face; frictional force between sliding chip and tool
$F_{\gamma N}$	force perpendicular to the rake face
f	feed rate, i.e. the displacement of the tool relative to the workpiece, in the direction of feed motion, per revolution of the workpiece or tool
f_m	feed per minute
f_{max}	maximum available machine feed
f_l	lower limit of feed
f_n	resonance of frequency
f_{nd}	natural damped frequency of the system
f_{opt}	optimum value of feed

f_u	upper limit of feed
f_z	feed per tooth
H_T	hardness of the tool material
H_W	hardness of the workpiece material
HRC	Rockwell hardness number (C scale)
HSC	High spot count (count(s))
h	uncut chip thickness
h_{ch}	chip thickness
h_D	nominal thickness of cut
h_{min}	minimum undeformed chip thickness
$Im[G]$	imaginary part of the FRF
K	constant for a machining operation; can be regarded as the distance travelled by the tool in relation to the workpiece during the machining time t_m
K_1–K_8	constant in LPM
$[K]$	global stiffness matrix
KB	distance from the cutting edge to the back crater contour
KE	radial displacement of the tool corner
KF	width of the land between the crater and cutting edge
KM	distance from the cutting edge to the deepest crater point
KT	crater depth; depth of groove backwall wear
K_{1C}	fracture toughness
k	shear stress in the slip-line field; constant in Stabler's formula; damping ratio; negative slope of the tool-life curve
k_c	specific cutting pressure
k_h	chip thickness compression ratio (also Λ_h)
L	tool length; cutting length; lay (surface texture)
l	land length in a grooved tool
l_c	natural tool-chip contact length
l_{ca}	length of the active cutting edge
l_{cr}	restricted tool-chip contact length
l_e	equivalent restricted contact length
l_m	length of machined surface
l_{nc}	natural contact length
l_p	length of the plastic contact
l_{sh}	length of shear plane (also l_{AB})
l_{sl}	sliding-contact length
l_{st}	sticking-contact length
l_t	length of tool
l_w	length of workpiece or hole to be machined; length of cut path or cut surface
M	total machine and operator rate (cost per unit time), including machine depreciation M_t, operator's W_o and machine and operator overheads; mean line (M) system
M_R	machinability rating
$Mr1$	upper material ratio (%)
$Mr2$	lower material ratio (%)
M_t	machine-tool depreciation rate (cost per unit time)
M_t'	machine-tool rate including overheads (cost unit time)
MT_1–MT_5	extreme finishing; finishing; semi-roughing, roughing and heavy roughing machining operations
m	slope of linear plastic stress/strain relation; relative shear stress in Rowe and Spick's model; mass of the vibration system; width of the contact zone
m_{avg}	average number of teeth in the cut

m_{ch}	mass of chip specimen
m_1	strain rate sensitivity exponent
N	number of teeth on the cutting tool; number of full waves; nose wear
N_b	batch size, i.e. the number of components in the batch to be machined
N_t	number of tools used in machining the batch of components
NL_1	notch wear length on main cutting edge
NL_2	notch wear length on secondary cutting edge
NW_1	notch wear width on main cutting edge
NW_2	notch wear width on secondary cutting edge
N_T	thermal number; number of tool changes necessary during the machining of a batch of components
n	strain-hardening index or exponent; constant in Taylor's tool-life equation; spindle rotation speed
n_{opt}	optimum value of rotational speed
n_s	rotational frequency of a machine-tool spindle
n_{s_c}	rotational frequency of a machine-tool spindle for minimum production cost
$n_{s_{ef}}$	rotational frequency of a machine-tool spindle for minimum efficiency (maximum profit rate)
n_{s_p}	rotational frequency of a machine-tool spindle for minimum production time
n_t	rotational frequency of the cutting tool or abrasive wheel
n_w	rotational frequency of workpiece
P	power
$\{P\}$	vector of all applied loads
Pc	local peak count (count/cm)
P_e	electrical power consumed by the machine tool during a machining operation
Pe_c	Peclet number
P_f	assumed working plane
P_{fe}	working plane
P_g	tool-face orthogonal plane
P_m	power required to perform the machining operation
P_n	cutting edge normal plane
P_o	tool orthogonal plane
P_p	tool back plane
P_{pe}	working back plane
P_r	tool reference plane, the rate of production
P_{re}	working reference plane
P_s	tool cutting edge plane
P_{se}	working cutting edge plane
P_{sh}	shear plane
p_A	hydrostatic pressure in point A at the free surface
p_s	specific cutting power, i.e. the work required to remove a unit volume of material
Q	total amount of heat generated in machining
Q_1	heat source due to plastic deformation
Q_2	frictional heat source
Q_3	heat source at the contact between the workpiece and the flank
Q_4	heat source from which a small part of heat is transferred to the subsurface layer
Q_W	volumetric material removal rate
q_c	heat flux flowing to the chip
q_t	heat flux flowing to the tool
q_w	heat flux flowing to the workpiece
$\dot{q}$	heat flow rate
R	thermal number; universal gas constant; surface roughness

$\{R\}$	load vector
R_a	arithmetical mean value of surface roughness (CLA)
R_c	Rockwell hardness number (C scale)
R_{ch}	heat partition coefficient, i.e. percentage of heat entering the chip
Rk	core roughness depth
Rku	kurtosis
R_{KF}	heat partition coefficient defined by Kato and Fujii
$R_{min}(\tau)$	minimum radius of up-curling
$Rmr(c)$	material ratio at depth 'c'
R_o	groove radius
Rp	maximum height of peaks
Rpk	reduced peak height
Rq	root mean square (RMS) average
R_R	heat partition coefficient defined by Reznikov
Rsk	skew (skewness)
Rsm	average peak spacing
R_{SH}	heat partition coefficient defined by Shaw
Rt	total height of the profile (obsolete R_{max})
Rv	maximum depth of valleys
Rvk	reduced valley depth
Rz	maximum height of the profile
Rzt	theoretical value of P-V parameter
$R\Delta a$	centreline average (CLA) slope (deg)
$R\Delta q$	RMS slope (deg)
$R\lambda a$	CLA wavelength
$R\lambda q$	RMS wavelength
r_{min}	radius at which cutting is taking place
r_c	cutting ratio
r_{chip}	radius of the chip curvature
r_n	radius of the cutting edge
r_s	side-curling radius
r_u	up-curling radius; chip curvature
r_{ui}	radius of initial chip curl
r_{uf}	radius of final chip curl
r_ε	corner radius, i.e. the radius of a rounded tool corner
S	tool major cutting edge; income per component
S_a	active cutting edge
S'	tool minor cutting edge
SD	depth of secondary face wear
SL	sampling length
SW	width of secondary face wear
s	lamellar spacing
T	temperature; absolute temperature; tool life, i.e. of the tool cutting edge while the cutting edge is engaged with the workpiece
$\overline{T}$	average tool life
T_e	economic tool life (also T_E)
T_m	melting temperature
T_o	reference temperature
T_{mod}	velocity modified temperature
T_p	tool life for maximum production rate (also T_Q)
T_R	reference tool life

T_r	room (ambient) temperature; tool life for a cutting speed of v_r
t	time
t_a	acceleration time
t_c	tool changing time, i.e. the average machine time to change a worn tool or to index (and, if necessary, replace) a worn insert
t_{cs}	interchange time
t_e	magazine indexing (travelling) time
t_d	deceleration time
t_l	non-productive time, i.e. the average machine time to load and unload a component and to return the cutting tool to the beginning of the cut
t_l	loading and unloading time
t_m	machining time, i.e. machine time to machine a component
t_{max}	maximum operation time
t_{pr}	production time, i.e. the average time to produce one component on one machine tool
t_r	transportation (approach) time per workpiece
t_x	rapid travel location time
$\{U\}$	matrix of nodal velocities
$\{u\}$	displacement vector
V_w	volume of tool material lost due to wear
VB_B	average width of flank wear land in the central portion of the active cutting edge
VB_{Bmax}	maximum width of flank wear land in the central portion of the active cutting edge
VB_C	width of flank wear at tool corner
VB_N	width of notch wear
V_m	volume of material removed in machining
VN	width of the flank wear land at the wear notch
VB'	wear of minor flank face
v_{ac}	mean cutting speed, i.e. the average value of v along the major cutting edge
v_c	cutting speed, i.e. the instantaneous velocity of the primary motion of the selected point on the cutting edge relative to the workpiece
v_{cc}	optimum cutting speed for minimum production cost
v_{ce}	cutting speed at minimum cost
v_{ch}	chip velocity
v_{cp}	optimum cutting speed for minimum production time
v_{cR}	reference cutting speed in tool-life equation for grooved tool
v_{cT}	cutting speed corresponding to tool life T
v_{cTmax}	cutting speed corresponding to maximum tool life T_{max}
v_e	resultant cutting speed, i.e. the instantaneous velocity of the resultant cutting motion of the selected point on the cutting edge relative to the workpiece
v_{ef}	cutting speed for maximum efficiency (maximum rate of profit)
v_f	feed velocity
v_{HSC}	cutting speed in HSC defined by Schultz
v_{max}	maximum cutting speed, i.e. maximum of v_c
v_{min}	minimum cutting speed, i.e. minimum of v_c
v_p	cutting speed for minimum production time
v_{po}	cutting speed when maximum power is used
v_r	cutting speed giving a tool life of T_r
v_s	shearing velocity; sliding velocity
v_{ch}^{sl}	chip velocity along the sliding region
v_{ch}^{st}	chip velocity along the sticking region
W	weight of workpiece; waviness
W_c	tool coating effect factor

W_g	chip-groove effect factor
w	width of cut
$X1$	coded value of speed in LPM
$X2$	coded value of feed in LPM
x	distance from the point of chip separation

Greek Symbols

α	alpha-phase, thermal diffusivity
α_e	thermal expansion coefficient
α_n	tool normal clearance
α_{n_e}	working normal clearance
α_T	thermal diffusivity of the tool material
α_W	thermal diffusivity of the workpiece material
β	proportion of heat conducted into the workpiece; beta-phase
χ	characteristic of contact length in Rowe and Spick's model
Γ	proportion of heat generated in primary deformation zone conducted into workpiece
γ	gamma-phase
γ_{AB}	strain on shear plane in Oxley's model
γ_{EF}	shear strain along the exit boundary *EF* in Oxley's model
γ_c	catastrophic shear strain
γ_e	effective rake angle (also γ_{ef} and γ_{eff})
γ_f	tool side rake angle
γ_{f1}	tool side rake angle in the land
γ_{f2}	tool side rake angle in the groove
γ_g	tool geometric rake angle (direction of the maximum slope of the rake)
γ_h	homogeneous shear strain
γ_n	tool normal rake
γ_{n_e}	working normal rake
γ_o	tool orthogonal rake
γ_p	tool back rake
γ_{sb}	total shear strain in the shear band (*sb*)
γ_{sh}	shear strain
$\dot{\gamma}_{sh}$	shear strain rate
∇T	local temperature gradient (Hamilton's vectoral operator-nabla) in Km^{-1}
Δx	thickness of the shear zone (band)
Δt	time elapsed for material element to travel a distance Δs
Δs	distance along the shear plane
Δs_2	thickness of the shear zone in Oxley's model
$\Delta\overline{\Theta}_f$	mean temperature rise due to friction
δ_u	response (deflection) in the u direction
δ_v	response (deflection) in the v direction
ε	uniaxial true strain; fraction of waves
ε_b	chip strain caused by bending
ε_{max}	chip strain at fracture
ε_p	the equivalent strain
ε^p	accumulated plastic strain
ε^p_{eff}	effective plastic strain
ε^p_o	reference plastic strain

$\dot{\varepsilon}_p$	equivalent strain rate
$\dot{\varepsilon}_o^p$	reference plastic strain rate
$\dot{\varepsilon}_p^0$	strain rate equal to $1.0\,s^{-1}$
ε_r	tool-included angle
ε_{r_e}	working included angle
η	resultant cutting speed angle, i.e. the angle between the direction of primary motion and the resultant cutting direction; angle between the texture line and the shear plane; contact length factor
η_b	chip back-flow angle
η_c	chip flow angle; angle of maximum slope of the rake angle
η_s	chip side-flow angle
θ	temperature, mean angle of friction on tool face; groove tangent angle
θ_{int}	temperature at tool/chip interface
θ_{fmax}	maximum temperature rise of material passing through the secondary deformation zone
θ_{max}	maximum interface temperature along the rake face (also t_{max})
θ_n	mean angle of friction measured in the normal plane
θ_{smax}	maximum shear-plane temperature (maximum temperature rise of material passing through the primary deformation zone)
θ_o	initial workpiece temperature
$\Theta(T)$	thermal softening factor
$\overline{\Theta}_s$	mean shear-plane temperature
$\overline{\Theta}_t$	average interface temperature
κ_r	tool cutting edge angle
κ_r'	tool minor cutting edge angle
κ_{r_e}	working cutting edge angle
κ_{re}'	working minor cutting edge angle
λ	thermal conductivity
λ_T	thermal conductivity of the tool material
λ_W	thermal conductivity of the workpiece material
λ_s	tool cutting edge inclination
λ_{s_e}	working cutting edge inclination
μ	coefficient of friction, viscosity
μ_a	adhesion component of coefficient of friction
μ_c	equivalent coefficient of friction
μ_{cmax}	maximum coefficient of friction
μ_m	mechanical component of coefficient of friction
$\mu Re[G]$	real part of the FRF
ν	coefficient of tool-life variability
ρ	density of work material
σ	uniaxial true stress
$+\sigma$	tensile residual stress
$-\sigma$	compressive residual stress
$\overline{\sigma}$	effective von Misses stress
σ_c	normal contact stress acting on the tool/chip interface
σ_{cmax}	maximum normal contact stress acting on the tool/chip interface
$\overline{\sigma}_c$	mean value of normal contact stress
σ_f	flow stress; fracture stress
σ_n	normal stress on the tool face
σ_{nmax}	maximum normal stress on the tool face
σ_o	initial yield stress at the reference temperature T_o; constant in uniaxial true strain relationship
σ_{sh}	normal stress on the shear plane (also σ_s)
σ_T	standard deviation

τ chip flow angle
τ_c shear contact stress acting on the tool/chip interface
$\overline{\tau}_c$ mean value of shear contact stress
τ_o shear flow stress at zero plastic strain in Oxley's model
τ_s shear stress on the shear plane
τ_{sh} shear flow stress of the work material
τ_{st} shear stress on the tool face in the sticking region
τ_{so} shear stress in the shear plane with zero normal stress applied
Φ shear angle
Φ_n shear angle in the normal plane *Pn*
Φ_o shear angle for unstrained (softer) material
Φ_T shear angle calculated from mechanical properties of the workpiece material
ω angular frequency of vibration; angle between the resultant cutting force and the shear plane
ω_f angular frequency of external harmonic force
ω_n natural angular frequency

ABBREVIATIONS

AC Adaptive control; air cooled
ACC Adaptive control constraint
ACO Adaptive control optimization
A/D Analogue-to-digital converter
ADF Amplitude distribution function
ADI Austempered ductile iron
AE Acoustic emission
AFM Abrasive-flow machining; atomic force microscopy
AFRP Aramid fibre reinforced plastic
AGV Automated guided vehicle
AI Artificial intelligence
AJM Abrasive-jet machining
ALE Arbitrary Lagrangian–Eulerian formulation
Al_2O_3 Aluminium oxide, white ceramics
AMPR Advanced Manufacturing Research Programme
AMZ Altered material zone
ANN Artificial neural network
ANSI American National Standards Institute
APL A programming language
APS Advanced process system
AR Autoregression
ARMD Area-restricted molecular dynamics
ASM American Society for Metals (now ASM International)
ATC Automatic tool changer

BAC Bearing area curve
BEM Boundary Element Method
BHN Brinell hardness number (see HB)
BUE Built-up-edge
bcc Body-centred cubic

CAD Computer-aided design; computer-aided drafting
CAM Computer-aided manufacturing

CAPP	Computer-aided process planning
CAT	Computer-aided testing
CAVE	Computer Automated Visualization Environment
CBGF	Circular thread-milling tool
CBN	Cubic boron nitride
CBN-HT	CBN hard turning
CCD	Charge-coupled device (camera)
CCI	Coherence correlation interferometry
CE	Concurrent engineering; control emulator
CIM	Computer-integrated manufacturing
CF	Cutting fluid
CFEST	Cutting Fluid Evaluation Software Testbed
CFRP	Carbon-fibre reinforced plastic
CGI	Compacted graphite iron
CIM	Computer-integrated manufacturing; Canadian Institute of Mining and Metallurgy
CIRP	International Institution for Production Engineering Research
CLA	Centreline average
CM	Communication medium
CMM	Coordinate measuring machine
CNC	Computer numerical control
COS	Computerized optimization system
CT	Cermet
CVD	Chemical vapour deposition
CVL	Copper vapour laser
DARPA	Defence Advanced Research Project Agency
DBGF	Direct circular thread-milling tool
DBTT	Ductile-to-brittle transition temperature
D_c	Diameter of cutter
DLC	Diamond-like coating
DLL	Dynamic link library
DM	Digital manufacturing
DN	Product of the spindle diameter in mm and the spindle speed in rpm
DNC	Direct numerical control; distributed numerical control
DPU	Data processing unit
DRIE	Deep reactive ion etching
DSC	Differential scanning calorimeter
DSP	Digital signal processing
DUV	Deep ultraviolet lithography
DVA	Dynamic vibration absorber
EBM	Electron-beam machining
ECG	Electrochemical grinding
ECM	Electrochemical machining
ECT	Effective chip thickness
EDM	Electrical discharge machining
EDG	Electrodischarge grinding
EDX	Energy dispersion X-ray
EL	Evaluating length
ELACM	Eximer laser-assisted chemical machining
ELID	Electrolytic in-process dressing
EMF	Electromotive force (also emf)

EP	Extreme pressure
Ew1;Ew2	Offsets in turn-milling operations
e-manufacturing	Electronic-manufacturing
e-work	Electronic-work
FDA	Finite different approach
FDM	Finite different method
FEA	Finite element approach (analysis)
FEM	Finite element method
FES	Fuzzy expert system
FFT	Fast Fourier transform
FIB	Focused ion beam (micromachining)
FMS	Flexible manufacturing system; Federation of Materials Societies
FOF	Factory of the future
FRF	Frequency response function
FRP	Fibreglass-reinforced plastic
FTP	File transfer protocol
fcc	Face-centred cubic
GAC	Geometric adaptive control
GFRP	Graphite-fibre reinforced plastic; glass-fibre reinforced plastic
GGG	Nodular cast iron (German equivalent to CGI; see CGI)
HB	Brinell hardness number
HEM	High efficiency machining
Hi-E	High efficiency machining (range)
HK	Knoop hardness number
HK_{100}	Knoop hardness using 100 g load
HM	Hard machining; hard milling
HMC	Horizontal machining centre
HMI	Human/machine interface
HPC	High pressure coolant (supply)
H-PCBN	High content PCBN
HPC	High performance cutting
HPDL	High power diode laser
HPM	High performance machining; hard part machining
HPMA	High precision motorized arm
HPPA	High precision pull-down arm
HPRA	High precision removable arm
HR	Rockwell hardness number, including scales such as HRA, HRB, HRC, etc.; hot rolled
HSM	High speed machining
HSC	High speed cutting
HSS	High speed steel
HSS-Co	Cobalt enriched high speed steels
HT	Hard turning
HTML	Hyper Text Markup Language
HV	Vickers hardness number
HVM	High velocity machining
hcp	Hexagonal close-packed
ICM	Iterative convergence method
ID	Inside diameter
IMM	Intelligent machining module

IMS	Intelligent manufacturing system; intelligent maintenance system
I/O	Input/output
IPM	Inductive probe module
IR	Infrared
IT	Information technology; intelligent tool
ITC	Intelligent thermal control
JC	Johnson–Cook material model
JIT	Just-in-time
KHN	Knoop hardness number (obsolete; see HK)
LAM	Laser-assisted machining
LAN	Local area network
LASER	Light amplification by stimulated emission of radiation
LBM	Laser-beam machining
LCD	Liquid crystal display
LFM	Laser flash method
LDF	Linear discriminant function
LIGA	Photo-lithography and electroplating method
LN	Liquid nitrogen
LODTM	Large optics diamond turning machine
LPM	Linear programming method
L-PCBN	Low content PCBN
MCD	Machine code data
MC-HT	Mixed ceramics hard turning
MCU	Machine control unit
MD	Molecular dynamics
MDB	Machinability database
MDC	Machinability Data Center
MDI	Manual data input
MEMS	Micro-electromechanical system
MES	Manufacturing execution system
MMC	Metal matrix composite
MO	Mineral oil
MQC	Minimum quantity cooling
MQL	Minimum (minimal) quantity lubrication
MQCL	Minimum quantity cooling lubrication
MRP	Material requirements planning
MRR	Material removal rate
MST	Microsystems technology
MTM	Multitasking machining
MVL	Minimum volume lubrication
mMT	Micro-/mezzo-scale machine tool
NC	Numerical control
NEMS	Nano-electromechanical system
NDT	Non-destructive testing; nil ductility transition
NGM	New generation manufacturing
NNI	National Nanotechnology Initiative
NPT	Non-productive times
OD	Outside diameter
OFHC	Oxygen-free, high conductivity (for copper)

PAC	Plasma-arc cutting
PACVD	Plasma-assisted CVD (coating deposition technique)
PAM	Plasma-assisted machining
PC	Personal computer; printed circuit; polycarbonate
PCB	Printed circuit board
PCBN	Polycrystalline cubic boron nitride
PCD	Polycrystalline diamond
PDZ	Primary deformation zone
PGI	Phase grating interferometer
PH	Precipitation hardenable
PKM	Parallel kinematic machine
PLC	Programmable logic controller
PLM	Product lifecycle management
P/M	Powder metallurgy
PSZ	Partially stabilized zirconia
PVD	Physical vapour deposition
QA	Quality assurance
QC	Quality control
RCF	Rolling contact fatigue
RCT	Restricted-contact tool
RF	Radio frequency
RMI	Radio machine interface
RMS	Root-mean-square (also rms)
RNS	Remote notification system
RP	Rapid prototyping
SDZ	Secondary deformation zone
SiC	Silicon carbide
Si_3N_4	Silicon nitride, nitride ceramics
SL or SLA	Stereolithography technique; sampling length
SLF	Slip-line field
SLS	Selective lased sintering
SMS	Short message service
SMART	Smart Assistant to Machinists
SPDT	Single-point diamond turning
SPM	Scanning probe microscopy
STM	Scanning tunnelling microscope
TAM	Thermally assisted machining
TCM	Tool condition monitoring
TDZ	Tertiary deformation zone
TFTs	Thin film thermocouple sensor
TiAlN	Titanium–aluminium nitride
TiC	Titanium carbide
Ti(C,N)	Titanium carbo-nitride
TiN	Titanium nitride
TMP	Total machining performance
TMS	Tool monitoring system
TQC	Total quality control
TQM	Total quality management
TRS	Tensile rupture strength

UAM	Ultrasonic-assisted machining
UCL	Upper control limit
UCT	Undeformed chip thickness
UF	Ultra fine (for example, carbide grade)
UHSM	Ultra-high speed machining
UM	Ultrasonic machining
UR	Unit removal
UTS	Ultimate tensile strength (also R_m)
UV	Ultraviolet
UVC	Ultrasonic vibration cutting
VED	Video edge detection
VLSI	Very large-scale integration
VM	Virtual manufacturing
VMC	Vertical machining centre
VR	Virtual reality
WAP	Wireless Application Protocol
WC	Sintered tungsten carbide
WEDG	Wire electro-discharge grinding
WIP	Work in progress
WWW	World Wide Web
XML	eXtensive Markup Language
Y	Yield strength
ZD	Zero defect (manufacturing)

Introduction

Contents

The best way to predict the future is to create it.
Peter Drucker

Manufacturing constitutes the economic backbone of an industrialized nation and, in general, the economic position of a country is based on the level of manufacturing activity. It is observed that the pressures of international competition serve as a catalyst for changes in manufacturing technologies and systems. During the past 10–20 years, new manufacturing technologies, automated systems, and numerous system innovations have been implemented to improve machine and operator efficiency and increase productivity to reduce the cost of manufacturing. Especially, manufacturing industry meets the current market challenge by continued development of machine tools with more advanced digital control, higher speeds, better accuracy and greater flexibility. More recent developments include incorporated probes, sensors and adaptive machining processes that represent the eyes and ears of an operator. Currently, highly sophisticated knowledge-based systems models are being developed and tested, which will support a completely automatic intelligent machining workstation. The trend toward mass customization in manufacturing has also inspired the adoption of various concepts and practices leading to the flexibility and agility of manufacturing operations (basically to accommodate a family-of-parts). A number of other systems innovations are available, such as artificial intelligence (AI), computer-integrated manufacturing (CIM), material requirements planning (MRP), total quality management (TQM), and others. However, it must be understood that they, either individually or collectively, cannot solve all manufacturing problems. In these aspects, the book tries to help the manufacturing engineers to use properly and effectively a huge number of well-proven technological processes and equipment, as well as proprietary innovations in the manufacturing environment.

Parts manufactured by casting, forming and other shaping processes often require further operations to be machined to its final shape and dimensions with one or more of the processes listed in Fig. 1.1. Machining is the most widespread metal-shaping process in mechanical manufacturing industry. Rapidly growing investments in metal-machining machine tools have continued despite perceived threats to machine volume, such as the replacement of metal by plastic products in the consumer goods sector and material saving by near-net (casting or forging) process substitution in the metal products sector. One reason is that metal machining is capable of high precision: tight part tolerances of 50 μm and surface finishes of 1 μm are readily achievable. Moreover, machining is very versatile: complicated free-form shapes with many features, including those of micro- and nano-scale, can be made more cheaply, quickly and simply by CNC multi-axis machine tools. Parts may have external and internal profiles, as well as sharp corners and flatness, which cannot be produced by forming and shaping processes. Special surface characteristics, such as mirror surfaces with very high reflectivity, can only be produced by machining with a single-crystal diamond tool. A third reason discussed above is that machining is able to sustain large changes, which guarantee the success on the market. In general, three types of theses changes are decisive: advances in machine tools (machine technology), in

Advanced Machining Processes of Metallic Materials
ISBN: 978-0-08-044534-2

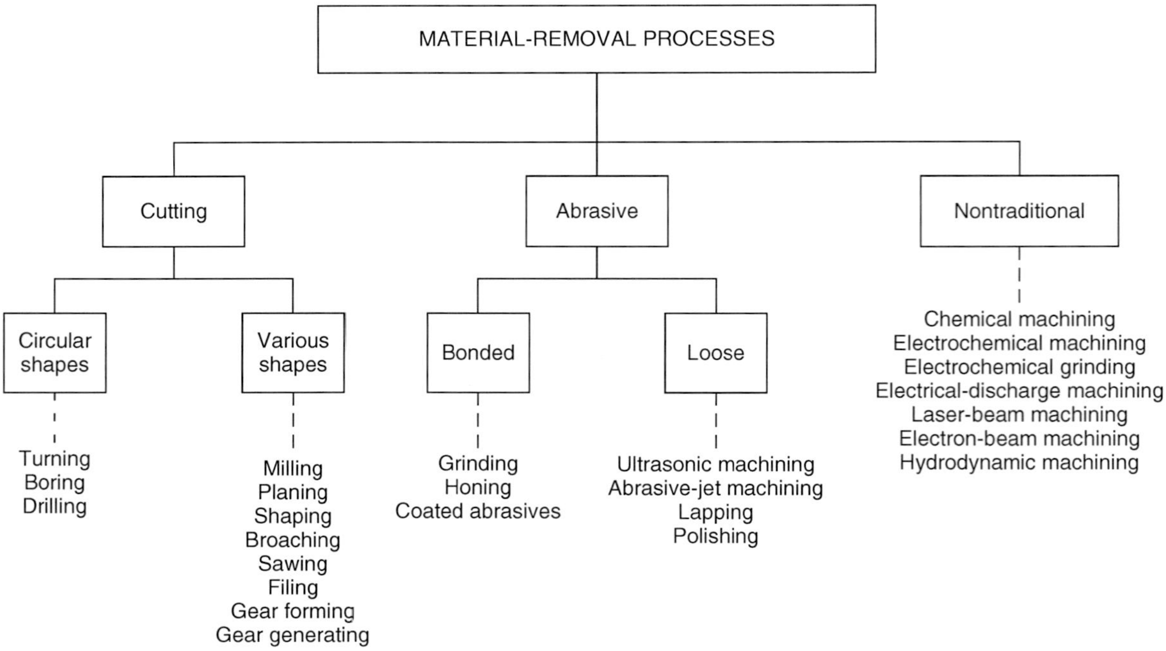

Fig. 1.1 A scheme illustrating a range of material-removal processes. After Kalpakjan [1].

the organization of machining (manufacturing systems) and in the cutting tools (materials technology) [2, 3]. It should be noted that each new improvement in one sector stimulates appropriate feedback from another.

In spite of some limitations related to mass-saving processes, such as material waste, more energy, capital and labour required, longer production time and adverse effects on the surface integrity, material-removal (mass-reducing) processes and machine tools are indispensable to modern manufacturing technology. As a result, today metal cutting is indeed a large segment of industrial activities. The automotive industry, precision engineering, electrical engineering, railways, shipbuilding, aircraft and aerospace industry, production of domestic equipment and the machine tool industry itself – all these branches have large machine shops with many thousands of employees engaged in machining. The author's primary intention is to address this book to all these people to be more active and creative in their common engineering practice. The book is essentially organized in three parts and 21 chapters, but sections in Chapter 15 can also be individual chapters.

Basic information generally needed for the proper utilization of basic science is provided in Chapters 2–4. Chapter 2 describes generically metal cutting operations and provides recommended terminology. Chapter 3 discusses new trends emerging in metal cutting theory and practice. Chapter 4 contains a useful *vade-mecum* for modelling and simulation techniques and their practical utilization. Chapter 5 covers a range of cutting tool materials, their properties and optimal areas of application. Chapter 6 focuses on mechanics of orthogonal and oblique cutting arrangements. Chapter 7 treats possible mechanisms of chip formation and practical methods of chip control. Chapter 8 deals with negative influences provided by cutting vibrations and methods of their control and prevention. Chapters 9 and 10 touch on thermal problems and the removal of heat by cooling. Chapters 11 and 12 concern overall tribological problems related to contact phenomenon and tool wear. Chapter 13 provides machinability rating for a majority of engineering materials, and Chapter 14 includes both economic and optimization considerations. Elaborated Chapter 15 characterizes several advanced machining processes which revolutionized manufacturing industry. It is completed by Chapters 16 and 17 which relate machining to micro- and nano-scale, respectively. Moreover, Chapters 18 and 19 provide some modern IT and sensor-based technology and design techniques. Chapter 20 covers cutting-edge measuring techniques and methods used in characterization of surface integrity. And finally, Chapter 21 provides some practical rules for typical troubleshooting. One appendix is provided to facilitate the comprehension of the physical phenomenon by numerous colour images.

According to Trent [4], progress in the technology of machining is achieved by the ingenuity and experimentation, intuition, logical thought and dogged worrying of many thousands of practitioners engaged in the many-sided arts of metal cutting. It is a belief of the author that texts involved should be informative in practical as well as in theoretical aspects.

Wit Grzesik
Poland
July 2007

References

[1] S. Kalpakjan, *Manufacturing Engineering and Technology*, Addison-Wesley, Reading, 1989.
[2] T.H.C. Childs, K. Maekawa, T. Obikawa, Y. Yamane, *Metal Cutting. Theory and Applications*, Arnold, London, 2000.
[3] G. Byrne, D. Dornfeld, B. Denkena, Advancing cutting technology, *CIRP Annals*, Vol. 52/2 (2003), 483–507.
[4] E.M. Trent, *Metal Cutting*, Butterworths, London, 1984.

CHAPTER TWO

Metal Cutting Operations and Terminology

Contents

2.1 Classification of Machining Processes

Modern manufacturing encompasses an ever-increasing variety of processes and the engineer's challenge is to select the most economic combination of processes to make a high quality product at the right price. Products range from high priced items made in low volume such as jet engines to everyday items such as razor blades which are made in their millions. To meet this challenge the manufacturing engineer needs to have a broad knowledge of the ways in which materials can be processed, and the shapes which can be formed by these processes.

Obviously, the term *machining* deals with any process in which material is removed gradually from a workpiece, including metal cutting with both single-point or multi-point tools (tools with geometrically defined cutting edges) and grinding with abrasive wheels which consists of a large number of micro-cutting edges, randomly shaped and oriented, i.e. tools with geometrically undefined cutting edges [2, 3]. In the engineering industry, the term machining is used to cover chip-forming operations resulting in mass reducing. The narrower term *cutting* is intended to include operations in which a thin layer of material, the *chip* or *swarf*, is removed by a wedge-shaped tool (case I in the mechanical category in Table 2.1) [3]. In general, material removal in machining processes can be based on four fundamental removal methods, which refer to the relationship between the imprinting of the information and the energy supply.

Table 2.1 shows the classification of mass-reducing processes in terms of the process category (the type of energy supplied) and the method of material removal. Group I in the mechanical category covers many popular conventional cutting processes, known also as machining processes with geometrically defined cutting edges. Other processes, which utilize mechanical, thermal and chemical energy (groups II and III which utilize highly concentrated energy beams or metal removal by dissolving or evaporating), are classified as non-conventional (non-traditional) machining processes. There is no sharp line separating chip-forming operations from others such as shearing, blanking and punching (case IV in the mechanical category) which are commonly classified as metal-forming operations. While metal cutting is commonly associated with big industries (automotive, aircraft, aerospace, home appliance, etc.) that manufacture big products, the machining of metals and alloys plays a crucial role in a range of manufacturing activities, including the ultraprecision machining of extremely delicate components (micro- and nano-machining). The three basic manufacturing technologies with characteristic ranges of part dimensions and achievable tolerances are presented in Fig. 2.1.

Table 2.2 illustrates the process class occurring in the machining process which creates subsurface features, including micro-structural, metallurgical and material property changes within the subsurface layer and

Advanced Machining Processes of Metallic Materials
ISBN: 978-0-08-044534-2

Table 2.1 Classification of machining processes based on the energy used to material removal proposed by Alting [1].

Category of basic process		Fundamental removal method	Examples of processes
Mechanical	I	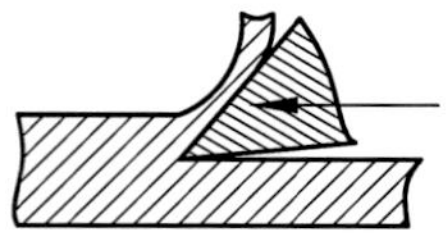	Cutting Turning Milling Drilling Grinding, etc.
	II	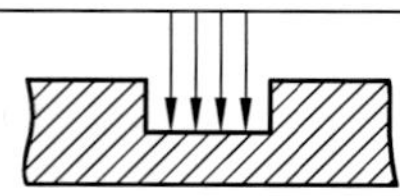	Water jet cutting Abrasive jet machining Sand blasting, etc.
	III	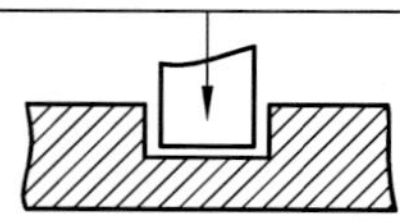	Ultrasonic machining
	IV	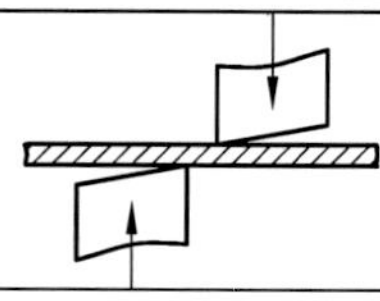	Blanking Punching Shearing
Thermal	II	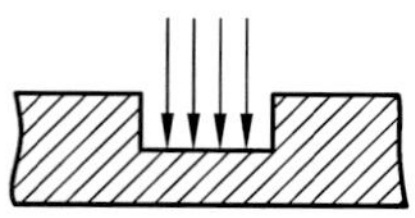	Thermal cutting (melting) Electron beam machining Laser machining
	III	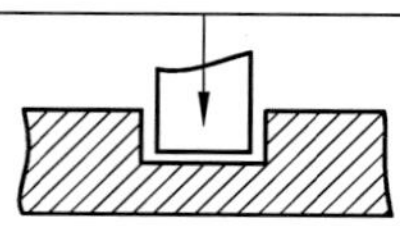	Electrodischarge machining (EDM)
Chemical	II	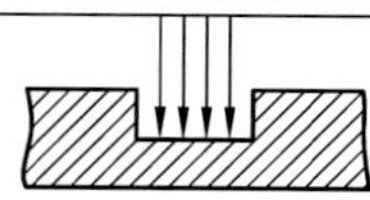	Etching Thermal cutting (combustion)
	III	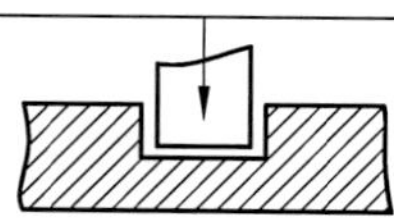	Electrochemical machining (ECM)

producing surface defects and residual stresses. For this purpose three types of generating mechanism, i.e. mechanical, thermal and chemical, were considered. Also of significance are the process operating conditions, i.e. gentle, conventional and abusive. It should be noted, based on the foregoing discussion, that from an engineering point of view the machining process must not only be capable of producing parts with required dimensional and form accuracy and surface quality/integrity, but also guarantee sufficient productivity and minimum costs.

The most commonly used criteria for selection of machining operations are the kinematical features and the capability of shaping appropriate surfaces, e.g. rotational, prismatic (planes), helical, threads, 3D free-form (sculptured) surfaces. In this content we distinguish three basic machining operations, i.e. turning operations (Fig. 2.2), drilling and related hole-making operations (Fig. 2.3) and milling operations (Fig. 2.4).

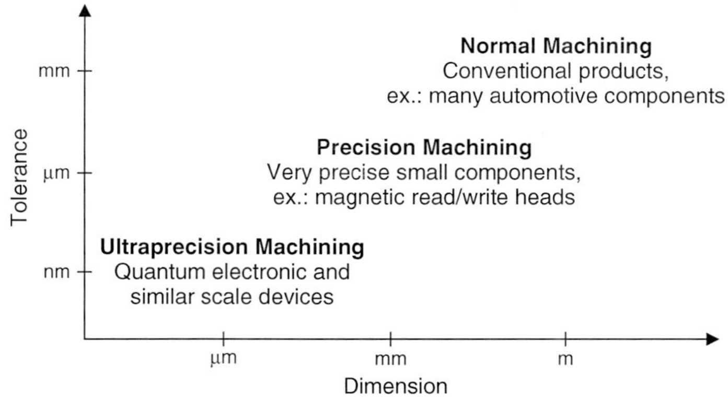

Fig. 2.1 Relationships between part dimension and dimensional tolerance for three groups of machining processes [4].

Table 2.2 Dependence of subsurface features on the type of energy used in machining processes [5]. A – abusive machining; C – conventional machining; G – gentle machining

	Manufacturing process energy class				
Subsurface features	*Chemical*	*Mechanical*	*Mechano-thermal*	*Thermo-mechanical*	*Thermal*
Untempered martensite	C	C	A	G&A	C
Overtempered martensite	C	C	A	G&A	C
Austenite	C	C	A	G&A	C
Intergranular attack	G&A	C	C	C	C
Plastic deformation	C	G&A	G&A	C	C
Burr	C	G&A	G&A	C	C
Cracks	G&A	C	C	C	C
Discontinuities	C	A	A	A	A
Tool fragments	C	G&A	G&A	G&A	G&A
Recast/redeposited material	C	G&A	G&A	G&A	G&A
Residual stress	C	G&A	G&A	G&A	G&A

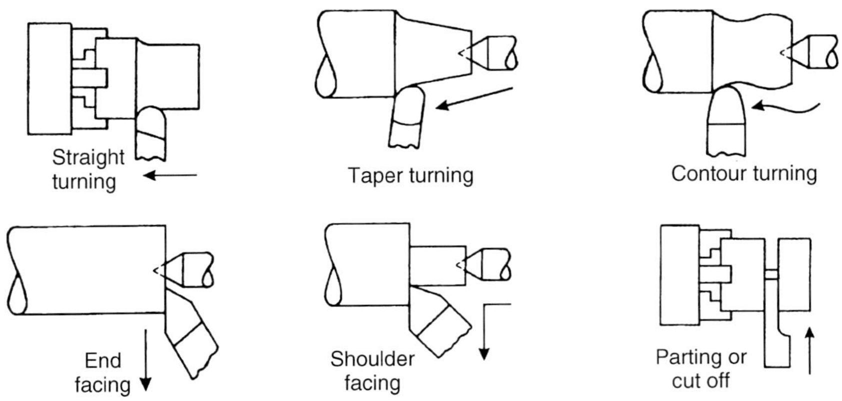

Fig. 2.2 Some typical turning operations.

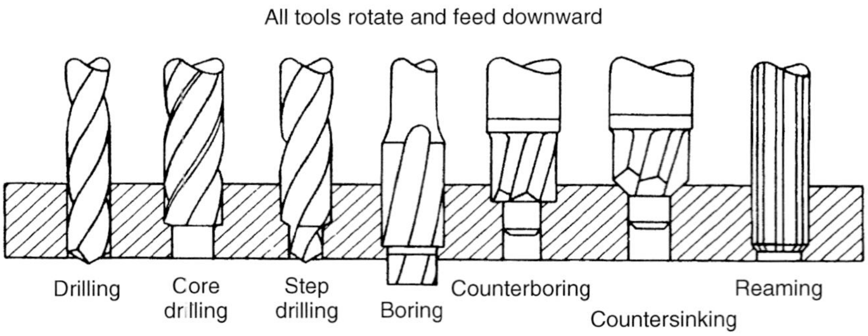

Fig. 2.3 Common drilling and related operations.

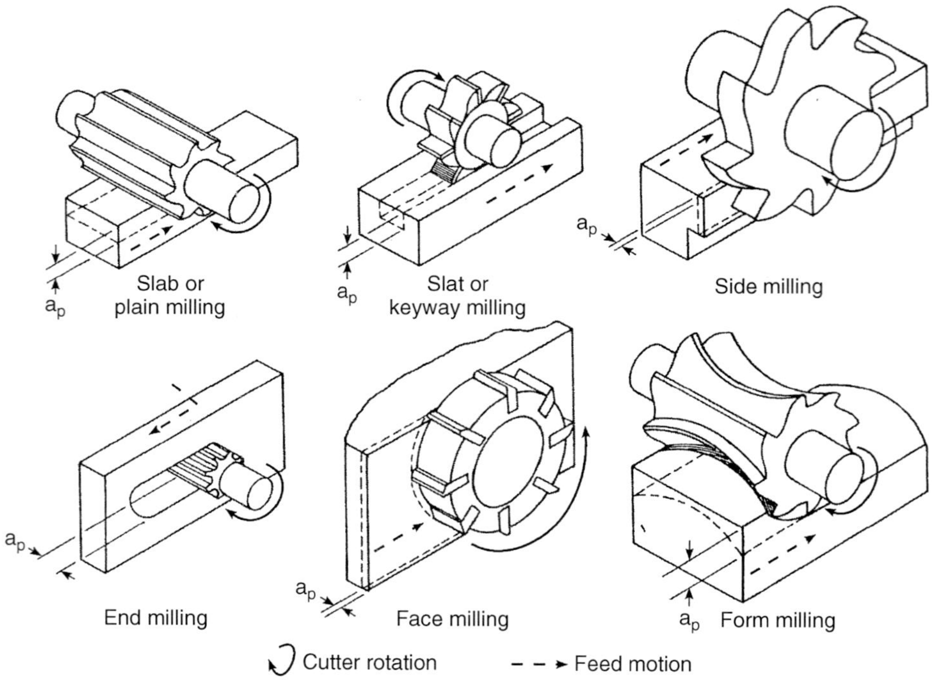

Fig. 2.4 Some typical milling operations.

As shown in Fig. 2.2, turning with single-point tools enables external and internal cylindrical surfaces (longitudinal turning) and taper surfaces, as well as 2D complex shapes using NC lathes equipped with contouring control systems, to be produced. It is also possible to perform facing, parting and cut-off and threading operations. Further, turning operations may be divided into two classes: those done with the workpiece between centres and those done with the workpiece chucked or gripped at one end with or without support, depending on the workpiece rigidity, at the outer end or sometimes between the chuck and the tailstock.

Hole-making operations based on drilling and allied operations are presented in Fig. 2.3. Drilling is the easiest way to make a hole in a solid metal with tolerances between IT 13 and IT 9(8). It is also applied to enlarge holes – a process called core drilling or counterdrilling. The operation of making the hole of two or more different diameters is called step drilling. Boring is the enlarging of a hole, typically resulting in obtaining higher accuracy and better surface finish than by drilling. Enlarging a hole to a limited depth is called counterboring. Producing an angular opening into the end of a hole is countersinking, also termed chamfering. Reaming is also a hole enlarging process, but its purpose is to produce a hole of accurate size and form (typically with IT 8(7) tolerances) and good surface finish (*Ra* parameter between 5 and 2.5 μm).

A range of milling operations shown in Fig. 2.4 can be used to generate flat surfaces (planes, sides and grooves) as well as 2D or 3D curved surfaces, but by using rotating cutters with multiple cutting edges or teeth. As indicated in Fig. 2.4, the same kind of surface can be milled in several ways. For instance, plane surfaces may be produced by slab milling, face or side milling. Peripheral (plain or slab) milling generates a surface parallel to the axis of rotation while face milling generates a flat surface (usually wider than 75 mm) normal to the axis of rotation. End milling, a type of face milling operation, is used for facing, profiling and slotting operations, i.e. to produce one-, two- or three-sided surfaces. Now, the use of inserted teeth milling cutters has minimized the problem of milling machine maintenance. Because of its versatility the milling machine, especially three- or five-axis vertical or horizontal CNC milling machines or machining centres, has become a workhorse in the parts fabricating industries. It should be noted that a machining centre is a multifunctional machine tool that can perform a wide variety of operations including milling, drilling, boring, reaming, tapping, etc. without changing the part setup.

Some other, less frequent or rare (like planing), machining operations are depicted in Fig. 2.5.

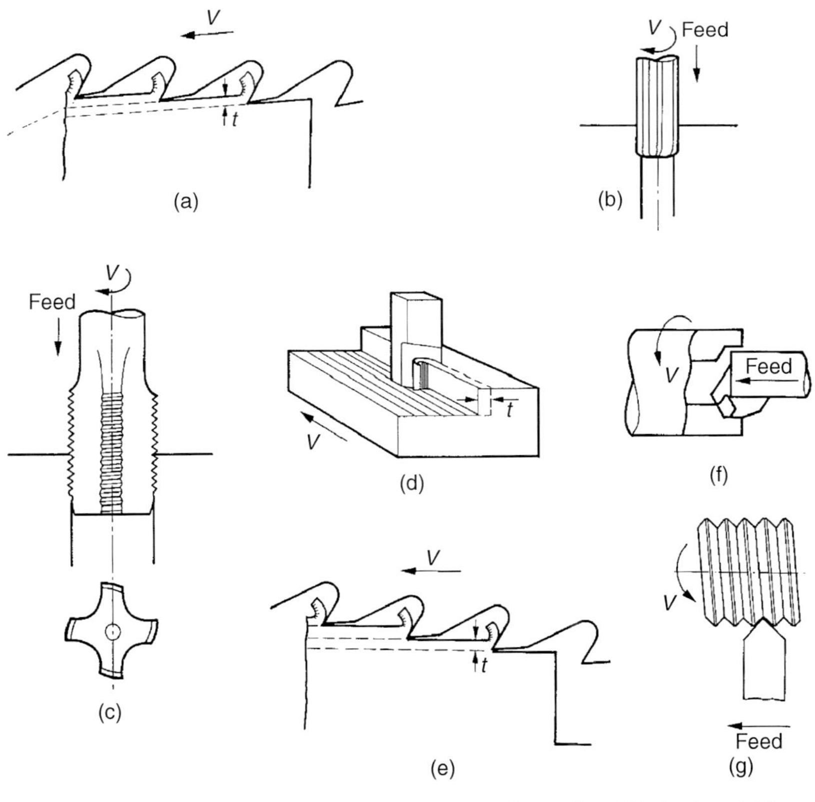

Fig. 2.5 Examples of cutting operations: (a) sawing, (b) reaming, (c) tapping, (d) planing, (e) broaching, (f) boring, (g) threading [6].

2.2 Kinematics of Cutting Process and Cutting Parameters

The principle used in all machine tools is to generate the machined surface by suitable relative motion between the cutting tool and the workpiece. In general, two kinds of relative motion must be provided by a metal-cutting machine tool in order to remove the material allowance planned. These motions are called primary motion and feed motion and are defined as follows:

1. The primary motion (rotational or linear) is the main motion provided to cause relative motion between the tool and the workpiece. Primary rotational motion is provided to the workpiece (turning, cylindrical grinding) or to the cutting tool (drilling, boring, milling) by spindle in cutting operations with stationary or rotating tools, respectively. In some cases such as broaching and planing the linear primary motion is applied.
2. The feed motion is a linear motion that is additionally provided to the tool or workpiece, leading to a repeated or continuous chip removal and the creation of a machined surface with the desired geometric characteristic including form accuracy, waviness and surface roughness. This motion may be realized stepwisely or most frequently continuously. To facilitate the description of machine tool motion it is useful to employ a system of machine tool axes recommended by the ISO (the *Z* axis of motion is arranged parallel to the axis of the machine spindle, the *X* axis is horizontal and parallel to the work-holding surface and the *Y* axis is selected to complete the right-hand coordinate system).

The kinematics of three basic cutting operations, i.e. turning, drilling and milling, are depicted in Fig. 2.6. The motion resulting from simultaneous primary and feed motions is called the resultant cutting motions.

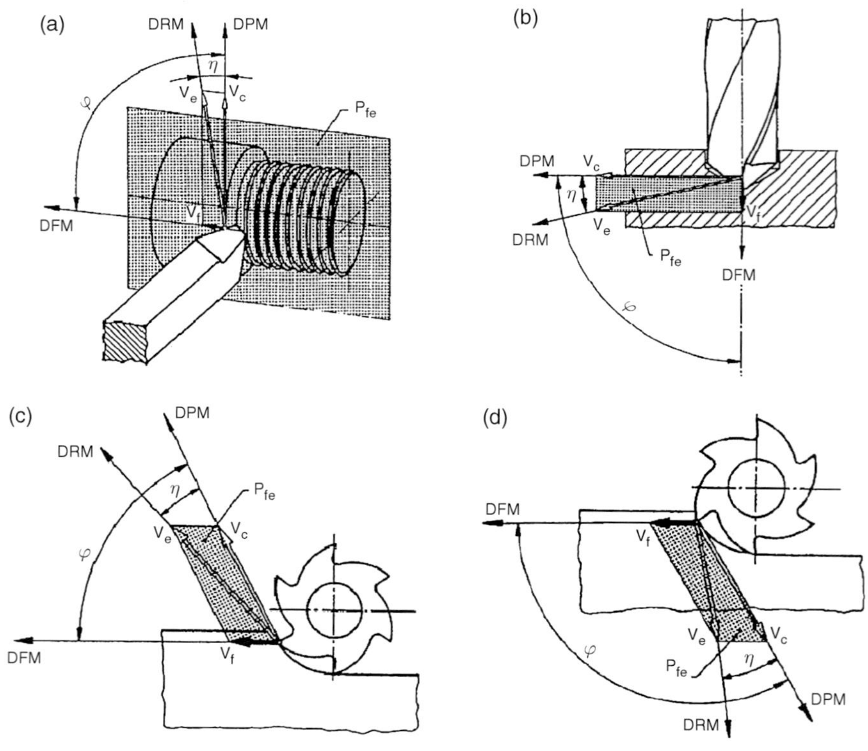

Fig. 2.6 Kinematics of typical machining operations: (a) turning; (b) drilling; (c) up-milling; (d) down-milling [7]. DPM-direction of primary (cutting) motion; DFM-direction of feed motion; DRM-direction of resultant cutting motion.

All three cutting motions are characterized by the appropriate velocities, i.e. the instantaneous velocity of the primary motion called the cutting speed v_c, the instantaneous velocity of the feed motion called the feed velocity v_f and the instantaneous velocity of the resultant cutting motion called the resultant cutting speed v_e. The angle η between the direction of primary and resultant motions is called the resultant cutting-speed angle. This angle is usually extremely small, especially for high speed machining with small feed velocity, and for most practical purposes can be assumed to be zero.

To characterize the magnitude of the feed motion, three different feed rates are used, namely:

1. The feed speed, also called the feed per minute (f_m in m/min), which is the machine feed allowing the tool or the workpiece to be moved in various directions.
2. The feed per revolution (f in mm/rev), which is the distance of the tool moving during one revolution. This is a key value in determining the quality of the machined surface in operations with a rotating workpiece, such as longitudinal turning.
3. Feed per tooth (f_z in mm/tooth) defined as the linear distance moved by the tool while one particular tooth is engaged in the cut. It is a key value in milling because the milling cutter is a multi-edged tool. In practice it is the distance covered by the table between the engagement of two successive cutting edges.

The cutting speed (v_c), the feed rate (f) and the depth of cut (a_p) create a set of technological cutting parameters which are of key importance in successfully performing any machining operation.

The depth of cut, properly termed the back engagement (a_p), is half the difference between uncut and cut surface diameters (Fig. 2.7a and b for turning and drilling) measured perpendicular to the working plane P_{fe} (that containing the directions of primary and feed motions). In addition, the feed engagement (a_f) is defined as the instantaneous engagement of tool cutting edge with the workpiece measured in the direction of feed motion. In milling operations (Fig. 2.7c and d) the working engagement of the cutting edge (a_e) is

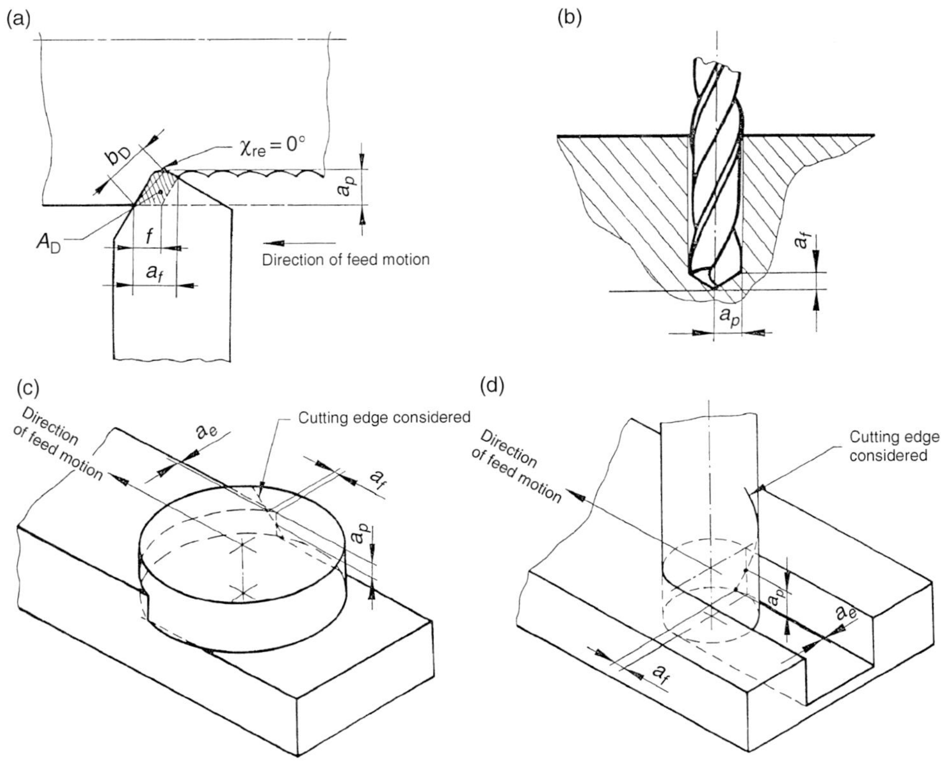

Fig. 2.7 Dimensions of removed material in: (a) turning; (b) drilling, (c) face milling and (d) end milling [8].

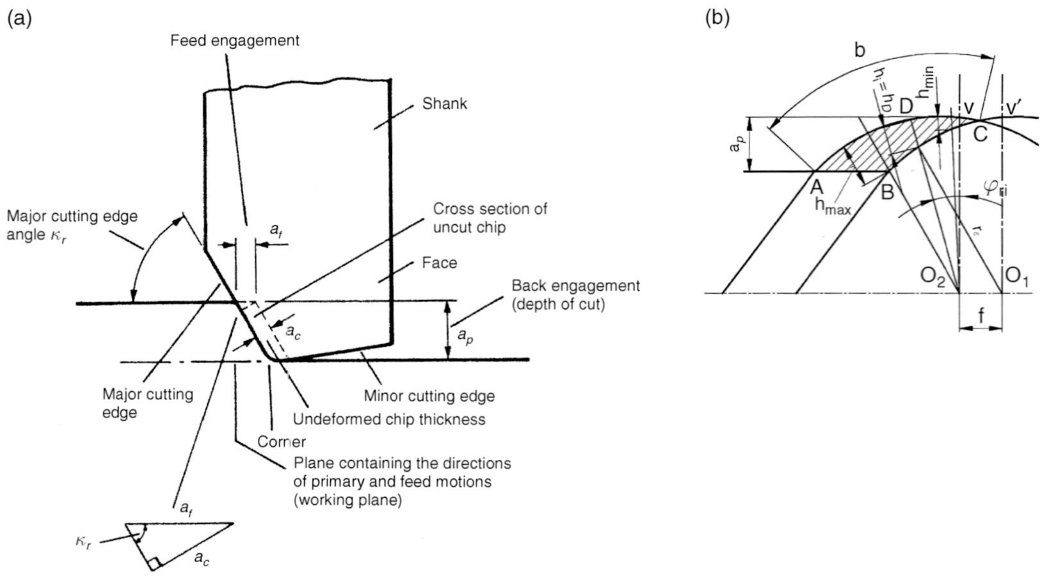

Fig. 2.8 Geometry of UCT in turning with large (a) and small (b) depth of cut.

defined as the engagement of the cutting edge with the workpiece measured in a direction parallel to the working plane and perpendicular to the direction of feed motion. All three kinds of tool engagement are defined in ISO-3002/3-1984(E).

The dimensions of the cut formed by a single action of a cutting part of the tool are shown in Fig. 2.8. For single-point cutting operations as in Fig. 2.8a the undeformed chip thickness (UCT) is given by

$$a_c = a_f \sin \kappa_r = f \sin \kappa_r \tag{2.1}$$

For small depth of cut the UCT varies along the rounded active cutting edge and its instantaneous value is given by

$$h_i = r_\varepsilon + f \sin\varphi_{ri} - \sqrt{r_\varepsilon^2 - f^2 \cos^2 \varphi_{ri}} \tag{2.2}$$

Also the nominal width of cut (b_D) and the nominal thickness of cut (h_D) are defined in the cut dimension plane, which is a plane perpendicular to the direction of primary motion in the reference point D dividing the length of the active major cutting edge into two equal parts, as shown in Figs 2.7a and 2.8b. The cross-sectional area A_c of the layer of material being removed (cross-sectional area of the uncut chip) is approximately given by

$$A_c = f a_p \tag{2.3}$$

2.3 Geometry of Cutting Tools

The geometry and nomenclature of cutting tools, even single-point cutting tools, are surprisingly complicated subjects. The ISO recommendation for the nomenclature of cutting tools [7] first establishes a system of planes that can be used to define the various angles which accurately localize the cutting edges, rake faces

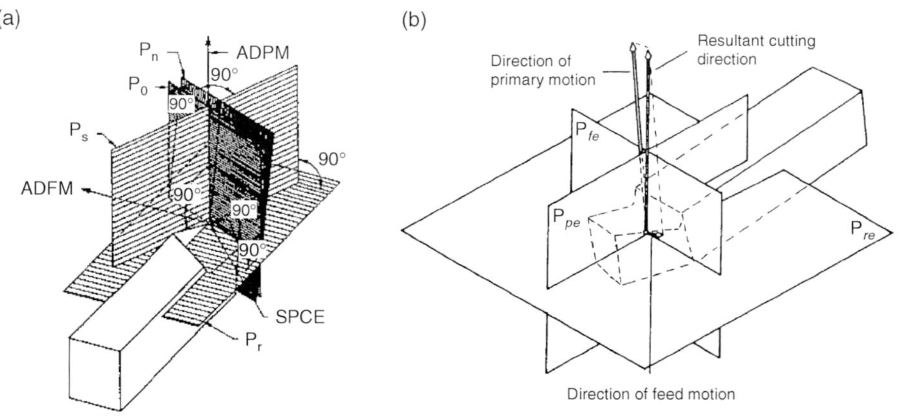

ADPM – assumed direction of primary motion;
ADFM – assumed direction of feed motion;
SPCE – selected point on cutting edge

Fig. 2.9 Reference planes in tool-in-hand system (a) and in tool-in-use system (b) according to ISO 3002-1 [7].

and flanks in the space. Two systems of planes and angles are proposed in the ISO standard. The first system is the *tool-in-hand system* of planes and angles (Fig. 2.9a) and refers to the direction of primary (cutting) motion and is used for the purpose of tool grinding (monolithic or brazed-tip tools) or manufacturing the tool holder (indexable-insert tools). The second system is the *tool-in-use system* (Fig. 2.9b) also called the working system, in which planes and angles are referred to the direction of resultant cutting motion like in a real machining operation. This is due to the fact that cutting tools can be held in a machine tool in various orientations, thereby altering the effective angles of the tool.

The following planes are defined in the tool-in-hand system:

1. The tool reference plane $\mathbf{P_r}$ parallel to the tool base (upper face of the shank for single-point tool as in Fig. 2.9a or rotation axis). Only for broachers, this plane is perpendicular to the tool axis.
2. The assumed working plane $\mathbf{P_f}$ perpendicular to the P_r plane and parallel to the ADFM.
3. The tool back plane $\mathbf{P_p}$ perpendicular to both P_r and P_f planes.
4. The tool orthogonal plane $\mathbf{P_o}$ perpendicular to the P_r plane and the projection of the major cutting edge on the P_r plane.
5. The tool-cutting edge plane $\mathbf{P_s}$ perpendicular to the P_r plane and tangential to the cutting edge S.
6. The cutting-edge normal plane $\mathbf{P_n}$ perpendicular to the cutting edge S.

Each plane is provided with a symbol consisting of the letter **P** with a subscript denoting the plane's identity. If the selected point lies on the minor cutting edge, the name of the plane and the relevant angle include the words *minor cutting edge* and the symbol would bear a prime (for example, P'_s and λ'_s). The symbol for a plane or angle in the tool-in-use systems bears an additional subscript, *e*, for *effective*. Figure 2.10 shows the various tool angles for a single-point tool with straight major cutting edge and their definitions along with some working angles are listed in Table 2.3. It should be noted that the planes and angles in the tool-in-hand system are generally prefixed by the word *tool* and those in the tool-in-use system are prefixed by the word *working*, except the normal angles because $P_n \equiv P_{ne}$.

Figure 2.11 illustrates more complex configurations of the tool wedge consisting of differently inclined faces on the rake and flank, along with several possible shapes of the cutting edge (*F*-sharp when the radius of the cutting edge (r_n) is less than 0.02 mm; *E*-rounded; *T*-chamfered (honed); *S*-rounded and chamfered; *K*-sharp and dually chamfered; and *P*-rounded and dually chamfered). Obviously, the rake faces of indexable inserts for turning, milling, drilling and boring tools are designed with 2D or 3D chip formers/chip breakers to achieve reliable chip control (see Chapter 7).

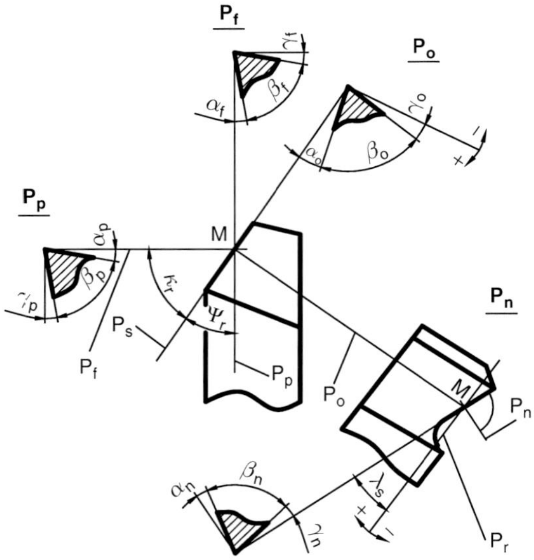

Fig. 2.10 Simplified tool geometry in tool-in-hand system.

Table 2.3 Tool and working angles, definitions.

Angle	*Is the angle between*	*and*	*Measured in plane*
Tool normal rake γ_n	A_γ	P_r	P_n
Working normal rake γ_{ne}	A_γ	P_{re}	$P_{ne}(\equiv P_n)$
Tool normal clearance α_n	A_α	P_s	P_n
Working normal clearance α_{ne}	A_α	P_{se}	$P_{ne}(\equiv P_n)$
Tool cutting-edge angle κ_r	P_s	P_f	P_r
Working cutting-edge angle κ_{re}	P_{se}	P_{fe}	P_{re}
Tool cutting-edge inclination λ_s	S	P_r	P_s
Working cutting-edge inclination λ_{se}	S	P_{re}	P_{se}
Tool-included angle ε_r	P_s	P'_s	P_r
Normal wedge angle $\beta_n(\equiv\beta_{ne})$	A_γ	A_γ	$P_{ne}(\equiv P_n)$

A_γ = tool face; A_α = tool flank; S = major cutting edge

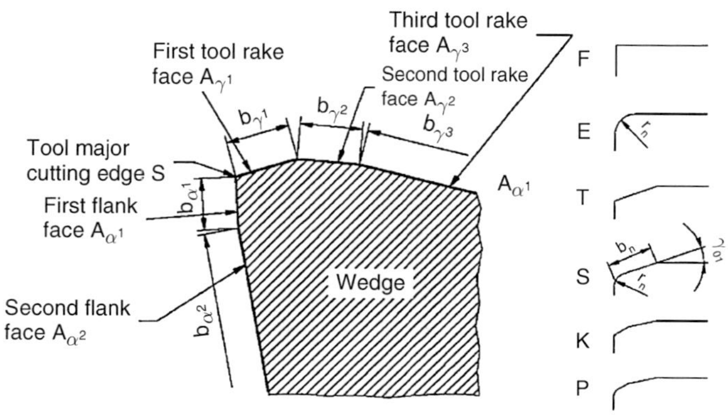

Fig. 2.11 General geometry of a cutting tool and variants of cutting edge shapes.

REFERENCES

[1] L. Alting, *Manufacturing Engineering Processes*, Marcel Dekker, New York, 1982.
[2] G. Boothroyd, W.A. Knight, *Fundamentals of Machining and Machine Tools*, Marcel Dekker, New York, 1989.
[3] W. König, F. Klocke, *Fertigungsverfahren 1: Drehen, Fräsen, Bohren*, Springer-Verlag, Berlin, 1997.
[4] E.M. Trent, P.K. Wright, *Metal Cutting*, Butterworth-Heinemann, Boston, 2000.
[5] B. Griffiths, *Manufacturing Surface Technology. Surface Integrity and Functional Performance*, Penton Press, London, 2001.
[6] M.C. Shaw, *Metal Cutting Principles*, Clarendon Press, Oxford, 1989.
[7] International Standard ISO 3003/1-1984(E), Basic quantities in cutting and grinding. Pt 1: Geometry of the active part of cutting tools.
[8] International Standard ISO 3003/3-1984(E), Basic quantities in cutting and grinding. Pt 3: Geometric and kinematic quantities in cutting.

CHAPTER THREE

Trends in Metal Cutting Theory and Practice

Contents

3.1 Evolution of Manufacturing Methods and Systems

Today, manufacturing is changing rapidly around the world due to many driving factors depicted in Fig. 3.1. The process, equipment and systems used to design and produce everything from automobiles to computer chips are undergoing dramatic changes in response to new customer needs, competitive challenges and emerging technologies. Recent advances in information systems, business practices, engineering techniques, and manufacturing science now enable companies to produce new and better products more quickly and at a lower cost than ever before.

Table 3.1 illustrates the evolution of previously and currently used manufacturing systems to 'next generation' manufacturing (NGM) systems by taking into account six factors such as production batch, costs, quality, flexibility, customer response (delivery time) and environmental demands. It is obviously known that the way to modern (lean and responsive) manufacturing enterprise is through the increasing level of production automation including intelligent control systems. In simple terms, intelligent control systems are those able to operate in uncertain environments to increase the probability of success. A logical extension of this definition is that intelligent control is realized through the integration of sensors, processing capability and actuators to assure the best product result from optimized processes. Seamlessly integrated suites of low cost, plug-and-play sensors will monitor process execution and provide real-time feedback to process controllers and actuators for continuous tuning of performance.

Flexible manufacturing equipment such as machining centres, manufacturing cells and linked systems is currently used for automatic machining of differently shaped parts. In particular, this type of automation permits fast adaptation to changing machining tasks. Reasons for the need for flexible manufacturing include small batch sizes with short production runs, reduction in stock and short-term delivery dates, under the constraints of high availability and high level-of-use of production equipment. In addition to the automatic setting of technological and geometrical data, CNC machines may perform other automated functions such as tool and workpiece handling and buffering, use of multiple tools (tool systems), measurement directly onto the machine, process monitoring and so on. Based on the scale of these issues performed it is possible

Advanced Machining Processes of Metallic Materials
ISBN: 978-0-08-044534-2

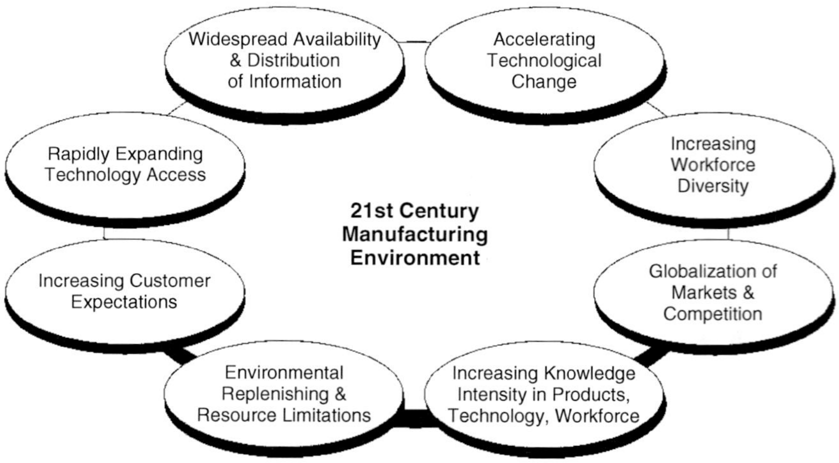

Fig. 3.1 Forces shaping the 21st century manufacturing environment [1].

Table 3.1 Evolution of manufacturing systems [1].

	Craftsmen and custom	*Mass production*	*Automated production*	*'Next generation' systems*
Timeframe	Up to present	1800s to Present	1950s to present	1990s and beyond
Lot sizes	Small	Very large	Moderate	Small
Unit costs	High	Low	Moderate	Low
Quality	Poor to good	Good	Good	Excellent
Delivery times	Long	Long	Moderate	Short
Flexibility	High	Low	Moderate	High
Environmental consciousness	Low	Low	Moderate	High

to distinguish between different manufacturing principles such as machining centres, machining cells, flexible manufacturing systems and production lines. Figure 3.2 links these machining concepts to the balance between flexibility and productivity which are contradictory objectives. The numerical values for quantities and numbers of parts shown on the X axis are solely examples because precise delimitation of the area of application of the different machining methods is rather impossible. The top line in this figure illustrates current demands on manufacturing/machining technology which may be met by flexible manufacturing methods.

For instance, Fig. 3.3 shows the structure of a flexible manufacturing cell consisting of a two-machine system comprising a machining centre and a CNC lathe linked together by a system for material transportation along with the machine-oriented control level containing the CNC and robot controls.

Moreover, a master cell control system is based on a control computer, so it superordinates to the CNC and PLC layers and carries out coordinating planning and control functions. In general, the two-machine system can contain complementary machines which execute different machining tasks or substituting machines providing the same machining facilities. The latter concept enables one machine tool to be replaced by another. Major paradigms in 21st century manufacturing are discussed in Chapter 19.

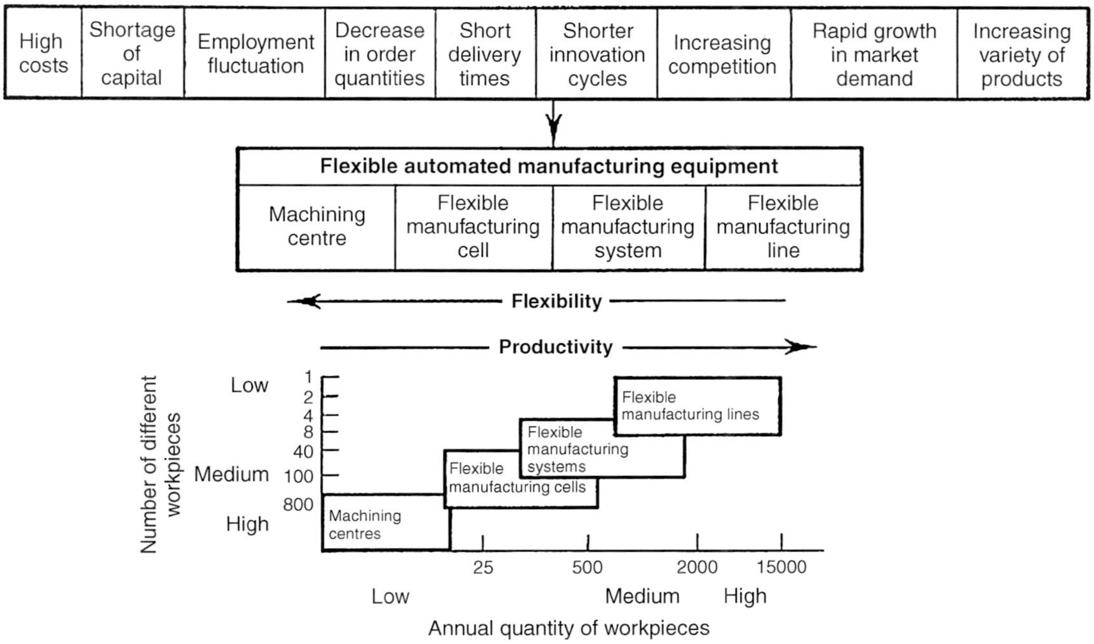

Fig. 3.2 Application areas of various manufacturing methods [2].

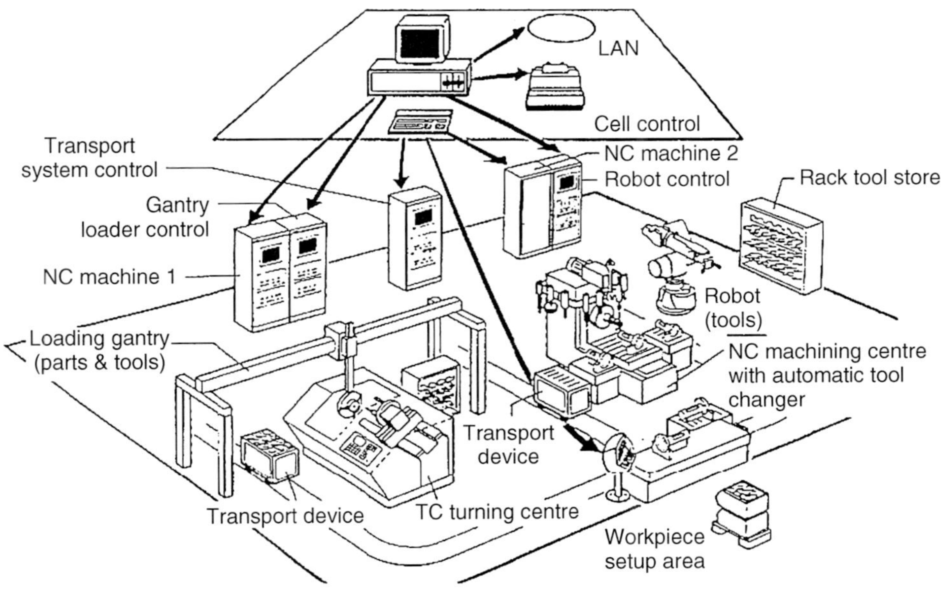

Fig. 3.3 Structure of a flexible manufacturing cell [2].

3.2 Driven Factors in Modern Machining Technology

Machining is a major manufacturing process, which plays the predominant role in the creation of a more competitive industry. In particular, its dynamic development and innovation allows for many new specific applications, including high precision machining, micro-machining and ultra precision machining [3]. Development of metal cutting technology results from evolutional and interactive improvements in machine tools, cutting tool materials, cutting tool designs and clamping systems. The scale of this development can be based on the fact that for a given operation the machining time was reduced from 100 min at the beginning of the 20th century to 1 min only when using modern CNC machine tools and appropriate tooling [4]. Currently, all machining processes should satisfy severe production requirements, including productivity, efficiency and surface quality, and additionally reliability, energy consumption and ecological standards. In case of automated manufacturing systems, new requirements such as flexibility of the product range and complete machining using one multifunctional machining centre have to be taken into account.

Possible influences of the machining technology on manufacturing processes can be visualized in the form of an external ring presented in Fig. 3.4. It is evident from Fig. 3.4 that the further increase in productivity and efficiency of the process can be achieved by the optimization of the operation sequence and cutting tool geometry and its design, and the optimal selection of cutting parameters. In consequence, some successfully implemented concepts of machining technologies, including *high speed machining* (HSM), *high velocity machining* (HVM) and *high performance machining* (HPM) are used in today's industry [6]. In contrast to high speed cutting, *high efficiency machining* (HEM) combines the advantages of HSM for finishing and high material removal rates in roughing by using medium cutting speeds (up to 2000 m/min when machining aluminium alloys) and high feeds per tooth up to 0.3 mm [7, 8]. In high efficiency machining, the most frequently used clamping devices are the hydraulic-expansion and the heat-shrinking chucks. Moreover, the tendency to eliminate cooling lubricants was a motivating factor for rapid developing near-dry (*minimal quantity lubrication* (MQL)) and *dry machining* technologies. It is extremely difficult to quantify the real costs

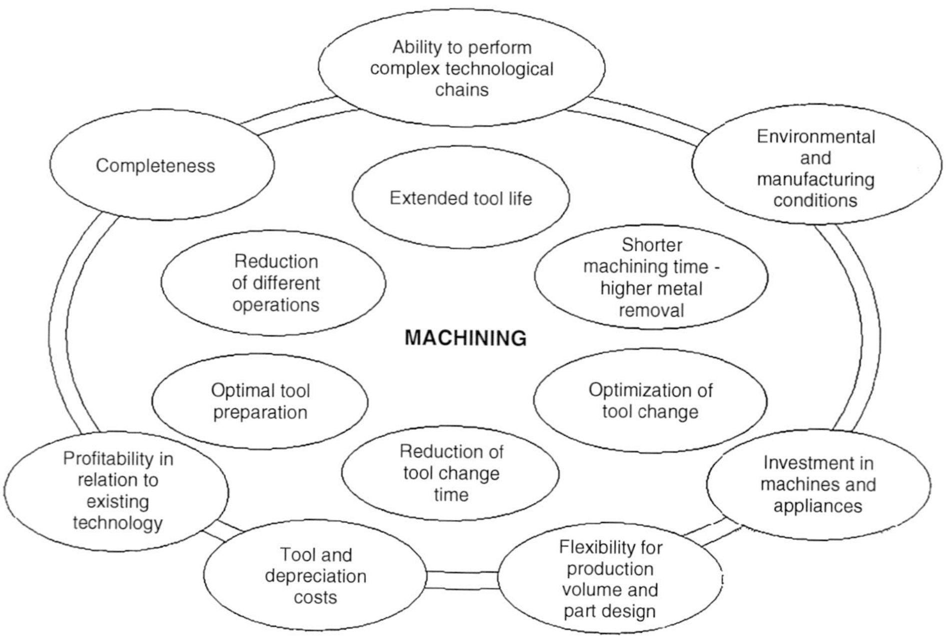

Fig. 3.4 Possible influences and importance of machining processes [5].

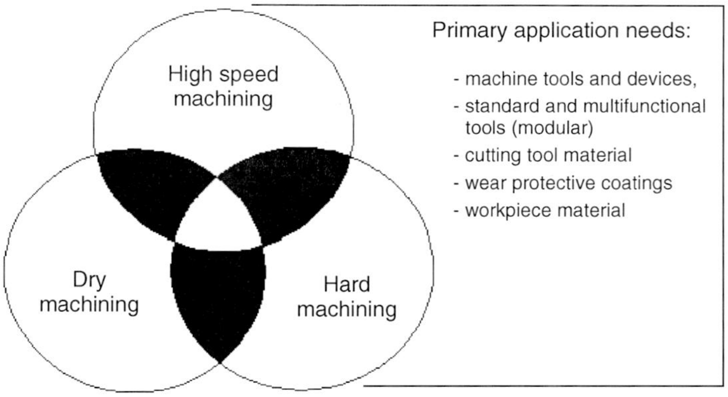

Fig. 3.5 Interactions between three advanced methods of machining.

associated with the use of cooling lubricants (maximum about 17%) but they significantly exceed tool costs (2.5–4%). It should be pointed out that in many cases some advanced machining technologies are performed jointly, as shown in Fig. 3.5.

It should be noticed that the reduction of the machining time and corresponding costs is also possible by implementing *near-shape technology*. As a result, the chip volume decreases by about 20% and the production of parts becomes also profitable in smaller batch sizes. All innovative cutting technologies cited in the technical literature are listed in Table 3.2 and their features are compared in Table 3.3. The range of applications of new highly efficient cutting processes depends mostly on the industrial tendencies in the use of particular workpiece materials. These tendencies are clearly observed in the automotive industry and in die and mould manufacture. For example, the automotive industry consumes currently more aluminium and Al-Si alloys, and ductile nodular and vermicular irons [10]. In contrast, die and mould manufacturing often requires the hardest steel materials (even 70 HRC) to be machined under the HSM conditions [11]. It can then be concluded that the progress in metal removal technology is stimulated interactively by new cutting tool materials, improved machining processes and government regulations. For example, Fig. 3.6 illustrates how the progress in materials engineering over last several decades allowed the cutting speed to be viably increased up to thousands of metres per minute when machining aluminium or its alloys using PCD tools, or grey cast

Table 3.2 Specification of currently used methods of machining.

Type of machining	*Acronym*
High speed machining	HSM
High velocity machining	HVM
High performance machining	HPM
High efficiency machining	HEM
Highly profitable cutting	HPC
Dry machining	DM
Minimum quantity lubrication	MQL
Hard machining	HM
Monolithic machining	MM
One-pass machining	O-PM
Complete machining	CM
Optical quality machining	OQM

Table 3.3 Basic features of machining strategies from Table 3.2 according to Seco Tools [9].

Machining strategies characteristics	*General machining*	*High speed machining*	*High performance machining*	*High feed machining*	*High speed precision machining*	*Micro-machining*
$a_e \times a_p$	$a_e \sim D_c$ $a_p \sim 0.4 \times D_c (40\% \ D_c)$	$a_e < D_c$ $a_p \sim D_c$	$a_e \sim D_c$ $a_p \sim D_c$	$a_e \sim 0.5 \times D_c$ $a_p < D_c$	$a_p < D_c$ $a_e < D_c$	$a_p < \sim D_c$ $a_e < D_c$
v_f (feed rate)	+ +	+ + + +	+ + +	+ + + + +	++	+
Required N (RPM)	+ +	+ + + + +	+ + +	+ + +	+ + +	+ + + + +
Volume (Q/min)		+ + +	+ + + + +	+ + + +	+	+
Cutting force (F)	+ +	+	+ + + +	+ + +	+	+
Power consumption	+ +	+	+ + + +	++	+	+
+ + + + + to + + + higher		+ + normal		+ lower		

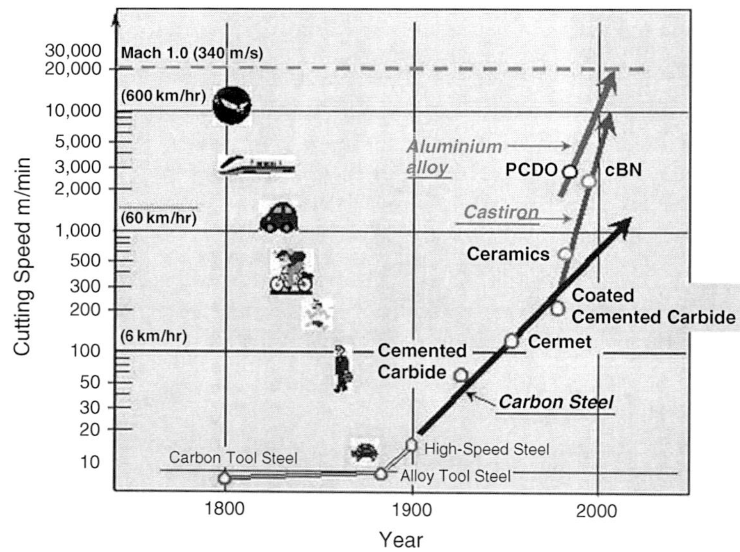

Fig. 3.6 Development of cutting tool materials for HSM. Source: Mitsubishi Carbide [9].

irons using PCBN tools. According to the forecast shown in Fig. 3.6, the future upper limit for the cutting speed will approach one Mach (340 m/s) which is comparable to speed ranges characteristic for high speed grinding with PCBN wheels.

The individual industrial branches drive the specific demands for new innovations using the different engineering materials. In particular in the automotive and the aerospace sectors, the use of light weight,

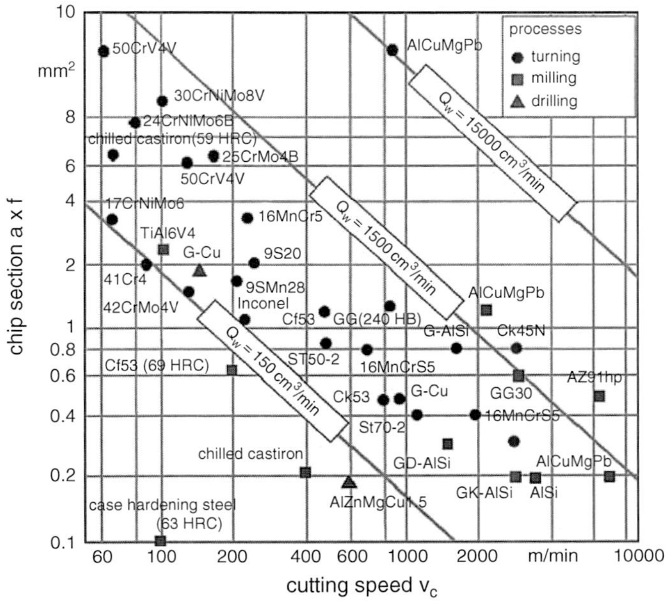

Fig. 3.7 Material removal rates for a range of materials and three most frequently used machining operations [12].

energy saving materials plays a vital role for structural components, housings and drive system components. The continuing generation of new material types feeds the demand for low density materials with high strength and easy manufacturability. Material removal rates for different processes and workpiece materials are summarized in double logarithmic form using chip section and cutting speed as coordinates in Fig. 3.7. The possibility of optimizing these process variables is strongly dependent on the combination of cutting technology used and machining task. As can be seen in Fig. 3.7, the material removal rates obtained in three basic machining operations (turning, milling and drilling) lie primarily in the range $Q_w = 150$ to $1500\,cm^3/min$ for the majority of materials. As can be expected, new machining processes have to shift the possibility of existing machining technologies toward higher material removal rates.

Machining processes which use the ultimate level of technology can be defined as super-processes [13]. A group of super-processes shown in Fig. 3.8 were found to competently show the way to technological and market leadership. The implementation of super-processes required extensive knowledge not only of the processes used in-house but also of competitive processes to support activities such as research and development, process simulation, rapid prototyping and information technology (IT). Typically, satisfactory use of one process leads to developing new technology, which in turn stimulates new innovative processes. Moreover, product and process lead and follow alternatively, i.e. the demand for high quality product initiates the development of new processes. In particular, hybrid machining processes can make the traditional manufacturing process of difficult-to-machine materials, including grinding, lapping or honing, more economical and environmentally conscious. Hybrid processes use a second form of energy that assists the original form, which is usually mechanical. Sometimes, the combination of the two processes, one of which applies the physical and chemical agents such as laser, ultrasound, liquid nitrogen, electrochemicals, water jets and microwaves, helps to overcome the original technical limitation (see section 15.6).

Cutting technology is a multidisciplinary manufacturing technique with an increasingly important role in economics. As a consequence, it has been suggested that manufacturing/machining technologies and management using IT, as, for example, to perform integration of the process planning and production planning, simulation of manufacturing systems, agile manufacturing, fast redesign of new products, modelling of manufacturing equipment performance, including the human operator, functional process analysis, virtual

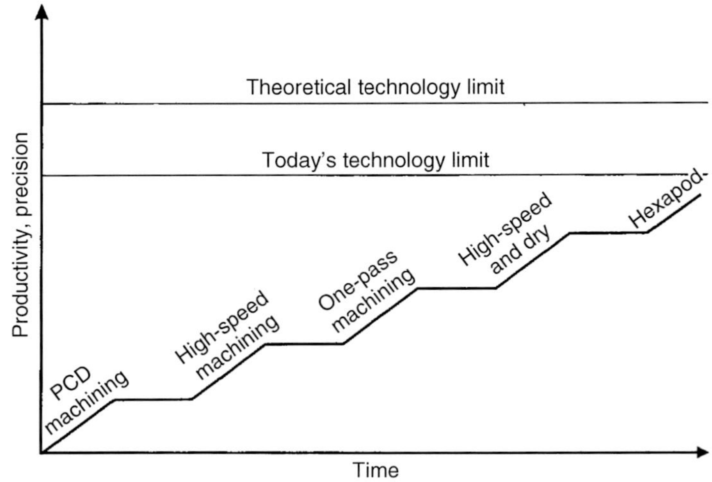

Fig. 3.8 Super-machining processes. Corrected version by Erdel [13].

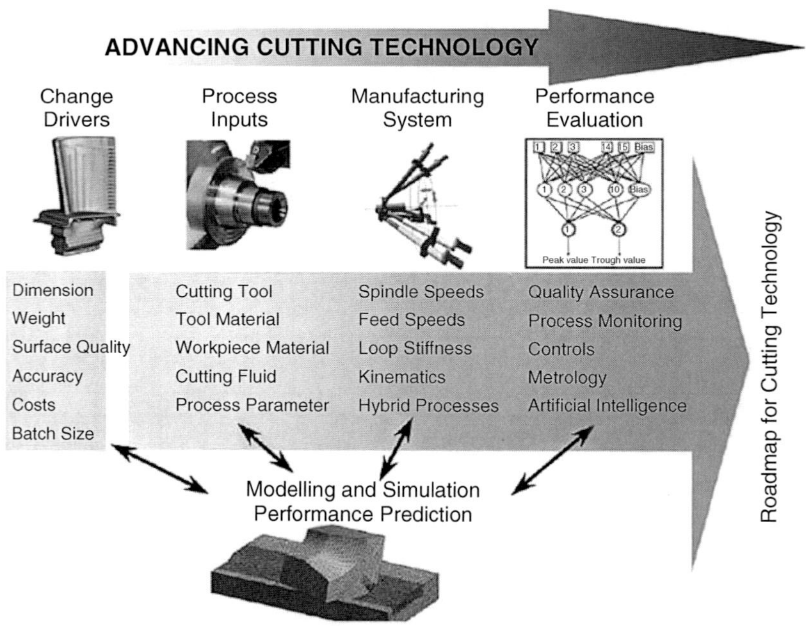

Fig. 3.9 Primary aspects associated with advancing cutting technology [12].

machining and inspection algorithms, etc., should be integrated. As shown in Fig. 3.9, the key drivers in the area of cutting technology include [12]: diminishing component size, enhanced surface quality, tighter tolerances and higher manufacturing accuracies, reduced costs, diminished component weight and reduced batch sizes. These change drivers have a direct influence on the primary inputs to the cutting process, namely, the cutting tool and tool material, the workpiece material and the cutting fluid. Further, these inputs are focused on the machine tool being a central element of the whole manufacturing system, which in

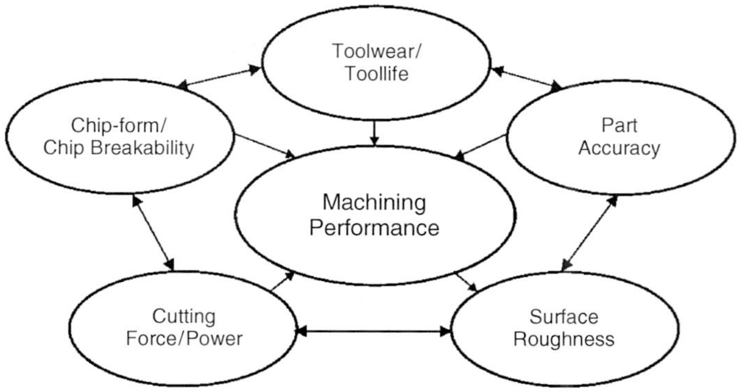

Fig. 3.10 Representation of the factors influencing machining performance for turning [12].

conjunction with the cutting technology makes a decision about the capability of the manufacturing process. Nowadays, the ability to predict/simulate and evaluate the cutting performance is embedded in the road map for cutting technology.

The cutting performance can be defined as an indication of the degree to which a set of production operations is optimized with respect to each single operation but also for the set of operations so that there is no negative influence of one operation on another [12]. This effect can ideally be achieved with the objective of shortening value added chains and with optimization of measurement activities by ensuring process quality control. For this purpose, a new manufacturing strategy termed *lean manufacturing* is developed. It aggregates a set of practices intended to remove waste (*muda*) from a manufacturing system, especially by reducing the cost of non-value-added activities. *Lean* encompasses various concepts such as just-in-time, Kaizen, Kanban, empowered teams, cycle time reduction, small lot manufacturing and flexible manufacturing [1]. A number of performance measures are commonly associated with cutting performance and those used in the case of turning operations are presented in Fig. 3.10.

Recently, predictive performance has been developed based on several modelling/simulation techniques, from which four primary categories have been evident over the past several decades, namely: analytical modelling, slip-line modelling, mechanistic modelling and finite element modelling (FEM). The application of these modelling techniques covers the range of cutting processes and cutting characteristics including cutting forces, power, tool wear and life, chip flow angle, chip curling and chip-breaking, built-up-edge, cutting temperatures, workpiece surface finish and integrity, burr formation, part distortion and accuracy, tool deflection, dynamic stability limits and thermal damage of the tool/workpiece. The processes modelled range from 2D orthogonal cutting, through 3D oblique cutting to non-free, V-shape cutting with triangular tools [14]. Cutting operations range from simple cases performed by single-point tools (turning) to more complex operations with multi-tooth tools (face and end milling and drilling).

3.3 The Future of Manufacturing

Based on the study of how the manufacturing environment might appear in 2020 and beyond, performed by the National Research Council, USA, a number of key technologies were revealed that will have a great impact on the future of manufacturing [15]. They are:

- Adaptable and reconfigurable manufacturing processes and systems.
- Extension of systems models for all manufacturing enterprise operations.
- Waste-free manufacturing.
- Synthesis and architecture technologies for converting information into desired knowledge.

- Processes and totally new customized materials with order of magnitude property improvements designed on the atomic scale (nanotechnology).
- Design methods and manufacturing processes for products than can easily be reconfigured with software or hardware.
- Direct machine/user interfaces that enhance human performance and promote intelligent input.
- Design methodologies that allow the process to enhance the range of product requirements by an order of magnitude.
- Biotechnology processes for manufacturing applications.
- Processes for rapid and cost-effective development, transfer and utilization of manufacturing technology.

During this decade, the industrial trend with two different prongs, namely technological and managerial, each of which generates specific evolving methodologies, is observed [16]. The most promising of the evolving technological methodologies are concurrent engineering (CE) and artificial intelligence (AI). On the other hand, the computer-integrated manufacturing (CIM) system concept and the formation of virtual companies are the most promising from the second group. All these concepts are briefly described in Chapters 18 and 19.

REFERENCES

[1] The NGM Project: A Key Element in the Government Manufacturing Infrastructure Strategy by B.M. Kramer, US National Science Foundation, www.nsf.gov.com.
[2] U. Rembold, B.O. Nnaji, A. Storr, *Computer Integrated Manufacturing and Engineering*, Addison-Wesley Publishing Ltd, 1993, p. 188.
[3] W. Grzesik, *Fundamentals of Machining of Metallic Materials* (in Polish), WNT, Warsaw, 1998.
[4] N.N., Coromant, Capto-system of modular quick-changed cutting tools, Sandvik Coromant, *World of Machining*, No. 1 (1999), 10–11.
[5] W. Grzesik, Developments in metal removal processes, *Proc. 4th Intln. Scientific Conference 'Development of Metal Cutting'*, Košice, Slovakia, 2002, pp. 103–110.
[6] B.P. Erdel, *New Dimensions in Manufacturing*, Hanser Gardner, Cincinnati, 1998.
[7] P. Andrae, High-efficiency machining, *Manuf. Eng.*, Vol. 125 (2000), No. 4, 82–96.
[8] C.J. Mege, Is there a HAM in your future?, *Manuf. Eng.*, Vol. 125 (2000), No. 1, 114–124.
[9] www.mitsubishi.carbide.com, www.secotools.com.
[10] D.B. Arnold, Trends that drive cutting tool development, Internet: www.kennametal.com.
[11] T. Altan, G. Lilly, Manufacturing dies and molds, *Annals of the CIRP*, Vol. 50/2 (2001), 1–23.
[12] G. Byrne, D. Dornfeld, B. Denkena, Advancing cutting technology, *CIRP Annals*, Vol. 52/2 (2003), 483–507.
[13] B.P. Erdel, *High-Speed Machining*, SME, Dearborn, 2003.
[14] E.J.A. Armarego, R.H. Brown, *The Machining of Metals*, Prentice-Hall, Englewood Cliffs, 1969.
[15] J.G. Bollinger, Visioneering manufacturing for the 21st century, *Trans. NAMRI/SME*, Vol. 26 (1998), 3–8.
[16] M.E. Merchant, Manufacturing in the 21st century, *J. Mater. Proc. Technol.*, Vol. 44 (1994), 145–155.

CHAPTER FOUR

Cutting Tool Materials

Contents

4.1 Classification and Properties of Cutting Tool Materials

Cutting tool materials constitute a special group of tool materials because they have to withstanding extreme process conditions specific for cutting, i.e. high temperature, high contact stresses, rubbing on the workpiece surface by fast moving chip. Consequently, cutting tool materials must have certain properties depending on the type of machining operations, workpiece material being machined and overall thermomechanical process conditions, namely:

1. High hardness at elevated temperatures (hot hardness) to resist abrasive wear.
2. High deformation resistance to prevent the cutting edge from plastic deformation under high stresses and temperatures arising during chip formation.
3. High fracture toughness to resist edge micro-chipping and breakage especially in interrupted cutting.
4. Chemical inertness (low chemical affinity or high chemical stability) with respect to workpiece material to protect against heat-affected wear types, i.e. diffusion, chemical and oxidation wear.
5. High thermal conductivity to reduce temperatures near the cutting edge (the cool edge of the tool).
6. High fatigue resistance for tools suffering from peaked mechanical loads.
7. High thermal shock resistance which naturally follows the mechanical shock.
8. High stiffness required to maintain accuracy.
9. Adequate lubricity (low friction) to increase welding resistance and to prevent seizure (built-up edge formation).

It should be noted that all properties deal only with an ideal tool material and in practice no single material exhibits all of the desirable properties for a tool material. Moreover, some of the desired properties are mutually exclusive, for example hardness vs toughness.

The main classes of tool materials currently in use include high speed steels (HSS) and cobalt enriched high speed steels (HSS-Co), sintered tungsten carbides (WC), cermets, ceramics (aluminas and silicon nitrides), super-/ultra-hard materials such as polycrystalline cubic boron nitride (PCBN) and polycrystalline diamond (PCD).

Advanced Machining Processes of Metallic Materials
ISBN: 978-0-08-044534-2

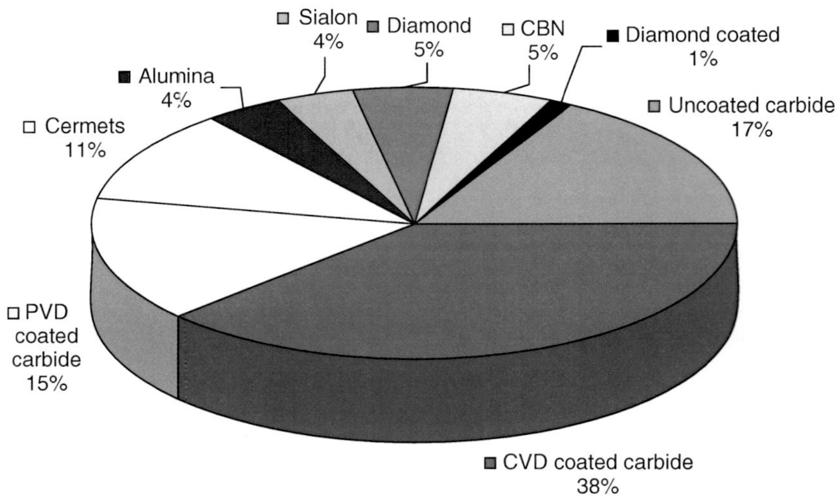

Fig. 4.1 Estimated global use of the main cutting tool materials in year 2005. Source: Kennametal, USA [1].

Figure 4.1 shows the global use of the main cutting tool materials based on a 'Technology Forecast 2005' by Kennametal, USA [1]. It can be seen in this pie diagram that currently CVD and PVD coated carbide tools are mostly used in the manufacturing sector (about 53% of the total usage).

As clarified below, the most important criteria of the suitability of cutting tool materials are the mechanical properties, such as hardness and toughness at elevated temperatures. The relationship between room temperature tool Vickers hardness and tensile rupture strength (TRS) data for some representative cutting tool materials is shown in Figs 4.2 and 4.3a. Figure 4.3a also comparatively records (in the column to the

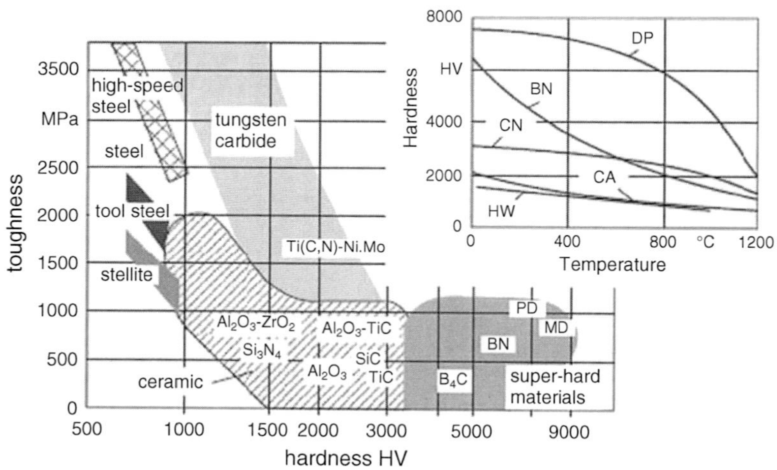

Fig. 4.2 Toughness and hardness of cutting tool materials [2]. HW – tungsten carbide, CA – alumina ceramics, CN – nitride ceramics, BN – boron nitride, DP – polycrystalline diamond.

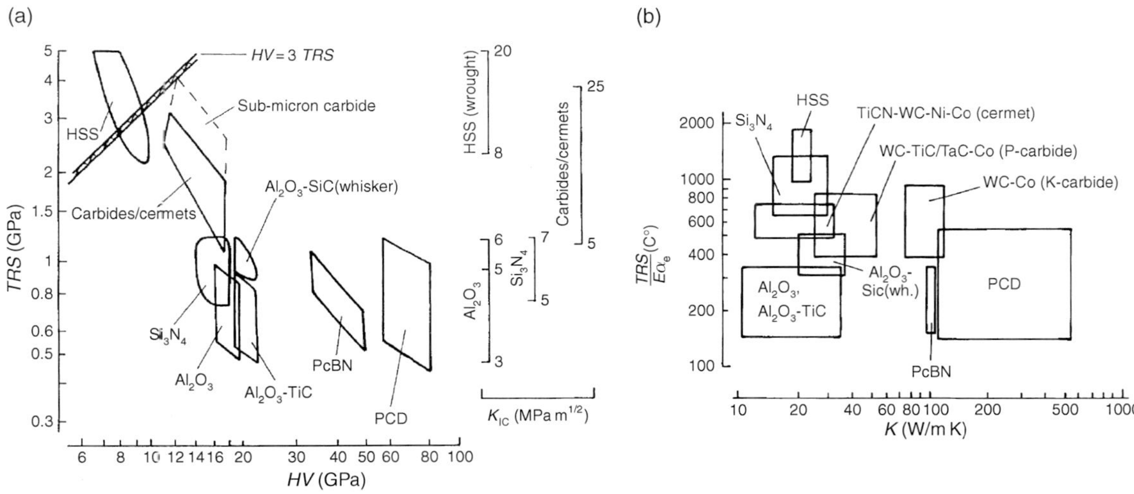

Fig. 4.3 Room temperature TRS and HV ranges of commercial uncoated cutting tool materials (a) and tool materials' characterization in terms of thermal conductivity vs shock resistance (b) [3].

right) the ranges of the fracture toughness K_{1C} vs fracture resistance (TRS) values. This figure includes the inclined line $HV = 3\ TRS$, above which a tool would be expected to show some ductile flow before fracture (the region below is of predominantly elastic fracture). Moreover, right upper insert provides the Vickers hardness as a function of temperature, referred to sometimes as hot hardness. As can be seen in Fig. 4.3a only the HSSs and ultra-fine grained carbides become predominantly or partly ductile before fracture, but at high temperatures they certainly reach this state. Thermal stability, i.e. changes of tool material properties with temperature, can be caused by unstable composition or microstructure above the critical temperature leading to rapid wear (usually oxidation), variation of hardness and resistance to fracture below this temperature and variation of thermal and elastic (elastic Young's modulus) properties with temperature. Table 4.1 summarizes some of the critical temperatures at which tool degradation occurs due to oxidation or structural changes.

The changes in tool thermal conductivity influence the heat partitioning (the rate of heat penetration into the tool) and the temperature rise in the cutting zone, as well as thermal shock resistance defined as the ratio of the TRS to $E\alpha_e$ (the product of the tool's elastic Young's modulus E and the thermal expansion

Table 4.1 Temperature ranges for oxidation and structural changes of tool materials [4].

Tool material	Temperature range (°C) for	
	Oxidation	Structural change (nature)
High speed steel	–	>600 (overtempering)
WC-Co carbide	>500	>900–950 (solution of WC in Co binder)
Mixed carbides/cermets	>700	–
Ceramics	–	>1350–1500 (intergranular liquids)
PCBN	–	>1100–1350 (transform to hexagonal form)
PCD	>900	>700 (change to graphite)

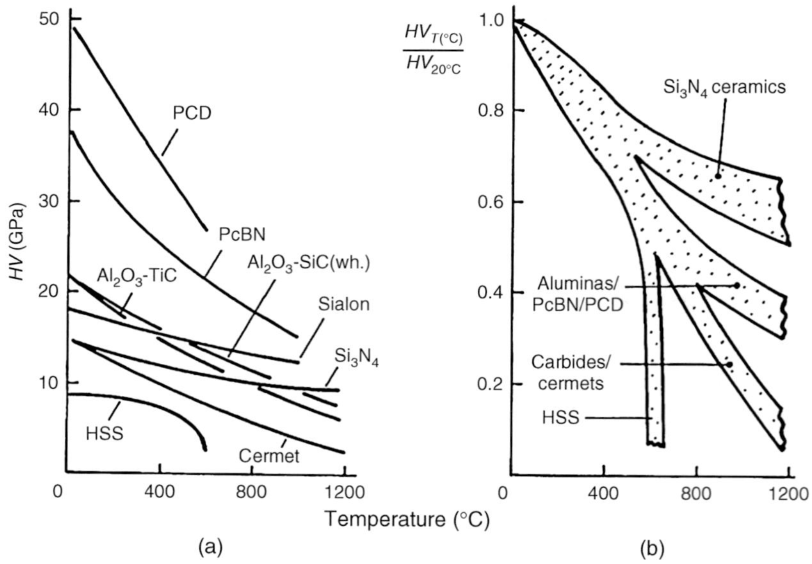

Fig. 4.4 Tool hardness changes with temperature: (a) representative values and (b) expressed relative to hardness at room temperature [3].

coefficient α_e), shown in Fig. 4.3b. It expresses the temperature range on cooling that would generate a tensile thermal stress equal to *TRS*, if the thermal strain were not allowed to relax. It is evident from Fig. 4.3b that three groups of tool material (WC-Co, PCBN and especially PCD) with high thermal conductivity reduce the temperature gradients in a tool during the cooling phase. In this arrangement, the position of Si_3N_4 ceramics does not correspond with relatively high values of *TRS* and K_{1C} parameters.

Figure 4.4 illustrates the changes in Vickers hardness with temperature (hot hardness) and the loss of hardness due to heating, respectively. As the first approximation, the relative hardness of all tool materials varies with temperature in the same way, up to about 500°C. At higher temperatures, depending on the rate of hardness reduction, four groups of tool material with similar hardness behaviour can be distinguished. High speed steels and silicon nitride base ceramics are the opposite groups of tool material in this ranking.

4.2 High Speed Steels and Cast-Cobalt Alloys

High speed steels are heavily alloyed, ferrous, conventionally cast tool materials that can be divided into four main categories: tungsten-, molybdenum-, molybdenum-cobalt-, and molybdenum-vanadium-based grades. They can be coated by PVD deposition method. In addition, P/M HSS materials made by powder metallurgy (P/M) are used. This type of steel has a uniform structure with very little carbide segregation, which can occur in conventionally cast materials. The influence of alloying components is as follows:

- Tungsten: increases tempering and wear resistance.
- Molybdenum: improves toughness.
- Vanadium: increases wear resistance of finishing tools.
- Cobalt: enables increased hardening temperatures and improves heat resistance.

Table 4.2 Grades of high speed steels (general ISO designation: S W-Mo-V-Co).

General description	*ISO/Germany/USA steel description*	*Material no. (steel code)*	*Range of application*
HSS	S 6-5-2 (DMo5/M2)	1.3343	Standard tool material for most common applications
HSS-Co	S 6-5-2-5 (EMo5Co5/M35)	1.3243	High heat resistance, especially suited for roughing or when coolant insufficient
HSS-E	S 6-5-3 (EMo5V3/M3)	1.3344	High friction resistance and cutting edge stability, especially important for reaming operations
HSS-E M 42	S 2-10-1-8 (M42)	1.3247	Increased heat resistance and hardness, suitable for difficult-to-machine materials
P/M HSS-E	S 6-5-3-9	1.3207	High hardness, heat resistance and cutting edge stability

All types of HSSs contain significant amounts of carbon, which reacts with the refractory metals to form hard carbides which increase wear resistance and hot hardness. Basic grades of high speed steels and their applications are summarized in Table 4.2.

Because of their high toughness and resistance to fracture, high speed steels are especially suitable for high positive rake angle tools and for machine tools with low stiffness. High speed steels are used in a wide variety of cutting operations requiring complex tool shapes such as drills, reamers, taps, broachers, milling cutters and gear cutters.

Sometimes, cast-cobalt alloys (Stellite tools) are classified between HSS and carbides. These molten alloys contain large amounts of carbide-forming elements such as chromium (30–33%) and tungsten (10–20%) in a cobalt matrix (38–53% Co) to produce a fine-grained, hard surface of 60–65 HRC made up of complex carbides. Now, their application is very limited to special, simple-shaped roughing tools because they will not be annealed at 815°C.

4.3 Sintered Tungsten Carbides

Cemented carbides are the most common cutting tool materials currently in use and are now the material of choice for more than half of all cutting tools produced worldwide. HSS accounts for about 40%; the remaining 10% or so are made up of all other materials. Cemented carbides are made by powder metallurgy methods using a number of different carbides and tough metals of the iron group as binders. The main carbides used are tungsten carbide (WC), titanium carbide (TiC), tantalum carbide (TaC) and niobum carbide (NbC). The binder is mostly cobalt (Co). However, the carbides are also soluble in each other and can form a cemented carbide without a separate metallic binder.

The percentage of hard particles in carbide tool materials can vary from about 60 to 90%. By adjusting the type, size and concentration of particles, producers can tailor properties of these tool materials to meet a wide variety of application requirements. The ISO system of classification uses the symbols P (blue colour), M (yellow colour) and K (red colour) to denote the application groups for long-chipping materials such as most steels and steel alloys, stainless steels and heat-resistant alloys, and short-chipping materials such cast iron, non-ferrous metals and non-metallic materials. These designations combine these letters with numbers

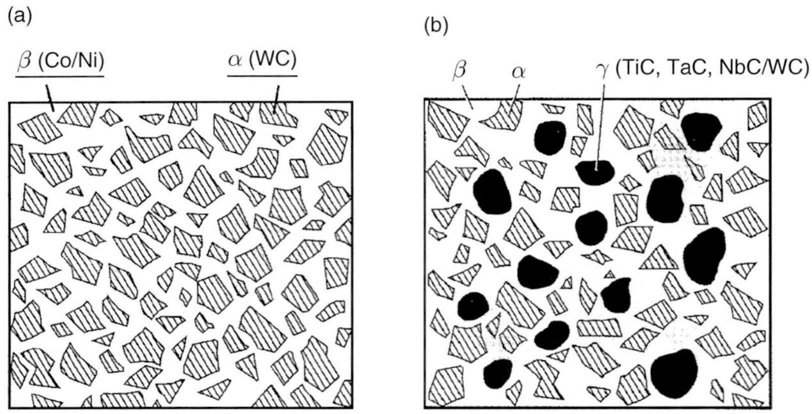

Fig. 4.5 The structure of two-phase and three-phase carbides [4].

indicating the suitability of the grade for application ranges from light finishing (01) to heavy roughing (50). According to Fig. 4.6, two groups of cemented carbides with different structure can be distinguished, namely, two-phase and three-phase cemented carbides. A two-phase cemented carbide (Fig. 4.5a) contains the hard WC phase called alpha-phase (α) and the cobalt binder called beta-phase (β). The affinity of carbon and austenite leads to the flow of carbon from the tool and creates a crater on the rake face. In order to improve the resistance to crater wear at high machining temperatures more chemically stable titanium and tantalum carbides are added to create the three-phase structure shown in Fig. 4.6b. The three-phase cemented carbides contain an additional gamma-phase (γ), represented by the TiC, TaC and NbC carbides. Two parameters, namely the ratio of Co to WC and the WC particle size, control the material properties. As shown in the diagrams in Fig. 4.6, the higher content of hard particles makes the cemented carbides more wear resistant because of the increase of both hardness (HV) and compressive strength (Cs). In addition, the elastic Young's modulus (E) increases. In contrast, a higher proportion of a binding metal makes the cemented carbide tougher and causes the bending strength (*Be*) to increase. Similar effects in improving the material toughness deal with an increase in the transverse rupture strength (*Trs*) and impact strength (*Imp*).

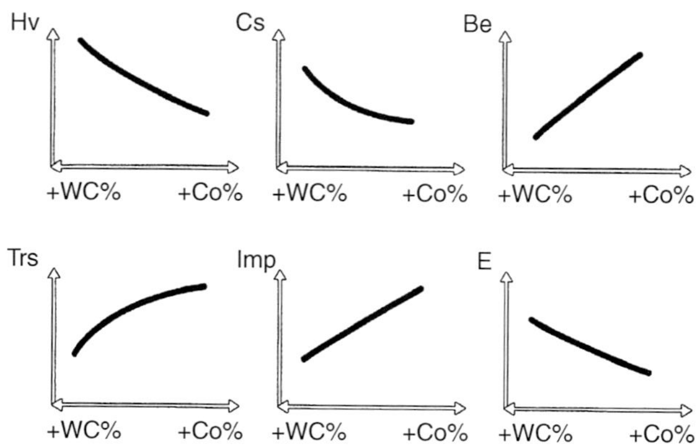

Fig. 4.6 Influence of WC and Co content on the mechanical properties of cemented carbides [4].

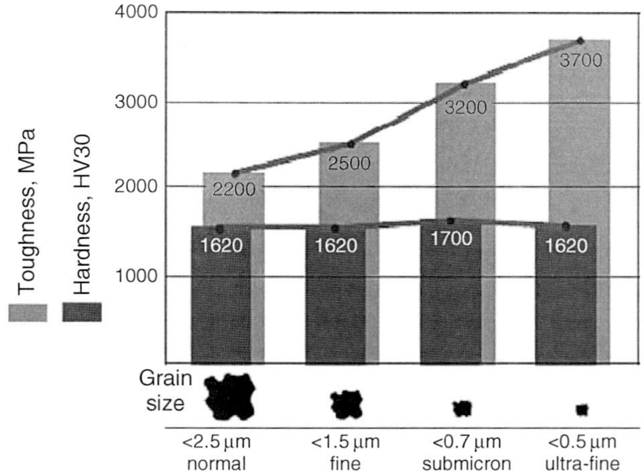

Fig. 4.7 Influence of grain size on the hardness and toughness of cemented carbides [7].

In general, grain size of carbide powders has shrunk from 10 µm to 1 µm. As in any P/M product, smaller grain sizes in carbides result in smaller voids and a denser finished material. The most significant possibility to increase hardness and toughness in the case of the WC-Co carbides are the submicron and ultra-fine (nano-phase) grained structures with WC grain sizes of 0.5–0.8 µm (Fig. 4.7) and 0.2 (0.1)–0.5 µm, with a respective content of 6–16% by mass. Interparticle distances in nano-phase carbides are so small that sintering temperatures can be lowered to inhibit grain growth that may occur at higher processing temperatures. Nano-phase carbides are being successfully examined for use as round tools, such as end-mills for exotic materials [5]. Other additions, such as vanadium carbide and chromium carbide, tend to inhibit growth of the submicron grains during processing and sintering.

Some of the latest developments in the field of carbides are cobalt enrichment of the surface layer [5] and a functional gradient in near-surface areas of cutting tools [1]. By increasing the concentration of cobalt binder phase near the surface, we can improve toughness within the layer of 0.015–0.025 mm in thickness, whereas the core material remains hard. A functional gradient is a specific, smoothly varying distribution of phases and/or element composition which provides a highly resistant surface region that withstands the cutting tool and workpiece interaction at high temperatures and which smoothly attains the microstructure of the carbide. Diamond dispersed cemented carbides with embedded diamond particles are potentially of great practical importance due to combining the hardness and wear resistance of diamond with the relative toughness of sintered hard metal. It was documented experimentally [6] that the diamond dispersed, ultra-fine grained tungsten carbide compacts have fracture toughness 50% greater and about a 25 times higher wear resistance than conventional low binder cemented carbides.

Cermet is the collective name for cemented carbides (CERamic plus METal) that have hard particles based on titanium carbide (TiC), titanium carbonitride (TiCN) and/or titanium nitride (TiN) rather than tungsten carbide (WC). Molybdenum carbide may also be used as part of the hard phase, and the metal binder material may be a mix of cobalt and nickel. Figure 4.8a–c shows ultra-fine (CT1), micro-grain (CT2) and conventional (CT3) structures of cermets. In comparison to cemented carbides, cermets have higher abrasive wear resistance, high hot hardness and better chemical stability, and low tendency for oxidation wear. On the other hand, strength and toughness are lower than cemented carbides, as is thermal shock resistance. The optimal range of application for cermets in turning and milling covers a very small area which combines low feed rates and high cutting speed (Fig. 4.8d); they even outperform coated carbides. Cermets denoted as CT1 and CT2 are recommended for finishing operations in turning and milling of steels and cast irons due to balanced fracture and wear resistance. Unfortunately, demanding profiling operations are not suitable applications for cermets.

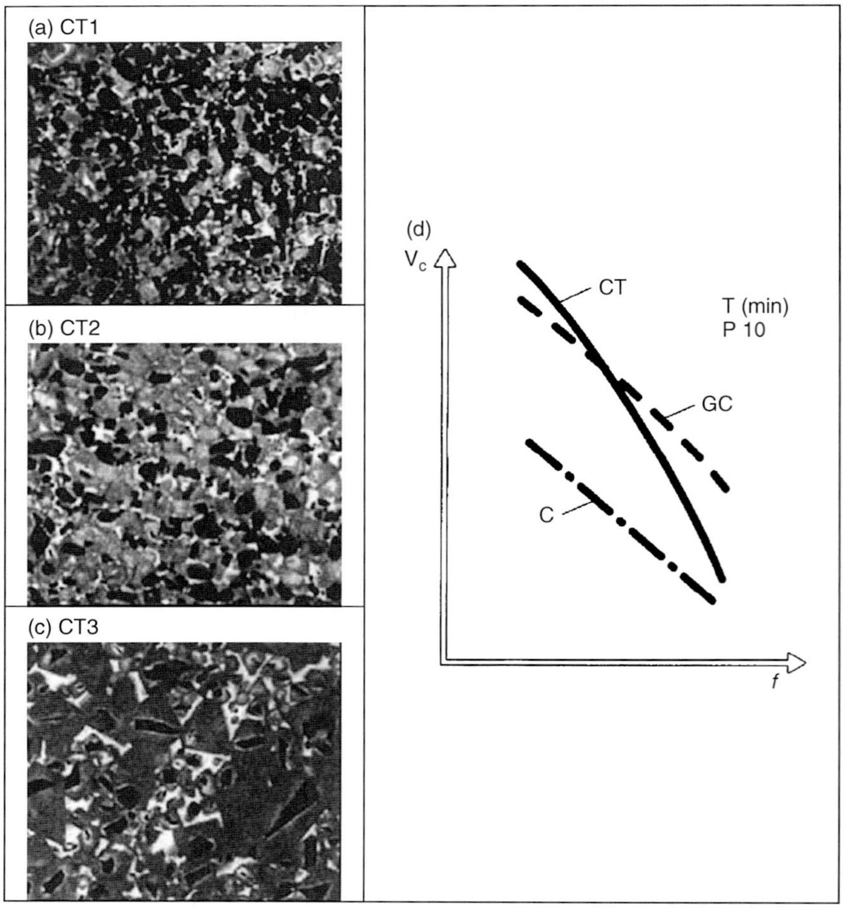

Fig. 4.8 Microstructures and application range of cermets [4, 7]. C – uncoated carbide; GC – coated carbide; CT – cermet.

4.4 CERAMICS

In general, ceramic tools have found applications principally in turning and milling cast irons and superalloys and in finishing hardened materials. For cutting tools, two kinds of ceramic composite materials are used, which can be differentiated according to the matrix materials. There are ceramics based on aluminium oxide (Al_2O_3) and silicon nitride (Si_3N_4). The key to successful application of ceramic cutting tools is to remember that they can take far more heat than carbides, because they soften in the range of 2200°C vs about 870°C for carbide tools. Aluminium oxide, commonly referred to alumina, possesses strong ionic interatomic bonding giving rise to its desirable material characteristics. It can exist in several crystalline phases which all revert to the most stable hexagonal alpha-phase at elevated temperatures Alumina-based cutting tool compositions (C1 in Fig. 4.9) include additions of zirconia (ZrO_2), titanium carbide, titanium nitride or silicon carbide (SiC) whiskers in order to change the mechanical properties of the matrix. Consequently, aluminium oxide-based ceramics are divided into pure, mixed and reinforced ceramics.

Alumina-zirconia (white ceramics) contains up to 10% ZrO_2 in order to improve toughness. White ceramic materials are especially effective in steel finishing operations. Alumina with additions of up to 40%

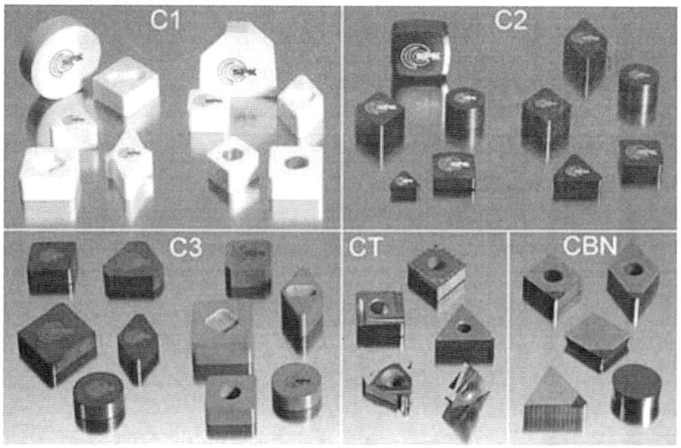

Fig. 4.9 A range of cutting tool materials: C1 – aluminium oxide (white) ceramics; C2 – Al_2O_3-TiC (black) ceramics; C3 – silicon nitride-based ceramics; CT – cermets; CBN – cubic boron nitride [7].

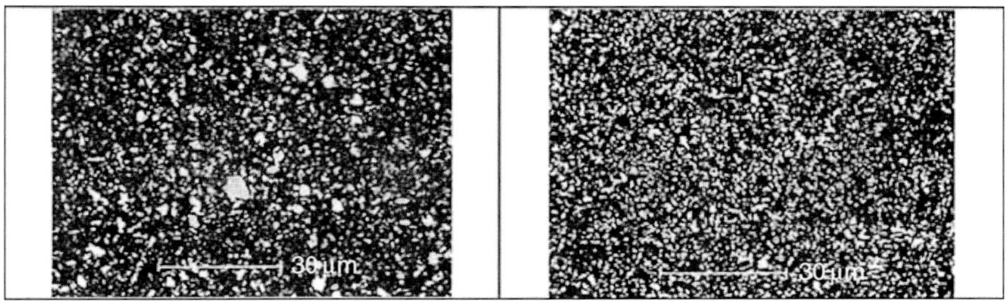

Fig. 4.10 Structures of mixed ceramics with conventional (left) and submicron grain size [8].

TiC or Ti(C,N) (black ceramics; C2 in Fig. 4.9 and Fig. 4.10) is exceptionally abrasion resistant and is used for machining chilled cast irons and hardened steels. Comparison of properties of mixed ceramics (Al_2O_3 + Ti(C,N)) with different grain sizes is made in Table 4.3. It should be noted that decreasing the grain size below 1 μm causes the fracture toughness K_{1C} to increase from 5 to 6.6 MPa $m^{1/2}$. Alumina reinforced with SiC whiskers is the toughest and most resistant to thermal shock of the Al_2O_3-based ceramics. SiC whiskers used for reinforcement are discontinuous, rod- or needle-shaped fibres in the size range of 0.1 to 1 μm in diameter and 5 to 100 μm in length (about 100 times longer than they are wide). The incorporation of SiC whiskers into the ceramic matrix results in the increase of strength, fracture toughness, thermal conductivity, thermal shock resistance and high temperature creep resistance. Because they are nearly single

Table 4.3 Comparison of properties of mixed ceramics with structures presented in Fig. 4.10.

Property	*Conventional ceramic*	*Submicron-grained ceramic*
Density, g/cm^3	4.28	4.33
Grain size	<2 μm	<1 μm
Hardness HV10	2100	2200
Fracture toughness K_{1C}, MPa × m$^{1/2}$	5.0	6.6

Table 4.4 Comparison of properties of aluminium-based and silicon nitride-based ceramics [4]. T – toughness, HV – Vickers hardness, TSR – thermal shock resistance; CS_{Fe}, CS_{Ni} – chemical stability against Fe and Ni. *Work exceptionally well when silicon nitride ceramics is paired with grey cast-iron.

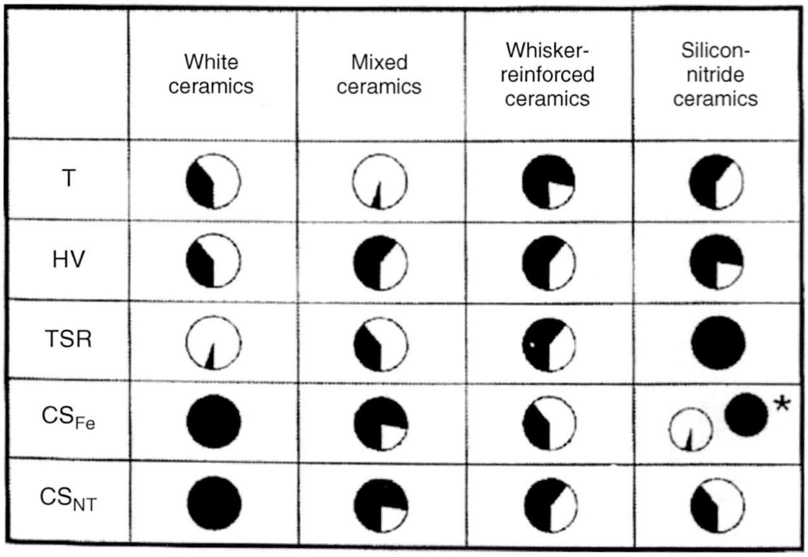

crystals, the whiskers typically have very high tensile strength on the order of 7 GPa and elastic Young's modulus values up to 550 GPa. Unlike other such materials, it can be run with coolant. High speed finishing of nickel/cobalt-based superalloys is a typical application for whisker-reinforced ceramic cutting tools.

Cutting tool materials based on silicon nitride include fully dense Si_3N_4 and SiAlON materials, which are solid solutions of alumina in Si_3N_4. Fully dense Si_3N_4 can have fracture toughness nearly as high as cemented carbides, high strength, and a low coefficient of thermal expansion, which result in very good thermal shock resistance. The silicon-based ceramics are excellent in maintaining hot hardness at temperatures higher than those suitable for cemented carbides, and are tougher than aluminium-based ceramics. Tools made with this material are excellent for turning grey cast iron at cutting speeds over 400 m/min, and are also used for milling and other interrupted operations on grey iron. Coolant can be used for tuning applications. SiAlONs are typically more chemically stable than Si_3N_4 but not quite as tough or resistant to thermal shock. They are mainly used in rough turning of nickel-based superalloys.

In summary, the main application areas for ceramic materials are grey cast iron, heat-resistant alloys, hardened steels, nodular cast iron and to some extent steel.

4.5 Superhard Materials

The development of superhard cutting tool materials began in the early 1970s with the introduction of polycrystalline diamond cutting tool materials. So far, two polycrystalline materials have been developed: polycrystalline cubic boron nitride (PCBN) and polycrystalline diamond (PCD). Although their current market share, at 1%, is almost insignificant in comparison with conventional materials, the trend is clearly towards the increasing use of both PCD and PCBN. CBN tool use is currently growing at a rate of 10–15% annually, driven mainly by increased use of hard machining [5]. Both superhard materials are manufactured using a high temperature, high pressure process in which individual diamond or CBN particles

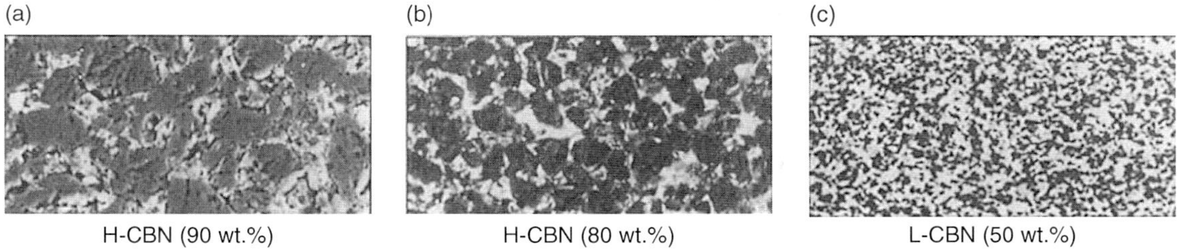

Fig. 4.11 Microstructure of different PCBN grades [9].

Fig. 4.12 A variety of PCBN inserts: (a) solid flat-faced inserts; (b) solid PCBN insert with chip-breaker; (c) brazed-tip inserts; (d) solid PCBN insert coated with copper coloured (Ti,C) Al layer [7]. Source: Seco Tools.

are consolidated in the presence of iron, nickel, and/or cobalt catalysts that promote grain consolidation into a solid mass (solid polycrystalline) or on a WC substrate (backed polycrystalline). Small pieces obtained from the polycrystalline compact are often brazed onto WC inserts as shown in Fig. 4.11. Some types of PCBN cutting tool inserts are shown in Fig. 4.12.

The properties of the CBN cutting tool material can be varied by altering the crystal size, content and the type of binder to make various grades, namely (Fig. 4.11):

1. The straight PCBN grade consisting of a high percentage of CBN (about 90 wt.%) and a metal second phase (Al).
2. High content composite CBN (CBN about or less than 80 wt.%) and a metal/ceramic second phase.
3. Low content composite CBN (CBN less than 70 wt.%) and ceramic (TiN or TiC) second phase.

The properties of some PCBN and PCD grades and, comparatively, of other cutting tool materials are listed in Table 4.5. It should be noted that the fracture toughness and thermal conductivity increase with increasing CBN content. It is thermally stable up to 1200°C and has good chemical wear resistance up to 1400°C. The grains in CBN are very small (even less than 1 μm in L-PCBN) and hard, and to avoid micro-chipping, the cutting edge must have appropriate chamfers and should be honed. CBN should be applied to workpiece materials harder than 45(48) HRC to avoid excessive wear. Steel forgings, hardened steel and cast iron, surface-hardened components, cobalt and iron-based powder metals, forming rolls, pearlitic cast iron and heat-resistant alloys are among the main types of application of CBN. There are limitations of PCBN, however, in the machining of austenitic steels and ferritic cast irons, whose more reactive γ and α ferrous phases result in increased PCBN edge wear.

PCD is produced by sintering together selected particles of diamond under conditions typically used for CBN synthesis. A secondary metal phase (Fig. 4.13a) facilitates controlled particle intergrowth of the polycrystalline mass. PCD can be therefore regarded as a composite material which combines the very high hardness of up to 8000 HV, high abrasion resistance and exceptional thermal conductivity in the range of 500 $Wm^{-1}K^{-1}$ with the toughness of tungsten carbides (see data in Table 4.5). The average starting grain size is about 25 μm. The diamond crystals are randomly oriented to eliminate any direction of crack

Table 4.5 Physical properties of various cutting tool materials [9].

	Al_2O_3	*Ceramics* Al_2O_3+TiC	*Sialon*	*Hard metal ISO K10*	*PCD*	*Polycrystalline materials* *H-PCBN (wt. 90%)*	*H-PCBN (wt. 80%)*	*L-PCBN (wt. 50%)*	*Mono-crystalline diamond*
Density (g/cm^{-3})	3.91	4.28	3.20	14.7	4.12	3.42	3.52	4.28	3.52
Compressive strength (GPa)	4.00	4.50	3.50	4.438	7.60	2.73	–	3.552	8.68
Fracture toughness (MPa m$^{1/2}$)	2.33	2.94	5.00	10.48	8.81	6.4	5.9	3.7	3.40
Knoop hardness (GPa)	16	17	13	17	50	31.5	30	27.5	57–104
Young's modulus (GPa)	380	390	300	593	776	680	649	587	1141
Thermal expansion ($10^{-6}K^{-1}$)	8.5	7.8	3.2	5.4	4.2	4.9	4.6	4.7	1.5–4.8
Thermal conductivity ($Wm^{-1}K^{-1}$)	8.4	9.0	20–25	100	540	100	85	44	500–2000
Wear coefficient	0.76	0.92	0.91	1.15	3.89	1.90	1.76	1.34	2.14–5.49

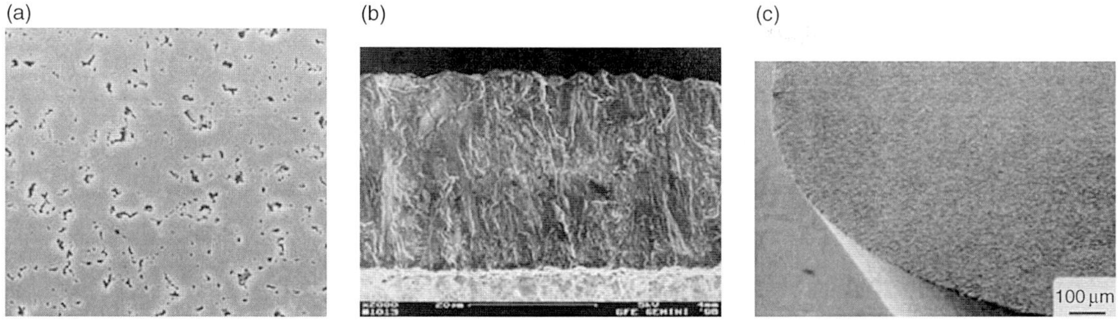

Fig. 4.13 Microstructures of PCD compact (a) and CVD PCD-coating morphology (b) deposited onto WC insert (c) [7, 10].

propagation. The small PCD cutting edges are bonded to cemented carbide inserts similar to PCBN brazed-tip inserts (Fig. 4.12c), adding strength and shock resistance. The main drawbacks of PCD materials are the upper thermal barrier of 600°C, chemical affinity to ferrous material and elimination of tough, high tensile workpiece materials. Today, PCD grades are used for turning, boring and milling operations, especially in finishing the abrasive silicon-aluminium alloys for which sharp cutting edges and positive rakes are essential, when surface finish and accuracy are predominant technological criteria. Because of the very brittle nature of PCD, very stable conditions, rigid tools and machines and high speeds are necessary for machining with PCD. Cutting fluids can be used, but generally for cooling. Today, many types of diamond materials are available for cutting tool applications. These also include diamond coatings deposited by various methods, as well as thick-film diamond (Fig. 4.13b and c), which is a thick (250 μm–1 mm) layer of pure diamond brazed to a carbide substrate. Another diamond product developed more recently is chemical vapour deposition (CVD) diamond, which is fully dense, free of metal binders and thus has hardness and thermal stability near that of natural diamond. As reported [10], adhesion of a deposited diamond film to the carbide tool has presented the greatest technological obstacle to wider commercial introduction of diamond coated tool inserts.

4.6 Cutting Tool Coatings

According to the prediction shown in Fig. 4.1, the use of CVD and PVD coated carbides recently approached 38% and 15%, respectively. Other production data assesses this level even at 70–75% [11, 12]. Today, a majority of cutting tool materials, i.e. HSS, HSS-Co, WC, cermets, various types of ceramic materials and CBN, are coated with differently structured hard/soft thin layers to increase tool life and can achieve higher cutting speeds to increase productivity. A variety of potential candidates for hard coatings or their components are given in Fig. 4.14.

In general, various borides, carbides, nitrides, and oxides, mainly transitional metals and elements of groups IV and VI in the periodic table, have been found to be applicable to hard coating materials. Among them, the binary and ternary phases such as titanium nitride (TiN), titanium carbide (TiC), titanium carbonitride (Ti(C,N)), titanium-aluminium nitride (TiAlN), aluminium oxide (Al_2O_3), chromium nitride (CrN), hafnium nitride (HfN) and CBN seem to be mostly used by cutting tool manufacturers. The reason for such rapid development of coating technology and growing popularity of coated tools in the metal-cutting industry is that the coatings can positively alter the cutting process performance as illustrated in the scheme from Fig. 4.15. Figure 4.15 shows two major ways in which a coating may influence tool wear, i.e. directly

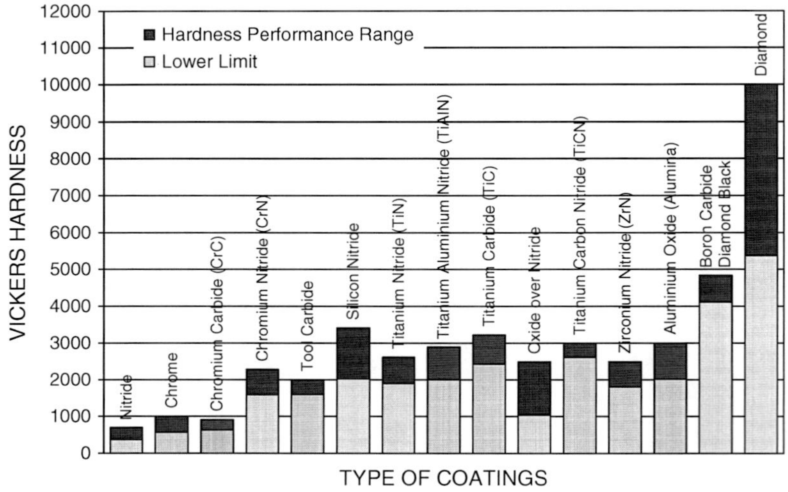

Fig. 4.14 Comparison of micro-hardness for several types of coatings [13].

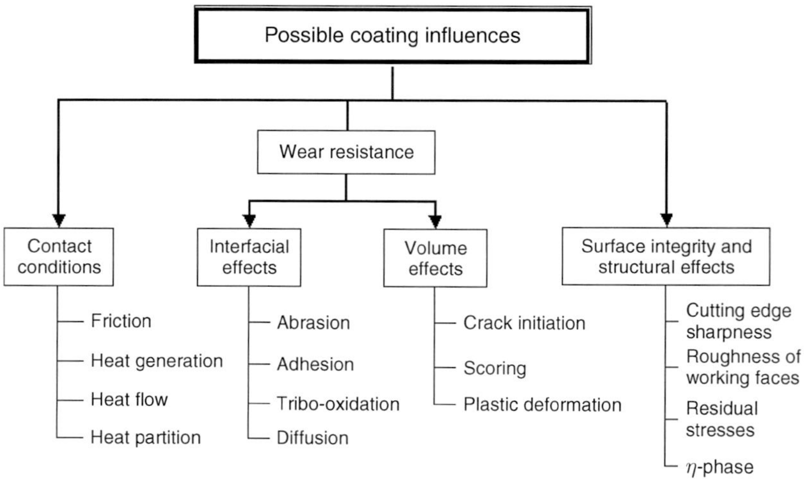

Fig. 4.15 The role of coatings in the alteration of cutting process performance. After Grzesik [11].

by increasing specific types of tool wear or indirectly by altering friction, heat generation or heat flow [14]. In particular, TiC provides superior abrasive (flank and crater) wear resistance, TiN reduces friction and resists adhesive wear and built-up edge formation, Ti(C,N) provides high fracture toughness and good abrasive wear resistance, TiAlN exhibits high hot hardness, ductility and thermal impact resistance, and Al_2O_3 provides excellent thermal isolation and oxidation resistance as well as abrasive wear and BUE resistance. In the light of multitude influences of the coatings on the contact and tribological phenomena, the choice of coating compound and its structure should be tuned to the specific tool/workpiece combination and process conditions. Figure 4.16 compares some important mechanical properties and the resistance against flank (VB) and crater wear (KT) of four selected coating materials (TiC, TiN, Ti(C,N) and Al_2O_3). In particular, an oxide ceramic layer is basically used as an intermediate layer in the multilayer/multiple structures due to lower

	HV	Br	Bo	CoF	VB	KT	T
TiC	3000	◐	●	●	●	◐	●
Al_2O_3	2300	●	◐	◐	◐	●	◐
TiN	2200	●	●	●	◐	◐	●
TiCN	2800	◐	●	◐	●	◐	●

Fig. 4.16 Comparison of mechanical properties and wear resistance of some coating compounds [3]. HV – hardness, Br – thermal barrier effect, Bo – bonding ability to the substrate, CoF – coefficient of friction, VB – flank wear, KT – crater wear resistance, T – toughness.

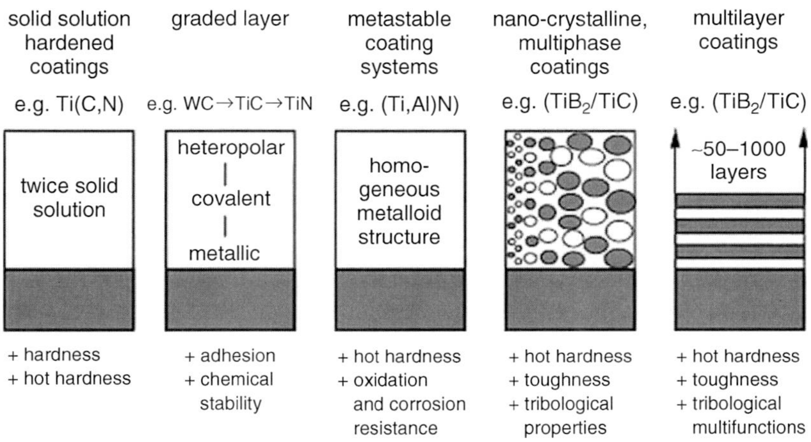

Fig. 4.17 Tailoring of coating properties in terms of structural concepts [14].

adhesive ability (Bo) to the carbide substrate. It can modify substantially the tool/chip interface behaviour and thus the dissipation of heat between sliding materials. In Fig. 4.17 the possibilities of enhancing the coating properties by deposition of differently structured coatings are presented.

The processes used to deposit coatings on cutting tools – regardless of whether the substrate is HSS, cemented carbide, cermet, ceramic, or a superhard material – are chemical vapour deposition (CVD) and physical vapour deposition (PVD), as shown in Fig. 4.18. The process selected depends on tool material composition and on the intended application of the tool. In the CVD process, the tools are heated in a sealed reactor to about 900°C (1000°C). Gaseous hydrogen and volatile compounds supply the metallic and non-metallic constituents of the coating material, which basically include TiN, TiC, TiCN and Al_2O_3. CVD is also used to produce diamond thin films ranging from ≈5 to 20 μm for graphite and non-ferrous cutting applications [11, 12]. Because the coating materials have a different coefficient of thermal expansion than carbide, tensile residual stress is induced into CVD coatings deposited on carbide substrates. One relatively new approach is to deposit pre-coating, to form a cobalt-enriched layer or induce compressive stress into the subsurface of the substrate by super-blasting. In order to avoid embrittlement and other microstructural changes characteristic of CVD coatings, the medium-temperature process (MT-CVD) was developed to allow coating deposition at lower temperatures ranging from 700 to 900°C, as shown in Fig. 4.19. Lower deposition temperatures lead to reducing residual stresses and improving the edge toughness. MT-CVD coated tools offer increased resistance to thermal shock and edge chipping, especially in interrupted

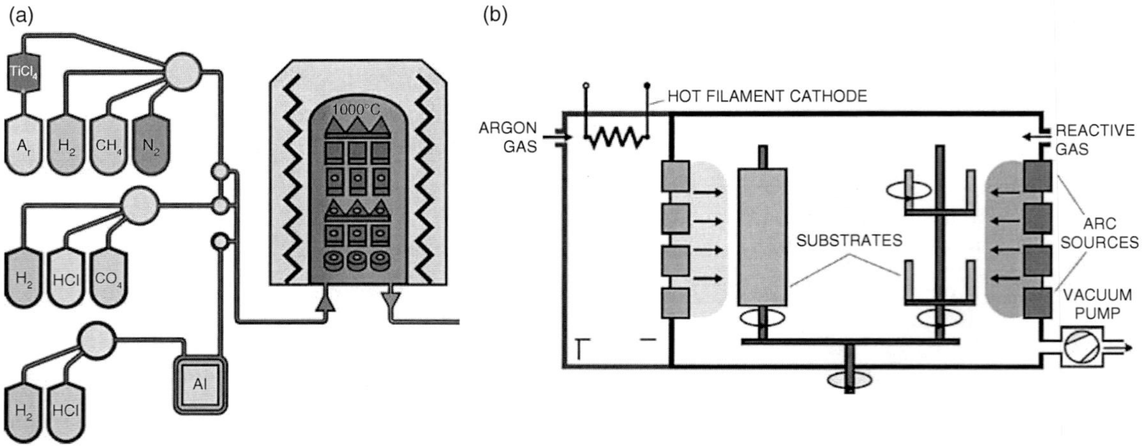

Fig. 4.18 Basic coating deposition methods: (a) CVD and (b) PVD [5].

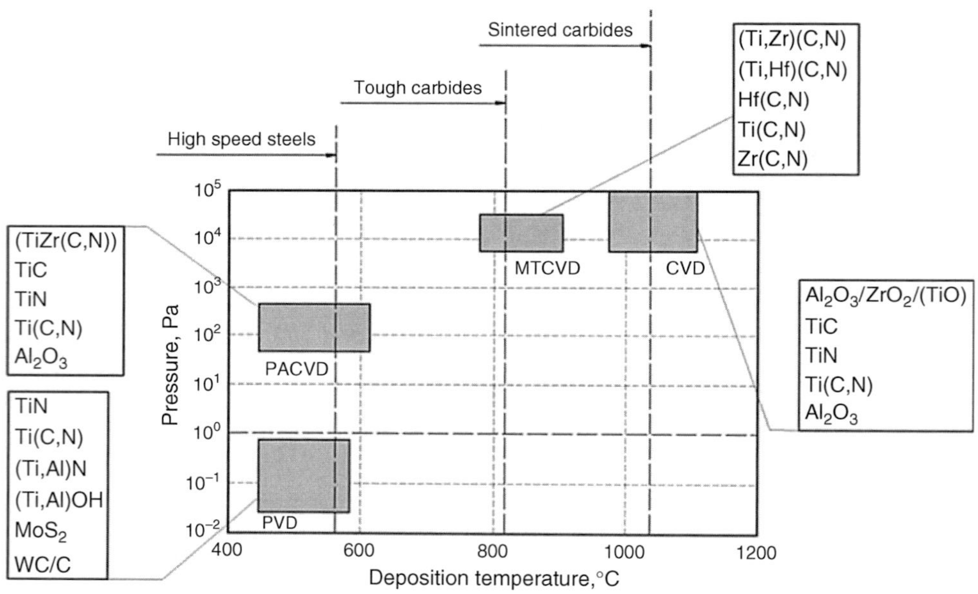

Fig. 4.19 Comparison of coating deposition methods [11].

cuts, in comparison to high temperature CVD coatings. A survey of coating deposition methods can be found in [11].

Physical vapour deposition (PVD) shown in Fig. 4.18b is the other major process used to produce cutting tool coatings. In contrast to the CVD method, in PVD the coating is deposited in a vacuum. The metal species of the coating, obtained via evaporation or sputtering, react with a gaseous species (nitrogen or ammonia) in the chamber and are deposited onto the substrate. Temperatures of about half those used in the CVD process, i.e. around 500°C, are used [11, 12]. The lower processing temperature results in a very

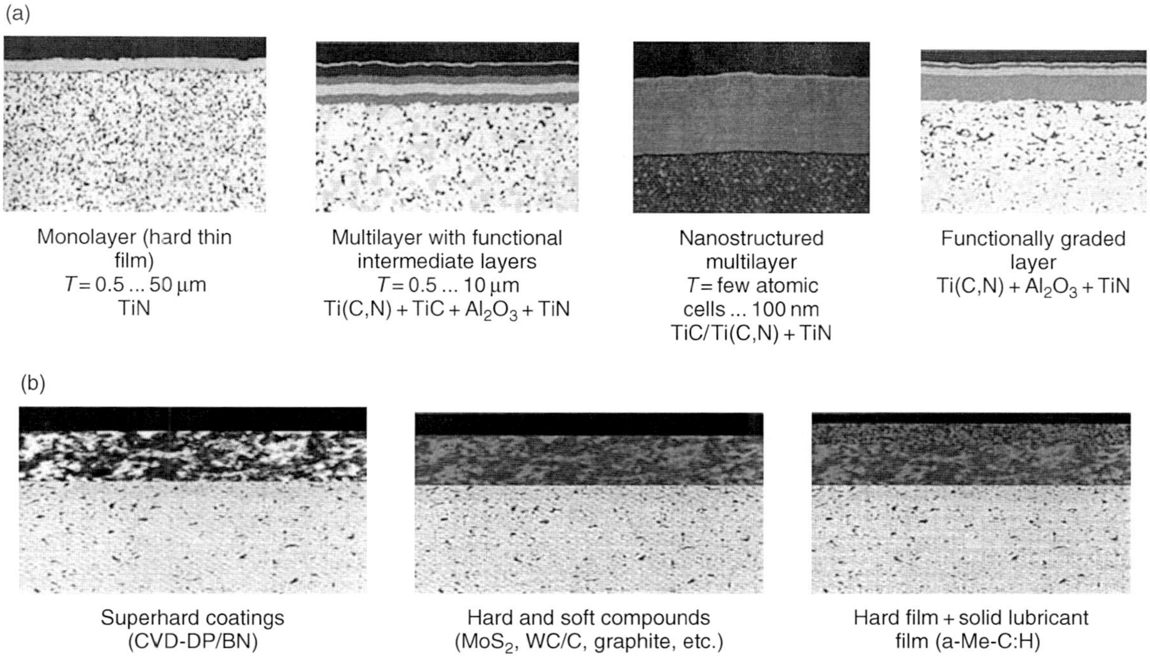

Fig. 4.20 Material-based classification of cutting tool coatings. (a) Ti-based mono- and multilayer coatings, (b) superhard and composite coatings [11]. (see Plate 1)

fine grain, free of the thermal cracks structure of the coating, very smooth, bright coating surface with a low coefficient of friction, and ability to coat sharp edges and complex chip-breaker surfaces (a PVD coating is thinner than a comparable CVD film). In addition, a brittle η-phase formation and associated substrate microporosity are minimized.

Figure 4.20 illustrates typical structures of some representative coatings selected from the groups selected above. From a material point of view, cutting tool coatings can be basically classified into the following groups [11]:

- Titanium-based coatings materials. This group includes TiN, TiC, TiB_2, Ti(C,N) and Ti(N,B) layers and materials in which the metallic phase can be supplemented by Al or Cr, as, for example, (Ti,Al)N coatings.
- Ceramic-based coatings materials like oxide ceramic Al_2O_3.
- Superhard coatings like CVD-DP/BN and DLC coatings, which compete with conventional, single-point polycrystalline diamond (PCD) tools.
- Solid lubricant coatings, such as amorphous metal-carbon, Me-C:H.
- Soft coatings deposited as a single layer or the top layer on a hard layer of TiAlN. Examples in this category are MoS_2, pure graphite and tungsten carbide/carbon (WC/C).

New developments of coatings, so-called the third generation coatings include a range of different types of complex coatings, such as multicomponent, multiphase, composite, graded, superlattice, nanocrystalline, etc. coatings, as show in Fig. 4.21.

Figure 4.22 shows characteristic examples of coatings produced by the four different deposition techniques, including: (a) CVD ($TiCN/Al_2O_3$); (b) plasma-assisted PACVD ($TiCN/Al_2O_3/TiN$); (c) PVD (TiN); and (d) MTCVD ($TiCN/Al_2O_3/TiN$). It is evident from Fig. 4.22 that their microstructures depend

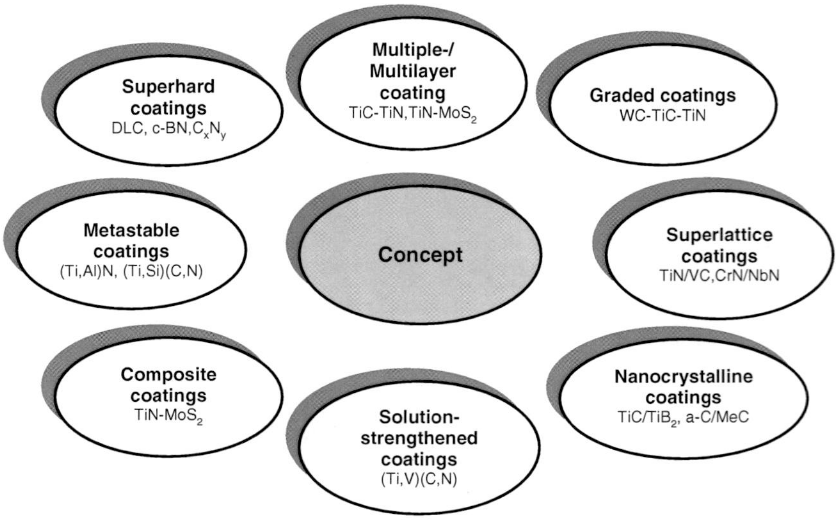

Fig. 4.21 Specification of possible concepts for hard and protective coatings [11].

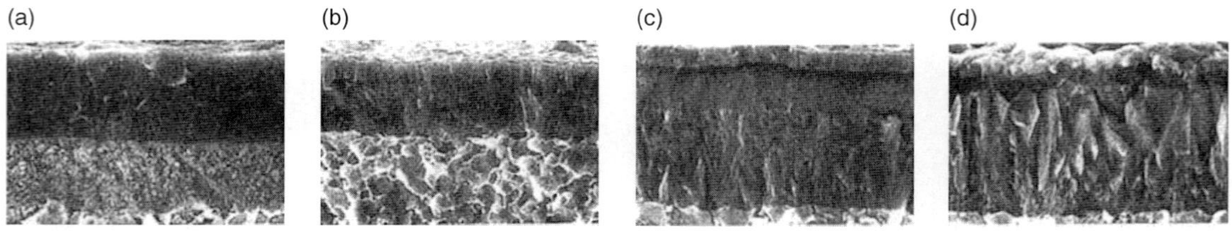

Fig. 4.22 Examples of microstructures obtained with different coating technologies [16].

on processing conditions. By adjusting process parameters, it is possible to modify the coating structure from a columnar to an equiaxed structure.

In general, proper and accurate data on the intrinsic properties of coating materials and their behaviour in exploitation is not available. The most complete databases can be found on the websites of two world class leaders in advanced coating technology, i.e. Gühring from Germany and Balzers from Liechtenstein. The most important properties and performance characteristics of the seven selected coatings along with their suggested application, i.e. recommended workpiece materials and machining operations, are listed in Table 4.6.

Some recently developed coatings for cutting tool materials other than sintered carbides are [7]:

1. PVD-TiN/TiCN/TiN of 18 microns thick on cermet substrate.
2. CVD-Al_2O_3/TiN of a total thickness 1 micron (also multiple film), MTCVD-TiCN/Al_2O_3 of a total thickness 16 microns or TiCN/Al_2O_3/TiN/TiN of a total thickness 18 microns on silicon nitride ceramic substrate. In addition, CVD-DLC/PCD coatings were deposited on β-Si_3N_4 ceramic substrates.
3. CVD-Al_2O_3/TiN on low content PCBN substrate to increase crater wear resistance when machining hardened steels.

Table 4.6 The properties of the coatings produced by Gühring [15].

Coating material	*TiN*	*TiAlN*	*TiCN*	*TiN-TiAlN*	*Diamond (D-CVD)*	*Diamond (D-ARC)*	MoS_2*-based (MOVIC®)*
Process	PVD (ion plating)	PVD (sputtering)	PVD (ion plating)	PVD (ARC)	CVD (low pressure diamond synthesis)	PVD-DLC	PVD (sputtering)
Maximum coating temperature, °C	450–500	350 (270)	450–500	450–500	1000	90	100–150
Substrates	HSS, carbides, cermet	HSS, carbides, cermet	HSS, carbides, cermet	HSS, carbides, cermet	carbides with <6% Co	HSS, carbides, cermet	HSS, carbides, cermet, nitrided, steam tempered
Structure	mono	mono	multi	multi	mono	nano-crystallic	mono
Layers (sublayers)	1	1 (2)	up to 7	up to 6	1	n/a	1
Colour	golden	black violet	violet	red violet	grey	black	dark olive
Thickness, μm	1.5–3	1.5–3	4–8	1–5	3–10	1.5 + 1	0.2–0.5
Hardness (HV 0.05)	2200	3300	3000	3600	10 000	6000	20–50
Friction coefficient (against steel)	0.4	0.3	0.25	0.3	machining steel is not possible	0.2	0.05–0.15
Heat transfer (kW/mK)	0.07	0.05	0.1	0.05	2	k.A.	<0.1
Maximum operating temperature (°C)	600	800	450	800	600	2600	800
Decoating	HSS only	HSS, carbide	in lab stage	carbide, HSS	no	yes	carbide, HSS
Typical machining operations	universal	turning, drilling	milling, drilling, tapping	universal	turning, drilling	drilling, tapping, reaming, milling	drilling, tapping, reaming, milling
Preferred machined materials	universal	cast iron, grey cast iron, Al-Si	steel	universal	graphite (Al-Si)	graphite (Al-Si)	Al, Al-Si, special alloys, steels
Characteristic	Cost effective	dry machining	interrupted cutting	dry machining, interrupted cutting	special coating	use instead of PCD tools with complicated geometries	dry machining, prevents built-up edge, water and oil resistant

4.7 Rules for Applications of Cutting Tool Coatings

In order to prevent technological difficulties and achieve a stable cutting process, the priority is to select tool grades corresponding to the workpiece properties and cutting conditions. As a general guide, Fig. 4.23 provides the choice of tool grades in terms of required properties of the tool material to achieve an optimal process performance. For example, to aggregate in one tool material such properties as hot hardness, abrasive wear resistance, plastic deformation resistance, high chemical stability and welding (attritional wear) resistance, alumina oxide ceramics will be the best choice. It is evident from this confrontation that a good competitor to ceramic tools would be relevant coated carbide grades. In sharp contrast high speed steel will be a tool material that will provided poor performance of the cutting process.

In practice, the selection of a tool grade should be made after considering a wider spectrum of factors, including at least:

- Machine tool characteristics.
- Cutting mode.
- Cutting edge geometry, tool overhang and its rigidity.
- Cutting condition.
- Shape, properties and rigidity of the workpiece.
- Required part accuracy, performance and cost.
- Use of cutting fluids.

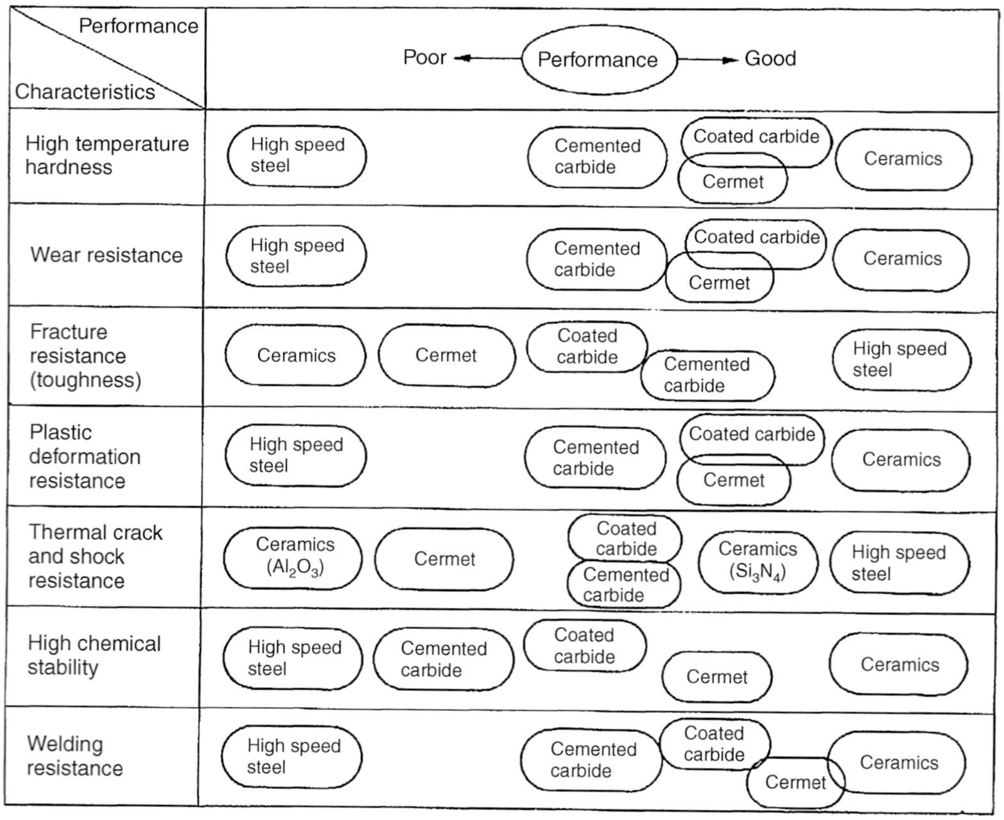

Fig. 4.23 Tool material characteristics [17].

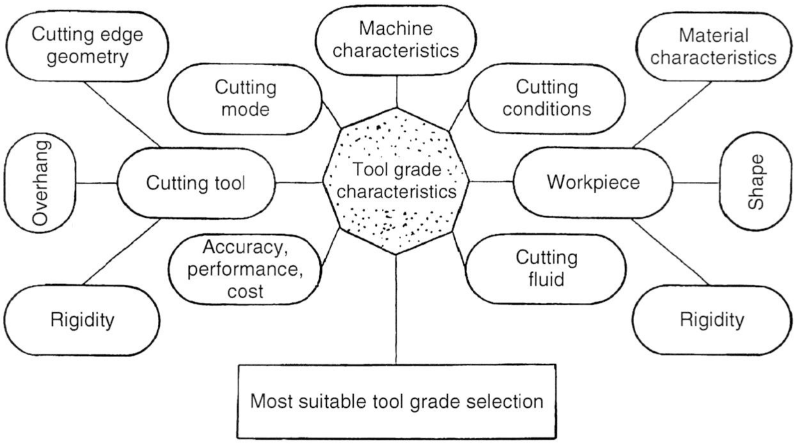

Fig. 4.24 A diagram for appropriate tool grade selection [17].

It seems to be cbvious, based on Fig. 4.24, that all these factors supply feedback on the most suitable tool grade selection.

As mentioned above, it is essential to select appropriate cutting conditions according to the workpiece, tool grade and machine tool. Figure 4.25 illustrates the relationship between workpiece material being machined, represented by its hardness, and the cutting speed range for a given tool grade.

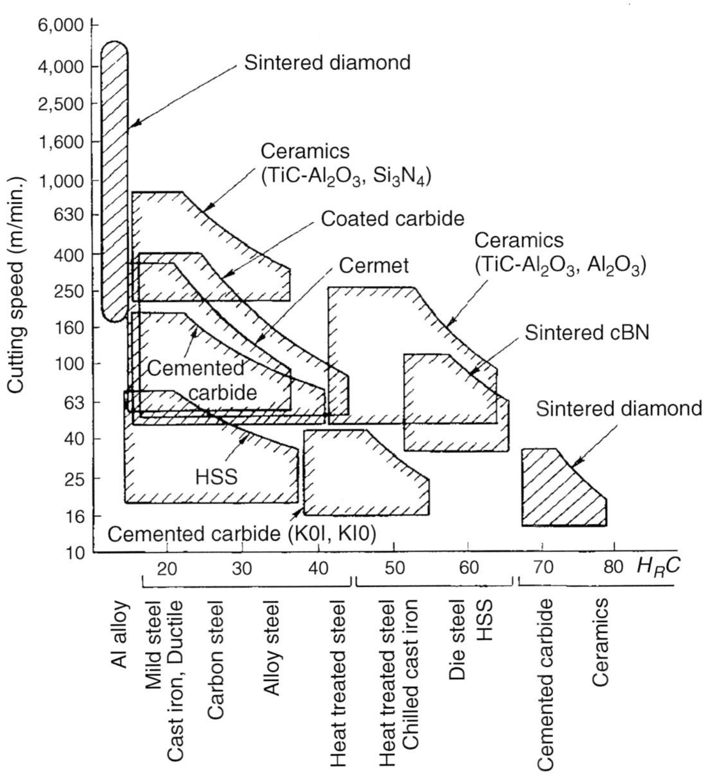

Fig. 4.25 Workpiece hardness and cutting speed application range for a range of tool grades [17].

REFERENCES

[1] M.F. Huston, G.W. Knobeloch, Cutting materials, tools, market trends in the USA, *VDI Berichte*, No. 1399 (1998), 21–45.
[2] G. Byrne, D. Dornfeld, B. Denkena, Advancing cutting technology, *CIRP Annals*, Vol. 52/2 (2003), 483–507.
[3] T.H.C. Childs, K. Maekawa, T. Obikawa, Y. Yamane, *Metal Cutting. Theory and Applications*, Arnold, London, 2000.
[4] *Modern Metal Cutting. A Practical Handbook*, Sandvik Coromant, Fair Lawn, 1996.
[5] J. Destefani, Cutting tools, Pt 1, *Manuf. Engng*, Vol. 129, No. 3 (2002), 57–69.
[6] News, *Int. J. Refract. Metals Hard Mater.*, Vol. 20 (2002), 151–167.
[7] Internet at www.cerasit.com; www.secotools.com.
[8] J. Schneider, G. Richter, Neue Entwicklungen und Anwendungen bei Schneidkeramik, *Werkstatt und Betrieb*, Vol. 132 (1999), 71–75.
[9] Introduction to De Beers PCD and PCBN cutting tool materials, De Beers Ind. Diamond Div., ABB 2000/98, England.
[10] W. Tillman, Trends and market perspective for diamond tools in the construction industry, *Int. J. Refract. Metals Hard Mater.*, Vol. 18 (2000), 301–306.
[11] W. Grzesik, *Advanced Protective Coatings for Manufacturing and Engineering*, Hanser Gardner, Cincinnati, 2003.
[12] J. Destefani, Cutting tools: coatings, Pt 2, *Manuf. Engng* (2002), www.sme.org.
[13] D.A. Stephenson, J.S. Agapiou, *Metal Cutting Theory and Practice*, Marcel Dekker Inc., New York, 1997.
[14] F. Klocke, T. Krieg, Coated tools for metal cutting – features and applications, *CIRP Annals*, Vol. 48/2 (1999), 515–525.
[15] T. Cselle, *Carbide Drills: At the Peak of Development?*, 3rd ed., Gühring, Albstadt, 1998.
[16] S. Söderberg, Advances in metal cutting tool materials, *Scand. J. Metall.*, Vol. 26 (1977), 65–70.
[17] K. Karino, *Trouble Shooting for Cutting*, Mitsubishi Materials, Ibaraki, Japan, 1998.

CHAPTER FIVE

Modelling and Simulation of Machining Processes and Operations

Contents

5.1 The Role of Modelling in Modern Production Systems

Manufacturing in the future will require much more powerful strategies for the control of processes in a highly automated manufacturing environment. Effective utilization of the newly developed manufacturing equipment depends heavily on the applications of information technology. Moreover, accurate predictions of the results of the manufacturing/machining operations are required. This approach relies strongly on the prediction of the accuracy of shape and dimensions, and the surface finish and properties of the subsurface layer of the parts produced, the machining times required and the costs of the operation performed. The rise in quality requirements with respect to precision, surface quality and subsurface integrity or internal stresses is the direct effect of increased focus on the component and its service properties. In the future, the range of surface specifications will be extended to surface microstructure and properties of subsurface zone (see Chapter 20). When successfully implemented, it can lead to minimizing the amount of scrap. In general, existing methods for the control of manufacturing operations, mainly based on the experience and craftsmanship of the manufacturing engineers/machinists, are becoming obsolete and have to be replaced by science-/knowledge-based methods. These are the main reasons for the development and introduction of CAD/CAM/CAPP systems to support process planning and control in flexible manufacturing/machining systems capable of performing a large variety of machining operations [1]. This is because, in a changing market situation, the manufacturing companies have to achieve not only productivity and quality, but also a high degree of flexibility for better satisfaction of individual customer achievements. These computer-aided systems require reliable models and simulation methods of all manufacturing/machining processes involved. A survey of highly advanced simulation methods including virtual modelling will be carried out in Chapter 19.

It is expected that future production will unquestionably require rapid and reliable adaptation to technological innovations, and these will be essential for a continued existence in the marketplace. Consequently, much more effort must be made to consider available process knowledge in order to improve component properties, and process parameters have to be based on product-related production requirements. One way

ISBN: 978-0-08-044534-2

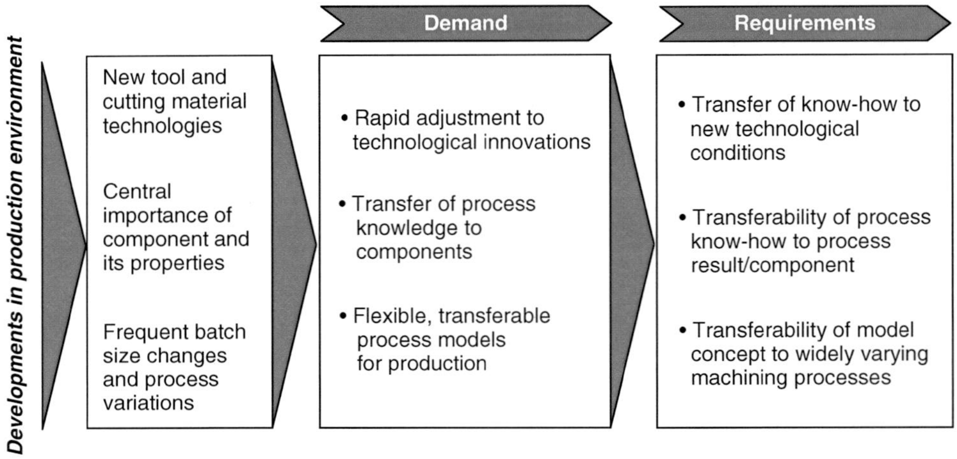

Fig. 5.1 Development trends in production, creating demand for modelling [2].

of meeting these numerous requirements is to use modelling, according to scheme shown in Fig. 5.1. The purpose of such modelling is as follows [2]:

- To find ways of optimizing processes quickly and realistically.
- To permit realistic predictions of process results.
- To obtain new knowledge of process steps and process design.
- To derive capabilities of process monitoring and control from this.

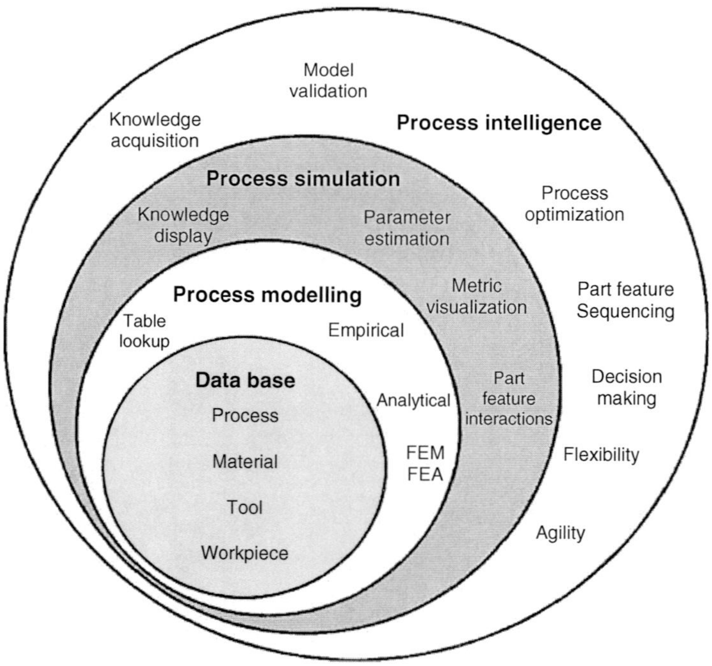

Fig. 5.2 Process data and models drive design and manufacturing decisions with increasing levels of complexity [3].

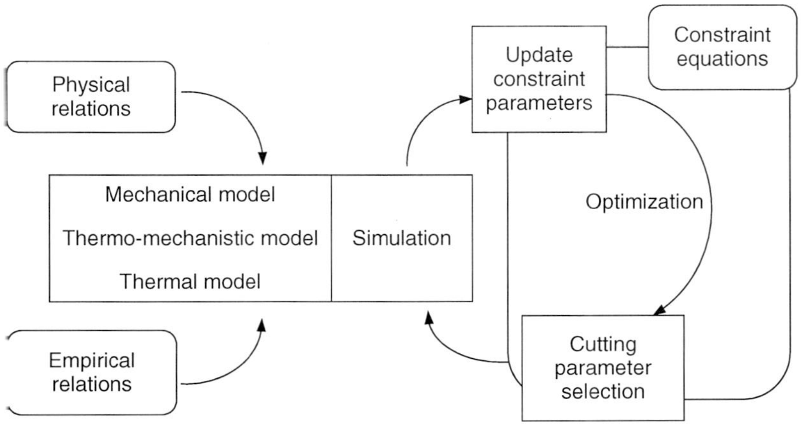

Fig. 5.3 Integration of simulation and parameter selection [4, 5].

According to Merchant *et al.* [3] manufacturing evolution, i.e. the design-to-manufacturing cycle shown in Fig. 5.2, is strongly driven by the database of information on process, tooling, materials and workpiece requirements. Evidently, process data and appropriate models can drive design and manufacturing decisions to achieve goals with increasing levels of complexity. Various models of the process, incorporating subtle details of the process capability, combined with sophisticated simulations tools and AI-based techniques, can extend the ability of the engineer even at the design stage to check his/her design effects in the process performance.

In recent years, simulation tools utilizing mechanistic and thermal models of machining processes have been proven valuable for the prediction of machining state variables over a wide range of operating parameters. However, such simulation packages are seldom an integral part of machining parameter optimization modules. Conventional approaches to machining operation planning typically rely on extensive empirical results or rule bases generated by human experts [4]. The question is, are the process simulation tools efficiently incorporated into a multivariable optimization procedure to select optimal machining parameters? As represented on the left side of Fig. 5.3, a thermo-mechanistic model combines both empirical and physical components. Simulations allow a more detailed prediction of the effects of various parameters on the cutting process than closed-form analytical predictions do. In order to directly address the planning issue the simulation tools are used as feedback to the parameter selection module, and to perform this issue an iterative optimization procedure has been developed. As a result, Fig. 5.3 schematically depicts the iterative relationship between simulation results and the optimization module when such a method is implemented. In particular, the model was tuned to fit simulation results by updating the constraint models from simulation feedback.

The key parameters that the day-to-day metal cutting practitioner finds valuable are [6, 7]:

1. Prediction of tool life.
2. Prediction of the accuracy of the component being machined.
3. Prediction of surface finish/surface integrity of the component being machined.
4. Prediction of chip control.
5. Prediction of the loads on the tool and/or workpiece and/or fixtures.

5.2 Classification of Models for Machining Processes

A model can be defined as an abstract system which is equivalent to the real system with respect to key properties and characteristics, and is used for investigations, calculations, explanation of demonstration purposes, which would otherwise be too expensive or not possible. A model permits general statement about elements, structure and behaviour of a section of reality (definition after Brockhaus, 1991) [43]. Simulation is an imitation of a dynamic process in a model in order to obtain knowledge which can be transferred to reality (definition after VDI Guideline 3663) [2]. In the machining environment, Ehmann *et al.* [6] proposed three analysis and modelling categories, i.e. analytical, experimental, and mechanistic and numerical methods to be distinguished. They noted that generally a combination of these methods is needed to obtain a working model since existing knowledge is insufficient to afford the superiority of any particular category of methods over the others.

Analytical methods were developed to relate the dynamic cutting force to: (1) the instantaneous uncut chip cross-section, (2) the shear plane area, and (3) non-linear mechanisms. In general, most of the models proposed do not predict the dynamic force accurately due to secondary non-linear effects related to the shearing mechanism, to the tool flank/workpiece interaction, and to the tool/chip interface. The plethora of experimental methods can be divided into three distinct categories: static methods, dynamic methods and time series methods. *The mechanistic modelling methods* view the machining process as a combination of the following components: the chip load/cutting force relationship, cutting tool geometry, cutting process geometry, workpiece geometry, machining conditions, tool/workpiece displacement due to cutting forces and displacement feedback which changes the cutting conditions and effective tool geometry. In addition, the mechanistic models consider the effects of process noise factors such as runout, tool wear, and inhomogeneity of the workpiece material on the process responses such as cutting forces and surface accuracy. Numerical methods will be overviewed in the second part of this section.

Van Luttervelt *et al.* [1] have given the present state-of-the-art of the complex field of modelling and simulation of metal cutting processes and operations based on various types of sources, e.g. academia, industry, literature sources, along with critical review on industrial applications of models. They considered the fact of complexity and extensiveness of modelling in the machining field. There are many different machining operations and for each of them different aspects can be modelled and different techniques for modelling can be used. Consequently, there exists a wide variety of models dealing in one way or another with machining

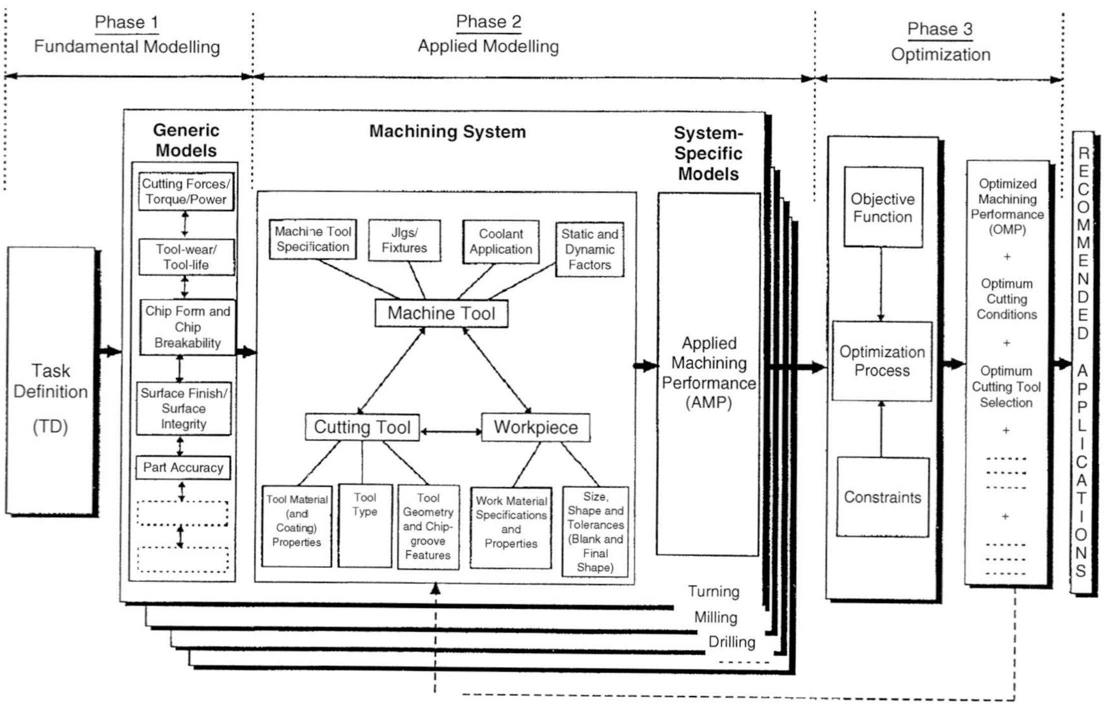

Fig. 5.4 Predictive models for machining operations in practical applications [1].

operations. In this report, models are classified according to six characteristics: type of operation, aspect, element, modelling technique, nature and capability. For example, according to the capability criterion, descriptive, predictive and learning models are selected [8].

Taking into account the modelling technique, the models can be classified as empirical, semi-empirical and physical. Predictive models, which are capable of predicting quantitatively the influence of the magnitude of one or more input parameters on the magnitude of one or more output parameters, are of special importance in the machining area. In general, among predictive models for machining operations, which require the need for experimental validation, analytical, numerical and AI-based modelling methods are the most commonly used methods. Analytical models were extensively developed in the past but their practical application is very limited due to consideration only of a few selected cutting conditions, tool geometries and work materials, as well as limited experimental validation. Moreover, further limitations result from the inaccuracies involved in the determination of the thermal and mechanical property variations of the work and tool materials at high cutting temperature and large strains and strain rates. The same kinds of limitations occur with the newer numerical and computational/simulation models and even with AI-based models.

As shown in Fig. 5.4, predictive modelling of machining operations for practical applications would consist of two phases:

Phase 1: The development of models for machining variables; and
Phase 2: The development of models for machining performance.

In the actual application, the third phase of optimization of machining conditions may follow the other two. Figure 5.5 shows the structural scheme of the cutting process combining different types of specific

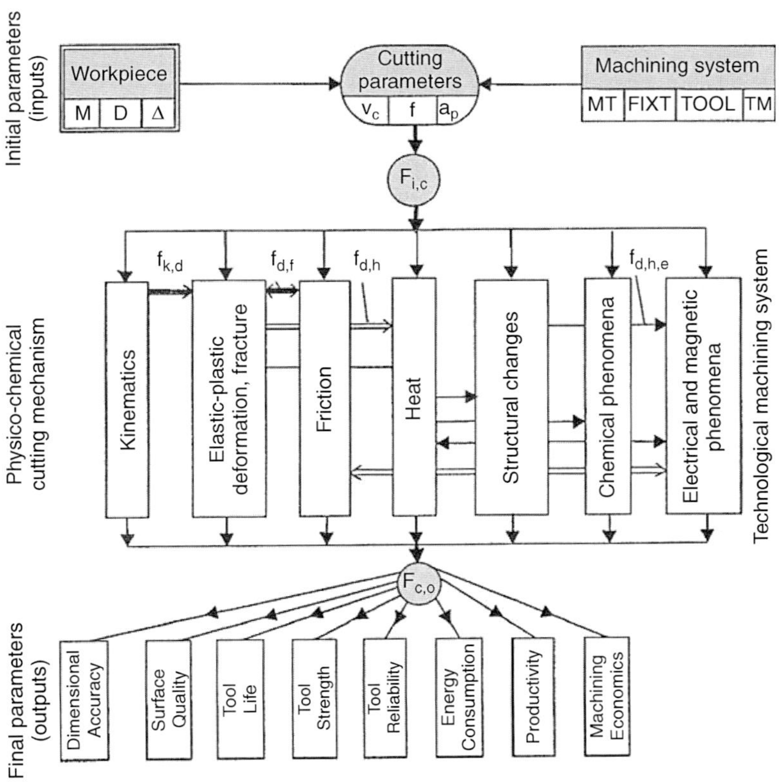

Fig. 5.5 Structural block scheme of machining process. Author version after Podurajev [2]. Element of machining system: MT – machine tool; FIXT – fixture; TOOL – cutting tool; TM – technological medium. Workpiece features: M – material grade; D – dimensions, Δ – machining allowance.

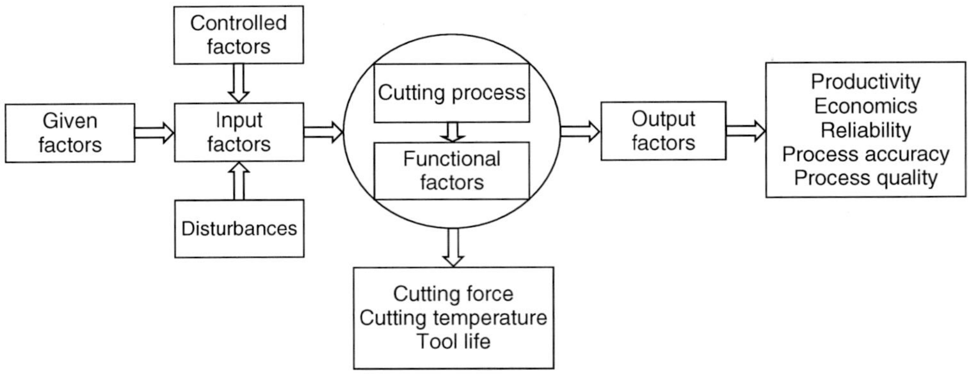

Fig. 5.6 Structural model of the cutting process. Author version after Starkov [2].

model defined by van Luttervelt [8], particularly predictive, empirical, physical and analytical models. The following relationships are distinguished in Fig. 5.5: one-sided (⇒) and two-sided (⇔); dominant (⇔), ordinary (⇔) and minor (secondary) (→) as well as direct and indirect (implicit) relationships. In general, the relations of the first type between input variables and cutting characteristics, and between the cutting characteristics and output variables ($F_{c,o}$), as well as the relations of the second type determined between different physical phenomena/mechanisms involved in the cutting process (for example, physical models such as $f_{d,h}$ between plastic deformation and the amount of heat or more complex indirect relationship $f_{d,h,e}$ which consider also the electric/thermoelectric effects occurring during the cutting process) are selected. It is also possible to aggregate ($F_{c,o}$) and ($F_{i,c}$) models in order to generate ($F_{i,o}$) types of models to establish, for instance, the function: workpiece material grade/cutting temperature/tool life. Implicit relations, similar in nature to mechanistic models, are very important for optimizing the complex dynamic influences, particularly the influence of cutting vibration on the tool life.

A simplified structural model of the cutting process is illustrated in Fig. 5.6. In this model, all factors influencing the cutting process are divided into given factors (work material grade, cutting operation, accuracy and quality requirements), controlled factors (cutting tool material, design and geometry of the cutting tool, machine tool type, cutting parameters, cooling method and the amount of cooling medium) and disturbances classified as systematic and random influences (for example, changes of static and dynamic features of the technological machining system).

5.3 Modelling Techniques for Machining Processes

For the past 50 years metal cutting researchers have developed many modelling techniques including analytical techniques, slip-line solutions, empirical approaches and finite element techniques. In this book modelling methods will be successively employed for description, prediction and simulations of the elements of machining processes/operations in five generic categories:

1. Empirical modelling typified by equations of cutting forces, cutting temperature, tool life and surface roughness.
2. Analytical and hybrid modelling represented by different solutions of the shear plane theory (see Chapter 7).
3. Mechanistic modelling typified by analyses of chip control and static and dynamic forces (Chapters 6, 7 and 8).
4. Numerical modelling using different modelling techniques (Chapters 6, 7, 9 and 12).
5. AI-based modelling (Chapters 14 and 19).

The first three modelling methods are discussed when considering the particular machining problems (AI-based techniques will be applied for optimization procedures in Chapter 14), and now a short characterization of numerical modelling/simulation techniques, which are widely used in the metal cutting modelling sector, will be provided.

According to the results of a survey covering the dissemination of modelling techniques for machining operations [2], about 18% of this specific research activity is engaged in numerical modelling. These statistics have distinguished three techniques which can be applied to solve the heat transfer problem in metal cutting, namely: the finite element (FEM), the finite difference (FDM) and the boundary element (BEM) methods. In recent years, the finite element method has particularly become the main tool for simulating metal cutting processes [9–11]. Finite element models are widely used for calculating the stress, strain, strain rate and temperature distributions in the primary, secondary and tertiary subcutting zones. In consequence, temperatures in the tool, chip and workpiece, as well as cutting forces, plastic deformation (shear angle and chip thickness), chip formation and possibly its breaking can be determined faster than by using costly and time-consuming experiments.

Typical approaches for numerical modelling of metal cutting processes are Lagrangian and Eulerian techniques, as well as a combination of both called the arbitrary Lagrangian–Eulerian formulation (denoted in the literature by the acronym ALE) [9, 12]. It should be noticed that all these methods are mathematically equivalent. Figure 5.7 illustrates these two options. The major feature of Lagrangian formulation used in this study is that the discretized mesh is 'attached to the workpiece' and the material model is elastic–plastic, only plastic, and viscoplastic or additive. In the Eulerian case, attention is drawn to how velocities vary from element to element (for example, elements 1 and 2) at the same time. In the Lagrangian case, attention is focused on how the velocity of a particular element varies with time.

Early finite element analyses were performed by Usui and Shirakashi [14], Iwata *et al.* [15] and Strenkowski and Carroll [16], who were the first to use Eulerian formulations for steady-state metal cutting simulations. On the other hand, Marusich and Ortiz [17] have developed a Lagrangian formulation in which the material model contains deformation hardening, thermal softening and strain-rate sensitivity tightly coupled with a transient heat conduction analysis appropriate for finite deformations. Most of the early investigations on the FEM modelling of orthogonal cutting were limited by the assumption of a perfectly sharp cutting edge which satisfies the fundamental 2D cutting model proposed by Ernst and Merchant. In recent years, a few trials were undertaken [18] to extend the FEM modelling technique (for example, the Lagrangian thermo-viscoplastic cutting simulation) to real non-sharp cutting tool geometries including round/hone and T-land/chamfer edges. It is especially important that FEM analysis can help to investigate some thermodynamical effects occurring in the cutting zone which, so far, cannot be measured directly [10]. An example for such effects is the influence of cutting tool coatings on the heat transfer and friction, and resulting cutting temperature distribution in the chip and the tool.

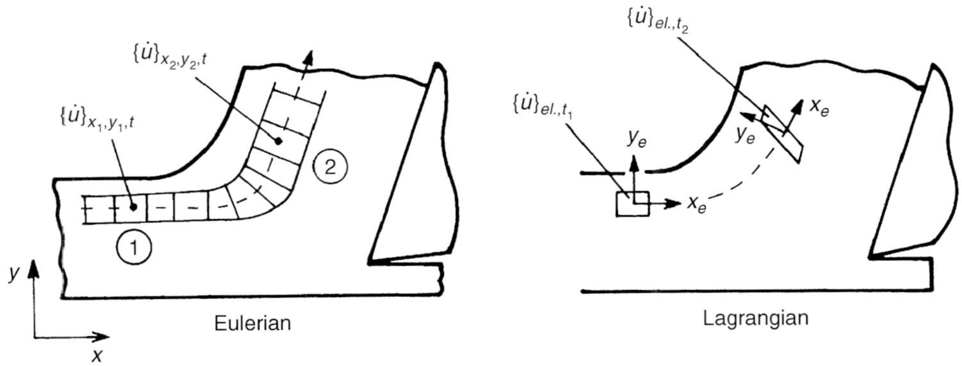

Fig. 5.7 Eulerian and Lagrangian representation of a plastic flow in the PDZ [13].

It should be noted that the majority of previous orthogonal metal cutting simulations is devoted to uncoated carbide tools and now the opposite trend to consider single and multiply coatings has been observed [10, 19–21]. In fact, the first completed work focusing on the evaluation of a predictive orthogonal cutting model for coated carbide tools with multiple coating layers using the FEM was presented by Yen *et al.* [20]. In this model the thermal properties of the three-layer TiC/Al_2O_3/TiN were implemented by considering individual coating layers and a composite monolayer coating with substitute (equivalent) thermal conductivity and diffusivity. Separate Lagrangian and Eulerian calculations were used to obtain the steady-state solution of temperature in the tool and chip/workpiece.

Usui and Shirakashi [13, 14] have developed the iterative convergence method (ICM) in which the shape of the pre-form changed until the generated plastic flow was consistent with the assumed shape. They also included realistic chip/tool friction conditions (from split-tool experiments), a temperature as well as a mechanical calculation, and material flow stress variations with strain, strain rate and temperature, measured from high strain rate Hopkinson bar tests. The ICM method is an updated Lagrangian elastic–plastic finite element analysis for predicting steady-state chip flows. Figure 5.8a shows an ICM mesh for two-dimensional machining with a single-point tool, in which the X and Y axes are set parallel and perpendicular to the cutting direction, in a rectangular Cartesian coordinate system. The tool is assumed to be stationary and rigid, while the workpiece moves towards it at the specified cutting speed. As can be seen in Fig. 5.8a the mesh is highly refined in the primary and secondary shear zones. The separation of material at the cutting edge is taken into account geometrically. Figure 5.9 compares some types of discretization of the cutting zone, available in some popular commercial FEM packages, such as DEFORM, ABAQUS, I-FORM2, FORGE2 and AdvantEdge.

In general, four different approaches to formulating and calculating the properties of individual elements (grids) can be applied in the FEM analysis, namely:

1. *Direct approach* based on a direct stiffness method of the structure.
2. *Variational approach* which relies on the calculus of variations and involves extremizing the functions.
3. *Weighted residuals approach* which uses the governing equations of the problem and proceeds without relying on a functional or a variational statement.
4. *Energy balance approach* which relies on the balance of thermal and/or mechanical energy of a system.

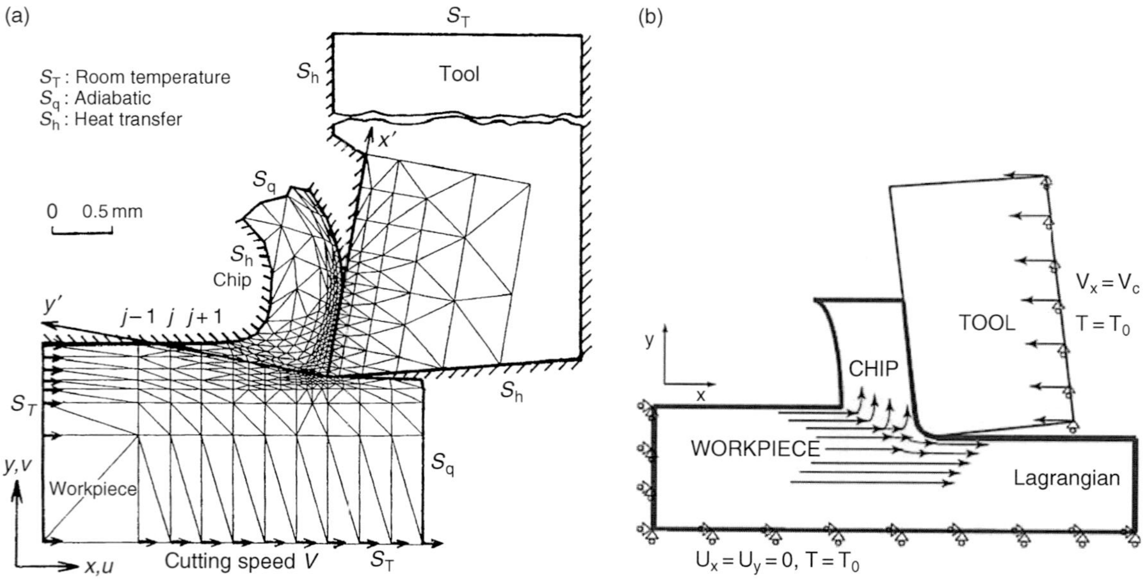

Fig. 5.8 Two-dimensional finite element assemblage with boundary conditions [13] (a) and simulation model used for ALE with Lagrangian boundary conditions [22] (b).

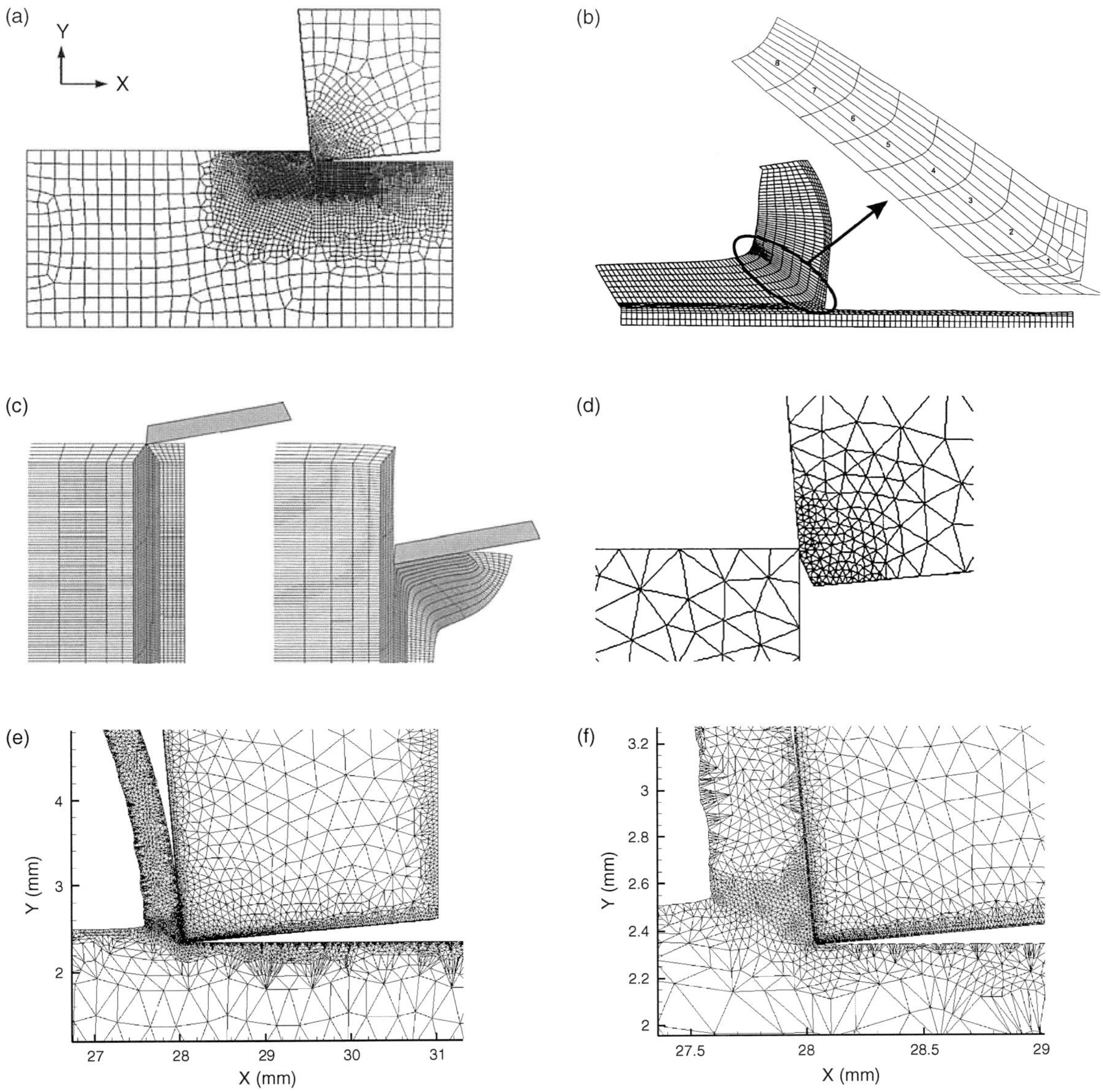

Fig. 5.9 Typical finite element meshes used in modelling the cutting zone, generated in some popular commercial FEM packages: (a) DEFORM-2D® by Yen *et al.* [23], Umbrello *et al.* [24] and Jawahir *et al.* [25], (b) ABAQUS™ by Shin *et al.* [26] and Kalhori [27], (c) I-FORM2 by Martins *et al.* [28], (d) FORGE2® by Ng *et al.* [29], (e) and (f) AdvantEdge by Kalhori [27] and Grzesik *et al.* [30].

In a Lagrangian formulation, the finite element analysis leads to a system equation for the nodal displacement vector $\{u\}$ of the system as follows:

$$[K]\{u\} = \{R\} \tag{5.1}$$

where $[K]$ is the global stiffness matrix and $\{R\}$ is the load vector which includes all its applied loads.

The solution of the system of the equation yields the nodal displacement vector $\{u\}$ from which the element strains and hence stresses can be calculated.

In an Eulerian formulation the constitutive law becomes a viscous flow equation relating flow stress to the instantaneous strain rates:

$$\{\sigma\} = [C\{(\mu)\}]\{\dot{\varepsilon}\} \tag{5.2}$$

where μ is the viscosity which is dependent on the strain rate.

The global matrix for all elements in the chip and workpiece is found by summing the contribution for each element. This results in the following equation:

$$[K]\{U\} = \{P\} \tag{5.3}$$

where $[K]$ is the global stiffness matrix, $\{U\}$ represents the nodal velocities and $\{P\}$ is the vector of all applied loads.

The flow chart in Fig. 5.10 illustrates the finite element procedure leading to the solution of the matrix equations for displacements.

The finite difference method/approach (FDM/FDA) can be used to solve the heat transfer problem in the cutting zone and generate the temperature fields in the chip and the selected area within the tool. It is assumed that for the idealized model of the cutting process the heat sources resulting from shearing at the shear plane and friction at the tool/chip interface are represented by one plane, uniformly distributed heat source at the tool/chip interface. For the two-dimensional model (the isothermal surface) the temperature T at the arbitrary selected point $P(x, y, t)$ will be a continuous function of coordinate axes x and y at time t.

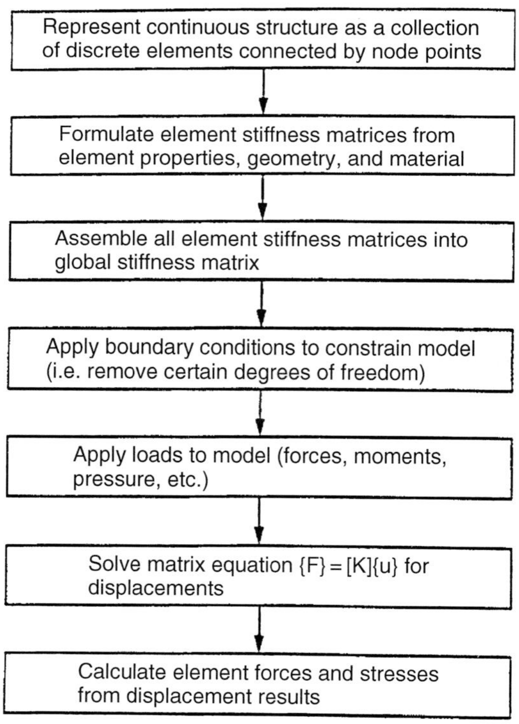

Fig. 5.10 Overview of finite element solution procedure [7].

Thus, the governing equation (Fourier–Kirchhoff equation) in Cartesian coordinates for a steady-state heat conduction can be expressed as [31]:

$$\frac{\partial T}{\partial t} = \alpha \nabla^2 T + \frac{\dot{q}_v}{c_p} \tag{5.4}$$

or

$$v_c \frac{\partial T(x, y, t)}{\partial x} = \alpha \left[\left(\frac{\partial^2 T(x, y, t)}{\partial x^2} \right) + \cdots \cdots + \left(\frac{\partial^2 T(x, y, t)}{\partial y^2} \right) + \frac{\dot{q}_v(x, y, t)}{\lambda} \right] \tag{5.5}$$

where $\dot{q}$ is the heat flow rate, ∇T(in Km^{-1}) is the local temperature gradient (Hamilton's vectoral operator-nabla), and λ is the thermal conductivity, $a = \lambda/(c_p \rho)$ is the thermal diffusivity of the substance, c_p is specific heat capacity and v_c is the chip velocity ($\frac{\partial T}{\partial t} = v_c \frac{\partial T}{\partial x}$).

In the case of steady heat flow for which the derivative $\frac{\partial T}{\partial t} = 0$ and the absence of the net heat flow into the element, Eqn (5.5) can be simplified to the Laplace's formula as:

$$\frac{\partial^2 T(x, y)}{\partial x^2} + \frac{\partial^2 T(x, y)}{\partial y^2} = 0 \tag{5.6}$$

Therefore, Eqn (5.6) considers the case of a solid through which heat is flowing, but within which no heat is generated (in Eqn (5.4) $\dot{q}_v = 0$). The circumstances encountered in metal cutting, especially when the Peclet number is greater than 10 [32], result in a relatively small $\partial T/\partial x$ and conduction in the direction of heat source motion can be neglected [33]. Then, a final analytical model for the transient heat transfer in the chip is [34, 35]:

$$\frac{\partial^2 T}{\partial y^2} - \frac{R}{a_c} \frac{\partial T}{\partial x} = 0 \tag{5.7}$$

where $R = c_p \rho v a/\lambda$ is the thermal number, ρ is the density in $\mathrm{kg\,m^{-3}}$, v is the cutting velocity, a is the undeformed chip thickness and a_c is the chip thickness.

The heat conduction rates in two orthogonal directions x and y ($\dot{q}_x$, $\dot{q}_y$) can be evaluated from Fourier's heat conduction law as [7]:

$$\dot{q}_x = -\lambda (dy) \frac{\partial T}{\partial x}, \dot{q}_y = -\lambda (dx) \frac{\partial T}{\partial y} \tag{5.8}$$

Therefore, the partial differential equation (5.6) can be rewritten for the mesh shown in Fig. 5.11 in the following finite difference form [35]:

$$\frac{T_{i-1,j} - 2T_{i,j} + T_{i+1,j}}{(\Delta x)^2} + \frac{T_{i,j-1} - 2T_{i,j} + T_{i,j+1}}{(\Delta y)^2} = 0 \tag{5.9}$$

The simplified difference equation (5.9) can be solved numerically using the boundary conditions described below. The uniqueness of the two-dimensional model is defined by three conditions, i.e. geometrical, thermophysical and limiting conditions. Further, among the last group we differentiate the following special conditions, namely:

The FDM model shown in Fig. 5.11 is built as a set of cells including adequate formulae and values, which finally represents both geometrical and physical features of the modelled process. Formulae are introduced according to the explicit method of the difference procedure. Moreover, the process conditions for zero intrinsic heat source $\dot{q}_v = 0$ and a steady two-dimensional heat flow problem were assumed.

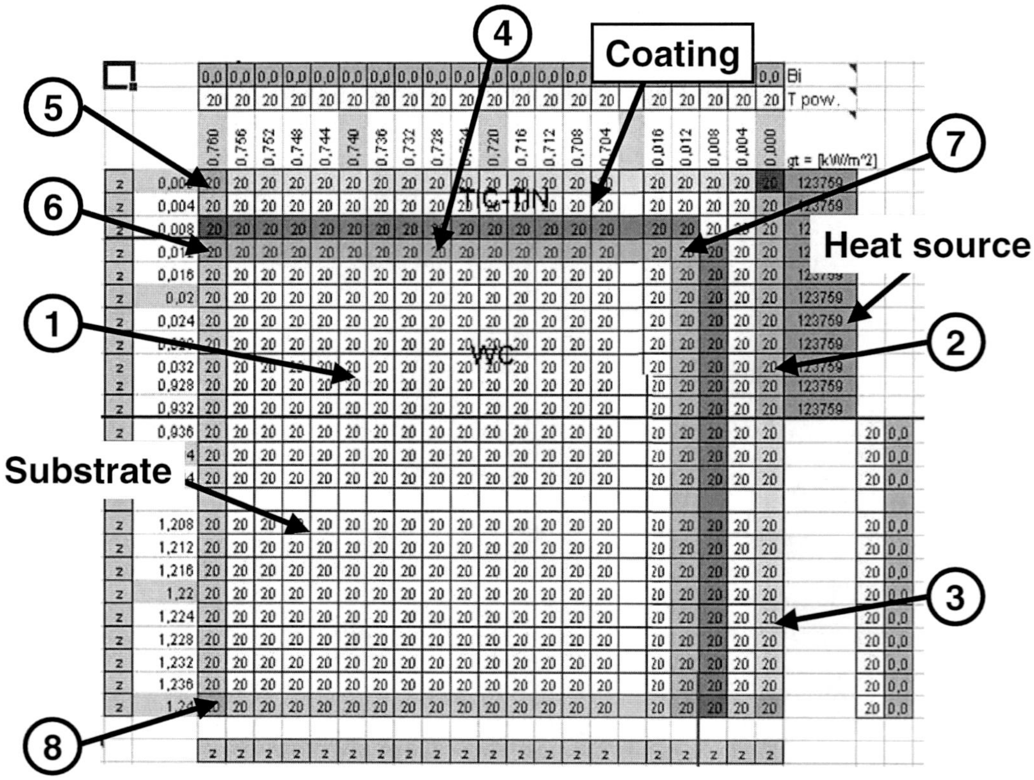

Fig. 5.11 FDM mesh model of heat transfer into coated tool with marked boundary conditions [36]. (see Plate 2)

In particular, the formulae used in the heat flow model from Fig. 5.11 are as follows:

1. Formulae inside the calculation field.
2. Linear boundary formulae defined by Neumann's condition (when heat source is localized at the tool/chip interface).
3. Linear boundary formula with the Fourier's condition (for the tool surface cooled by surrounding air).
4. Linear boundary formula with the condition of IVth type (it is assumed that an ideal thermal contact $R = 0$ exists on the boundary between adjacent coatings).
5. Corner boundary formula with Fourier/Neumann's condition (two sides of coating contact surrounding air).
6. Corner boundary formula with the condition of Neumann/IVth type (boundary cell being in contact with surrounding air and coating/substrate from the same side).
7. Corner boundary formula with the condition of mixed IVth/IVth type (internal corner of the contact between two different materials – sintered carbide and coating).
8. Corner boundary formula with mixed Neumann/Neumann's condition (two sides of carbide substrate being in contact with surrounding air).

All boundary conditions specified above are schematically illustrated in Fig. 5.11. It should be noted that for all places where the discrete linear boundaries of the model exist we can use linear formulae.

5.4 DATA NEEDED FOR MODELLING OF MACHINING PROCESSES

In order to obtain reliable results from numerical (FE, FD, BE) simulations, it is necessary to determine, as input, the properties of the workpiece and tool materials, as well as the characteristics of the tool/chip interface. These input data include physical and thermal data, friction and heat transfer, and, most importantly, the flow stress of the workpiece material under high strain, strain rate and temperature conditions that proximately exist during the process.

For example, to allow performing a successful simulation of different machining processes and operations, a material property database termed *MA*terial *DA*tabase for *M*achining *S*imulation (MADAMS) has been developed at the Engineering Research Center for Net Shape Manufacturing, the Ohio State University, USA [37]. In addition, the author and coworkers have developed at Opole University of Technology, Poland, a thermal property database termed *T*hermal *P*roperty *D*atabase for *M*achining *S*imulation (TPDMS), oriented on the thermal properties of both workpiece and tool materials, especially on cutting tool coating components.

Table 5.1 specifies the mathematical forms of the most commonly used constitutive equations to represent flow stress data. The accuracy of the results from FE simulation strongly depends on the accuracy of flow stress

Table 5.1 Mathematical expressions for material constitutive models [38].

Material model	*Equivalent material flow stress*
Johnson–Cook	$$\sigma_{eq} = \Big(\underbrace{A + B\varepsilon_p^n}_{\text{I}}\Big)\left(\underbrace{1 + C\ln\left(\frac{\dot{\varepsilon}_p}{\dot{\varepsilon}_p^0}\right)}_{\text{II}}\right)\left(1 - \underbrace{\left(\frac{T - T_r}{T_m - T_r}\right)^m}_{\text{III}}\right) \quad (5.10)$$ where: I – elastic–plastic term; II – viscosity term; III – softening term; ε_p – the equivalent strain, $\dot{\varepsilon}_p$ – the equivalent strain rate; $\dot{\varepsilon}_p^0$ – the strain rate equal to $1.0\,s^{-1}$, T – the absolute temperature; T_r – the room (ambient) temperature, T_m – the melting temperature of a specific metal or alloy, A, B, C, n, m – five material constants
Zerilli–Armstrong	I. For bcc (body-centred-cubic) metals: $$\sigma_{eq} = C_o + C_1 \exp\left(-C_3 T + C_4 T \ln\left(\tfrac{\dot{\varepsilon}}{\dot{\varepsilon}_0}\right)\right) + C_5\varepsilon^n \quad (5.11)$$ II. For fcc (faced-centred-cubic) metals: $$\sigma_{eq} = C_o + C_2\varepsilon^{0,5} \exp\left(-C_3 T + C_4 T \ln\left(\tfrac{\dot{\varepsilon}}{\dot{\varepsilon}_0}\right)\right) \quad (5.12)$$ where: ε, $\dot{\varepsilon}$, $\dot{\varepsilon}_0$ are the same as in Eqn (5.10), C_0, C_1, C_2, C_3, C_4, C_5, n – seven material constants
Oxley	$$\sigma_{eq} = \sigma_1\varepsilon^n \quad (5.13)$$ $$T_{\text{mod}} = \left(1 - \nu \cdot \log\tfrac{\dot{\varepsilon}}{\dot{\varepsilon}_0}\right) T \quad (5.14)$$ where: σ_1 – the material flow stress at $\varepsilon = 1,0$ and n is the strain-hardening index. Values of σ_1 and ε depend on the velocity modified temperature $T_{mod,}$, ν, $\dot{\varepsilon}_o$ – two material constants for a given material and range of testing conditions

(Continued)

Table 5.1 (Cont.)

Material model	*Equivalent material flow stress*
Maekawa *et al.*	$\sigma_{eq} = A\left(\frac{\dot{\varepsilon}}{1000}\right)^{M} e^{aT}\left(\frac{\dot{\varepsilon}}{1000}\right)^{-m}\left[\int_{D_\varepsilon} e^{-aT/N}\left(\frac{\dot{\varepsilon}}{1000}\right)^{-m/N} d\varepsilon\right]^{N}$ (5.15) For a constant strain rate and temperature path, Eqn (5.6) simplifies to $\sigma_{eq} = A\left(\frac{\dot{\varepsilon}}{1000}\right)^{M} \varepsilon^{N}$ (5.16) where: D_ε– strain path, A, M, N, a, m – five material constants
Marusich (Thirdwave AdvantEdge)	$\overline{\sigma} = \sigma_f(\varepsilon^p)\cdot\left(1+\frac{\dot{\varepsilon}^p}{\dot{\varepsilon}^p_o}\right)^{\frac{1}{m_1}}$ (5.17) where: $\overline{\sigma}$ is the effective von Misses stress, σ_f is the flow stress, ε^p is the accumulated plastic strain, $\dot{\varepsilon}^p_o$ is a reference plastic strain rate, m_1 is the strain rate sensitivity exponent $\sigma_f = \sigma_o\cdot\Theta(T)\cdot\left(1+\frac{\varepsilon^p}{\varepsilon^p_o}\right)^{1/n}$ (5.18) where: σ_o is the initial yield stress at the reference temperature T_o, ε^p_o is the reference plastic strain, n is the hardening exponent and $\Theta(T)$ is the thermal softening factor ranging from 1 at ambient to 0 at melt

data. To be useful in metal cutting simulation, flow stress data must be obtained at high strain rates (up to $10^6\,s^{-1}$), strains up to 4 and temperatures up to 1000°C [37]. Today, flow stress data can be experimentally obtained mainly using five methods: high speed compression tests, Split-Hopkinson's Pressure Bar (SHPB) tests, practical machining tests, integration of conventional tests at low strain rates and machining tests (see section 6.4), and inverse analysis using the FEA technique. It is obviously known that this data cannot be obtained reliably with conventional tensile and compression tests.

The production of photomicrographs of the chip formation zone is usually limited to low cutting velocities. With the aid of the modified SHPB test, it is possible to perform an analogous test allowing for pictures of chip roots at speeds up to 2000 m/min. The production of such photomicrographs of the chip root at high cutting velocities is of fundamental importance for the investigation of the chip formation process during high speed cutting, since the knowledge of the mechanical behaviour of metallic materials at conventional cutting velocities cannot be transferred directly to the high speed domain. Figure 5.12 shows the principal setup of the Split-Hopkinson Pressure Bar apparatus with quick-stop device [37]. The Split-Hopkinson bar consists of two long rods. An elastic mechanical wave is induced in the input bar using an accelerated mass. At the end of the input bar, the test piece is fixed. On the output bar, two cutting inserts are mounted. When the elastic wave reaches the end of the input bar, the test piece is moved into the cutting inserts and two chips are formed. The movement is stopped in a few microseconds by two stoppers which are connected with the input bar.

With strain gauges on the input and output bar, the cutting force can be measured. By the use of high pressure gun and faster heat rate generated by an induction coil, it is possible to obtain material data at strain rates up to $2\times10^3\,s^{-1}$, by further modifying the sample geometry (see circled insert) the strain-rate boundary can be elevated up to $10^4\,s^{-1}$. However, these strain-rate values are still too low to satisfy those of up to $10^6\,s^{-1}$, occurring in the majority of HSM and abrasive processes.

Figure 5.13 shows the comparison, with strain ε, strain rate $\dot{\varepsilon}$, and temperature T varying within a very wide range ($\varepsilon = 0.8$–3.8, $\dot{\varepsilon} = 10^2$–$10^5\,s^{-1}$ and $T = 100$–1000°C). It is seen from Fig. 5.13 that predictions from the Johnson–Cook (JC) model are very close to those obtained from the Oxley model in most cases. However, unlike the Oxley model, the JC model does not consider the dynamic strain-aging (blue-brittleness) effect during a certain range of temperature variations in the plastic deformation of AISI 1045 carbon steel, as

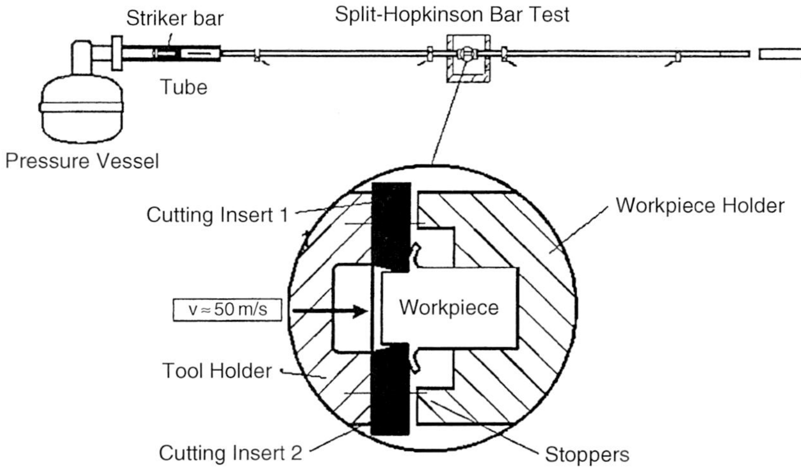

Fig. 5.12 Split-Hopkinson's Pressure Bar apparatus [37].

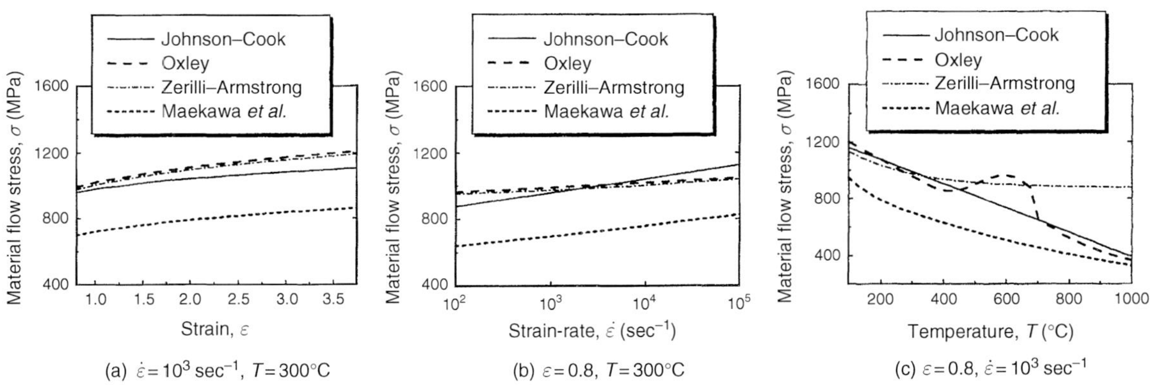

Fig. 5.13 Comparison of four material constitutive models for AISI 1045 carbon steel [38].

depicted in Fig. 5.13c. It should be noted, based on the sensitivity analysis performed for 18 engineering materials by Fang [38], that the first predominant factor governing the material flow stress is either strain hardening or thermal softening, depending on the specific material employed and the varying range of temperatures.

Childs has noted [39] that despite differences in their structure, the material constitutive laws give similar numerical values of flow stress under conditions in the primary shear zone, but show large differences in the secondary shear zone. This is due to the fact that in the secondary shear zone, there is not only further pure strain hardening at high strain rate, but external heating due to intensive friction. Maximum temperatures in this region, when machining steels, can easily rise to 1000°C and above. The heating occurs in milliseconds, so thermal equilibrium microstructures do not develop. In consequence, flow stresses predicted by extrapolating observations from lower temperatures vary strongly with the assumed flow stress model (Oxley model, JC model and Zerilli and Armstrong model). This discrepancy suggests a need for more accurate flow stress as well as friction input data for FE modelling. Later, Childs has documented [39] by numerous numerical experiments that varying strain-rate hardening and particularly thermal softening characteristics do not lead to changed friction and shear angle relationships. The effect of strain hardening is to reduce $(\Phi + \Theta - \gamma_o)$. In particular, the normal stress on the shear plane peaked once $(\Phi + \Theta - \gamma_o) > 51^0$.

In general, modelling of tool/chip friction in FE simulation is based on Coulomb's friction law defined by a pair of equations:

$$\bar{\tau} = \mu\sigma_n, \mu\sigma_n \leq \bar{\tau}_{max} \quad \text{(sliding region)} \tag{5.19a}$$

$$\bar{\tau} = \bar{\tau}_{max}, \mu\sigma_n \geq \bar{\tau}_{max} \quad \text{(sticking region)} \tag{5.19b}$$

where $\bar{\tau}$ is the shear flow stress of the chip, μ is the friction coefficient and σ_n is the normal stress along the tool rake face. It is assumed in commercial FE codes that the mean value of friction coefficient equals to 0.2 (ABAQUS) or 0.5 (AdvantEdge).

The third group of input data for numerical simulations involves thermal properties of the workpiece and cutting tool material, such as the thermal conductivity (λ) and the thermal diffusivity (α). It was revealed by many researchers including author [30, 40, 41] that the most successful simulation results were obtained by using the temperature-dependent values of λ and α. Recently, the author and coworkers have applied to this principle the so-called Laser Flash Method (LFM) developed by Netzsch, Germany [42].

A scheme of laser flash apparatus along with the sample holder system is shown in Fig. 5.14. The sample is mounted on a carrier system which during the measurements is located in a furnace. After the sample reaches a predetermined temperature, a burst of energy emanating from a pulsed laser is absorbed on the front face of the sample, resulting in homogeneous heating. The relative temperature increase on the rear face of the sample is then measured as a function of time by an IR detector. First, the thermal diffusivity is computed by software using the time/relative temperature increase data. For adiabatic conditions, α is determined from the simple equation:

$$\alpha = 0.1388\frac{l^2}{t_{0.5}} \tag{5.20}$$

where l is the sample thickness in mm and $t_{0.5}$ is time (in sec) at 50% of the temperature increase.

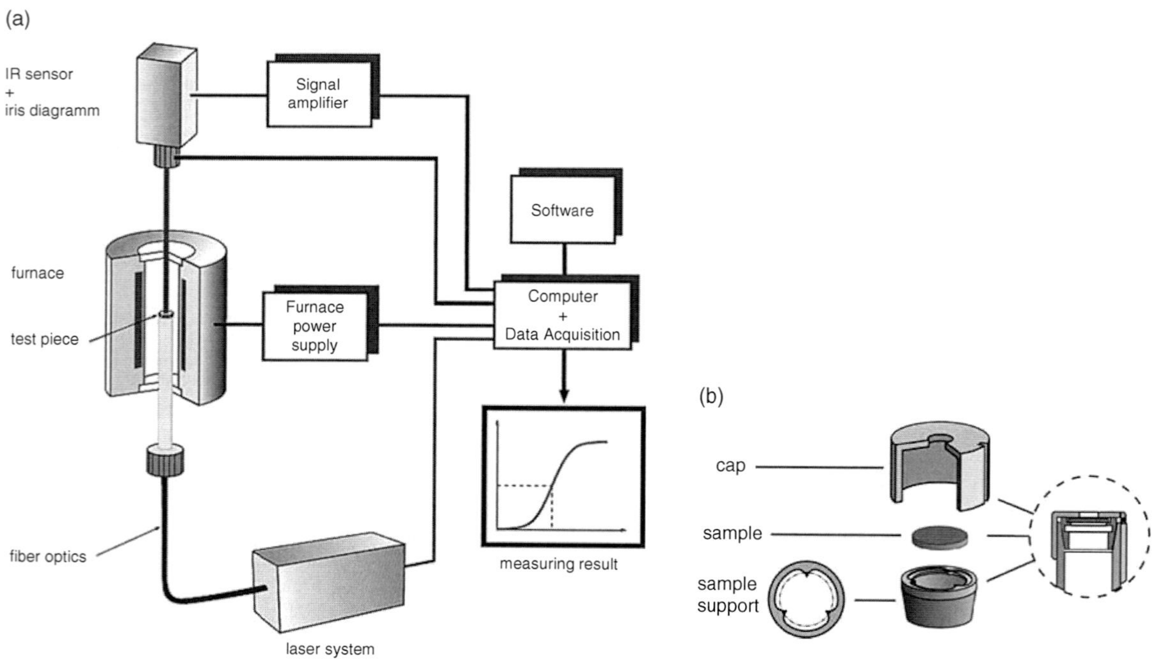

Fig. 5.14 Laser flash apparatus for thermal diffusivity (a) and holder system for solid sample (b) [42].

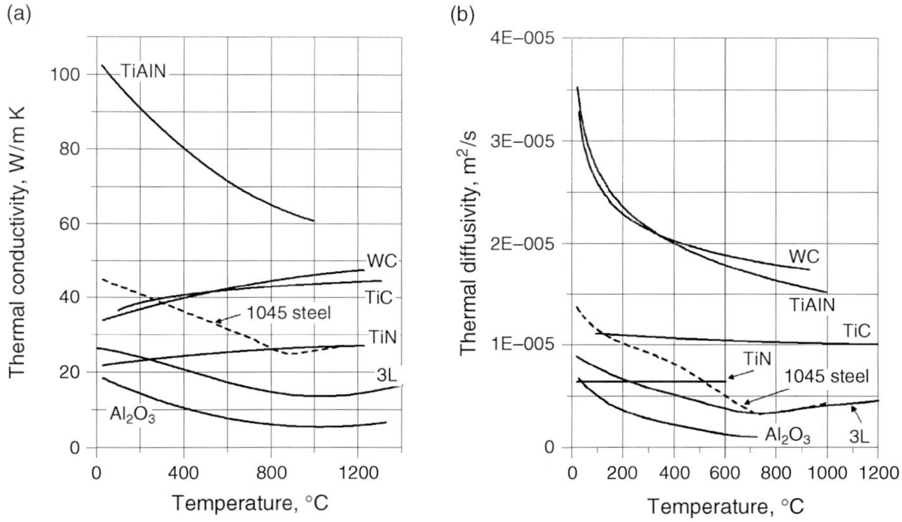

Fig. 5.15 Influence of temperature on thermal conductivity and thermal diffusivity. Courtesy of P. Nieslony.

On the other hand, the temperature-dependent thermal conductivity $\lambda(T)$ was determined by prior measurement of the thermal diffusivity (α), the specific heat (c_p) by means of a differential scanning calorimeter (DSC), and density (ρ), all as functions of temperature. That is:

$$\lambda(T) = \alpha(T)\rho(T)c_p(T) \tag{5.21}$$

The diagrams in Fig. 5.15 provide temperature-dependent values of the thermal conductivity and thermal diffusivity for AISI 1045 carbon steel as a reference work material, and some frequently used cutting tool materials including coatings or their components. In particular, the thermal properties of single layer TiAlN coating and multilayer TiC/Al_2O_3/TiN coating (denoted as 3L) were determined using the LF technique.

Table 5.2 lists possible sources of computational errors arising frequently in metal cutting numerical computations, based on the author's own experiences. The problem of heat partitioning between the chip and the tool and the distribution of the heat source at the tool/chip interface, as well as measurements of cutting temperature, will be discussed in Chapter 9.

Table 5.2 Possible sources of errors in FDM-based computing.

	Input data/computation procedure/validation of results	*Comments*
1	Heat flux value/Heat partitioning	Heat partition coefficient R_{SH}, R_{KF}, R_R
2	Heat flux distribution	*Constant,* triangular, trapezoid, parabolic
3	Heat flux reflection	Workpiece temperature rise/heat reflection coefficient
4	Tool/chip contact length	Correct measurement – LOM, *digital image processing*

(Continued)

Table 5.2 (Cont.)

	Input data/computation procedure/validation of results	*Comments*
5	Dependence of thermal properties on temperature	*Corrected (temperature-dependent) values for* λ *and* α
6	Material constitutive law	Johnson–Cook material model is preferred
7	Number of iterations	*Based on own experiences*
8	Calculation method of average interface temperature	*Based on own experiences*
9	Replacement of differential equation. with its finite difference representation	Euler scheme (forward difference method)
10	Reference method of temperature measurement (experimental vs modelled temperatures)	Tool/work thermocouple Thermocouple-insert Infrared thermovision/microscopy Thermosensitive coatings or paints

References

[1] C.A. van Luttervelt, T.H.C. Childs *et al.*, Present situations and future trends in modelling of machining operations. Progress report of the CIRP working group 'Modelling of Machining Operations', *CIRP Annals*, Vol. 47/2 (1998), 587–626.

[2] W. König, F. Klocke, M. Rehse, Prospects for the use of process models in metal cutting, *VDI Berichte*, No. 1399 (1998), 235–249.

[3] M.G. Merchant, D.A. Dornfeld, P.K. Wright, Manufacturing – its evolution and future, *Transactions of NAMRI/SME*, Vol. 33 (2005), 211–218.

[4] J.A. Stori, P.K. Wright, C. King, Integration of process simulation in machining parameter optimization, *J. Manuf. Sci. Engng*, Vol. 121 (1999), 134–143.

[5] W. Grzesik, Developments in metal removal processes, *Proc. 4th Intln. Scientific Conference 'Development of Metal Cutting'*, Košice, Slovakia, 2002, pp. 103–110.

[6] K.F. Ehmann, S.G. Kapoor, R.E. deVor, I. Lazoglu, Machining process modeling: a review, *J. Manuf. Sci. Engng*, Vol. 119 (1997), 655–662.

[7] E.M. Trent, P.K. Wright, *Metal Cutting*, Butterworth Heinemann, Boston, 2000.

[8] C.A. van Luttervelt, Typology of models and simulation methods for machining operations, *Mach. Sci. Technol.*, Vol. 5 (2001), 415–428.

[9] S.M. Athavale, J.S. Strenkowski, Finite element modelling of machining: from proof-of-concept to engineering applications, *Mach. Sci. Technol.*, Vol. 2(2) (1998), 317–342.

[10] F. Klocke, T. Beck, S. Hoppe, T. Krieg *et al.*, Examples of FEM application in manufacturing technology, *J. Mat. Proc. Technol.*, Vol. 120 (2002), 450–457.

[11] W. Zhang, O.W. Dillon, I.S. Jawahir, A finite element analysis of 2-D machining with a grooved tool, *Trans. NAMRI/SME*, Vol. 29 (2001), 327–334.

[12] M.R. Movahhedy, M.S. Gadala, Y. Altintas, Simulation of chip formation in orthogonal metal cutting process: an ALE finite element approach, *Mach. Sci. Technol.*, Vol. 4(1) (2000), 15–42.

[13] T.H.C. Childs, K. Maekawa, T. Obikawa, Y. Yamane, *Metal Cutting. Theory and Applications*, Arnold, London, 2000.

[14] E. Usui, T. Shirakashi, Mechanics of Machining – from Descriptive to Predictive Theory. *On the Art of Cutting Metals – 75 Years Later*, Vol. 7, ASME PED, 1982, pp. 13–55.

[15] K. Iwata, K. Osakada, Y. Teresaka, Process modeling of orthogonal cutting by a rigid-plastic finite element method, *J. Eng. Mat. Technol.*, Vol. 106 (1984), 132–138.

[16] J.S. Strenkowski, J.T. Carroll III, A finite element model of orthogonal metal cutting, *J. Eng. Ind.*, Vol. 107 (1985), 347–354.
[17] T.D. Marusich, M. Ortiz, Modelling and simulation of high-speed machining, *Int. J. Numer. Methods in Engng.*, Vol. 38 (1995), 3675–3694.
[18] Y.-Ch. Yen, A. Jain, T. Altan, A finite element analysis of orthogonal machining using different tool edge geometries, *J. Mater. Proc.Technol.*, Vol. 146 (2004), 72–81.
[19] J. Rech, A. Kusiak, J.L. Battaglia, Tribological and thermal functions of cutting tool coatings, *Surf. Coat. Technol.*, Vol. 186 (2004), 364–371.
[20] Y.-Ch. Yen, A. Jain, P. Chigurupati, W.-T. Wu, T. Altan, Computer simulation of orthogonal cutting using a tool with multiple coatings, *Proceedings of the 6th CIRP Int. Workshop on Modeling of Machining Operation*, McMaster University, ON, Canada, 20 May 2003, pp. 119–130.
[21] J. Leopold, M. Meisel, R. Wohlgemuth, J. Liebich, High performance computing of coating – substrate systems, *Surf. Coat. Technol.*, Vol. 142–144 (2001), 916–922.
[22] T. Özel, E. Zeren, Finite element method simulation of machining of AISI 1045 steel with a round edge cutting tool, *Proc. 8th CIRP Int. Workshop on Modeling of Machining Operations*, Chemnitz, 2005, pp. 533–541.
[23] Y.-C Yen, A. Jain, P. Chigurupati, W.-T. Wu, T. Altan, Computer simulation of orthogonal cutting using a tool with multiple coatings, *Mach. Sci. Technol.*, Vol. 8/2 (2004), 305–326.
[24] D. Umbrello, R. M'Saoubi, J.C. Outeiro, The influence of Johnson–Cook material constants on finite element simulation of machining of AISI 316L steel, *Int. J. Mach. Tools Manuf.*, Vol. 47 (2007), 462–470.
[25] K.C. Ee, O.W. Dillon Jr, I.S. Jawahir, Finite element modeling of residual stresses in machining induced by cutting using a tool with finite edge radius, *Int. J. Mech. Sci.*, Vol. 47 (2005), 1611–1628.
[26] S. Lei, Y.C. Shin, F.P. Incropera, Thermo-mechanical modeling of orthogonal machining process by finite element analysis, *Int. J. Mach. Tools Manuf.*, Vol. 39 (1999), 731–750.
[27] V. Kalhori, Modeling and simulation of mechanical cutting, Doctoral thesis, Lulea University of Technology, 2001.
[28] P.A.R. Rosa, P.A.F. Martins, A.G. Atkins, Revisiting the fundamentals of metal cutting by means of finite elements and ductile fracture mechanics, *Int. J. Mach. Tools Manuf.*, Vol. 47 (2007), 607–617.
[29] E.-G. Ng, D.K. Aspinwall, D. Brazil, J. Monaghan, Modelling of temperature and forces when orthogonally machining hardened steel, *Int. J. Mach. Tools Manuf.*, Vol. 39 (1999), 885–903.
[30] W. Grzesik, M. Bartoszuk, P. Nieslony, Finite element modelling of temperature distribution in the cutting zone in turning processes with differently coated tools, *J. Mater. Proc. Technol.*, Vol. 164–165 (2005), 1204–1211.
[31] H.S. Carslaw, J.C. Jaeger, *Conduction of Heat in Solids*, Oxford University Press, 1980.
[32] A.J.R. Smith, E.J.A. Armarego, Temperature prediction in orthogonal cutting with a finite difference approach. *Annals of the CIRP*, Vol. 30(1) (1981), 9–13.
[33] I. Lazoglu, Y. Altintas, Prediction of tool and chip temperature in continuous and interrupted machining, *Int. J. Mach. Tools Manuf.*, Vol. 42 (2002), 1011–1022.
[34] G. Boothroyd, W.A. Knight, *Fundamentals of Machining and Machine Tools*, Marcel Dekker, New York, 1989.
[35] J. Tlusty, *Manufacturing Processes and Equipment*, Prentice-Hall Intl, NJ, 1999.
[36] W. Grzesik, M. Bartoszuk, P. Nieslony, Finite difference analysis of the thermal behaviour of coated tools in orthogonal cutting of steels, *Int. J. Mach. Tools Manuf.*, Vol. 44 (2004), 1451–1462.
[37] P. Sartkulvanich, T. Altan, J. Soehner, Flow stress data for finite element simulation in metal cutting: a progress report on madams, *Mach. Sci. Technol.*, Vol. 9 (2005), 271–288.
[38] N. Fang, A new quantitative sensitivity analysis of the flow stress of 18 engineering materials in machining, *Trans. ASME J. Eng. Mater. Technol.*, Vol. 127 (2005), 192–196.
[39] T.H.C. Childs, Material property needs in modeling metal machining, *Proc. 1st CIRP Int. Workshop on Modeling of Machining Operations*, Atlanta, 1998, pp. 193–202.
[40] W. Grzesik, P. Nieslony, A computational approach to evaluate temperature and heat partition in machining with multilayer coated tools. *Int. J. Mach. Tools Manuf.*, Vol. 43 (2003), 1311–1317.
[41] W. Grzesik, The role of coatings in controlling the cutting process when turning with coated indexable inserts, *J. Mater. Proc.Technol.*, Vol. 37 (1998), 133–143.
[42] Laser Flash Apparatus LFA 427 for the measurement of thermal diffusivity. www.ngb.netzsch.com.
[43] Der Brockhaus in fünfzehn Bänden, 1997, F.A. Brockhaus, Wiesbaden.

CHAPTER SIX

Orthogonal and Oblique Cutting Mechanics

Contents

6.1 Geometrical and Kinematical Characterization

In metal cutting theory all operations performed with the wedge-shaped tool when a layer of metal is removed in the form of continuous or discontinuous chip can be conventionally divided into two general cases termed orthogonal and oblique cutting. These two characteristic cutting arrangements are illustrated in Fig. 6.1 and comparison of their specific features is made in Table 6.1.

The wedge-shaped cutting tool basically consists of two surfaces intersecting to form the straight cutting edge. Functionally, the surface along which the chip flows is termed the *rake face* and the surface which contacts a new or machined workpiece surface is known as the *flank face*. Since orthogonal cutting represents a two-dimensional rather than a three-dimensional mechanical problem, it leads to a relatively simple cutting process which is therefore widely used in theoretical and experimental work. In particular, it enables the adoption of the process model assuming the plane state of both plastic deformation and stresses in the cutting zone. The vectors of the tool, the deformed material and the chip velocities for orthogonal and oblique cutting are shown in Fig. 6.2a and b, respectively. It should be noted that for oblique cutting (Fig. 6.2b) the chip formation process consists of two parallel subprocesses including the plane plastic flow of the work material in the plane perpendicular to the cutting edge, similarly to orthogonal cutting (plane *Omn* in the *Olmn* coordinate system), and lateral plastic flow along the cutting edge (axis l parallel to the cutting edge S). The lateral material flow causes the chip to flow at the angle τ relative to the axis m_1, i.e. it is deviated from the direction perpendicular to the inclined cutting edge.

It is evident from the vector diagram of velocities shown in Fig. 6.2a that by using the rule of continuity, the velocity of shearing v_s along the shear plane inclined at the shear angle Φ is the vectorial sum of the chip

Advanced Machining Processes of Metallic Materials
ISBN: 978-0-08-044534-2

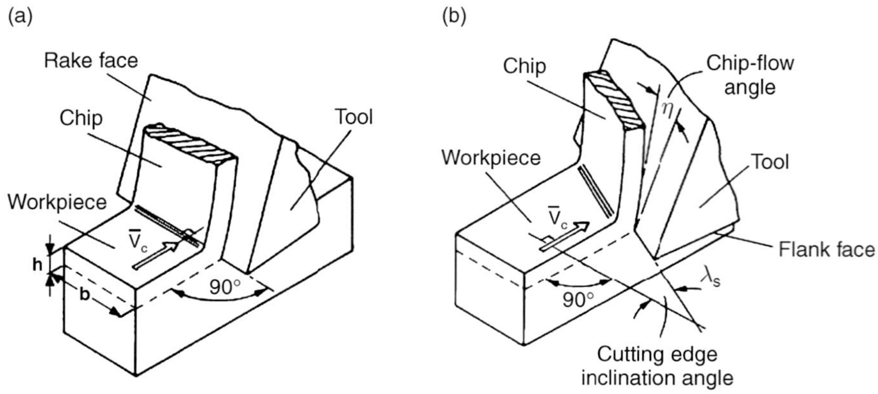

Fig. 6.1 Geometries of orthogonal (a) and oblique cutting processes with straight cutting edge (b) [1, 2].

Table 6.1 Comparison of two basic cutting processes [2].

	Characteristic feature	*Type of cutting*	
		Free orthogonal	*Free oblique*
1	Number of active cutting edges	one (main)	
2	Shape of the cutting edge	straight	
3	Position of the active cutting edge vs direction of relative tool-work motion	$\angle \bar{v}_e, S_a = 90°$	$\angle \bar{v}_e, S_a = 90° + \lambda_s$
4	Tool cutting edge angle	$\kappa_r = 90°$	$\kappa_r \neq 90°$
5	Cutting edge inclination angle	$\lambda_s = 0$	$\lambda_s \neq 0$
6	Length of the cutting edge	$1_c > b$	
7	Length of the active cutting edge (S_a)	$1_{ca} = \mathrm{B}$	$1_{ca} = S_a / \cos \lambda_s$
8	Width of the cutting edge vs width of workpiece	$B = \mathrm{B}$	$B = 1_{ca}$
9	Width vs thickness of the undeformed chip	$B >> h$	
10	Character of plastic flow in the chip formation zone	plane (2D)	spatial (3D)
11	Direction of chip flow	$\tau = 0$	$\tau \geq \lambda_s$ or $\tau < \lambda_s$
12	Direction of chip material movement at individual points along the active cutting edge	$\tau =$ const (for oblique cutting $\tau_i \neq$ const)	
13	Characteristic shape of a continuous-type chip	logarithmic spiral	helix line

velocity v_{ch} (the sliding velocity at the tool/chip interface) given by Eqn (6.1) and the cutting velocity v_c. It can be determined from Eqn (6.2).

$$v_{ch} = v_c \frac{\sin \Phi}{\cos(\Phi - \gamma_0)} \tag{6.1}$$

$$v_s = v_c \frac{\cos \gamma_0}{\cos(\Phi - \gamma_0)} \tag{6.2}$$

where Φ is the shear angle and γ_0 is the orthogonal rake angle.

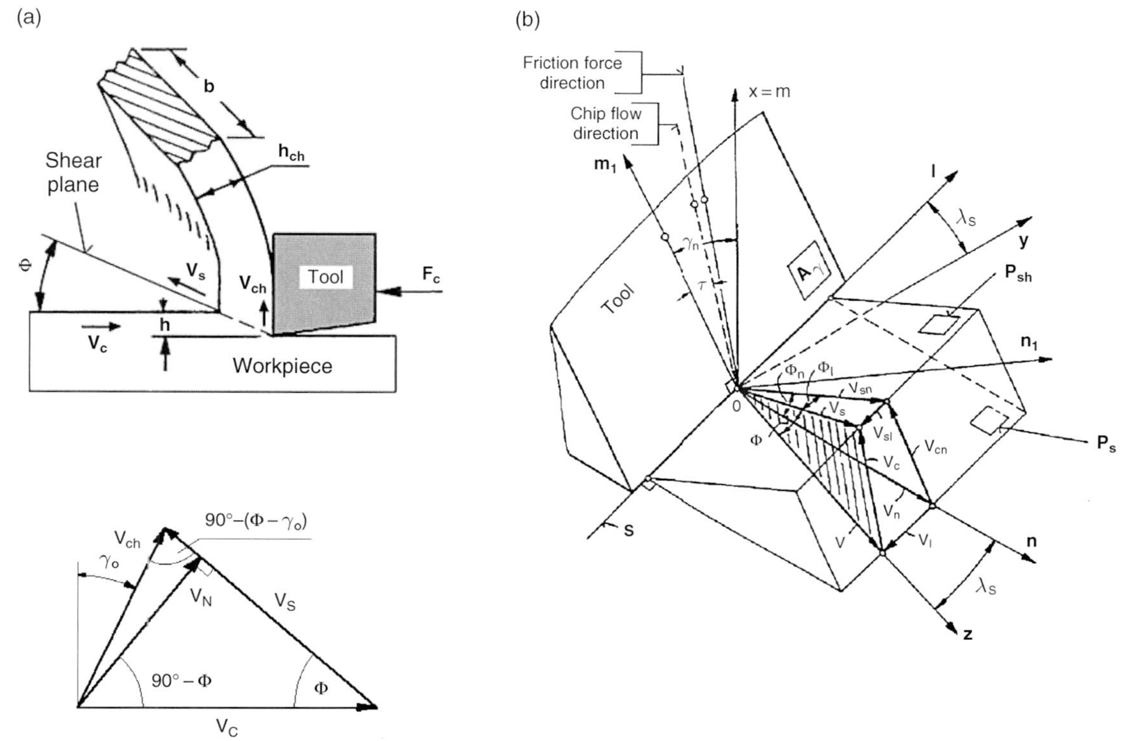

Fig. 6.2 Velocity resolution for orthogonal (a) and oblique (b) cutting [2].

6.2 Forces in the Cutting Zone

In the general case, the force system exerted by a cutting tool is three dimensional, i.e. the resultant (total) cutting force F has three mutually perpendicular components shown in Fig. 6.1a. Typically, the resultant cutting force (RCF) is resolved into geometrical and physical components. In the first case the coordinate system is oriented based on the directions of primary and feed motions, i.e. the total force is resolved by perpendicular projection to these two directions. As a result, the cutting force and feed force denoted by symbols F_c and F_f are distinguished. Moreover, the back (passive) force F_p tending to push the tool away from the work in a radial direction and perpendicularly to the working plane is defined. This force is usually ignored for orthogonal cutting and the total cutting force is reduced to the active force F_a. As can be seen in Fig. 6.3 in orthogonal cutting the entire force system lies in a single plane and appropriate force components can be easily calculated by drawing the Merchant circle diagram. The force component in the direction of relative tool travel F_c determines the amount of work required to move the cutting tool a given distance. The feed force F_f determines the power required to feed drives installed onto a machine tool.

Geometrical resolution of the total cutting force is done by assuming that the chip is a body in stable mechanical equilibrium under the action of the forces exerted on it at the rake face and at the shear plane as shown in the right upper detail.

The force components acting on the shear plane are F_{sh} and F_{shN}. The shearing force F_{sh} represents the force required to shear the material on the shear plane. Force F_{shN} acts perpendicularly to the shear plane

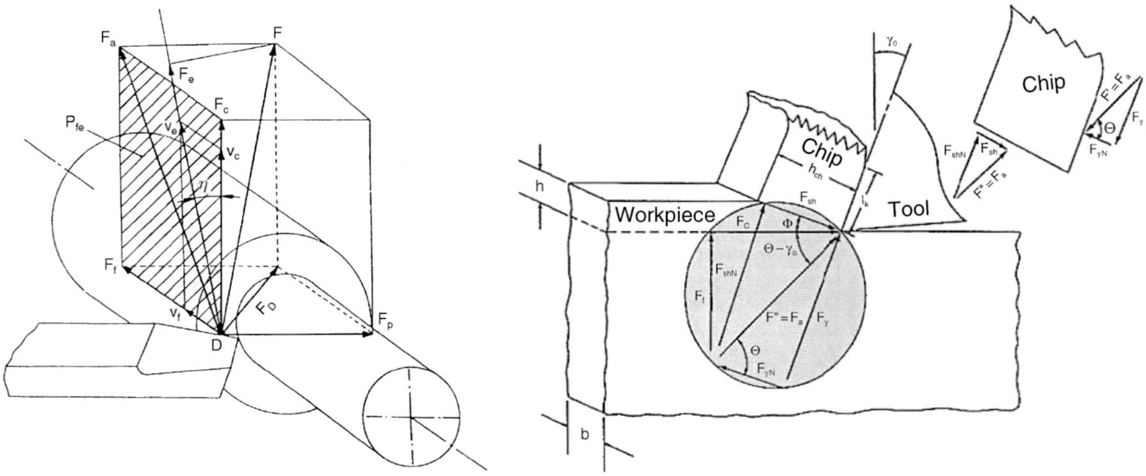

Fig. 6.3 Geometrical (a) and physical (b) force resolution in the cutting zone for orthogonal cutting with a continuous-type chip [3].

and results in an additional compressive stress being applied on the shear plane. They can be calculated from the following equations:

$$F_{sh} = F_c \frac{\cos(\Theta + \Phi - \gamma_0)}{\cos(\Theta - \gamma_0)} \tag{6.3}$$

$$F_{shN} = F_c \frac{\sin(\Theta + \Phi - \gamma_0)}{\cos(\Theta - \gamma_0)} \tag{6.4}$$

At the tool/chip interface, the force components F_γ and $F_{\gamma N}$ act parallel and perpendicularly to the face. F_γ, known as friction force (the tangential force at the rake), represents the frictional resistance met by the chip as it slides over the tool rake face.

The friction force on the chip/tool interface:

$$F_\gamma = F_c \frac{\sin\Theta}{\cos(\Theta - \gamma_0)} \tag{6.5}$$

where Θ is the friction angle.

The normal force on the chip/tool interface:

$$F_{\gamma N} = F_c \frac{\cos\Theta}{\cos(\Theta - \gamma_0)} \tag{6.6}$$

The ratio of F_γ to $F_{\gamma N}$ is the apparent coefficient of friction between chip and tool, namely:

$$\mu_\gamma = tg\Theta = \frac{F_\gamma}{F_{\gamma N}} = \frac{F_c \sin\gamma_0 + F_f \cos\gamma_0}{F_c \cos\gamma_0 - F_f \sin\gamma_0} \tag{6.7}$$

It was experimentally proven that in cases such as cutting with a small undeformed chip thickness (UCT), cutting of materials with high elastic (Young's) modulus and when tool wear at the flank progresses, also the forces acting on the flank/workpiece contact should be considered in the total cutting force. It can be

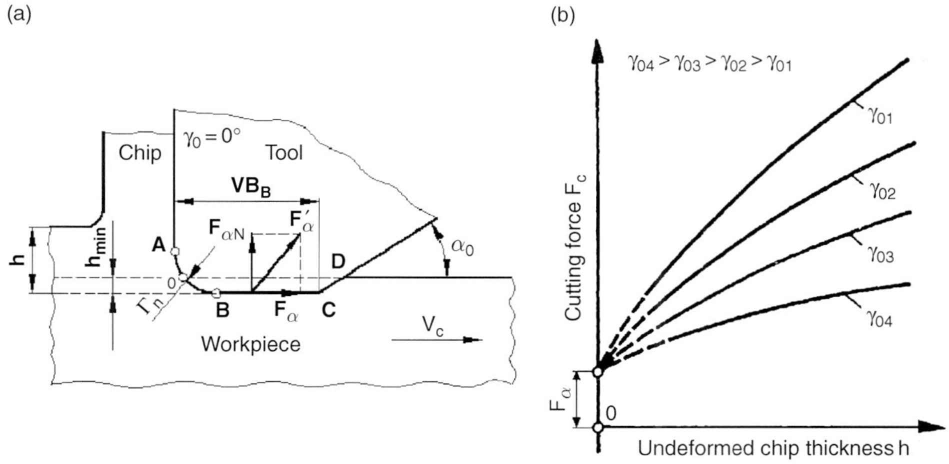

Fig. 6.4 Force distribution on the workpiece/flank contact (a) and projection of the cutting force on zero undeformed chip thickness (b) [2].

assumed that the force acting on the flank wear land VB_B (Fig. 6.4) can be resolved into two components F_α and $F_{\alpha N}$ which are independent of the rake angle value. These forces can be determined using the method of extrapolation of F_c and F_f components on zero undeformed chip thickness (Eqn (6.8)) proposed by Zorev [5]. As illustrated in Fig. 6.4b the Zorev method needs a few F_c/UCT curves to be obtained for different rake angles.

From the definition

$$F_\alpha = F_{c(h\to 0)} \ i \ F_{\alpha N} = F_{f(h\to 0)} \tag{6.8}$$

In addition, forces acting on the flank face can be calculated by solving the problem of penetration of a rigid punch into a plastically deformed half-space using the slip-line method [2].

The normal force on the worn flank face

$$F_{\alpha N} \approx c\sigma_p A_\alpha \tag{6.9}$$

where coefficient $c = 2.6$–4, σ_p is the flow yield stress and A_α is the flank/workpiece contact area.

The tangential force on the flank face and

$$F_\alpha = \tau_{c\gamma}\, A_\alpha = (0.5 \div 0.6)\,\text{UTS}\, A_\alpha \tag{6.10}$$

where $\tau_{c\gamma}$ is the main shear stress on the rake face and UTS is ultimate tensile strength.

As depicted in Figs 6.5 and 6.6, the cutting forces are influenced by the cutting parameters as well as by tool geometry. Figure 6.5 a shows a linear increase of all components of RCF due to relevant increase of the cross-sectional area of the uncut chip. On the other hand, force/feed curves are non-linear because increasing the feed results in decreasing the specific cutting pressure k_c [4]. The kink in the force/cutting speed curves in the medium speed range illustrates the effect of a built-up edge (BUE). With steels, a BUE disappears when the speed is raised. The cutting forces are also influenced by tool geometry, the most important parameter being the rake angle. An increase in the rake angle lowers the force components but at the same time reduces the strength of the tool wedge and may lead to its fracture. The strongest tool edge is achieved with negative rake angle tools, and these are frequently used for the harder grades of carbide and for ceramic and CBN tools which lack toughness. Tools with high positive inclination angles (Fig. 6.6b)

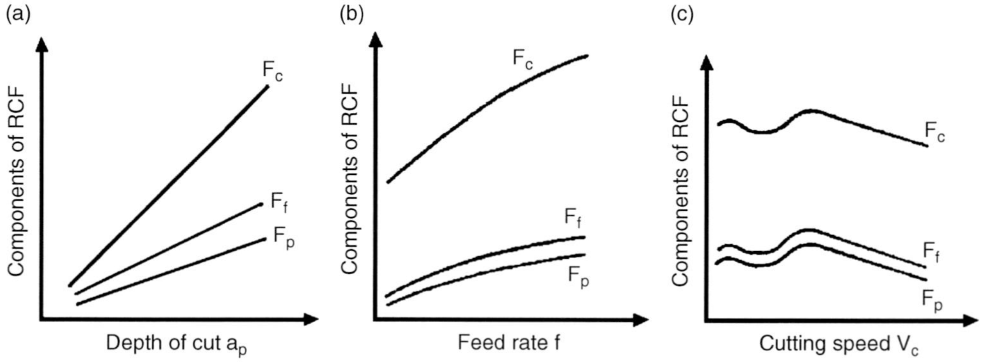

Fig. 6.5 Influence of cutting parameters on the components of resultant cutting force (RCF) [6].

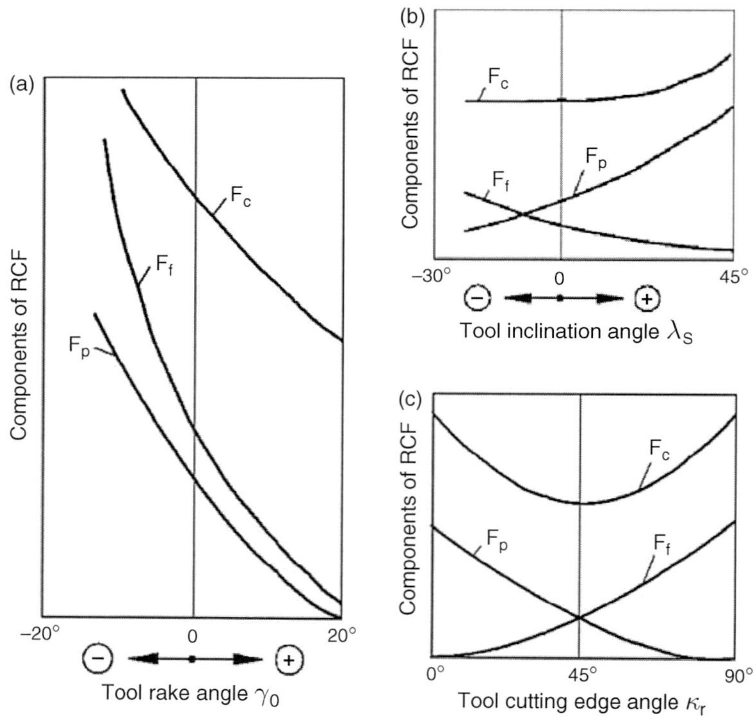

Fig. 6.6 Influence of cutting tool angles on the components of RCF [2].

generate high thrust forces causing the deflection of the workpiece. In contrast, this force is very small when using tools with a 90° cutting edge angle (Fig. 6.6c) and machining of small diameter workpieces is possible. The tool material and deposited coatings can also influence the cutting forces mainly by changes in the area of seized contact between the chip and the tool [7]. Finally, the contact length and cutting forces may be greatly influenced by cutting lubricants or special soft low friction coatings. In the speed range used in most machine shop operations, it is not possible to prevent seizure near the cutting edge but liquid or gaseous lubricants, by penetrating from the periphery, can restrict the area of seizure to a small region. This phenomenon will be discussed in Chapter 11.

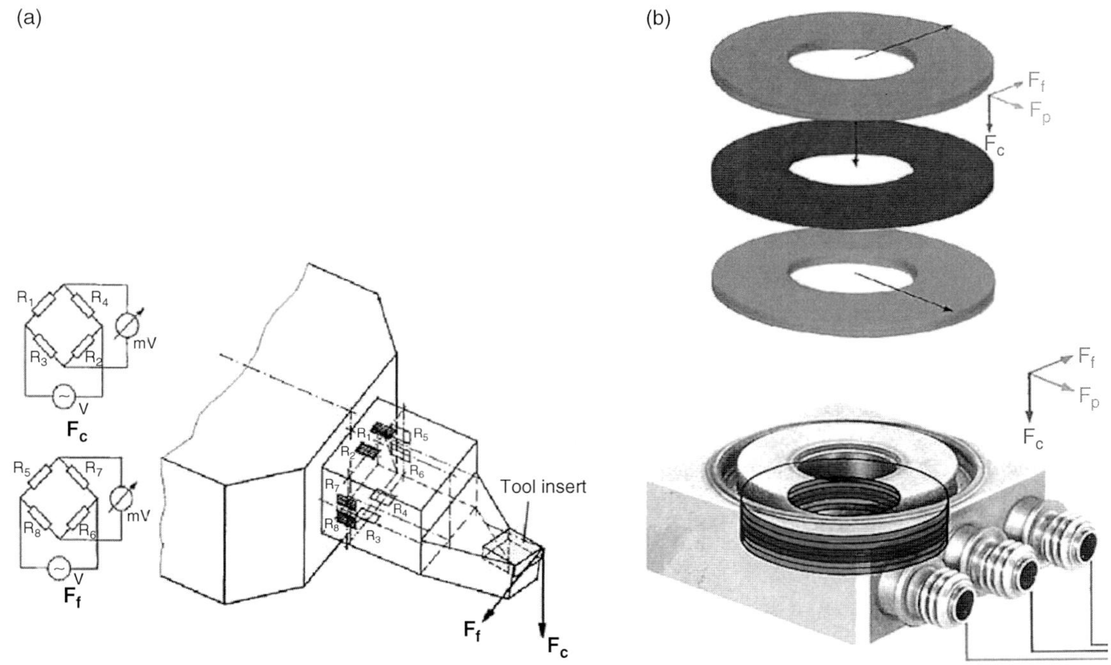

Fig. 6.7 Measurement of cutting forces using biaxial strain-gauge (a) and piezoelectric (b) dynamometers [2, 9].

Measurements of cutting forces are carried out using strain-gauge and piezoelectric dynamometers. The principles of both types of tool dynamometers are shown in Fig. 6.7. In general, a tool dynamometer should have high sensitivity, high rigidity, high frequency response, high linearity and low drift [8]. A common type of dynamometer, a biaxial device shown in Fig. 6.7a, uses strain gauges to sense elastic strains caused by cutting forces. Typically, Wheatstone bridge circuits for each force component, consisting of two pairs of tension and compression gauges and amplifier system, are used to measure strain changes in the wire-type gauges. Before (prior to), the measurements force dynamometer is calibrated to find the relationship between voltage signal from the bridge and the actual value of a force. Figure 6.7b shows the piezoelectric type of dynamometer capable of measuring all three components of the RCF. Commercial machining dynamometers are available with natural frequencies from 2 kHz to 5 kHz, depending on size. The basic rule is that each force component is detected by a separate pure quartz crystal oriented relative to the force or torque in its piezoelectric sensitive direction. Quartz rings are used in pairs (low view in Fig. 6.7b) with a common electrode between them to yield double sensitivity. In Fig. 6.7b two pairs of shear quartzes for F_f and F_p and one pair of compression quartzes for F_c, assembled in a single housing, constitute a three-component force transducer. Two-, three- and four-component piezoelectric dynamometers and force plates for measuring cutting forces and torques in various machining operations (turning, milling, drilling, etc.) are produced by world leading manufacturer Kistler, Switzerland [9].

6.3 Cutting Energy

The total work done by the cutting tool to remove metal, determined by the value of the force component F_c, is dissipated in the following forms for a sharp tool: (1) as work done in shear deformation, (2) as work done to overcome chip/tool interfacial friction and (3) as work done to generate a new surface in cutting, but the third form is often neglected. Practically, all of the energy generated in metal cutting operations is dissipated in plastic deformation and friction.

In general, the energy required to perform primary and feed motions can be calculated using Eqns (6.11) and (6.12), respectively. As mentioned above, Eqn (6.13) defined two basic components of the cutting energy required to perform plastic deformation (E_{sh}) and overcome friction (E_f):

$$E_c = \int_o^t F_c v_c dt \tag{6.11}$$

$$E_f = \int_o^t F_f v_f dt \tag{6.12}$$

$$E = E_p + E_f = E_{sh} + E_{f\gamma} + E_{f\alpha} \tag{6.13}$$

Often, equivalent specific (volumetric) cutting energy (SCE) and its components are used as defined by Eqns (6.14)–(6.17), where V denotes the volume of the layer removed and A_c is cross-sectional area of the uncut chip:

$$\mathrm{e_c} = \frac{\mathrm{E_c}}{\mathrm{V}} = \frac{\mathrm{F_c v_c}}{\mathrm{bhv_c}} = \frac{\mathrm{F_c}}{\mathrm{A_c}} = \mathrm{k_c} \tag{6.14}$$

$$e_c = e_{sh} + e_{f\gamma} \tag{6.15}$$

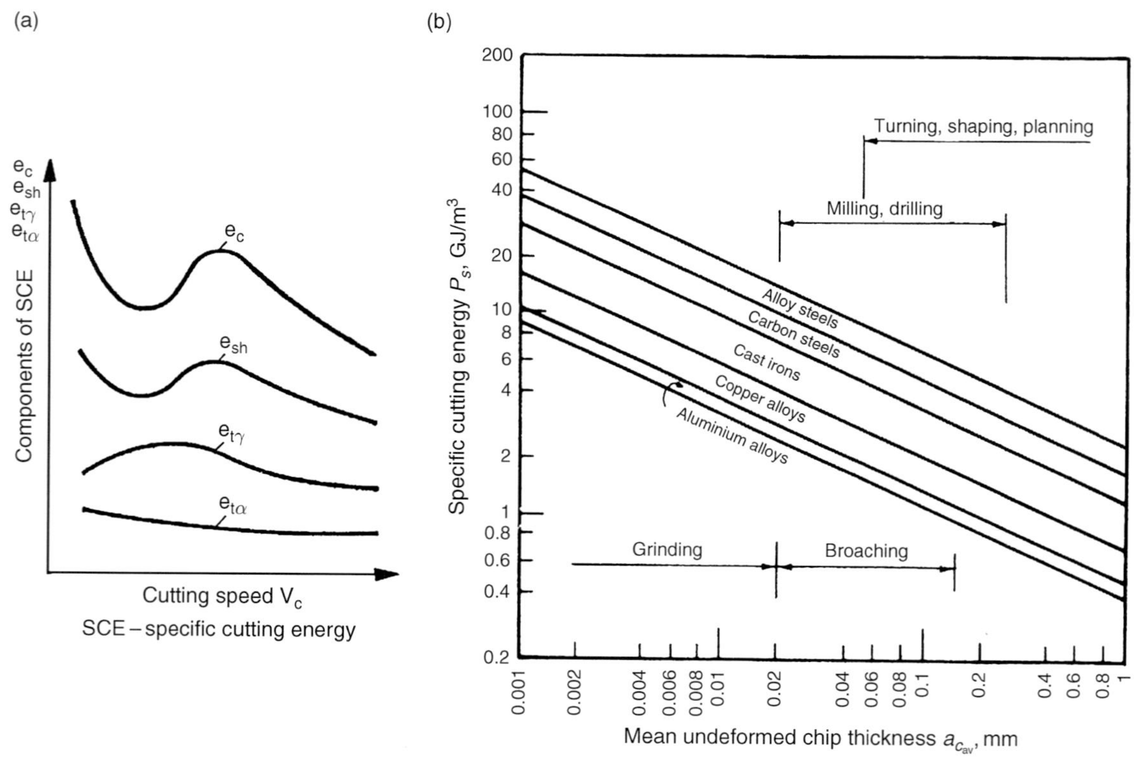

Fig. 6.8 Influence of cutting speed on the components of specific cutting energy (a) and approximate values of the cutting energy for various materials and operations (b) [1, 2].

It is evident from Eqn (6.14) that the specific cutting energy is equal to the specific cutting pressure [4]. Using Eqns (6.3) and (6.4), components of specific cutting energy are equal:

$$e_{sh} = F_c \frac{\cos(\Phi + \Theta - \gamma_o)\cos\gamma_o}{A_c \cos(\Phi - \gamma_o)\cos(\Theta - \gamma_o)} = \tau_{sh}\gamma_{sh} \tag{6.16}$$

$$e_{f\gamma} = F_c \frac{\sin\Theta}{A_c k_h \cos(\Theta - \gamma_0)} \tag{6.17}$$

where τ_{sh} is shear stress of the material in the shear plane, γ_{sh} is the shear strain and k_h is the chip thickness compression ratio. All these quantities will be defined in section 6.5.

Figure 6.8a illustrates changes to the components of SCE with cutting speed, and Fig. 6.8b shows the influence of the undeformed chip thickness relative to typical machining operations and work materials. The shape of SCE/speed curve is the same as in Fig. 6.5c. It is important to note that all finishing operations, especially performed on hard materials like grinding, are more energy consuming due to removing a very small layer of material.

6.4 Stresses on the Shear Plane

Measurement of the forces and the chip thickness make it possible to estimate the stresses on the shear plane for orthogonal cutting conditions, where a continuous chip is formed without built-up edge. With the assumption of uniform stress distribution and for known forces acting on the shear plane, the shear stress τ_s and normal stress σ_s are found to be

$$\tau_s = \frac{F_{sh}}{A_{sh}} = F_c \frac{\cos(\Theta + \Phi - \gamma_0)\sin\Phi}{bh\cos(\Theta - \gamma_0)} \tag{6.18}$$

$$\sigma_s = \frac{F_{shN}}{A_{sh}} = F_c \frac{\sin(\Theta + \Phi - \gamma_0)\sin\Phi}{bh\cos(\Theta - \gamma_0)} \tag{6.19}$$

where A_{sh} is the shear plane area ($A_{sh} = bh/\sin\Phi$), b is the width of cut and h is the uncut chip thickness.

The ratio of the normal stress to the shear stress is

$$\frac{\sigma_s}{\tau_s} = tg(\Theta + \Phi - \gamma_0) \tag{6.20}$$

The distribution of stresses along the shear plane is shown in Fig. 6.9a. The shear stresses are uniformly distributed but the normal stresses are constant only for zero or negative rake angle as indicated in case b. On the contrary, for positive rake angle (case a) normal stress decreases in the vicinity of the cutting edge and becomes tensile. Consequently, the stress ratio given by Eqn (6.20) is not constant.

The rake angle is also a predominant factor controlling the distribution of elastic stresses in the subsurface layer. As shown in Fig. 6.10, the compressive zone expands visibly when cutting with large negative rake angles (Fig. 6.10c) or with rounded cutting edges (Fig. 6.10d). As a consequence, in such cases the work-hardening effect will probably be more intensive than for cutting tools with positive rake angles.

In general, the shear strength of metals and alloys in cutting has been found to vary only slightly over a wide range of cutting speeds and feeds [7]. The problem is to correlate the shear stress magnitude in cutting with relevant values of shear yield strength determined in tensile, compression, torsion or hardness tests. In general, the shear yield strength in cutting exceeds the shear yield strength of a given material by a factor of 2–4, depending on the intensity of plastic deformation in the primary deformation zone (PDZ). For example,

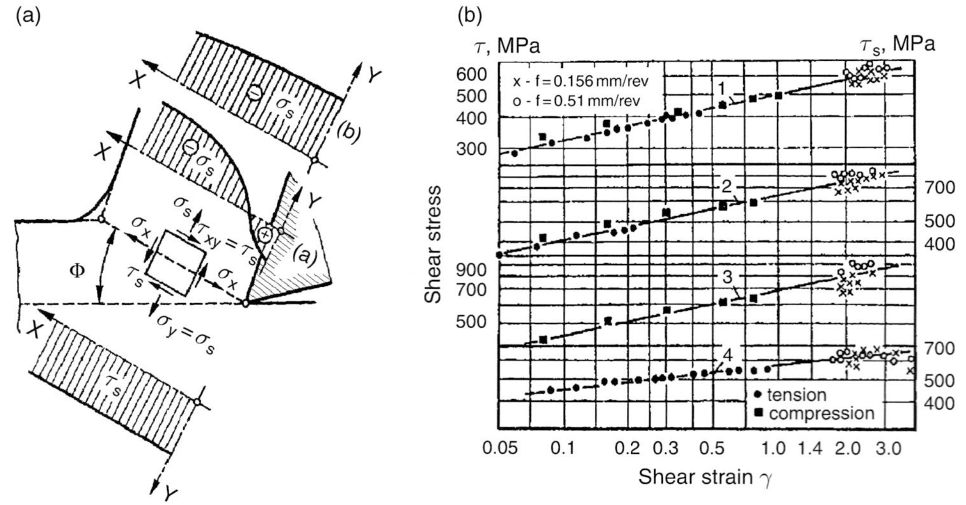

Fig. 6.9 2D model of stresses acting on the shear plane (a) and estimation of shear stress using simple tension and compression tests [2, 5].

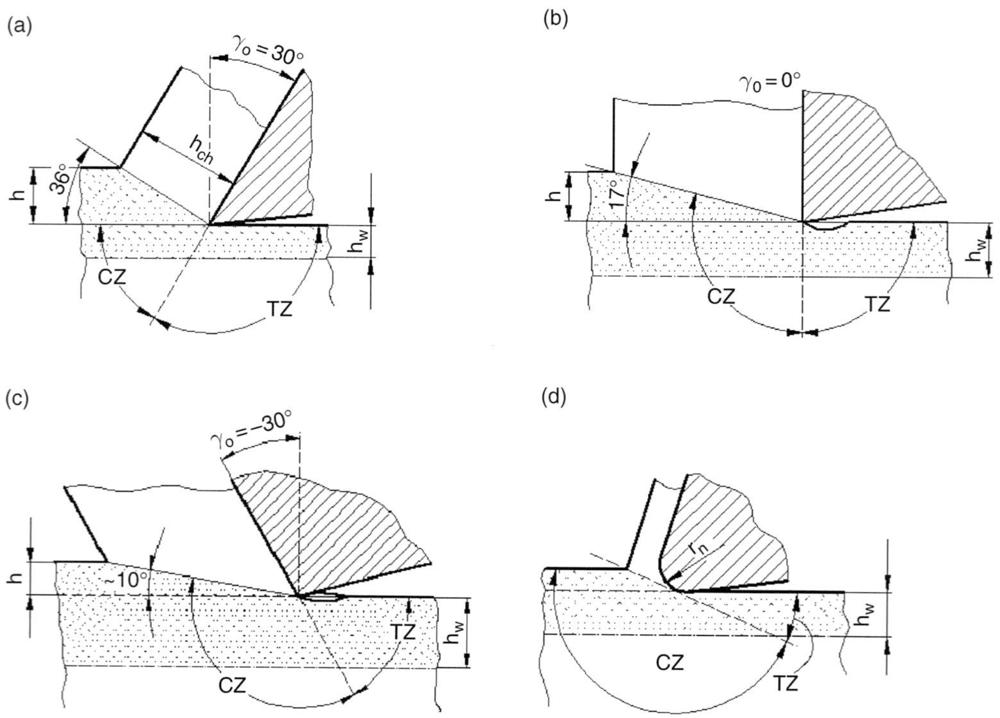

Fig. 6.10 Distribution of stresses in the subsurface layer [2]. CZ – compressive zone, TZ – tensile zone.

Table 6.2 Mechanical properties of different work materials [2].

Work material	*Shear stress* τ_s *(MPa)*	*Ultimate tensile strength (UTS) (MPa)*	*Fracture stress* σ_f *(MPa)*	*Ratio* $M = \tau_s/UTS$
Carbon steel 0.45%C	485	655	830	0.740
Alloy steel 0.4%C, 1%Cr	563	615	715	0.910
Die steel 1%C, 1.4%W, 1.1%Cr	603	630	830	0.960
Austenitic stainless steel 0.12%C, 18%Cr, 9%Ni	775	628	925	1.230
Heat-resistant alloy (Nimonic) 20.5%Cr, 2.5%Ti, 4%Fe, 0.75%Al	1175	1072	1460	1.100
Duralumin AlCu4Mg1	430	474	575	0.910
Aluminium alloy AlCu2Si1Mg	200	264	295	0.760
Copper	280	243	310	1.150
Leaded brass CuZn40Pb1	490	465	640	1.050
Aluminium bronze CuAl9Fe3	735	577	870	1.270
Titanium alloy (β) TiAl6Mo3	718	990	1050	0.725

Fig. 6.9b compares the shear stresses obtained for cutting of different steels with stresses determined in both tensile and compression tests. Based on this comparison, Zorev [5] concluded that the agreement exists only for very high shear strains, approximately in the range of $\gamma_{sh} = 2.5$. Some other trials correlate the shear stress measured in cutting with the UTS or fracture stress. Table 6.2 lists all the above-mentioned mechanical properties of both ferrous and non-ferrous materials.

6.5 Plastic Deformation in the Cutting Zone

In metal cutting, a chip formation area consisting of three characteristic zones I–III can be schematically presented as in Fig. 6.11a. The extensive primary deformation (shear) zone I (PDZ) is distinguished within the *OAB* area by a number of slip lines including the *OFA* line (the entry boundary) on which plastic deformation begins and the *OCQB* line (the exit boundary) on which chip material is entirely work hardened. A material particle during deformation moves along the *FQ* path but the deformation process does not take place uniformly but it is more intensive between the shear band of Δx thick and the upper boundary *OB*. Then, the material adjacent to the tool/chip interface of *OD* length is subsequently deformed in depth Δ_1 due to intensive interfacial friction. The relevant *OCD* region is called the secondary deformation (shear) zone II (SDZ). Additionally, the tertiary deformation zone III (TDZ) localized below the cutting edge in depth Δ_2 is separated. This zone is characterized in Chapter 20.

The engineering approach to the characterization of plastic deformation in the cutting zone is based on three basic simplifications (b1–b3 in Fig. 6.11) including replacing curvilinear boundaries by straight lines *OP* and *OK* ('fan'-type or pie-shaped shear zone model proposed by Zorev [5]), parallel-sided shear band inclined at angle Φ to the tool motion (model with parallel boundaries proposed by Oxley [10]) and one shear surface being a plane P_{sh} (the shear plane model proposed by Merchant [11]. It should be noted that in the last case the work material is assumed to be perfectly plastic (incompressible), i.e. work-hardening effect

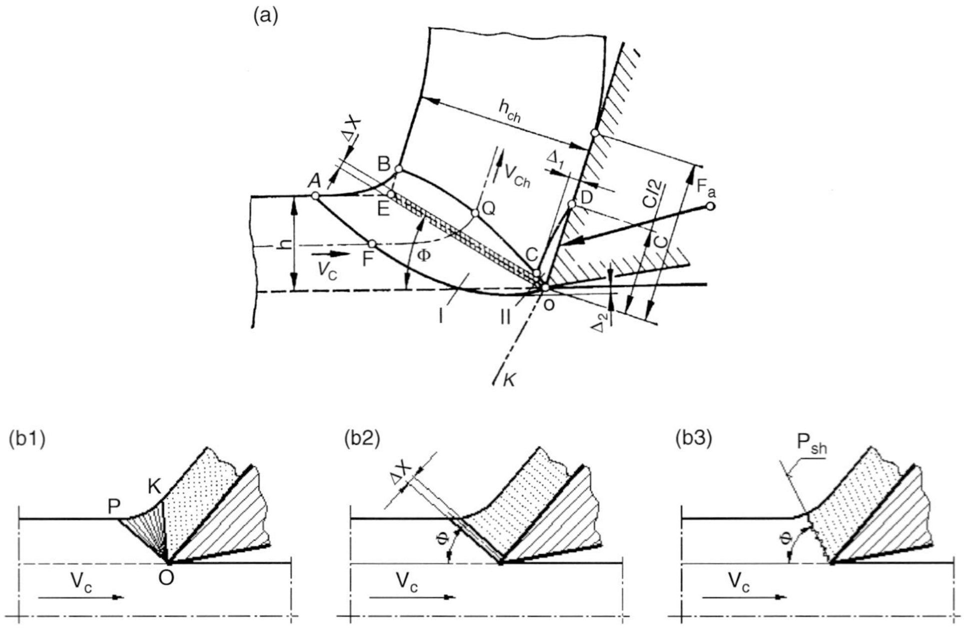

Fig. 6.11 Schematic (a) and simplified models (b) of the chip formation zone [2].

does not occur. Moreover, it ignores BUE, which may be present, as well as chip curl, and depicts tool face friction as elastic rather than plastic.

According to International Standard ISO 3002/4, the chip compression ratio (Λ_h or k_h) is the ratio of the idealized chip thickness to the nominal thickness of the cut. It can also be determined using the weight method, as follows:

$$k_h = \frac{h_{ch}}{h} = \frac{m_{ch}}{\rho b l h} \tag{6.21}$$

where m_{ch} is mass of the chip specimen, ρ is density of the workpiece material, b, l and h are dimensions (width, length and thickness) of the uncut chip. In experimental work, the chip thickness can be easily measured with a ball-ended micrometer.

From the geometry of cut (Fig. 6.12) it can be shown that the shear plane (AB) length is

$$l_{AB} = \frac{h}{\sin \Phi} = \frac{h_{ch}}{\cos(\Phi - \gamma_0)} \tag{6.22}$$

The shear angle is derived as a function of the chip compression ratio and the rake angle:

$$tg\Phi = \frac{\cos \gamma_o}{h_{ch}/h - \sin \gamma_o} = \frac{\cos \gamma_o}{k_h - \sin \gamma_o} \tag{6.23}$$

To derive an expression for the shear strain, the deformation can be idealized as a process of block slip, as shown in Fig. 6.12b. From definition, the shear strain for pure shearing is the ratio of absolute deformation Δs to the nominal distance Δx between the deformed and undeformed planes, namely

$$\gamma_{sh} = \frac{\Delta s}{\Delta x} = \frac{AB'}{CF} = \frac{FB' + AF}{CF} = \frac{FB'}{CF} + \frac{AF}{CF} = \frac{CF\ ctg\Phi}{CF} + \frac{CF\ tg(\Phi - \gamma_0)}{CF} \tag{6.24}$$

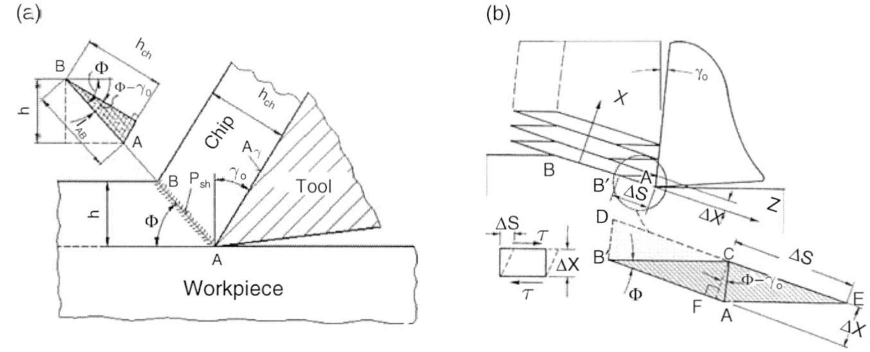

Fig. 6.12 Geometry of cut (a) and the shear-strain model [2, 7, 11].

After simple trigonometrical rearrangements three formulae for the shear strain can be obtained. Thus,

$$\gamma_{sh} = ctg\Phi + tg\,(\Phi - \gamma_0) \tag{6.25a}$$

$$\gamma_{sh} = \frac{\cos \gamma_0}{\sin \Phi \cos (\Phi - \gamma_0)} \tag{6.25b}$$

$$\gamma_{sh} = \frac{k_h^2 - 2k_h \sin \gamma_0 + 1}{k_h \ \cos \gamma_0} \tag{6.25c}$$

The strain may also be expressed in terms of the shearing velocity (see Fig. 6.2):

$$\gamma_{sh} = \frac{v_s}{v_c \sin \Phi} \tag{6.26}$$

It is interesting to note that the shear strain defined by Eqn (6.25c) will be equal to zero only for $k_h = 1$ and $\gamma_0 = 90°$. For example, for the rake angle $\gamma_0 = 0°$ the strain $\gamma_{sh} = 2$, and covertly $\gamma_{sh} = 1$ when for rake angle $\gamma_0 = 36°52'$.

The shear strain rate in cutting is defined as

$$\dot{\gamma}_{sh} = \frac{\gamma_{sh}}{\Delta t} = \frac{\Delta s}{\Delta x \, \Delta t} = \frac{v_s}{\Delta x} = \frac{\cos \gamma_0}{\cos (\Phi - \gamma_0)} \frac{v_c}{\Delta x} \tag{6.27}$$

where Δt is the time elapsed for the material element to travel a distance Δs along the shear plane, and Δx is the thickness of the shear zone.

The final formula is derived using Eqns (6.25b) and (6.2). Since the shear plane is usually very thin, as shown by photomicrographs (exemplarily in Fig. 6.13), it is expected that very few slip lines will be active at any one time. A reasonable mean value for the spacing of successive slip planes (Δx) would appear to be about 25 μm [11]. By assuming $\gamma_0 = 0°$, $\Delta x = 0.02$ mm, $\Phi = 30°$, the shear strain rate is equal to $2.3 \times 10^5\,s^{-1}$ [2]. In many practical operations, peak shear stain rates are of the order of $10^4\,s^{-1}$, but for high speed machining and grinding processes it can reach a value of 10^5–$10^7\,s^{-1}$.

One piece of the experimental evidence of chip formation by shearing and existence of both PDZ and SDZ zones is characteristic deformation texture of the chip observed on the etched chip roots (Fig. 6.13b)

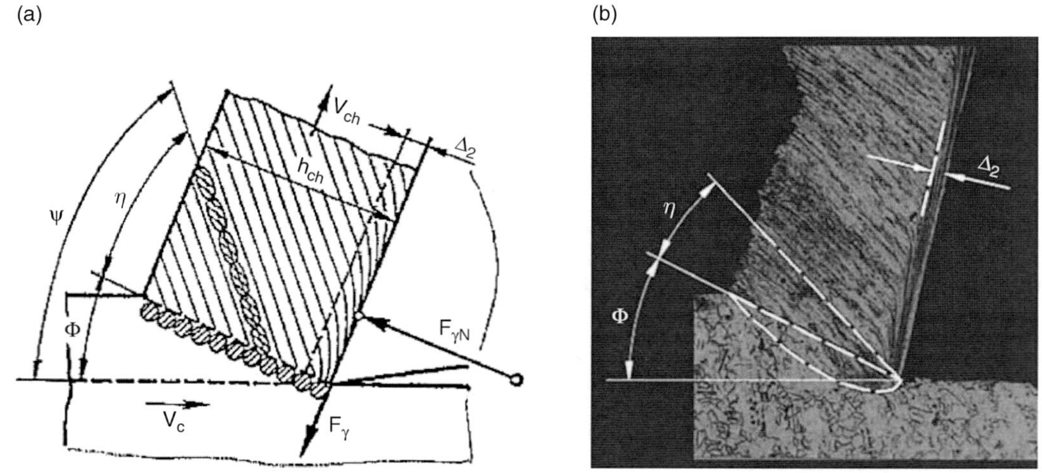

Fig. 6.13 Deformation texture of the chip [2]: preferred orientation of grains (a) and micrograph of the chip root (b).

obtained by means of a special quick-stopping device. The angle η between the chain of preferably oriented grains and the shear plane, shown in Fig. 6.13a, is given by

$$ctg\eta = \frac{\gamma_{sh}}{2} + \sqrt{1 + \frac{\gamma_{sh}^2}{4}} \tag{6.28}$$

In the parallel-sided shear zone model (Fig. 6.14a) proposed by Oxley and Welsh [10], elaborated for higher cutting speeds, the shear plane is assumed to open up so the entry boundary *CD* between the shear zone and work and the exit boundary *EF* between the shear zone and chip are parallel to and equidistant from

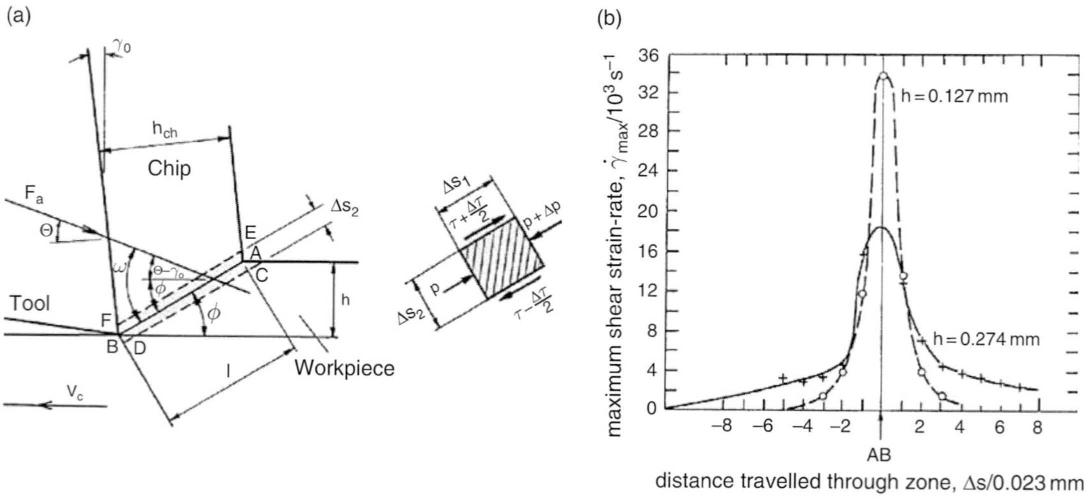

Fig. 6.14 Parallel-sided shear zone model (a) and distribution of maximum shear strain in chip formation zone for $\gamma_0 = 10°$ and $v_c = 120\,m/min$ (b). After Oxley [10].

AB and like AB are slip lines. The overall geometry of the shear model is practically the same as for the shear plane model with AB and Φ but the resultant force F_a will not in general pass through the mid-point of AB as in Fig. 6.11a. Oxley proposed [10] the solution for the shear angle to be a function of variations of the ratio of hydrostatic pressure in point A at the free surface to the shear flow stress on the shear plane AB, i.e. p_A/τ_{AB} which is close to the upper boundary predicted by the slip-line field theory [8].

$$\frac{p_A}{\tau_{AB}} = 1 + 2\left(\frac{\pi}{4} - \Phi\right) \tag{6.29}$$

Then

$$\tan(\Phi + \Theta - \gamma_0) = 1 + 2(\pi/4 - \Phi) - \frac{\Delta\tau}{2\tau_{AB}}\frac{l}{\Delta s_2} = 1 + 2(\pi/4 - \Phi) - Cn \tag{6.30}$$

Where the term Cn (c is the strain-hardening index) may be considered as a correction to the value p_A/τ_{AB} that $\tan(\Phi + \Theta - \gamma_0)$ would have in the absence of any strain-hardening effects.

The expression for the shear strain along the exit boundary EF is the same as for the shear strain as material crosses the shear plane:

$$\gamma_{EF} = \frac{\cos\gamma_0}{\sin\Phi\cos(\Phi - \gamma_0)} \tag{6.31}$$

The flow shear stress at AB is calculated by assuming that half of the total strain γ_{EF} will have occurred at AB. τ_0 is the shear flow at zero plastic strain (for the machining model used $\tau_0 = \tau_{CD}$) and m is the slope of linear plastic stress/strain relation:

$$\tau = \tau_{AB} = \tau_o + \frac{1}{2}m\gamma_{EF} \tag{6.32}$$

The thickness of the shear zone is given by

$$\Delta s_2 = \frac{h}{10\sin\Phi} \tag{6.33}$$

In the tests performed by Oxley the strain rate increased with increase in cutting speed and had a maximum value which occurs in the region defined by the plane AB, as shown in Fig. 6.14b. From the results in Fig. 6.14b and others discussed in [10] the line of best fit passing through the origin is

$$\max(\dot{\gamma}_{\max}) = 2.59\frac{v_s}{h} \tag{6.34}$$

where the unit of v_s/h is s^{-1}.

It can be seen in Fig. 6.14b that plastic deformation starts well in advance of AB and continues well beyond it. The chip formation zone is therefore of substantial thickness, which changes only slightly with cutting speed but significantly with undeformed chip thickness.

REFERENCES

[1] G. Boothroyd, W.A. Knight, *Fundamentals of Machining and Machine Tools*, CRC Press, Boca Raton, 2006.
[2] W. Grzesik, *Fundamentals of Machining of Metallic Materials* (in Polish), WNT, Warsaw, 1998.
[3] Ed. by T.J. Drozda, C. Wick, *Tool Manufacturing Engineers Handbook*, Vol. 1, *Machining*, SME, 1983.

[4] J. Kaczmarek, *Principles of Machining by Cutting, Abrasion and Erosion*, Peter Peregrinus Ltd, Stevenage, 1976.
[5] N.N. Zorev, *Metal Cutting Mechanics*, Pergamon Press, Oxford, 1966.
[6] W. König, F. Klocke, *Fertigungsverfarhen. Drehen, Fräsen, Bohren*. Springer, Berlin, 1997.
[7] E.M. Trent, P.K. Wright, *Metal Cutting*, Butterworth-Heinemann, Boston, 2000.
[8] T.H.C. Childs, K. Maekawa, T. Obikawa, Y. Yamane, *Metal Cutting. Theory and Applications*, Arnold, London, 2000.
[9] Optimize metal cutting by measuring cutting forces. www.kistler.com.
[10] P.L.B. Oxley, *Mechanics of Machining. An Analytical Approach to Assessing Machinability*, Ellis Horwood, Chicester, 1989.
[11] M.C. Shaw, *Metal Cutting Principles*, Clarendon Press, Oxford, 1989.

CHAPTER SEVEN

Chip Formation and Control

Contents

7.1 Chip Classification

In machining with a flat-faced tool when a chip leaves the chip formation zone undisturbed, one of the three possible basic forms of free chips can be achieved (Fig. 7.1) [1]:

1. A straight line form on the tool rake face inclined at an angle indicating the chip flow direction (Fig. 7.1a), or
2. A circular form on the plane tangential to the tool rake face, called side-curling (Fig. 7.1b), or
3. A circular form on the plane perpendicular to the tool rake face, called up-curling (Fig. 7.1c).

All real shapes of chips formed without any artificial obstacles can be constructed by combination of these three shapes. The three geometrical factors influencing the entire chip shape, i.e. the angle of chip side flow (η_s), the side-curling radius (r_s) and the up-curling radius (r_u), depend on the dimension of cut, the geometry of the tool, the work and tool materials. In practice, chip shapes are additionally controlled by special provisions on the rake face, such as the chip former steps or grooves, to obey chip-breaking effect. Details of chip-breaking phenomena are given in section 7.4.

ISO 3685-1977 gives a comprehensive classification of chip forms that are generally produced in metal machining operations based on their sizes and shapes. This classification system shown in Fig. 7.2 consists of eight descriptive shape groups, with each of these groups being subdivided into further 1–4 subgroups defining the size (e.g. long, short, etc.) and the physical conditions (e.g. snarled, connected, loose, etc.). Moreover, it also includes a third digit numerical code for certain chip shapes, to characterize the direction of chip flow (numbers 1–4) and the chip-breaking mode (numbers 5–8).

The third possible classification of chips shown in Fig. 7.3 is based on the mechanisms of material deformation and fracture resulting from the properties of the work material and process conditions. This classification generally involves spalled or torn-off chips, usually called discontinuous [5] and shear types of chips shown in Fig. 7.3a and Fig. 7.3b–d, respectively. The latter group of chips can be subdivided into elemental, segmented (also called serrated or saw-tooth type) and continuous chips, which indicate different levels of material continuity.

Discontinuous chips (Fig. 7.3a) are generally formed when machining brittle materials or materials containing points of stress concentration such as the graphite flakes in cast iron or the manganese sulphide inclusions in free-machining steels. They may also be produced when machining ductile materials at very low speeds and

Advanced Machining Processes of Metallic Materials
ISBN: 978-0-08-044534-2

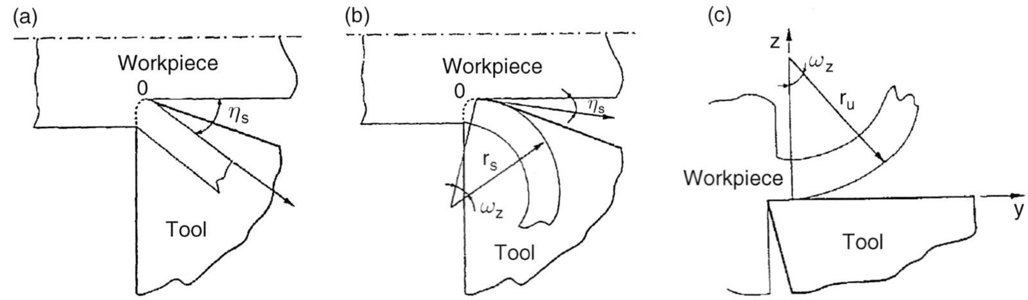

Fig. 7.1 Three basic forms [1]: (a) straight chip, (b) side-curling, (c) up-curling.

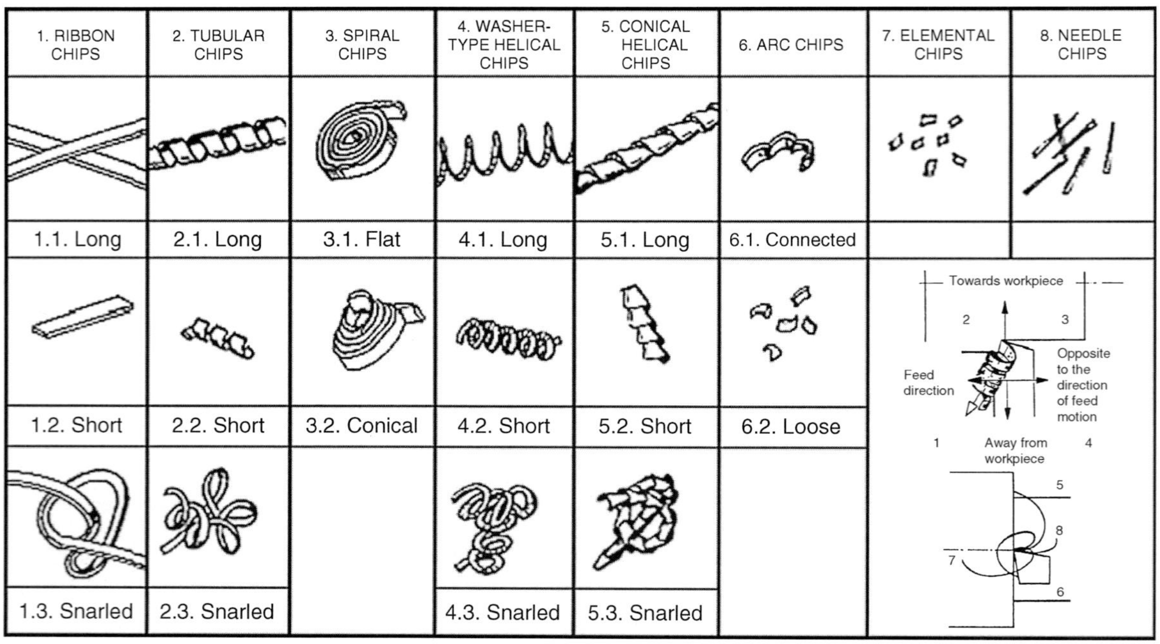

Fig. 7.2 Classification of chip forms according to ISO 3685-1977 (E).

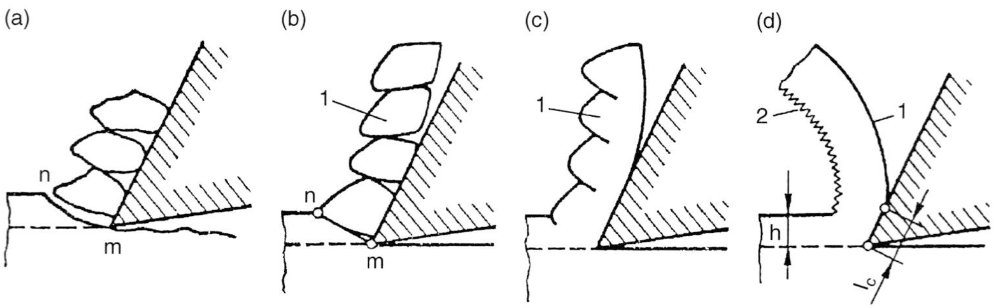

Fig. 7.3 Classification of chip types [3]: (a) discontinuous, (b) elemental, (c) segmented, (d) continuous.

high feeds. Fracture occurs when the chip does not undergo plastic deformation and a crack propagates from the cutting edge (point *m*) to the free surface (point *n*). As a result, the chip breaks up into small segments in a cyclic manner.

Continuous chips (Fig. 7.3d) are common when most ductile materials, such as wrought iron, mild steel, copper, and aluminium, are machined at relatively high cutting speeds. The reason is ductile metals and alloys do not fracture on the shear plane. It is noteworthy that cutting under these conditions is a steady-state process. The formation of a continuous chip takes place in the extended zone as described in section 6.5. Continuous chip formation in ductile materials is associated with reduced cutting forces and generally results in good surface finish. As shown in section 7.1 continuous chips may adopt many shapes – straight, tangled or with different types of helix. Often they have considerable strength due to intensive plastic deformation, and control of chip shape is one of the practical problems challenging machinists and tool designers. Usually, continuous chips with built-up edge (BUE) are additionally distinguished at speeds at which the temperature at the chip/tool interface is relatively low and friction is so great that the chip material welds itself to the tool face. A BUE tends to grow until it reaches a critical size and then passes off with the underside of the chip and the new workpiece surface. A cyclic variation in size of the BUE gives rise to a variation in depth of cut which deteriorates dimension accuracy and surface finish. The effect of BUE formation on cutting-tool wear will be discussed in Chapter 12.

Continuous and discontinuous chips are not two sharply defined categories and some gradations between them occur as shown in Fig. 7.3b and c. In particular, under certain conditions, strain in the chip is very inhomogeneous even though the chip seems to be continuous. High speed and hard machining discussed in Chapters 15.1 and 15.3 are representative examples of such behaviour. It is observed under these conditions that the process of chip formation tends to become periodic with corresponding variations in chip thickness. The basic types of such chips are *elemental* (chips with separated segments), *segmented* and *wavy* (partially segmented) chips [3, 4]. *Wavy chips* are sometimes observed as a transition from continuous chip formation to segmented chips as cutting speed is increased.

Segmented chips (Fig. 7.3c) are characterized by regions of intense shear separated by regions of material with relatively lower deformation. Such chips are produced for most materials as cutting speeds are increased to high levels, but for materials with low thermal conductivity and low thermal capacity such as titanium alloys and austenitic stainless steels, segmentation takes place at more moderate speeds. In the latter case, cutting with a thermo-plastic instability, called adiabatic shear, occurs. As speed increases, the separation of the segments becomes more severe until complete separation of the segments (Fig. 7.3b). As a consequence, if the heat generated is not dissipated quickly enough, the catastrophic shear becomes localized and narrow segments formed are separated by relatively fewer undeformed regions.

There is a transition from one type of chip formation to another as speed is increased as shown in Fig. 7.5 for AISI 4340 steel with different hardness values. It can be observed that for more hardened material the speed at which segmental chips are formed, and especially when complete chip separation occurs as in Fig. 7.4b, is lower.

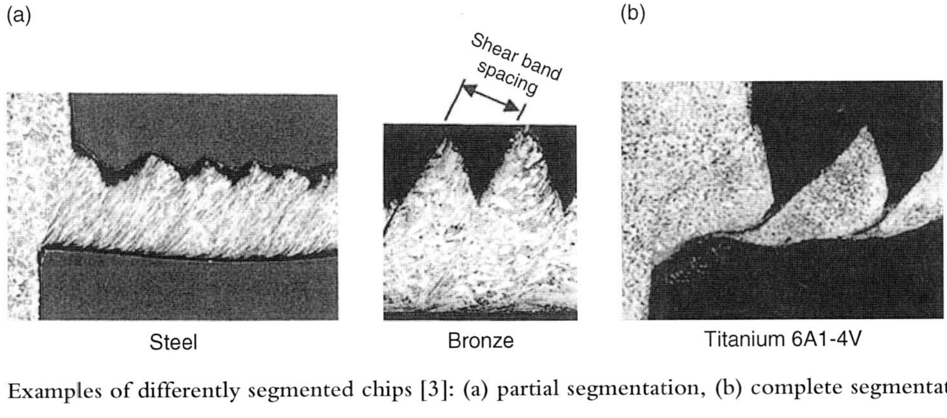

Fig. 7.4 Examples of differently segmented chips [3]: (a) partial segmentation, (b) complete segmentation.

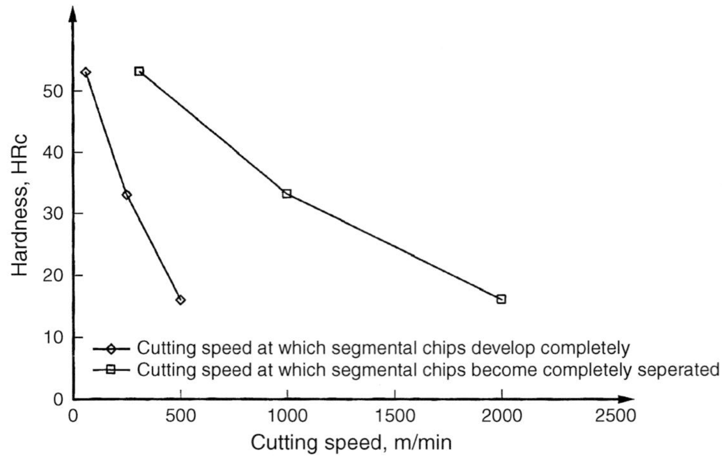

Fig. 7.5 Segmentation of chips from AISI 4340 with different hardness values vs cutting speed [3].

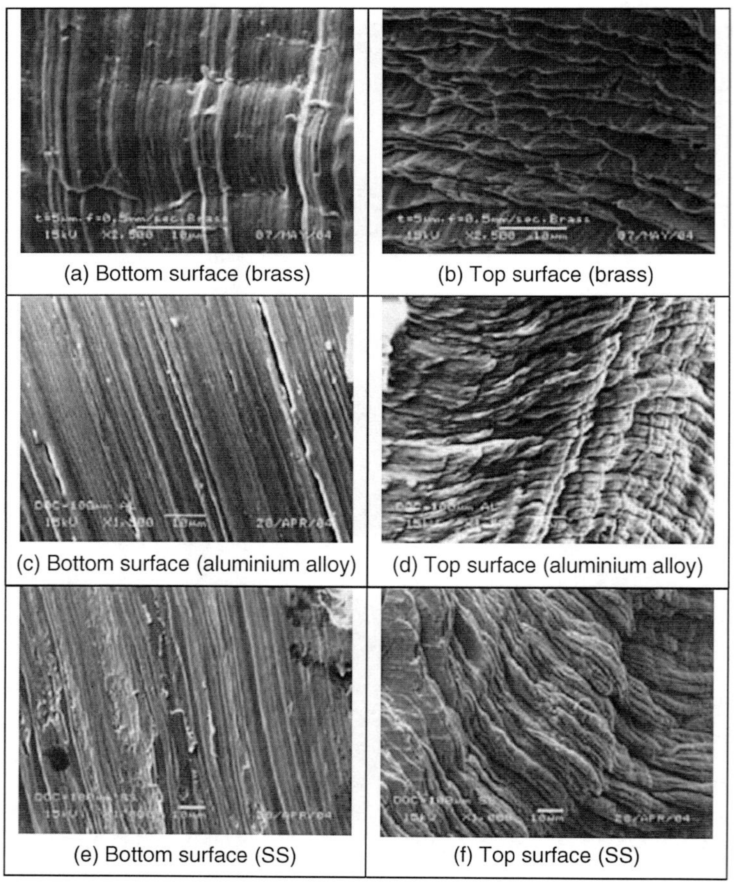

Fig. 7.6 SEM images of free side and underside of chips produced from different work materials: (a) brass; (b) aluminium alloy; and (c) stainless steel [8].

It is observed that even with orthogonal cutting, the cross-section of the chip is not strictly rectangular. First, in free cutting the chip tends to spread sideways and the maximum width is sometimes greater than the nominal depth of cut. Second, the chip thickness is greatest near the middle, tapering off somewhat towards the sides. As shown in Fig. 7.6, the upper surface of the chip is always rough, usually with minute corrugations and steps. Even with a ribbon, continuous chip cracks are often observed, breaking up the top surface into a series of thin segments. Also, the underside of the chip is not smooth due to intensive friction and transferring wear products, including BUE particles when they occur.

7.2 Chip Formation Mechanisms

The formation of torn-off and shear-type chips can be explained in terms of material behaviour during the deformation process using the appropriate stress/strain curves and relevant fracture mechanisms shown in Figs 7.7 and 7.8, respectively. In Fig. 7.7 continuous chips are related to curve 1 with large plastic flow of the material and characteristic neck of a tensile specimen developed in region II. This phenomenon is termed by Shaw [4] as negative strain hardening in the engineering stress/strain curve.

If the plastic flow is not so intensive and the shear strain exceeds slightly the critical value of strain γ_{so} the onset of chip segmentation is observed (partially segmented chips vs curve 2). When the entire shear strain further decreases, the chip segmentation starts to develop and the separation of chip segments begins (elemental chips vs curve 3). In this case, the material necks before fracture but at relatively lower shear stress. It should be noted that all three types of chips formed by shearing mechanisms can be related, in general, to the plastic deformation of the ductile work materials. In contrast, brittle materials undergo only elastic deformation (curve 4) which leads to producing discontinuous chips.

All previously selected chip types can be also quantitatively related to the types of material fracture based on the Fridman diagram, shown in Fig. 7.8. This diagram utilizes the Tresca criterion for ductile

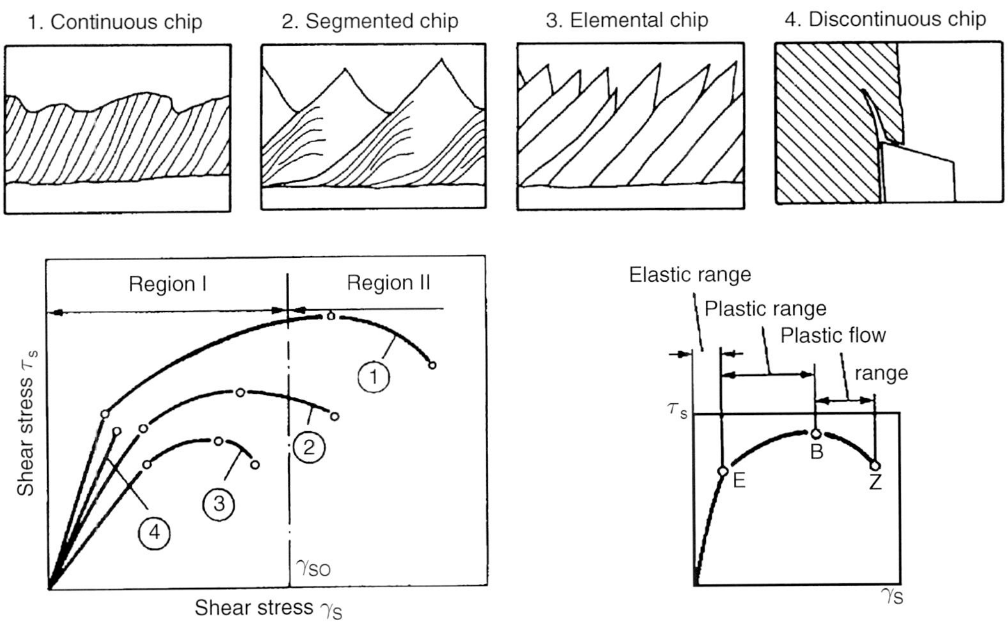

Fig. 7.7 Chip formation in terms of stress/strain curves. After Weber and Loladze [6, 7].

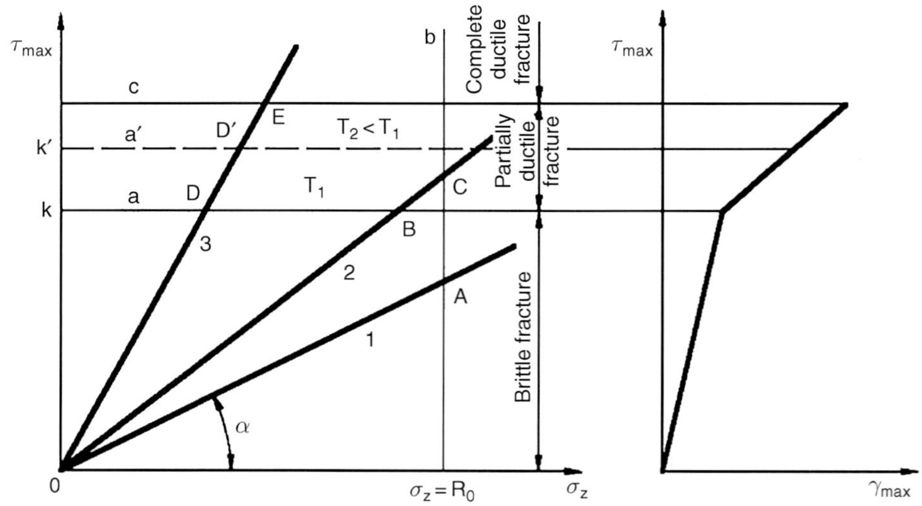

Fig. 7.8 Chip formation in terms of fracture mechanisms [6].

fracture (maximum shear stress on the *Y* axis) and the maximum tensile stress (Saint Venant) criterion on the abscissa. Lines *b*, *a* and *c* represent adequately the brittle fracture strength (b), the yield point in shearing (a) and the yield point in shearing for perfectly ductile material (c). Additionally, line *a*′ represents some specific case when the shear stress is artificially lowered by intensive cooling by supplying gaseous or liquid coolants. Inclined lines started from the origin of such a coordinate system represent simple tension (1), simple compression (2) and simple torsion (3). Consequently, three types of material fracture can be distinguished, namely: brittle fracture in point *A*, partially ductile fracture in point *C* because of the linear work-hardening effect (see τ_{max}–γ_{max} diagram on the right) and perfect ductile fracture in point *E*. It can be assumed that these mechanical material states are equivalent to those responsible for the formation of discontinuous, segmented and continuous chips.

It was suggested by Shaw that the real distribution of stresses on the shear plane is not uniform but their variations can be observed as in Fig. 7.9. In region I (nearest the cutting edge) τ_s is independent of normal stress on the shear plane (σ_s), while in region II adjacent to the free surface of the chip, the ratio of τ_s/σ_s is approximately constant corresponding to Amontons's law of friction. These two regions should be connected by a transition region, where the density of microcracks increases to the point of intercommunication. Due to such stress distribution, the mechanism of plastic flow involving the rewelding of microcracks would be expected to pertain over part I of the shear plane, where $\sigma_s > \tau_s$.

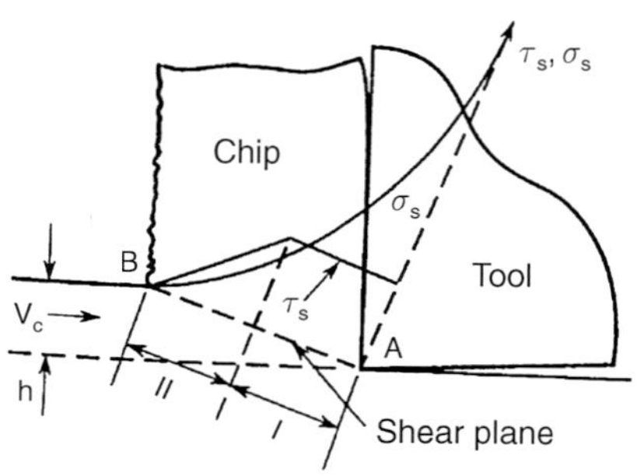

Fig. 7.9 Variation of shear and normal stresses across the shear plane. After Shaw [4].

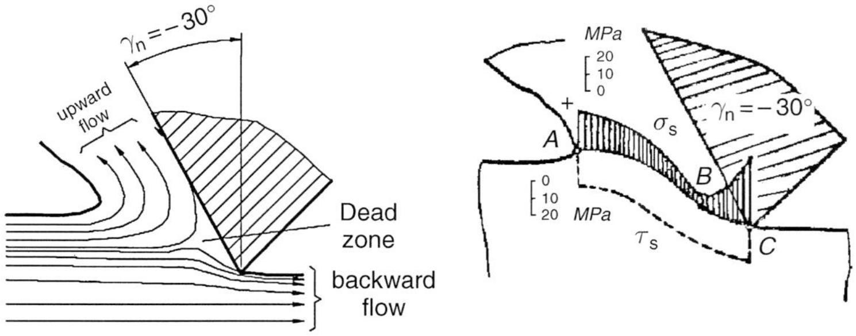

Fig. 7.10 Material plastic flow for a tool with large negative rake angle [9].

The microcrack theory of plastic flow has later been confirmed by investigation of plastic flow in cutting of lead with tools having large negative rake angles, reported in [9]. As can be seen in Fig. 7.10 the stagnant (dead) zone of the range depending on the value of negative rake angle appears between upward flow and backward flow zones. As a result, point *B* corresponding to the inflexion point on the normal stress distribution curve will hypothetically represent the initiation of perfect ductile fracture and consequently the onset of continuous chip formation.

The adiabatic shear theory suggests that the root cause of saw-tooth chip formation is a catastrophic thermo-plastic instability where the decrease in flow stress due to thermal softening associated with an increase in strain is more pronounced than the associated strain hardening. Figure 7.11a is a schematic diagram of the shear localized chip formation process proposed by Komanduri [10, 11]. In general, two stages occurring simultaneously are involved in the formation of shear localized chips but the stress state and strain distribution are different in the two regions. This means that actually two shear planes, the auxiliary and primary shear planes, operate simultaneously [13]. One stage comprises plastic instability and strain localization in a narrow band (i.e. along line (5)) in the primary deformation zone. The other stage involves gradual build-up of the segment with negligible deformation by upsetting the inclined work material by the advancing tool. During this stage intense shear (even shear strain about 60) between the segment just formed and the undergoing upsetting, takes place in a narrow band. A thermo-plastic instability occurring along dashed line (5) extends from the tool tip, curving upward toward the free surface. As the tool advances, the material above line (5) is extruded into the shape (2, 3, 4, 5), line (5) becoming surfaces (3) and (4). At the same time, inclined

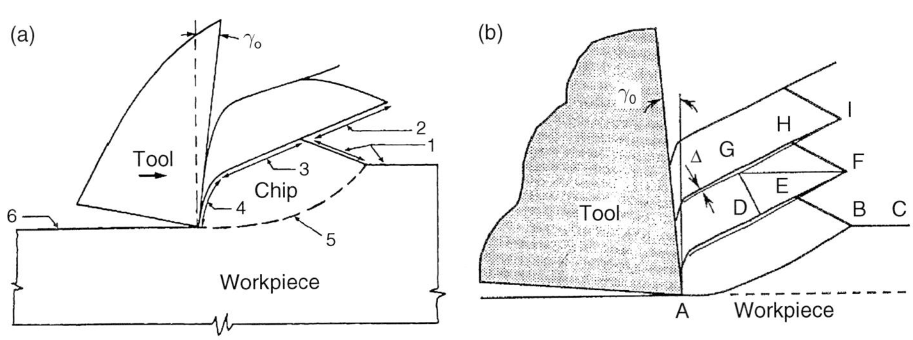

Fig. 7.11 Schematic of the shear localized chip formation process (a) [10, 11] and schematic diagram for calculating shear strains (b) [12]. 1 – undeformed surface; 2 – part of the catastrophically shear failed surface separated from the following segment owing the intense shear; 3 – intense shear band formed by catastrophic shear during the upsetting stage of the segment being formed; 4 – intensely sheared surface of a segment in contact with the tool and subsequently slid on the tool face; 5 – intense localized deformation in the primary shear zone.

surface (1) becomes surface (2). In this model the final chip shape is completed after one cycle and then the process starts again. With increase in cutting speed the extent of contact between the segments decreases until the speed at which completely isolated segments of the chip will be formed. The chip formation by the catastrophic (adiabatic) shear was found to occur with Inconel 718 at a cutting speed above 100 m/min, hardened AISI 4340 steel at high speed of 260 m/min, Ti-6Al-4V titanium alloy at all speeds [10], case carburized steel of HRC 62 [11] and AISI 321 stainless steel at a cutting speed of 50 m/min [13]. It should be noted that the frequency of crack formation is very high in hard turning operations [11], hence a very rigid machining system is necessary in hard turning to accommodate shock occurring at the beginning and end of cut.

Calculations of the shear strains occurring within the lamellae and along the shear fronts can be performed from geometric considerations of the shear strains illustrated in Fig. 7.11b. The shear strain within the segments is termed the homogeneous shear strain γ_h and the shear associated with the formation and propagation of shear bands is designed as the catastrophic shear strain γ_c. The maximum amount of homogeneous shear strain applied in the first stage is given by

$$\gamma_h = \frac{GH}{GD} \tag{7.1}$$

and the amount of catastrophic (localized) shear is given by

$$\gamma_c = \frac{HI}{\Delta} \tag{7.2}$$

where Δ is the thickness of a thin shear band.

Hence, the total shear strain γ_{sb} in the shear bands is given by the sum of both components, i.e.

$$\gamma_{sb} = \gamma_h + \gamma_c \tag{7.3}$$

Chip formation in micro-scale depends strongly on the minimum undeformed chip thickness (h_{min}) due to the fact that the chip may not be formed when the depth of cut is less than h_{min}. As a result, simple scaling of the chip thickness variation model, assuming complete removal of the undeformed chip thickness (UCT) from the workpiece and generating the chip, cannot be valid. Therefore, finding the minimum UCT has received much attention for several decades. In practice, the small UCT causes a large negative effective rake angle (γ_e) due to the small feed rate and the action of the rounded cutting edge, and, as a result, ploughing phenomenon occurs leading to the elastic recovery (also termed springback effect) or elastic–plastic deformation of the subsurface layer. The concept of the minimum chip thickness is that the feed rate or depth of cut must be over a certain critical value before the chip is formed. This transition from ploughing to chip formation is depicted in Fig. 7.12.

When $h < h_{min}$ (Fig. 7.12a) only elastic deformation occurs but as the UCT approaches h_{min} chip formation by shearing becomes a predominant effect over elastic deformation (the removed depth of material layer is less than the nominal depth of cut). However, when $h > h_{min}$, the elastic deformation decreases significantly

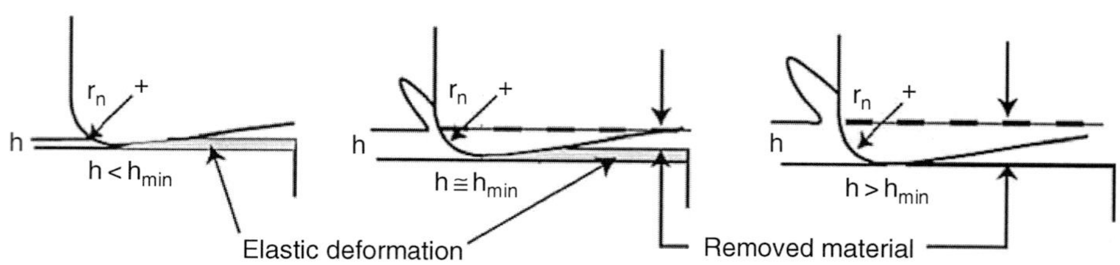

Fig. 7.12 Schematic of the effect of the minimum chip thickness on chip formation [19].

and the entire depth of cut is removed as a chip. Approximately, it can be assumed that under dry or semi-dry friction conditions the ratio of h_{min} to the radius of the cutting edge is about 0.1, i.e. $h_{min} \approx 0.1r_n$ [6].

More accurate prediction of the beginning of micro-cutting can be made based on the molecular-mechanical theory of dry friction, which considers the work-hardening effect and adhesion at the contact. It should also be noted in Fig. 7.13 (after Komanduri *et al.* [18]) that the ratio of the cutting force to the thrust force decreases from 2 to values considerably below 0.5 when transition from conventional machining performed with positive or slightly negative ($\gamma_e = -5°$) rake tools to large negative rake angle cutting tools takes place. Large negative rake angle tools are represented in Fig. 7.13 by simple abrasive grains in grinding (b), ultraprecision machining (c) and, for comparison, by sliding of the rounded indenter.

As an example, the simulation of nano-scale cutting using atomic models of the mono-crystalline copper workpiece and a diamond tool is shown in Fig. 7.14a. The workpiece has the size of $6a \times 34a \times 2a$, where 'a' denotes the lattice constant of copper. The completely rigid tool with an extremely sharp diamond cutting edge has the 6.3° rake and 5° clearance angles. The two-dimensional cutting with the 1 nm depth of cut was performed by prohibiting atomic motion in the Z direction and maintaining the relative XY position of different Z-coordinate atoms. The simulated results shown in Figs 7.14b and c include the chip formation from the viewpoint of the minimum thickness of cut and the distribution of energy dissipation in plastic deformation, respectively. It is evident in Fig. 7.14b that the points imitating plastic deformation are densest around the shear zone and also within the chip. The energy distribution represented by a pie graph shows that a great amount of the total energy is consumed on plastic deformation in front of the tool (about 52% in region I) and under the tool (about 42% in region II). Especially, the second energy fraction is considerably larger compared with that in macro-scale cutting, although the energy dissipated on new surface generation (about 6%) is relatively low. This fact suggests that in submicron diamond cutting a significant fraction of mechanical energy is not dissipated on material removal but rather in the redundant plastic work resulting from ploughing at the tool edge and sliding of the flank face which is elastically loaded at the machined surface [20]. Figure 7.14c confirms the significance of the minimum chip thickness in nanometric chip removal on Cu and Al based on molecular simulation data. The chip formation shown in Fig. 7.14c was simulated when the UCT of about 1 atomic layer was successively removed with the cutting edge radii of 0.5, 1.5 and 5 nm. It is seen in Fig. 7.14c that the Al atoms localized on the surface atomic layer of about 0.3 nm thick tend to move underneath the cutting edge when its radius increases. Ultimately, when the cutting edge radius was

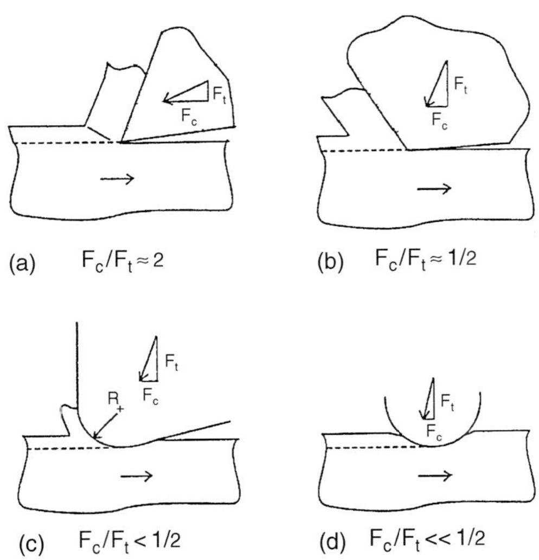

Fig. 7.13 The progression of tool/workpiece interaction which results in the rotation of the resultant force anticlockwise downward towards the workpiece surface: (a) conventional cutting, (b) grinding, (c) ultra-precision machining at small depths of cut, and (d) indentation sliding [18].

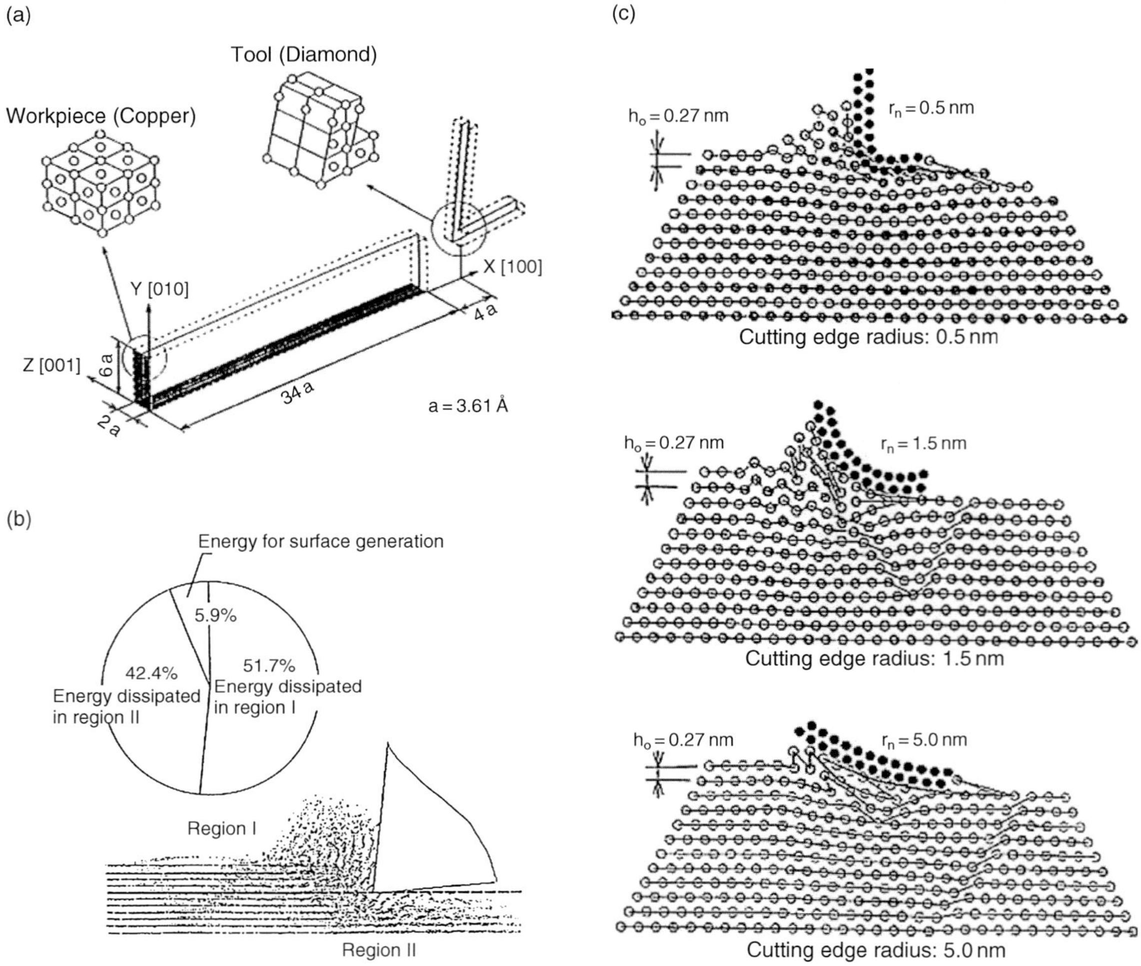

Fig. 7.14 Mechanics of nano-scale cutting [14, 15]: (a) cutting model with 1 nm depth of cut; (b) distribution of energy dissipated; (c) chip formation on Cu cutting with different edge sharpness.

larger then 5 nm, the chip cannot be produced. These results evidently indicate that the minimum UCT in diamond cutting of Al and Cu is on the order of 1/10 of the cutting edge radius, similar to the adequate conditions for conventional cutting provided below. Moreover, the influence of relative cohesive energy for the Al/diamond junction was found to be of secondary importance.

Figure 7.15a shows a nanometric cutting model for the 2D molecular dynamics (MD) simulation of Cu and Al being cut by rigid diamond cutting edge. The work material is divided into three different zones, namely, the moving zone, the peripheral zone and the boundary zone. The motion of the atoms in the moving zone is determined by the forces produced by the Morse interaction potential. In addition, the dislocations generated near the cutting edge are supposed to move in the zig-zag way, as shown in Fig. 7.15b. The chip formation from one atomic layer is possible by successively passing the dislocations through the shear zone to the free surface (no dislocations remain in the chip). Figure 7.15c presents some results of the area-restricted molecular dynamics (ARMD) simulation with reduced computation time, applying to orthogonal machining of a (111) plane of a copper single crystal by a diamond tool with 3 nm edge radius, where the UCT and the cutting speed were set at 1 nm and 20 m/s, respectively. Friction between the

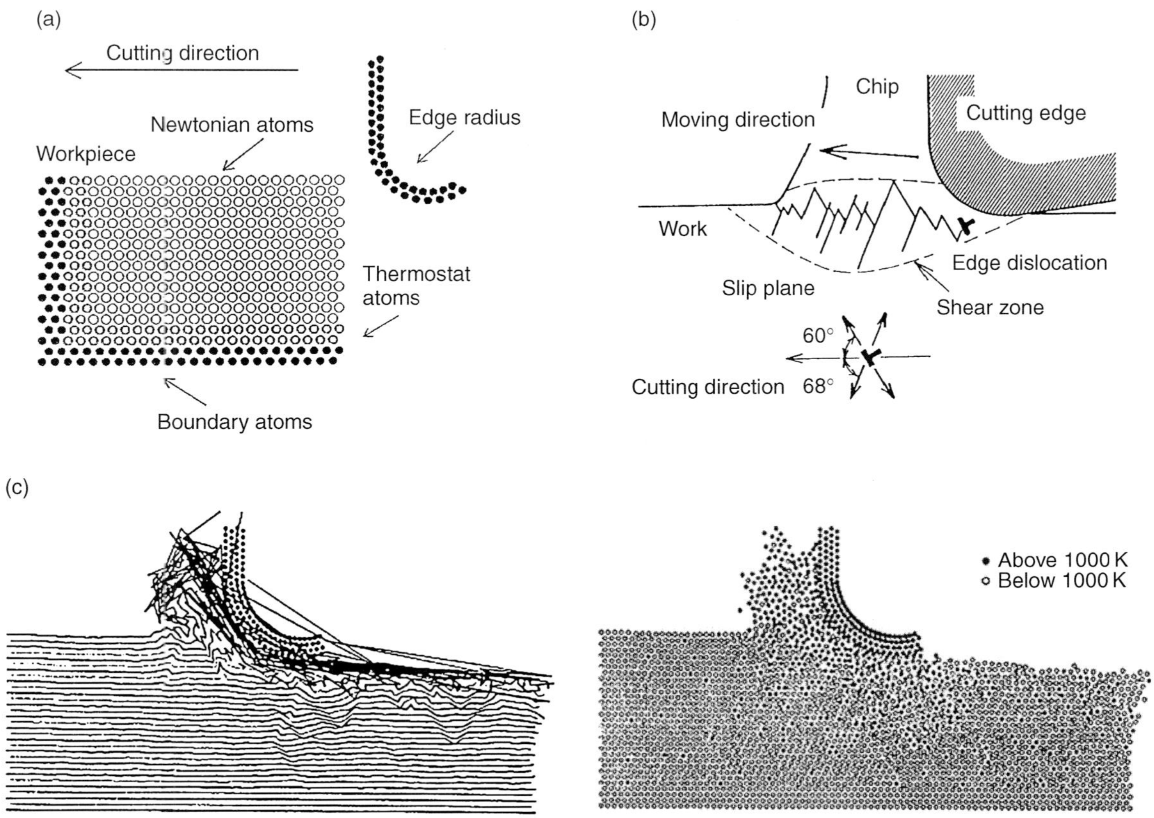

Fig. 7.15 Modelling of nanometric (micro) cutting using molecular dynamics simulation [16–18]: (a) schematic model for molecular dynamics simulation; (b) schematic model of dislocation movement in shear zone; (c) behaviour of work atoms and distribution of temperature for given potential shape.

tool and workpiece in the nano-scale machining was approximated by changing the cohesion energy of the interatomic potential function [17]. Figure 7.15c shows both the trajectories of the work atoms (left) and the instantaneous temperature of each atom at 60 000 time steps (equivalently after a cut distance of about 12 nm). It was found that similar to macro-scale machining all friction-depending quantities such as the chip volume, the contact length, the number of atoms above 1000 K and the degree of the atomistic disturbance below the machined surface decrease with reducing cohesion energy. This approach can also be applied to model tool wear in nano-scale machining. The ultimate surface roughness attainable in Cu and Al machining with a cutting edge radius of 5 nm, a UCT of 0.5 nm and a cutting speed of 200 m/s was equal to 0.5 and 0.8 nm, respectively [16].

7.3 Modelling of Chip Formation

The prediction of shear angle which quantifies macroscopically the state of chip formation is one of the pioneering solutions in metal cutting theory. As mentioned in Chapter 6, in the first model of the cutting process proposed by Ernst and Merchant, the shear in chip formation was confined to the shear plane oriented by the shear plane to the movement of the tool. In shear angle theories many equations were proposed but it

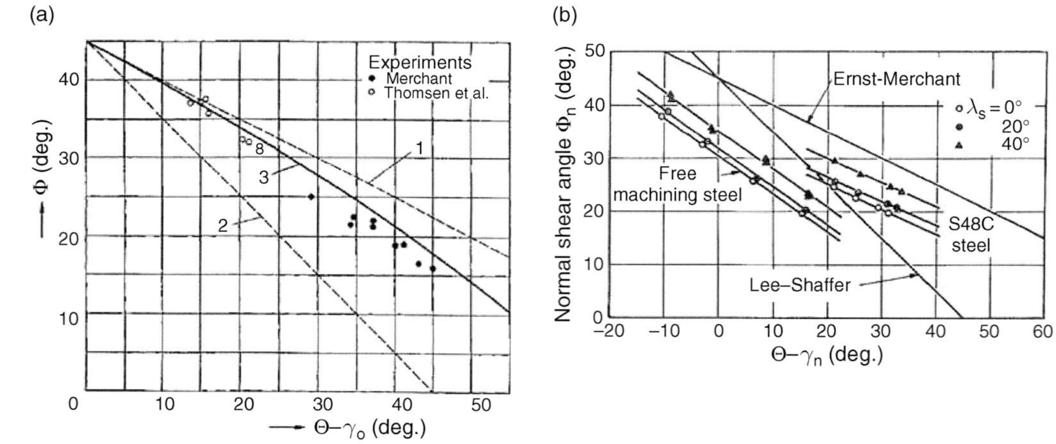

Fig. 7.16 Shear angle relations for orthogonal (a) and oblique (b) cutting [6, 20].

is still not clear which equation is most suitable, because of the many influencing factors. Ernst and Merchant and Lee and Schaffer in their early studies of orthogonal machining, proposed to express the shear angle (Φ) as a function of the difference between the mean friction angle and the rake angle ($\Theta - \gamma_o$) and ignore any effects of work hardening. The computation problem is that a unique function of ($\Theta - \gamma_o$) for the shear angle does not exist. Merchant proposed to derive the shear angle by minimizing the cutting energy (the same as minimizing the cutting force) or maximizing the shear stress on the shear plane. The well-known equation represented by line 1 in Fig. 7.16a results in:

$$\Phi = \frac{\pi}{4} - \frac{1}{2}(\Theta - \gamma_o) \tag{7.4a}$$

This optimum value of Φ gives $Fa = 2kh \cot \Phi$, the normal pressure p on the shear plane being $k \cot \Phi$.

On the other hand, Lee and Schaffer proposed a simple slip-line field to describe the material flow in the triangular plastic zone ABC above the shear plane AB with slip lines parallel and perpendicular to AB (Fig. 7.17a). For force equilibrium of the free chip, it requires that the pressure on the primary shear plane

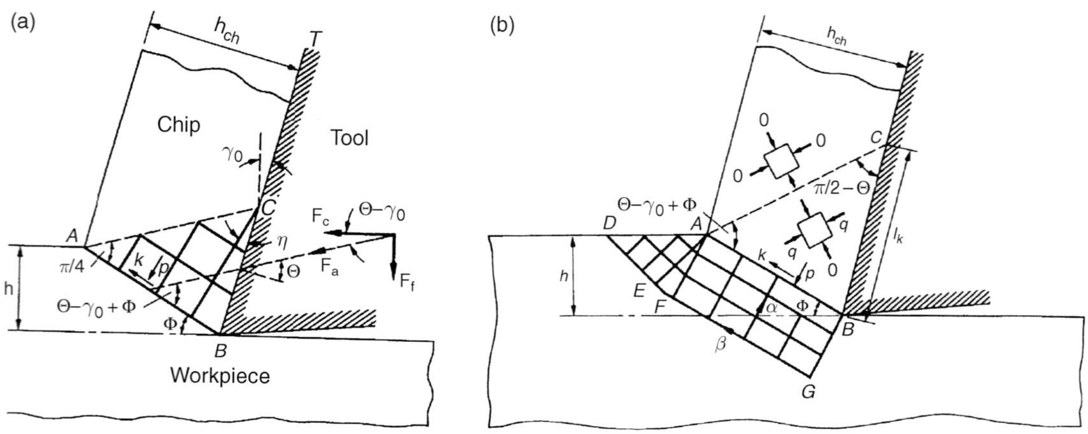

Fig. 7.17 Slip-line fields for a rigid perfectly plastic material proposed by Lee and Schaffer (a) and for linear work-hardening effect after Chakrabarty [24].

is constant along the length of the shear plane and equal to k. Assuming $(p/k) = 1$ and $\Delta k = 0$, Lee and Shaffer's solution, expressed by line (2) in Fig. 7.16a, is obtained as

$$\Phi = \pi/4 - (\Theta - \gamma_o) \tag{7.4b}$$

This means that AC is parallel to the line of action of the resultant cutting force, which is equal to $F_a = kh(1 + \cot\Phi)$. As shown in Fig. 7.17a, neither Eqn (7.3) nor Eqn (7.4) is supported by the experiment. Childs *et al.* [25] suggest that in plotting the results as a graph of Φ against $(\Theta - \gamma_o)$, the process-centred view, i.e. $(\Theta - \gamma_o)$ is the complement of the angle between the shear plane and the tool rake face, is a better choice. This is due to the fact that almost a single relation is observed between the friction coefficient and $(\Theta - \gamma_o)$.

Moreover, Fig. 7.16b presents both the classical solutions for oblique cutting with the inclination angle ranging from 0 to 40°. It is clear that Lee and Schaffer's equation coincides better with the qualitative tendency in the experimental data obtained for both the carbon and free-machining sulphurized steels. However, the quantitative agreement was not achieved, similarly to the orthogonal cutting process.

According to Shaw [4], when the various shear angle equations are tested against experimental data, the points gathered for various rake angles lie along a straight line whose slope indicates that

$$\Phi = C - \Theta + \gamma_o \tag{7.5}$$

The problem is the value of the constant C in Eqn (7.5) is not always the same.

The next step in predicting shear angle includes slip-line modelling with strain-hardening effect and variation of the normal contact stress on the tool face. It should be noted that the straightness of the slip lines results in a constant normal stress along the chip/tool contact and a sliding velocity everywhere equal to the chip velocity. An overview of slip-line fields for different material models and process variants, e.g. with restricted contact tools, chip breaking, the presence of work hardening, changing of hydrostatic stress on the shear plane, stagnant zone, variable flow stress, friction factor, etc., is given by Childs *et al.* [25]. Slip-line field modelling with variable flow stress behaviour proposed by Oxley was shortly reviewed in Chapter 6. As an example, Fig. 7.17b illustrates the slip-line field obtained for the assumption that the linear strain-hardening occurs when material is cut. For the shear angle Φ equal to or less than $\pi/4$, the slip-line field $BADEFG$ consists of two regions with constant hydrostatic stresses, separated by a fan of angular span $\pi/4 - \Phi$. In the triangular segment ADE, the slip lines meet the stress-free surface AD at $\pi/4$, while in the region $ABGF$ the slip lines are parallel and perpendicular to AB. Consequently, the normal stress acting on the shear plane AB is given by

$$p/k = (1 + \pi/2 - 2\Phi) \tag{7.6}$$

The solution for the shear angle can be obtained by substituting the ratio of p/k in the general $p/k = \tan(\Phi - \Theta - \gamma_o)$ relationship. It is represented by straight line (3) in Fig. 7.16a.

The third group of shear angle relationships was derived based on a simple upper bond model of orthogonal cutting [4]. Figure 7.18a shows Rowe and Spick's model with two discontinuities, which accounts for the evidence of sticking friction and expresses friction at the tool face in terms of the relative shear stress (m). The general solution for the shear angle is obtained by differentiating the equation expressing the total work done according to the shear angle, as follows [26, 27]:

$$m \sin^2\Phi = \frac{1}{\chi}\cos\gamma_o \cos(2\Phi - \gamma_o) \tag{7.7}$$

where the ratio $m = \tau_c/\tau_s$ represents the friction condition as in strip rolling and χ characterizes contact length.

The Eqn. (7.7) can be solved graphically for shear angle by plotting separately two families of curves corresponding to the left and right sides of the equation. Intersection of curves gives the value of shear

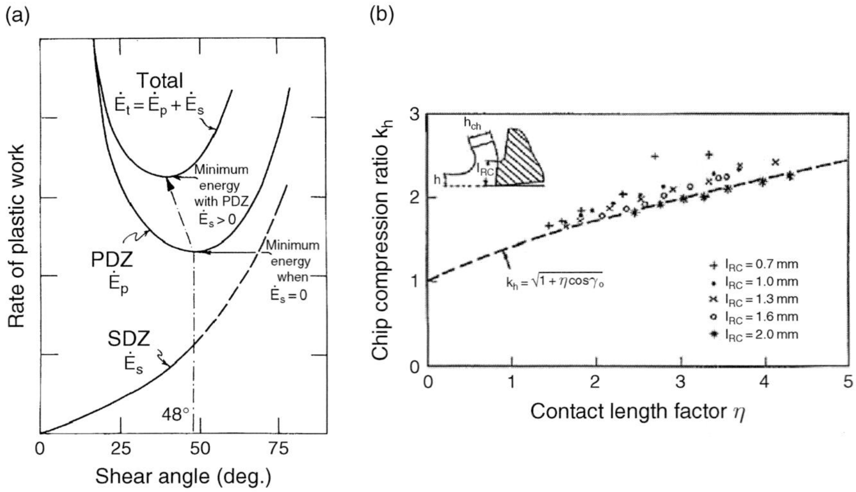

Fig. 7.18 Shear angle (a) and chip compression ratio (b) relationships using upper bound models [26, 27].

plane angle. For example, for $m = 1$ (the case of full sticking friction and contact length equal to UCT) the solution is identical to that proposed by Merchant (7.4).

In the new formulation proposed by Wright [26] the predicted shear angle utilizing the lowest upper bound to cause shear on severely work-hardened material of shear strength k_1 with defined power input is given by

$$\cos(\Phi_o - \gamma_o)\sin\Phi_o = k_o/k_1\,[\cos(45° - \gamma_o'2)\sin(45° + \gamma_o)] \tag{7.8}$$

where shear angle Φ_o corresponds to the unstrained (softer) material with the shear stress of k_o.

De Chifree [28] has also used an upper bound approach, but instead of using a friction angle he introduced the contact length factor $\eta = l_c/h$ or l_{rc}/h for restricted contact tools (Fig. 7.18b). Applying the minimum energy principle and assuming sticking friction conditions ($m = 1$) in Eqn (7.7) the following relationship for the chip compression ratio can be derived:

$$k_h = \sqrt{1 + \eta}\cos\gamma \tag{7.9}$$

Also the corresponding shear angle relationship is as follows.

When turning tests were run on an aluminium alloy with HSS tools ($\gamma_o = 12°$, v_c =50 m/min) the results shown in Fig. 7.18b for various reduced contact lengths are in excellent agreement with Eqn (7.9).

$$p/2k = \frac{k_h}{\cos\gamma_o} - \tan\gamma_o = \cot\Phi \tag{7.10}$$

It should be noted that Eqn (7.10) is identical to the Merchant equation and this has proven that the simple shear plane model belongs to the types of upper-bound solutions.

The fourth group of modelling approaches utilizes computation machining methods, especially finite element modelling (FEM), which is rapidly developed to simulate and visualize the continuous and discontinuous chip formation. The background of the FEM technique was previously described in Chapter 4. This section shortly surveys developments of the finite element approach applied to chip formation.

The results of simulation of continuous chip formation in a high speed orthogonal cutting process by means of an explicit, Lagrangian, elastic–plastic, finite element method is shown in Fig. 7.19. The work

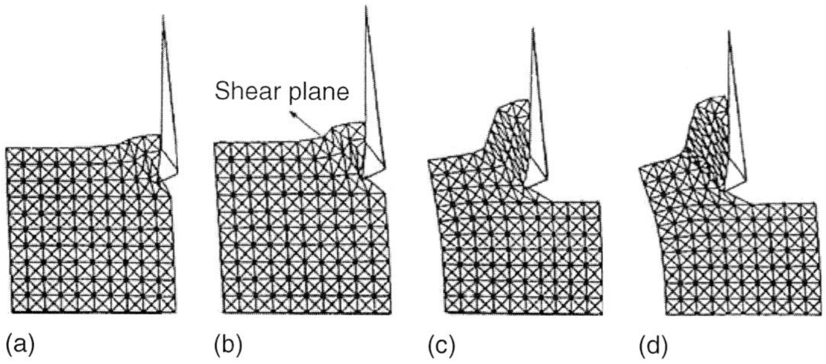

Fig. 7.19 Different stages of the cutting process associated with continuous chip formation [21].

material was an aluminium alloy with 27.6 GPa shear modulus, 517 MPa yield strength and 76.6 GPa bulk modulus, modelled as elastic–perfectly plastic. A special fracture algorithm was implemented together with the sliding surface algorithm to simulate the separation of the chip from the workpiece. Simulations were performed for a cutting speed of 25 m/s and rake angle of 10°. A fracture criterion based on the critical value of effective plastic strain ε^p_{eff} equal to 0.6 was adopted. Four characteristic stages showing continuous chip formation are illustrated in Fig. 7.19a–d. Figure 7.19a shows the initiation of the chip formation process by shearing ($\varepsilon^p_{eff} = 0.29$) which progresses from the tool tip up to the free surface. In the second stage (Fig. 7.19b) the plastic deformation in the PDZ is visibly higher than in other regions of the workpiece and the SDZ is formed. Chip separation (Fig. 7.19c) begins at $\varepsilon^p_{eff} = 0.34$ and in the final stage of chip formation the computed values of ε^p_{eff} in the PDZ and SDZ were equal to 0.45 and 1.13, respectively.

Additionally, Fig. 7.20 shows the full development of a continuous chip removed from the AISI 1045 steel workpiece using AdvantEdge commercial FEM package. As can be seen in Fig. 7.20, the chip formation process includes the indentation of the tool into the workpiece (a), the beginning of chip forming (b), and final chip curling (c) when it is hitting the workpiece ahead of the cut.

On the other hand, Fig. 7.21 shows the formation of a discontinuous chip at different cut distances l in the orthogonal cutting of 60% Cu and 40% Zn β-brass at a low cutting speed of 0.217 mm/s. The simulations were performed using an updated Lagrangian, elastic–plastic formulation and a ductile fracture criterion keeping the fracture strain at 0.62. As can be seen in Fig. 7.21, a shear-type discontinuous chip is simulated, with a crack initiation periodically at tool side of the chip, within a highly deformed workpiece, and propagating towards the free surface side. The crack penetrates through the chip with a sharp drop in the forces. As depicted in Fig. 7.21f, four chip segments are formed periodically at intervals of the cutting distance of about 1.25 mm, thus the second, third and fourth segments have almost the same fan-type shape,

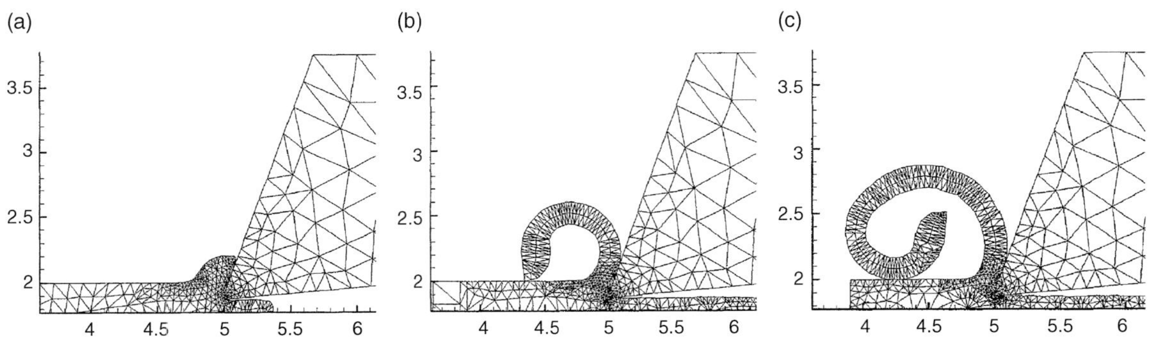

Fig. 7.20 Simulation of continuous chip using AdvantEdge FEM modelling software [29].

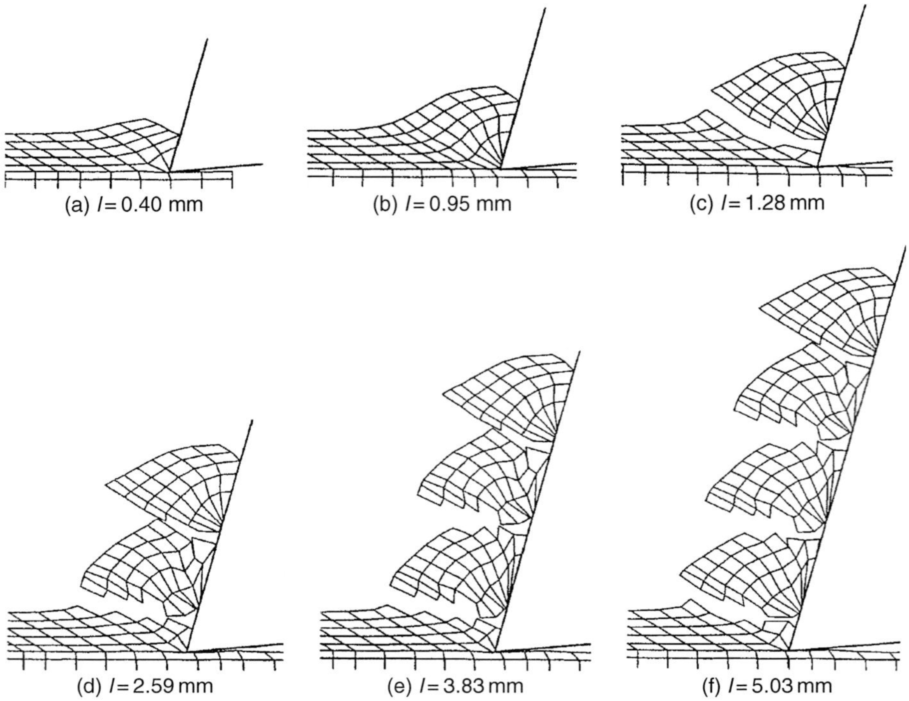

Fig. 7.21 Variation of the discontinuous chip formation of β-brass by changing the cutting length obtained by FEA [22, 25].

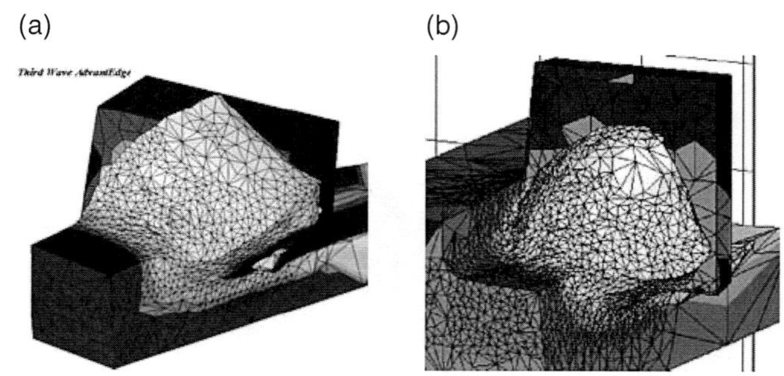

Fig. 7.22 Three-dimensional FEM simulation of oblique cutting with flat rake face tool (a) and nose turning with non-constant chip load modelled (b) [29]. (see Plate 3)

size and strain. The maximum values of equivalent plastic strain in the underside of the chip segment were computed to be of 3.0.

Figure 7.22 shows the simulated images of 3D chip flow obtained in free oblique cutting performed with a flat rake face tool (a) and in non-free (nose) turning using an explicit dynamic, thermo-mechanically coupled finite element modelling software termed *AdvantEdge*. In this case, the FEM-based modelling was run for the AISI 4130 workpiece material using a power-hardening law model with thermal softening.

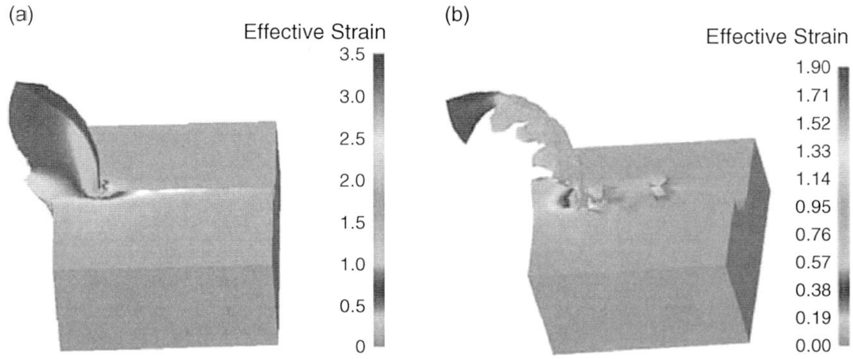

Fig. 7.23 Result with only continuous adaptive remeshing (a) and result with modelling of fracture (b) [30]. (see Plate 4)

In order to simulate the segmented chip formation, a 3D thermo-mechanically coupled finite element model of dry turning operations was developed using commercial FE code, Deform-3D™, v.5.1 [30]. A normalized Cocroft–Latham model was applied as the damage criterion for chip erase. Figure 7.23 shows two specific results of simulation. In the first case (Fig. 7.23a) no modelling for chip segmentation was achieved because only a continuous adaptive remeshing technique was employed. In contrast, with the help of fracture modelling distinct chip segmentation was simulated, more severe in comparison to modelling with shear localization resulting from thermal softening of the material (when a brief segmentation was achieved).

7.4 Chip Flow

The knowledge about chip flow direction across the rake face is necessary for determining the effective tool rake angle under more complex cutting arrangements, such as semi-orthogonal cutting with round-nosed tools and 3D oblique cutting, and primarily for planning the process in terms of efficient chip control. As shown earlier in Chapter 6, for ideal, sharp-nosed tools the chip flows in the direction perpendicular to the main cutting edge. For the same type of tools but in free-oblique machining, involving three-dimensional chip formation, the local chip flow direction follows Stabler's chip flow rule, that the chip flow angle measured from the normal to the cutting edge is equal to the tool inclination angle, i.e. $\eta_c = \lambda_s$ or $\eta_c = k\lambda_s$ where k is between 0.9 and 1 [2, 3]. This means that Stabler's rule is invariable with the cutting speed and it neglects side flow by assuming the chip width is the same as the undeformed chip width. Better agreement with experimental data can be achieved by considering the normal rake angle ($\tan \eta_c = \tan \lambda_s \cos \gamma_n$) or the cutting speed ($\eta_c = \lambda_s v_c^{-0.08}$) as proposed by Russel and Brown and Zorev [6]. According to the rule of minimum cutting energy, in free-orthogonal cutting the chip should flow along the line with maximum slope of the rake face (in the tool face orthogonal plane P_g), i.e. can be geometrically defined as

$$\tan \eta_c = \frac{\tan \gamma_f}{\tan \gamma_p} \tag{7.11}$$

where γ_f is the tool side rake angle and γ_p is the tool back rake angle.

In consequence, the effective rake can be calculated as

$$\sin \gamma_{ef} = \sin \eta_c \sin \lambda_s + \cos \eta_c \cos \lambda_s \sin \gamma_n \quad (7.12a)$$

or after considering Stabler's rule

$$\sin \gamma_{ef} = \sin^2 \lambda_s + \cos^2 \lambda_s \sin \gamma_n \quad (7.12b)$$

More complex equations, which in addition to given cutting conditions require an extensive knowledge of factors which must either be predicted from machining theory or measured experimentally, have also been proposed. For example, Armarego and Brown [32] using an equivalent shear plane model of chip formation in oblique cutting showed geometrically that

$$\tan \eta_c = \frac{\tan \lambda_s \cos \gamma_n}{\tan(\Phi_n + \Theta_n)} + \tan \lambda_s \sin \gamma_n \quad (7.13)$$

where Φ_n is the shear angle measured in the normal plane and Θ_n is the mean friction angle at the tool/chip interface measured in the same plane.

For machining with nose radius tools, Colwell suggested a simplified geometrical method (Fig. 7.24) which assumed that the direction of chip flow over the rake face would be normal to the chord joining the extreme points of the engaged cutting edge or to the major axis of the projected area of cut. However, the predicted results of chip flow direction satisfied only the cutting conditions with zero tool inclination and rake angles. The equation for the predicted chip flow angle obtained from this method can be expressed in three different ways depending on the cut geometry. For the case in which the tool cutting edge $\kappa_r = 90°$ and depth of cut a_p is greater than the tool nose radius r_ε (Fig. 7.24a) the chip flow angle is

$$\tan \eta_c = \frac{a_p}{r_\varepsilon + f/2} \quad (7.13)$$

When the depth of cut is small enough in relation to the nose radius, so that only the curved edge is engaged (Fig. 7.24b), then

$$\eta_c = \frac{a_p}{\left(2r_\varepsilon a_p - a_p^2\right)^{1/2} + f/2} \quad (7.14)$$

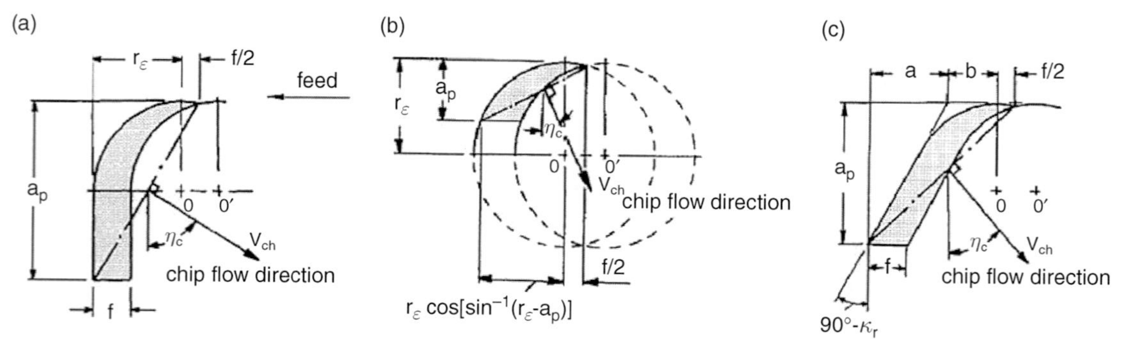

Fig. 7.24 Colwell's method of determining the chip flow angle for nose radius tools [31].

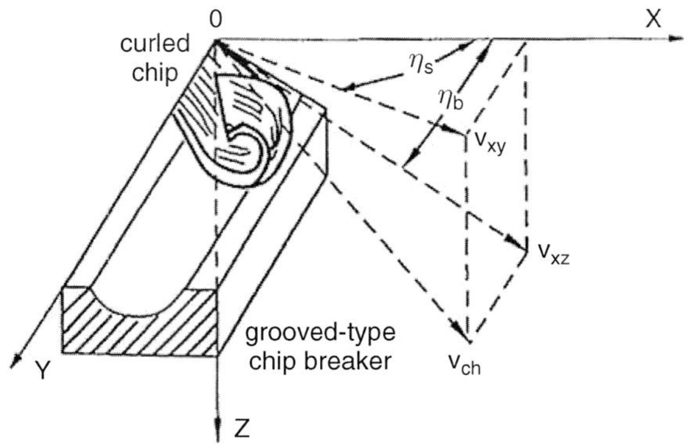

Fig. 7.25 Three-dimensional (3D) chip flow with a grooved chip breaker [35].

For the more general case, where the tool cutting edge $\kappa_r = 90°$ and the depth of cut is greater than the nose radius, as shown in Fig. 7.24c, a more involved relationship is obtained, namely

$$\tan \eta_c = \frac{a_p}{a_p \tan \kappa_r + r_\varepsilon \tan(\pi/2 - \kappa_r/2) + f/2} \tag{7.15}$$

When using chip-grooved inserts, the chip-groove geometry alters the chip flow from its natural state that would normally occur in machining with a flat-faced tool. In particular, under the same cutting conditions and tool geometry, the change in the chip-groove geometry will decisively result in different chip-flow patterns and therefore in different force distribution [36]. Chip flow for a grooved tool occurs as a combination of chip side- and back-flow. Chip side-flow in a grooved tool is measured on the horizontal tool face plane and is similar to that of a flat-faced tool, while the chip back-flow determines the degree of chip streaming into the groove.

As documented by an extensive investigation performed by Jawahir [35], the chip flow in practical oblique machining operations with grooved tools having varying tool geometries is decisively more complex than that described below for the flat-faced tools. In the case of a grooved chip breaker with varying tool restricted contact length shown in Fig. 7.26, the effect of 3D chip flow can be presented by two components, i.e. chip-side flow angle η_s and chip-back flow angle η_b. According to the sketch from Fig. 7.26, η_s is the angle formed by the chip-side flow on the tool face projected onto the (X,Y)-plane and η_b is the angle formed by the chip down flow (X,Z)-plane.

In 3D oblique cutting, the effectiveness of grooved chip breakers is reduced due to the effect of chip-side flow angle η_s. As shown in Fig. 7.26a, when the chip-side flow angle is larger than zero, the equivalent tool restricted contact length l_e and the equivalent groove width B_e measured in the direction parallel to chip flow are larger than adequate constructive dimensions l and B measured directly on the insert. Hence $he = h/\cos \eta_s$ and $Be = B/\cos \eta_s$. Consequently, the chip-groove profile will be useless if the equivalent restricted contact length l_e is larger than the natural contact length l_n. Also, when η_s is large, say more than 75°, both dimensions l_e and B_e will increase rapidly even with a small increase in η_s. As a result, the direction of chip-side flow will be nearly parallel to the cutting edge and therefore such a chip breaker will be no longer usable for chip curling under these process conditions.

The chip-back flow angle η_b plays a key role in chip-groove utilization, which determines the chip breakability and power consumption rates. Figure 7.26b shows a typical chip curling pattern in the case of a grooved chip breaker with restricted contact length, i.e. $l < l_n$. It is apparent from this figure that the smaller is the up-curl radius r_u, the greater is the chip-groove utilization. Optimally, the chip may fully utilize the groove profile when the chip-curl radius r_u is equal to the groove radius $R_o = B/2 \sin \Theta$. In other words, this case is observed when the chip-back flow angle η_b is equal to the groove tangent angle Θ. On the contrary, if $\eta_b > \Theta$, an additional friction between the chip and the groove surface occurs, thus consuming

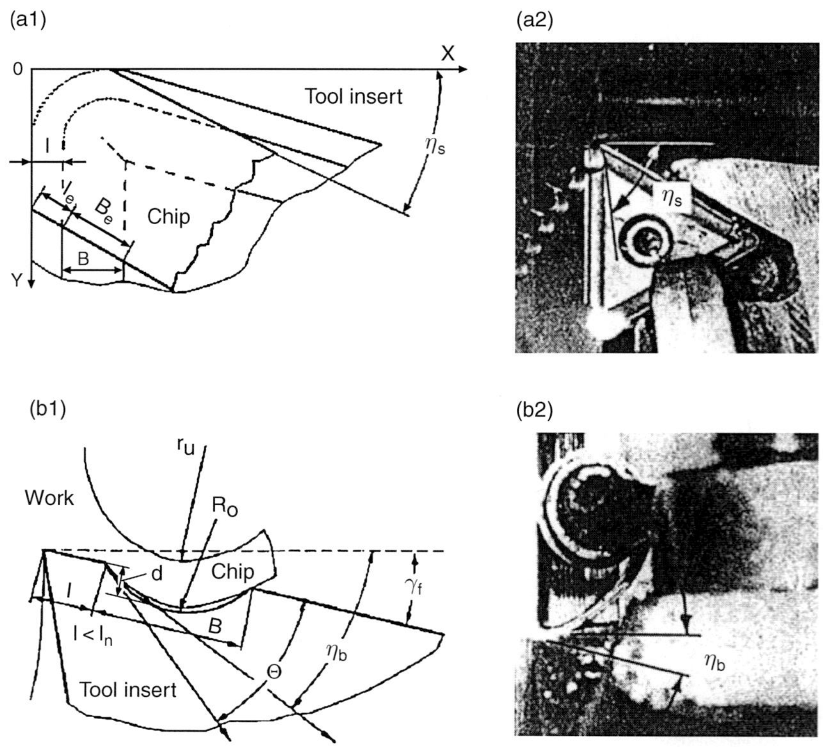

Fig. 7.26 Effects of chip-side flow (a) and chip-back flow (b) angles on 3D chip flow and curling [35]. Sections a2 and b2 present relevant images obtained by high speed filming.

more power. Therefore, in designing grooved inserts, $(\eta_b - \gamma_f)/\Theta = 1/2$ could be taken as the norm in order to guarantee the optimal performance of the groove profile.

In general, taking into account a fine and uniform lamellar shear on an unmachined free surface in continuous chips, the chip curvature (curl radius) can be expressed as [34]

$$r_u = \frac{a_p}{s \cos \gamma_o \sin \Phi} \tag{7.16}$$

where s is the lamellar spacing, γ_o is the rake angle and a_p is the depth of cut.

Usually, groove-type, restricted-contact tools work with $r_u > R_o$ and in such cases the effective rake angle γ_{eff} should be considered instead of the orthogonal rake angle, as shown in Fig. 7.27. Its value depends on the position of chip cross-section relative to the insert corner (Fig. 7.27a and b) and ranges between the side rake γ_{f1} in the land and the side rake γ_{f2} in the groove. Consequently, the up-curling tendency can be expressed in terms of chip flow direction and the up-curling radius varies according the following function:

$$r_u = R_{\min}(\eta_b) \frac{\sin(\gamma_{f2} - \gamma_{f1})}{\sin(\gamma_{eff} - \gamma_{f1})} \tag{7.17}$$

where $R_{min}(\tau)$ is the minimum radius of up-curling defined in the narrower cross-section of the chip-former in chip flow direction. It was found that the function (7.17) can be approximated by a cubic spline as shown in Fig. 7.27c.

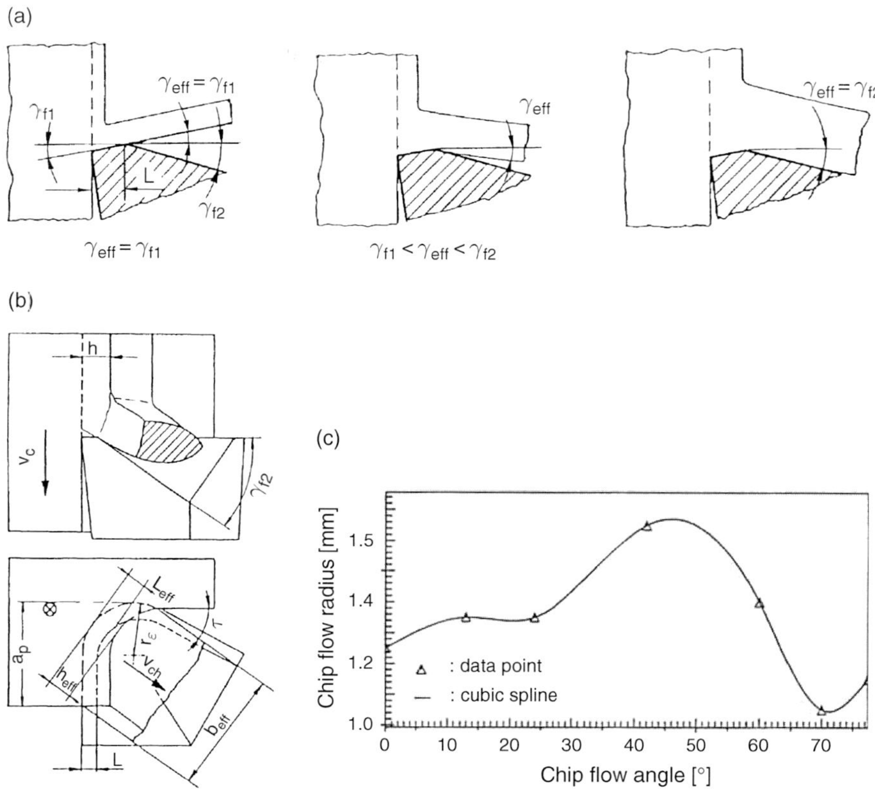

Fig. 7.27 Chip up-curling in three-dimensionally shaped chip formers [37].

7.5 Chip Breaking

When planning machining operations for automated manufacturing systems, it is absolutely essential to select such cutting conditions and tool geometries (including chip-groove parameters) that result in efficient chip breaking, i.e. producing small, handleable and well-broken chips, for the given tool/work combinations. In machining practice, the *chip breakability* has been recognized as a significant machinability factor among the other traditionally used factors such as tool wear, surface roughness, cutting power and dimensional accuracy [1]. The four most common elements of the machining system that have profound influence on chip sizes/shapes are: cutting tool, machine tool, cutting tool and work material properties. Relative to cutting tool, the most significant features are the tool material properties, tool geometry and chip former/chip breaker design parameters, namely the groove and cutting edge configurations.

It has been observed that there are three most common modes of chip breaking, as shown in Fig. 7.28: they are as follows (see also right insert in Fig. 7.2):

1. Chip breaking by contact with the surface to be machined (caused by chip side-curling).
2. Chip breaking by contact with the transient surface (caused by up-curling).
3. Chip breaking by contact with the tool flank (three-dimensional chip breaking).

Nowadays, a great variety of inserts which combine the effect of restricted contact length with complex grooved chip breakers exists. Their geometries are well suited to the cutting conditions in order to enhance

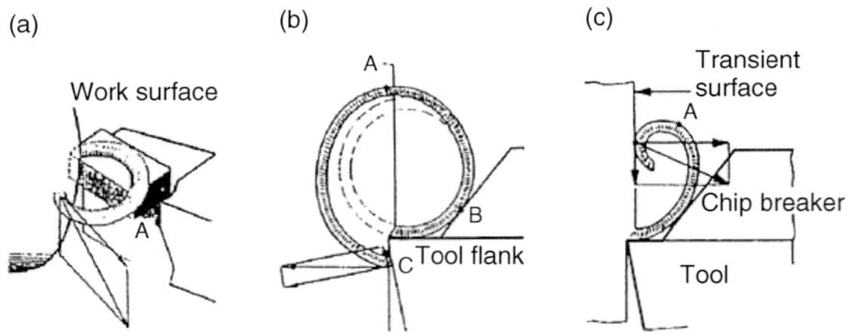

Fig. 7.28 Three common modes of chip breaking [40].

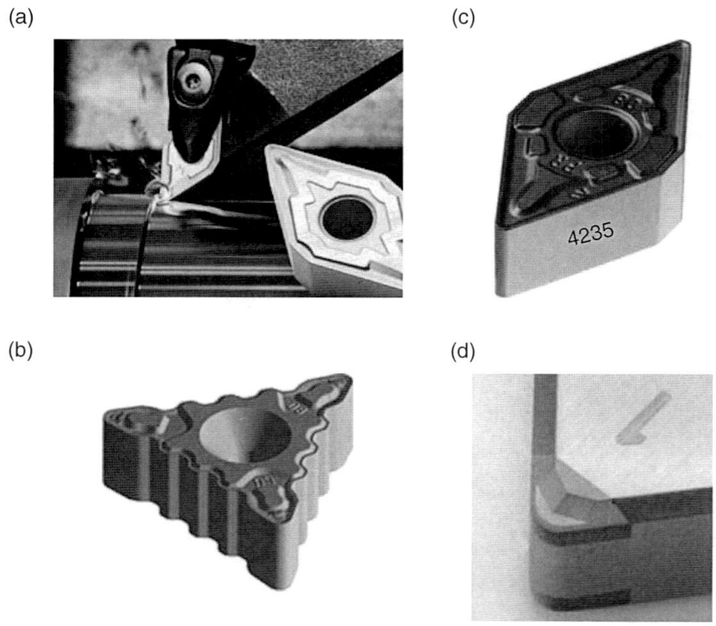

Fig. 7.29 New designs of chip breakers for coated carbide (a–c) and CBN (d) inserts [33]. (see Plate 5)

chip breaking. Figure 7.29 presents some examples of newly designed cutting tool inserts with complex chip breaker geometries recommended for machining carbon and alloy steels by leading cutting tool manufacturers, namely:

1. 55° rhomboidal, TP2000 insert coated with Ti(C,N)/Al_2O_3/Ti(C,N)/ TiN for general turning of steels and alloyed steels (Seco Tools, Sweden).
2. 55° rhomboidal T-REX insert with six cutting edges for machining carbon and alloy steel (Sumitomo Electric, Japan); this design features both dimples in and pimples on the rake face as well as interrupted cutting edges.
3. 55° rhomboidal, T-Max P insert coated with MTCVD TiCN/Al_2O_3 for rough turning of steels (Sandvik Coromant, Sweden).
4. CBN BreakMaster insert with integral chip breaker (Sumitomo Electric, Japan). It should be noted that in this case an obstruction type of chip breaker was applied.

Manufacturers of tool inserts offer specially designed insert types for given combinations of work material and cutting conditions, usually for rough (R), medium (M) and finish (F) machining operations or more universal geometries, such as F and M or M and R. However, fields of their applications are specified by chip-breaking diagrams, known also as a_p-f diagrams, shown in Fig. 7.30. In addition, Fig. 7.30 illustrates how the localization of chip-breaking field in the a_p-f coordinates is influenced by the cutting parameters, mechanical properties of work material, both cutting tool and chip breaker geometries, as well as coolant supplied.

Prediction of 2D chip breaking is based on the assumption that fracture of the chip modelled as a curved beam occurs when the tensile strain on the outer surface of the chip during bending will reach the ultimate tensile strain ε_{max} of the chip material. The relevant chip-breaking criterion, proposed by Nakayama [1, 41], can be expressed as

$$\varepsilon_{\max} \leq \frac{h_{ch}}{2}\left[\frac{1}{r_{ui}} - \frac{1}{r_{uf}}\right] = \varepsilon_b \tag{7.18}$$

where h_{ch} is the chip thickness, ε_b is the chip strain caused by bending, r_{ui} is the radius of initial chip curl and r_{uf} is the radius of final chip curl prior to fracture. This criterion is valid for chip-breaking occurring by chip/tool flank surface contact (case 'b' in Fig. 7.28) in an obstruction-type chip breaker.

Assuming, after Worthington *et al.* [1], as a first approximation that for machining steels at high cutting speeds $h_{ch} \approx 2h$ (h is undeformed chip thickness) and the chip is effectively broken if it expands to twice its original radius, the chip-breaking criterion can also be expressed by equation

$$\varepsilon_{\max} \approx \frac{h}{r_{ui}}\left(\frac{c-1}{c}\right) = \frac{h}{r_{ui}}A \tag{7.19}$$

where for carbon steels and feeds ranging from 0.3 to 0.6 mm/rev, constant 'A' varies from 0.166 to 0.5, i.e. a ratio of 1:3.

It should be noted that chip breaking is a very complex phenomenon depending strongly on the materials and metallurgical aspects of the cutting process [1], such as chip segmentation due to the adiabatic shear instability (the chip breaking is easier) and specific thermal effects caused by tool coatings with low thermal conductivity (due to a thermal bi-metallic effect, additional thermal stresses cause the chip to curl to a smaller radius).

Figure 7.31 shows three major stages in the development of chip up-curl radius for a spiral chip. As seen at stage (a), a new chip breaking cycle begins with initial up-curl radius r_{ui} (the same as in Eqn (7.18)). Stage (b) shows a higher degree of chip straightening by chip/work surface contact forces. It can be remembered that the chip straightening process is a plastic deformation, in which the material yields severe plastic deformation leading to its fracture. It is assumed that the chip at this stage has reached its final up-curl radius r_{uf}. Stage (c) shows a development of chip fracture from the outer profile of the chip (or from the inner profile of the curled chip). At this stage, the previous (departing) and new (incoming) chips are shown with r_{uf} and r_{ui} being retained. The high filming of many transient positions within this cycle shows the continuous variation of r_{ui} leading towards r_{uf}, with variation in the chip/contact point C (Fig. 7.32).

As mentioned above, Fig. 7.32 presents some selected pictures obtained at recorded intervals of time, additionally showing the contact forces acting in the contact points A and C. Moreover, linear measurements were made from the cutting edge B to the respective contact points C based on the screen of the film analyser [39].

The results of individual chip breaking cycle revealed a regular pattern of variation in chip streaming, with a rapid increase in r_u with time at a certain interval of time (between 15 and 20 ms in Fig. 7.33a) and then a sudden drop just prior to the chip fracture. On the other hand, Fig. 7.33b illustrates a variation in the chip/work contact position on the moving work surface. As has been observed, the rapid increase in distance BC increases r_u by an action of chip straightening at the chip root due to increased contact forces resulting from the chip/workpiece contact at the free end of the chip. The fracture for both chip breaker

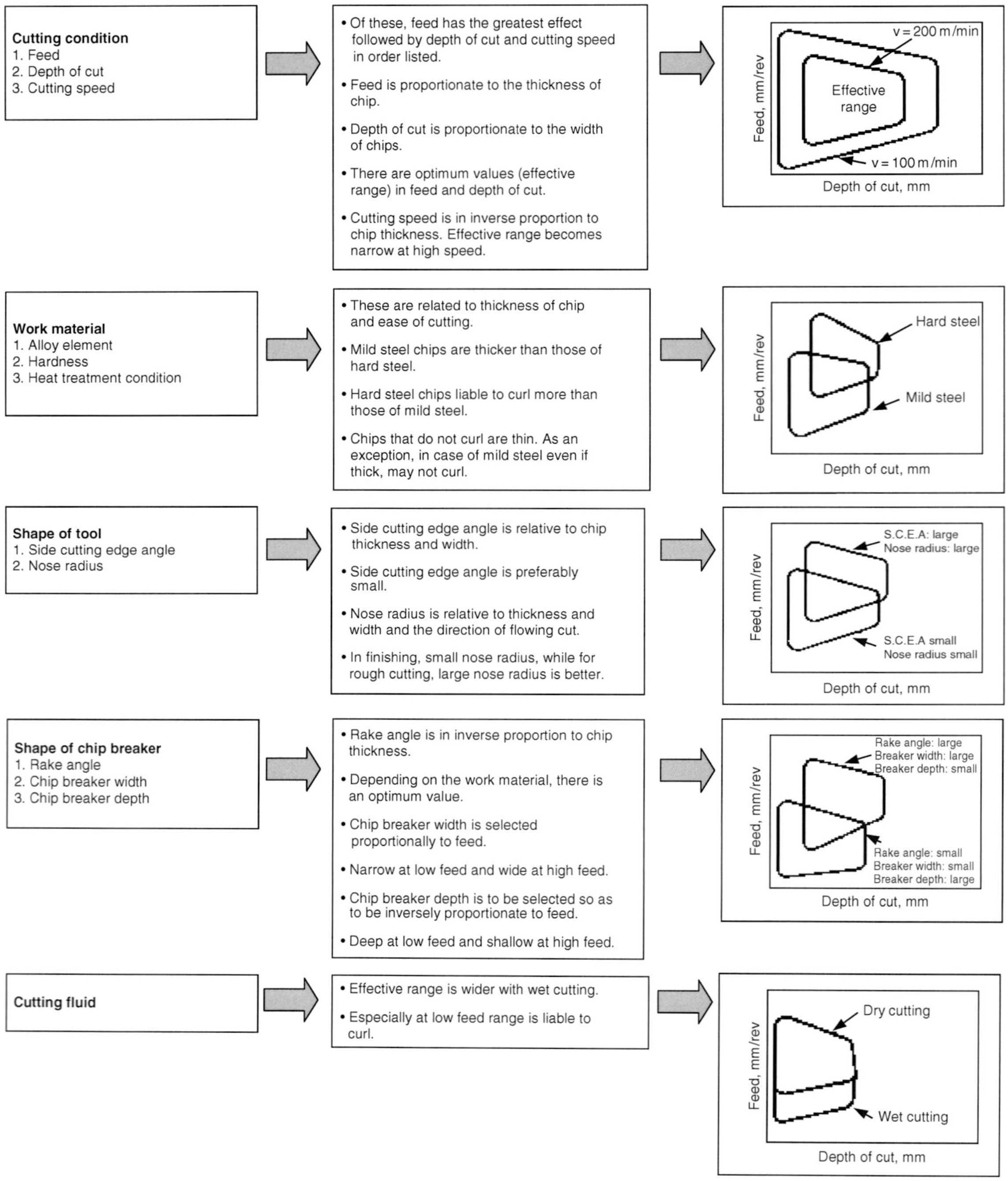

Fig. 7.30 Localization of chip-breaking fields vs process variables. After Toshiba Tungaloy [33].

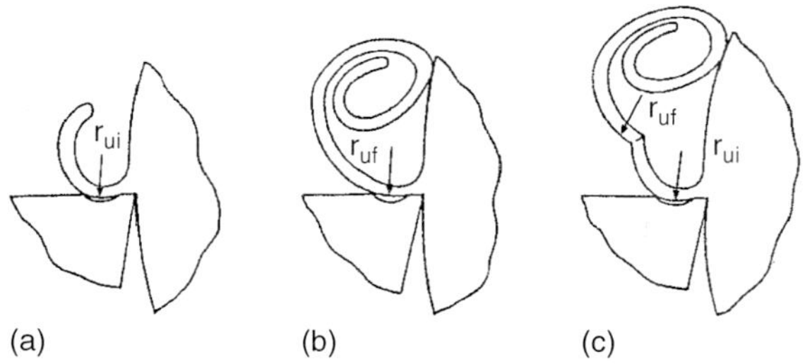

Fig. 7.31 Development of chip up-curl radius before chip fracture [1]: (a) initial chip curl, (b) chip curl just before fracture, (c) chip curl at fracture.

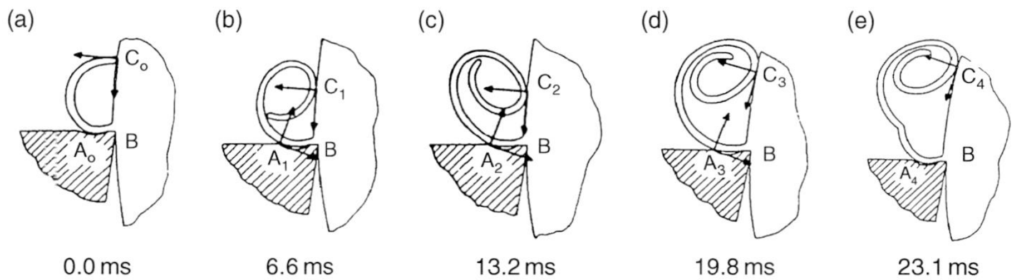

Fig. 7.32 Changes of contact forces within a chip-breaking cycle [39].

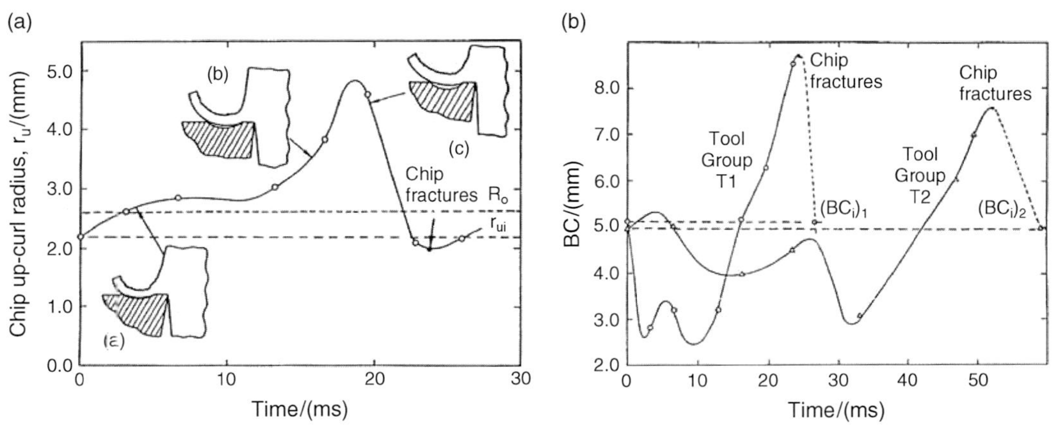

Fig. 7.33 Variation in chip up-curl radius (a) and chip/work surface contact point (b) within a chip-breaking cycle [39]. Cutting conditions: orthogonal cutting, $v_c = 100$ m/min, $h = 0.2$ mm, $b = 3$ mm, restricted contact (RC) tools T1 – $B = 2$ mm, $l = 0.3$ mm, $R_o = 2.6$ mm, $d = 0.2$ mm; T2 – the same chip and RC dimensions apart from the flat land length $l = 0.2$ mm. All dimensions as in Fig. 7.26b1.

configurations occurs when the distance BC is maximum, but for the tool with larger RC length the distance reaches its maximum substantially quicker. At this point the chip curvature is the highest and $r_u < R_o$ with the chip being lifted from the tool groove. Therefore, no contact is observed between the groove back wall and the chip. During this interval of time a certain elastic recovery occurs as depicted in Fig. 7.33a. It was also documented by Grzesik *et al.* [42] that the transition from unacceptable chip forms to the chip-breaking

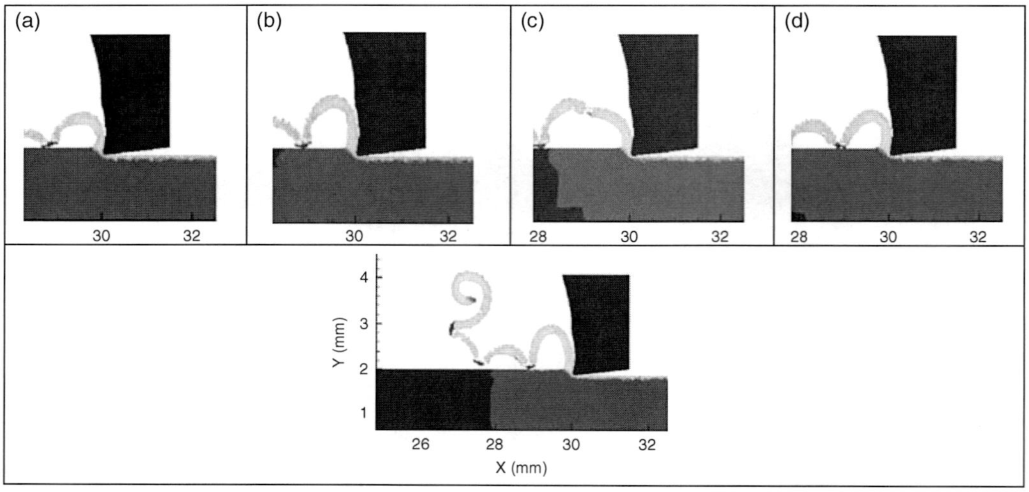

Fig. 7.34 Simulation of chip breaking when turning AISI 1045 carbon steel using FEM AdvantEdge package. Chip breaker with land length $l_{RC} = 0.08$ mm, width $B = 0.8$ mm and depth $d = 0.1$ mm, tool coating type TiN/Al_2O_3/TiC+Ti(C,N), cutting speed $v_c = 220$ m/min, depth of cut $a_p = 1$ mm, feed $f = 0.14$ mm/rev. (see Plate 6)

area can easily be correlated with a critical value of the specific cutting energy (substantially lower than in the case of long, continuous chips), which quantitatively support all experimental results obtained by Jawahir and coworkers [1, 39].

Chip breaking can be modelled analytically incorporating the effects of strains, strain rates and temperature from Oxley's predictive machining theory and appropriate rigid-plastic slip-line model for machining with grooved tools [31, 43] or by using the finite element model [29]. In the case shown in Fig. 7.34, chip breakage was modelled as a tensile fracture event when the chip curls over and contacts the workpiece ahead of the tool. The resulting bending moment induced in the chip due to contact causes areas of high tensile stress, in which their values are compared with a maximum normal stress fracture criterion. If this criterion is exceeded, the resulting local loss in strength becomes the dominant defect where the chip eventually breaks. Evidence of the influence of chip up-curling on the cutting force can be seen in Fig. 7.35 for coated and uncoated tools. Similar to the variations shown in Fig. 7.33 distinct patterns in periodical force fluctuations corresponding to chip breakage were computed. It is interesting to note that uncoated insert did not exhibit chip breakage, while the coated insert did.

Efficient chip breaking and effective chip disposal are the two functional elements of 'total' chip control, shown in Fig. 7.36. Figure 7.36 shows the various factors that affect the process of chip breaking, indicating the very complex nature of research on chip breaking. Chip control is essential to assure reliable machining operations and chip form monitoring is one of the basic tasks in automated machining systems. Several systems for chip form detection have been developed during the last two decades [1, 2, 40, 44]. The physical principles, which are used to detect the chip form, are:

1. The dynamic components of cutting forces owing to chip breakage and the detection technique based on the fast Fourier transform (FFT).
2. The chip/tool contact length determined by an ultrasonic method.
3. The variation on chip/tool contact conditions measured by means of the chip/tool thermocouple method.
4. The chip flow direction by electric contacts touched by the chip.
5. The monitoring of chip flow using a radiation thermometer.

All these sensor systems can be used only in medium to roughing operations and, furthermore, they are not suitable for industrial practice. Therefore, two other physical effects were chosen as a basis for chip form

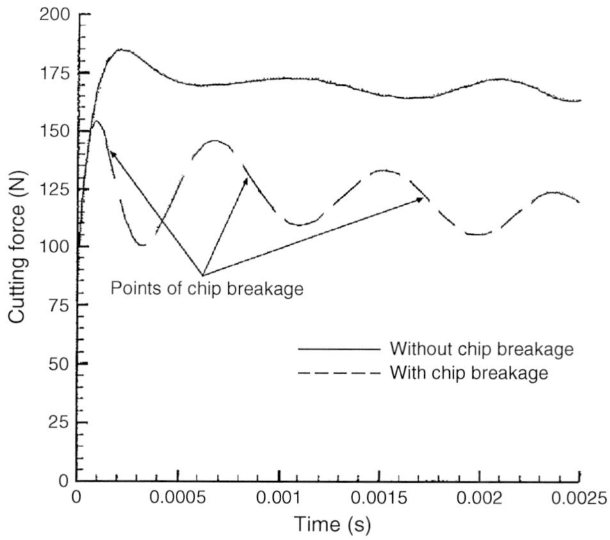

Fig. 7.35 Force comparison between a simulation with chip breakage and one without [29].

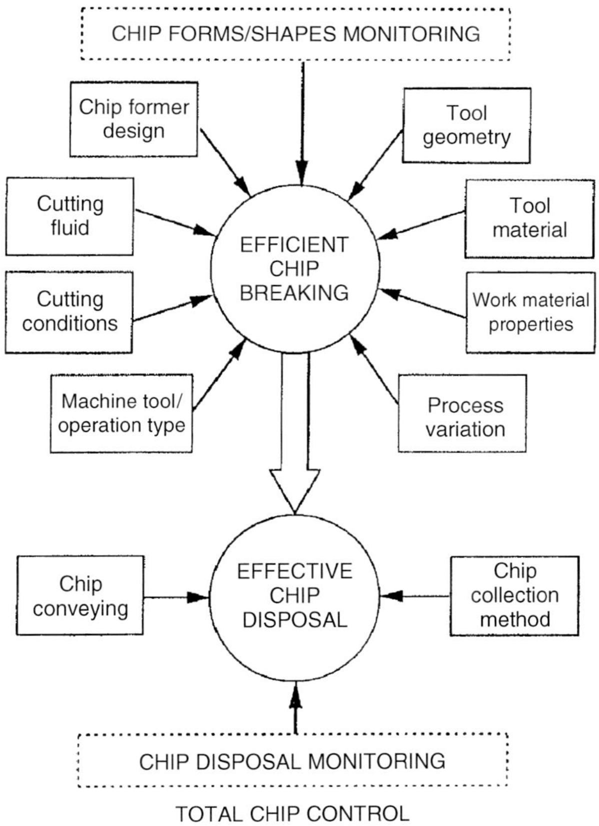

Fig. 7.36 The total chip control system [1].

detection. These are the infrared emission of the hot chip and the acoustic emission of the cutting process. Also, more recently, a new chip form detection method using the frequency analysis of the cutting force has been introduced. At the same time, advanced techniques involving a sensor fusion method based on a neutral network approach and fuzzy logic and neural network methodologies have been developed.

References

[1] I.S. Jawahir, C.A. van Luttervelt, Recent developments in chip control research and applications, *CIRP Annals*, Vol. 42/2 (1993), 659–693.
[2] W. Kluft, W. König, C.A. van Luttervelt, K. Nakayama, A.J. Pekelharing, Present knowledge of chip control, *CIRP Annals*, Vol. 28/2 (1979), 441–455.
[3] G. Boothroyd, W.A. Knight, *Fundamentals of Machining and Machine Tools*, CRC Press, Boca Raton, 2006.
[4] M.C. Shaw, *Metal Cutting Principles*, Clarendon Press, Oxford, 1989.
[5] J. Kaczmarek, *Principles of Machining by Cutting, Abrasion and Erosion*, Peter Peregrinus Ltd, Stevenage, 1976.
[6] W. Grzesik, *Fundamentals of Machining of Metallic Materials* (in Polish), WNT, Warsaw, 1998.
[7] H. Weber, T.N. Loladze, *Grundlagen des Spanens*, VEB Verlag Technik, Berlin, 1986.
[8] M.A. Rahman, M. Rahman, A.S. Kumar, H.S. Lim, CNC microturning: an application to miniaturization, *Int. J. Mach. Tools Manuf.*, Vol. 45 (2005), 631–639.
[9] Y. Kita, M. Ido, A study of metal flow ahead of tool face with large negative rake angle. *Trans. ASME J. Eng. Ind.*, Vol. 104 (1982), 319–324.
[10] R. Komanduri, T.A. Schroeder, On shear instability in machining a nickel-iron base superalloy, *Trans. ASME J. Eng. Ind.*, Vol. 108 (1986), 93–100.
[11] A. Vyas, M.C. Shaw, Mechanics of saw-tooth chip formation in metal cutting, *Trans. ASME J. Eng. Ind.*, Vol. 121 (1999), 163–172.
[12] D.M. Turley, E.D. Doyle, Microstructural behaviour – its influence on machining, in *On the Art of Cutting Metals – 75 Years Later*, PED – Vol. 7, ASME, 1982, 99–118.
[13] B.M. Manyindo, P.L.B. Oxley, Modelling the catastrophic shear type of chip when machining stainless steel, *Proc. Instn. Mech. Engrs*, Vol. 200 (1986), No. C5, 349–358.
[14] T. Inamura, N. Takezawa, Y. Kumaki, Mechanics and energy dissipation in nanoscale cutting, *CIRP Annals*, Vol. 42/1 (1993), 79–82.
[15] N. Ikawa, Sh. Shimada, H. Tanaka, G. Ohmori, An atomistic analysis of nanometric chip removal as affected by tool-work interaction in diamond turning, *CIRP Annals*, Vol. 40/1 (1991), 551–554.
[16] S. Shimada, H. Tanaka, G. Ohmori, J. Uchikoshi, Feasibility study on ultimate accuracy in microcutting using molecular dynamics simulation, *CIRP Annals*, Vol. 42/1 (1993), 91–94.
[17] K. Maekawa, A. Itoh, Friction and tool wear in nano-scale machining – a molecular dynamics approach, *Wear*, Vol. 188 (1995), 115–122.
[18] R. Komanduri, N. Chandrasekaran, L.M. Raff, Effect of tool geometry in nanometric cutting: a molecular dynamics simulation approach, *Wear*, Vol. 219 (1998), 84–97.
[19] J. Chae, S.S. Park, T. Freiheit, Investigation of micro-cutting operations. *Int. J. Mach. Tools Manuf.*, Vol. 46 (2006), 313–332.
[20] N. Ikawa, R.R. Donaldson, R. Komanduri *et al.*, Ultraprecision metal cutting – the past, the present and the future, *CIRP Annals*, Vol. 40/2 (1991), 587–594.
[21] E. Usui, A. Hirota, Lee-Shaffer's type solution for oblique cutting, *Bull. J. Soc. Prec. Eng.*, Vol. 12 (1978), No. 2, 72–110.
[22] J. Hashemi, A.A. Tseng, P.C. Chou, Finite element modeling of segmental chip formation in high-speed orthogonal cutting, *J. Mater. Eng. Perform.*, Vol. 3 (1994), No. 5, 712–721.
[23] T. Obikawa, H. Sasahara, T. Shirakashi, E. Usui, Application of computational machining method to discontinuous chip formation, *J. Manuf. Sci. Engng*, Vol. 119 (1997), 667–674.
[24] J. Chakrabarty, *Theory of Plasticity*, McGraw-Hill Book Co., New York, 1987.
[25] T.H.C. Childs, K. Maekawa, T. Obikawa, Y. Yamane, *Metal Machining. Theory and Applications*, Arnold, London, 2000.
[26] P.K. Wright, Predicting the shear plane angle in machining from workmaterial strain-hardening characteristics, *Trans. ASME, J. Eng. Ind.*, Vol. 104 (1982), 285–292.

[27] E.M. Trent, P.K. Wright, *Metal Cutting*, Butterworth-Heinemann, Boston, 2000.
[28] L. De Chiffre, What can we do about chip formation?, *CIRP Annals*, Vol. 34/1 (1985), 129–132.
[29] T.D. Marusich, C.J. Brand, J.D. Thiele, A methodology for simulation of chip breaking in turning processes using an orthogonal finite element model, *Proc. 5th CIRP Int. Workshop on Modeling of Machining Operations*, West Lafayette, 2002, pp. 139–148.
[30] J.C. Aurich, H. Bil, 3D finite element modelling of segmented chip formation, *CIRP Annals*, Vol. 55/1 (2006), 47–50.
[31] P.L.B. Oxley, *Mechanics of Machining*, Ellis Horwood, Chichester, 1989.
[32] E.J.A. Armarego, R.H. Brown, *The Machining of Metals*, Prentice-Hall, New Jersey, 1969.
[33] Internet at: www.secotools.com (Seco Tools); www.sumitomotool.com (Sumitomo Electric); www.coromant.sandvik.com (Sandvik Coromant).
[34] J.G. Horne, A new model for initial chip curl in continuous cutting, *Int. J. Mech. Sci.*, Vol. 20 (1978), 739–745.
[35] I.S. Jawahir, A knowledge-based approach for designing effective grooved chip breakers – 2D and 3D chip flow, chip curl and chip breaking, *Int. J. Adv. Manuf. Technol.*, Vol. 10 (1995), 225–239.
[36] K.C. Ee, A.K. Balaji, I.S. Jawahir, Progressive tool-wear mechanisms and their effects on chip-curl/chip-form in machining with grooved tools: an extended application of the equivalent toolface (ET) model, *Wear*, Vol. 255 (2003), 1404–1413.
[37] Ch. Nede, B.W. Hintze, Characteristic parameters of chip control in turning operations with indexable inserts and three-dimensional shaped chip formers, *CIRP Annals*, Vol. 38/1 (1989), 75–79.
[38] A.K. Balaji, G. Sreeram, I.S. Jawahir, E. Lenz, The effects of cutting tool thermal conductivity on tool-chip contact length and cyclic chip formation in machining with grooved tools, *CIRP Annals*, Vol. 48/1 (1999), 33–38.
[39] I.S. Jawahir, On the controllability of chip breaking cycles and modes of chip breaking in metal machining, *CIRP Annals*, Vol. 39/1 (1990), 47–51.
[40] C.A. van Luttervelt, Neue Entwicklungen in der Spanüberwachung, *VDI Berichte*, No. 988 (1993), 431–477.
[41] K. Nakayama, Chip control in metal cutting, *Bull. J. Soc. Prec. Engng*, Vol. 18 (1984), 97–103.
[42] W. Grzesik, E. Kwiatkowska, An energy approach to chip-breaking when machining with grooved tool inserts, *Int. J. Mach. Tools Manuf.*, Vol. 37 (1997), 569–577.
[43] N. Fang, I.S. Jawahir, An analytical predictive model and experimental validation for machining with grooved tools incorporating the effects of strains, strain-rates, and temperatures, *CIRP Annals*, Vol. 51/1 (2002), 83–86.
[44] W. Grzesik, P. Bernat, An investigation of the cutting process for chip breaking monitoring in turning of steels, *J. Manuf. Sci. Engng*, Vol. 120 (1998), 555–562.

CHAPTER EIGHT

Cutting Vibrations

Contents

8.1 Classification of Cutting Vibrations and their Sources

The structural system consisting of the machine tool, the cutting tool and the workpiece have very complicated dynamic characteristics and as a consequence vibrations can frequently occur under certain machining conditions. As in all types of machinery, the three classes of mechanical vibrations can be distinguished in such a structural system: free (natural), forced and self-excited vibrations, as shown in Fig. 8.1 and defined by the differential equation of a damped motion, Eqn (8.1).

$$m\ddot{x}(t) + k\dot{x}(t) + cx(t) = \begin{cases} 0 & \text{Free vibrations} \\ F(t) & \omega = \omega_0,\ x_0 \to 0 \\ & \text{Forced vibrations} \\ F[x(t)] & \omega = \omega_F,\ x_0 = \text{const} \\ & \text{Self-excited vibrations} \\ & \omega \approx \omega_F,\ x_0 \to \infty \end{cases} \tag{8.1}$$

where m is mass of the vibration system (kg), k is the damping ratio (Ns/m), c is the rigidity constant (N/m), ω is the angular frequency of free vibrations (rad/s), ω_0 is the angular frequency of self-excited vibrations (rad/s), ω_F is the angular frequency of forced vibrations (rad/s), and x_0 is amplitude of vibration.

Free, natural or *transient* vibrations result from impulses transferred to the structure through its foundation, from rapid reversals of reciprocating masses such as machine tables, or from the initial engagements of the cutting tool. The result is a motion with an amplitude decaying and a frequency equal to the natural damped frequency f_{nd} of the system (Fig. 8.1a).

Forced vibrations result from external, periodic force $F(t)$ acting on the system, such as unbalanced rotating shafts or due to the periodic component of the cutting force during the intermittent engagement of multi-tooth milling cutters, or transmitted through the foundations from nearby machinery. The resulting motion (Fig. 8.1b) reaches a steady state in which amplitude A is constant and the frequency of vibration f is equal to the frequency of the exciting force. For $f = f_n$, a *resonance* is obtained at which the amplitude of vibrations is maximum. Forced vibrations are present in all kinds of machinery where periodic forces are generated. Forced vibration can produce waviness of the machined surface in finish grinding and boring as well as the

ISBN: 978-0-08-044534-2

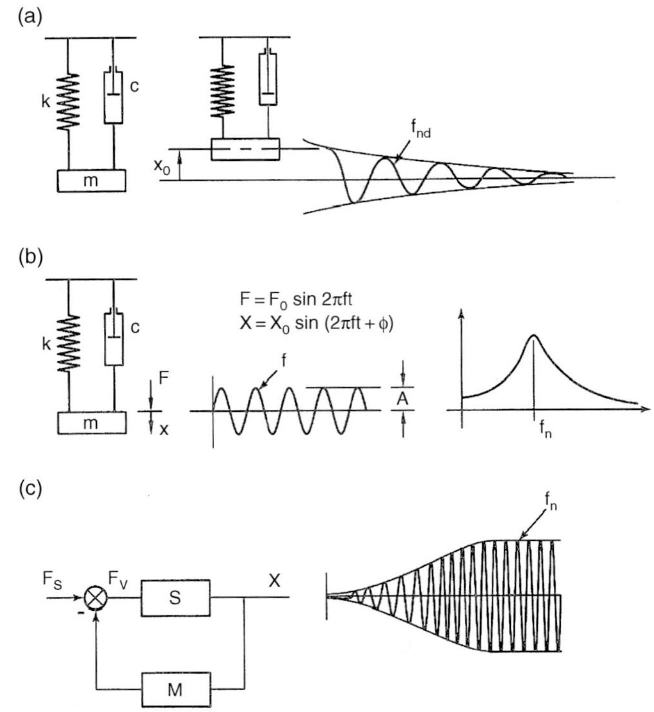

Fig. 8.1 The three classes of mechanical vibrations [1].

shift of the machined surface (over- or undercut effect) in slab milling, and non-flatness of the machined surface in end milling with helical cutters.

Self-excited vibrations usually result from a dynamic instability of the cutting process and they develop due to the built-in (or intrinsic) up mechanism providing for a closed-loop relationship in the system. This mechanism is capable of modulating a steady, non-periodical energy source and generating a periodic force F_v through the vibration system S (Fig. 8.1c), since it sustains the vibration. The characteristic features of self-excited vibrations are: (1) the amplitude increases until it stabilizes on a constant value; (2) the frequency of the vibration is equal to or close to the natural frequency of the system; (3) there is no independent, external periodic force; (4) there is a steady energy force from which the system derives a periodic force through its vibration. This phenomenon is commonly referred as *machine tool chatter* and, typically, if large tool/work engagements are attempted, oscillations suddenly build up in the structure. They effectively limit metal removal rates and the conditions for *the limit of stability* are of decisive interest.

8.2 Forced Vibrations in Milling Operations

As mentioned earlier, forced vibrations in machine tools are most often caused by cyclic variations in the cutting forces. Such variations will occur in end milling, which is a very significant operation in the aerospace industries, and in face milling where the frequency of the forced vibration equals the product of the tool rotational frequency and the number of teeth on the milling cutter.

It is obvious that similar to the single-degree-of-freedom system, the behaviour of a complex structure with many degrees of freedom, characteristic for the machine tools, can also be illustrated by the *harmonic response locus* of the system [2]. This is due to the fact the periodic forces are not sinusoidal but have significant

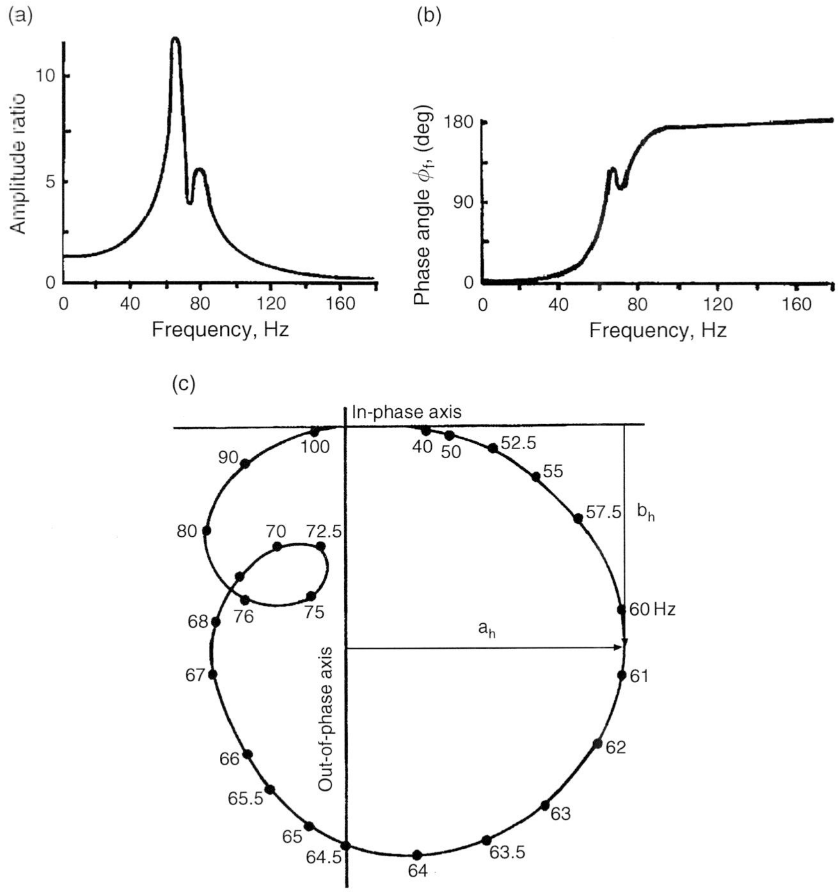

Fig. 8.2 Frequency (a, b) and polar (c) response of a machine tool structure [2].

harmonics (similar to that shown in Fig. 8.3a) and the tool has the same frequency as the periodicity of the force. Figure 8 2 shows the resonance (a) and phase (b) curves and the relevant harmonic response locus in polar coordinates (c). Because a complex structure exhibits several different resonances or natural frequencies, the frequency response curve will also have several peaks corresponding to each resonance (two peaks at approximately 60 Hz and 80 Hz in Fig. 8.2a and b). Moreover, each resonance produces a separate loop of approximately circular form in the polar harmonic response locus shown in Fig. 8.2c.

The surface error produced in end milling with cutters having four helical teeth is shown in Fig. 8.3. Every tooth starts through a surface starting at line *A–A* and ending at line *B–B*. As the cutting edge *CE* rotates, it climbs on line *A–a* and imprints the sinusoidal vibration exiting by the force (b) on the surface *S* through the succession of points 1, 2, ..., 6. As a result, the surface error is produced as in detail (d). In practice, each resonance has a corresponding mode of vibration or mode shape that the structure adopts or, in other words, the amplitude and relative phases of various points on the structure will be different. For example, the modes of vibration which result in relative displacement of the tool and workpiece, causing a detrimental effect on the surface finish or surface error of the workpiece, can be determined by special computing algorithms or structural analysis programs based on the finite element method [1, 2]. In the case of a three-dimensional structure of a machine tool the exciting forces applied at certain points and direction result in the displacements at another points in different directions, so the resulting frequency response curve will be referred to as a cross-frequency response.

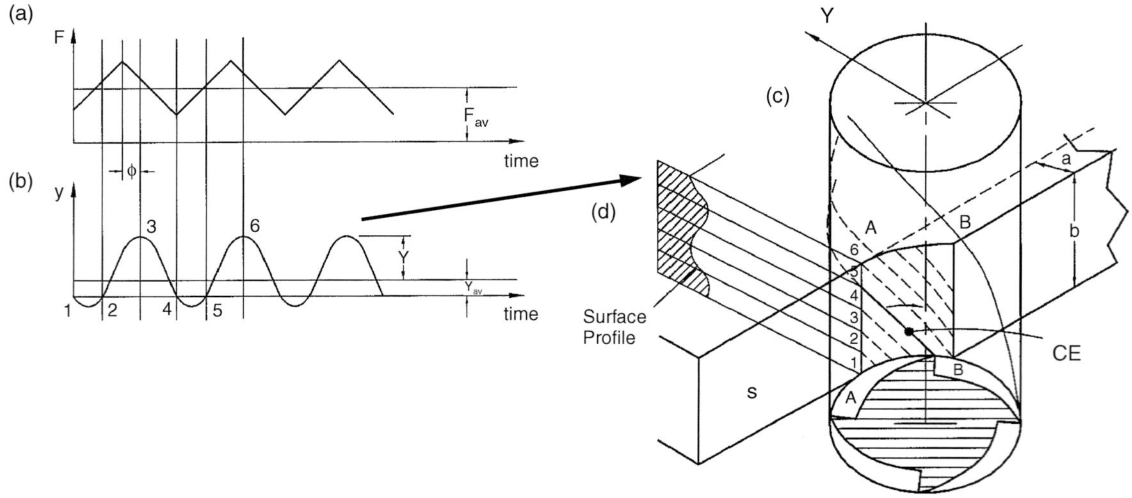

Fig. 8.3 Development of surface error by end-milling cutter with helical teeth and non-regenerative approach according to Tlusty [1]: (a) force variation; (b) corresponding vibration of the tool in the *Y* direction shown in (c); and (d) imprints the vibration on surface S.

8.3 Mechanisms of Self-excitation in Metal Cutting

The basic cause of chatter is the dynamic interaction of the cutting process and the machine tool structure which behaves as a closed-loop feedback system. In such a situation, a deflection of the structure caused by an accidental disturbance in the cutting process may alter the undeformed chip thickness, and then the cutting force, and initiate self-sustaining vibrations. In general, chatter occurs in metal cutting if the chip width is too large with respect to the dynamic stiffness of the system. The cutting force becomes periodically variable, reaching considerable amplitudes, the machined surface becomes undulated, and the chip thickness varies in the extreme so much that it becomes dissected. Chatter is easily recognized by the noise associated with these vibrations and by the chatter marks on the machined surface. Such unstable cutting with large peak values of the variable cutting force is mostly unacceptable because of tool breakage or a machine part failure. Moreover, the limit of the chip width can substantially reduce the metal removal rate. In practice, the most significant cutting parameter for generating chatter is the width of cut b (width of chip). The limiting value of b_{lim} depends on the dynamic characteristics of the structure, the workpiece material, cutting speed and feed, and the geometry of the tool. Two major effects causing the instability of the closed-loop cutting process structural system are:

1. The mode coupling effect (mode coupling for short) shown in Fig. 8.4.
2. The regenerative effect (or simply regeneration of waviness) shown in Fig. 8.5.

Mode coupling is a mechanism of self-excitation that can only be associated with a situation where the relative vibration between the tool and the workpiece can exist simultaneously in at least two directions in the plane of the orthogonal cut. Obviously, mode coupling instability occurs when successive passes of the tool do not overlap, as in screw cutting. Simultaneous vibrations in the X and Y directions with the same frequency and a phase shift between the two vibrations result in an elliptical motion as shown in Fig. 8.4a. The resultant cutting force F_r does not work during this periodic motion of the tool from A to B because it acts against the motion and takes energy away. During the motion from B to A the force drives the tool and imparts energy to its motion and a larger value of the cutting force causes the surplus of energy sustaining the vibrations against damping losses.

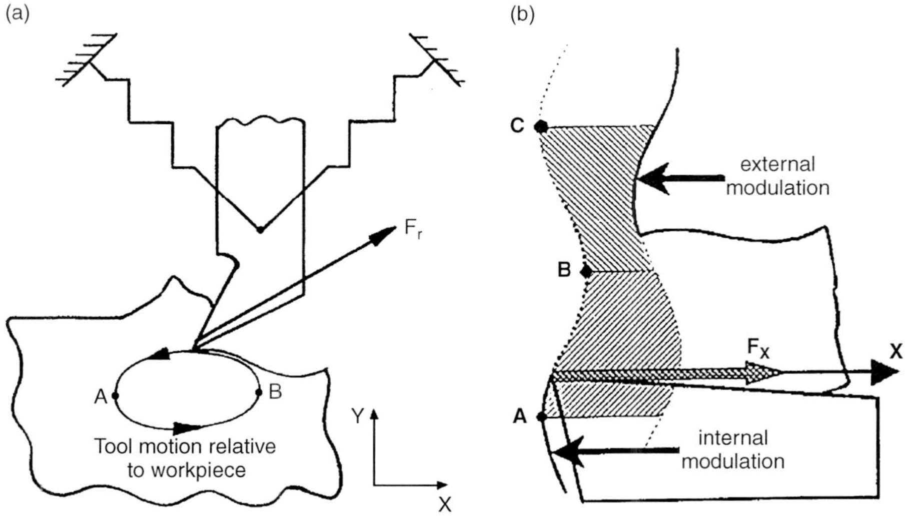

Fig. 8.4 Mode coupling instability [2, 3].

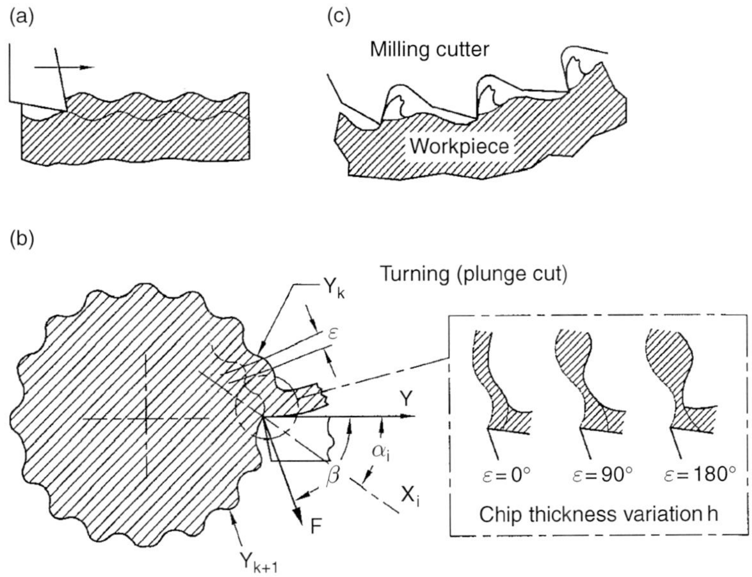

Fig. 8.5 Regeneration of surface waviness: (a) shaping; (b) turning; (c) milling [1].

Regeneration of waviness shown in Fig. 8.5 is possible because in almost all machining operations the tool removes the chip from the surface which was produced in the preceding pass. In other words, regenerative instability occurs when successive passes overlap. The tool in the next pass (next revolution in turning or next tooth in milling) encounters a wavy surface and removes a chip with periodically variable thickness (Fig. 8.5c). The newly created surface is again wavy and in this way the waviness is continually regenerated.

In turning, as in diagram (b), the phase ε between the subsequent undulations is determined by the relationship between spindle speed n and frequency of chatter f. The number of waves between subsequent cuts is

$$N + \frac{\varepsilon}{2\pi} = \frac{f}{n} \tag{8.2}$$

where f is the frequency in Hz and n is the spindle rotational speed in rev/sec. In practice, there are N full waves and a fraction equals $\varepsilon/2\pi$. The chip thickness variation can be either zero for $\varepsilon = 0$ or maximum for $\varepsilon = \pi$. In milling the geometric constraint is much more important because the regeneration of waviness is produced by subsequent teeth of the cutter and Eqn (8.2) is changed to

$$N + \frac{\varepsilon}{2\pi} = \frac{f}{nm} \tag{8.3}$$

where m is the number of teeth of the milling cutter. It may easily be computed that there are two and a fraction (three and a fraction) waves between subsequent teeth. Moreover, a change of ε by 2π, for $N = 2$, requires a 50% change of frequency or the cutting speed. This causes the process stability to be importantly influenced by the geometric conditions.

8.4 Stability of Chatter

As discussed in the previous section, in turning changing spindle speed affects the level of stability very little, but in milling, especially in high speed milling, a change of spindle speed may be a very effective means for increasing the limit depth of cut and, correspondingly, the metal removal rate (MRRR) [1].

In practice, stability of a machine tool can be represented graphically in the form of a special chart called the *stability lobe diagram* which depicts the effect of the width of cut, drill diameter and so on, and the rotational speed of the tool or workpiece. An example of a stability chart for an end-milling operation with characteristic unstable ranges shown as lobed areas is presented in Fig. 8.6. It is shown at the top of this figure how in milling the waviness, which is cut into the surface during chatter vibrations by a tooth, gets recut by the subsequent tooth. Cases (a1), (a2) and (a2) were recorded for increased spindle and it can be seen that for higher spindle speed (a3) only one and a small fraction of a wave occur. In addition cases (b1) and (b2) illustrate that a substantial variation in chip thickness (twice the vibration amplitude) occurs with one and half waves between the teeth, for the same amplitude. In contrast, no chip thickness and force variations were obtained with exactly one wave between the teeth being in phase.

In the stability diagram shown in Fig. 8.6, the vertical coordinate is the ratio $q = b_{lim}/b_{cr}$, where b_{cr} is the lowest b_{lim} obtained for phasing most favourable for chatter generation, and the horizontal scale expresses the value of the number p being the ratio of the tooth frequency over the natural frequency of the system. It should be pointed out that the individual 'lobes' in the diagram correspond to a different integer N in Eqn (8.3). The practical interpretation of the graph is to consider the envelope of all the lobes as the boundary between the stable field below the envelope and the chatter field (shaded areas) lying below the envelope. The upturn of the stability boundary on the left end of the horizontal scale is the effect of process damping. In contrast, at the high speed end, on the right, gaps of increased stability occur. The highest stability, permitting the highest value of stable depth of cut, is obtained with the spindle speed at which the tooth frequency equals the natural frequency of the system (for $p = 1$). It can be seen in Fig. 8.6a that peaks of stability are close to values of $p = 1(N + 1)$. This means that for low spindle speeds the peaks of stability are not very high and localized close to each other. But as the spindle speed approaches the values of $n = 0.5f_n/m$ (and mainly $n = f_n/m$), which denotes one wave between the subsequent teeth, substantial increase of the stability may be achieved by exact selection of the right speed. In practice, it is necessary to have steplessly variable spindle speed to be able to select the most stable speed. Figure 8.7 shows how

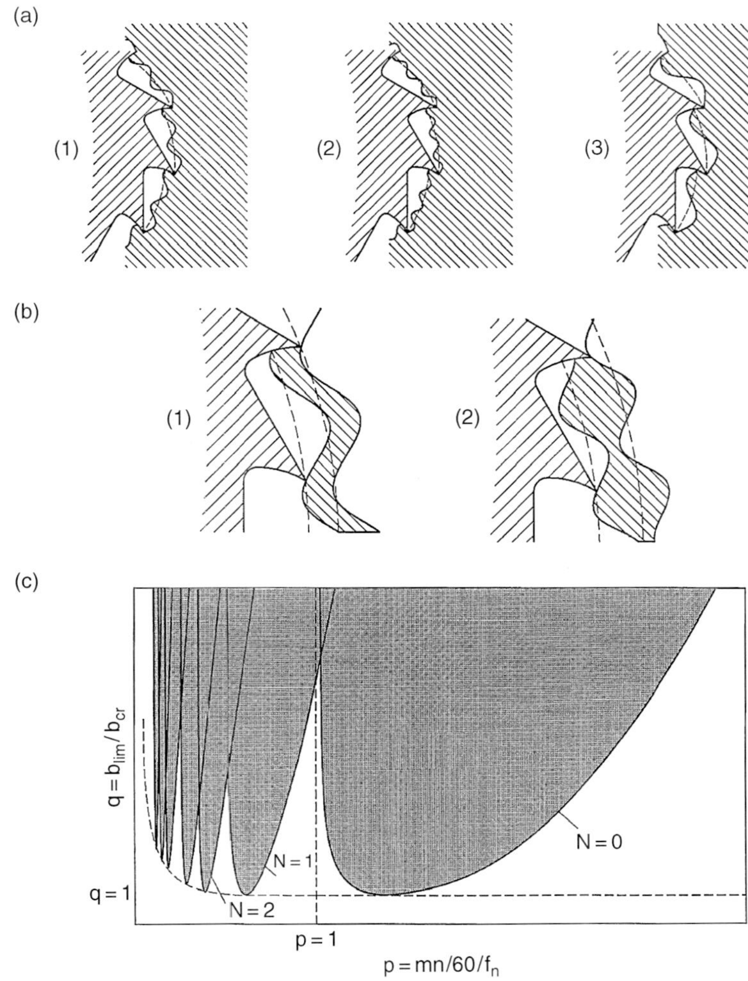

Fig. 8.6 Stability lobe diagram for a milling operation [1].

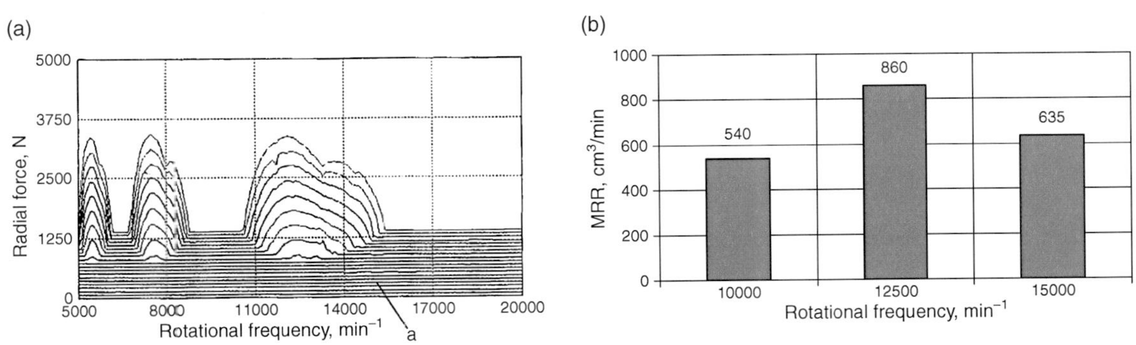

Fig. 8.7 Optimized stability chart (a) and corresponding values of MRR (b) [4]. *a* – depth of cut change in every step by 0.5 mm.

the MRRR can be increased by dynamical optimization of the interface between the spindle unit and tool holder. As a result, when using an end-milling cutter of 19 mm in diameter at a most stable spindle speed of 15 000 m^{-1}, it was possible to decrease the radial force down to 1250 N and to increase the material volume removed up to 900 cm^3/min.

Most of the techniques for predicting and controlling machining chatter are based on the three stability lobe equations in milling as [5]:

$$b_{\text{lim}} = \frac{-1}{2K_s m_{avg} \mu \text{Re}[G]} \tag{8.4}$$

where b_{lim} is the limiting stable axial depth of cut, K_s is the material specific power, m_{avg} is the average number of teeth in the cut, and $\mu\text{Re}[G]$ is the real part of the oriented frequency response function (FRF) measured at the tool tip.

The second is Eqn (8.3) and the third is as follows:

$$\varepsilon = 2\pi - 2\tan^{-1}\left(\frac{\text{Re}[G]}{\text{Im}[G]}\right) \tag{8.5}$$

where Im[G] is the imaginary part of the FRF measured at the tool tip.

The measurements of the FRF are most commonly based on the impact testing when an instrumented hammer is used to strike the tool tip and the force of the impact is measured or the 'pop-off' device in which the impact is created by using an explosive device. Another method is the impact excitation technique in which a magnet and the rotation of the spindle are used to provide the excitation [5]. Once the measured FRF is available, the stability lobe diagram may be computed using a set of Eqns (8.3)–(8.5). The methods of determining stable cutting regions without the requirement of an FRF measurement involve recording the sound spectrum during the cutting process or a special device which uses the rotating tool and a non-contact actuator to generate the excitation force [6]. The displacement of the tool tip is measured in response to the excitation at the tool pass frequency and by utilizing once-per-revolution sampling of the displacement at each tachometer impulse, the superimposed real part of the FRF (see Eqn (8.5)) is constructed. Each of these real parts corresponds to the regions where the tooth passing frequency matches the chatter frequency, and therefore corresponds to the most stable regions of the stability diagram. Commercially available software exists [1, 5] which can be used to simulate the chatter and calculate the cutting power from the MRR (it is proportional to the spindle speed and the axial depth of cut). Moreover, such runs can be automatically programmed over ranges of spindle speeds and depths of cut, resulting in comprehensive graphs of vibration and force amplitudes for all combinations of the axial and radial depth of cuts and spindle speeds. Finally, these lobes can be used for NC programming of optimum machining operations but the dynamic characteristics of all cutting tools used have to be measured.

8.5 Methods for Improving Machine Tool Stability

During the operation of a machine tool, particularly under NC control, it is not always possible to avoid the machining conditions at which chatter may build up. Therefore, in such a situation, improving the stability of a machine tool is necessary. The possibilities of improving machine tool stability are as follows [1, 2]:

1. Structural alterations dealing with adding more damping and/or stiffness to the structure in order to reduce the magnitude of the in-phase component of harmonic receptance a_h on the operative response locus (see Fig. 8.2c). For example, it is possible to alter the principal directions of the major modes of vibration in such a way that they will not coincide with the cutting forces and/or the directions normal to the cut surface. As shown in Fig. 8.8, damping is cumulative owing to a number of joints in a machine tool.

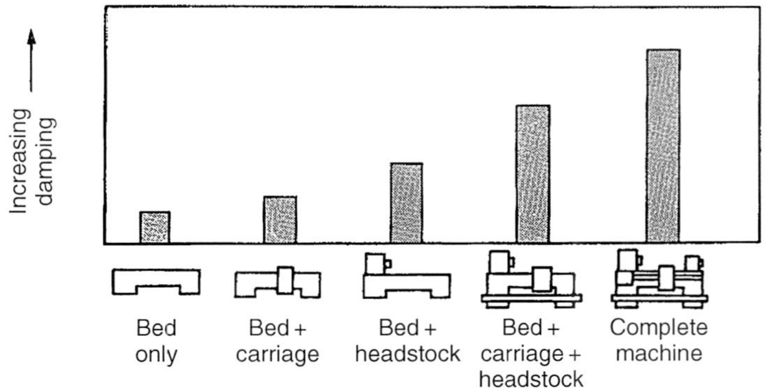

Fig. 8.8 Damping of vibrations as a function of the number of components on a lathe [9].

2. Application of a vibration absorber being the mass-spring system added to structure and tuned so that its natural frequency coincides with that of the major vibration mode. Such solution can be especially effective for machine tools with a single dominant mode of vibration.
3. Modification of the regenerative effect which utilizes disturbing the phasing between successive waves cut on the surface by a vibrating tool. This effect can be achieved in several ways.

   - The use of special milling cutters with non-uniform tooth spacing, cutters that use teeth with alternating helix (the average chip thickness variation is very small), cutters that either have interrupted or undulated edges (they divide a wide chip into narrower and shorter pieces), and variable-pitch cutters.
   - Chatter suppression by the effect of time-varying cutting parameters – width of cut, the spindle speed and mainly time varying feed rate [7]. In the latter case the stability is higher because a negative feedback loop is introduced into the system.
   - Active force control by means of active feedback control. For instance, the effect of providing damping force can be generated by a piezoelectric actuator with an inertial mass attached to the controlled cutting tool [8]. The actuator resonance can be tuned over a wide frequency range by adjusting the size of the inert mass, so that the actuator can provide an extremely large damping force to suppress undesired forced or self-excited vibration of the cutting tool at the resonance frequency of the actuator.

Vibration from machining creates noise and sometimes it is necessary to use damped tools to keep it within the maximum noise level permitted in a workshop. Currently used damped tooling includes tools (Fig. 8.9) that are pre-tuned to the correct frequency in relation to the tool length that is ordered. This is basically required for the extremely damped boring bar in order to sustain the high material removal rate, as shown in Fig. 8.9b.

In general, several techniques are known for enhancing dynamic stiffness and stability (chatter resistance) of long cutting tools and increasing the allowable overhang. The four most widely used and most universal approaches are [10, 11]:

1. Use of anisotropic bars with specifically assigned orientations of the stiffness axes. This approach utilizes the development of chatter during cutting by an intermodal coupling in the two-degrees-of-freedom system referred to the orthogonal plane to the bar axis and passing through the cutting zone. A specific orientation of the stiffness axis relative to the cutting force leads to a significant increase of the dynamic stability.
2. Use of high Young's modulus and/or high damping materials, such as sintered tungsten carbides and machinable sintered tungsten alloys with an added 2–4% of copper or nickel. Solid bars made of these materials allow stable cutting with ratios $L/D > 7$.
3. Use of passive dynamic vibration absorbers (DVA) with an inertia mass, called pre-tuned bars. In boring bars, the DVA has to be placed in an inertial cavity of much smaller diameter than the bar diameter, inside the tool structure.

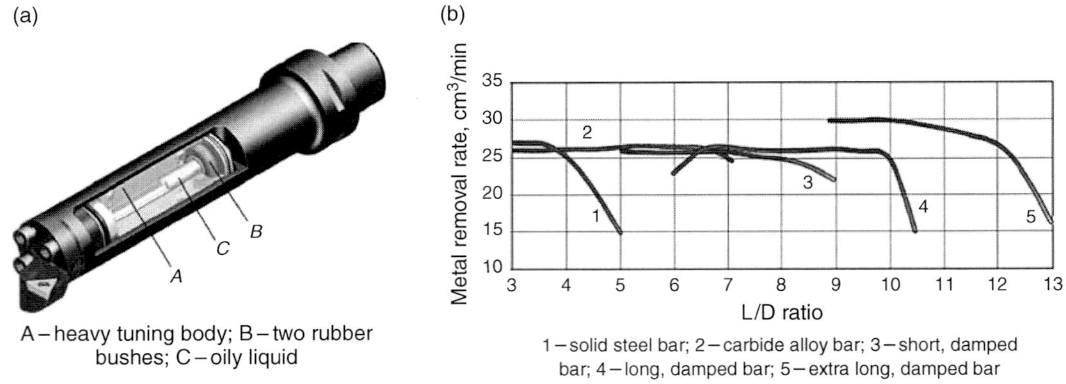

Fig. 8.9 Damped boring bar (silent boring bar) (a) and obtainable MRR for different length-to-diameter ratio *L/D* (b) [10].

4. Use of active vibration control with active vibration dampers. This approach requires vibration sensors and actuators generating forces opposing the deflection of the tool when chatter occurs. The most frequently used are active systems with cavities in the mandrel body filled with variably pressurized oil. In this case, the ultimate *L/D* ratio increases up to 10–12.

To measure and control vibrations, advanced machine systems must rely on accurate sensors. A number of direct and indirect sensors have been used to measure also vibrations, including contact strain gauges/accelerometers, capacitive probes, inductive probes, fibre-optic displacement sensors, laser Doppler vibrometers and acoustic pick-up devices (microphones). An alternative method involves detection of tool vibration from an indirect measurement, via measured fluctuations in tool rotation speed or spindle power requirements, for example. Figure 8.10 presents the principle of the active vibration monitor for high speed machining which utilizes a non-contact ultrasonic sensor.

In the vibration monitoring system shown in Fig. 8.10, the sensor head consists of a piezoelectric transducer and a fluid system that directs a column of fluid to the tool (similar to supplying cooling or lubrication media) under measurement. An ultrasonic carrier propagates toward the tool through the fluid

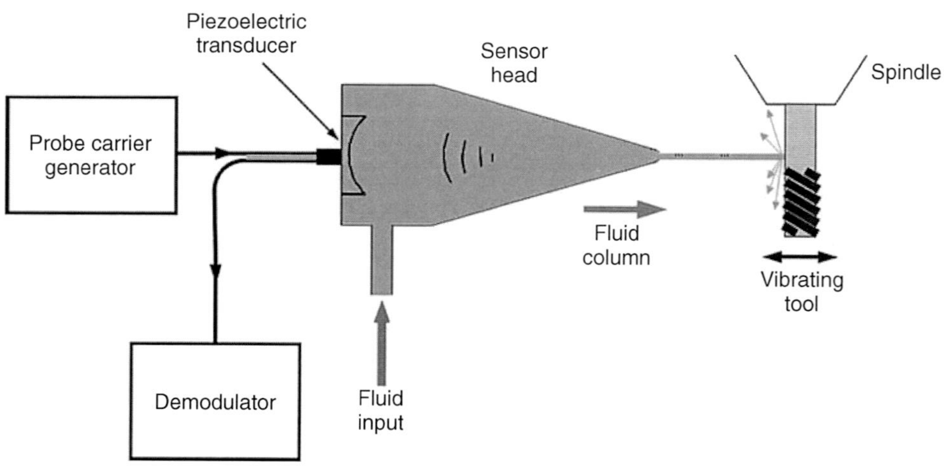

Fig. 8.10 The principle of the active vibration monitoring [12].

column and is reflected back to the sensor by the surface of the cutting tool. If the cutting tool does not vibrate, the reflected ultrasonic wave has the same frequency as the incoming ultrasonic wave. On the other hand, if the cutting tool vibrates, the frequency spectrum of the ultrasonic wave is frequency shifted according to the Doppler effect. The tool vibration is obtained from the reflected wave using an electronic demodulator. The sensor output can be used to improve the cutting performance or control the machine during the cutting process through real-time feedback of critical information.

References

[1] G. Tlusty, *Manufacturing Processes and Equipment*, Prentice Hall, Upper Saddle River, New Jersey, 2000.
[2] G. Boothroyd, W.A. Knight, *Fundamentals of Machining and Machine Tools*, Marcel Dekker, New York, 1989.
[3] W. Grzesik, *Fundamentals of Machining of Metallic Materials* (in Polish), WNT, Warsaw, 1998.
[4] H. Voll, K. Gebert, Motor-spindeln für hohe Spanvolumina, *Werkstatt und Betrieb*, Vol. 132 (1999), 50–52.
[5] S. Smith, Techniques for predicting and controlling chatter, Invited paper on the topic of machine tools presented at the 29th NAMRC, University of Florida, 2001.
[6] J.P. Snyder, J.R. Pratt, M.A. Davies, S. Smith, A new stable speed test apparatus for milling, *Trans. NAMRI/SME*, Vol. 29 (2001), 153–156.
[7] F. Yang, Bi Zhang, J. Yu, Chatter supression through time-varying feedrate in cutting process, *Trans. NAMRI/SME*, Vol. 25 (1997), 87–92.
[8] J.L. Jang, Y.S. Tarng, A study of the active vibration control of a cutting tool, *J. Mater. Proc. Technol.*, Vol. 95 (1999), 78–82.
[9] S. Kalpakjian, *Manufacturing Engineering and Technology*, Addison-Wesley Publishing Company, 1998.
[10] How to reduce vibration in metal cutting, *Silent Tools*, Sandvik Coromant, Sandviken, Sweden, 2006.
[11] Minimizing vibration tendencies in machining, C-5000.458 ENG, Sandvik Coromant, 2003.
[12] M. Klein, J. Eichenberger, T. Delio, Noncontact vibration monitoring for high-speed machine tools, *Machining Technology*, 4th Quarter 2005, Vol. 16, No. 4.

CHAPTER NINE

HEAT IN METAL CUTTING

Contents

9.1 HEAT SOURCES IN METAL CUTTING AND CUTTING TEMPERATURE

During metal cutting, high temperatures are generated because the power consumed for plastic deformation is largely converted into heat near the tool cutting edge (length 1–2 in Fig. 9.1), and these temperatures have a controlling influence on the rate of tool wear and the friction at the tool/chip interface. Because of the effects of friction and wear, considerable attention is paid to the determination of the heat amounts and temperatures in the tool, chip and workpiece. Figure 9.1 shows four heat sources existing in metal cutting, i.e. heat source due to intensive plastic deformation (Q_1 on the shear plane 1–2–3–4), frictional heat sources Q_2 localized at the too/chip interface (1–2–5–6) and Q_3 at the contact between the workpiece and the flank (1–2–7–8), and an additional source Q_4 from which a small part of heat is transferred to the subsurface layer. Usually, in the thermal analysis of the cutting process, the heat source Q_3 becomes important for worn tools and the heat source Q_4 is comparatively small and is neglected. Hence, in practice conversion of cutting energy into heat occurs in the primary and secondary deformation zones.

The heat fluxes flowing to the chip (q_c), the tool (q_t) and the workpiece (q_w) can be schematically aggregated as in Fig. 9.2a. The three ongoing fractions of the total heat amount depend on the thermal properties of cutting tool and workpiece materials, cutting parameters with predominant effect of the cutting speed and cooling method applied. The fact that the cutting speed greatly influences not only the temperature but also the heat distribution between the chip, the tool and the workpiece has been practically utilized in machining practice. As shown in Fig. 9.2b, as cutting speed increases, a larger proportion of the heat generated is carried away by the chip, and a little heat goes into the workpiece. When cutting speed increases, the time in which the heat fluxes q_{1w} and q_{2t} flow into the workpiece and the tool is shortened. For example, at the cutting speed of 150 m/min, 75–80% of heat is transported by the chip, 10–15% is conducted into the tool, and the remaining 5–10% is conducted into the workpiece. As a result, at extremely high cutting speeds characteristic for high speed machining, the majority of the generated heat is evacuated with chips and the cutting temperature diminishes substantially (see Chapter 15.1).

In real machining processes complex temperature fields occur but the maximum temperature is observed at the rake face near the middle of the tool/chip interface. Because the temperature is not distributed uniformly, the term *cutting temperature*, which denotes the average tool/chip interface temperature, is popularly used in metal cutting theory and practice.

Advanced Machining Processes of Metallic Materials
ISBN: 978-0-08-044534-2

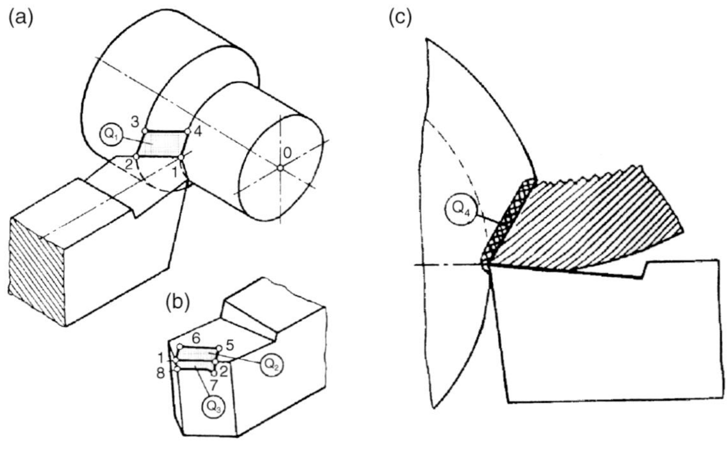

Fig. 9.1 Sources of heat generation in metal cutting.

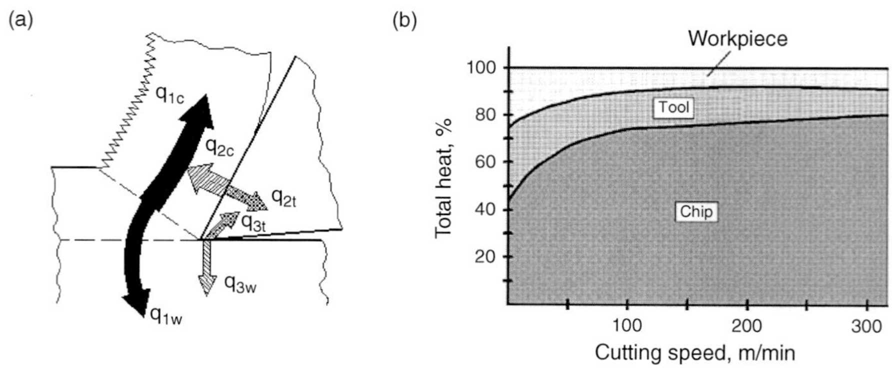

Fig. 9.2 A scheme of heat flow (a) and percentage of the heat generated going into the workpiece, tool, and chip, as a function of cutting speed (b) [1].

9.2 Heat Flow and Distribution in the Cutting Zone

Most thermal analyses of the cutting process are based on the moving heat source theory [2] for which it is necessary to know the fraction of the total cutting energy conducted as heat to the tool in order to estimate the interface temperature. In this theory, the chip is the body with a moving heat source and the heat partition coefficient (R_{ch}) defines the percentage of the heat entering the moving chip. Hence, the fraction $(1 - R_{ch})$ provides the percentage of the dissipated energy going to the tool, i.e. the member that is stationary relative to the heat source. In general, three different methods of calculations of the heat partition coefficient are used, namely those proposed by Shaw [3] – R_{SH}, Fujii and Kato [4] – R_{KF}, and Reznikov [5] – R_R.

$$R_{SH} = \frac{1}{1 + \left[\left(0.754(\lambda_T/\lambda_W)\right) / \left(A_a\sqrt{N_T}\right) \right]} \tag{9.1}$$

In Eqn (9.1) thermal conductivities of work (λ_W) and tool (λ_T) materials, thermal number (N_T) and area shape factor (A) describe the thermal and geometrical features of the interface.

The thermal number is given by

$$N_T = v_{ch} l / 2\alpha_W \tag{9.2}$$

where v_{ch} is the chip (sliding) velocity, $l = l_{nc}$ is the natural contact length and α_W is the diffusivity of work (chip) material.

The shape factor (A) is determined as follows [3]:

- for estimating the average temperature the corresponding average value of this factor is equal to

$$A_a = \frac{2}{\pi}\left[\sinh^{-1}\left(\frac{m}{l}\right) + \left(\frac{m}{l}\right)\sinh^{-1}\left(\frac{l}{m}\right) + \frac{1}{3}\left(\frac{m}{l}\right)^2 + \frac{1}{3}\left(\frac{l}{m}\right) - \frac{1}{3}\left\{\left(\frac{l}{m}\right) + \left(\frac{m}{l}\right)\right\}\left\{1 + \left(\frac{m}{l}\right)^2\right\}^{0.5}\right] \tag{9.3a}$$

- the maximum value of factor A is

$$A_m = \frac{2}{\pi}\left[\sinh^{-1}\left(\frac{m}{l}\right) + \left(\frac{m}{l}\right)\sinh^{-1}\left(\frac{l}{m}\right)\right] \tag{9.3b}$$

In Eqns (9.3a) and (9.3b) width of the contact zone $m = a_p$ (a_p is the depth of cut) and the ratio m/l is the aspect ratio of the contact area. Kato and Fujii [4] originally defined heat partition for conventional surface grinding and for the thermal analysis in cutting a modified version is proposed [6]:

$$R_{KF} = \frac{1}{1 + \sqrt{(c_p\rho\lambda)_T / (c_p\rho\lambda)_W}} \tag{9.4}$$

It should be noted that in Eqn (9.4) triple products, also called *heat transmission ratios*, for work (W) and tool (T) materials are introduced.

The third formula uses both Peclet ($\mathrm{Pec} = v_{ch} l_c / \alpha_{ch}$) and Fourier ($\mathrm{Fo} = \alpha_w / (l_c v_{ch})$) numbers in order to consider the velocity and duration of the frictional heat source [7]. The relevant Eqn (9.5) includes equivalently thermal conductivities (λ_T and λ_W) and diffusivities (α_T and α_W) of both matching materials.

$$R_R = \frac{1}{1 + \left[(3\lambda_T / 2\lambda_W)\sqrt{\alpha_W / \alpha_T}\right]} \tag{9.5}$$

Figure 9.3 shows changes of the heat partition coefficient computed for multilayer tool coatings (courses 3L and 4L) which were replaced by equivalent composite monolayers, when machining AISI 1045 carbon steel using R_R, R_{KF} and R_{SH} partition coefficients. It was found that the heat partition coefficient calculated from Shaw's version (case C) is higher than 0.9 when using multilayer coated tools, similar to uncoated carbide tools, and it increases slightly with the cutting speed rise up to the maximum value of about 200 m/min. In contrast, when using the R_R value (course A), 0.55–0.6 of the dissipated heat flows to the chip. This implies that the use of multilayer coated tools can cause, at most, about 40% of heat generated during the cutting process at higher cutting speeds to be transferred into the tool body due to the heat isolation effect. These findings were schematically illustrated in Fig. 9.4, in which two workpiece materials (carbon AISI 1045 steel and AISI 304 stainless steel) were machined with differently coated tools. As a result, heat partition obtained in these cases is quite different.

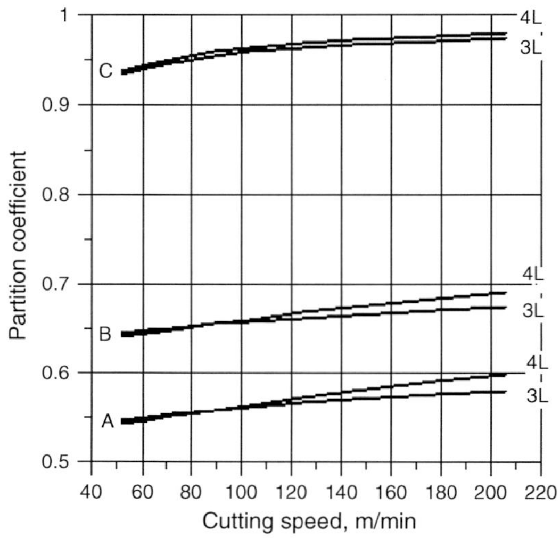

Fig. 9.3 Influence of cutting speed on the heat partition coefficient for composite layer [8]. A after Reznikov, B after Fujii and Kato, C after Shaw. 3L-TiC/Al_2O_3/TiN coating, 4L-TiC/TiCN/Al_2O_3/TiN coating.

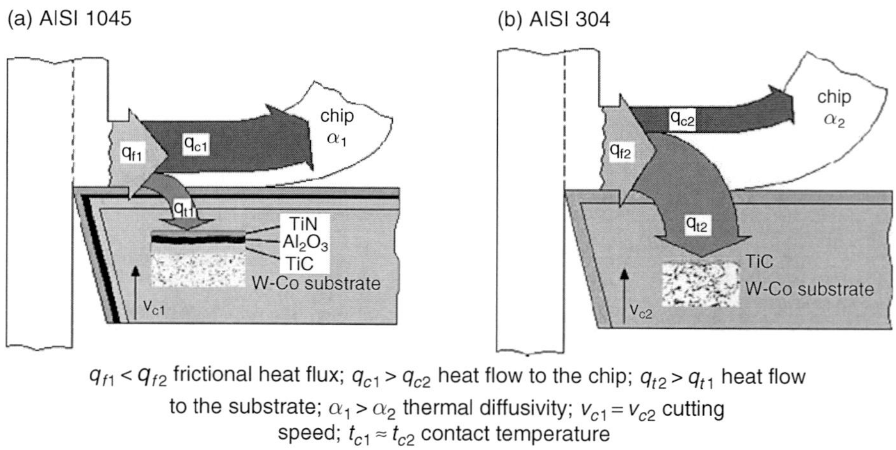

Fig. 9.4 Distribution of heat generated for material with high (a) and low (b) thermal diffusivity. (see Plate 7)

Boothroyd (after Wiener) [10] proposed the proportion of heat conducted into the workpiece β to be calculated using a unique function of R tan Φ (where Φ is the shear angle). The thermal number R is expressed as follows:

$$R = \frac{c_p \rho \cdot v_c \cdot h}{\lambda_W} \tag{9.6}$$

where v_c is the cutting speed, λ_W is the thermal conductivity of work material and undeformed chip thickness $h = f(f = \text{feed rate})$.

The graphical representation of the relationship between heat partition β and R tan Φ is shown in Fig. 9.5. Equation (9.6) is used to calculate temperatures in the primary deformation zone where heat is

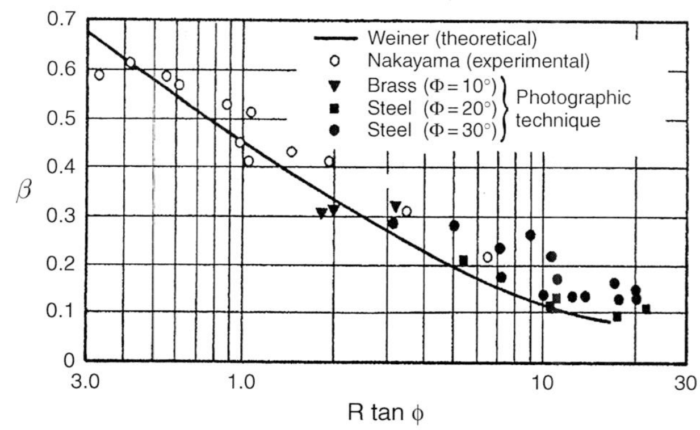

Fig. 9.5 Experimental curve for the division of shear zone heat between chip and workpiece [9, 10].

generated due to plastic deformation (shearing) required for chip formation. Theoretical calculation and experimentally determined values show that the coefficient β may be as high as 50% for very low rates of metal removal, materials with high conductivity like aluminium and copper alloys, and small shear plane angles. However, for high MRRs and machining steel it is of order of 10–15% [9].

9.3 Prediction and Modelling of Temperatures in the Cutting Zone

Typically, analytical methods allow the temperature rises for the primary and secondary deformation zones to be calculated based on the appropriate fractions of heat conducted into the workpiece and the chip calculated by means of a set of Eqns (9.1)–(9.6).

9.3.1 Calculation of temperature rise due to plastic deformation in the PDZ

The computation scheme used consists of three steps. In the first one, the thermal number R is calculated using Boothroyd's formula (9.6). The next two steps of computations are based on the theory of similarity elaborated for metal cutting purposes by Silin [6]. This enables the estimation of the maximum and mean temperatures occurring at the shear plane. It should be noted that the theory of similarity enables calculating this temperature more accurately than Oxley's theory because more calculation formulae are proposed depending on the value of the product $R \tan \Phi$.

The maximum temperature at the shear plane can be computed as:

$$\Theta_{s\max} = \frac{\tau_{sh}}{c_p \rho \tan \Phi} erf \sqrt{\frac{R \tan \Phi}{4}} \tag{9.7}$$

where τ_{sh} is the shear flow stress and Φ is the shear angle.

Then, the mean temperature at the shear plane is equal to:

- for $R \tan \Phi \leq 5$

$$\bar{\Theta}_s = 0.685 \, (R \tan \Phi)^{0.07} \, \Theta_{s\max} \tag{9.8a}$$

- for $5 \leq R \tan \Phi \leq 20$

$$\bar{\Theta}_s = 0.620\,(R \tan \Phi)^{0.13}\,\Theta_{s\max} \tag{9.8b}$$

- for $R \tan \Phi \geq 20$

$$\bar{\Theta}_s = 0.820\,(R \tan \Phi)^{0.04}\,\Theta_{s\max} \tag{9.8c}$$

9.3.2 Calculation of average and maximum interface temperatures

In general, the computing concept is based on the well-known principle of the simultaneous action of two independent heat sources, which suggests that the total heat flux is generated by aggregation of the plastic deformation and sliding friction effects. Hence, the average interface temperature is defined as the sum of the mean shear-plane temperature ($\overline{\Theta}_s$) and the mean temperature rise due to friction ($\Delta\overline{\Theta}_f$), namely

$$\overline{\Theta}_t = \overline{\Theta}_s + \Delta\overline{\Theta}_f \tag{9.9}$$

According to Shaw [3] the temperature increment due to the action of the frictional heat source (l_c is the tool/chip contact length) can be determined as

$$\Delta\overline{\Theta}_f = \frac{0.377 R q l_c}{\lambda_W \sqrt{N_T}} \tag{9.10}$$

By analogy to Eqn (9.9) the maximum interface temperature is the effect of proper temperature peaks, namely:

$$\Theta_{\max} = \Theta_{s\max} + \Theta_{f\max} \tag{9.11}$$

Using the methodology described above, the average temperature at the tool/chip interface was predicted for several different cutting speeds as shown in Fig. 9.6. It can be seen that a higher temperature is reached along the tool/chip interface when the uncoated tools were used. As documented in Fig. 9.6, substantially lower

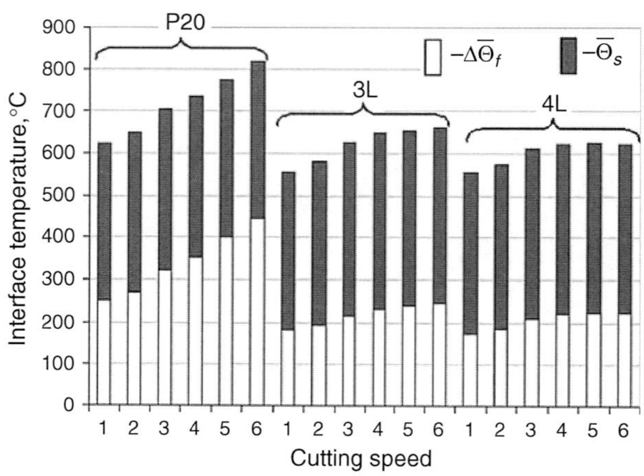

Fig. 9.6 Components of average interface temperature vs cutting speed [11]: 1 – 90 m/min; 2 – 105 m/min; 3 – 145 m/min; 4 – 180 m/min; 5 – 205 m/min; 6 – 235 m/min. 3L – $TiC/Al_2O_3/TiN$ coating, 4L – $TiC/TiCN/Al_2O_3/TiN$ coating.

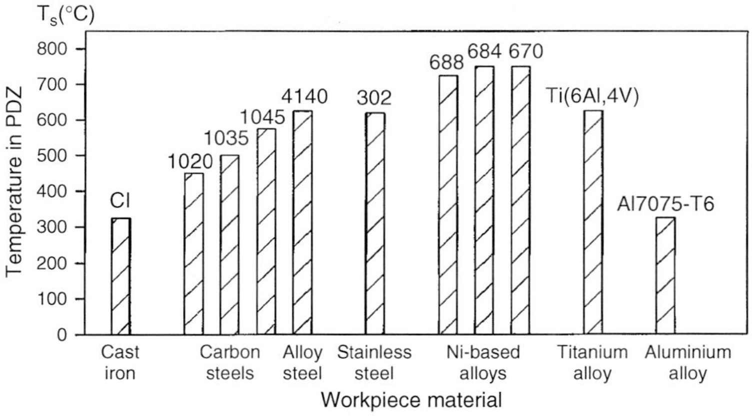

Fig. 9.7 Temperature on the shear plane for various workpiece materials [11].

temperature rises generated by friction at the secondary deformation zone (SDZ) are observed for multilayer coatings. Under experimental conditions, adequate differences dealing with the reduction of friction are in the range of 50–175°C depending on the cutting speed. On the other hand, the temperature rises due to shearing in the primary deformation zone (PDZ) were found to be, in general, slightly higher than for uncoated tools.

Additionally, Fig. 9.7 compares the temperatures in the shear plane for a broad range of workpiece materials. In this analytical prediction the following average values of the chip formation parameters were chosen: shear angle $\Phi = 28°$, friction angle $\Theta = 17°$, chip compression ratio $\lambda_h = 2$ ($r = 0.5$) and rake angle $\alpha_o = 0°$. Moreover, it is assumed that at low cutting speed, the shear-plane temperature is primarily determined by the mechanical and thermal properties of the workpiece material only.

9.3.3 FEM and FDA prediction of cutting temperature

As an example of the finite element model, the Lagrangian technique and explicit dynamic, thermo-mechanically coupled modelling software with adaptive remeshing were applied to simulate the plane-strain orthogonal metal cutting. This means that the initial mesh becomes distorted after a certain length of cut as shown in Fig. 9.8b and is remeshed in this vicinity to form a regular mesh again. The upper part of the

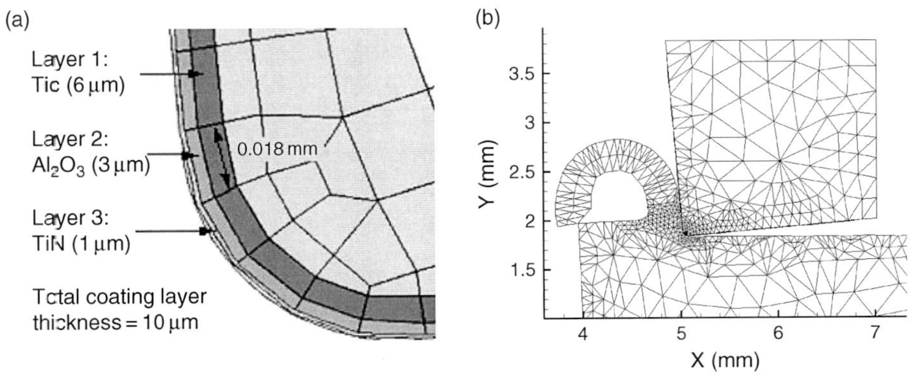

Fig. 9.8 The mesh model for a $TiC/Al_2O_3/TiN$ coated tool (a) and shape of the deformed chip after a tool path of 4.0 mm (b).

mesh, which constitutes the removed workpiece material, is finer, to enable the stress, strain, strain rate and temperature in the chip and the tool tip regime to be accurately predicted. In the model applied, constant thickness coating layers are added to the tool substrate to create a multiple coating as shown in Fig. 9.8a. For the dimensions of the rest a coarser mesh is sufficient. In the case shown in Fig. 9.8b, the workpiece consists of about 1340 six-noded plane-strain triangular elements and 1460 nodes. Dimensions of the element size can range from a minimum value of 0.02 mm to a maximum of 0.1 mm. On the other hand, the tool model consists of the adequate number of node planar heat-transfer elements, because heat transfer analysis is carried out for the thermal model including the tool. The lower half of its mesh, expected to be in contact with the chip, is modelled with a finer mesh, in order to be able to predict the temperature field developed in the tool.

The FEM simulations have revealed that cutting tool coatings influenced distinctly the performance and intensity of the thermal interactions when turning C45 carbon steel. In particular, coatings change both the heat transfer and its distribution in the cutting zone as shown in Fig. 9.9b. In comparison to an uncoated P20 carbide tool (Fig. 9.9a), the coating applied caused the fraction of heat which flows into the tool to decrease. On the other hand, it can be observed that more heat is transferred to the chip and workpiece. Moreover, coatings cause areas with the maximum temperatures to be localized near the chip and workpiece. In consequence, the maximum interface temperature exists in the vicinity of the cutting edge, i.e. in the first part of the tool/chip contact. Furthermore, it was revealed, based on temperature distributions illustrated in Fig. 9.9, that temperature developing on the workpiece surface increases by about 50°C in comparison to the conventional tool used. It is evident that this effect can be related to differences in the thermal properties of the tool materials. In particular, the thermal conductivity of Al_2O_3 ceramic layer in the $TiC/Al_2O_3/TiN$ coating decreases distinctly and it is apparent that at higher contact temperatures the carbide substrate is partly thermally insulated by the coating.

Figure 9.10 shows the selected segments of tool temperature maps located closer to the tool/chip interface obtained for the two identical material/tool pairs, as in Fig. 9.9, during FDA-based simulations. In particular, Fig. 9.10b depicts that temperature decrease through the coating of 0.01 mm total thickness is only about 30°C. Using these isothermal maps the relevant temperature distribution curves along rake and flank faces of the tool can be plotted [12]. Comparable temperature gradients between the hot point and the unloaded part of insert are generated in the machining trials performed. For instance, the maximum temperatures drop from 740(685)°C to about 620(555)°C at the end of chip contact for uncoated and coated tools, respectively.

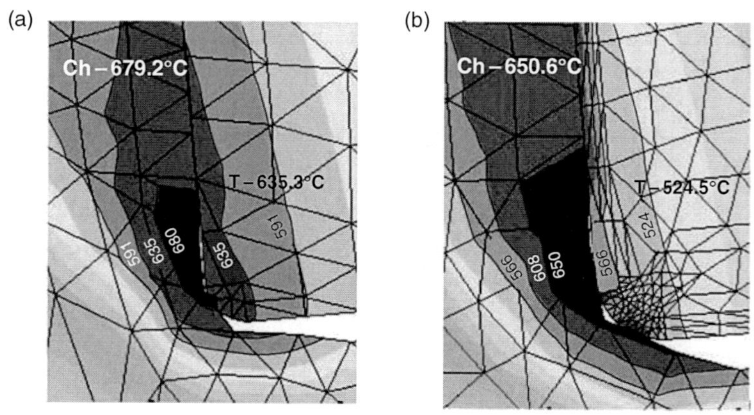

Fig. 9.9 Magnification of temperature distribution in the vicinity of the cutting edge. (a) P20 uncoated carbide; (b) $TiC/Al_2O_3/TiN$ coated carbide [8]. Workpiece – AISI 1045 steel. (see Plate 8)

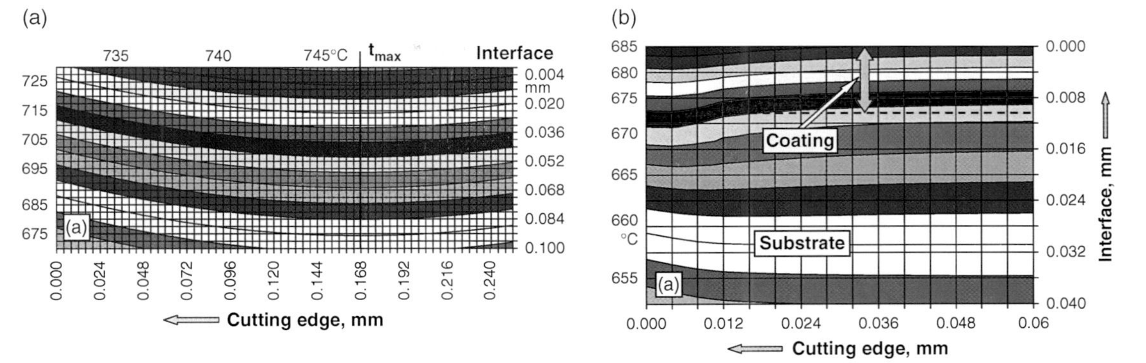

Fig. 9.10 Temperature field for a P20-AISI 1045 (a) and for a TiC/Al_2O_3/TiN-AISI 1045 pair (b) [12]. Heat flux into the tool: $q_T = 29.08\,\mathrm{MW/m^2}$ (a) and $q_T = 22.23\,\mathrm{MW/m^2}$ (b). The reference cutting conditions: $v_c = 145\,\mathrm{m/min}$, $f = 0.16\,\mathrm{mm/rev}$ and $a_p = 2\,\mathrm{mm}$. (see Plate 9)

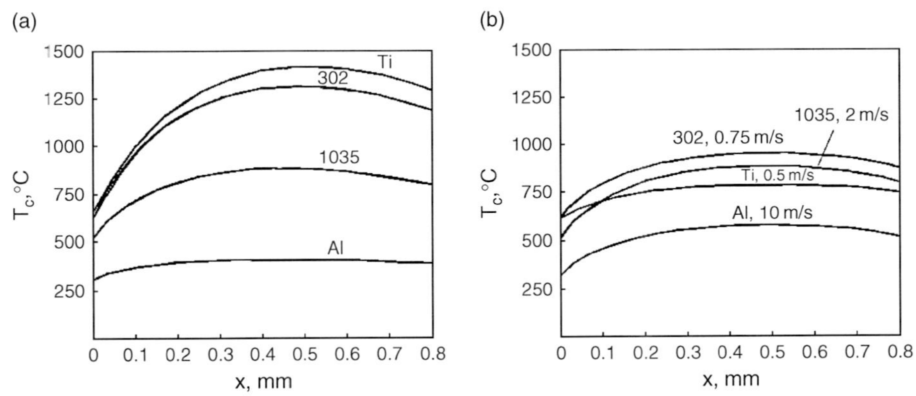

Fig. 9.11 Temperature distribution along the tool-chip contact for four different work materials at constant speed of 2 m/s (a) and different speeds (b) obtained by FDM modelling. Tool material: cemented carbide [11].

Figure 9.11 presents the computed distributions of the tool/chip interface temperatures for a constant speed of 2 m/s keeping the same geometry of cut (h, Φ, Θ) for each work material (a), and for different cutting speeds taking into consideration the machinability rating for the four materials applied. It should be noted based on Fig. 9.7 that a substantial rise of the contact temperature over the shear-plane temperature is documented. The low temperature for Al results from its low specific cutting pressure ($k_c = 850\,\mathrm{N/mm^2}$) and its very high thermal conductivity and diffusivity. In contrast, extremely high temperature obtained for a titanium alloy is due to substantially higher $k_c = 2000\,\mathrm{N/mm^2}$ and extremely low thermal properties, which cause the heat to be cumulated in a thin subsurface layer at the chip/tool contact. When using different cutting speeds, as shown in Fig. 9.11b, the temperature for the stainless steel and titanium alloy drops from about 1300°C and 1400°C down to about 970°C and 900°C, respectively, and the increase of the maximum temperature over the shear-plane temperature is much smaller. When high speeds ($v_c = 10\,\mathrm{m/s}$) are applied to aluminium machining, the maximum contact temperature reaches approximately the melting point of about 620°C. This comparison clearly indicates the large differences in the machinability of various materials.

9.4 MEASUREMENTS OF TEMPERATURES IN THE CUTTING ZONE

The interfacial temperatures play a major role in the performance of machining processes and a general move toward dry machining increases the importance of understanding how machining temperatures are affected by the process variables involved (cutting parameters, cutting tool materials, etc.) and by other factors such as plastic deformation, friction and tool wear. Based on this fact, a number of various measuring techniques of cutting temperatures were developed in the past and now new ones are developing successively. The main two groups of temperature measuring techniques (contact/non-contact or conduction based/radiation based) adapted to machining are schematically illustrated in Fig. 9.12.

Thermocouples have always been a popular transducer used in temperature measurement. Thermocouples are very rugged and inexpensive and can operate over a wide temperature range. According to the Thomson–Peltier law, a thermocouple is created whenever two dissimilar metals touch and the hot contact point produces a small open-circuit voltage (EMF – electromotive force) as a function of temperature. If these two dissimilar materials are the cutting tool and the workpiece materials, as shown schematically in Fig. 9.13a,

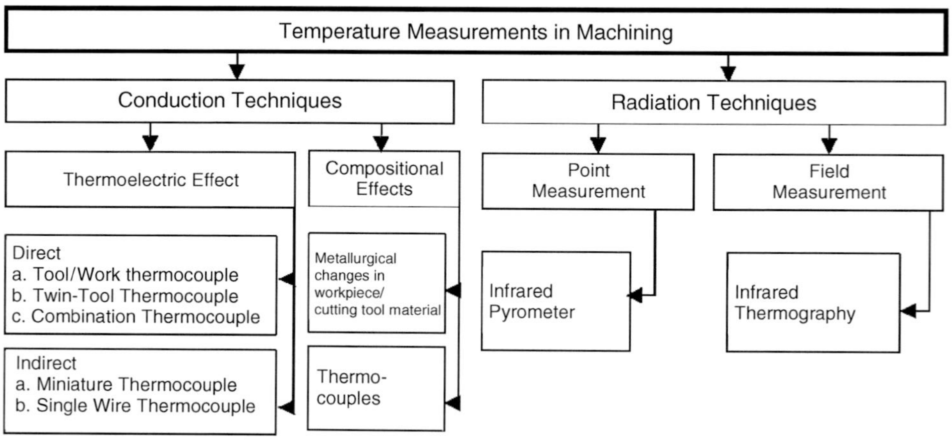

Fig. 9.12 Temperature measurement techniques in machining [13].

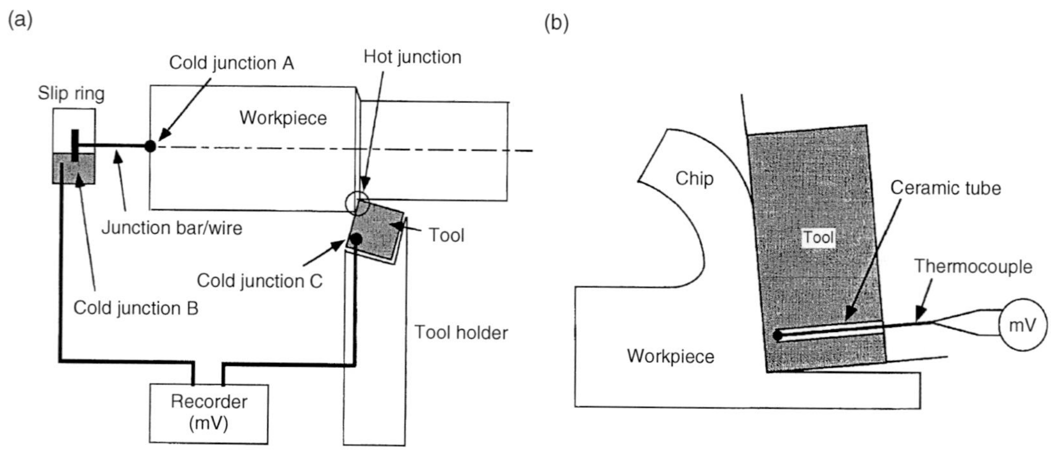

Fig. 9.13 A tool/work thermocouple circuit (a) and inserted thermocouple (b) [14].

then this thermocouple is called a tool/work (chip) natural thermocouple or dynamic thermocouple. Some wires are required to make a complete circuit between the recorder and the tool and the recorder and the work, as well as a special mercury slip-ring device due to transmission of the EMF signal from the rotating workpiece. As a rule, both the tool and workpiece are electrically insulated from the post and the chuck, respectively. The EMF measured in cutting must be converted to temperature, hence prior to measurement the tool/work thermocouple is calibrated using the same materials as in the cutting tests and the reference thermocouple, for example the chromel/alumel standard. Each different type of tool and work materials used must be calibrated individually. It is possible to calibrate separately, for example, the tool/chromel and work/chromel junctions and in such a case the tool/work EMF vs temperature relation is the difference between the foregoing relations [14]. The hot junction is created by rapid heating using an infrared heating furnace equipped with a high power halogen lamp or standard TIG welding apparatus. It should be noted that generally, the EMF/temperature relation for tool/work thermocouples is non-linear. Errors arising from uncertain calibration of the thermocouple can be partially eliminated by using two different tool materials (for example, high speed steel and tungsten carbide) to cut the same bar of work material, simultaneously, under the same conditions.

A standard thermocouple or one wire embedded (inserted) in the cutting tool (Fig. 9.13b) or workpiece material can be used to measure the temperature at a single point or at different locations to establish the temperature distribution in the tool. They can also be positioned at the interface between an indexable insert and the tool holder. In particular, embedded thermocouples have been found to provide a good indication of the transient changes in frictional heat generation that accompany contact area changes. A small diameter hole has been made in the tool and insert (often by EDM) in a precisely determined position close to the cutting edge. This must be repeated many times with holes in different positions to map the temperature gradients at the tool/chip interface (optionally the rake and the clearance faces of the tool may be progressively ground away). If the hole is positioned accurately, this may be a satisfactory method for comparing the tool temperature when cutting different materials.

Within the conduction techniques some innovation techniques were developed such as those utilizing the melting points of thermo-sensitive materials, thin film sensors and thin film thermocouple sensors. In the PVD film method, various PVD deposited films of different materials with known melting points (Ge 1211 K; Sb 904 K; Te 723 K; Pb 601 K; Bi 545; In 429 K) are used as thermal sensors [15]. The boundary between the melted film zone and the unmelted film zone determines directly and clearly the isotherm due to the difference in optical reflectivity of these two zones. By subsequently depositing different films on the inner surface of the split tool the temperature distribution in the tool was mapped.

The structure of a thin film thermocouple sensor (TFT) built in the cutting tools is shown in Fig. 9.14. For instance, the TFT can be made of Pt and Pt-13%Rh or Ni and Ni-Cr (80:20 in mass %) layers deposited on Al_2O_3 tool substrate [16]. The TFT layers of approximately 0.5 μm thick were insulated by a PVD hafnium oxide (HfO_2) and, then, a TiN was deposited to protect the built-in TFT against the wear and

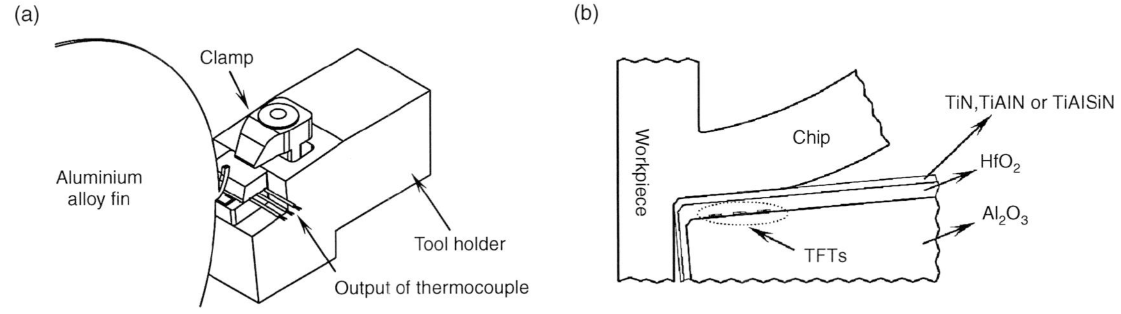

Fig. 9.14 Concept of tool with a built-in thin film thermocouple: (a) application of TFT tool to turning and (b) view of longitudinal cross-section [15].

high stresses in high speed machining. In addition, a chromel/alumel thermocouple was used to measure the terminal temperature of the cold junctions (the EMF is due to the temperature difference between the hot junction and terminals of the TFT). The new TFT sensors were successfully tested in the machining of A6061-T6 aluminium alloy at very high cutting speeds up to 16 m/s.

Recently, many efforts have been made for the development of non-contact radiation techniques (infrared thermometry) for the measurement of cutting temperature. The radiation techniques can be divided into pyrometry which measures the temperature of a single spot and thermography which provides the temperature distribution of a thermographic image. The development of an infrared detection method was spread out because the significant brittleness and electrical resistance of some tool materials such as ceramics make it difficult to use contact-type sensors such as embedded tool/work thermocouples for measuring temperature at the tool/work interface. An infrared (IR) pyrometer with fibre optics is a non-contact method of measuring temperature of a body based on its emitted thermal energy. Thermal imaging (thermographic) methods, measuring the radiation from the outside surfaces of the tool, the workpiece and the chip, have a number of advantages, if surface temperatures are of interest. However, much care must be taken, as real materials are some fraction of the black body because the emissivity (or emittance) varies with surface roughness, state of oxidation, cleanness and other factors. Emissivity is defined as the ratio of the energy radiated by an object at a given temperature to the energy emitted by a perfect radiator, or 'black body' at the same temperature. As a result, calibration of the IR camera under the same conditions as cutting is necessary because the infrared emission of the chip is measured while the chips are produced. For instance, the IR camera can be calibrated for each typical chip roughness and oxidation colour by reheating the chip with a thermocouple attached to one point on its surface, in a special vacuum chamber [17]. The use of an IR-CCD camera allows the data to be converted into a high resolution, colour, thermographic image with the complete temperature distribution in the vicinity of the cutting edge, as shown in Fig. 9.15.

In order to eliminate the problem of emissivity effect, which strongly influences the evaluated temperature measurements, a two-colour pyrometer technique was developed [18, 20]. For a two-colour pyrometer the measured temperature is independent of the surface emissivity when grey-body behaviour can be assumed. As shown in Fig. 9.16a, an optical fibre of 2 m length is inserted into a thin plastic pipe and in a fine 0.6 mm hole, and is fixed when the distance between the incidence face of the fibre and the rake surface approaches 1 mm. The calibration is obtained by sighting on radiating surfaces of known, uniformly distributed temperature. As can be seen in Fig. 6.16b, a specimen is inserted inside the cylindrical siliconit-heating element and a Pt/Pt-Rd embedded thermocouple is used for monitoring the surface temperature. The fibre

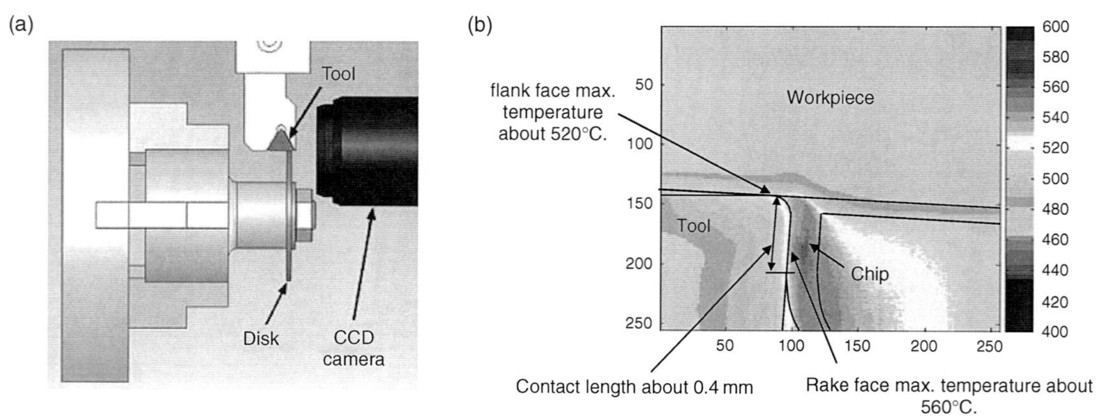

Fig. 9.15 Thermal map obtained by means of IR CCD camera. Courtesy of J. Rech. Work material: INOX 316L steel, cutting tool: carbide insert TPKN coated with TiN; cutting conditions in orthogonal cutting: v_c = 60 m/min, f = 0.1 mm/rev. (see Plate 10)

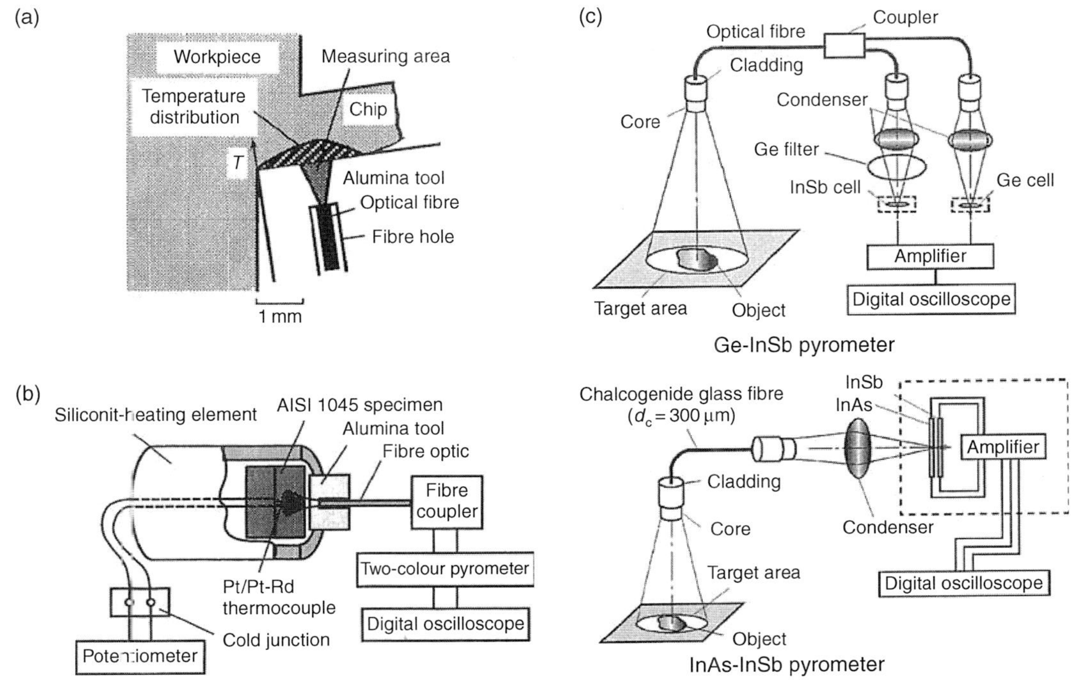

Fig. 9.16 Measurement of cutting temperature using two-colour pyrometer [18, 19]. (a) illustration of optical fibre setting; (b) scheme of calibration setup; (c) two structures of two-colour pyrometers.

optic that captures infrared radiation from the specimen is inserted in the same manner as in the alumina ceramic insert.

The measurements are performed when infrared rays radiated from the tool/chip interface are received by an optical fibre, then divided into two fibres by a fused fibre coupler. Each fibre leads the infrared rays to two infrared detectors which have different spectral sensitivity. The electric signals converted from infrared energies by these cells are amplified and stored into digital memory. The temperature is obtained from the ratio of the output voltage of the two cells. In the upper scheme (Fig. 9.16c) germanium (Ge) and indium antimonide (InSb) are used as detector and quartz (Si02) fibre is selected to measure temperature higher than 500°C. The Ge filter is added to the InSb cell to reduce the short wavelength of infrared rays and improve the sensitivity of the apparatus at high temperature range. In the lower scheme, the infrared energy is received by a chalcogenide optical fibre and led to a two-colour detector consisting of InAs and InSb detectors (the first one responds to incident radiation from 1 to 3 μm and transmits waves larger than 3 μm, whereas the second one responds to radiation from 3 to 5 μm). It was proven experimentally that the temperature measured by the two-colour pyrometer was the maximum temperature in the target area of the optical fibre by an accuracy of at least 90% [19].

REFERENCES

[1] W. Grzesik, *Fundamentals of Machining of Metallic Materials* (in Polish), WNT, Warsaw, 1998.
[2] F.P. Incropera, D.P. De Witt, *Fundamentals of Heat and Mass Transfer*, Edward Arnold, London, 1990.
[3] M.C. Shaw, *Metal Cutting Principles*, Clarendon Press, Oxford, 1989.

[4] T. Kato, H. Fujii, Energy partition in conventional surface grinding, *ASME Trans. J. Manuf. Sci. Eng.*, Vol. 121 (1999), 393–398.
[5] A.N. Reznikov, *Thermophysical Aspects of Metal Cutting Processes* (in Russian), Mashinostroenie, Moscow, 1981.
[6] W. Grzesik, P. Nieslony, Physics based modeling of interface temperatures in machining with multilayer coated tools at moderate cutting speeds, *Int. J. Mach. Tools Manuf.*, Vol. 44 (2004), 889–901.
[7] W. Grzesik, P. Nieslony, A computational approach to evaluate temperature and heat partition in machining with multilayer coated tools, *Int. J. Mach. Tools Manuf.*, Vol. 43 (2003), 1311–1317.
[8] W. Grzesik, Composite layer-based analytical models for tool-chip interface temperatures in machining medium carbon steels with multi-layer coated cutting tools, *J. Mat. Proc. Technol.*, Vol. 176 (2006), 102–110.
[9] E.M. Trent, P.K. Wright, *Metal Cutting*, Butterworth-Heinemann, Boston, 2000.
[10] G. Boothroyd, W.A. Knight, *Fundamentals of Machining and Machine Tools*, Marcel Dekker, New York, 1989.
[11] G. Tlusty, *Manufacturing Processes and Equipment*, Prentice Hall, Upper Saddle River, New Jersey, 2000.
[12] W. Grzesik, Analytical models based on composite layer for computation of tool-chip interface temperatures in machining steels with multilayer coated cutting tools, *CIRP Annals*, Vol. 54/1 (2005), 91–94.
[13] D. O'Sullivan, M. Cotterell, Temperature measurement in single point turning, *J. Mat. Proc. Technol.*, Vol. 118 (2001), 301–308.
[14] T.H.C. Childs, K. Maekawa, T. Obikawa, Y. Yamane, *Metal Machining. Theory and Applications*, Arnold, London, 2000.
[15] T. Kato, H. Fujii, PVD film method for measuring the temperature distribution in cutting tools, *J. Engn. Ind.*, Vol. 118 (1996), 117–122.
[16] A. Basti, T. Obikawa, J. Shinozuka, Tools with built-in thin film thermocouple sensors for monitoring cutting temperature, *Int. J. Mach. Tools Manuf.*, Vol. 47 (2007), 793–798.
[17] S.P.F.C. Jasper, J.H. Dautzenberg, D.A. Taminiau, Temperature measurement in orthogonal metal cutting, *Int. J. Adv. Manuf. Technol.*, Vol. 14 (1998), 7–12.
[18] M. Al Huda, K. Yamada, A. Hosokawa, T. Ueda, Investigation of temperature at tool-chip interface in turning using two-color pyrometer, *J. Manuf. Sci. Engng*, Vol. 124 (2002), 200–207.
[19] T. Ueda, A. Hosokawa, K. Yamada, Effect of oil mist on tool temperature in cutting, *J. Manuf. Sci. Engng*, Vol. 128 (2006), 130–135.
[20] B. Müller, Temperature measurements with a fiber-optic two-colour pyrometer, in: H. Schulz (ed.), *Scientific Fundamentals of HSC*, Carl Hanser Verlag, Munich, 2001.

CHAPTER TEN

Cutting Fluids

Contents

10.1 Basic Categories of Cutting Fluids

Cutting fluids (CFs) are applied to the chip formation zone to improve the cutting performance by cooling and/or lubrication effects. In general, from a thermal point of view, the lubricating action is more important at low cutting speeds, whereas the cooling is rather the predominant effect at higher cutting speeds due to more intensive heat generation. Different types of cutting fluids available can be classified according to widely varying criteria, although a somewhat unified system of terminology is used in the different standards. For example, German DIN 51385 proposes cutting fluids to be primarily divided into non-water miscible and water miscible CFs, as shown in Fig. 10.1. Non-water miscible, also known as oil-based cutting fluids, straight or neat oils, are supplied as ready-to-use products. Water mixed cutting fluids are also supplied as ready-to-use, while water miscible products are supplied as a concentrate which must be diluted before application. Once mixed, the constituents can form either a solution or an emulsion.

Popularly, there are four basic categories of cutting fluids [2]:

1. Straight or neat oils which are used undiluted and are usually mineral oils but they often include other lubricants such as fats, vegetable oils and esters, together with extreme-pressure (EP) compounds based on chlorine, sulphur and phosphorus. Neat cutting fluids provide excellent lubricity but are relatively poor coolants.
2. Mineral-soluble oils (emulsions) which consist of oil with emulsifiers that allow the oil to disperse in water. These oils are used in diluted form with a typical concentration range of 3 to 15%. These fluids are widely used in industry and are generally the least expensive cutting fluids available.
3. Semi-synthetic fluids (or micro-emulsions) are essentially a combination of synthetic and soluble oil fluids and, as a result, exhibit characteristics of both types of compounds. They consist of natural and synthetic emulsifiers, oil droplets and clear emulsions. They offer good corrosion resistance, lubrication and tolerance to contamination. The definition of semi-synthetics can cause some difficulty but the best definition is this: 'A semi-synthetic forms a translucent emulsion and contains mineral oil' [3].
4. Synthetic fluids, which are mineral oil-free solutions formulated from alkaline inorganic and organic compounds, usually with the addition of corrosion inhibitors. They are used diluted with water, with a concentration range from 3 to 10%. According to industrial practice, these fluids often provide the best cooling performance.

Figure 10.2 illustrates the equilibrium between the cooling and lubrication properties of the most commonly applied cutting fluids. As mentioned above, the boundary properties represent water and mineral oils.

Advanced Machining Processes of Metallic Materials
ISBN: 978-0-08-044534-2

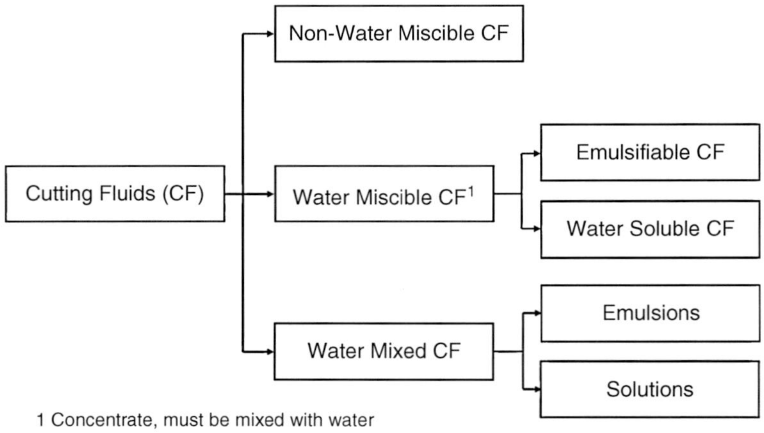

Fig. 10.1 Classifications of cutting fluids according to DIN 51385 [1].

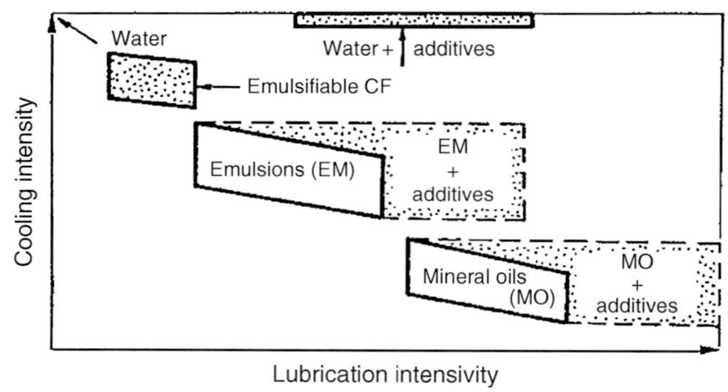

Fig. 10.2 Comparison of cooling and lubrications effects for different cutting fluids.

As mentioned above, to preserve a separation film between the tool and workpiece for difficult-to-machine materials or high cutting speeds, oil soluble EP additives that contain chlorine, phosphorus and/or sulphur are added to neat cutting fluids and micro-emulsions. At higher temperatures and pressures, EP additives react chemically with the metal surface to form chlorides, phosphates or sulphides, thereby acting much like a solid lubricant at the workpiece/tool interface. Active temperature regions for different types of EP additives, covering the temperature range from 180°C up to 1000°C, are shown in Fig. 10.3.

Single point turning, planning and shaping, and drilling of shallow holes are among the operations where simple water-based coolants may be the only fluids required. Lubricating requirements are most exacting with difficult operations such as broaching, threading, reaming, trepanning of deep holes and the hobbing of gears. According to a statistic gathered from the US manufacturing industry, around 80% of cutting fluids used are

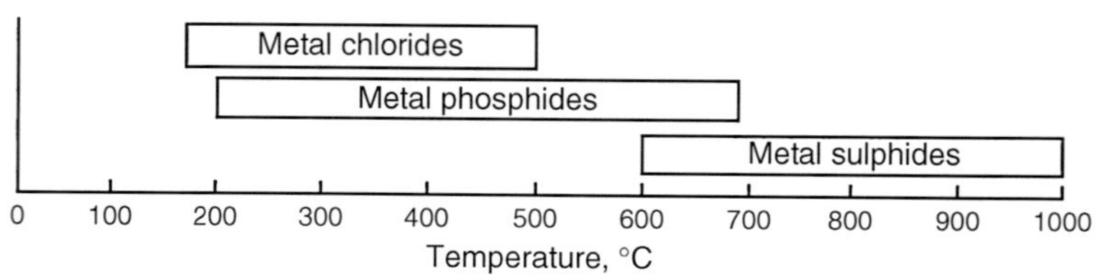

Fig. 10.3 Active temperature regions for extreme-pressure additives [1].

water-based emulsions of oils and/or synthetic fluids. On the other hand, the European automotive industry assesses the cost of cutting fluids at about 20%, whereas for comparison the tooling costs are only around 7.5%.

10.2 Functions and Action of Cutting Fluids

It is well known that cutting fluids perform a very important role in cutting operations and many operations cannot be efficiently carried out without the correct fluid. Correct selection, application and control of the cutting fluid will result in its becoming an integral part of a cost-efficient, well-organized and safe machine shop. In recent years, coolant systems that enable high pressure coolant supply and internal coolant-type cutting tools have been implemented to facilitate highly efficient coolant functions. As a result, high speed, high efficiency, long tool life, high cutting accuracy and improved machinability of difficult-to-machine materials have been established. As shown in Fig. 10.4, cutting fluids are essential for solving

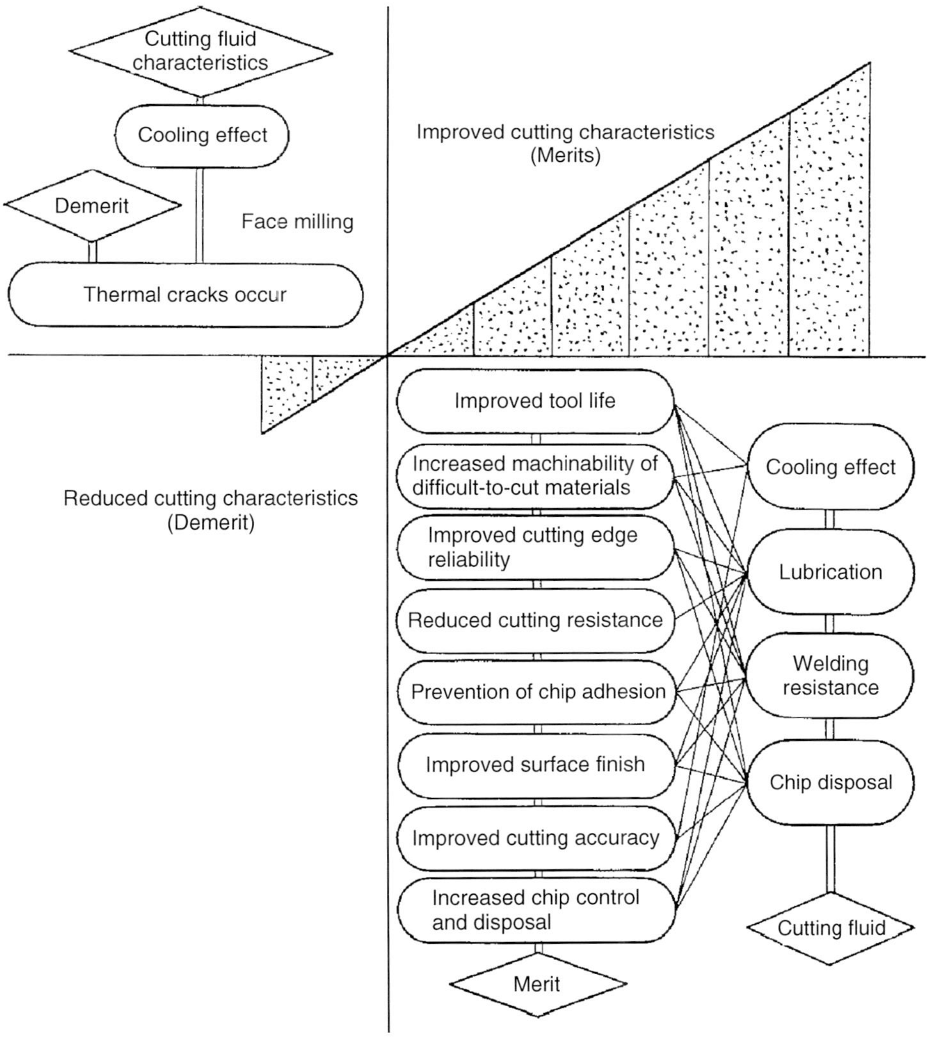

Fig. 10.4 Standard characteristics and function of cutting fluids [4].

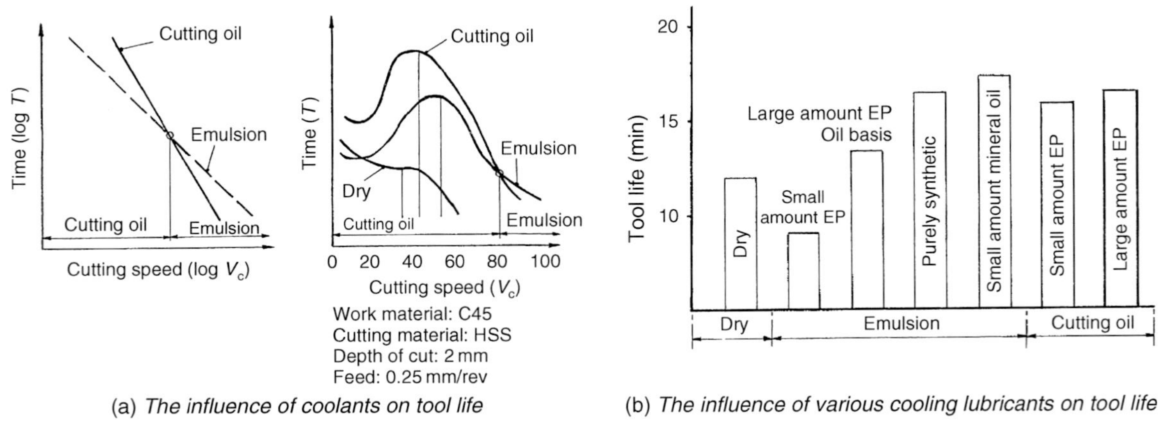

Fig. 10.5 Influence of coolants (a) and cooling lubricants (b) on tool life [3]. Work material: C45 steel; tool material: HSS; depth of cut: 2 mm; feed rate: 0.25 mm/rev.

problems that aim at high productivity. The basic functions of cutting fluid are cooling and lubrication effects and welding resistance, chip disposal (transport) and cleaning the workpiece. All other positive functions (merits) are listed in the right lower part of Fig. 10.4.

However, some negative aspects exist, as, for example, tool damage due to supplying unsuitable cutting fluid, but they are small compared to the positive aspects of applying cutting fluid. In particular, application of cutting fluid improves tool life by aggregating the cooling effect, lubrication, welding (built-up edge) resistance and easier chip disposal. Furthermore, it enables high speed and highly efficient cutting. The cooling effect prevents normal wear and plastic deformation. Welding resistance and chip disposal prevent abnormal tool damage. Lubrication prevents both normal and abnormal tool damage. Influence of different cutting fluids on tool life, in comparison to dry cutting, is shown in Fig. 10.5. It is evident from Fig. 10.5a that a lubricant fulfils many other functions besides simply reducing wear. Because it lowers the friction and temperature, the extent of temperature-induced wear types is reduced. In general, an adequate emulsion-type coolant or a lubricant (oil) should be selected depending on the cutting speed applied. On the other hand, Fig. 10.5b confirms the significant role of EP additives in the prolongation of tool life for both cooling and lubrication predominant effects. In particular, extreme pressure additives and other individual ingredients can extend the life of the emulsions to cover a wider range of cutting conditions.

10.3 Application of Cutting Fluids and Other Cooling/Lubrication Media

The cutting fluid can perform its functions if it is effectively supplied to the cutting zone. Generally, fluids with predominant lubrication function must be delivered to form a wear protected film on the sliding surfaces, while coolants must penetrate to the tool cutting edge. Four main fluid application methods are used in practice for general cutting operations: manual, flood, high pressure jet and mist applications (see Chapter 15.2). Usually, external and internal (through the tool) supplying methods are distinguished, as presented in Fig. 10.6 for end-milling operations. It is worth noticing that due to differently cooled cutting zone, as shown in Fig. 10.6, the cutting temperature in milling can be changed by ±50%.

The most common method of fluid application is to flood the tool, workpiece and cutting zone with fluid delivered from a low pressure nozzle directed appropriately, as in Fig. 10.6d. It allows a continuous

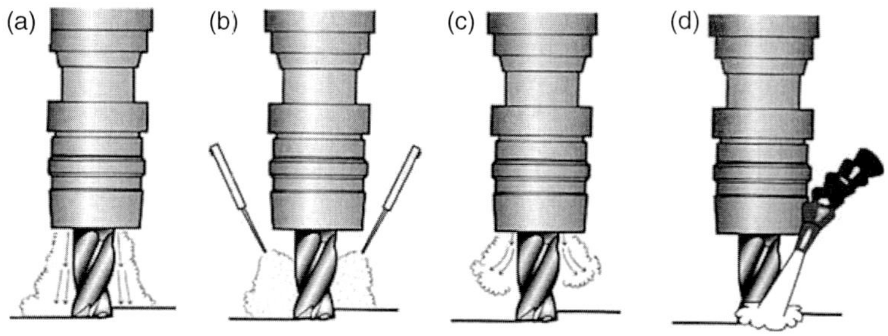

Fig. 10.6 Different methods of cooling in milling: (a) compressed air, (b) oil mist, (c) internal cooling, (d) external cooling [5]. Source: Sandvik Coromant.

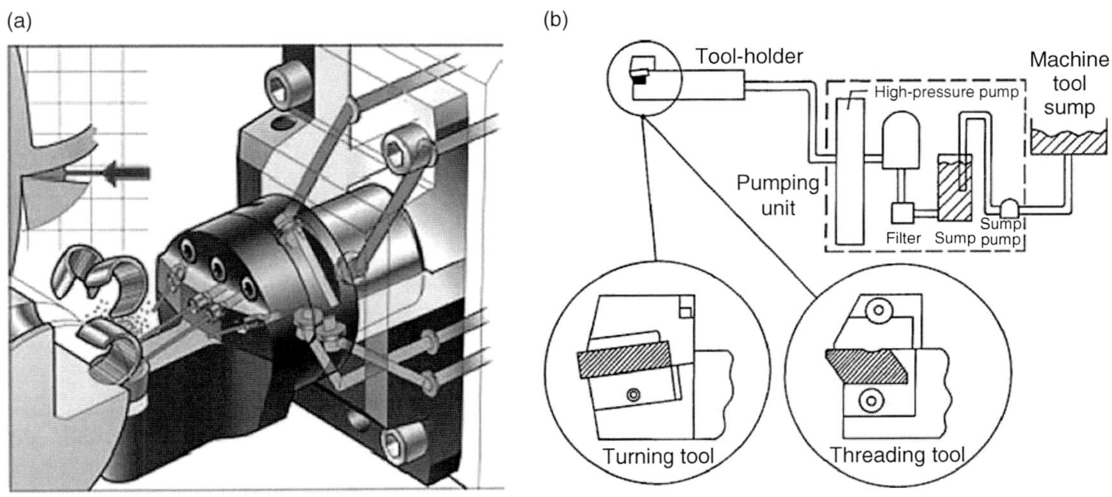

Fig. 10.7 High pressure coolant (HPC) supply systems called Jetbreak (a) and Hyperson (b). After Sandvik Coromant (a) [5] and Kennametal [3].

flow of fluid to the cutting zone and removes chips from the cutting zone (prevents recutting of the chips), but its effectiveness depends on the geometry of the cutting process performed.

A high pressure and hence higher velocity jet of fluid is directed at the cutting zone. In the solution shown in Fig. 10.7a, the coolant supply at a high pressure up to 1000 bars (for example, 200 bars is the maximum requirement for titanium machining) shortens the tool/chip contact length and provides improved chip control. The Jetbreak system, giving a specifically directed nozzle, can be coupled with the appropriate tooling system. As a result, the coolant supplied through the cone of the coupling into the nozzle jet is directed towards the cutting edge to form a hydraulic wedge between the chip and the rake face of the tool insert. Most applications are on single spindle turning machines such as vertical turning centres and turn-milling centres.

Figure 10.7b shows the commercial system called the Hyperson which provides a high velocity of 122 m/s and a high pressure flow of coolant at 110 bar to the interface between the tool and chip. The system consists of a pumping unit, the piping necessary to transport the fluid, and a tool-holder through which the coolant can pass and which can focus the high pressure flow of coolant on the interface. In addition a programmable control unit which adopts the needs of users and the machine tool is installed.

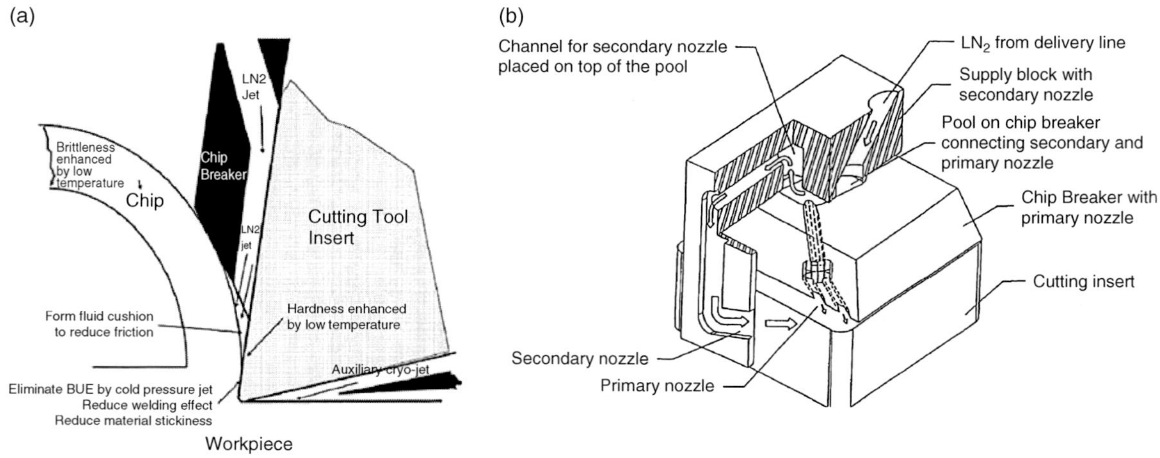

Fig. 10.8 A schematic of the cryogenic machining approach (a) and the design implementation of two cryogenic nozzles (b) [6].

Predominantly, this system works efficiently in the turning of titanium alloys and the threading on alloy steels. As well as its chip-breaking and chip disposal abilities, improvement of tool life by a factor of between 30% and 100% over a range of difficult-to-machine materials was achieved [3]. Another method which is also very effective and clean is the use of CO_2 as a coolant. CO_2 at high pressure is supplied through a hole in the tool and allowed to emerge from small channels under the tool tip as close as possible to the cutting edge. The expansion of the CO_2 lowers the temperature and the tool close to the jet is kept below 0°C. A further extreme potential use of coolants involves cryogenic machining, when both tool and work material are cooled at the temperature of liquid nitrogen (−196°C).

Figure 10.7a presents the cryogenic machining cooling concept using a flat cutting insert with an obstruction-type chip breaker. The micro-nozzle supplying a minimum amount of liquid nitrogen (LN2) is located between the tool face and chip breaker because the chip breaker allows liquid nitrogen to reach and cool the highest temperature spot, i.e. the tool/chip interface. The liquid nitrogen absorbs the heat, evaporates it quickly and forms a fluid/gas cushion between the chip and the face that functions as a lubricant. Consequently, the coefficient of friction and deformation in the SDZ are reduced, and both the lubrication and cooling effects reduce the tool temperature which, in turn, effectively reduces both crater and flank wears. An auxiliary cryogenic nozzle may be added as in Figs 10.8a and b, to cool the flank face for further reduction of the flank wear. In the case from Fig. 10.8b the secondary nozzle is integrated with the LN2 supply block. The system causes tool life to increase up to five times versus conventional emulsion cooling and by application of two nozzles, the highest temperature is reduced about 320°C in comparison to emulsion cooling [6].

10.4 Maintenance and Disposal of Cutting Fluids

The significant cost associated with cutting fluid application places particular importance on good maintenance to extend useful life. The concentration and composition of cutting fluid alters with use due to water evaporation, contamination, bacterial growth, reaction when leakage oil mixes with the cutting fluid, etc. Water-soluble cutting fluids require more maintenance than cutting oils. Therefore, it is important to check the concentration of the mix regularly. When the concentration needs to be changed, mixed cutting fluids are added. In addition, a bacteria killing agent, known as biocides, must be added when cleaning.

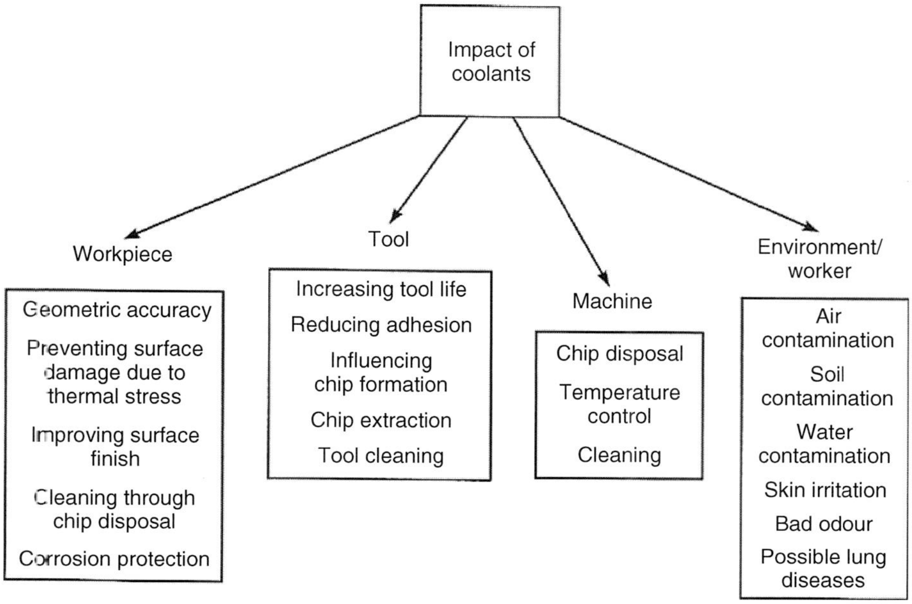

Fig. 10.9 Impact of coolants on manufacturing environment [7].

If pollutants and contaminated cutting fluid remain, fresh cutting fluid which is added will soon become contaminated. Sedimentation of undissolved particles using the force of gravity is the easiest way to remove pollutants from the cutting fluid. To ensure that the coolant system is working properly, periodic testing of the pH levels should be regularly determined and the oil concentration values should be determined using refractometers. Modern systems for the reclamation of cutting fluid should have automatic equipment for measuring, filling up and mixing supplementary cutting fluid.

The lifecycle of a cutting fluid within a machining facility involves four stages: storage and handling, mixing of concentrate with water, cutting fluid in-process use and cutting fluid disposal. With long-term use, the fluid condition deteriorates until the coolant ultimately fails and requires disposal. So, proper monitoring of fluid degradation and fluid recycle leads to significantly prolonged fluid life. As shown in a diagram from Fig. 10.9, the decision to dispose of cutting fluids can involve the evaluation of corrosion, dermal initiation, air, water and soil contamination, reduced tool life and surface finish, colour and odour. Bacterial growth and the release of small droplets (aerosols) into the atmosphere are the two main environmental concerns with cutting fluids and there are stringent national recommendations of the permissible levels [2].

Rejected cutting oil and oil emulsion should not be released into the public sewage system. Emulsions are treated so that oil and water are separated. In order to break down emulsion, ferrous sulphate, salts and, in certain cases, strong acids are used. If acids are used the water is neutralized before it is released into the sewage system. In general, straight oils are relatively simple to treat prior to disposal, requiring only removal of water and solid contaminations. The oil is then either burnt for energy recovery, or recycled into a virgin product. The easiest fluids to separate are emulsions, followed by semi-synthetics. With water-diluted coolants, the water can be removed and the remaining solids disposed of separately. The most common types of secondary treatment include thermal separation (evaporation), chemical treatment and ultrafiltration. For example, evaporation can reduce the disposed mass by 90%. The primary stage involves removing the free oil (skimming) and suspended solids (filtration and settling). This is followed by ultrafiltration (solids greater than 0.01 μm can be rejected) and chemical treatment to separate fluids. The final stage uses carbon filtration or reverse osmosis for final purification of the water that can then be reused or discharged.

Despite the significant environmental and health impacts of cutting fluids, there is a lack of knowledge regarding proper selection and application of fluids in various machining operations. In most cases, cutting fluid selections are based mostly on their perceived process performance benefits, cost and marketing. As a result, overuse or misuse of cutting fluids in machining operations are common. The waste streams associated with cutting fluid usage may be reduced to a great extend or eliminated by proper selection and application of cutting fluids based on a fundamental understanding of cutting fluid properties, their impact on the process and their environmental and health effects. To address this deficiency, an Internet-based cutting fluids evaluation software testbed (CFEST) was developed by the Machine Tool-Agile Manufacturing Research Institute [8]. This software is structured in several modules to provide, via the Internet using a web browser, quantitative information for manufacturing practitioners on cutting fluid performance (heat generation, heat transfer, mist generation and chip carry-off characteristics), environmental impact, health and safety hazards, and costs for cutting fluid comparison and selection.

References

[1] P.S. Sheng, S. Oberwalleney, Life-cycle planning of cutting fluids – a review, *J. Manuf. Sci. Engng*, Vol. 119 (1997), 791–800.
[2] G. Boothroyd, W.A. Knight, *Fundamentals of Machining and Machine Tools*, CRC Press, Boca Raton, 2006.
[3] G.T. Smith, *Advanced Machining*, Springer-Verlag, Berlin, 1989.
[4] K. Karino, *Trouble Shooting for Cutting*, Mitsubishi Materials, Ibaraki, Japan, 1998.
[5] *Titanium Machining*, Sandvik Coromant, No. C-2920:22 ENG/01, 2004.
[6] S.Y. Hong, I. Markus, W.-Ch. Jeong, New cooling approach and tool life improvement in cryogenic machining of titanium alloy Ti-6Al-4V, *Int. J. Mach. Tools Manuf.*, Vol. 41 (2001), 2245–2260.
[7] B.P. Erdel, *High-speed Machining*, SME, Dearborn, 2003.
[8] J.W. Sunderland, P. Sheng, R.E. DeVor *et al.*, CFEST: an Internet-based cutting fluid evaluation software testbed, *Trans. NAMRI/SME*, Vol. 25 (1997), 243–248.

CHAPTER ELEVEN

Tribology of Metal Cutting

Contents

11.1 Tribological Characterization of the Cutting Zone

Tribology is the common term used to cover the performance of real surfaces in terms of lubrication, friction and wear [1, 2]. The cutting process is performed in the specific tribo-mechanical system consisting of three basic elements: the work (chip) material, the tool and the environment which can change the conditions between the two sliding surfaces by controlled supply of solid, liquid or gaseous lubricants. Specific tribological features of the cutting process result from the following causes [3]:

1. Freshly generated sliding surfaces of the tool and the chip are extremely active chemically.
2. The contact load is concentrated on a very small area and, as a result, very high normal and shear stresses acting on the interface with maximum values of $\sigma_c = (900\text{–}1600)$ MPa and τ_c of about $(30\text{–}60\%)\sigma_c$ are recorded.
3. Plastic deformation of the chip in the secondary deformation zone (SDZ) is performed under upper boundary conditions with an additional influence of high temperatures, often causing welds between the asperities or sometimes melting the high points of mating asperities (so-called 'flash' points) at extremely high sliding velocities (for example, in high speed machining).
4. Friction at relatively high speeds causes cutting fluid not to easily penetrate to the contact zones and form the solid lubrication films.
5. Associated wear phenomenon causes alteration of the cutting tool geometry which influences friction conditions.

Taking into consideration all the above-mentioned tribological conditions, the tribological model of the cutting process can be schematically presented as in Fig. 11.1. In this tribo-model with heavily loaded sliders, two/three characteristic regions can be distinguished, namely [3–5]:

1. The seized (sticking or plastic frictional contact) region A placed on the parts of both rake and flank faces adjacent to the tool cutting edge, in which the ratio A_r/A_a (real contact area to apparent contact area), approaches unity and continuous refreshing of the sliding surfaces takes place. In this region physically dry friction with chemically clean mating active surfaces exist.
2. The sliding (elastic frictional contact) region B extends from the end of the sticking region to the point where the chip loses contact with the tool (narrow region C). In contrast to the sticking region, in the region B dry technical sliding friction exists when supplying the cutting fluid boundary lubrication (the lubrication film is not completed) or fluid film lubrication. More accurate models consider the transition zone A–B (Fig. 11.1b) with mixed friction or lubrication conditions.

Advanced Machining Processes of Metallic Materials
ISBN: 978-0-08-044534-2

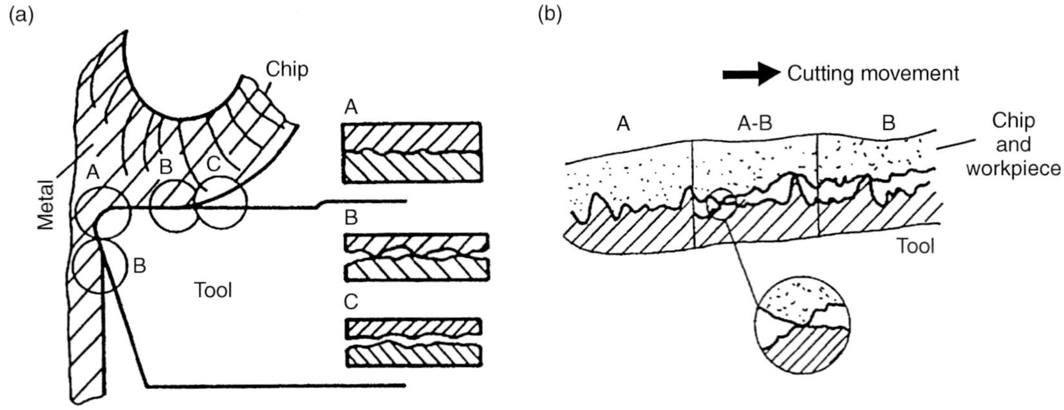

Fig. 11.1 Tribological characteristics of the cutting zone [3]: (a) three zone model; (b) model with transition *A–B* zone.

The evidence of the sticking mode of frictional contact was examined on the undersurface of the chip after the cutting action was suddenly stopped by using a quick-stop device. After cutting a wide range of materials, low magnification photomicrographs of the rake faces of the tools showed that there was generally less transfer near the cutting edge, particularly for soft work materials. It was observed that in the region adjacent to the tool cutting edge, the grinding marks on the tool rake face were imprinted on the undersurface of the chip, indicating also that no relative motion between the chip and the tool occurred and that the real and apparent areas of contact are equal in this region. These macroscopic observations were confirmed optically using transparent (photoelastic) sapphire tools and a high speed (fast-frame) photography technique [4] in machining steels and brass.

The evidence of two different frictional contact modes' existence was experimentally confirmed by Loladze [6] on the pictures of material plastic flow in the SDZ, as shown in Fig. 11.2. It is evident from Fig. 11.2 that in the region adjacent to the cutting edge of l_{st} in length, the texture line No. 10 is parallel to the rake face and the velocity of a thin interfacial material layer v_{ch}^{st} approaches zero. This means the formation of the stagnation (dead) zone resulting from very strong adhesion between fresh undersurface of the chip and the rake face. The shear strain in this region reaches unusually high values of $\gamma_{sh} = 30–40$. On the other hand, in the sliding region with uniform thickness Δ_2 of the SDZ, classical dry sliding friction occurs with the sliding velocity of v_{ch}^{sl}, which is proportionally lower than the cutting speed v_c.

Based on experiments, Abuladze [7] proposed the sticking contact length l_{st} to be expressed in the form:

$$l_{st} = l_p = h\left[k_h(1 - \tan\gamma_0) + \frac{1}{\sin\gamma_o}\right] \tag{11.1}$$

where h is the undeformed chip thickness, k_h is the chip thickness ratio and γ_o is the rake angle.

In practice, the ratio l_{st}/l_{sl} ranges between 0.3 and 0.8 but for steels it can be kept approximately equal to 0.5, which means that the total contact length can be divided into two equal parts belonging to frictional sticking and sliding contacts. According to Eqn (11.1), the contact length increases with the increase of the UCT (equivalently with feed rate) and chip compression ratio, but decreases for high positive rake angles. It was also documented that the tool/chip contact length decreases for high cutting speeds and when friction is modified by coatings deposited on the cutting tools.

Figure 11.3 specifies the alterations caused by coatings, which were classified as surface (superficial) and volume effects. The fact that coatings can modify the contact conditions between the chip and the tool has

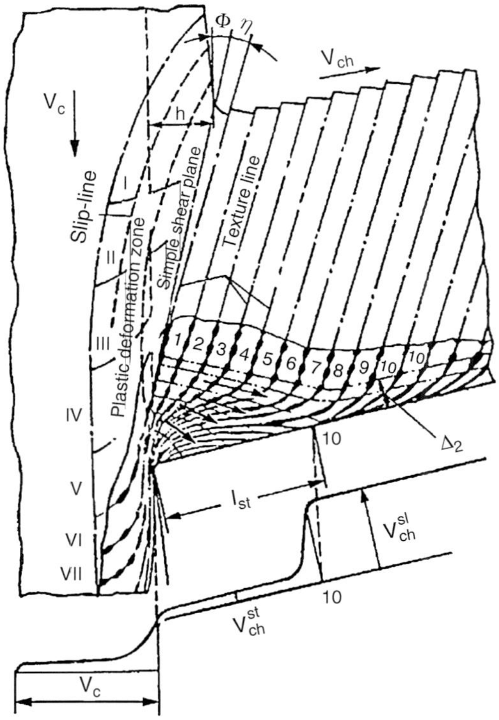

Fig. 11.2 Scheme of material plastic flow in the SDZ. After Loladze [3, 6].

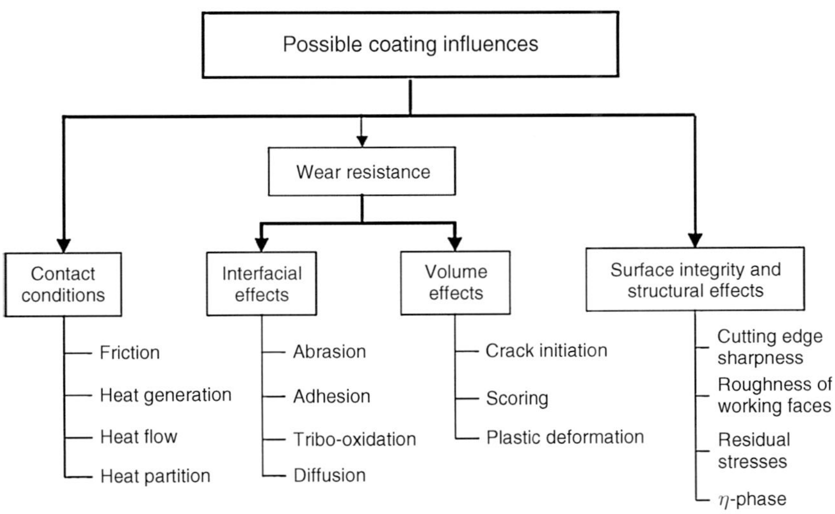

Fig. 11.3 The role of coatings in the alteration of cutting process performance [8].

been practically utilized for several decades. In order to adapt the desired properties (including tribological ones) of coating materials to a specific machining task, more attention should be focused on the contact zone including integral friction, heat and wear.

11.2 Distribution of Stresses in the Tool/Chip Interface

Since the maximum stress acting on any part of the tool/work interface is the most critical stress determining the requirements for the tool material and the strength of the tool, it is essential to know the distribution of the compressive (normal) and shear (tangential) stresses. In metal cutting theory, both calculated and experimentally determined distribution of stresses have been attempted.

The most famous model for the stress distribution proposed by Zorev [7] shown in Fig. 11.4 is valid for the case in which, in the absence of a built-up edge, sticking or seizure occurs at the interface near the tool edge and sliding takes place beyond the sticking region. Zorev's model assumes that the normal stress distribution satisfies the power function $\sigma_c = qx^n$, where x is the distance from the point B where the chip breaks contact with the tool. Because the normal stress is distributed with the maximum at the cutting edge, falling to zero in the chip separation point, the expression for σ_c is

$$\sigma_c = \sigma_{c\,max}(x/l_c)^n \tag{11.2}$$

where $\sigma_{cmax} = \sigma_c$ for $x = l_c$.

The shear stress shows a lower maximum at the boundary between the sticking and sliding regions, and a plateau across the sticking region. Due to this dualism, shear contact stress for the model from Fig. 11.4 can be determined as follows:

$$\tau_c = \begin{cases} \mu\sigma_{c\,max}(x/l_c)^n & \text{for } 0 \le x \le l_{sl} \\ \mu\sigma_c & \text{for } l_{sl} \le x \le l_c \end{cases} \tag{11.3}$$

where l_c is the tool/chip contact length and μ is the coefficient of friction.

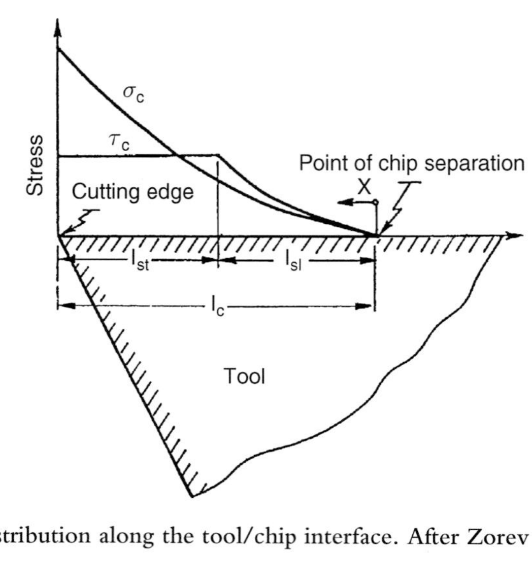

Fig. 11.4 Simplified stress distribution along the tool/chip interface. After Zorev [7].

By using Eqns (11.2) and (11.3) average forces $F_{\gamma N}$ and F_{γ} acting on the rake face of the tool can be determined:

$$F_{\gamma N} = \int_{0}^{lc} \sigma_c \, b_c dx = \frac{\sigma_{c\,max} b_c l_c}{1+n} \tag{11.4}$$

$$F_{\gamma} = \int_{0}^{lc} \mu \sigma_{c\,max} (x/l_c)^n \, b_c dx + \int_{l_{sl}}^{lc} \tau_c b_c dx = \tau_c b_c \left[l_{st} + l_{sl}/(1+n)\right] \tag{11.5}$$

where b_c is the contact width and n is the exponent.

Based on the slip-line theory Loladze [6] proved that the maximum contact stress is equal to:

- for intensive friction, when $\mu \to \mu_{max}$

$$\sigma_c \approx 2\tau_{sh} (1.3 \pm \gamma_0) \tag{11.6}$$

where τ_{sh} is the shear stress on the shear plane and γ_0 is the rake angle in rad.
- for frictionless condition, when $\mu \to 0$

$$\sigma_c \approx 2\tau_{sh} (1 \pm \gamma_0) \tag{11.7}$$

Sign (+) is valid for the rake angle $\gamma_0 < 0$, whereas sign (−) should be used for $\gamma_0 > 0$.

As an example, assuming $\gamma_0 = -30°$ ($-\pi/6$ rad) we obtain that $\sigma_c = 3.65\tau_{sh}$ and $\sigma_c = 05\tau_{sh}$, respectively.

The mean values of contact stresses are given by

$$\overline{\sigma}_c = \frac{F_{\gamma N}}{l_c b_c} = \frac{F_{\gamma N}}{A_c} \tag{11.8}$$

$$\overline{\tau}_c = \frac{F_{\gamma}}{l_c b_c} = \frac{F_{\gamma N}}{A_c} \tag{11.9}$$

where A_c is the contact area.

Examples of estimated values of the mean compressive stress are specified in Table 11.1 to give a comparison between the behaviours of pure metals and alloys. In fact, high forces when cutting pure metals such as iron, copper and titanium, are an indirect result of a large contact area, and then mean normal stress on the tool is relatively low in comparison to medium carbon steel.

Table 11.1 Values of mean normal contact stress. After Trent and Wright [4].

Work material	*Cutting force (N)*	*Contact area (mm²)*	*Average normal stress (Mpa)*
Iron	1070	3.1	340
Copper	4150	13.5	310
Titanium	455	0.77	570
Medium carbon steel	490	0.65	770
Brass 70/30	500	12.2	420
Lead	323	22.5	14

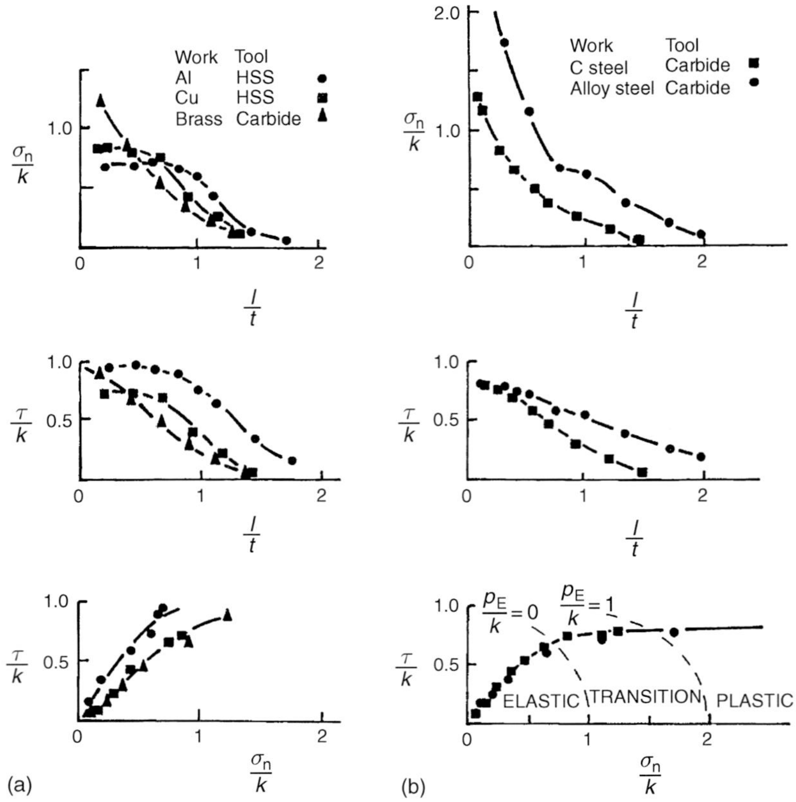

Fig. 11.5 Derived rake face stresses for non-ferrous (a) and ferrous (b) work materials [9]. Literature sources: Al/HSS and Cu/HSS after Kato *et al.* [10]; brass/carbide and C-steel/carbide after Shirahashi and Usui [11]; low alloy steel/carbide after Childs and Maekawa [12].

The real situation at the tool/chip interface is much more complex than that of the simplified model adopted for calculation, and experimental determination of stress distribution is therefore of more interest. Measurements of compressive and shear stresses will be discussed later. Several possible stress distributions shown in Fig. 11.5 were obtained using split tool data available in [10–12]. The contact stresses have been normalized by the shear stress k (τ_{sh}) acting on the primary shear plane, and by analogy the distance l from the cutting edge has been normalized by the chip thickness t (h_{ch}). In most cases, the normal stress rises to a peak at the cutting edge. However, in two cases (for aluminum and copper), the rise in normal stress towards the cutting edge is capped by a plateau. Peak normal stresses range from $0.7k$ to $2.5k$. Friction (shear) stress also rises towards the cutting edge, but is always capped at a value $\leq k$. When friction stress is replotted against normal stress (bottom panels of the figure), these stresses seem to be proportional at low stress levels (in the sliding region) but at high normal stresses (near the cutting edge) the friction stress becomes independent of normal stress. Consequently, the characteristic zones were labelled as elastic, transition and plastic (by analogy the two regions in Fig. 11.1 were also proposed to be termed elastic and plastic). It should be concluded that the experimental data confirmed, in general, Zorev's model of stress distribution along the tool/chip contact.

Today, a large portion of the tools used in the turning process have restricted contact (RC) lengths by a groove or an attached chip-former (in [13] this portion was assessed to be about 65–75%). It is also denoted as artificial contact length in contrast to the natural tool/chip contact length for tools with a plane rake face. It is assumed that the optimum contact length for restricted-contact tools (RCT) is in the range of 55–65% of the natural contact length. The reduction of tool/chip contact can provide several benefits for the machining process, such as a substantial reduction in power consumption, lower tool/chip interface temperature, improved surface integrity of the machined surface, an increased tool life and promoting chip control. As shown in Fig. 11.6, the effect of RC can also result in altering the distribution of contact stresses. In general, for large UCT (tool/chip contact length increases if the feed increases) which provides an optimum RC contact length, as in Fig. 11.6c, the average stresses on the tool rake face increase by producing flatter stress distributions.

It is evident that tool/chip contact length is one of the most important parameters controlling the cutting process and its estimation is based on several experimental, analytical and numerical methods such as experiments with split tools, slipe-line theory, and FEM simulation. Table 11.2 contains some important

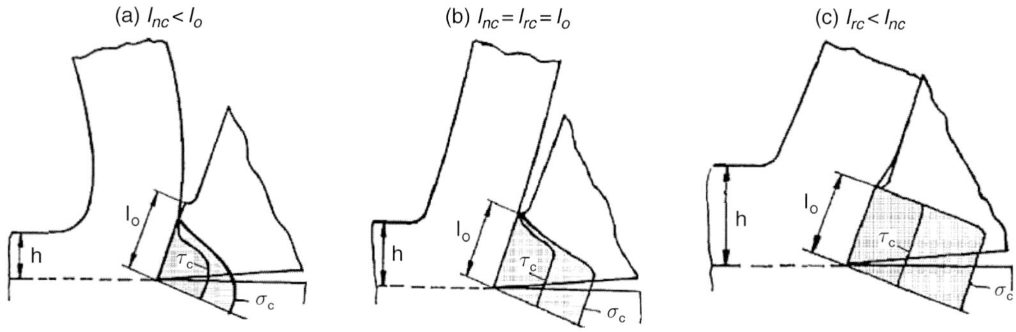

Fig. 11.6 Change of stress distribution for restricted-contact tools. After De Chiffre [14].

Table 11.2 Equations for calculation of the chip/tool contact length.

Equation for relative tool/chip contact length	*Author(s)*
$\frac{l_c}{h} = \frac{\sqrt{2}}{\sin\left[\tan^{-1}\left(\frac{\cos\gamma_0}{k_h - \sin\gamma_o}\right) \times \sin\left[45^0 + \tan^{-1}\left(\frac{\cos\gamma_0}{k_h - \sin\gamma_o}\right) - \gamma_o\right]\right]}$	Lee and Schaffer [15]
$\frac{l_c}{h} = \frac{\sin\omega}{\cos\Theta\sin\Phi}\left\{1 + \frac{Cn}{3[1+2(1/4\pi - \Phi)] - Cn}\right\}$	Oxley [16]
$\frac{l_{st}}{h} = \frac{\sin\pi/4}{\sin\Phi_T\sin[\pi/4 + \Phi_T - \gamma_0]};\ \frac{l_{st}}{l_c} = 0.4 - 0.56$	Vinogradov [17]
$\frac{l_c}{h} = \left(\frac{h_{ch}}{h}\right)^{0.1}\left[\frac{h_{ch}}{h}(1 - \tan\gamma_o) + \frac{1}{\cos\gamma_0}\right]$	Abuladze [7]
$l_c = \frac{2h\cos(\gamma_o - \Phi)}{\sin\Phi}$ or $\frac{l_c}{h} = 2\left(\frac{h_{ch}}{h}\right)$ or $\frac{l_c}{h} = 2.05\left(\frac{h_{ch}}{h}\right) - 0.55$ (for $1 \le k_h \le 10$)	Kachanov and Poletika [18]

where Φ is the shear angle, Φ_T is the shear angle calculated from mechanical properties of the workpiece material, Θ is the friction angle, ω is the angle between the resultant cutting force and the shear plane, h is the UCT, h_{ch} is the chip thickness, k_h is the chip thickness coefficient ($k_h = h_{ch}/h$) and γ_o is the rake angle.

equations for calculation of the chip/tool contact length (in reality, the non-dimensional parameter l_c/h, termed as relative tool/chip contact length). This is based on the suggestion that the length of sticking region can be assumed to be nearly equal to the chip thickness [10].

In general, the most important factor influencing l_c is the undeformed chip thickness (the feed) and this relationship is almost linear when machining carbon steels with cutting speeds of 150–300 m/min. The tool/chip contact length is also affected by the cutting speed (lower l_c values are recorded for higher cutting speeds) and depth of cut (higher l_c values are recorded for higher depths of cut). Contact length is also a function of both the mechanical properties of the workpiece material and the tool material including coatings. Particularly, contact length decreases as the hardness of the workpiece materials increases.

11.3 Characterization of Friction at the Tool/Chip Interface

In metal machining the frictional processes are very complex because of existence of very high normal loads. An increase in the normal load across the contacting surfaces produces an increase in the real contact area (A_r) until it is capable of supporting the applied load. Because a newly generated chip having a highly chemically activated surface slides over the rake face, adhesion (welding of asperities) is a predominant phenomenon at the interface and the frictional force stems from shear fracture of the bonded materials. The equivalent coefficient of friction is given by

$$\mu_c = \frac{F_t}{N_t} = \frac{\tau_t A_p}{\sigma_t A_p} = \frac{\overline{\tau}_c}{\overline{\sigma}_c} \tag{11.10}$$

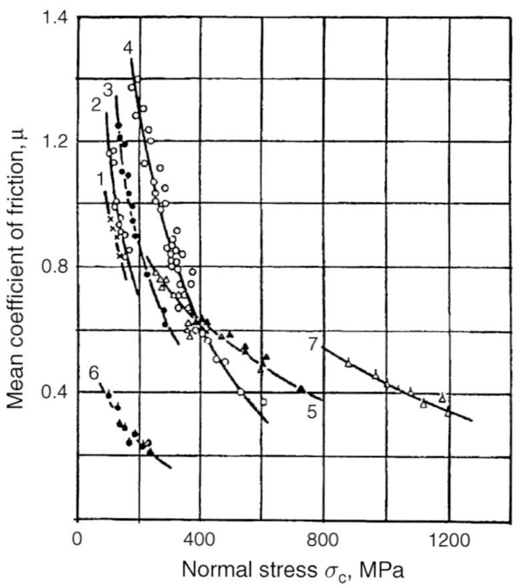

Fig. 11.7 Influence of normal stresses on the mean coefficient of friction for different workpiece materials. After Poletika [3]. 1 – copper, $\gamma_0 = 10°$; 2 – copper, $\gamma_0 = 25°$; 3 – Armco iron, $\gamma_0 = 10°$; 4 – stainless steel 12HN3; 5 – bronze, HB 110 and 200, $\gamma_0 = 10°$; 6 – bronze, HB 320, $\gamma_0 = 10°$; 7 – cadmium, $\gamma_0 = -30° \div 10°$.

where A_p is the apparent area of contact and τ_t and σ_t are the shear strength and yield pressure of the softer (chip) material.

In a lightly loaded condition (region of sliding contact) where A_p/A_r is very small, the Coulomb–Amonton laws of dry sliding friction hold. In case of relative motion between the highly loaded surfaces (region of sticking friction close to region of sliding friction) the frictional force is no longer linearly related to the normal force. At extremely high normal loads (region of sticking friction close to the cutting edge) the real contact can approach, or become equal to, the apparent area of contact. With these conditions, the coefficient of friction reaches a limiting value of 0.577, if it applies the von Mises plastic flow rule [19]. For $A_p/A_r = 1$, seizure occurs and the frictional force is independent of the normal force, and the coefficient of friction decreases with a further increase in normal load, as shown in Fig. 11.7. In metal cutting a simple method for calculating the mean normal $\overline{\sigma}_c$ and frictional stresses $\overline{\tau}_c$ from the shear plane model and then the mean coefficient of friction (Eqn (7.10)) is used. By substituting this formula with Eqn (11.6), the maximum value of friction coefficient is given by

$$\mu_{c\,\max} = \frac{1}{2\,(1.3 \pm \gamma_0)} \qquad (11.11)$$

For example, for $\gamma_o = \pm\pi/6$ values of μ_{cmax} are equal to 0.64 and 0.27. The first value of 0.64 is close to the limiting value of 0.577 satisfying the von Mises plastic flow rule.

The total coefficient of friction on the rake face can also be estimated by summing the adhesion and deformation (mechanical) terms [20], namely:

$$\mu_c = \mu_a + \mu_m \qquad (11.12)$$

The interfacial coefficient of friction μ_a can be expressed as the ratio of the shear strength of the adhesive junction τ_a between local asperities of dissimilar materials, and the yield stress of the plastically deformed material obtained from the true stress–true strain curve. On the other hand, the mechanical term μ_m can be easily derived from Lee and Schaffer's solution for the shear angle (see Chapter 7). Under orthogonal cutting of AISI 1045 carbon steel and AISI 304 stainless steel with uncoated carbide tools (cutting speeds of 40–270 m/min; feeds of 0.08–0.27 mm/rev) the adhesive term was found to be higher than the mechanical term of friction [20]. The maximum value of μ_a was equal to 0.7 for carbon steel and 0.4 for stainless steel, whereas the total coefficient of friction μ_c was equal to 1.1 and 0.82, respectively.

During metal cutting, it has generally been observed that the mean coefficient of friction between the chip and the tool can also vary considerably and is affected by changes in cutting speed (partly by thermal softening effect), feed (by controlling normal load), rake angle (by controlling plastic deformation in the SDZ) and so on, as well as by low friction coatings. Occasionally, in machining processes, values of the mean coefficient of friction reach a level of 1.5(2), such as in aluminium machining with HSS tools having large-positive rake angles. Normally, in machining steels, copper and brass μ ranges from 0.6(0.7) to 1 [19]. In contrast, μ can approach values close to zero when machining with tools having large negative rake angles (see Chapter 6). For extremely high sliding velocities, up to 60 m/s, obtained by a special air gun setup, the reduction of μ is observed especially at velocities higher than 20 m/s and the effect of the normal pressure is still remarkable [21].

The influence of cutting speed and normal load on the mean coefficient of friction for machining carbon and austenitic stainless steels with uncoated and differently coated carbide tools is shown in Fig. 11.8. One can observe a substantial decrease in the value of μ with increase of both cutting speed and feed rate. The similar effects can be explained in terms of the thermal softening phenomenon (Fig. 11.8a) and the rise of the normal load when varying the feed rate as in Fig. 11.8b. It is evidently documented in Fig. 11.8a that the workpiece material (stainless steel) with lower thermal conductivity, generating higher contact temperatures, tends to produce higher gradients of μ. This evidence is mostly pronounced for uncoated tools.

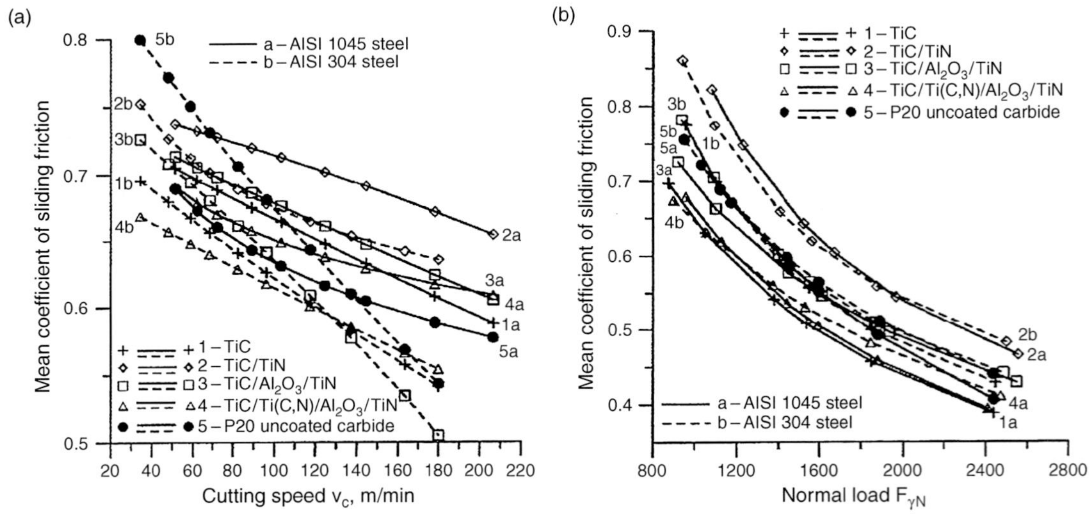

Fig. 11.8 The mean coefficient of friction vs cutting speed (a) and normal load (b) [22].

11.4 Measurements and Predictions of Contact Stresses and Friction

The following experimental and simulation techniques can be used to measure or predict the distribution of normal and shear stresses over the tool/chip contact length:

1. In early experiments commercially pure lead was cut with photoelastic polymeric (epoxy resin) tools or later steel and brass specimens were cut with sapphire tools [4]. Fast-frame photography allowed a series of images to be collected from which colour images provided isochromatics (for normal stresses) and black and white provided isoclinics (for shear stresses). It was found that the variation of the boundary stress distribution was more pronounced with respect to feed rate rather than to cutting speed.
2. The most commonly applied method uses a split cutting tool shown in Fig. 11.9a. This method employs a special HSS or carbide tool in which the split is parallel to the cutting edge and the length (l) of tool 1 (RCT) is much shorter than the natural tool/chip contact length (l_c). Two segments of a tool are separately mounted, at least one part (usually the outer part) on a force measuring platform, with a small gap between them of width g. This gap must be small enough for the chip flowing over the rake face not to extrude into the gap and large enough for the two parts of the tool not to touch as a result of any deflections caused by the acting forces.

   Since the area corresponding to the length l can be varied from a practical minimum of about 0.1 (0.12) mm to the full contact length l_c, then a series of force measurements give the stress distribution on the tool rake face (Fig. 11.9b). Split tool data were shown above in Fig. 11.5.
3. In tribology, many experimental arrangements employing several types of wear testing apparatus (*tribometer* or *tribotester*) have been used to study sliding friction and wear. In metal cutting applications, the asymmetric system employing a pin pressed against a flat disc (called *pin-on-disc* tribometer) can be of special interest [24]. In order to partially reproduce the contact conditions in machining, the system should be capable of continuous introduction of a fresh surface of work material (disc). Two designs of the modified tribotester with such possibilities are shown in Fig. 11.10. The horizontal version of the pin-on-disc device shown in Fig. 11.10a employs a Kistler three-axis piezoelectric dynamometer for measuring the normal F_f and tangential F_c forces ($\mu = F_c/F_f$), and an infrared camera for recording thermal images of the contact zone.

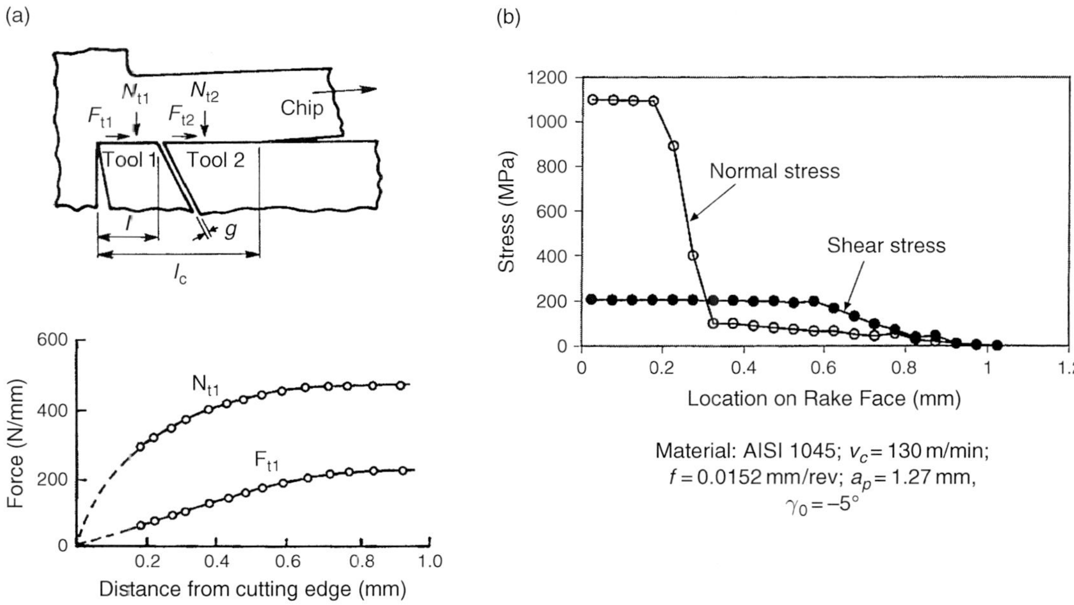

Fig. 11.9 Schematic split tool with measured specific contact forces [9] (a) and representative stress distribution obtained in [23] (b).

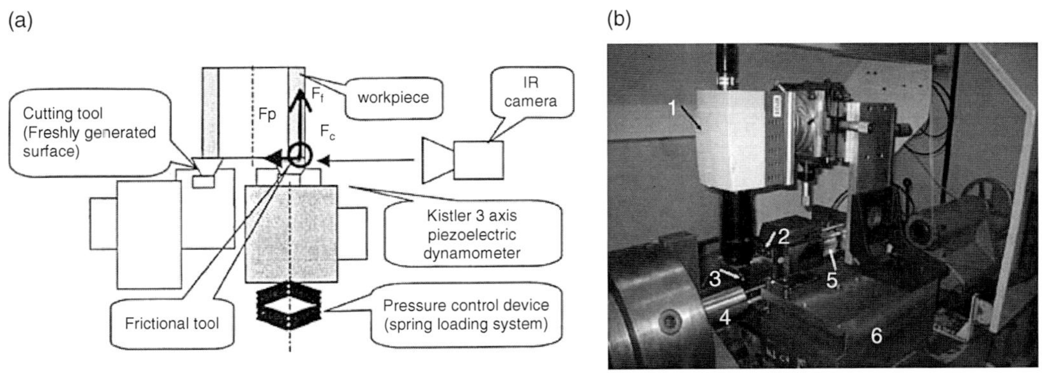

Fig. 11.10 Modified vertical [25] (a) and horizontal (b) pin-on-disc tribotester. 1 – IR camera; 2 – three-axis piezoelectric dynamometer; 3 – coated pin; 4 – thin-walled tube; 5 – pneumatic loading system; 6 – rigid base to avoid chatter. Picture (b) with permission of Prof. J. Rech.

The flat surface of friction on the tool (area of 1–2 mm^2) was prepared by a precision diamond saw and an additional cutting insert with the cutting edge parallel to the friction surface was clamped on the base. In the design from Fig. 11.10a, a spring loading system was applied to control the normal load, whereas the second similar solution (Fig. 11.10b) applied a pneumatic controller of the normal load. Both devices enable correlating friction and thermal phenomena in order to calculate the temperature-dependent value of heat flux, which can be of special interest in FEM-based simulation.

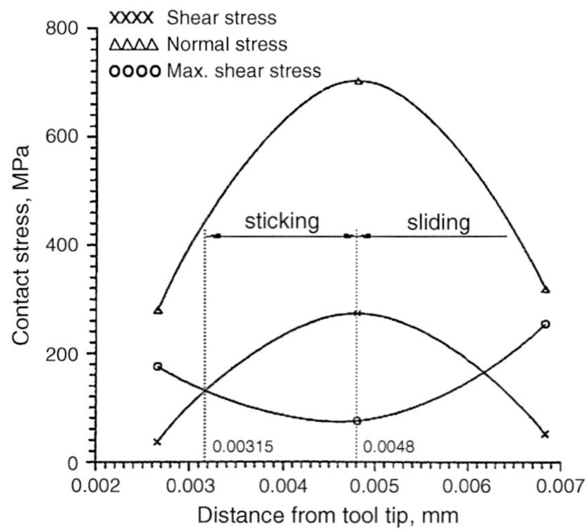

Fig. 11.11 Relationship between stresses on the tool/chip interface when sticking occurs at a diamond tool in machining OFHC copper [26].

4. Numerical simulation using FEM technique which satisfies, in general, the stress distribution obtained by means of the split-tool technique shown in Fig. 11.9b [27]. To model the sticking phenomenon on the rake face for high precision cutting when a single-crystal diamond tool is coupled with an oxygen-free high conductivity (OFHC) copper, the effect of high strain rate and high temperature was combined in a thermo-elastic–plastic FEM model to realize a pseudo-friction coefficient concept. The pseudo-friction coefficient $\mu_p = 0.22$ was introduced instead of Coulomb's friction coefficient $\mu = 0.13$ when the frictional shear stress in Eqn (11.10) is greater than the allowed maximum shear stress of the chip, corresponding to the onset of seizure. Figure 11.11 shows such a case, along with gradual increase of the normal stress, over the distance of 3.15–4.8 μm from the tool tip. After passing this distance, gradually decreasing the normal stress indicates that Coulomb's sliding friction exists.

References

[1] M.C. Shaw, *Metal Cutting Principles*, Clarendon Press, Oxford, 1989.
[2] I.M. Hutchings, *Tribology: Friction and Wear of Engineering Materials*, Edward Arnold, London, 1992.
[3] W. Grzesik, *Fundamentals of Machining of Metallic Materials* (in Polish), WNT, Warsaw, 1998.
[4] E.M. Trent, P.K. Wright, *Metal Cutting*, Butterworth-Heinemann, Boston, 2000.
[5] G. Boothroyd, W.A. Knight, *Fundamentals of Machining and Machine Tools*, CRC Press, Boca Raton, 2006.
[6] T.N. Loladze, *Strength and Wear Resistance of Cutting Tools* (in Russian), Mashinostroenie, Moscow, 1982.
[7] N.N. Zorev, *Development of Metal Cutting Science* (in Russian), Mashinostroenie, Moscow, 1967.
[8] W. Grzesik, *Advance Protective Coatings for Manufacturing and Engineering*, Hanser Gardner, Cincinnati, 2003.
[9] T.H.C. Childs, K. Maekawa, T. Obikawa, Y. Yamane, *Metal Machining. Theory and Applications*, Arnold, London, 2000.
[10] S. Kato, K. Yamaguchi, M. Yamada, Stress distribution at the interface between tool and chip in machining, *J. Eng. Ind.*, Vol. 94B (1972), 683–689.
[11] T. Shirakashi, E. Usui, Friction characteristics on tool face in metal machining, *J. Jap. Soc. Prec. Eng.*, Vol. 39 (1973), 966–972.
[12] T.H.C. Childs, K. Maekawa, Computer aided simulation of chip flow and tool wear, *Wear*, Vol. 139 (1990), 235–250.

[13] M.I. Sadik, B. Lindström, The effect of restricted contact length on tool performance, *J. Mater. Proc. Technol.*, Vol. 48 (1995), 275–282.

[14] L. De Chiffre, *Metal Cutting. Mechanics and Applications*, Technical University of Denmark, 1990.

[15] E.H. Lee, B.W. Schaffer, The theory of plasticity applied to a problem of machining, *Trans. ASME, J. Appl. Mech.*, Vol. 18 (1951), 405–413.

[16] P.L.B. Oxley, *The Mechanics of Machining*, Ellis Horwood, Chichester, 1989.

[17] A.A. Vinogradov, *Physical Foundations of the Process of Drilling Difficult-to-Machine Materials using Carbide Drills* (in Russian), Naukova Dumka, Kiev, 1985.

[18] A. Toporov, S.-L. Ko, Prediction of tool-chip contact length using a new slip-line solution for orthogonal cutting, *Int. J. Mach Tools Manuf.*, Vol. 43 (2003), 1209–1215.

[19] J.A. Bailey, Friction in metal machining – mechanical aspects, *Wear*, Vol. 31 (1975), 243–275.

[20] W. Grzesik, Experimental investigation of the influence of adhesion on the frictional conditions in the cutting process, *Tribol. Inter.*, Vol. 32 (1999), 15–23.

[21] S. Philippon, G. Sutter, A. Molinari, An experimental study of friction at high sliding velocities, *Wear*, Vol. 257 (2004), 777–784.

[22] W. Grzesik, The influence of thin hard coatings on frictional behaviour in the orthogonal cutting process, *Tribol. Inter.*, Vol. 33 (2003), 131–140.

[23] D. Buryta, R. Sowerby, I. Yellowley, Stress distributions on the rake face during orthogonal machining, *Int. J. Mach. Tools Manuf.*, Vol. 34 (1994), 721–739.

[24] W. Grzesik, Z. Zalisz, P. Nieslony, Friction and wear testing of multilayer coatings on carbide substrates for dry machining applications, *Surf. Coat. Technol.*, Vol. 155 (2002), 37–45.

[25] M. Meiller, J.L. Lebrun, M. Touratier, D. Ryckelynck, Friction law for tool/workpiece contact area in dry machining, *Proc. Int. Workshop on Friction and Flow Stress in Cutting and Forming*, Paris, 2000, pp. 101–109.

[26] S.-P. Lo, Y.-Y. Lin, An investigation of sticking behaviour on the chip-tool interface using thermo-elastic-plastic finite element method, *J. Mater. Proc. Technol.*, Vol. 123 (2002), 285–292.

[27] B. McClain, S.A. Batzer, G.I. Maldonaldo, A numeric investigation of the rake face stress distribution in orthogonal machining, *J. Mater. Proc. Technol.*, Vol. 123 (2002), 114–119.

CHAPTER TWELVE

Tool Wear and Damage

Contents

12.1 Types of Tool Wear

Chapter 11 has demonstrated that the cutting tool environment features much higher shear and normal stresses, usually along with higher localized temperature, than in classical tribological systems. Moreover, the tool may experience repeated contact loads during interrupted cuts, and the workpiece may chemically interact with the tool material. The response of a tool material to the above tribological conditions dictates its performance.

The damages of a cutting tool are influenced by the stress state and temperature at the tool surfaces, which in turn depend on the cutting mode – for example, turning, milling or drilling, the cutting conditions and the presence or not of cutting fluid and its type. In machining, the tool damage mode and the rate of damage are very sensitive to changes in the cutting operation and the cutting conditions. To minimize machining cost, it is necessary not only to find the most suitable tool and work material combination for a given machining operation, but also to reliably predict the tool life. Tool damage can be classified into two groups, wear and fracture (Figs 12.1 and 12.2), by means of its scale and how it progresses [3]. Unfortunately, in practice these two groups of tool damage are not clearly distinguished. Wear is loss of material, which usually progresses continuously, on an asperity or micro-contact, or smaller scale, down to molecular or atomic removal mechanisms. As for fracture (failure), on the other hand, there is a continuous spectrum of damage scales from micro-wear (for example, micro-chipping) to gross fracture (catastrophic failure), as illustrated in Fig. 12.2. As shown in Fig. 12.1, the progressive wear of a tool takes place mainly in two distinct ways [4]:

1. Wear on the rake face characterized by the formation of a crater and/or built-up edge resulting from the action of the chip flowing along the face.
2. Wear on the flank face, including often regions adjacent to the major and minor cutting edges and tool nose, where characteristic land is formed from the rubbing action of the newly generated workpiece surface.

Crater wear (Fig. 12.1a), observed on the rake face of cutting tools, generally occurs during machining of relatively soft steels and ductile irons at high speeds when continuous chips are formed. It is primarily caused

Advanced Machining Processes of Metallic Materials
ISBN: 978-0-08-044534-2

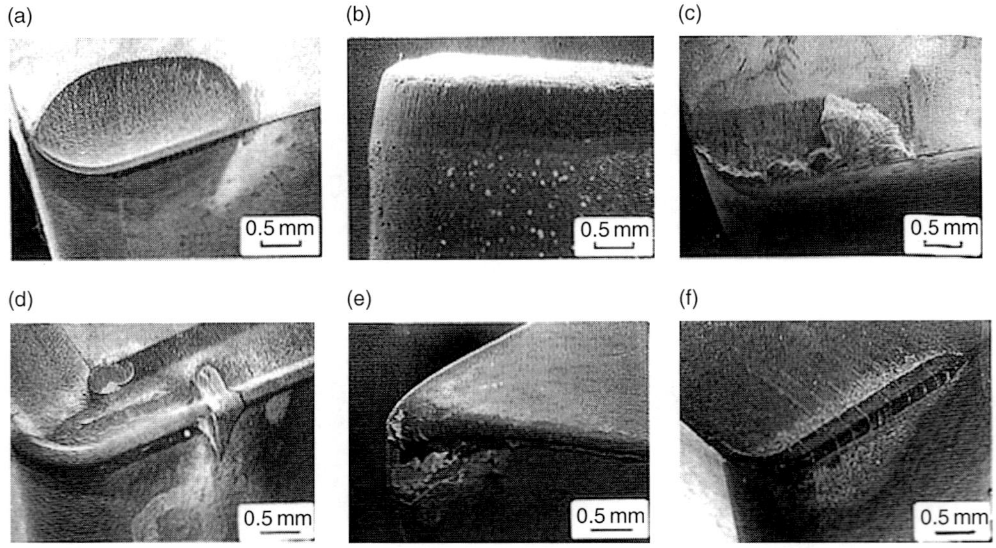

Fig. 12.1 Tool wear forms: (a) crater wear on a cemented carbide tool produced during the machining of plain carbon steel; (b) abrasive wear on the flank face of a cemented carbide tool produced during the machining of grey cast iron; (c) built-up edge produced during low speed machining of a nickel-base alloy; (d) depth-of-cut notching on a carbide tool produced during the machining of a nickel-base superalloy; (e) carbide insert showing nose wear (tool-tip blunting) due to insufficient deformation resistance of the tool; (f) thermal and mechanical cracks in a carbide insert after interrupted cutting of low alloy steel [1].

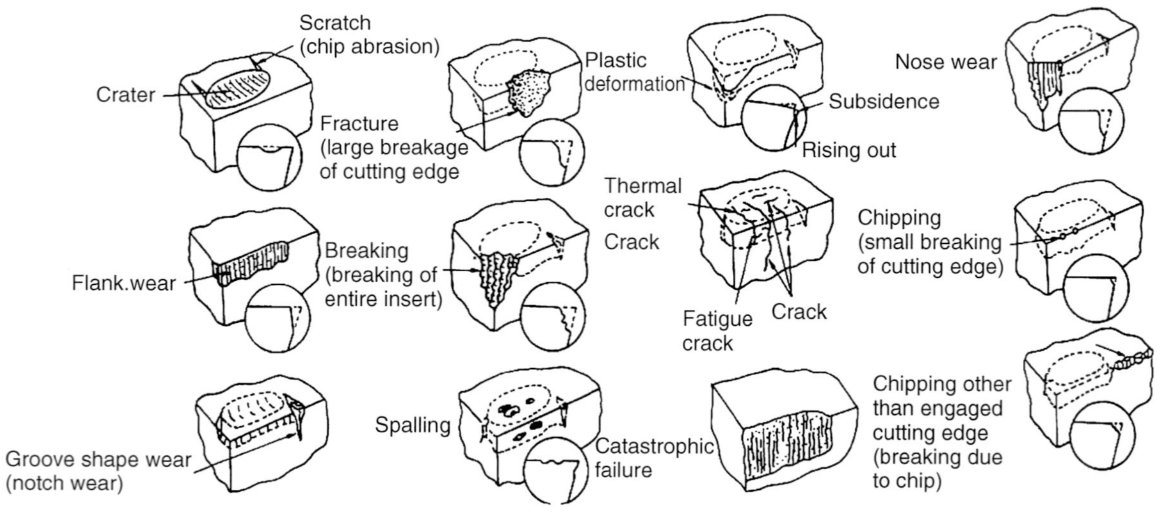

Fig. 12.2 Representative types of tool failure [2].

by a chemical interaction between the rake face of a tool insert and the hot chip. Many thermally activated phenomena such as adhesion, diffusion or dissolution may be involved in the wear process.

Flank wear (Fig. 12.1b), on the other hand, is observed on the flank or clearance flank as a result of abrasion by the hard constituents of the workpiece material. This failure mechanism is commonly observed during machining of steels and cast irons where the abrasive particles are mainly Fe_3C cementite and non-metallic inclusions.

Built-up edge (BUE), shown in Fig. 12.1c, is referred to when welding the chip to the tool tip. It typically occurs when soft low carbon steels or ductile stainless steels, and non-ferrous materials (mainly aluminium, nickel, cobalt and titanium alloys), are machined at low cutting speeds and light feed rates. This typically results in severe attrition wear if small fragments of tool material are carried away as the BUE breaks off by every cycle.

Notch wear (depth-of-cut notching), shown in Fig. 12.1d, is often attributed to the oxidation of the tool material from the sides of major and minor cutting edges, or to abrasion by the hard, saw-tooth outer edge of the chip (for example, in hard machining). Notching is a serious technological problem with workpiece materials that tend to have high work-hardening and generate high tool-tip temperatures, such as austenitic stainless steels and nickel-based superalloys. Notch wear can obviously lead to tool fracture and can be minimized by applying tools with chamfered edges, rounded inserts and avoiding small depths of cut.

Nose wear (Fig. 12.1e), termed also tool-tip blunting, results from insufficient deformation resistance of a tool material in a given machining operation.

Thermal cracks (Fig. 12.1f) develops when the repeated heating and cooling associated with interrupted cutting (thermo-mechanical fatigue), such as milling, creates high temperature gradients at the cutting edge. With prolonged time, lateral cracks may appear parallel to the cutting edge. The thermal and lateral cracks may joint together and cause small fragments of tool material to break away.

The preferred mode of tool wear is flank wear, because it progresses gradually and can be easily monitored for tool-changing protocol in NC programs. In practice, some directly measured dimensional characteristics of typical wear patterns, i.e. crater and flank wear, and notch wear at the depth-of-cut extremities, for HSS, carbide and ceramics tools, are standardized in ISO 3685, as shown in Fig. 12.3. As shown in Fig. 12.3a,

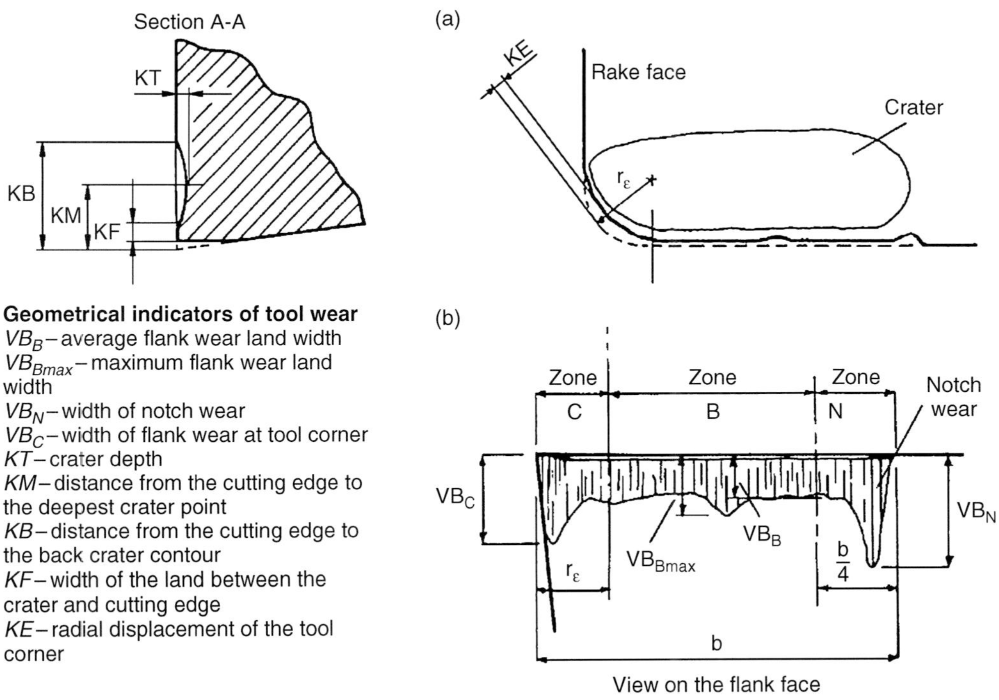

Geometrical indicators of tool wear
VB_B – average flank wear land width
VB_{Bmax} – maximum flank wear land width
VB_N – width of notch wear
VB_C – width of flank wear at tool corner
KT – crater depth
KM – distance from the cutting edge to the deepest crater point
KB – distance from the cutting edge to the back crater contour
KF – width of the land between the crater and cutting edge
KE – radial displacement of the tool corner

Fig. 12.3 Typical wear pattern of a carbide tool according to ISO 3685.

the crater wear grows both in depth (*KT*) and width (*KF-KB*). Primarily to characterize the progress of cratering, the crater depth *KT* is measured at the deepest point of the crater (section *A–A*) localized at the distance *KM* from the cutting edge. In particular, as *KF* decreases and the crater boundary approaches the cutting edge, the effective rake angle γ_{eff} can be found using the equation:

$$\gamma_{eff} = \gamma_o + \tan^{-1}\left(\frac{KT}{KM}\right) \tag{12.1}$$

where γ_o is the nominal rake angle.

It can be seen that flank wear is usually greatest at the extremities (tool corner and point corresponding to maximum depth of cut) of the cutting edge. As indicated in Fig. 12.3b, the active major cutting edge of length b is divided into three zones for the purpose of wear measurements: (1) zone C of the length r_ε which includes the rounded part of the cutting edge at the corner; (2) zone N of a length $b/4$ at the opposite end of the active cutting edge where a groove or notch wear is formed; and (3) the central zone B including the remaining part of the cutting edge, between zone A and zone N. In the central zone B, the wear land is usually fairly uniform and the relevant average wear land is designated VB_B. However, to allow for variations that may occur, the maximum wear land width is denoted by VB_{Bmax}.

It should be noted that the ISO recommendations were established for flat-faced tools only, but taking into account the most frequent use of grooved inserts some additional wear indicators are needed, as shown in Fig. 12.4. In general, the tool-wear pattern on a worn grooved insert is influenced by the three-dimensional chip flow and complex chip groove configuration. For example, it was found that quite often some grooved tools fail long before the major flank wear VB_B reaches its failure criterion [5]. This confirms that the tool wear/tool life is significantly affected by the combined effects of cutting conditions and the chip-groove configurations.

The criteria recommended by the ISO to define the effective tool life for HSS tools are:

1. $VB_B = 0.3$ mm if the flank is regularly worn in zone B, or
2. $VB_{Bmax} = 0.6$ mm if the flank is irregularly worn, scratched, chipped, or badly grooved in zone B, or
3. Catastrophic failure.

For sintered-carbide tools, one of three criteria is recommended:

1. $VB_B = 0.3$ mm, or
2. $VB_{Bmax} = 0.6$ mm if the flank is irregularly worn, or
3. $KT = 0.06 + 0.3f$, where f is the feed rate.

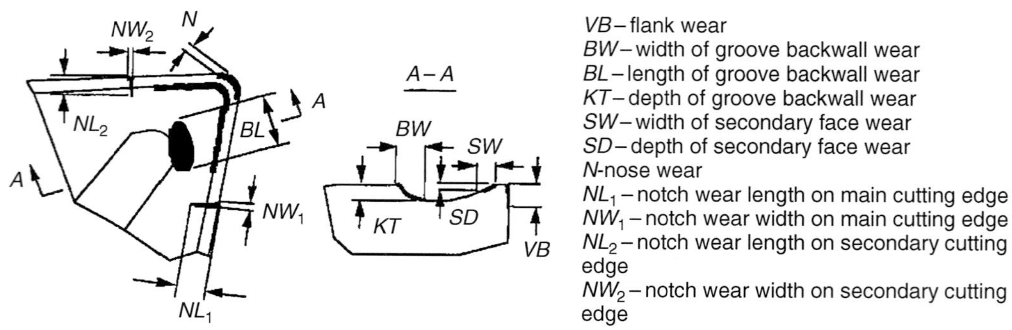

Fig. 12.4 Measurable tool-wear parameters in a grooved tool [5].

12.2 Physical Mechanisms of Tool Wear

In the context of cutting tool wear three groups of causes can be qualitatively identified: mechanical, thermal and adhesive (Fig. 12.5). Mechanical types of wear, which include abrasion, chipping, early gross fracture and mechanical fatigue, are basically independent of temperature. Thermal causes with plastic deformation, thermal diffusion and oxygen corrosion as their typical forms, increase drastically at high temperatures and can accelerate the tool failure by easier material removal by abrasion or attrition.

Abrasive wear occurs when hard particles slide against the cutting tool, primarily on the flank surface. The hard particles come from either the work material's microstructure (for example, hard carbide inclusions and cementite in steels or free silicon carbide particles in aluminium-silicon alloys), or are broken away from the cutting edge by brittle fracture. Moreover, they can also result from a chemical reaction between the chips and cutting fluid when machining PM steels or cast irons alloyed with chromium. Abrasive wear reduces the harder the tool is relative to the particles in high temperatures, and generally depends on the machining distance. This can primarily be achieved by coating the tool with hard, highly abrasive-resistant coatings having high hot hardness. If the hardness of the tool material reduces with temperature, the thermal softening effect can lead to plastic deformation of the cutting edge. The primary cause of plastic lowering of the cutting edge is high temperature creep. Deformation of the edge of HSS, high cobalt content cemented carbide and cermet tools lead to insufficient tool geometry and premature tool failure. *Adhesive* or *attritional wear* and accompanying BUE formation were discussed in the previous section. Adhesive and abrasive wear are the most significant types of wear at lower cutting speeds. At high cutting speed, temperature-activated wear mechanisms including *diffusion (solution wear)*, chemical wear (oxidation and corrosion wear), and thermal wear (superficial plastic deformation due to thermal softening effect) occur. Sometimes, thermal electro-motive force (EMF) wear is distinguished [3].

As shown in Fig. 12.6, in diffusion wear, which according to Loladze [8] begins for carbide tools at temperatures of 800–850°C, constituents of the tool material diffuse into or form solid solutions with the chip material. Because the first diffuses carbon (mobility of carbon atoms is 70×10^3 higher than iron atoms), and more slowly tungsten, cobalt and titanium, the tool subsurface containing Fe_3W_3C or more complex carbides weakens and a crater wear develops rapidly. Severe crater wear ultimately leads to tool failure due to breakage. Figure 12.7a illustrates the changes of the ratio between hardnesses of a Cr-Ni-Co alloy steel and WC carbide (1) and the cutting distance to tool failure (2), as functions of temperature. The transition point at about 900°C, after which the hardness ratio increases and tool life decreases drastically, can be considered as the region of intensive diffusion wear. This evidence is confirmed in Fig. 12.7b where the transition zone between adhesion and diffusion was found for three different steels to occur at 1170 K (~900°C). This means

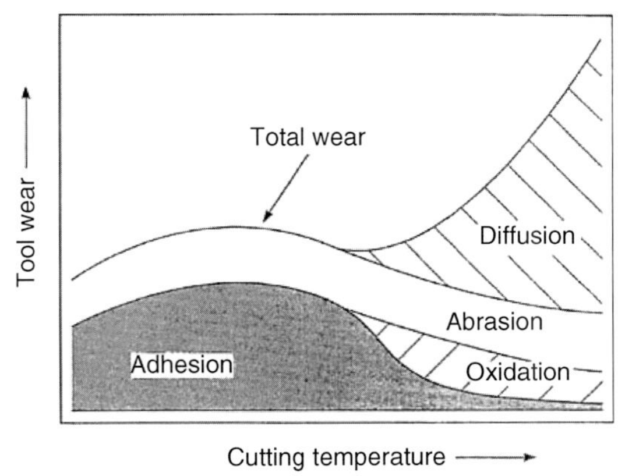

Fig. 12.5 Wear mechanism for machining as a function of cutting temperature [4].

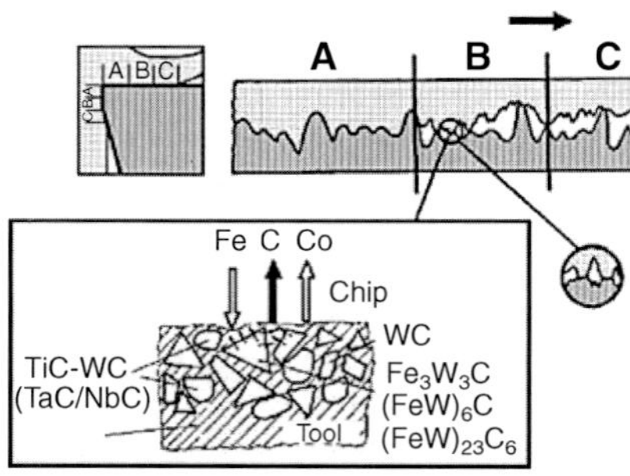

Fig. 12.6 Mechanism of diffusion wear for carbide tools [6].

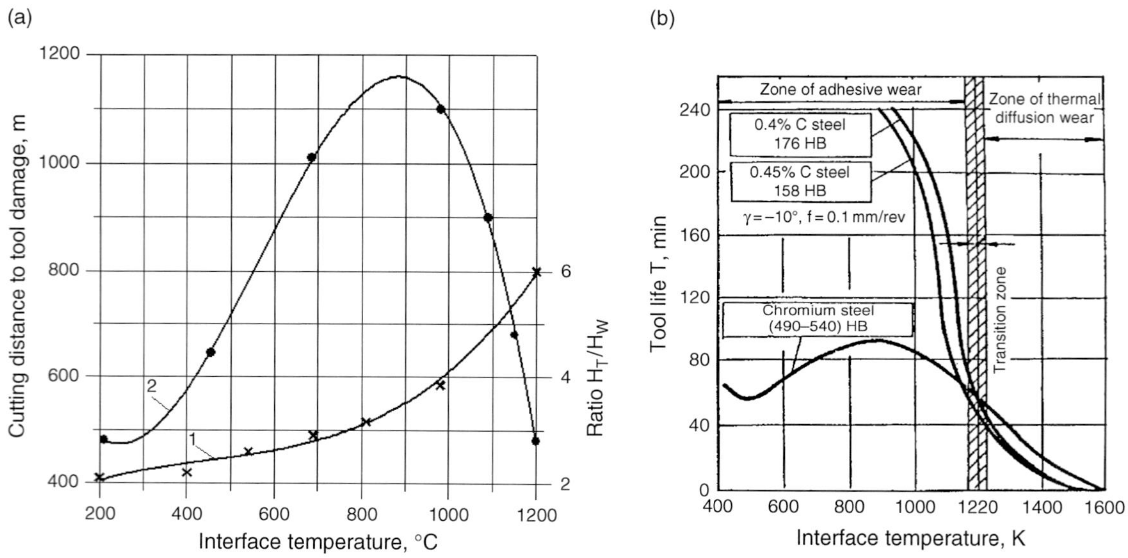

Fig. 12.7 Influence of interface temperature on the onset of diffusion wear [6, 7]. H_T – hardness of tool material; H_W – hardness of workpiece material.

that from the diffusion point of view, tool life is independent of the mechanical properties of workpiece material and can be estimated based on chemical compositions of sliding materials, the contact temperature and the cutting speed. Gekonde and Subramanian [9] revealed that for machining ferritic ductile iron and AISI 1020 steel with uncoated cemented carbide tools, the maximum depth of crater formed from dissolution wear at a cutting speed of 200 and 300 m/min corresponds to the local tool/chip interface temperature of 835°C and 840°C, respectively. Thus, the diffusion hypothesis for crater wear of cutting tools proposed by Trent [10] applies under seizure conditions.

Oxidation, which is one of the chemical reactions in the cutting zone, occurs at temperatures around 800°C when constituents of the tool material (especially the Co binder) react with atmospheric oxygen. It most often occurs near the free surface of the part, where the hot portion of the tool in and around the tool/chip contact region is exposed to the atmosphere. The tool surface, directly weakened either by oxidation or by reaction with oxidized work material, becomes porous and brittle and is then easily carried away by the chip. Oxidation results in severe depth-of-cut notch wear or in the production of hard oxide particles which increase abrasive wear. A typical example of oxidation wear is that occurring on the rake face of cemented carbide tools (may be possible for alumina ceramic and cermet tools) in high speed milling/turning of steel. *Corrosion wear* is also classified as chemical reaction but is caused primarily by free sulphur or chlorine EP additives in a cutting fluid. The surface layer of the tool is changed due to the reaction product, which is typically soft and wears immediately by abrasion.

12.3 Tool Life

With the progress of tool wear, the surface roughness and dimensional accuracy of the machined part deteriorates, which in consequence leads to changing the worn tool. In practice, the most important practical consideration in selecting cutting tools and cutting conditions is the tool life (T) which defines the time corresponding to the prevailing tool-life criterion (see section 12.2). In other words, tool life is determined when the measured wear levels exceed a limit. A standard measure of tool life is the time to develop a flank wear length VB_B of 0.3 mm or VB_{Bmax} of 0.6 mm obtained from tool-life curves, as shown in Fig. 12.8a. Typical wear curve, termed also the curve of natural wear, consists of three characteristic parts: the initial (running-in) period (I), the longest uniform (progressive) wear period (II) and accelerated wear period (III) leading often to catastrophic failure. As shown in Fig. 12.8b, generally similar shapes of wear curve were recorded in finish turning ($v_c = 115$ m/min; $f = 0.1$ mm/rev; $a_p = 0.5$ mm) of an AISI 4140 steel for grooved tool inserts when considering major flank wear (VB_B), crater depth (KT) and minor flank wear VB'. Only nose wear VB_N indicated approximately linear relation vs time. However, it should be noted in Fig. 12.8b that the accelerating stage of minor flank wear VB' (at 10 min) occurs earlier than those of major flank wear (at 15 min) and crater wear (at 15 min).

It is well known that tool life depends on a number of factors which characterize the performance of machine/tool/work system, including those related to the tool (material, geometry, surface integrity and internal defects), the work (hardness, strength, chemical composition, etc.), the cutting parameters (speed, feed and doc (depth of cut)), the cutting mode and the machine tool (mainly stiffness and state of maintenance). Hence, it is impossible to derive a universal tool-life criterion for these all quantitative and qualitative variables. In practice, the problem for a given cutting regime is reduced to determining effects of cutting parameters (Fig. 12.9), but primarily to the influence of the cutting speed limited to its values higher than v_{c23} (region III – shaded area in all figures). It should be noted in Fig. 12.9 that in all three cases

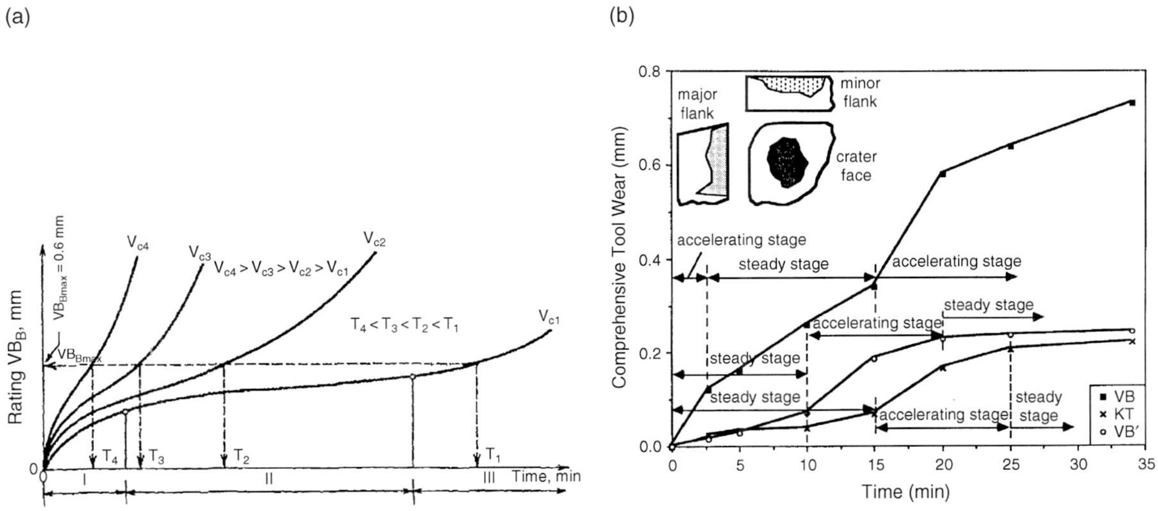

Fig. 12.8 Development patterns of tool wear for flat-faced [6] (a) and grooved [11] (b) tools.

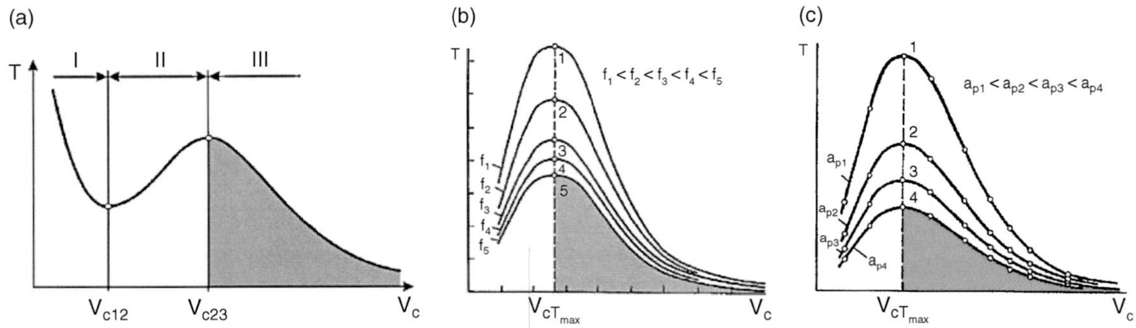

Fig. 12.9 Influence of cutting parameters on tool life: cutting speed (a); feed (b); depth of cut (c) [6].

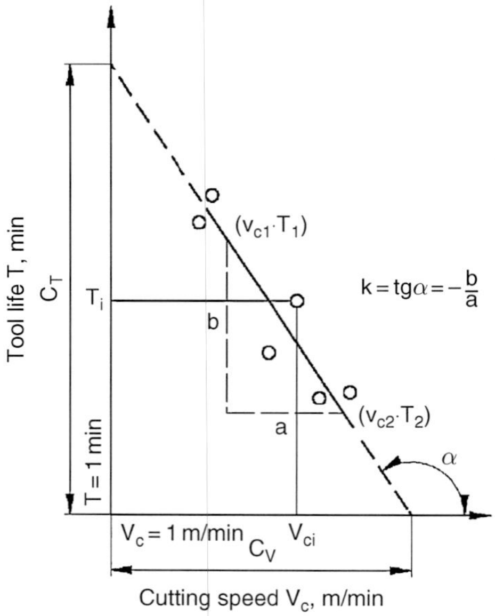

Fig. 12.10 A Taylor tool-life curve in a log–log scale.

maximum tool life corresponds to the optimum cutting speed v_{cTmax}. The mostly used constructing tool-life curve is based on the ISO turning test performing for two different cutting speeds v_{c1} and v_{c2}. Using linear regression method and a log–log scale, the plot $T - v_c$ in Fig. 12.10 shows a linear variation between tool life and the cutting speed over limited ranges of v_c.

The most widely used tool-life model is the Taylor tool-life equation (in original form it uses English units for cutting parameters), which has the form [12, 13]:

$$v_c T^n = C_t \tag{12.2}$$

where T is tool life, C_t is the cutting speed which yields on-minute tool life and n is the exponent depending primarily on the tool material (0.1–0.17 for HSS tools; 0.2–0.25 for uncoated WC tools; 0.3 for TiC and TiN coated WC tools; 0.4 for Al_2O_3 coated WC tools; 0.4–0.6 for solid ceramic tools [14]).

For a linear $T - v_c$ relationship shown in Fig. 12.10, the Taylor equation has the other form:

$$T = C_T v_c^k \tag{12.3}$$

where C_T is the constant equal to T for $v_c = 1\,\text{m/min}$ and k is the negative slope of the curve and equal in magnitude to the inverse of the exponent n in Eqn (12.2). The slope k can be determined as

$$\log T = \log C_T + k \log v_c \tag{12.4a}$$

$$k = tg\alpha = -\frac{b}{a} = \frac{\log T_2 - \log T_1}{\log v_{c2} - \log v_{c1}} \tag{12.4b}$$

Mamalis *et al.* [15] proposed a new general tool-life equation for superhard cutting tools (especially CBN tools) to reflect the real, non-monotonic shape of the $T - v_c$ curve in a wide range of cutting speeds (see Fig. 12.9a). This relationship is:

$$T = \frac{C_{T1}}{v_c^3 + C_{T2}v_c^2 + C_{T3}v_c} \tag{12.5}$$

where values of three constants C_{T1}, C_{T2} and C_{T3} are determined experimentally.

Sometimes, an inverse tool-life equation is used to determine the appropriate cutting speed, namely:

$$v_c = C_v\ T^{\frac{1}{k}} \tag{12.6}$$

where C_v is the constant equal to the cutting speed when tool life was assumed to be only 1 min.

$$\frac{\log C_T}{\log C_v} = -k \Rightarrow C_v = C_T^{-\frac{1}{k}} \tag{12.7}$$

Some characteristic changes of tool-life curve caused by different wear mechanisms of carbide tools, resulting in its shifting and/or rotation, are shown in Fig. 12.11. As a rule, several wear mechanisms act simultaneously. When low cutting speeds are utilized, there is a greater chance that a BUE condition will occur. At the other extreme, using high cutting speeds, there is a high probability that the insert will be plastically deformed. At intermediate positions on the graph, where the 'optimum' cutting speed is likely to be selected, several wear mechanisms can occur simultaneously.

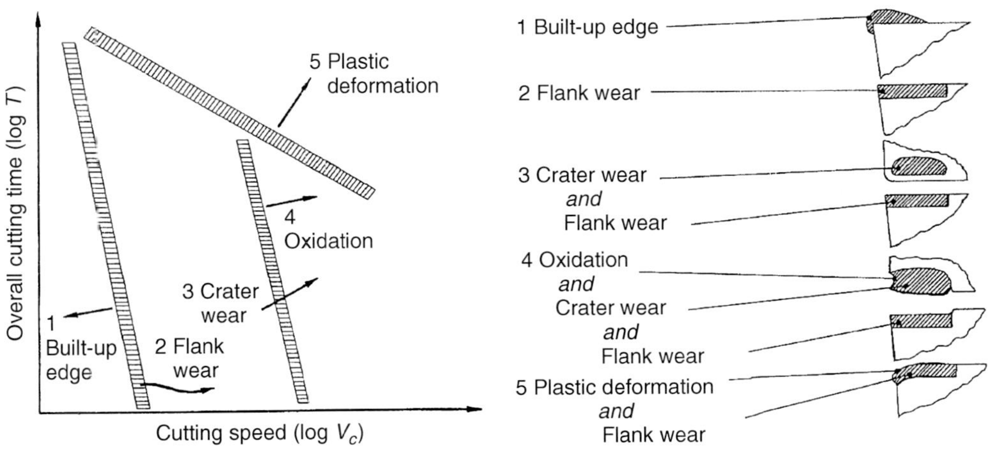

Fig. 12.11 The wear and failure criteria for carbide inserts [16].

According to Fig. 12.9, the partial tool-life models for all cutting parameters can be formulated as $T = C_{T1}/v_c^{x_T}$, $T = C_{T2}/f^{y_T}$ $T = C_{T3}/a_p^{z_T}$ which, in turn, leads to the extended Taylor equation reflecting not only the dominant influence of cutting speed, but also smaller independent (but significant) effects of the feed rate and the depth of cut. Hence,

$$T = \frac{C_T}{v_c^{x_T}\, f^{y_T}\, a_p^{z_T}} K \tag{12.8a}$$

where constant K takes into account other influences, such as the hardness of workpiece material.

$$v_c \cdot T^n \cdot f^{n_1} \cdot a_p^{n_2} = C \tag{12.8b}$$

An alternative equation proposed by Colding [14] combined the influence of f and a_p by assuming tool life to be dependent on the effective chip thickness (ECT) which is equivalent to the average chip thickness ($T = A \times v_c^m \times ECT^p$). Moreover, first and second order mathematical (linear regression) models for tool life can also fit to experimental data in order to select optimum economic cutting conditions involving a single tool/work material combination. The extended tool-life relationship which in parallel considers a chip-groove effect factor and a tool coating factor can be written as [17]:

$$T = T_R W_g \left(\frac{v_{cR}}{v_c}\right)^{W_c \frac{1}{n}} \tag{12.9}$$

where T_R and v_{cR} are the reference tool life and reference cutting speed, v_c is the cutting speed, n is Taylor's tool-life exponent, W_g is the chip-groove effect factor and W_c is the tool coating effect factor.

The tool-life equations described above treat tool life as a deterministic property. In reality, tool life exhibits significant scatter and is probabilistic in nature. Sources of tool-life variability in practice include: (a) variation in the workpiece material hardness, microstructure, composition, and surface characteristics; (b) variation in the cutting tool material, geometry and sharpening technology; and (c) vibrations due to the compliance of the machine tool structure, workpiece, fixture, tool-holder and tool body. Under these circumstances, tool life can be determined more reliably using probability theory and the introduction of hazard or failure rate function. Taking into account that tool life is a random variable with characteristic logarithmic-normal distribution (the Gaussian distribution is determined for log T), the coefficient of tool-life variability ν, defined as the ratio of the standard deviation to the average tool life can be determined ($\nu = \sigma_T/\overline{T}$). Minimizing the coefficient ν leads to a more stable machining performance and reliable tool life in automated production. A methodology for the quantification of tool-life measurement uncertainties in the extended Taylor's equation, including uncertainty of tool flank wear measurements, cutting data variability and uncertainty of cutting time measurements, was proposed by Axinte *et al.* [18].

12.4 Modelling of Tool Wear

A number of predicting models for the rate of tool wear due to specific mechanisms occurring in cutting has been developed and some of them are characterized in Table 12.1.

The best known model of the tool wear rate was developed by Usui *et al.* [20] based on the ideas of contact mechanics and wear. In particular, they assumed that wear should increase with normal contact stress. A differential Eqn 2 in Table 12.1 relating the incremental volume of material worn away at a given point of the tool dV, and the incremental sliding distance, $dL_s = v_s\,dt$. As shown in Fig. 12.12a a plot of wear rate against $1/\Theta$ gives a straight line, the slope of which is $-B$. The constants C_1 and C_2 can be

Table 12.1 Some important models for tool wear rate.

Tool wear rate model	*Author(s)*
1. $\frac{dW}{dt} = v_c(\Theta, f)A + B\exp(-E/R\Theta)$ where A and B are constant, E is the process activity energy, R is the universal gas constant and Θ is cutting temperature	Takeyama and Murata wear model considering abrasive wear and diffusive wear [19]
2. $\frac{dW}{dt} = C_1\sigma_n v_s \exp(-C_2/\Theta)$ where σ_n is the normal stress, v_s is the sliding velocity, Θ is the interfacial temperature, and C_1 and C_2 are constants	Wear model considering adhesive wear by Usui *et al.* [20]
3. $dW' = A_r \frac{c}{b} Z\, dL$ where dW' is the wear volume for sliding distance dL, Ar is the real area of contact, c is the height of the postulated plate-like wear particle, b is the mean spacing of the asperities and Z is the probability of producing a wear particle per asperity encounter (Holm's probability).	An Archard type of equation for adhesive wear by Shaw and Dirke [13]
4. $\frac{dW}{dt} = K\left[-D\frac{dc}{dy} + Cv_y\right]_{y=0}$ where K is the ratio of the molar volumes of the tool material and workpiece material, D is the diffusivity of the slowest diffusing tool constituent into the work material, c is the concentration of the tool material in the chip, C is the equilibrium concentration or solubility of the tool material in the work material and v_y is the bulk velocity of the work material normal to the rake face	Kramer and Suh's model for crater wear rate due to diffusion [21]
5. $dW_r = \frac{\Delta V}{F_n l}$ where ΔV is the wear volume, F_n is the mean normal cutting force and l is the cutting distance	Thermo-chemical wear rate of the diamond tip in nanometric cutting [22]

estimated from a series of orthogonal cutting tests. Usui's equation yields accurate results when crater wear is developed in cutting low carbon steels with an uncoated carbide tool. The model is useful because it incorporates the influence of the tool hardness (through constant C_1) and interfacial temperature on the crater tool wear. It should be noted that in Fig. 12.12a, the experimental data appears to scatter in a range of low interfacial temperatures. Figure 12.12b shows experimental results for both the crater and flank depth wear rates (wear volume W in the differential equation was replaced by wear depth h) when machining a 0.25%C steel and 0.46%C steel with a P20 grade carbide tool. Two linear regions with the characteristic steep line, belonging to flank wear (lower line with smaller slope) and crater wear (upper line with larger slope), are seen in Fig. 12.12b. In this case the boundary is at $1/\Theta \approx 8.5 \times 10^{-4}\text{K}^{-1}$ (or $\Theta = 1175\,\text{K}$). The slope of the higher temperature data ($\Theta > 1175\,\text{K}$) is typical of the diffusion process between steels and cemented carbides, whereas the smaller slope at lower temperatures is typical of a temperature dependent mechanical wear process such as abrasion. The results of predicting wear progress curves for crater wear and wear progress lines, based on the Usui's models, are very satisfactory.

Recently, partially successful efforts have been made to integrate tool wear models with FEM calculations to predict the evolution of tool wear for uncoated tools [19]. It is postulated here that the growth of tool wear can be evaluated at discrete points in time, although in reality it is a continuous process. As shown in Fig. 12.13, the procedure for wear rate prediction involves four phases which make one simulation cycle that

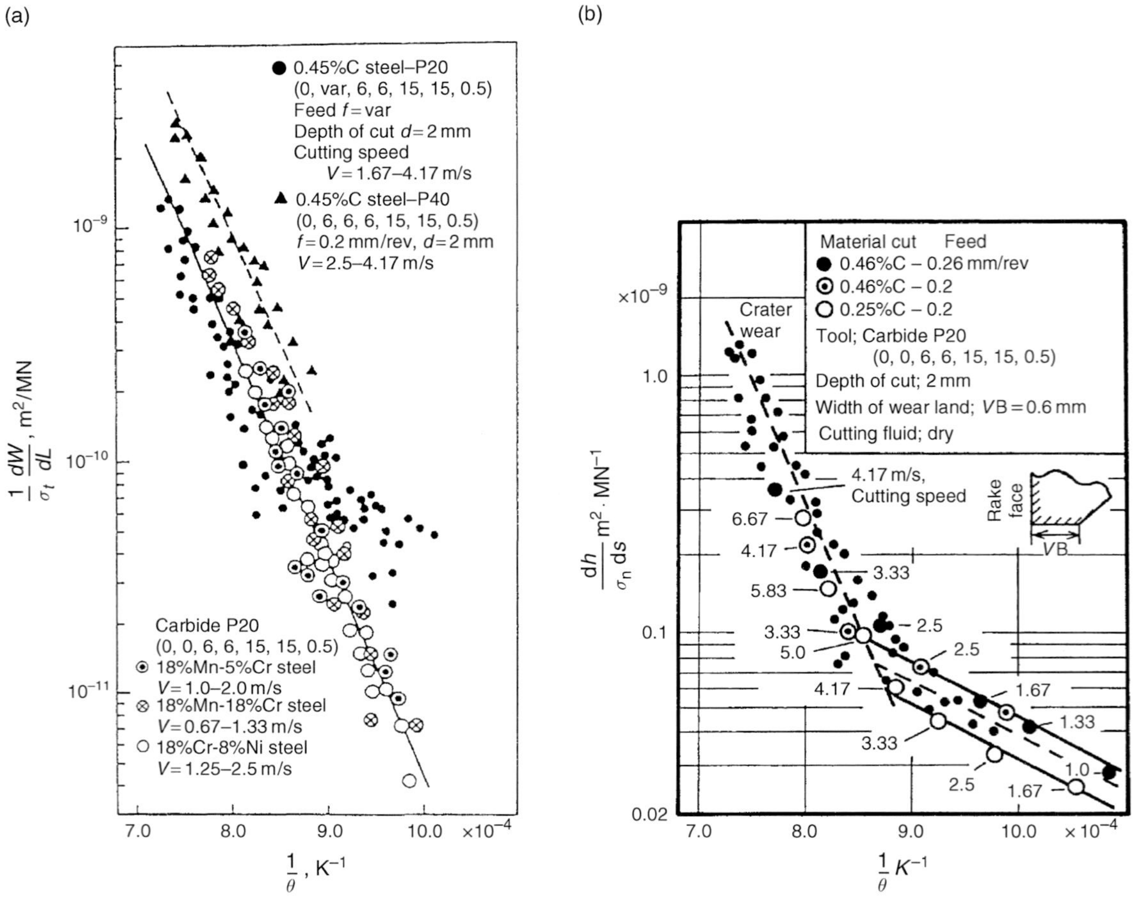

Fig. 12.12 Crater (a) and flank (b) wear characteristics for carbon steels turned by P20 or P40 carbide tools [20].

corresponds to one data point on the flank wear (*VB*) vs cutting time curve. The simulation was based on a coupled thermo-viscoplastic Lagrangian formulation with isotropic strain-hardening model, and a special simulation module, 'Konti-Cut', was implemented to generate near-steady-state chip geometry. Usui's tool wear model with updated values of constants was adopted. Figure 12.14 shows some predicted results of gradual crater wear growth and its distribution for tools with a predefined wear land of 0.06 mm on the tool flank face (sharp edge was replaced by initially worn cutting edge. It can be seen from the figure that the wear rates of crater wear and flank wear are equal on order and hence the increase of both types of tool wear is quite consistent. Moreover, the location of the maximum crater wear coincides well with that of the maximum cutting temperature and a region of very low wear rates is observed close to the tool radius on the rake face side.

Tool wear in nanometric cutting can be simulated using a molecular dynamics (MD) simulation approach, described previously in Chapter 5. Figure 12.15 summarizes the results obtained by MD simulation of tool wear in nanometric cutting performed on a single crystal silicon plate with a 30 nm single diamond tip of an atomic force microscope (AFM), similar to the test described in Chapter 17. The wear mechanism was investigated by the computation of the temperature, the stress in the diamond tip, and the relationship

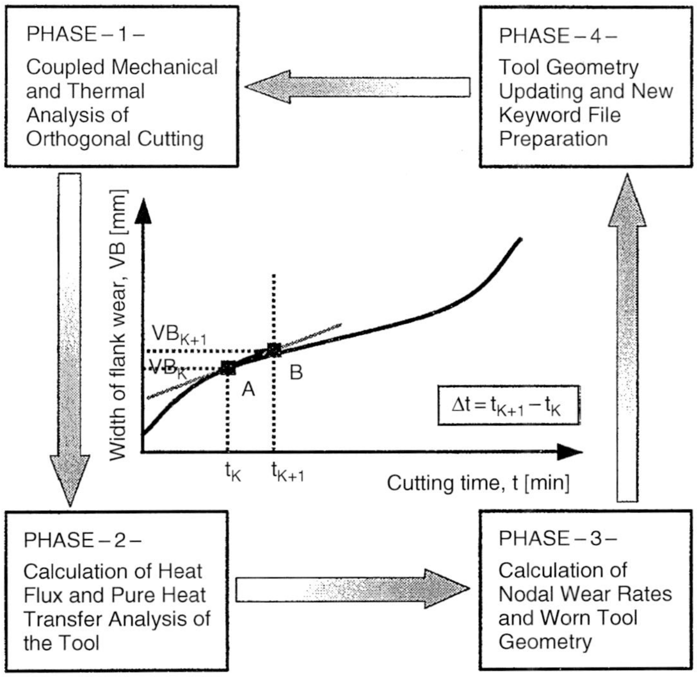

Fig. 12.13 Simulation procedure for tool wear prediction using FEM model [19].

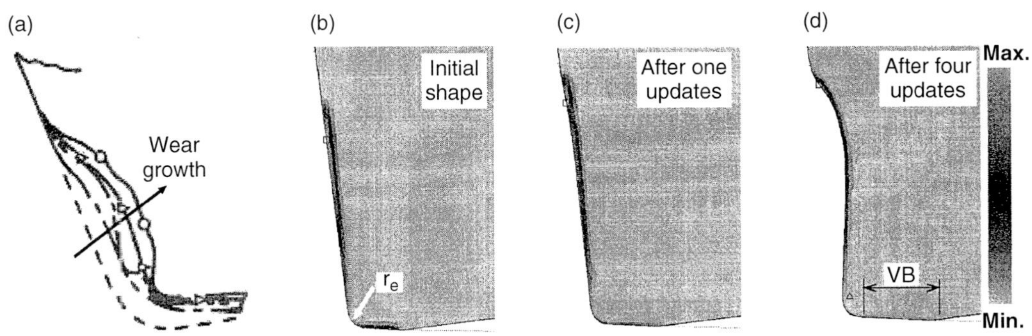

Fig. 12.14 Predicted results of the wear rate distribution and the updated geometries of crater wear and flank wear for the tool with an initial flank wear land of 0.06 mm (AISI-1045 workpiece vs uncoated carbide, $v_c = 300$ m/min, $f = 0.145$ mm/rev, the length of VB in (d) ≈ 0.12 mm, Δ – location of maximum wear rate) [19].

between the temperature and sublimation energy of the diamond and silicon atoms. The wear rate was expressed by formula 5 in Table 12.1. Figure 12.12b shows the image of the diamond tip before and after 20 times nanometric scratching. The flat wear land and a characteristic concavity, consistent with the MD simulation (Fig. 12.12a), were found be the major wear patterns. The results of MD simulation and AFM scratching tests show that the thermo-chemical wear mechanism dominates in nanometric scale-diamond cutting.

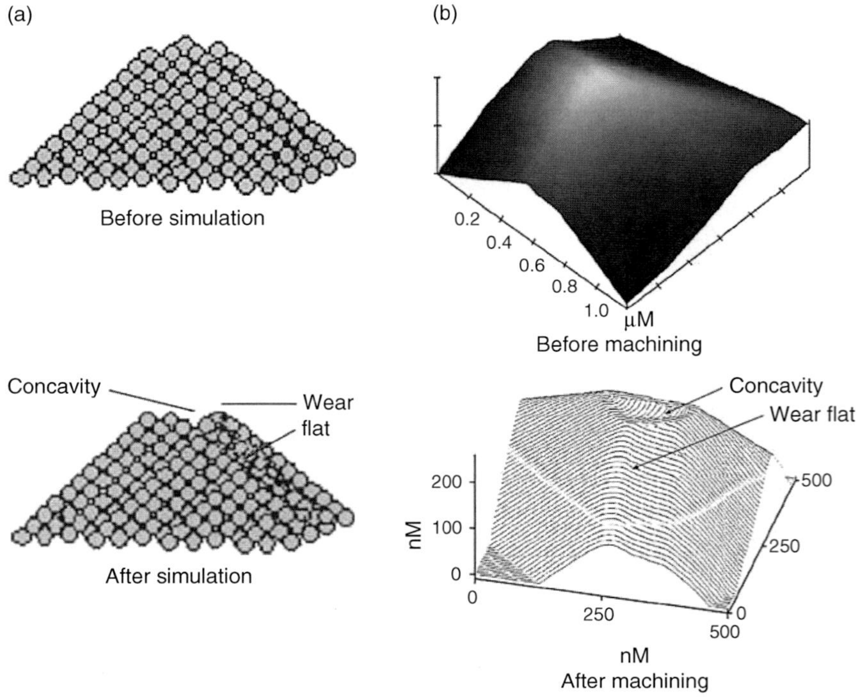

Fig. 12.15 The wear of diamond tip in nanometric cutting [22]: (a) MD simulation; (b) AFM scratching test.

12.5 Advanced Method of Tool Wear Identification and Measurement

Basically, the scanning electron microscope (SEM), stereo microscope with digital camera attachment, computer vision systems, surfcom (a surface roughness instrument) and coordinate measuring machine (CMM) can be used to measure tool wear parameters. In addition, more advanced measuring techniques using white light interferometry or confocal microscopy can be of interest when the analysis in nano-scale is necessary. It is generally desirable to follow the examination of wear scars with low power optical microscopy, such as can be done with stereo- and toolmaker's microscopes. Frequently, magnification in the range of 30 to 50 times (or higher at 100 times) is often most useful. Comparison of wear scars at this macro level can be facilitated by the use of photographs or microphotographs. Scanning electron microscopy (SEM) with magnification in the range of several hundred to several thousand times is frequently used to supplement the macro examinations. Similarly, it is often useful to augment the SME topographical examination with some form of elemental mapping, scanning or profilometer tracing. Comparison of wear depth with surface roughness may be important in assessing the significance of initial surface conditions in the observed wear or distinguishing between initial wear and long-term wear. With a wear analysis with elemental analysis techniques, such as Energy Dispersive X-Ray Analysis (EDX), it is generally necessary to determine the depth of wear by means of profilometers. Techniques used to examine wear scars are not limited to those mentioned above. For instance, in order to identify special wear regions (e.g. sticking and sliding regions) a typical micro-analysis technique such as Energy Dispersive X-ray Spectroscopy (EDS), Backscattering Electron Spectroscopy (BSE) or Atomic Force Microscopy (AFM) techniques can be used.

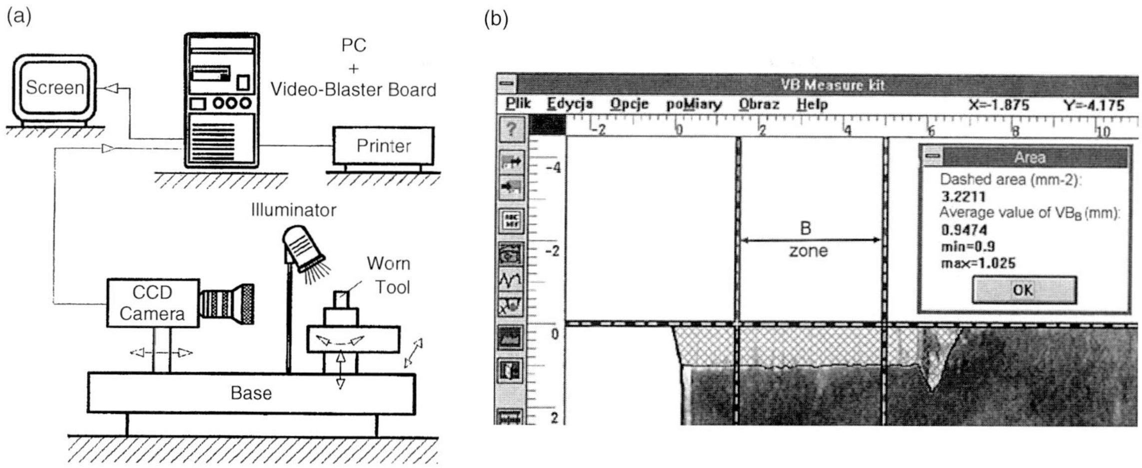

Fig. 12.16 Vision system for tool wear measurement (a) and a typical window for measurement of the land width VB_B (b). Courtesy of Dr J. Balamucki.

An example of a computer vision system for tool flank wear measurements is shown in Fig. 12.16. The well-known image principle in which the CCD camera sees the light source reflected in the plane flank surface under the cutting edge was employed. The other surfaces of the insert will normally have other orientations, so that the reflected light from these surfaces will have a smaller intensity than that reflected by the flank surface. Using this principle it is possible to measure any dimension of the flank area under the main cutting edge. Figure 12.16 shows a typical window with the magnified worn flank area, available in special computer software, named 'VB-measure'. In general, other wear indicators specified in the ISO Standard, such as VB_{Bmax}, VB_C and VB_N, can be easily measured.

The integral effect of both SEM microphotograph and X-ray map in differentiating between the sticking and sliding regions across the tool/chip interface is shown in Fig. 12.17. In this case, the BUE layer can be evidently localized with the Ni-distribution, whereas the W distribution (substrate material) and Ti distribution (TiAlN coating layer) show the abrasive wear effects. Moreover, the white light interferometry technique can be employed to generate 3D colour topographies of the wear patterns on the cutting tools. It

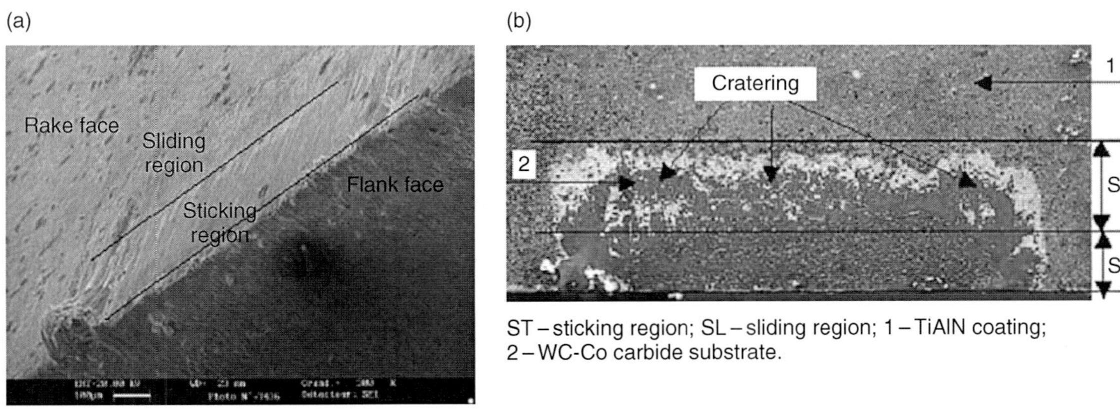

Fig. 12.17 SEM image of the rake face after machining of Inconel 718 with AlTiN coated tools (a); and analysis of the tool/chip contact area using Electron Diffraction Spectrometry (EDS) technique. Courtesy G. Poulachon.

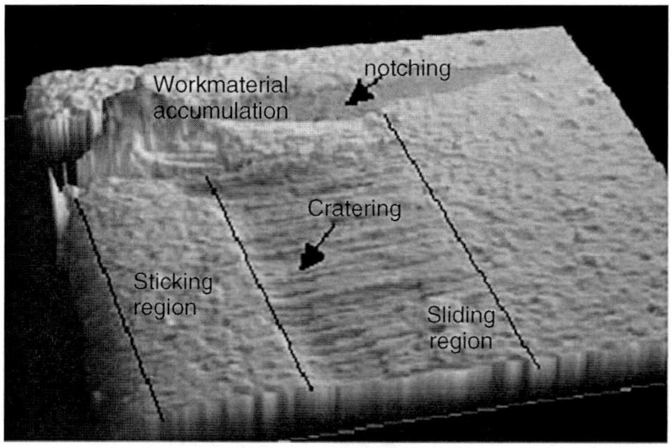

Fig. 12.18 A typical wear pattern and types of wear on the 3D rake face of a carbide insert in machining of Inconel 718 obtained by a Veeco NT 100 (Wyko®) optical profiler. Courtesy G. Poulachon. (see Plate 11)

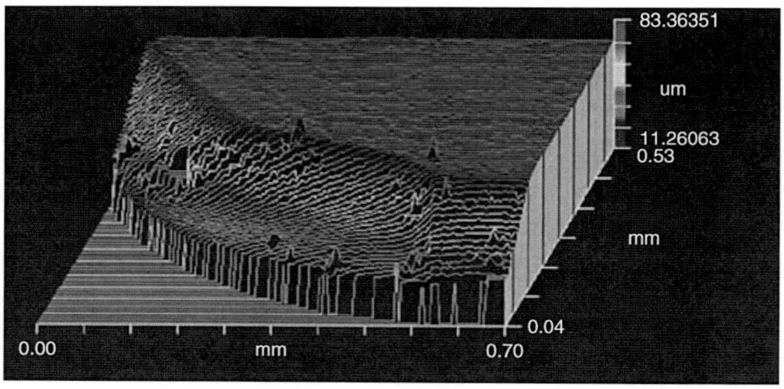

Fig. 12.19 Worn CBN cutting tool data from a white light interferometer [23]. (see Plate 12)

allows not only the sticking and sliding regions to separate but also other wear mechanisms to be observed, as, for example, notch wear apart from the predominant crater wear shown in Fig. 12.18.

A 3D image of a lightly worn CBN-tipped tool obtained by reflected light interference microscopy (RLIM) using a Zygo *New View* 200 microscope, is shown in Fig. 12.19. The Zygo *NewView* 200 is a scanning white light interferometer which uses frequency domain analysis (FDA) to generate quantitative 3D images of surfaces [24]. This 3D data can be used to determine the volumetric wear loss in the crater and flank regions. The technique proposed by Dawson and Kurfess [23] relies on the comparison of 3D data obtained by white light interferometry with an ideal representation (CAD model) of unworn tools.

Recently, the applicability of Confocal Laser Scanning Microscopy (CLSM) to the characterization of tool wear topography has been tested [25]. In this investigation, a Zeiss LSM 210 confocal microscope with appropriate software was used in reflection mode to obtain the top views and the $z(x, y)$ matrices which contain the height information on the surface (z position) as a function of pixel position (x, y). As a result, Fig. 12.20 a shows an evolution of the crater wear up to 720 s obtained from the z matrices. It should be noted in this figure that the crater depth expands rapidly after 300 s of cutting test. A top view of the worn rake face of a multilayer (8.6 μm TiCN/7 μm Al_2O_3/3 μm TiN) coated tool with distinguished six zones (numbered successively from A to F) is illustrated in Fig. 12.20b. Zone A is an area where AISI 1045 steel

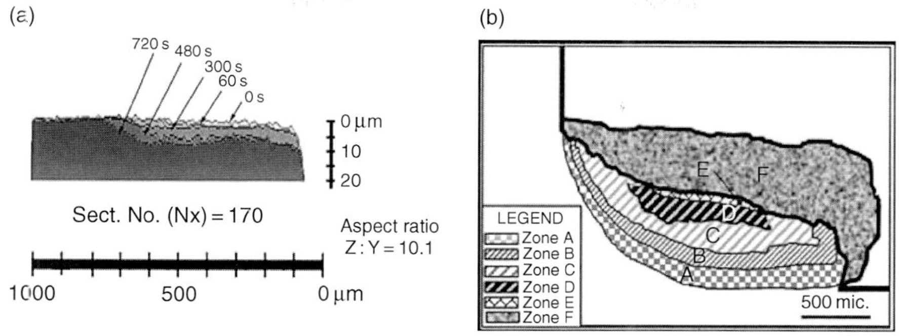

Fig. 12.20 Evolution of crater profiles from CLSM topography (a) and zones of the worn face (b) [25].

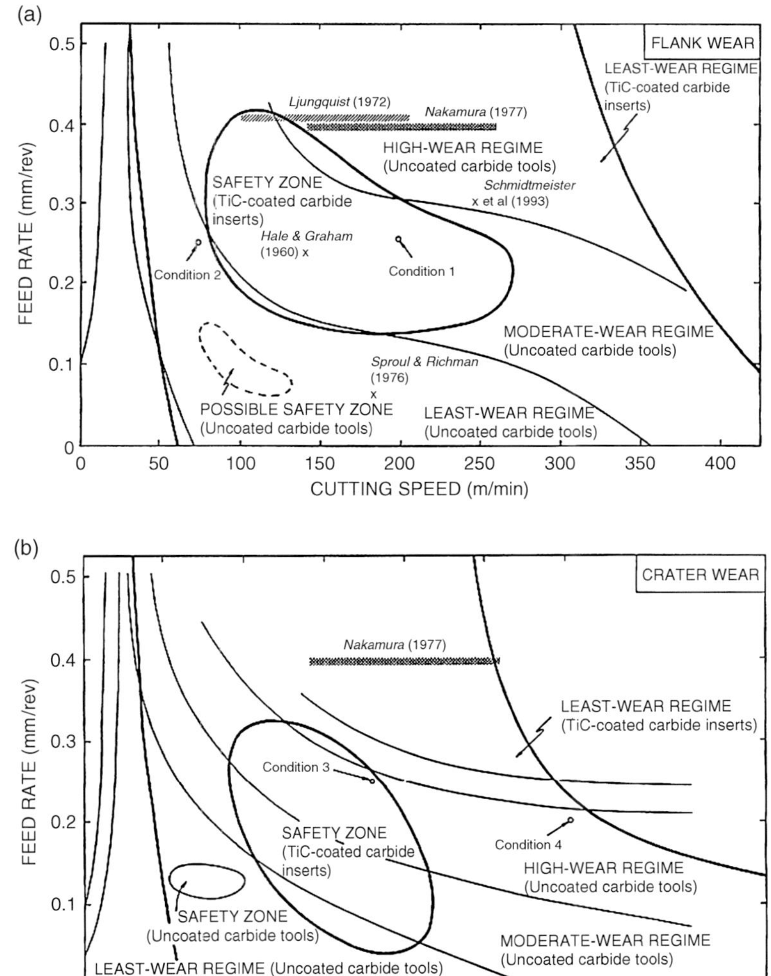

Fig. 12.21 The effect of TiC coating on the flank wear (a) and the crater wear (b) of carbide tools [27].

is adhered to TiN layer, zone *B* is a TiN layer with slight scoring marks, zone *C* is the transition from TiN layer to Al_2O_3 layer, zone *D* is oxide alumina layer with scoring marks, zone *E* is a narrow region where a high concentration of steel in alumina layer was revealed, and zone *F* is a top TiN layer with partially adhered steel. This suggests that the CSLM technique is capable of imaging the crater wear topography of multilayer coated tools.

One way of exploring many mechanisms of tool wear is to use the wear-mapping technique to construct *wear-mechanism maps*, which summarize data and models for wear, showing how the mechanisms interface, and allowing the dominant wear mechanism to be selected, for any given set of process conditions. In application to cutting tool wear, wear maps representing the rates of specific tool wear (for example, crater and flank wear) in a two-dimensional space defined by the machining conditions can be generated [10, 26].

In general, wear maps showing the predominant wear mechanisms occurring for a given combination of cutting speed and feed rate, or wear-rate maps which indicate trends in the development of selected wear types that can be generated. In order to develop the predictive wear-rate mapping system for crater and flank wear rates, theoretical wear models are modified based on experimental data in terms of workpiece material properties, tool geometry and cutting conditions. Figure 12.21 presents the predicted wear maps for flank (a) and crater wear when machining AISI 1045 carbon steel with uncoated ISO P10 and P30 grades and Ti coated carbide tools.

REFERENCES

[1] *ASM Handbook*, Surface engineering Vol. 5, ASM International, USA (1999).

[2] Toshiba Tungaloy Cutting Tools, catalogue No. 99905E1, 1999/2000.

[3] T.H.C. Childs, K. Maekawa, T. Obikawa, Y. Yamane, *Metal Machining. Theory and Applications*, Arnold, London, 2000.

[4] W. König, F. Klocke, *Fertigungsverfahren. Drehen, Fräsen, Bohren*, Springer-Verlag, Berlin (1997).

[5] K.C. Ee, A.K. Balaji, I.S. Jawahir, Progressive tool-wear mechanisms and their effects on chip-curl/chip-form in machining with grooved tools: an extended application of the equivalent toolface (ET) model, *Wear*, Vol. 255 (2003), 1404–1413.

[6] W. Grzesik, *Fundamentals of Machining of Metallic Materials* (in Polish), WNT, Warsaw, 1998.

[7] T.N. Loladze, *Strength and Wear Resistant of Cutting Tools* (in Russian), Mashinostroenie, Moscow, 1982.

[8] T.N. Loladze, Of the theory of diffusion wear, *CIRP Annals*, Vol. 30/1 (1981), 71–76.

[9] H.G. Gekonde, S.V. Subramanian, Tribology of tool-chip interface and tool wear mechanisms, *Surf. Coat. Technol.*, Vol. 149 (2002), 151–160.

[10] E.M. Trent, P.K. Wright, *Metal Cutting*, Butterworth-Heinemann, Boston, 2000.

[11] X.D. Fang, Experimental investigation of overall machining performance with overall progressive tool wear at different tool faces, *Wear*, Vol. 173 (1994), 171–178.

[12] G. Boothroyd, W.A. Knight, *Fundamentals of Machining and Machine Tools*, CRC Press, Boca Raton, 2006.

[13] M.C. Shaw, *Metal Cutting Principles*, Clarendon Press, Oxford, 1989.

[14] D.A. Stephenson, J.S. Agapiou, *Metal Cutting Theory and Practice*, Marcel Dekker, New York, 1997.

[15] A.G. Mamalis, J. Kundrak, M. Horvath, Wear and tool life of CBN cutting tools, *Int. J. Adv. Manuf. Technol.*, Vol. 20 (2002), 475–479.

[16] G.T. Smith, *Advanced Machining*, Springer-Verlag, Berlin, 1989.

[17] I.S. Jawahir, P.X. Li, R. Gosh, E.L. Exner, A new parametric approach for the assessment of comprehensive tool wear in coated grooved tools, *CIRP Annals*, Vol. 44/1 (1995), 49–52.

[18] D.A. Axinte, W. Belluco, L. De Chiffre, Reliable tool life measurements in turning – an application to cutting fluid efficiency evaluation, *Int. J. Mach. Tools Manuf.*, Vol. 41 (2001), 1003–1014.

[19] Y.-C. Yen, J. Söhner, H. Weule, J. Schmidt, T. Altan, Estimation of tool wear of carbide tool in orthogonal cutting using FEM simulation, Proc. *5th CIRP Int. Workshop on Modeling of Machining Operations*, West Lafayette, 2002, pp. 149–160.

[20] E. Usui, T. Shirakashi, T. Kitagawa, Analytical prediction of cutting tool wear, *Wear*, Vol. 100 (1984), 129–151.

[21] B.M. Kramer, N.P. Suh, Tool wear by solution: a quantitative understanding, *ASME J. Eng. Ind.*, Vol. 102 (1980), 303–309.

[22] K. Chang, X. Luo, R. Ward, R. Holt, Modeling and simulation of the tool wear in nanometric cutting, *Wear*, Vol. 255 (2003), 1427–1432.
[23] T.G. Dawson, T.R. Kurfess, Quantification of tool wear using white light interferometry and three-dimensional computational metrology, *Int. J. Mach. Tools Manuf.*, Vol. 45 (2005), 591–596.
[24] What is frequency domain analysis?, R&D Technical Bulletin, Zygo Corporation, No. 9/23/93.
[25] J.A. Olortegui-Yume, P.Y. Kwon, Tool wear evolution in multilayer coated inserts using topographic imaging, *Trans. NAMRI/SME*, Vol. 34 (2006), 405–412.
[26] X.P. Li, H.H. Ng, S.C. Lim, A predictive mapping systems for tool wear in metal cutting, *J. Mater. Proc. Technol.*, Vol. 89–90 (1999), 279–286.
[27] C.Y.H. Lim, S.C. Lim, K.S. Lee, Wear of TiC-coated carbide tools in dry turning, *Wear*, Vol. 225–229 (1999), 354–367.

CHAPTER THIRTEEN

Machinability of Engineering Materials

Contents

13.1 Definition and Machinability Criteria

The popular, standardized property *machinability* refers to the ease or difficulty with which a given material (or group of materials of similar machinability) can be machined, meaning how easy or demanding it is to shape the workpiece with a cutting tool. As a result, all engineering materials are often divided into *easy-to-machine* or *difficult-to-machine* materials. As shown in Fig. 13.1, machinability cannot be simply related to the unique work material property but rather is considered as a resultant 'property' of the machining system which is affected, directly or indirectly, by the work and tool materials, machine tool, part, fixture, cutting fluid, and cutting conditions. All components of the machining system affect physical processes such as tool wear or surface finish generation which, as discussed below, are generally used as machinability criteria. It should be noted, however, that machinability rating can only serve as a guide to starting values for further optimization.

Some commonly appearing causes of difficulties in effective/optimal performing of the machining process are specified in Fig. 13.2. In this case, the source of the general problem triggers troubles, and workpiece characteristics that decrease machinability magnify the problem. For instance, the metallurgy, chemistry, mechanical properties, microstructure, heat treatment, additives, inclusions, surface skin, etc. affect the machinability. There are many cases in which unsuitable tool grades and cutting conditions applied result in worsening the original machinability level of the workpiece.

Machinability cannot be uniquely defined in quantitative terms and may have varying meanings in different contexts. A material may have good machinability by one criterion, but poor by another. Also, relative machinability may change when a different type of operation is being carried out – turning versus milling – or when the tool material is changed [3]. In such cases, the term machinability is often used for comparison or ranking purposes. Despite the lack of a universal metric, machinability may be assessed by one or more of the criteria classified as basic, substitutive and auxiliary criteria, shown in Fig. 13.3.

Machinability data is generally gathered from production experience and summarizes machining conditions which yield acceptable tool life and part quality under specified operating conditions. Consequently, the machinability rating is predominantly determined from normalized tool life test results as for roughing

Advanced Machining Processes of Metallic Materials
ISBN: 978-0-08-044534-2

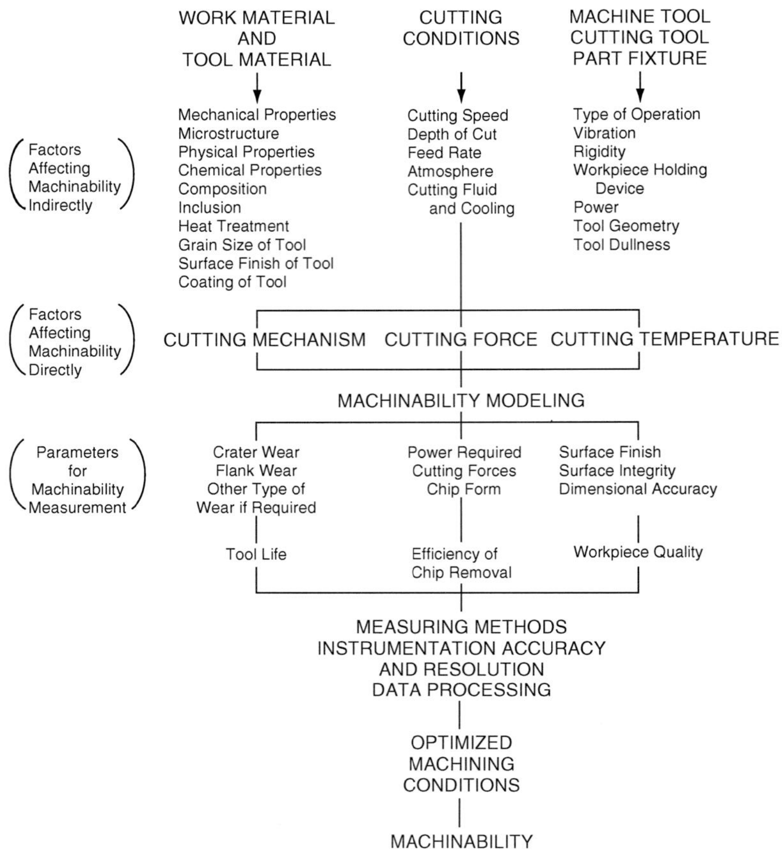

Fig. 13.1 Factors affecting machinability [1].

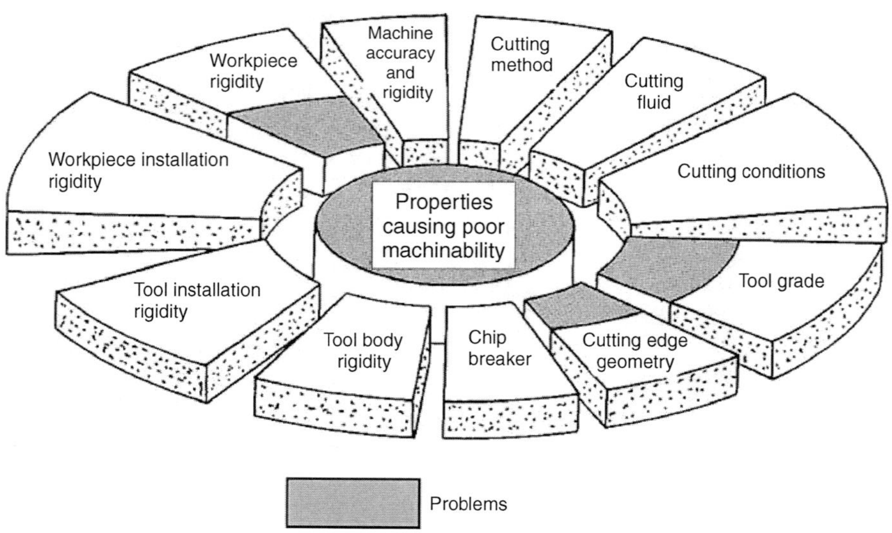

Fig. 13.2 Causes for poor machinability of difficult-to-cut materials [2].

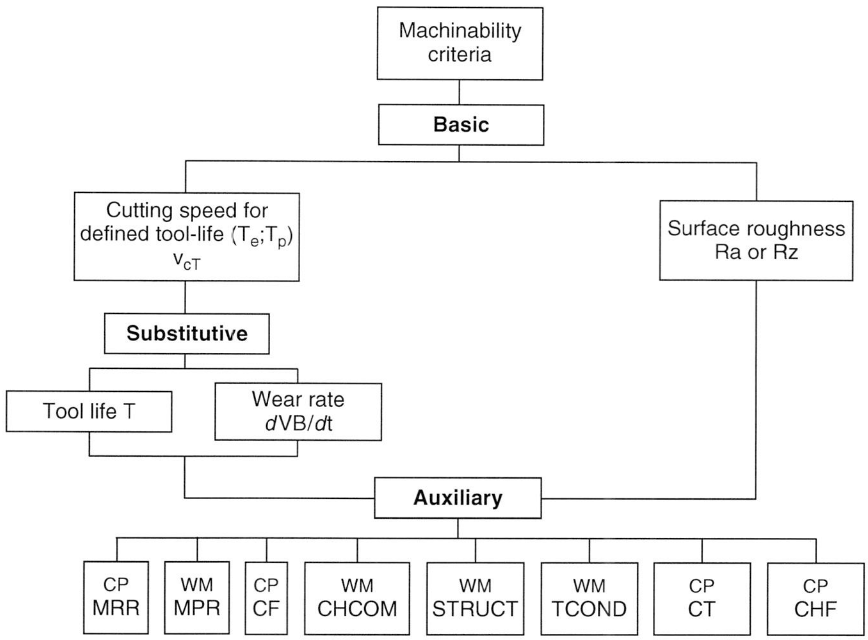

Fig. 13.3 Three-layer division of machinability criteria. After Grzesik [4]. MRR – material removal rate at which the part can be machined for a standard tool life; MPR – mechanical properties of work material (WM); CF – cutting forces (power); CHCOM – chemical composition of WM; TCOND – thermal conductivity of WM; CT – cutting temperature; CHF – chip type and form.

machining operations or achievable surface finish as for finishing operations. Usually, from a practical point of view, the cutting speed v_{cT} corresponding to the economic tool life (T_e) or tool life for maximum production rate (T_p) is a basic machinability criterion for roughing operations. In principle, a range of auxiliary criteria related to work material (WM) or process characteristics (CP) can be used. Extensive current machinability data, which follows new developments in tool materials and machine tool capabilities, is available from laboratories specializing in machinability testing, and from cutting tool manufacturers in the form of insert grade advisers or professional guides. General machining practices and machinability concerns deal with several material groups denoted by specific ISO codes and characteristic colours, specified in Fig. 13.4.

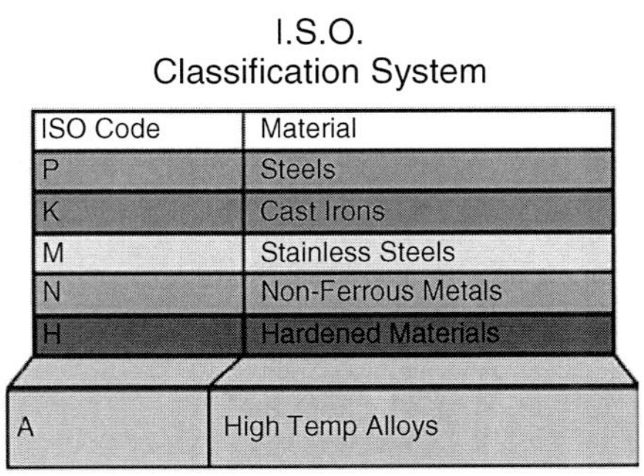

Fig. 13.4 ISO classification of basic groups of construction materials. (see Plate 13)

13.2 MACHINABILITY RATING

In general, relative and absolute machinability rating is utilized in machining practice. For cost estimating purposes, there is an interest in the introduction of absolute or quantitative measures for machinability. The most commonly used relative measure is the machinability rating (M_R) or index (I_m) defined by

$$M_R = \frac{(v_{cT})_{mat}}{(v_{cT})_{ref}} \times 100\% \qquad (13.1)$$

where $(v_{cT})_{mat}$ is the cutting speed at which the material being rated yields the defined tool life ($T = 90$-, 60-, 20- or 15-minute tool life) for a specified feed rate, depth of cut, tool material, and tool geometry, and $(v_{cT})_{ref}$ is the cutting speed at which the reference material with a machinability rating of 100 yields the defined tool life under the same conditions.

The idea of the comparative machinability rating for workpiece materials of different hardness is shown in Fig. 13.5. In the US, SAE B1113 sulphurized free machining steel has traditionally been used (AISI 1018 steel is also used) as the reference material and a 60-minute tool life is preferred. On the other hand, ISO Standard ISO 683-1 prefers a hot rolled C45 E4 carbon steel of 750 MPa in UTS as the reference material for steels and 15(20)-minute tool life speeds (v_{c15}). For irons, a 25 grey cast iron grade with a 100% perlitic matrix and flat graphite flakes is used.

In order to implement an absolute metric of machinability, it is proposed by Seco Tools [5] to select some material grades within specific material groups, for example 7 for mild and alloy steels (ISO P), 4 for stainless steels (ISO M), 4 for cast irons (ISO K) and 5 for non-ferrous materials and superalloys (ISO A and N), based on UTS values for carbon and alloy steels or specific cutting pressure $k_{c1.1}$ for other materials. So, the choice of suitable initial values of cutting data is performed for these 22 material groups.

In practice, it is necessary to compensate for the change in a recommended cutting speed or feed rate, or when workpiece material has another hardness, in relation to the starting value of tool life. This can be done by the appropriate compensation factors provided by cutting tool manufacturers, as, for example,

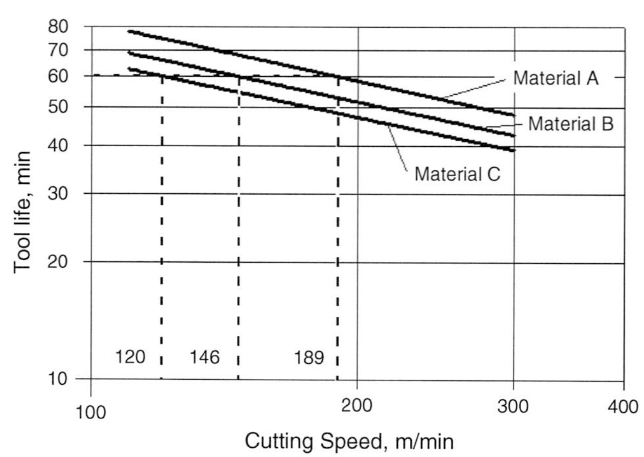

Fig. 13.5 Determination of the cutting speed v_{c60} for three different materials A, B and C.

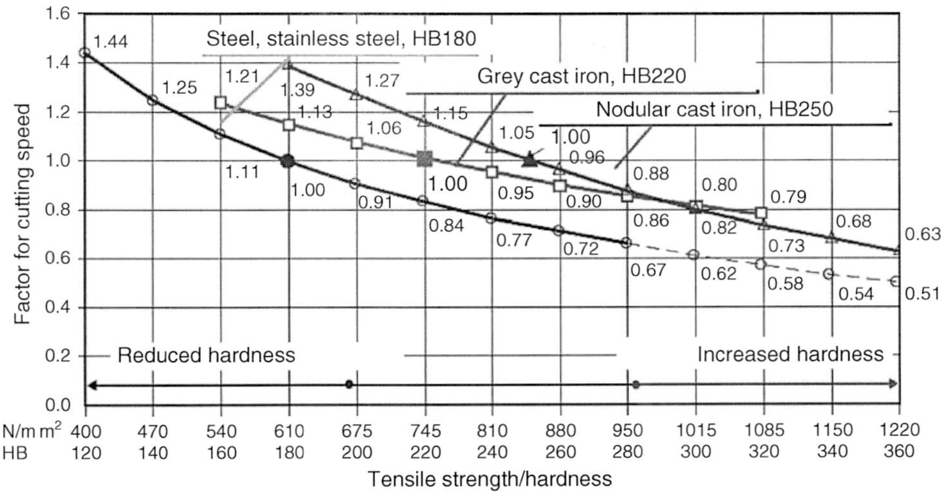

Fig. 13.6 Correction of cutting speed for different hardness of workpiece material classified in ISO P, M and K groups [5].

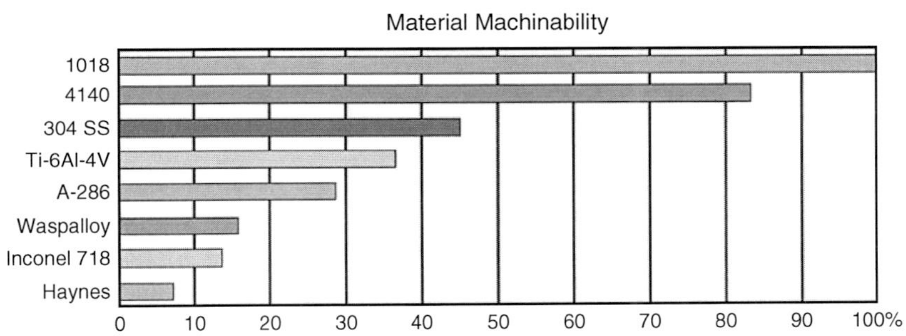

Fig. 13.7 Comparison of the machinability ratings of some popular materials [5].

in Fig. 13.6 [5]. The relationship between recommended and corrected cutting speed and depth of cut for different values of workpiece hardness is [1]:

$$v_{c2} = v_{c1}\left(a_{po}/a_p\right)^p \left(RC_o/RC\right)^q \tag{13.2}$$

where v_{c1} is the recommended cutting speed for a depth of cut a_{po} and Rockwell hardness RC_o.

For example, if a workpiece material from ISO P group has a higher hardness of 240 HB, relative to the given HB 180, the cutting speed adjusted to HB 240 is equal to $305 \times 0.77 \approx 235\,\text{m/min}$ ($a_p = 3\,\text{mm}$ and $f = 0.4\,\text{mm/rev}$) [5].

Figure 13.7 compares the relative machinability indices for some popular representatives of ISO groups of construction materials. It can be seen in this figure that superalloys (Waspalloy, Inconel 718 and Haynes), which belong to the group of extremely difficult-to-cut materials, have the machinability rating of about 8–15% in comparison to an AISI 1018 low carbon steel.

Figure 13.8 shows how the machinability rate varies for different iron grades starting from MR = 1 for conventional easy-to-machine grey cast iron (A) up to MR = 0.3 or even less for ductile (D) and white (E) irons, respectively. According to ASTM data [7], relative machinability ratings for nodular irons depend on their hardness and for the hardness of 183, 200 and 230 BHN they are equal to 0.6, 0.5 and 0.4, respectively.

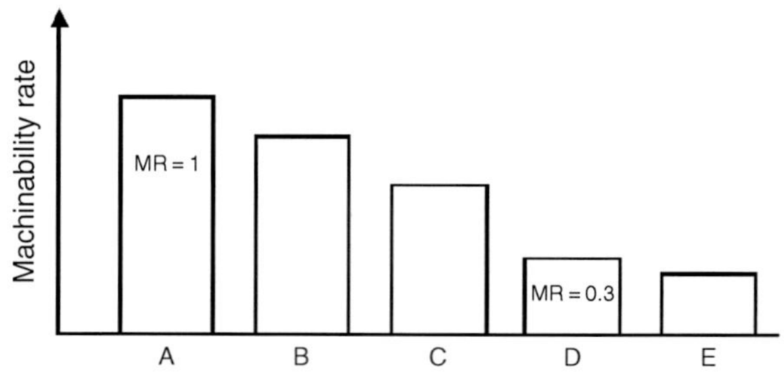

Fig. 13.8 Qualitative comparison of various irons: (A) grey cast, (B) annealed, (C) vermicular (CGI), (D) ductile (nodular), (E) white (chilled) [6].

13.3 Machinability Data Systems

The practical problem arising is that the analytical methods for determining recommended cutting conditions and estimating machining costs required data on machinability of a wide range of materials for the effective use of theses procedures. Establishing the tool-life equation for a particular set of conditions requires lengthy cutting tests and wear measurements which must be performed for the whole range of tool and work materials under consideration. The extent of this procedure can be substantially reduced by utilizing data found in handbooks or in computer-based machinability data systems. Two different types of machinability data systems can be developed: data-based systems and mathematical model systems.

Machinability database (MDB) systems were developed to assist numerically controlled (NC) machines not only in selecting the recommended cutting speed, feed rate and depth of cut, but also the tool material, appropriate tool geometry from available inventory, tool-holder and other elements of automatic modular quick-change tooling system, sequence of operations, and optimum cutting conditions based on economic or productivity requirements. In order to accomplish these functions, the MDB system may use data storage and retrieval algorithms, generalized empirical equations, mathematical models, or a combination of these methods [1]. In particular, generalized empirical equation systems limit the amount of data to be stored for each of the many available combinations of materials, tools and operations. Because empirical equations available in handbooks are often useless due to large number of the relevant inputs, *mathematical model* systems can be more effective in model building using a multiple regression analysis routine. Based on stored machinability data, the constants for the appropriate tool-life equations are determined and the optimum feeds and speeds are then determined using appropriate economic or production rate equations for the particular operation being considered. An example of the MDB system consisting of three modules – manufacturing data, model building and manufacturing analysis and optimization – is presented in Fig. 13.9. In particular, the model building module helps to develop mathematical models relating the required machinability responses, such as tool life, surface finish, horsepower, cutting forces or tool breakage, to the various input variables. The optimization module contains the specific algorithm to carry out the optimization using the techniques discussed in Chapter 14. In general, the sources of professional information for an MDB system include handbooks, laboratory experiments and production experience but extensive data collection sheets for the machinability response should be elaborated by machining information centres.

Recently, an online fuzzy expert system (FES) for machinability data-on-demand, which can be easily incorporated into the internet and intranet environment (see Chapters 14 and 19) and can be virtually accessed at any place, is proposed [8]. The expert system is created as a dynamic link library (DLL) and an ActiveX Control interface is developed to deliver dynamic documents via the Web.

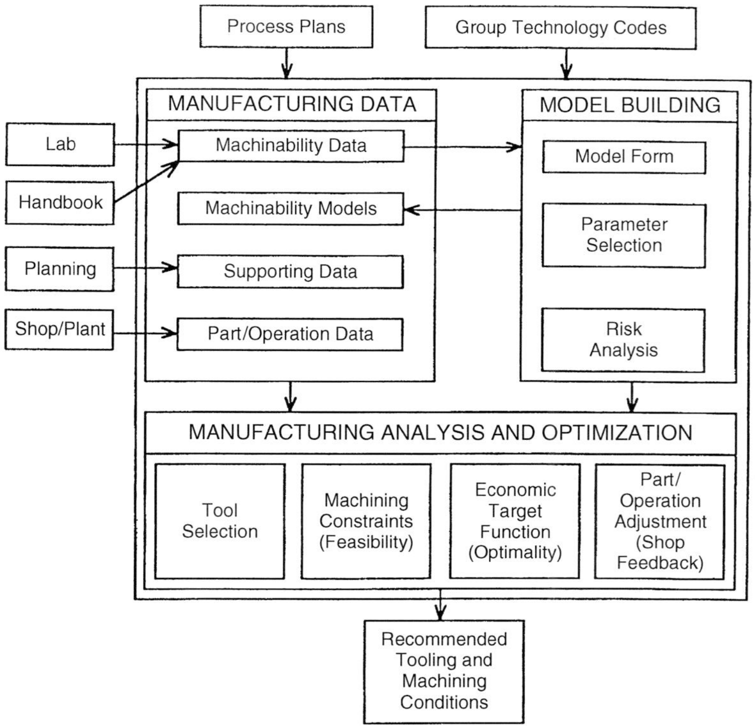

Fig. 13.9 Machinability database system consisting of three basic modules [1].

13.4 Survey of Machinability of Engineering Materials

This section provides only some new trends observed in improving machinability of six basic ISO groups of engineering materials and relatively new materials such as light metals (mainly intermetallics and magnesium alloys), metal matrix composites, and fibre-reinforced composite materials (FRCM), as well as refractory metals which systematically replace traditionally used construction materials in most advanced sectors of industry. More fundamental information concerning the machinability of easy-to-machine materials, mainly steels and cast irons, can be found in [7, 9, 10].

13.4.1 Carbon/unalloyed steels

Steels vary greatly in chemical content and microstructure. Traditionally, these materials are classified according to their carbon content:

- Low-carbon steel with carbon content lees than 0.25 (0.3)%C;
- Medium carbon steel with 0.25–0.55(0.6)%C; and
- High carbon steel with 0.55–0.8%C.

The factors affecting the machinability of carbon/low alloy steels are well recognized and the main practical machining problem deals with increasing productivity. With regards to machinability of low carbon steels, the low hardness and high ductility is often a negative factor with considerable tendency for smearing and BUE which reduce tool life and give poor surface texture. A higher carbon content improves machinability in such a way that hardness is increased moderately and ductility decreased. Machinability within this group of materials varies considerably with the addition of free machining additives (sulphur, lead, bismuth, tellurium and selenium often in combination with sulphur), conditions of rolling or drawing process, and post-process treatment (annealed, normalized, tempered conditions). Some manganese is often present in these steels (not regarded as an alloying element unless it exceeds 2%), especially to produce planes of low strength when manganese sulphides in free machining steels are plastically deformed. Free machining steels with sulphur and selenium additions are common in carbon and stainless steels – ferritic and martensistic, as well as austenitic.

Carbon steels are most commonly turned, bored and milled with grooved carbide tools coated with single (TiN, TiC) or multilayer films to reduce cratering due to diffusion. Moreover, carbon steels can be finished by turning with cermets, turned and bored with alumina-based ceramic tools, and also turned, bored and milled with PCBN tools. Machining of hardened steels is discussed in Chapter 15.3.

13.4.2 Alloyed steels

When the various alloying elements are equal to or less than 5%, the steel is rated as low alloy steel and above 5% is considered as high alloy. Adding alloying elements in a precisely controlled way to steel generally increases the strength and hardness, resulting in lower machinability and reducing the cutting speeds and feed rates which produce acceptable tool life. Alloying elements alter mechanical properties such as strain-hardening ability. They are added in various proportions for different reasons: to modify the alloy structure, to change eutectoid point, to affect phase transformation, hardness, strength, wear resistance and corrosion resistance. Some elements (refractory metals such as Cr, Mo, V, W) combining with carbon form very hard, abrasive carbides which reduce tool life. Nickel (N), chromium (Cr) and molybdenum (Mo) are the most typical additives, but vanadium (V), tungsten (W), cobalt (Co), silicon (Si) and manganese (Mn) occur in many steels. The major reason for poorer machinability of alloy steels is that they are generally machined in harder and stronger conditions, as, for example, die and tool steels. Tempered alloy steels produce better surface finish than annealed steels because the tempered structure reduces or eliminates BUE formation. Alloyed steels represent a large group of materials that are machined extensively, recently often as near-net-shape blanks produced by precision forging and casting techniques. In such cases more demanding chip control and other tool wear parameters for maintaining close tolerances are required. Also hardened, highly alloyed components are machined to an increasing extent with cermet, ceramics and PCBN tools. Typically, machinability tables to asses the tool life to be expected from specific high alloy grades in comparison to plain carbon grades, and formulas for computing an equivalent carbon content, are used.

13.4.3 Stainless steels

Stainless steels contain chromium with a content of 12–25% Cr which is responsible for corrosion resistance and a few tens to 25% Ni (austenitic and maraging steels) to produce an austenitic structure, which leads to extreme high work-hardening rates. Stainless steels are usually classified into four categories depending on their primary constituent of the matrix: ferritic, martensistic, austenitic and duplex (combined ferritic/austenitic). Other classifications that distinguish precipitation-hardenable vs non-precipitation hardenable grades or free machining versus non-free machining are used in practice. Stainless steels are considered to be difficult to machine due to their high tensile strength, high ductility, high work hardening rate, low thermal conductivity and abrasive character. This combination of properties often results in high cutting forces and temperatures, and tool wear rates, as well as a susceptibility to notch wear, difficulties with chip breakability, BUE formation and poor surface finish. Ferritic steels are generally more machinable and their machinability decreases with increasing chromium content. The machinability of martensistic stainless steels is influenced by carbon content (increasing hardness), nickel content and metallurgical structure (free ferrite content) in such a way that higher content of carbon and nickel reduces machinability. The austenitic stainless steels represent the largest group of stainless steels in use, making up 65–70% of the total for the past several years. The most common grade of the AISI 300 series of austenitic stainless steels with a basic level of corrosion resistance is the 18/8 type (AISI 304) with 18% Cr and 18% Ni. To obtain acid-resistant stainless steel, 18/8 + 2% Mo, molybdenum is added. They are typically more difficult to machine than ferritic and martensitic stainless steels. Duplex alloys have a chemistry similar to austenitic stainless steels but are generally harder to machine due to their high annealed strength.

The following basic grades of coated carbides are recommended for machining austenitic stainless steels and other alloys [5]: TiCN-TiN (4 μm) PVD coated M15 fine-grain carbide for finishing operations, TiCN-Al_2O_3-TiN (total 5.5 μm) CVD coated M15 carbide for finishing and semi-finishing operations, the latter coated grade but with M25 substrate, and TiAlN-TiN (total 4 μm) PVD coated M25 carbide for medium and roughing operations. The last two grades can also be used for machining duplex steels. General guidelines for machining stainless steels include [10]: use of lower cutting speeds and MRRs than for carbon steels, use of rigid tooling and fixture to avoid chatter, maintain feed above a minimum level to avoid poor surface integrity, maintain doc (depth of cut) above the depth of work-hardened layer to avoid excessive notch wear, use of high sharp cutting edges (note that all grades recommended above are PVD coated carbides) to avoid BUE and reduce work-hardening effect, use of proper cutting fluids with sufficient flow rates for heat removal.

13.4.4 Cast irons

In recent years the development of new cast iron materials has been seen to offer greater competition to other materials and make cast iron a contender for components not traditionally manufactured from this material [10]. In particular, the two newer materials with significantly higher strengths creating important impact into these recent developments are compacted graphite iron (CGI) and austempered ductile iron (ADI). Their excellent mechanical and fatigue strengths result from lamellar graphite and globular graphite structures, respectively [11, 12], shown in Fig. 13.10. As an example, the increasing use of vermicular/compact graphite irons as a suitable material for engine blocks in high performance diesel engines is reported [12, 13]. The trends observed in the application of various irons in the automotive industry along with recommended cutting tool materials are outlined in Table 13.1. Basically, they originated from the necessity to design car engines with higher power trains, lighter and dimensionally smaller in comparison to engines manufactured using traditional materials such as Al-Si alloys or grey cast irons [11].

The importance of improving the machinability of nodular irons is sourced in the promising statistics from the automotive industry which reported substantially growing use of cast irons, not only previously mentioned compacted graphite iron (CGI) and austempered ductile iron (ADI), but also ductile nodular iron. On the other hand, machining the range of grey and ductile irons is complicated by the fact that materials can vary in machinability so greatly, often within the same batch [14].

Titanium alloys and nickel-based alloys best satisfy the criteria for aero-engine alloys and therefore are the most widely used in the aerospace industry (Fig. 13.11). Unfortunately, these alloys represent the most difficult to machine materials, so the highest degree of fracture resistance and chemical-abrasive wear resistance is demanded of the cutting tools applied. The same elements (Ni, Co, Cr, Ti) which give these materials their high strength and corrosion resistance also give serious difficulties when machining. It is

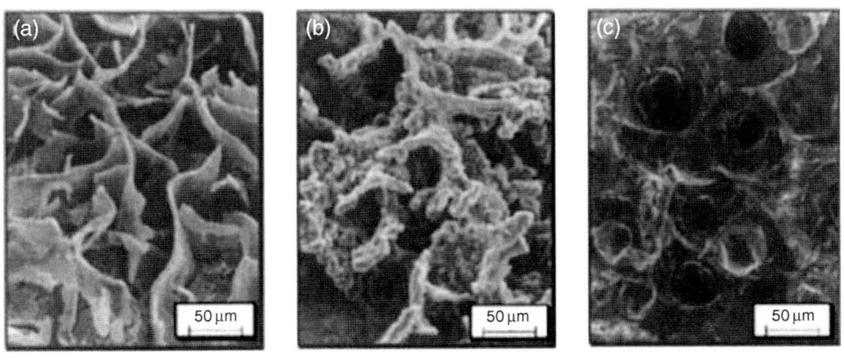

Fig. 13.10 Possible graphite shapes in grey cast (a), CGI (b), and nodular (c) irons [11].

Table 13.1 Application tendencies of cast irons and relevant cutting tool materials in the automotive industry [13].

Iron grade	*Trend*	*Typical parts*	*Typical cutting tool materials*
Grey cast EN-GJL	downward	Cylinder blocks, cylinder heads and brake drums and disks	Carbide coated with Al_2O_3, silicon nitride Si_3N_4 ceramics, CBN
Vermicular CGI	upward	Cylinder blocks of diesel engines, retainers and fasteners	According to process requirements
Nodular EN-GJS	upward	Crankshafts, camshafts, gear wheels, cases and housings	Multilayer MT-CVD and PVD coated $TiCN/Al_2O_3/TiN$ and $TiCN/Al_2O_3/TiN/TiCN$ carbides, cermets, mixed ceramics, CBN

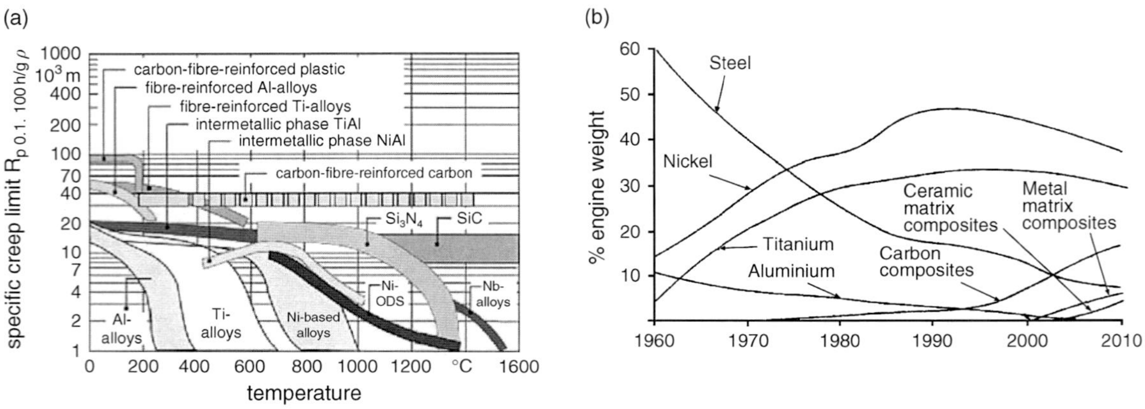

Fig. 13.11 A range of materials in turbine (a) and aero-gas turbine (b) manufacture [12, 15].

expected that cutting speed for these materials can be increased to the level of 300–450 m/min when a new generation of whisker-toughened alumina ceramic tools will be launched.

13.4.5 Titanium and its alloys

Titanium has a very high strength-to-weight ratio, a high melting point, and very good corrosion resistance [9, 10]. Its density is only 57% the density of iron, while the strength of Ti alloys is comparable to that of alloy steels. Titanium alloys are generally divided into three groups: alpha (α), alpha-beta (α-β) and beta (β). These mechanical and exploitation properties are decisive in their applications as structural materials for supersonic aircraft and for spacecraft. Non-aerospace applications include automotive, chemical/energy, medical and sporting goods. Commercially pure (CP) titanium and α alloys are used mainly in cryogenic applications whereas α-β alloys (Ti-6Al-4V is the most common alloy) are predominantly used in the aerospace industries (in airframes and engine components). β alloys find high strength applications due to good forgeability. On the other hand, these materials indicate low thermal conductivity resulting in a machinability rating of 0.3–0.5 in comparison to a reference C45 carbon steel. According to many production sources, the machinability decreases starting from CP titanium and α-type alloys and ending at β-type alloys due to its high ductility. Moreover, the rapid oxidation process of the machined surface results in a high tendency to react with the tool materials. At the same time the chips produced tend to stick adhesively to the cutting edges leading to built-up edge. Taking into consideration all serious difficulties observed in machining titanium-based materials, many large companies (for example, Boeing and General Electric) invest large sums of money in new machining technologies and the leading cutting tool manufacturers (for example, Sandvik Coromant) offer special selection systems for both tools and cutting parameters, contributing to minimizing machining costs and extended tool life. Recently, several vital components for the Airbus A320 and the new Airbus A380 were successfully produced by machining in solid titanium blocks using machining centres equipped with tooling by Seco Tools, Sweden [16]. For example, the first roller track beam of the A320 is machined with highly positive, sharp tools and inserts, which generate minimum cutting forces. Fine grain carbides (F15M, F40M) are used for milling operations, such as face milling, contouring, slot milling, pocket milling, and chamfer and copy milling. All tools have internal coolant troughs directed on the cutting edges. The cutting speeds for titanium are between 50 and 70 m/min, providing tool lives between 30 and 60 minutes.

General machining points to be considered by technologists are as follows [5, 17]:

- Use sharp, positive cutting edges with sufficient clearance angles.
- Machine under stable conditions with tight workpiece clamping.
- Optimize the feed rate.
- Apply copious amounts of coolant correctly for the operation performed.
- Limit wear progress to minimize heat generation.
- Minimize vibration tendencies.

According to Seco, Sweden [5], the machinability rating for Ti-6Al-4V titanium alloy ranges, depending on the treatment used, from 30 to 35% in comparison to C45 carbon steel. On the other hand, the machinability rating of CP titanium semi-products (tubes, plates, bars, rings and forges) is assessed to be of 45–60%. It is roughly evaluated that the typical metal removal rate is often only around 25% of the rates achieved in steels and stainless steels [17].

Detailed selection system for tool geometry, cutting parameters, cutting cooling and machining strategy for milling, drilling and turning operations can be found in an application guide published recently by Sandvik Coromant [17] or in other advisory systems available on the internet [5]. It should be noted that the cutting parameter that has the predominant damaging effect is cutting speed.

Recently, Sandvik Coromant [5] offered a special material group denoted by an apricot colour and the S symbol. So, for example, it recommends a special H10F (S30) grade with ultra fine grains and very high abrasive resistance of the flank for milling aerospace alloys. Last year, an innovative *X*cel insert configuration which combines the advantages of 90° and 45° approach angle for semi-roughing turning was launched. Seco Tools [5] recommends the hardest uncoated grade 890 (H15) with fine grain structure and inserts with positive geometry, sharp and precisely ground cutting edges for turning. These features minimize cutting forces, temperatures, built-up edge and metallurgical surface damage. In drilling with indexable drills an H13A (S15–S25) grade is proposed to be suitable for both inner and outer inserts because it provides sharp cutting edges, resistance against abrasive wear and dynamic loads. In the case of turning the Ti-6Al-4V alloy, the process can be easily performed while using diamond and cubic boron nitride tools with suitable cutting conditions [5]. As an example, Table 13.2 provides optimum process conditions for milling, drilling and turning operations.

Table 13.2 Recommended start cutting data for three basic machining operations [17].

Type of operation	*Application*	*Type of tool*	*Max. depth of cut a_p/hole length (mm)*	*Cutting speed v_c (m/min)*	*Feed, mm/rev or f_z (mm/tooth)*
Milling	Low to medium a_p	Cutter with round inserts	4	40–60	0.18
Drilling	Drill dia. 14–20 mm	Solid drill with internal cooling	$7.1 \times D_c$	20–60	0.16–0.30
Turning	Finishing (LSM)	RCMT round inserts	0.25–1.0	80–120	0.2–0.3

13.4.6 Nickel-based alloys

Nickel-based alloys, such as commercially available grades Inconel, Nimonic, Rene, Udimet and Pyromet, are the most widely used superalloys, accounting for about 50 wt.% of materials used in an aerospace engine, mainly in the gas turbine compartment [15]. Nickel-based superalloys (especially Inconel 718 accounting for as much as 45% of wrought nickel-based alloys) are generally known to be one of the hardest materials to machine because of their high hardness, high strength at high temperature, strong affinity to react with tool materials and low thermal conductivity. Their pronounced tendency to BUE formation and strain hardening, and the abrasive effect of carbides and intermetallic phases, results in exceptionally high mechanical and thermal stresses on the cutting edge during machining. The usual range of turning speeds for uncoated carbides of the ISO K10/K20 group on Inconel 718 and Waspaloy is 20–35 m/min [18]. Alternatives to carbides for lathe tools are ceramics and PCBN. Oxide ceramics are unsuitable for machining nickel-based alloys owing to intensive notch wear. Sialon materials were mostly used for roughing operations with cutting speeds of 100–200 m/min but currently turning operations rely on whisker-reinforced alumina ceramics ($Al_2O_3+SiC_w$). On the other hand, PCBN tools can also be suitable for finishing operations on Inconel 718 and Wespaloy with cutting speeds of 300–600 m/min. Fine-grained PCBN grades with a TiC- or TIN-based binder and a binder fraction of 30–50 vol.% were found to be a first choice.

13.4.7 Lightweight materials

Aluminium and aluminium alloys, in the form of cast and wrought alloys, belong to the most machinable materials in small-to-medium volume production and mass production. Due to low density and excellent recycling potential, the demand for aluminium in the transport industry is increasing. Technically, the most important alloying elements are silicon, magnesium, zinc and copper. The most commonly machined cast aluminium alloys by volume are cast aluminium-silicon (Al-Si) alloys in which a silicon content of 12.5% defines the eutectic point. Instead of the traditional materials used to strengthen aluminium (silicon, magnesium), new patented DSC aluminium contains more evenly dispersed ceramic particles, which are smaller than the ceramic grains used previously [19]. Machinability of various categories of aluminium alloys, such as wrought alloys, hypo-eutectic and hyper-eutectic cast alloys, varies significantly depending on the alloy composition and the associated grain structure [1, 10, 12]. For some alloys, the high tendency of aluminium to adhere during cutting presents a significant risk that can lead to tool breakage. Long, ductile chips, that complicate the machining process, are formed regardless of the cutting tool geometry chosen. The cutting action needs a sharp, positive cutting edge and is usually performed with tools specially designed for aluminium. Diamond tipped and diamond coated tools are developed for high productive machining of these alloys. Another strategy used is to develop sharp, uncoated carbide tools with large positive rake angles and polished rake faces. On the other hand, machining temperatures are generally low and high cutting speeds can be used, for example for the eutectic Al-Si alloys speeds up to 450 m/min and as high as 5000 m/min are achievable with carbides and PCD tools, respectively. The presence of hard free SiC particles in hypereutectic alloys lowers (decreases) machinability (cause accelerated abrasive wear of carbide tools and chipping of PCD tools) and due to this fact cutting speeds are typically limited to about 100 m/min in turning with carbide tools and to 1000 m/min when milling with PCD tools. Moreover, a malleable aluminium alloy (AlZnMgCu1.5) for aircraft parts and an aluminium casting alloy (AlSiMg) for automotive engine parts can be effectively machined dry using specific coating systems, e.g. partially crystalline hard coatings with amorphous (softer) parts such as WC/C, CrC/C or $TiN+MoS_2$, and superhard PCD coatings [20].

Aluminium alloys are the most critical materials for dry drilling and tapping operations. Figure 13.12 shows substantial differences in the operational behaviour of the different coatings tested in dry drilling an AlSi9Cu3 cast alloy and with MQL. In general, tool life was longer and surface roughness was better than those with dry cutting. Related to the tool life the comparable results were obtained with the three coatings tested, i.e. TiCN, TiAlCN + Me-C:H and $TiAlCN+MoS_2$ coatings, as shown in Fig. 13.12a. On the other hand (Fig. 13.12b), the smallest surface roughness was achieved with TiAlCN + Me-C:H and $TiAlCN+MoS_2$ coatings when using the MQL system. During these tests fatty synthetic ester with lubricant quantity of about 10 ml/h was fed to the tool on a stream of compressed air (air pressure 2 bar).

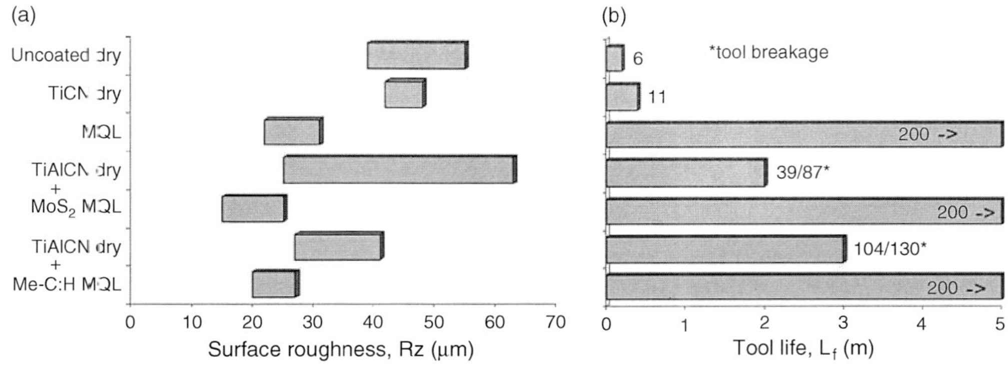

Fig. 13.12 Comparison of surface roughness and tool life in dry and near-dry (MQL) drilling operations of AlSi9Cu3 cast aluminium alloy [21].

Also, hypereutectic aluminium-based metal matrix composites (Al-MMCs) are of interest for automotive, aerospace and train companies for components such as brake discs and drums, suspension arms, callipers, pistons and cylinder liners. The machinability tests performed in the Fiat Research Centre on Al + 20% SiC composites with a cutting speed of 500 m/min and feed rate of 0.4 mm/rev, revealed that PCD-inserted tools perform better than PACVD diamond coated tools due to rapid detachment of the coating [22].

13.4.8 Magnesium alloys

Magnesium is the lightest metal used in construction and therefore offers the greatest potential for weight reduction and, as a result, reducing energy consumption. Its density of 1.74 g/cm^3 means that a weight reduction of up to 25% is possible in comparison to aluminium. Especially, the reduction of car weight is based on such lightweight materials as plastics but light metals such aluminium or magnesium are easier to machine and superior to recycling (lifecycle of a product is not regarded as the primary factor). There are two basic concerns in machining magnesium: the risk of fire when heated to its melting temperature (typically less than 650°C) and BUE formation. In general, magnesium is easy to machine and tool life is roughly five times longer than when machining aluminium wet under comparable dry conditions. Low hardness, low cutting forces, segmented chips, all mean that machining the alloy AZ91 HP with PCD coated tools is successful at a high cutting speed of about 2400 m/min [23]. This is because, flank BUE due to adhesion can occur when machining magnesium dry at a cutting speed of 900 m/min with uncoated and Ti coated carbide tools. PCD tools are not preferred due to best resistance against abrasion but, on the other hand they reduce the chip temperature and therefore the danger of chip ignition is limited in dry machining. A magnesium-based MMC is more difficult to machine due to combining the adhesive effects with the abrasive action of reinforced components such as Al_2O_3 or SiC in the form of fibres or particles. For example, cutting speeds of 100 m/min are achievable when machining an SiC reinforced, magnesium-based MMC with PCD coated tools.

13.4.9 Composite materials

Composites have superior high strength-to-weight ratio and resistance to corrosion and fatigue and are widely used in strength demanding applications. Exact numbers are not available, but estimates are that 50–60% of the material used in the new Boenig 787 Dreamliner and Airbus A380 will be composite [24]. In general, composites can be segmented into three types: fibre reinforced, particle reinforced and dispersion strengthened. The materials used for metal cutting are glass fibre reinforced plastics (GFRP) and carbon fibre reinforced plastics (CFRP), rarely aramid fibre reinforced plastics (AFRP), matrixed in an organic polymer, epoxy or phenolic resin. A typical structure involves the use of carbon fibre for the outer surfaces and a honeycomb in the middle [10]. Like metals, composites have different machining characteristics because material behaviour in cutting is non-homogeneous. The type of fibre and matrix and the fibre content ratio to the matrix are important factors in machinability. In the case of GFRPs and CFRPs, it is the cutting tool material that dominates the tool selection. In the case of AFRPs, the tool geometry is the most significant factor in the choice of cutting tool. Composites should be cut cleanly with a sharp cutting edge. Excessive wear can cause fibres to be broken away, rather than cut, leading to delamination. Very fine grained cemented carbide and PCD tools are recommended. From non-traditional methods, the most effective seems to be very high pressure abrasive water-jet machining [24].

Metal matrix composites (MMCs) are metals (aluminium, magnesium, titanium and steel) reinforced with particles, fibres or whiskers of harder materials such as boron, silicon carbide (SiC), aluminium oxide, graphite, tungsten and various proprietary fibres [25]. Due to the intensity of abrasive wear and adhesion, aluminium MMCs can be machined using diamond-based cutting materials, i.e. PCD, diamond tipped or CVD diamond coated tools. In general, aluminium MMCs reinforced with alumina are more difficult to machine than SiC reinforced materials. Holemaking operations such as drilling, reaming and tapping are typically the most critical cutting operations with composite materials. Magnesium MMCs have a machinability similar to aluminium MMCs, but titanium MMCs are more difficult to machine and are often machined using lasers or other non-traditional methods.

13.4.10 Refractory metals

These alloys (niobium + hafnium + titanium) have high strength in temperatures up to 2700°F (1480°C), making them attractive for a variety of rocket-propulsion components and industrial-process assemblies. They are relatively impervious to exposure to high temperature propellant gases and have low ductile-to-brittle transition temperatures for withstanding high frequency vibration [26]. Unfortunately, these materials impose new constrains on cutting tool design, coolants, and speeds and feeds, compared to more conventional materials. Typical machining operations involve turning, drilling, boring, milling and threading fairly thick-walled body sections. Because of the high cutting forces and the tough, stringy chips produced, chip breakers are completely ineffective. Turning HSS, cobalt alloy and carbide tools with high positive rake angle are able to machine refractory alloys at 250 fpm (76 m/min) with carbide tools and 75 fpm (23 m/min) [26]. Similarly, milling is accomplished with standard HSS end-mills with feeds no higher than 0.1 mm/tooth and speeds of 23 m/min. The best fluid selected was a water-soluble coolant with high lubricity and good wetting characteristics that cleans up using water-based cleaning processes.

References

[1] D.A. Stephenson, J.S. Agapiou, *Metal Cutting Theory and Practice*, Marcel Dekker, Inc., New York, 1997.
[2] K. Karino, *Trouble Shooting for Cutting*, Mitsubishi Materials, Ibaraki, Japan, 1998.
[3] E.M. Trent, P.K. Wright, *Metal Cutting*, Butterworth-Heinemann, Boston, 2000.
[4] W. Grzesik, *Fundamentals of Machining of Metallic Materials* (in Polish), WNT, Warsaw, 1998.
[5] Internet: www.coromant.sandvik.com, www.secotools.com, www.kennametal.com.
[6] S. Söderberg, Fortschritte beim Grauguß-Drehen, *Werksatt und Betrieb*, Vol. 132, No. 5 (1999), 145–151.
[7] *Tool and Manufacturing Engineers Handbook*, Vol. I, *Machining*, SME, Deaborn, 1983.
[8] S.V. Wong, A.M.S. Hamouda, A fuzzy logic based expert system for machinability data-on-demand on the Internet, *J. Mater. Proc. Technol.*, Vol. 124 (2002), 57–66.
[9] B. Mills, A.H. Redford, *Machinability of Engineering Materials*, Applied Science Publ., London, 1985.
[10] *Modern Metal Cutting. A Practical Handbook*, Sandvik Coromant Publ., Fair Lawn, 1996.
[11] H. Schulz, U. Reuter, Verschleißmechanismus geklärt: GGV-Motoren reif für die Großserie, *Werkstatt und Betrieb*, Vol. 134, No. 7–8 (2001), 80–82.
[12] G. Byrne, D. Dornfeld, B. Denkena, Advancing cutting technology, *CIRP Annals*, Vol. 52/2 (2003), 483–507.
[13] F.J. Momper, Neue Schneidstoffe für neue Werkstoffe, *Werkstatt und Betrieb*, Vol. 131, No. 5 (1998), 390–394.

[14] D. Graham, Machining cast iron, *Manufacturing Engineering*, Vol. 136, No. 2 (2006).
[15] E.O. Ezugwu, J. Bonney, Y. Yamane, An overview of the machinability of aeroengine alloys, *J. Mater. Proc. Technol.*, Vol. 134 (2003), 233–253.
[16] High tech at low cost, *The New Edge*, internet magazine for the Seco Tools Group, No. 1-2005, 2–3 (internet: www.secctools.com).
[17] *Titanium Machining – Application Guide*, Sandvik Coromant, Sweden, 2004.
[18] I.A. Choudhury, M.A. El-Baradie, Machinability of nickel-based super alloys: a general review, *J. Mater. Proc. Technol.*, Vol. 77 (1998), 278–284.
[19] C. DeJong, How to machine light metals, *Automotive Design and Production*, internet at www.autofieldguide.com.
[20] M. Lahres, P. Müller-Hummel, O. Doerfel, Applicability of different hard coatings in dry milling aluminum alloys, *Surf. Coat. Technol.*, Vol. 91 (1997), 116–121.
[21] F. Klocke, G. Eisenblaetter, Dry cutting-state of research, *VDI Berichte*, No. 1399 (1998), 159–188.
[22] S. Durante, G. Rutelli, F. Rabezzana, Aluminum-based MMC machining with diamond-coated cutting tools, *Surf. Coat. Technol.*, Vol. 94–95 (1997), 632–640.
[23] H.K. Tönshoff, J. Winkler, The influence of tool coatings in machining of magnesium, *Surf. Coat. Technol.*, Vol. 94–95 (1997), 610–616.
[24] R.B. Aronson, Cutting composites, *Advanced Technologies Supplement*, 2006, www.sme.org.
[25] R. Teti, Machining of composite materials, *CIRP Annals*, Vol. 51/2 (2002), 611–634.
[26] J.D. Petrush, S. Olivares, Machining refractory metals, *Manuf. Eng.*, Vol. 132 (2004), No. 4, 89–99.

CHAPTER FOURTEEN

Machining Economics and Optimization

Contents

14.1 Machining Economics

Economic considerations are obviously important in planning a machining process and the economics or cost play a vital role in determining the rate or speed of each machining operation. Taylor in his classical paper on the *Art of Cutting Metals* [1] pointed out that too slow a cutting speed (sufficiently slow as not to cause softening of the tool material) is uneconomic. Similarly, tools that run at too high a cutting speed (which produces rapid wear in a short time) unfeasible for daily shop use for the same reason. This, therefore, suggests that for a given tool/work material combination, the most economical cutting speed should lie somewhere between these two limits. It is obviously practised that before the most economical cutting speed is selected, maximum depth of cut should be used (consistent with the part/machine tool/cutting tool/system requirements) followed by maximum feed consistent with the strength of the tool and part finish requirements. Hence, a more efficient methodology for finding the most productive method of machining is needed for producing a particular part based on different approaches. This methodology should enable the identification of the optimum machine tool, cutting tool materials and geometries, cutting fluids and cutting conditions for each approach, and to rate the different manufacturing chains to determine the best solution.

Apart from the cost of material used, the cost per part is made up of four components: the machining cost, tool cost, tool change and its setup costs, and handling cost. As shown in Fig. 14.1 these costs vary significantly with the cutting speed but at a certain cutting speed the costs reach a minimum which corresponds to the optimum cutting speed.

In general the economics of machining considers three factors, namely: the production cost per component (the unit cost), the production time per component (equivalently the production rate as a reciprocal of the unit time) and the profit rate. Although in practice a high production would probably mean low production costs, it should be pointed out that these two factors must be considered separately because the manufacturing conditions giving maximum production rate will not be identical to those conditions giving minimum cost of production [2]. In this analysis, the *production time* is defined as the average time taken to produce one component, and the *production cost* is defined as the total average cost of performing the machining operation on a component using one machine tool. In general, the production of a component will involve several machining operations using a variety of machine tools (often called *setups*). Consequently, the total cost will be the sum of production costs for each machine setup. Hence, the total manufacturing costs, apart from the cost of the material, involve many items. In case of an NC machine for which the tool

Advanced Machining Processes of Metallic Materials
ISBN: 978-0-08-044534-2

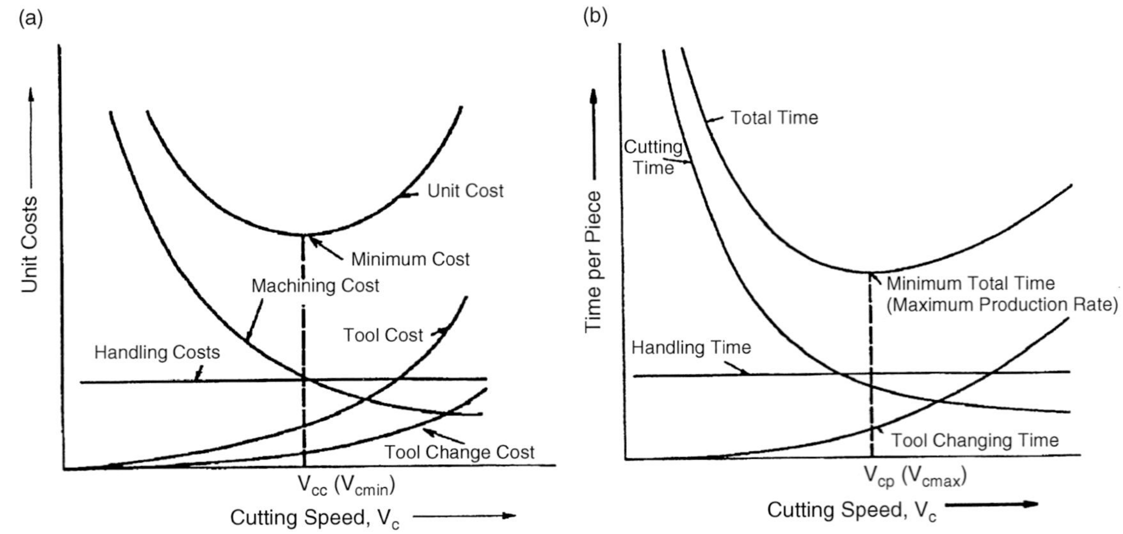

Fig. 14.1 Machining cost (a) and production time (b) for one component vs cutting speed.

change time between consecutive operations is not constant, the total production time for a given machining operation, such as turning, milling, drilling, etc., is given by [3]:

$$t_{pri} = t_{mi} + \frac{t_{mi}}{T_i} t_{li} + t_{cs} + t_{ei} + t_{ri} + t_{ai} + t_{di} + t_{xi} \tag{14.1}$$

where t_{mi} is the cutting time, T_i is the life of the tool used in ith operation and t_{li}, t_{cs}, t_{ei}, t_{ri}, t_{ai}, t_{di} and t_{xi} are the tool's loading/unloading, interchange, magazine indexing (travelling), approach, acceleration, deceleration, and rapid travel location times, respectively, for the ith operation or pass.

In a constant-cutting-speed operation, for example in cylindrical turning, the time spent by the operator and the machine in producing a batch of components N_b can be separated into three items [2]:

1. The non-productive (lead) time, given by the product $N_b t_l$, where t_l is the time taken to load and unload each component and to return the tool to the beginning of the cut.
2. The total machining time, given by the product $N_b t_m$, where t_m is the machining time for the component.
3. The total time involved in changing worn tools, given by $N_t t_c$, where t_c is the tool changing time and N_t is the number of tools used.

By setting the tool costs equal to zero, the following equation for unit time is obtained

$$t_{pr} = t_m + t_l + t_c \frac{N_t}{N_b} = t_m + t_l + \frac{t_m}{T} \tag{14.2}$$

where t_m is the total machining time, t_l is the loading/unloading time, t_c is the tool changing time, N_t is the number of tools used in machining the batch of components N_b ($N_t = N_b t_m / T$). It should be noted that the reciprocal of the production time is termed as the rate of production ($P_r = 1/t_{pr}$).

In order to reduce manufacturing costs, the direct concern should be towards the improvement of machine utilization and the reduction of part lead time. Structurally, machine time and part lead time are composed as follows:

$$\textit{Machine time} = \textit{setup time} + \textit{loading/unloading time} + \textit{cutting time} + \textit{idle time}$$

$$\textit{Part lead time} = \textit{setup time} + \textit{loading/unloading time} + \textit{cutting time} + \textit{waiting time} + \textit{moving time}$$

Consequently, machine utilization and lead time efficiency can be defined respectively:

$$\textit{Machine utilization} = \frac{\textit{Cutting time}}{\textit{Machine time}}$$

$$\textit{Lead time efficiency} = \frac{\textit{Cutting time}}{\textit{Lead time}}$$

Much effort has been put into improving machine utilization and lead time efficiency through the improvement of both software and hardware. Software improvements include better scheduling and CNC control strategies, but are relatively difficult to implement in case of long setup times, batch production environments, and machine flexibility or capability limitations. Hardware improvements, including the development of new machines or machining systems with better versatility and flexibility, can be used to improve both machine utilization and lead time efficiency. Some solutions, termed generally as high performance machining (HPM) or multitasking machining, are discussed in Chapters 15.4 and 15.5. Reducing the setup time, loading/unloading time and cutting time for a part reduces not only the machine and lead time, but also the waiting time for the other parts. In fact, the waiting time for job shop part applications typically accounts for more than 90% of the lead time. For instance, queuing theory indicates that the reduction of some fraction of the part processing time requires the reduction of a higher fraction of the part waiting time. Therefore, it is especially important to assess preliminarily work-in-progress or WIP inventory cost. Hence, economic analysis of all operations of the machine(s) should be incorporated into investment planning.

The cost to produce each component in a batch is given by

$$C_{pr} = C_o t_m + \frac{t_m}{T}(C_t + C_o t_l) + C_o t_c \tag{14.3}$$

where C_o and C_t are the operating cost and tool cost.

Assuming M as the total machine and operator rate (including overheads), the total machine and operator costs can be expressed as the product $M(N_b t_l + N_b t_m + N_t t_c)$ [2]. Thus, in order to calculate the unit cost, the cost of the tools used, given by the product $N_t C_t$, must be added. The average production cost C_{pr} for each component can be expressed as:

$$C_{pr} = Mt_m + Mt_l + M\frac{N_t}{N_b}t_c + \frac{N_t}{N_b}C_t \tag{14.4a}$$

or

$$C_{pr} = Mt_m + Mt_l + M\frac{t_m}{T}t_c + \frac{t_m}{T}C_t \tag{14.4b}$$

In Eqn (14.4b) M can also be defined as the machine operator's wage plus the overhead cost of the machine, Mt_l represents *non-productive* costs which vary depending on loading and fixturing tasks, Mt_m denotes actual costs of cutting operation, and Mt_c is the tool replacement cost shared by all the parts machined. This cost is divided among all the components because each one uses up t_m minutes of total tool life (T), and is allocated t_m/T of Mt_c [4].

The third consideration is the profit rate (P_r) which can be given by the following equation [2]:

$$P_r = \frac{S - C_{pr}}{t_{pr}} \tag{14.5}$$

were S denotes the amount of money which the machine shop receives for each component (income per component), C_{pr} is the average production cost for each component and t_{pr} is the average production time for one component. The profit per component is given by the difference $S - C_{pr}$.

14.2 Optimizing Cutting Speed and Feed Based on Tool-life Equation

This exercise is usually done for a single-tool, single-pass turning operation and the cutting parameters can be optimized from two different points of view:

1. Minimum cost of the operation, and
2. Minimum time of the operation.

The optimum cutting speed for minimum production cost can be derived by substituting the tool life (T) in Eqn (14.4b) by the cutting speed from the Taylor tool-life equation, and differentiating it with respect to the cutting speed and equating the derivative to zero ($\partial C_{pr}/\partial v_c = 0$). Hence [2]:

$$v_{cc} = v_{c\,min} = v_r \left(\frac{n}{n-1} \frac{MT_r}{Mt_c + C_t} \right)^n \tag{14.6}$$

where T_r denotes measured tool life for a given cutting speed v_r and n is Taylor's tool-life exponent.

To find the cutting speed giving maximum production rate (or equivalently minimum production time), it is necessary to follow a similar procedure to that described above for the v_{cc}. The resulting expression for the cutting speed v_{cp}, for minimum production time ($\partial t_{pr}/\partial v_c = 0$) is given by

$$v_{cp} = v_{c\,max} = v_r \left(\frac{n}{n-1} \frac{T_r}{t_c} \right)^n \tag{14.7}$$

Comparison of Eqns (14.6) and (14.7) suggests that both the criteria provide different solutions and the final choice of cutting speed requires a compromise. The cutting speed for maximum production rate (v_{cp}) is always higher than that related to minimum cost (v_{cc}). In Fig. 14.2, the speed range from v_{ce} (the economic cutting speed) to v_{cp} is referred to as the high efficiency machining range or *Hi-E* range. This means that depending on the process demand one could choose any cutting speed within this region and justify it on economic grounds. In practical machining operations, it is convenient to use appropriate expressions for the optimum tool life for minimum cost (T_c), called the economic tool life (T_e), and the optimum tool life for maximum production rate (T_p). The values of T_c and T_p accepted by cutting tool users are equal to 20 and 15 min, respectively. Thus, the value of v_{ce} can be obtained from the $T - v_c$ graph shown in the upper left side in Fig. 14.2, when the economical cutting edge life (T_e) has been determined. Alternatively, T_e can be calculated for a specific feed by means of a modified version of the Taylor formula attached to this graph. When a number of components produced per hour (the production rate calculated from Eqn (14.8)) are plotted against cutting speed, a typical curve with a maximum corresponding to the cutting speed v_{cQ} is found.

$$P_r = \frac{60}{t_p} \left(1 - \frac{t_c}{T} \right) \tag{14.8}$$

where T is tool life, t_c is tool-changing time and t_p is total time per component.

The procedure for optimizing cutting speed and feed for a multi-tool operation typical for the transfer line or a multi-spindle lathe is described in [6]. In general, the optimization can be performed with different strategies for changing tools, i.e. basically whenever the tool life ends for every tool working for the whole cycle time or rarely by changing several tools at once whenever one of them arrives at its tool life. In minimizing production cost C_{pr}, it is usually convenient first to determine all feeds by applying appropriate constraints and next express all cutting speeds by one spindle speed only. In cases where the production rate is fixed, as in transfer lines or dial machines, the economic tool life or maximum production rate should not be used. Instead, the tool cost should be kept as low as possible, i.e. by selecting the cutting conditions

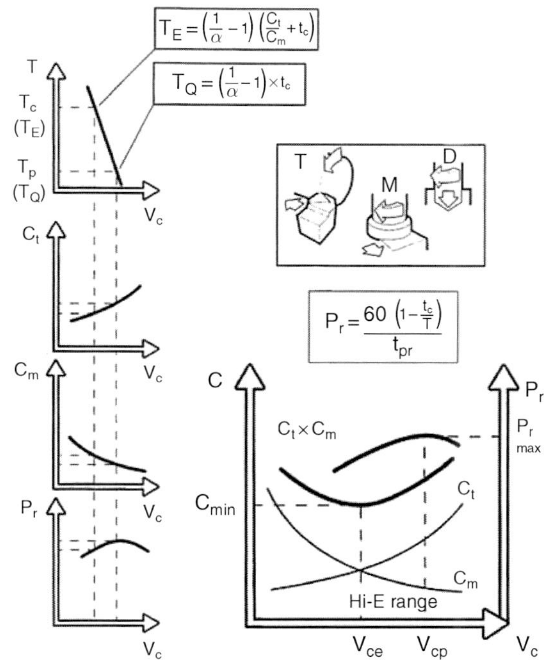

Fig. 14.2 Relationship between cost factors, machining costs and productivity [5]. C – machining cost; C_m – machine charge per minute, i.e. machine, labour and related overhead costs per minute; C_t – cutting tool cost per cutting edge; T – tool life; t_p – total time per piece including machining and downtime related to each piece; t_c – tool changing time for the operation in question; T_E – economic tool life in minutes; T_Q – tool life at maximum production rate; α – slope of $T - v_c$ curve measured from the graph.

most suitable for the required production rate (cycle time). An objective function based on production cost does not consider the influence of production time. Maximum production objectives identify the cutting conditions which best balance the material removal rate (MRR) and tool life to produce the highest output. Minimum cost objectives identify the cutting conditions which best balance the MRR and tool life for the lowest cost. As a result, maximum rate of profit criteria (or maximum efficiency criteria) have been proposed by Wu and Ermer [7] to include and eventually achieve a balance between the contributions of both minimum production cost and total time criteria into the objective function (Eqn (14.5)). However, this criterion predicts the condition for maximum efficiency always lying between the conditions for minimum cost and maximum production time, and that when the profit is zero, the conditions for maximum efficiency become equal to the condition for maximum cost. It also is proven that unless the profit is very high, the optimum condition for maximum profit lies close to the minimum-cost condition.

In fact, both the feed rate and depth of cut affect the tool life and the effect of feed is more important in single cut operations. By differentiating the production cost with respect to the cutting speed and feed, and equating the derivatives to zero one can calculate the optimum cutting speed for any one value of feed, as shown in Fig. 14.3a. As can be seen in Fig. 14.3a, the production cost is smaller for a higher value of feed. Therefore, maximum feed, which gives the desired surface finish and power, may be used. Figure 14.3b provides a methodology for minimum cost per component, when various restrictions apply: requirement of surface finish and dimensional accuracy, available horsepower in machine tool and available speed and feed steps on the machine tool. The minimum possible cost per component is tentatively set at point 1, with the maximum available machine feed f_{max} satisfying the restriction dealing with the maximum feed and maximum speed. With the power restrictions introduced, the conditions are moved to point 2 and later to

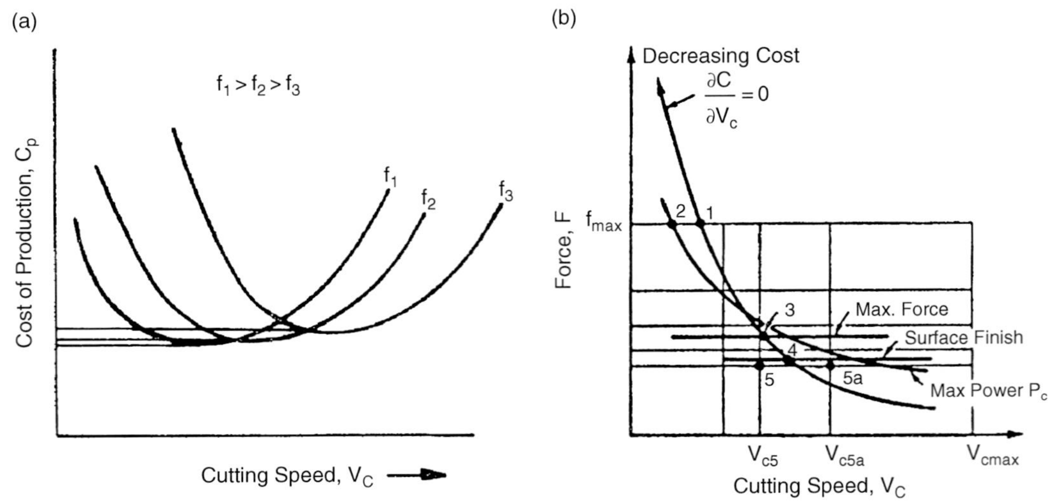

Fig. 14.3 Effect of feed and cutting speed on cost of production (a) and selections of minimum-cost-per-component conditions when various restrictions apply (b) [8, 9].

point 3 to account for the force restriction. The surface finish restriction moves the condition further to point 4, and since the machine tool does not have the required steps for point 4, the next lower feed step is used (point 5) and the available speed steps checked for lower cost (point 5a).

14.3 Advanced Methods of Optimization

The complexity of the optimization problem to be solved depends on the type of machining operations and different optimization techniques are used for single-pass and multi-pass operations. This is because in the first case the problem is aimed at the two variables – the cutting speed and feed – and thus can be represented in two-dimensional space, allowing for an easy graphical illustration. On the other hand, a multi-pass operation, particularly multi-pass turning application, is a four-variable problem in which the number of passes and depth of cut (doc) for each pass must be determined through a dynamic programming procedure, while the optimum cutting speed and feed for each pass is determined using single-pass optimization methods [3]. For a single-pass operation, the objective function is generally non-linear and the various constraints could be linear or non-linear. Several optimization techniques are proposed, namely:

1. *Linear programming method* (LPM) which requires a linear objective function and several linear constraints. Because process constraints are typically expressed as power functions, the mathematical relations are linearized by logarithms. Some special type of LPM is *goal programming* which was developed to solve problems involving complex and usually conflicting multiple objects.
2. *Non-linear programming* methods, classified as deterministic or probabilistic approaches. In the first case the *sequential unconstrained minimization technique* (SUMT), which is a penalty function method, and a logarithmic transformation is applied. A probabilistic approach assumes the tool life to be of a probabilistic nature and the optimum cutting condition can be estimated using a tool-life probability distribution function. For example, one of the probabilistic methods known as *change-constrained programming* proposed by Iwata [3] employs normal distributions for the constants in the constraints formulated. Moreover, *polynomial geometric programming* applying an extended Taylor tool-life equation can be involved into these optimization techniques.

3. *Fuzzy optimization* in which the selection of optimal machining conditions is based on fuzzy set theory. In particular, different uncertainties presented in the empirical models of various machining responses or experimental data (e.g. for tool life, surface finish, cutting forces, chip breakability, etc.) in terms of machining variables can be described and evaluated by the theory of fuzzy sets.
4. *Z map modelling (algorithm)* as a simulation alternative to the solid model using advanced computer graphics [10]. In a conventional Z map, part model or cutting tool immersion volume is detected by regular sampling and the simulation is performed by Boolean operation of the tool movement envelope from the workpiece model. The Z map algorithm instantaneously calculates cutting force using force buffer algorithm, and optimizes the cutting conditions subjected to several constrains. Sometimes, this algorithm can be used for verifying NC programmes by correcting errors or inefficiency of tool paths.
5. *AI-based modelling and expert systems.* This group of approaches includes artificial neural network (ANN), genetic algorithm (GA) and an artificial intelligence (AI) approaches [11]. Artificial intelligence is an area of computer science that attempts to recreate, to some extent, human intelligence 'inside the computer'. Thus, for metal cutting, an AI-based model attempts to recreate the skills and expertise of a manufacturing craftsperson. The desired outcome is that a computer controlled CNC machine tool can be set up and then operate itself without human intervention. In other words, the AI-based modelling asks the simple question [11]: *How does a skilled craftsperson set up a machine tool, choose the cutting conditions, and then monitor the operation for a successful outcome?*

The graph illustrating the methodology of linear programming with the numerous constraints and the feasible region of machining conditions (dashed area) in $v_c(X_1) - f(X_2)$ plane is shown in Fig. 14.4.

Technological constraints included in the mathematical model of single-pass machining operation can be listed in the relevant logarithmic forms (they can be presented graphically as straight lines 1–8 in Fig. 14.4) as follows:

1. Cutting speed and feed rate from the Taylor tool-life equation

$$\ln n_1 + \gamma_T \ln f_1 \leq \ln K_1 \tag{14.9a}$$

2. Cutting power which should be lower than the available power of spindle drive

$$\ln n_2 + \gamma_c \ln f_2 \leq \ln K_2 \tag{14.9b}$$

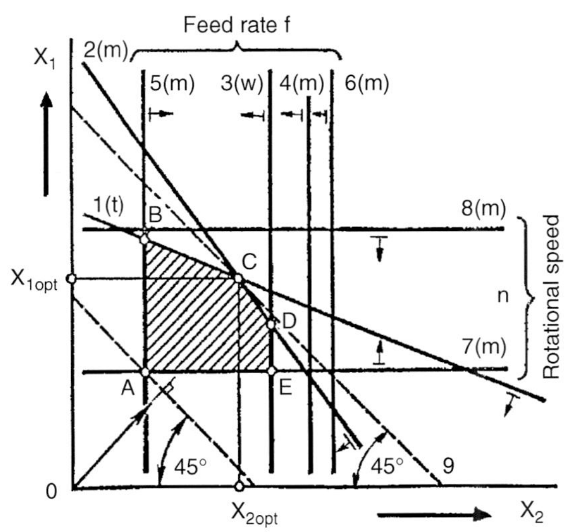

Fig. 14.4 Optimization of cutting conditions using LPM method [12]. m – machine tool; t – tool; w – workpiece.

3. Surface roughness

$$\ln f_3 \leq K_3 \tag{14.9c}$$

4. Strength of feed drive

$$\gamma_f \ln f_4 \leq K_4 \tag{14.9d}$$

5. Minimum feed on the machine tool

$$\ln f_5 \geq K_5 \tag{14.9e}$$

6. Maximum feed on the machine tool

$$\ln f_6 \leq K_6 \tag{14.9f}$$

7. Minimum rotational speed of spindle

$$\ln n_3 \geq K_7 \tag{14.9g}$$

8. Maximum rotational speed of spindle

$$\ln n_4 \leq K_8 \tag{14.9h}$$

The objective function minimizes production cost (machining time $t_m = L/(f \times n)$ in such a way that

$$F_o = \ln n + \ln f \rightarrow \max \tag{14.9i}$$

$$F_o = X_1 + X_2 \rightarrow \max \tag{14.9j}$$

After decoding, one can obtain optimum values of speed and feed corresponding to point C in Fig. 14.4, as below:

$$n_{opt} = e^{X_{1opt}},\ f_{opt} = e^{X_{2opt}} \tag{14.10}$$

where e is a base of natural logarithm.

A realistic optimization should consider the numerous technological and practical constraints, which limit the feasible domain for the selection of optimum cutting conditions. In modern machining systems there may be many constraints, different than those selected in Fig. 14.4. For example, the model-based economic optimization procedure shown in Fig. 14.5 includes the following set of constraints on the optimization: chip breakability (C1, C1′ and C2″), tool geometry and stock allowance (C2–C4), surface finish (C5), chatter limits (C6), maximum operation time per part t_{max} (C7 and C7′), maximum rotational speed N_{max} (C8), maximum spindle motor power P_{lim} (C9), and force limits (C10). Similarly to the LPM, a feasible region from which optimal operating conditions may be selected can be established and finally an optimal set of conditions can be chosen. The feasible feed, depth of cut and cutting speed space for a particular cutting operation is the space that satisfies all the constraints selected. When the cutting speed and the depth of cut are specified, the feasible domains in the f–a_p or v_c–f planes are given inside and on closed lines as shown in Figs 14.5a and b, respectively. Each segment of the closed lines represents a limit due to one constraint. The point of tangency M_{opt} between the surface of constant cost and the boundary of feasible space will locate at a coordinate of large depth of cut ap_{opt}, large feed f_{opt} and intermediate cutting speed vc_{opt}. In Fig. 14.5a, this point is placed at the upper right corner $M_v(f_{opt}, ap_{opt})$ and at a point $M_d(vc_{opt}, f_{opt})$ on the upper boundary of the respective feasible domain. On the other hand, the condition of minimum cost for v_c–f space (Fig. 14.5b)

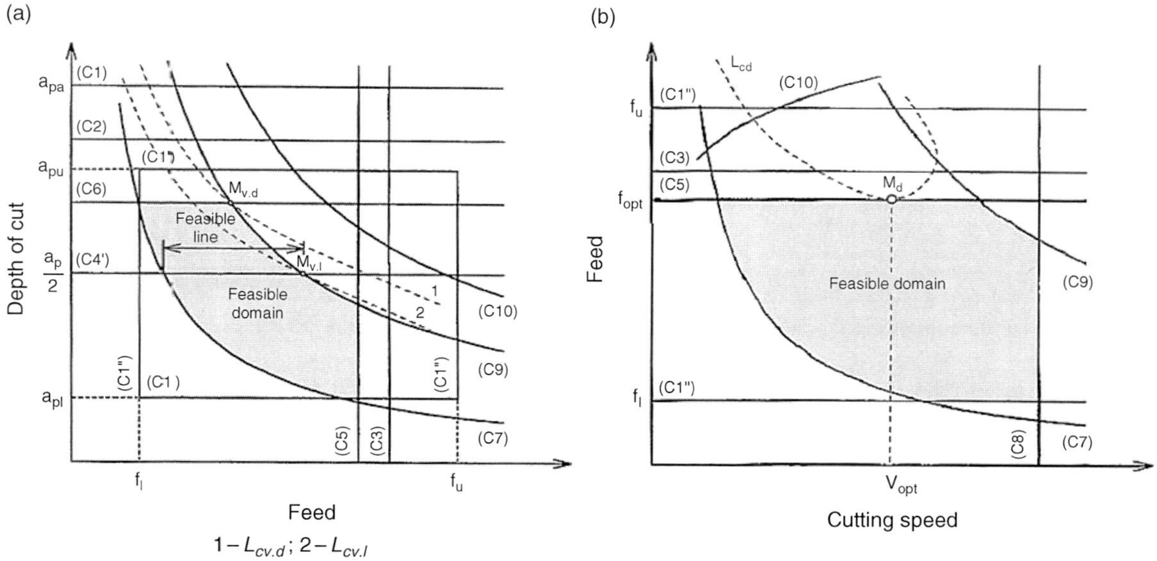

Fig. 14.5 Constraints and feasible regions of machining conditions in (a) (f–a_p) and (b) (v_c–f) planes [13, 14]. a_{pl} and a_{pu} – lower and upper limits of doc; f_l and f_u – lower and upper limits of feed.

corresponds to the optimum point $M_d(vc_{opt},f_{opt})$. The lines of minimum cost L_{cv} and L_{cd} are respectively drawn in both figures.

The selection of optimal cutting parameters in multi-stage rough and finish operations for maximizing production rate or minimizing production cost can be performed using a machining model based on a polynomial network. The polynomial network is a self-organizing adaptive modelling tool with the ability to construct the relationships between input variables and output feature spaces [13]. As a result, the polynomial network can learn the relationships between the three cutting parameters (v_c,f, a_p) and cutting performance characterized by surface roughness, cutting force and tool life using, for instance, a sequential quadratic programming method (being the most effective non-linear programming method) as an optimization algorithm. The flow chart for the optimization of multi-stage turning operations is shown in Fig. 14.6. As can be seen in Fig. 14.6 this method required geometrical drawing of a part with the area removed by repetitive rough and finish cycles, and experimental data on surface roughness, cutting force and tool life to train the polynomial networks and construct a machining model.

Nowadays, machining data selection is an important component of the CAD/CAM system and plays a very important role in the efficient utilization of machine tools and influences the product quality. For this purpose, a rule-based expert system and materials database have been incorporated into many CAD/CAM systems in order to obtain optimal machining parameters. It is suggested [15] that fuzzy logic is an efficient tool to describe the strategy and action of the skilled operator when selecting the metal cutting data. Generally, a fuzzy logic system accepts input or some inputs and then passes it/them to a process called fuzzification. In the fuzzification process, the input data undergoes some translation into linguistic quantity such as *very soft, soft, medium, hard, very hard* in Fig. 14.7a. The translated data will be sent to an inference mechanism that will apply the predefined rules and will generate the outputs in linguistic form. The linguistic output will go through a defuzzification process and then turn out again to be in numerical form. One of the first fuzzy models for metal cutting application was developed to evaluate the relationships between material hardness (input) and the corresponding recommended cutting speed and feed rate (output). After establishing the fuzzy rules (for example, *if material is very hard, then speed is very slow*) and developing the control algorithm, the second step is to partition the discretized universes of the input and output using a triangular shape

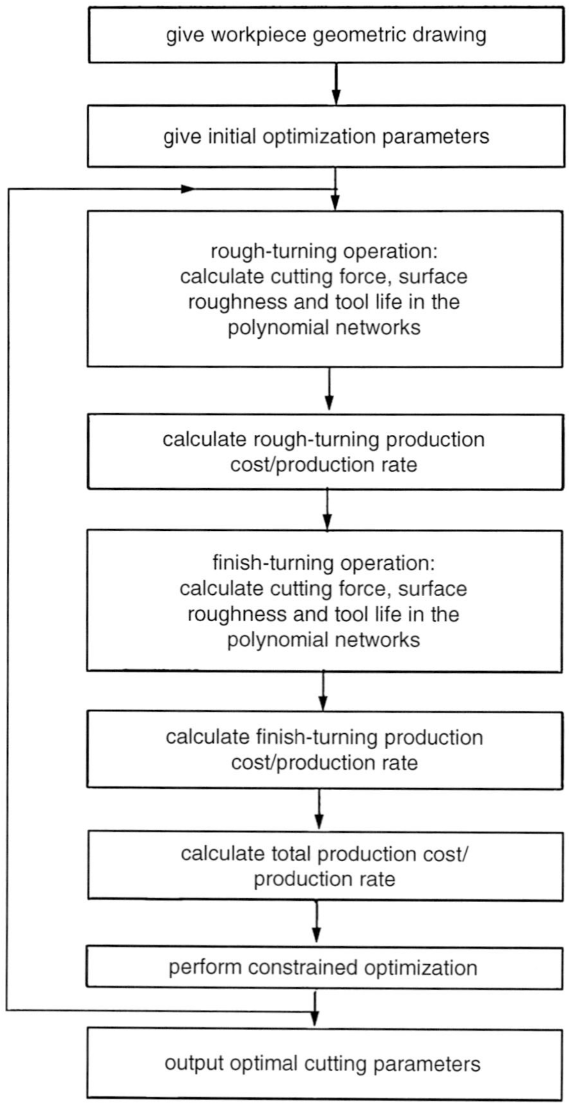

Fig. 14.6 Flow chart for the optimization of multi-stage turning operations [7].

(in Fig. 14.7 in the range 0–8). Figure 14.7a and b shows the material hardness membership (input) and the cutting speed membership (output) functions μ. As an example, the results of fuzzy logic-based optimization of the cutting speeds for different cutting tools when machining a carbon steel with variable hardness is shown in Fig. 14.8.

A fuzzy expert system for the design of turning operations, named SMART (Smart Assistant to Machinists), consisting of four modules: a database, a tool selection module, a cutting condition design module and a learning module, is shown in Fig. 14.9. The database consists of four data files: work material data file, machine tool data file, machining plan data file, which defines desirable MRR, surface finish and cutting parameters, and cutting tool data file. The cutting tool selection module is developed based on fuzzy logic, in which the tool selection is performed in three steps: in the first the input information is 'fuzzyized', in the

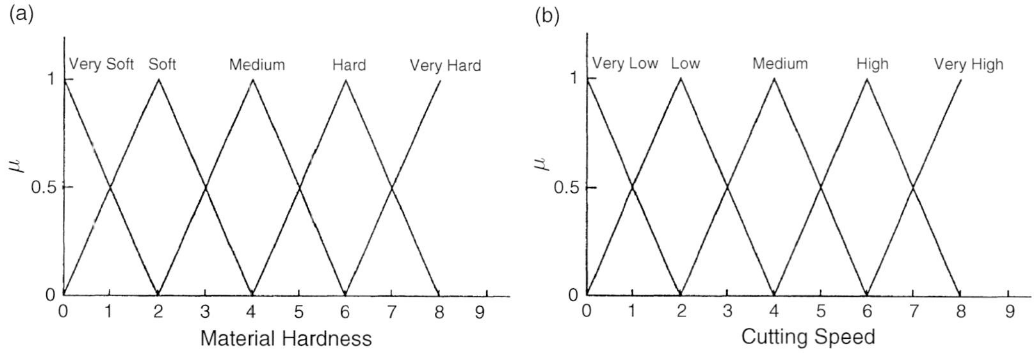

Fig. 14.7 Membership functions for (a) material hardness and (b) cutting speed [15].

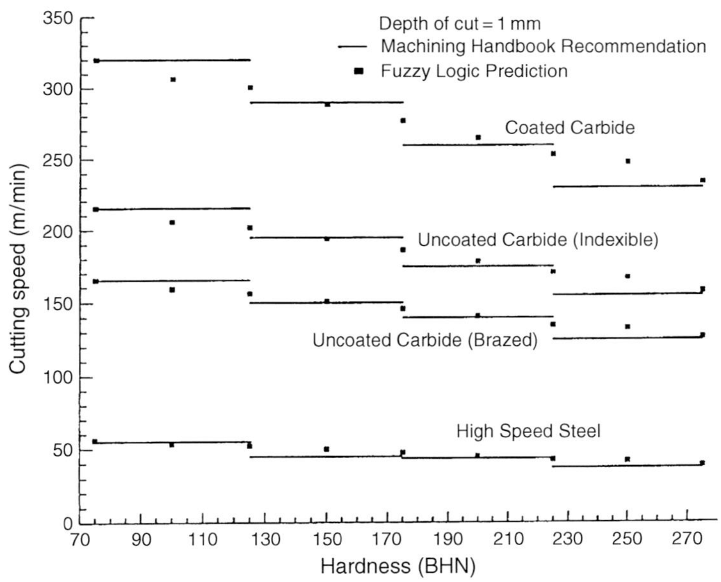

Fig. 14.8 Fuzzy logic model predicted cutting speeds for a carbon steel at 1 mm depth of cut [15].

second cutting tool material grades and cutting tools are selected from appropriate product catalogues (for example, the world-leading tool manufacturers – Sandvik Coromant and Kennametal) using the fuzzy correlation functions, and then the selected tool materials and tools are searched against the tool data file to check the ability. The cutting condition design module is developed on the base of fuzzy non-linear programming and user interactive graphics. The learning module allows the user to fine-tune the fuzzy functions and to further improve the system performance. For instance, for tool selection, when a qualitative term is input, such as 'finishing' for machining type, a fuzzy membership function is assigned after the manner:

$$\mu(MT_2) = 0.8/MT_1 + 1.0/MT_2 + 0.8/MT_3 + 0.4/MT_4 + 0.0/MT_5$$

where MT_1 is extreme finishing, MT_2 is finishing, MT_3 is light finishing (semi-roughing), MT_4 is roughing and MT_5 is heavy roughing type of machining operation. In SAM's (Smart Assistant to Machinists) system,

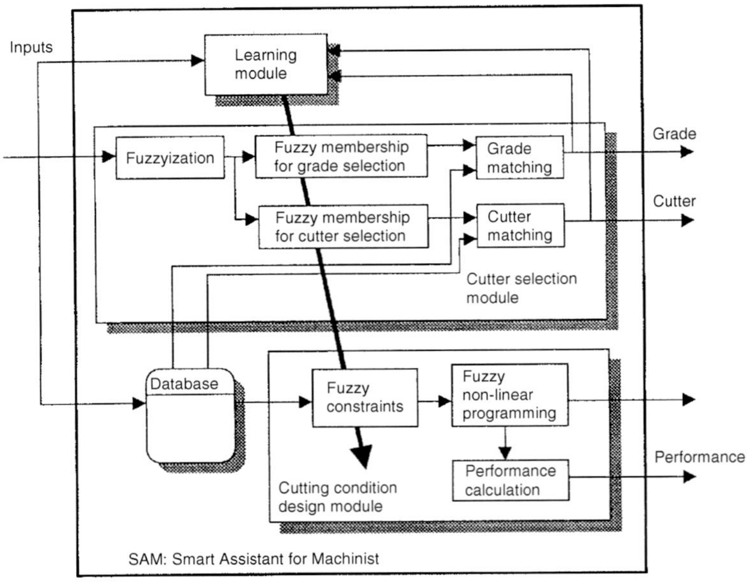

Fig. 14.9 A fuzzy expert system for the design of cutting operation [16].

over 100 functions of an element applicability to input variables are defined, based on metal cutting principles and various tools manuals, handbooks and technical reports.

Fang and Jawahir [17] elaborated a methodology for predicting machining performance in finish turning, termed the total machining performance (TMP), using integrated TMP knowledge base and integrated fuzzy-set models for six machining variables, including surface finish, tool wear rate (*VB*, *VB′*, *KT*, *N* wear indicators), dimensional accuracy, cutting power and chip breakability.

It was considered that all elements of the TMP (except the chip breakability which is already a membership function) need to be assigned to a membership function, ranging from 0 to 1, in order to quantify their variations in terms of the relevant fuzzy linguistic values specified in Fig. 14.10b. Assuming the minimum, average and maximum values of R_a roughness parameters for a given cutting tool under typical finishing conditions used in turning, membership functions for other elements of the TMP system are shown in Fig. 14.10a. A fuzzy exponential function was selected to quantify the variations of the membership values for each element of machining performance. Further, a series of modified membership values is calculated in order to consider the differences between the input machining conditions (cutting parameters, tool geometry, workpiece hardness, carbon content and chip-breaker type) and those collected in the machining reference database. They were transferred back to the actual TMP values through the corresponding membership functions, and finally a set of optimal TMP values was predicted.

The manufacturing database of a typical CIM system includes a large database (termed as a computerized optimization system (COS)), which not only stores machinability data but also allows the optimization to be performed at different levels [3]. The first level can be utilized to select the optimum cutting condition based on economic criteria, select appropriate cutting tools, determine the type of operation and select the most suitable machine tool from the available inventory. The second level is capable of optimizing the manufacturing process and sequence of operations involved on a single or multi-stage machining system. It uses detailed part geometry, machine motion and tooling data files to calculate individual machine motions and cutting times based on the optimum cutting conditions identified on the first level of the COS. The third level is used to determine the optimum combination of all machining parameters including the machine tool, cutting tool material and geometry, and cutting fluid. So far, such approaches as an expert system and artificial neural networks (ANNs) can support the activities involved in third-level COSs. Figure 14.11 presents the

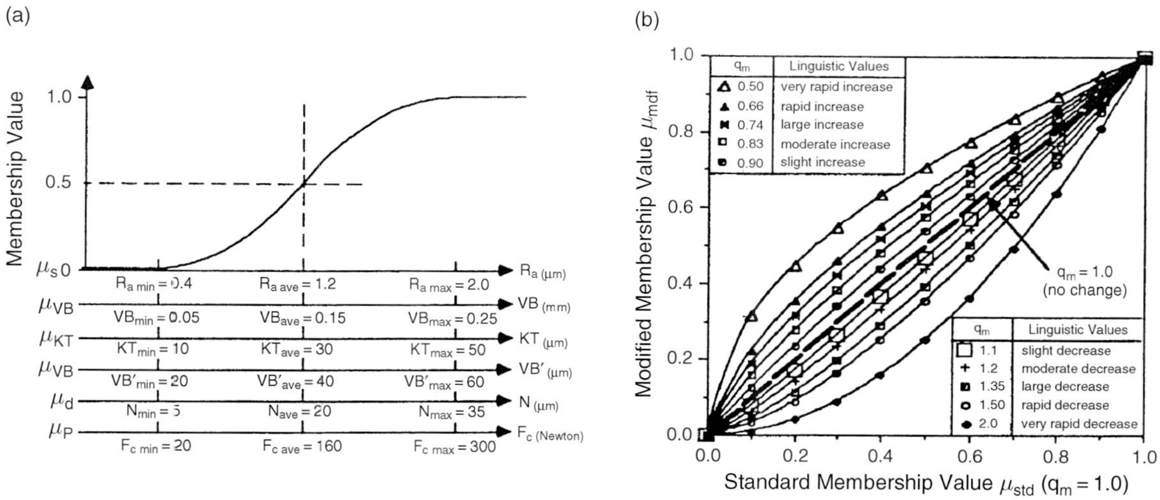

Fig. 14.10 Membership functions for the TMP (excluding the chip breakability) used in finish turning (a) and the linguistic value function (b) [17].

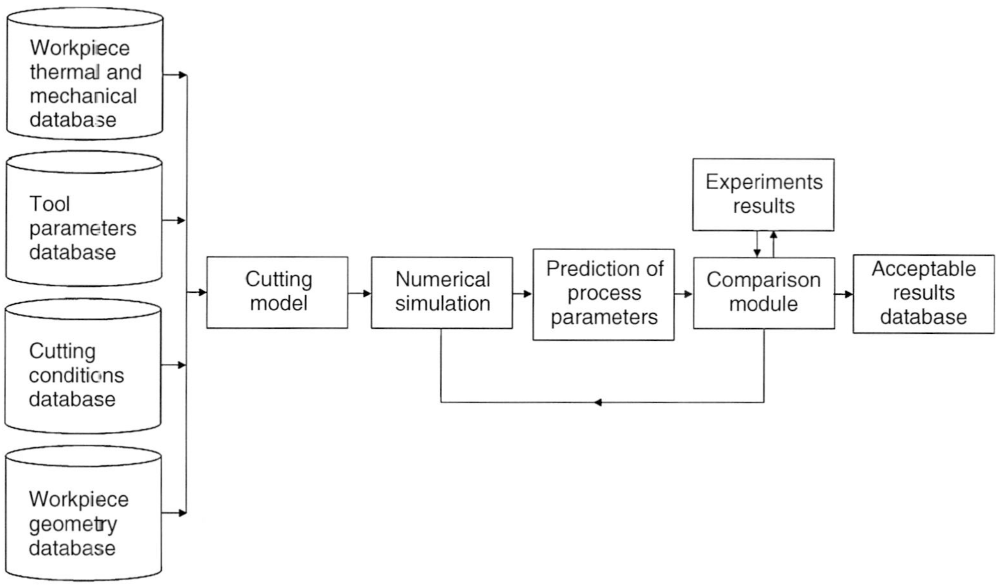

Fig. 14.11 The structure of the computer aided modelling and database systems [18].

structure of the computer aided cutting process evaluation and parameter selection which can be classified as the first level of COS. The computer selection system has an integrated database for work materials, machine tools, cutting conditions and cutting parameters, and workpiece geometry modules, but without any constraints. In particular, the computer simulation models included allow for interactive prediction of some important machining characteristics, such as cutting forces, cutting energy, cutting temperature and comparing them with experimentally obtained values for high speed milling in the aerospace industry.

References

[1] E.G. Thomsen, F.W. Taylor – A historical perspective, in *On the Art of Cutting Metals – 75 years later*, *PED*, Vol. 7 (1982), 1–12.
[2] G. Boothroyd, W.A. Knight, *Fundamentals of Machining and Machine Tools*, CRC Press, Boca Raton, 2006.
[3] D.A. Stephenson, J.S. Agapiou, *Metal Cutting Theory and Practice*, Marcel Dekker, New York, 1997.
[4] P.K. Wright, *21st Century Manufacturing*, Prentice Hall, Upper Saddle River, 2001.
[5] *Modern Metal Cutting. A Practical Handbook*, Sandvik Coromant, Fair Lawn, 1996.
[6] G. Tlusty, *Manufacturing Processes and Equipment*, Prentice Hall, Upper Saddle River, 1999.
[7] D.S. Ermer, A century of optimizing machining operations, *J. Manuf. Sci. Engng*, Vol. 119 (1997), 817–822.
[8] B.L. Juneja, G.S. Sekhon, *Fundamentals of Metal Cutting and Machine Tools*, John Wiley & Sons, New Delhi, 1987.
[9] E.J.A. Armarego, R.H. Brown, *The Machining of Metals*, Prentice Hall, Inc., New Jersey, 1969.
[10] M. Kang, S.-K. Lee, S.-L. Ko, Optimization of cutting conditions using enhanced Z map model, *CIRP Annals*, Vol. 51/2 (2002), 429–432.
[11] E.M. Trent, P.K. Wright, *Metal Cutting*, Butterworth-Heinemann, Boston, 2000.
[12] W. Grzesik, *Fundamentals of Machining of Metallic Materials* (in Polish), WNT, Warsaw, 1998.
[13] B.Y. Lee, Y.S. Tarng, Cutting-parameter selection for maximizing production rate or minimizing production cost in multistage turning operations, *J. Mater. Proc. Technol.*, Vol. 105 (2000), 61–66.
[14] T.H.C. Childs, K. Maekawa, T. Obikawa, Y. Yamane, *Metal Machining. Theory and Applications*, Arnold, London, 2000.
[15] M.A. El Baradie, A fuzzy logic model for machining data selection, *Int. J. Machine Tools Manuf.*, Vol. 37 (1997), 1353–1372.
[16] Y. Chen, A. Hui, R. Du, A fuzzy expert system for the design of machining operations, *Int. J. Machine Tools Manuf.*, Vol. 35 (1995), 1605–1621.
[17] X.D. Fang, I.S. Jawahir, Predicting total machining performance in finish turning using integrated fuzzy-set models of the machinability parameters, *Int. J. Prod. Res.*, Vol. 32 (1994), 833–849.
[18] A. Kaldos, I.F. Dagiloke, A. Boyle, Computer aided cutting process parameter selection for high speed milling, *J. Mat. Proc. Technol.*, Vol. 61 (1996), 219–224.

CHAPTER FIFTEEN

Advanced Machining Processes

Contents

15.1 High Speed Machining

15.1.1 Definition of basic features of HSM

High speed machining (HSM) or more precisely high speed cutting (HSC) has been of special interest to both manufacturing and academia sectors for many years. One of the first most systematic researches on HSM was the AMRP (Advanced Manufacturing Research Program) which was started in 1979 in the USA sponsored by the Defense Advanced Research Project Agency (DARPA) [1]. In general, these research activities were focused on turning and milling of aluminium, nickel-based, titanium and ferrous alloys. In the last decade, there have been many significant developments in high speed machining relative to machining centres and controls, tooling and CAD/CAM systems. These advances in cutting technology are helping manufacturing companies lower costs, shorten delivery times, manufacture complex, high quality parts and accelerate product development cycles.

High speed machining is a difficult term to define due to a number of decisive factors including the properties of the material being machined, the availability of suitable tool materials, ranges of speeds and feeds of available machine tools. Basically, four different options in defining HSM can be distinguished, namely: cutting operation-, work material-, process stability- and cutting force- or equivalently cutting energy-oriented ones [2–4], as illustrated in Figs 15.1, 15.2 and 15.3, respectively.

The material removal rate is a function of both undeformed chip cross-section and cutting speed as the specific characteristics of the cutting technology and the workpiece material properties. Figure 15.1 shows estimated cutting speed ranges for seven typical machining operations (a) and a variety of construction materials of different machinability rating. From Fig. 15.1, it is evident that the highest attainable machining speeds are recorded for easy-to-machine materials, such as aluminium and fibre-reinforced plastics and they

Advanced Machining Processes of Metallic Materials
ISBN: 978-0-08-044534-2

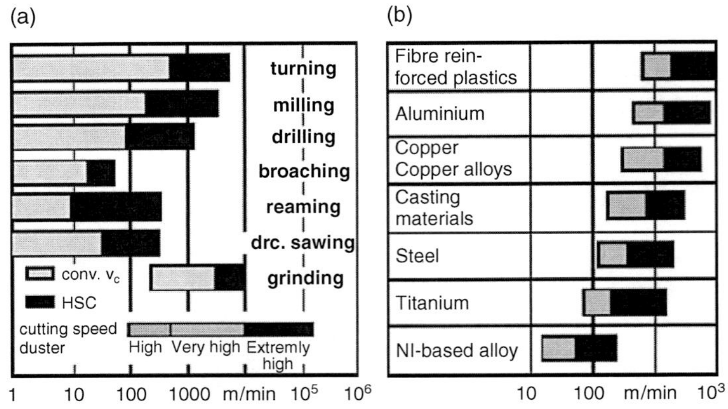

Fig. 15.1 Achievable cutting speed depending on the type of machining operation (a) and work material (b) [2].

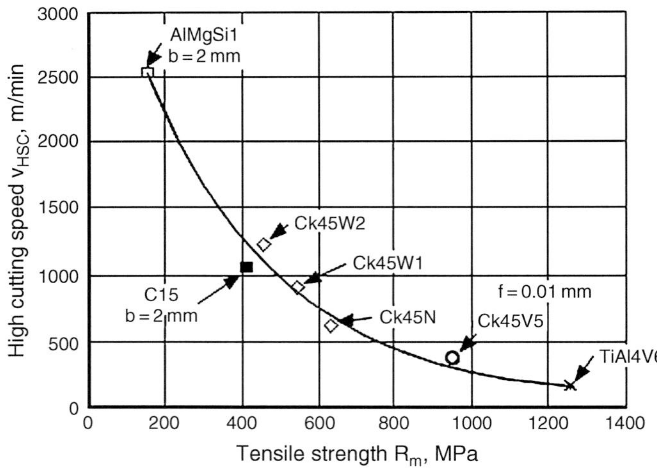

Fig. 15.2 The magnitude of the high cutting speed versus the ultimate tensile strength (UTS) for different work materials [3]. Cutting conditions: $b = 3$ mm; $f = 0.1$ mm/rev; cutting tool: HC P30–P40 coated with Ti(C,N)/Al_2O_3; tool geometry: $\alpha_{ff} = 6°$, $\gamma_{ff} = -6°$, $\varepsilon_r = 90°$.

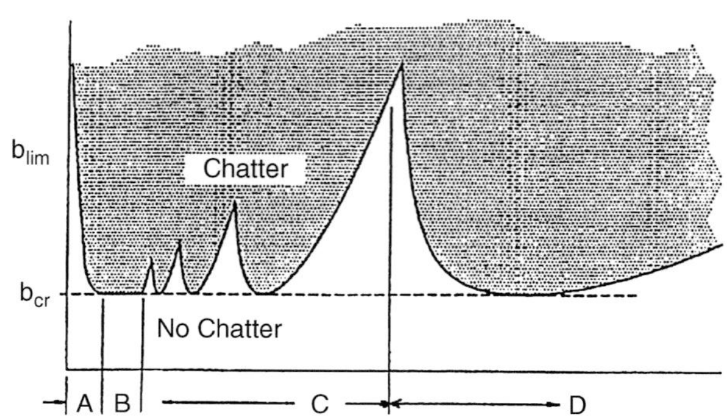

Fig. 15.3 Stability lobe diagram with spindle speed ranges [4].

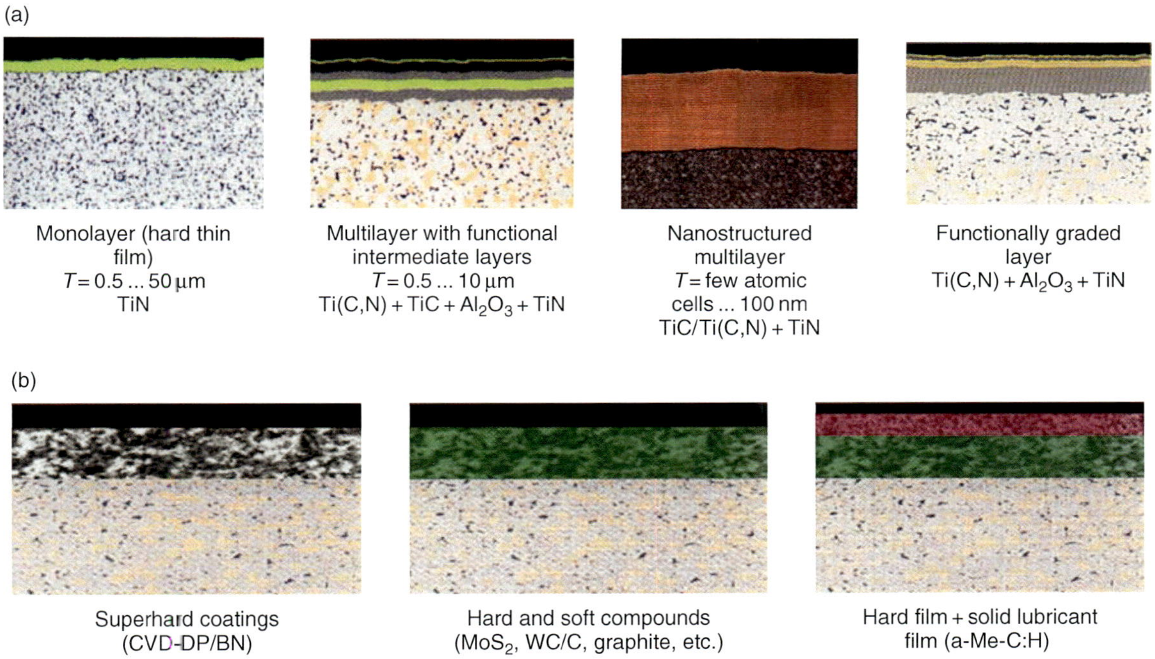

Plate 1 Material-based classification of cutting tool coatings. (a) Ti-based mono- and multilayer coatings, (b) superhard and composite coatings [11]. (see Fig. 4.20, p. 43)

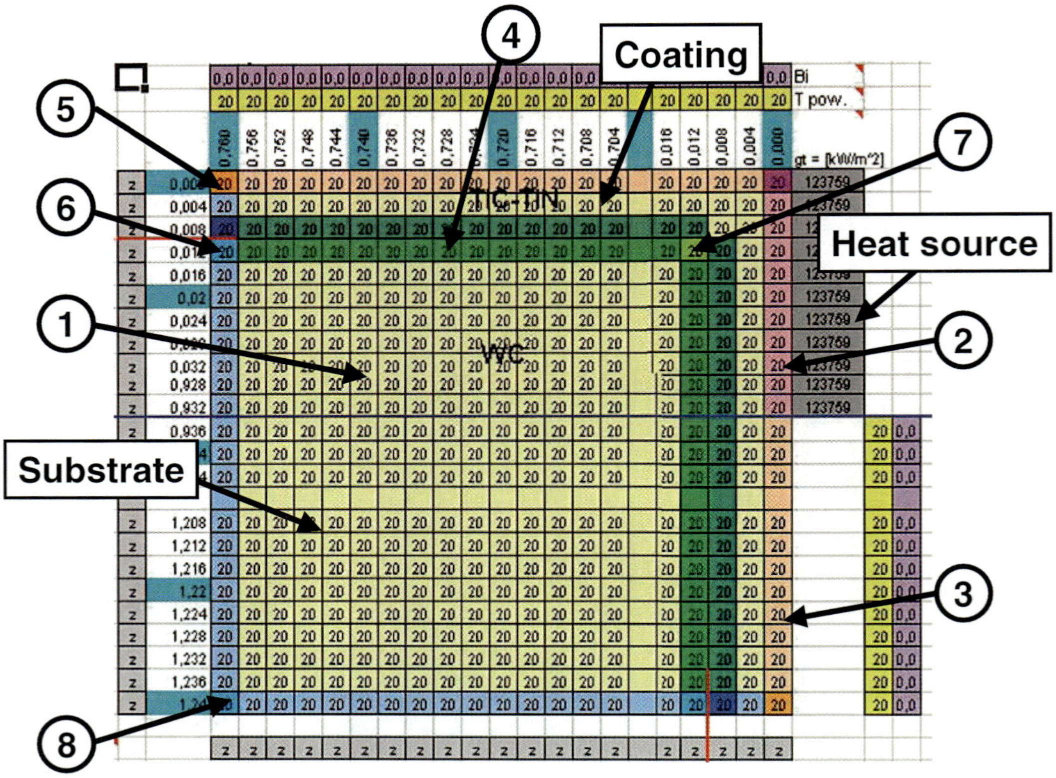

Plate 2 FDM mesh model of heat transfer into coated tool with marked boundary conditions [36]. (see Fig. 5.11, p. 60)

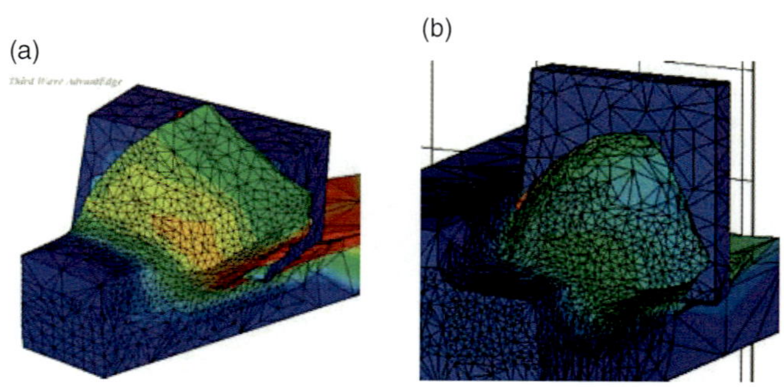

Plate 3 Three-dimensional FEM simulation of oblique cutting with flat rake face tool (a) and nose turning with non-constant chip load modelled (b) [29]. (see Fig. 7.22, p. 100)

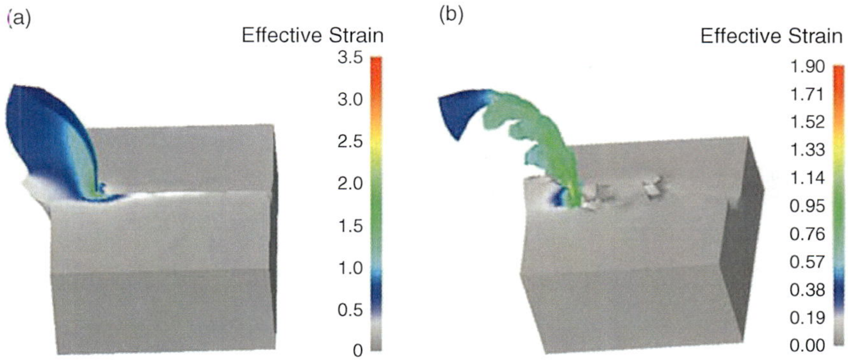

Plate 4 Result with only continuous adaptive remeshing (a) and result with modelling of fracture (b) [30]. (see Fig. 7.23, p. 101)

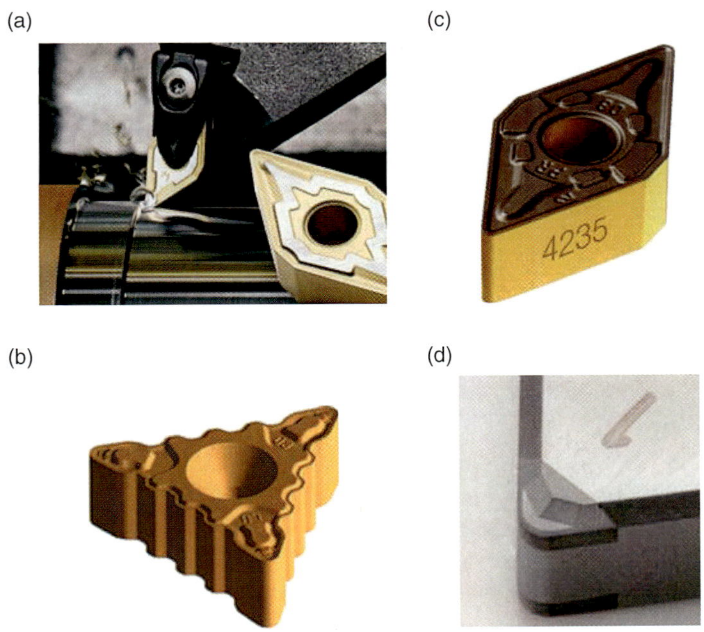

Plate 5 New designs of chip breakers for coated carbide (a–c) and CBN (d) inserts [33]. (see Fig. 7.29, p. 106)

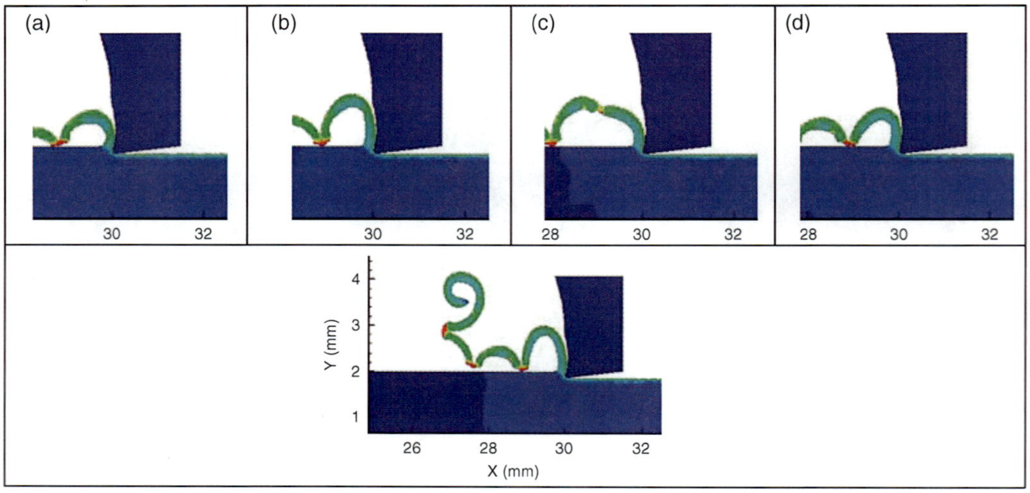

Plate 6 Simulation of chip breaking when turning AISI 1045 carbon steel using FEM AdvantEdge package. Chip breaker with land length $l_{RC} = 0.08\,\text{mm}$, width $B = 0.8\,\text{mm}$ and depth $d = 0.1\,\text{mm}$, tool coating type TiN/Al_2O_3/TiC+Ti(C,N), cutting speed $v_c = 220\,\text{m/min}$, depth of cut $a_p = 1\,\text{mm}$, feed $f = 0.14\,\text{mm/rev}$. (see Fig. 7.34, p. 110)

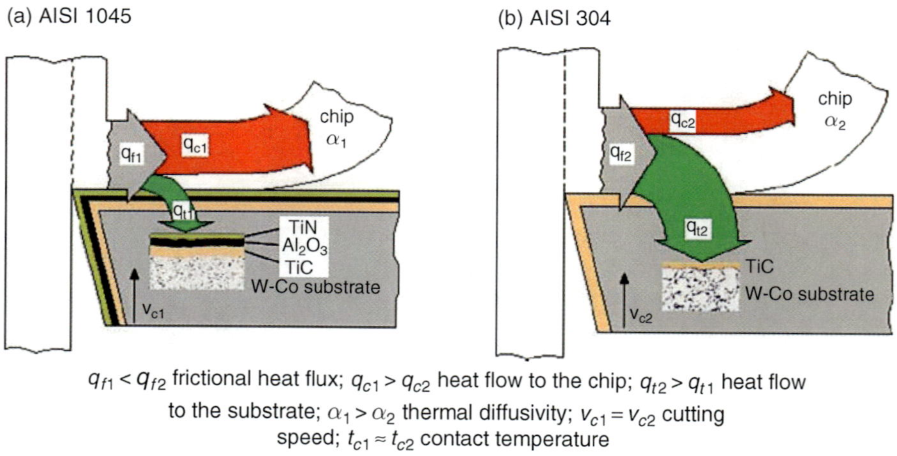

Plate 7 Distribution of heat generated for material with high (a) and low (b) thermal diffusivity. (see Fig. 9.4, p. 130)

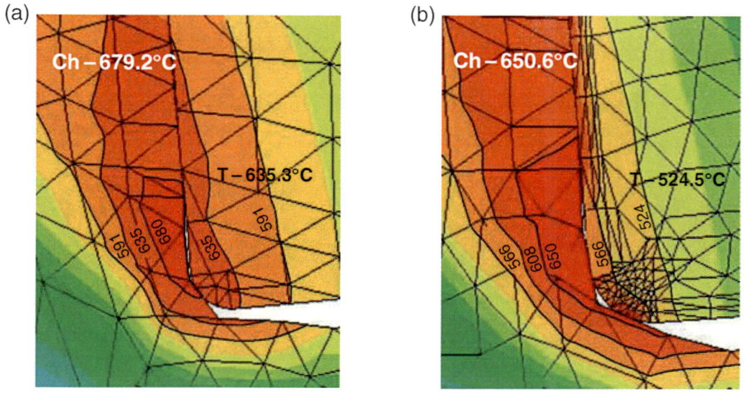

Plate 8 Magnification of temperature distribution in the vicinity of the cutting edge. (a) P20 uncoated carbide; (b) TiC/Al_2O_3/TiN coated carbide [8]. Workpiece – AISI 1045 steel. (see Fig. 9.9, p. 134)

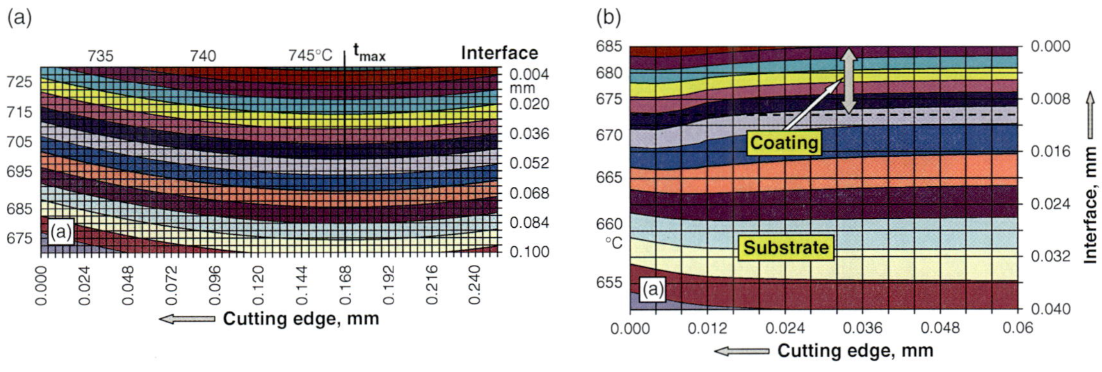

Plate 9 Temperature field for a P20-AISI 1045 (a) and for a TiC/Al_2O_3/TiN-AISI 1045 pair (b) [12]. Heat flux into the tool: $q_T = 29.08\,\mathrm{MW/m^2}$ (a) and $q_T = 22.23\,\mathrm{MW/m^2}$ (b). The reference cutting conditions: $v_c = 145\,\mathrm{m/min}$, $f = 0.16\,\mathrm{mm/rev}$ and $a_p = 2\,\mathrm{mm}$. (see Fig 9.10, p. 135)

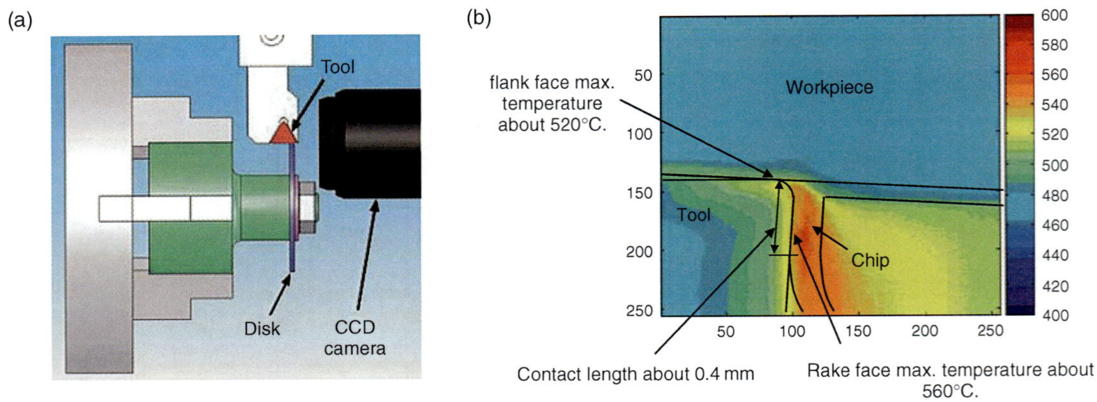

Plate 10 Thermal map obtained by means of IR CCD camera. Courtesy of J. Rech. Work material: INOX 316L steel, cutting tool: carbide insert TPKN coated with TiN; cutting conditions in orthogonal cutting: $v_c = 60$ m/min, $f = 0.1$ mm/rev. (see Fig. 9.15, p. 138)

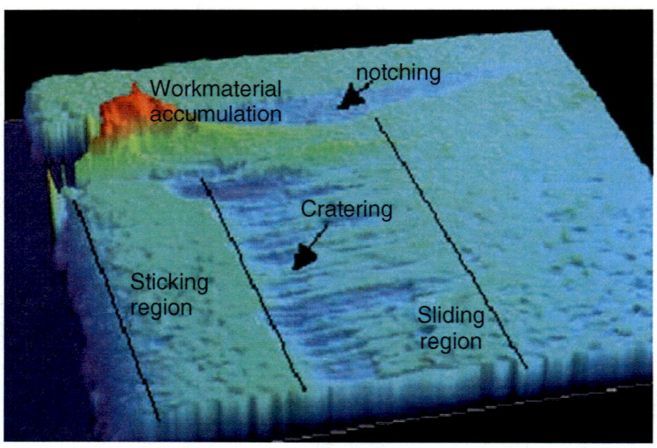

Plate 11 A typical wear pattern and types of wear on the 3D rake face of a carbide insert in machining of Inconel 718 obtained by a Veeco NT 100 (Wyko®) optical profiler. Courtesy G. Poulachon. (see Fig. 12.18, p. 178)

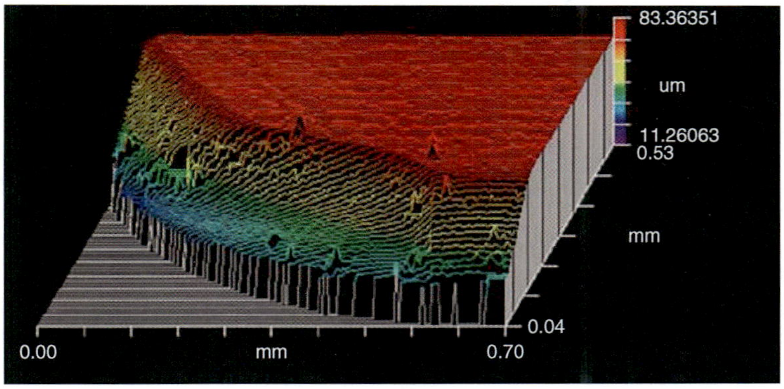

Plate 12 Worn CBN cutting tool data from a white light interferometer [23]. (see Fig. 12.19, p. 178)

I.S.O.
Classification System

ISO Code	Material
P	Steels
K	Cast Irons
M	Stainless Steels
N	Non-Ferrous Metals
H	Hardened Materials
A	High Temp Alloys

Plate 13 ISO classification of basic groups of construction materials. (see Fig. 13.4, p. 185)

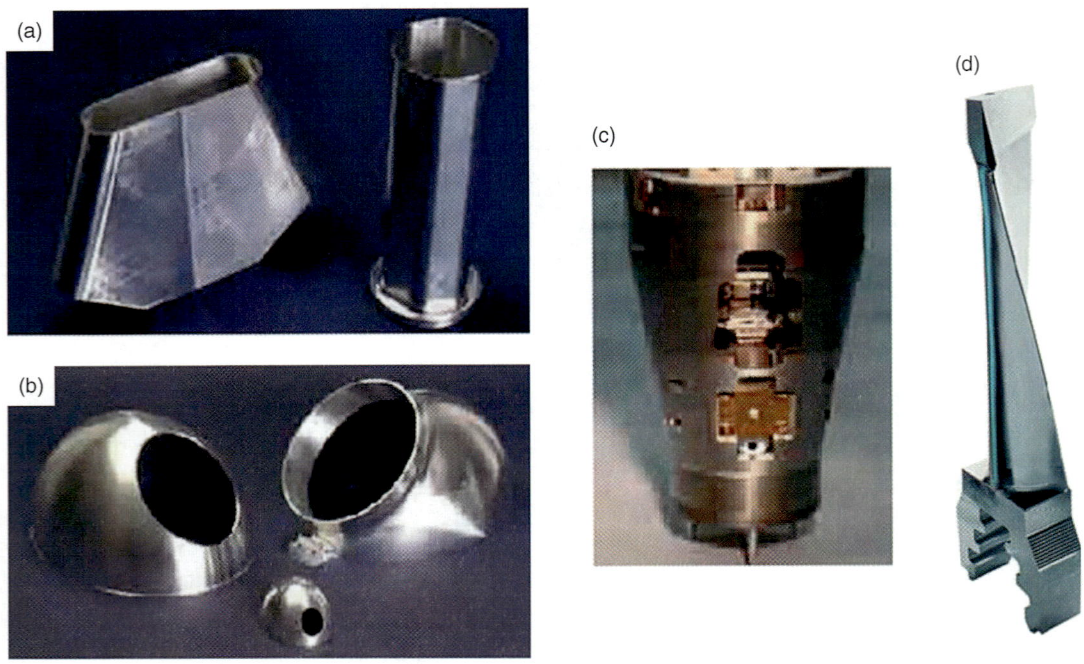

Plate 14 Examples of monolithic components machined at high speeds [21, 22]: (a) deep-pocketed aluminium parts; (b) thin-walled aluminium pressure vessels for neutron reactor; (c) the STM microscope body made of pure molybdenum; (d) turbine blade made of titanium. (see Fig. 15.17, p. 224)

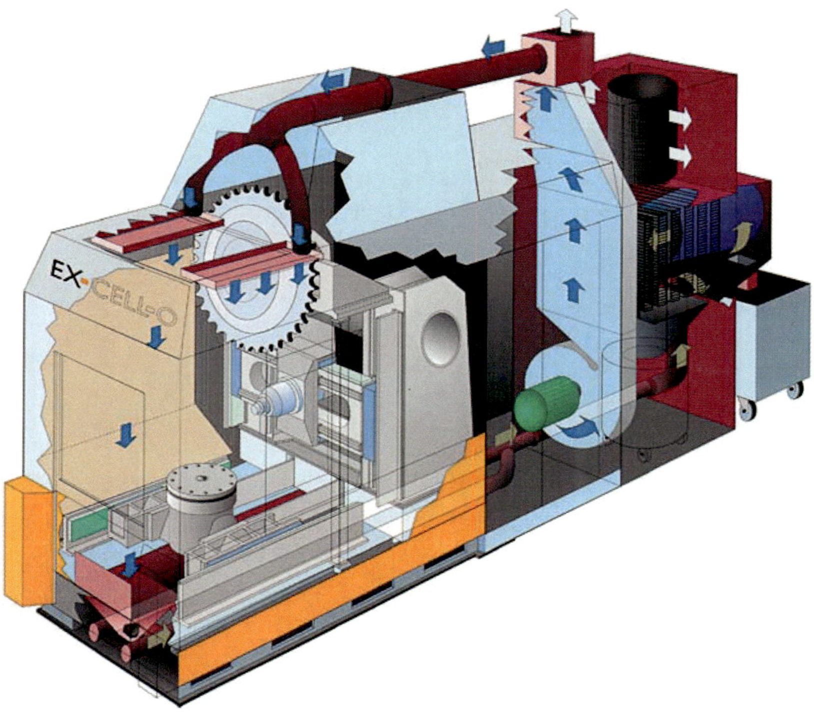

Plate 15 Horizontal machining centre equipped with dry and MQL machining systems [3, 6]. (see Fig. 15.23, p. 230)

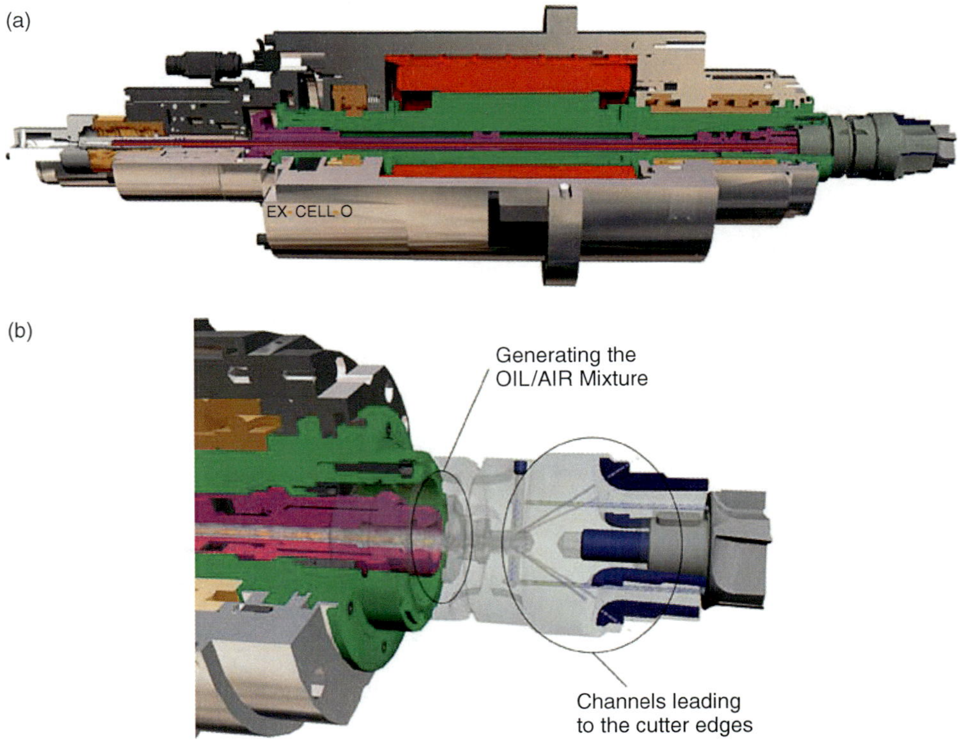

Plate 16 Construction details of spindle with two-channel supplying system [3]: (a) cross-section of the spindle, (b) near-dry machining with MQL channels. (see Fig. 15.24, p. 231)

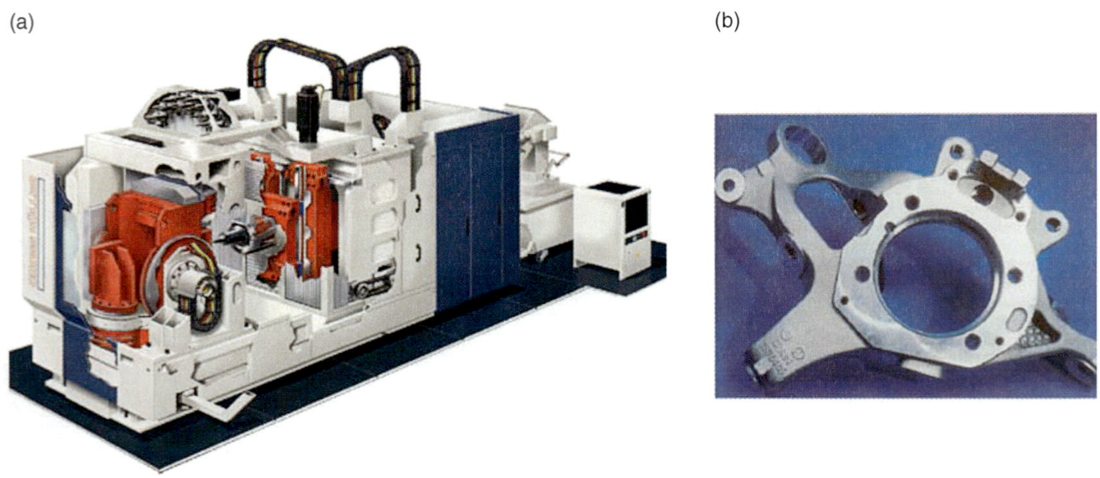

Plate 17 Dry machining centre with MQL supply (a) and steering knuckle machined on it (b) [3, 14]. (see Fig. 15.39, p. 241)

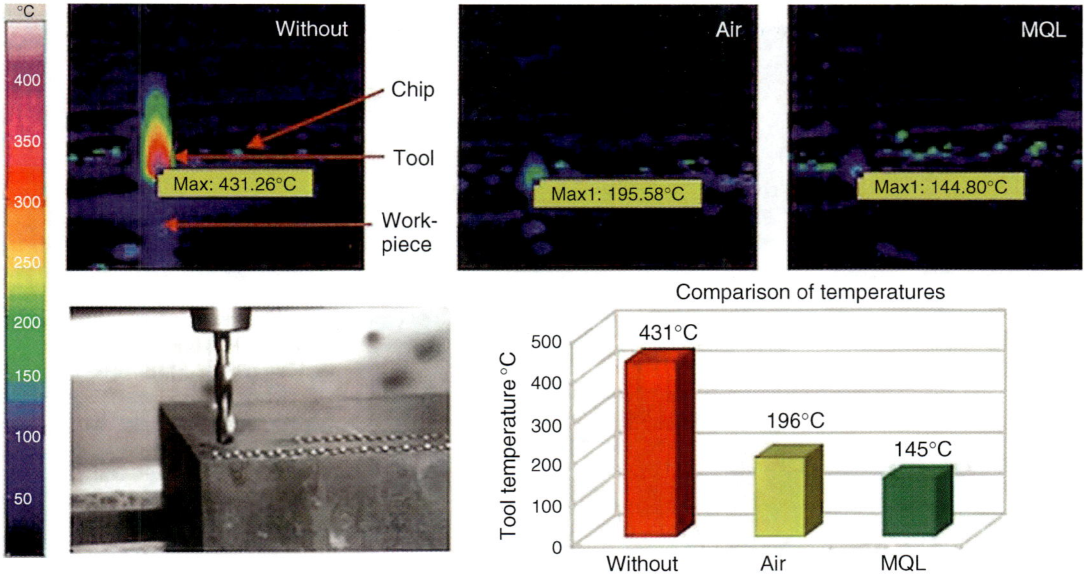

Plate 18 Comparison of cutting temperatures recorded by thermographic camera in drilling at different cooling systems [9]. Workpiece material: GGG40 nodular iron, drill diameter of 8.5 mm, hole length of 42 mm, cutting conditions: $v_c = 130$ m/min, $f = 0.26$ mm/rev. (see Fig. 15.40, p. 242)

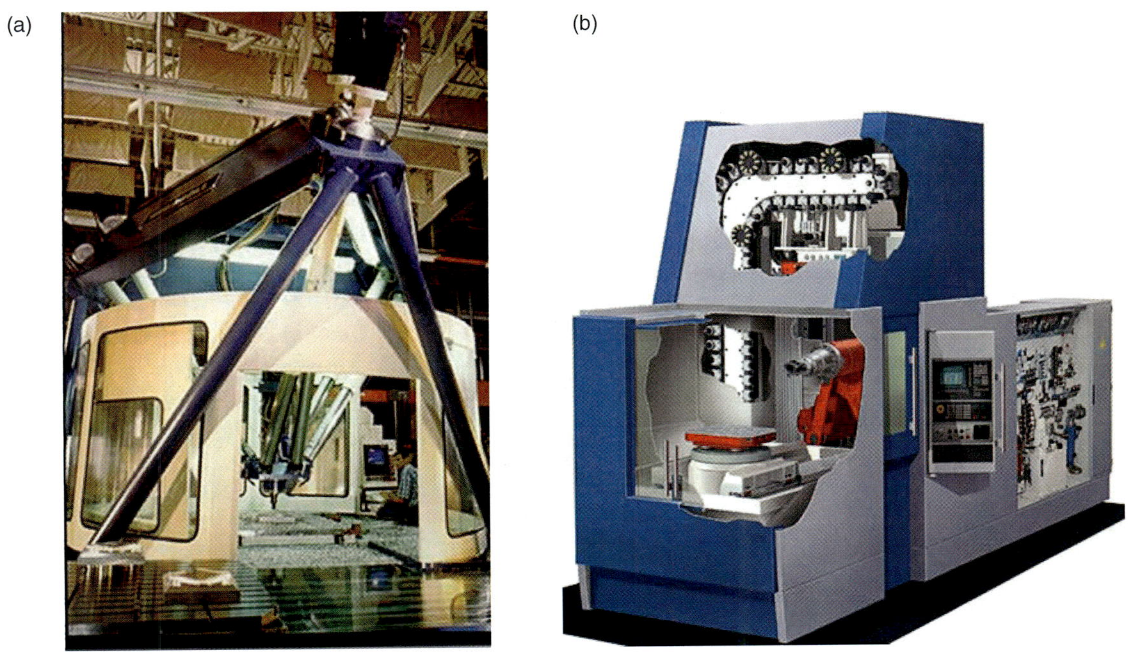

Plate 19 Examples of parallel kinematic machines [7]: (a) octahedral hexapod machine installed at NIST, (b) five-axis HMC with PKM integrated with spindle motion actuation. (see Fig. 15.74, p. 266)

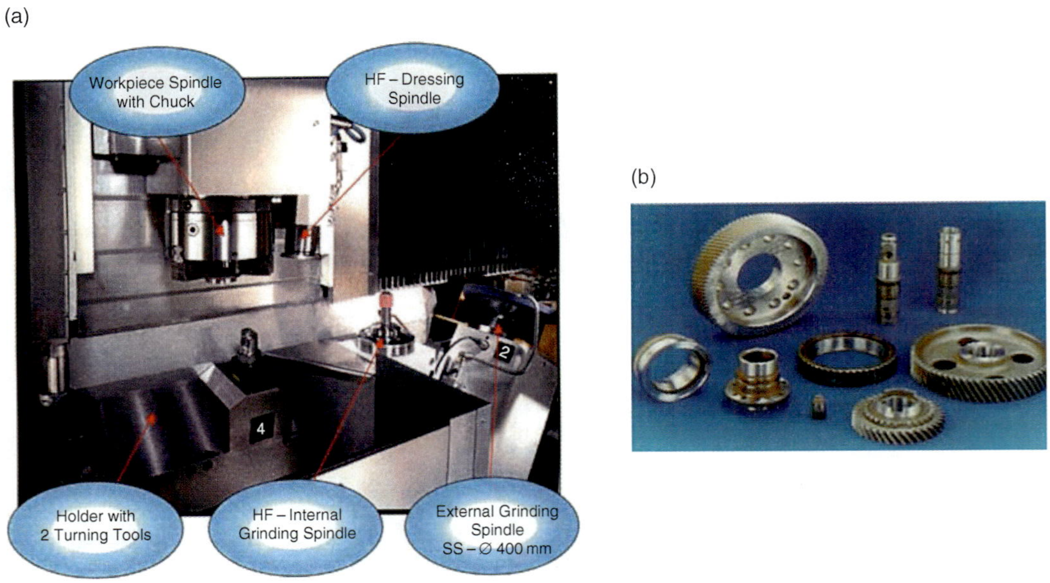

Plate 20 Turning and grinding machining centre: (a) working area with two separate stations, (b) a variety of turn/grind parts [3]. (see Fig. 15.96, p. 283)

Plate 21 From left to right: top being cut, one side filled in, second side cut with wax removed [14]. (see Fig. 16.30, p. 325)

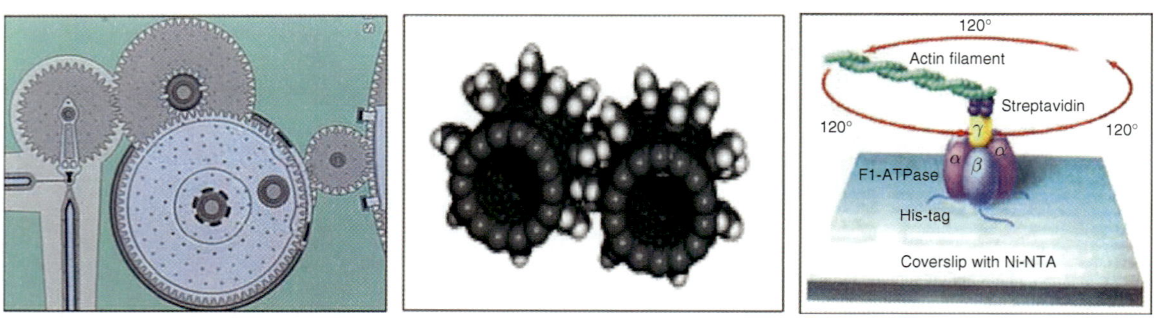

Plate 22 Comparison of micro- and nanoproducts of different generations: (a) micro-scale gear, (b) nano-scale gear and (c) biological nanomotor [5]. (see Fig. 17.3, p. 334)

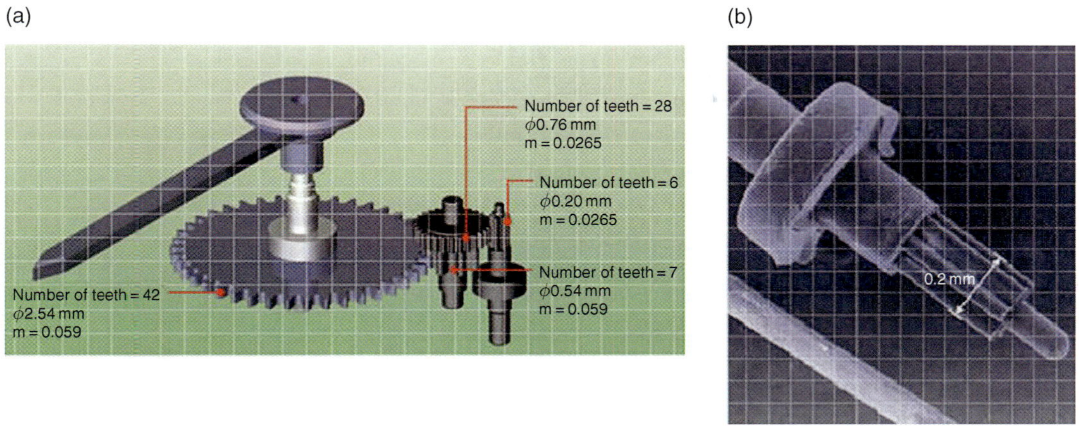

Plate 23 Exemplary nanomanufacturing products: (a) nanogear, (b) world's smallest carbon nanofibre composite-made wheel [6]. (see Fig. 17.4, p. 334)

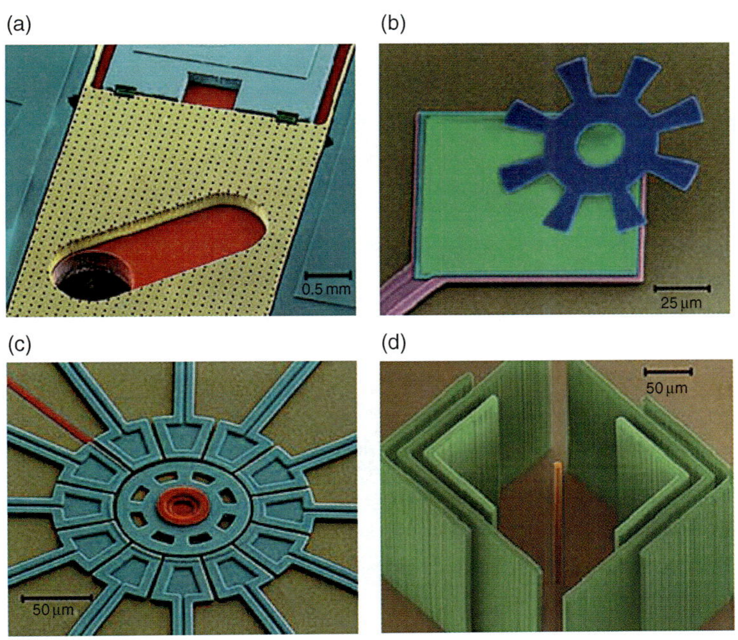

Plate 24 Examples of nanoproducts: (a) a high-aspect ratio MEMS device, (b) SEM of mechanical gear, (c) SEM of a salient-pole electrostatically actuated micromotor, (d) SEM of high-aspect ratio microstructure with central post [27]. (see Fig. 17.25, p. 349)

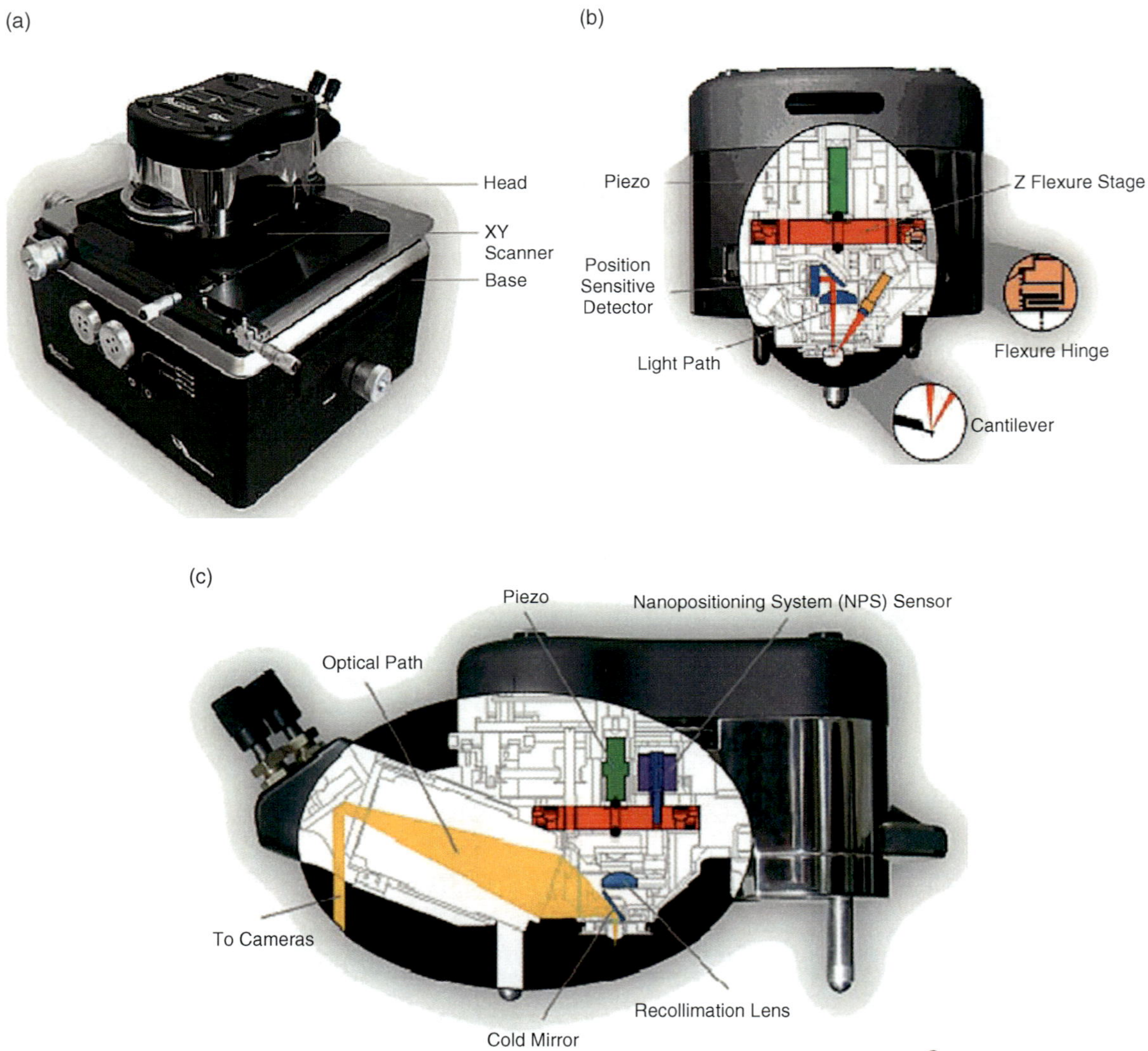

Plate 25 Atomic force microscope (AFM): (a) general view, (b) head, (c) XY scanner. Source: Asylum Research [28]. (see Fig. 17.26, p. 350)

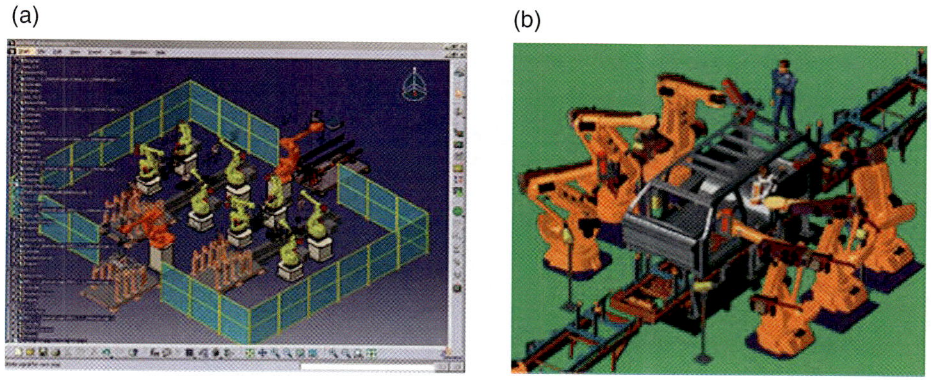

Plate 26 Examples of virtual robotized workcells: (a) DELMIA V5 virtual workcell [12], (b) automotive workcell's 3D simulation from Tecnomatix [14]. (see Fig. 19.10, p. 393)

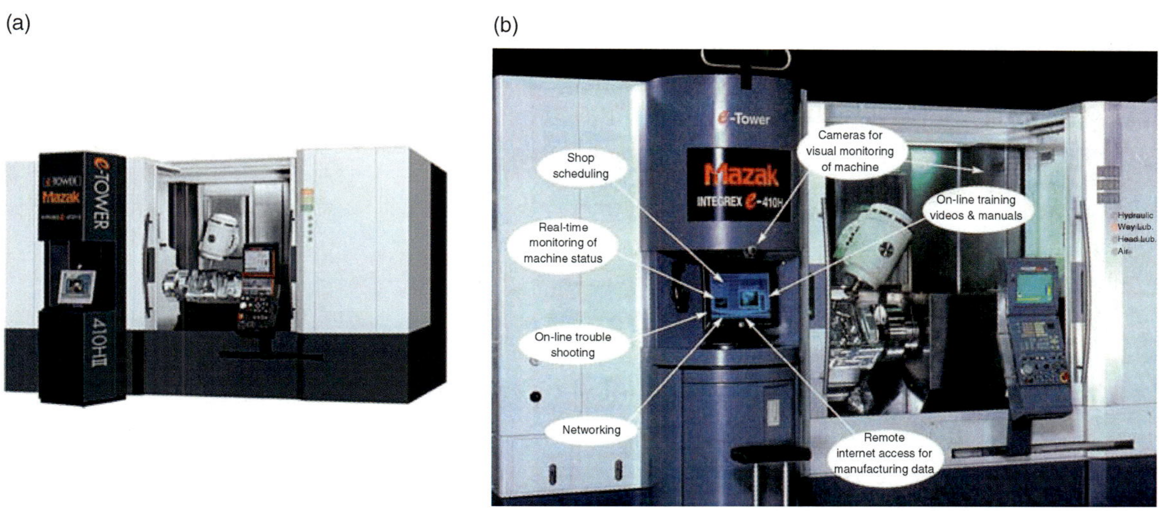

Plate 27 Multitasking machining centre Integrex e-410H II (Mazak Corp.) equipped with an internet information terminal (e-Tower) [19]: (a) general view, (b) functions of information terminal. (see Fig. 19.15, p. 398)

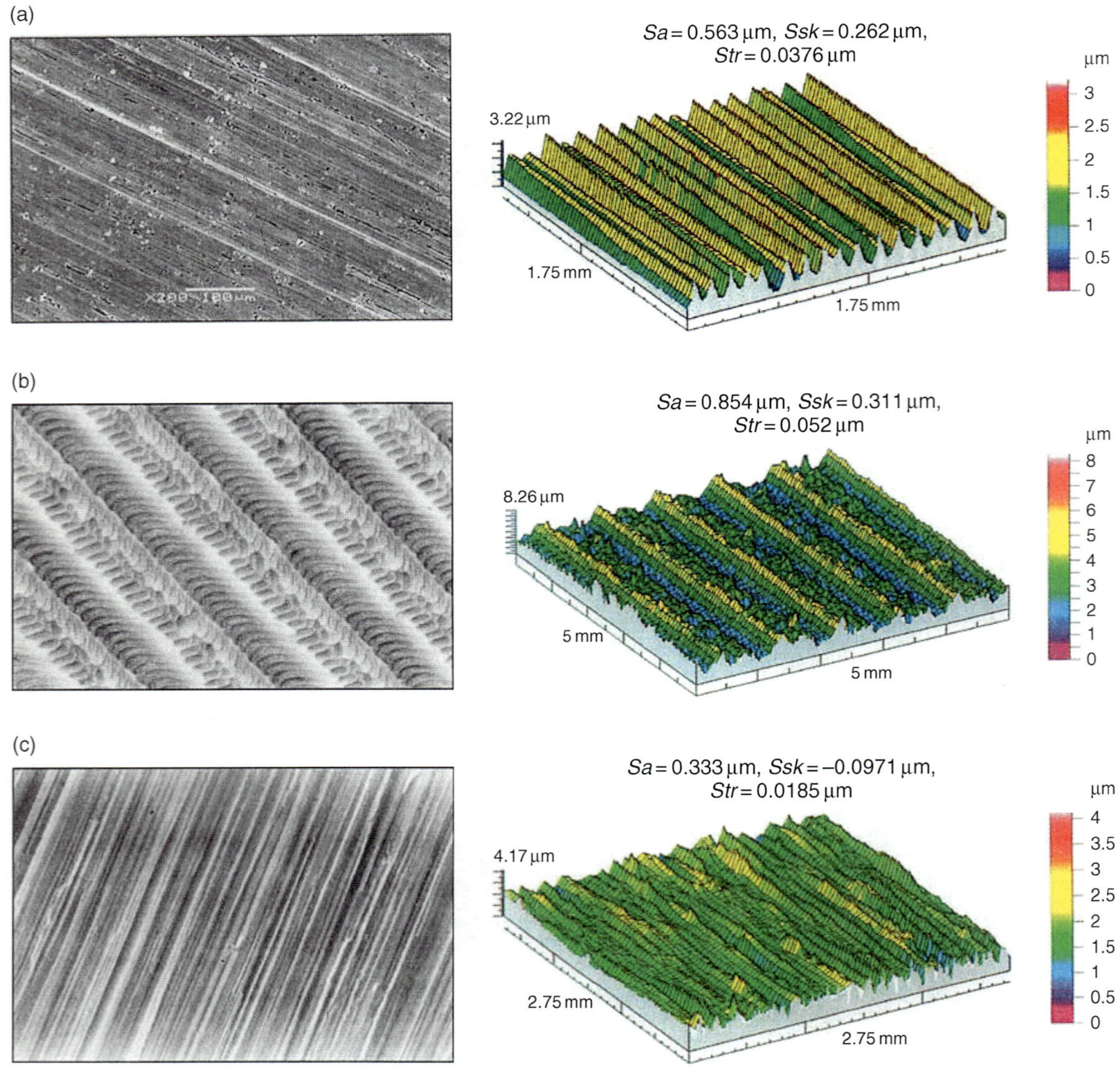

Plate 28 Examples of 3D surfaces after: precision diamond turning of aluminium alloy, $v_c = 6.6\,\text{m/s}$, $f = 0.1\,\text{mm/rev}$ (a), ball-nose end milling of tool steel, $v_c = 1.8\,\text{m/s}$, $a_e = 1\,\text{mm}$ (b), plain peripheral grinding of alloy steel, $v_c = 22.1\,\text{m/s}$, $v_{ft} = 0.4\,\text{m/s}$ (c) [20]. (see Fig. 20.12, p. 416)

are limited by the speeds of available machines. On the other hand, machining speeds on the group of difficult-to-cut materials such as titanium, nickel-based alloys and hardened steels are limited by the availability of suitable cutting tool materials. Moreover, turning, milling and drilling operations (besides grinding where achievable speeds with CBN wheel reach 200 m/s, i.e. ultra-high speed range) are mostly suitable for HSC equivalents. As a result, there is an ongoing, worldwide trend towards improving the characteristics of tool materials, workpiece materials and machine tools.

From the cutting mechanics point of view the high cutting speed coincides with the asymptotic tendency on both the friction and deformation force components. Hence, the high cutting speed v_{HSC} indicates the beginning of the range in which the friction and practical deformation energies remain constant (for example, at $v_c = 4000$ m/min for Ck45N carbon steel of UTS about 650 MPa). The value of v_{HSC} can be calculated from Eqn (15.1) based on measured cutting force (F_c) values, as follows [3]:

$$F_c(v_c) = F_{c\infty} + F_{dyn} e^{-\left(\frac{2v_c}{v_{HSC}}\right)} \tag{15.1}$$

where $F_{c\infty}$ is an asymptotic value of F_c force due to friction, F_{dyn} is the force component due to chip deformation, and v_c is the cutting speed.

When the high cutting speed model is extended over different materials producing continuous chips, as shown in Fig. 15.2, its values can be matched with the appropriate values of ultimate tensile strength as an exponential function (15.2). The relevant expression for UTS-depending v_{HSC} is [3]:

$$v_{HSC} = 56 e^{\frac{-Rm}{400\,\mathrm{MPa}}} \tag{15.2}$$

According to Smith and Tlusty [4] the definition of HSM can also be related to the cutting tool and spindle dynamics. In general, they postulated using the DN number (the spindle diameter in mm multiplied by the spindle speed in rpm) instead the absolute spindle speed only. It should be noted that any speed greater than 8000 rpm is usually defined as high speed. There are a number of commercial machines available with DN numbers ranging from 1.5 million to 2 million, and some of them exceed 2 million. The definition proposed depends on the natural frequency of the dominant (most flexible) mode of vibration and is based on the selection of four categories of the entire spindle speed range, as shown in Fig. 15.3. This figure illustrates the well-known stability lobe diagram which relates stable axial depth of cut to spindle speed for defined radial depth of cut. For example, if the dominant natural frequency is 1000 Hz and the tool used is a two flute end mill of 25 mm diameter (classical referring tool) these categories are as follows:

1. The low speed machining below about 2300 rpm, as indicated by section A in Fig. 15.3.
2. The midrange machining in the range between 2300 rpm and 7500 rpm, which corresponds to section B.
3. The high speed machining which occurs in section C, free of chatter. This range is extended from 7500 rpm to approximately 45 000 rpm. Due to this fact high velocity machining is used as a synonym for HSM.
4. Ultra-high speed machining at spindle speeds greater than about 60 000 rpm, i.e. when the tool passing frequency is greater than about 2 to 3 times the dominant natural frequency. This speed range is usually not feasible with current spindle bearing technology.

15.1.2 Physical aspects of HSM

From a historical point of view the idea of HSM originated from Salomon's theory patented in 1931 in Germany. It assumed that at a certain cutting speed of 5–10 times higher than in conventional machining the chip removal (chip/tool interface) temperature will start to decrease. Unfortunately, so far, this theory has not been verified to its full extent and the relative decrease of temperature observed in Fig. 15.4 starts at different speeds for non-ferrous and ferrous materials. The reduction of cutting temperature with appropriate increase of cutting speed is more pronounced for aluminium and other non-ferrous metals (for example, for magnesium). In sharp contrast, it is marginal for steel and cast iron.

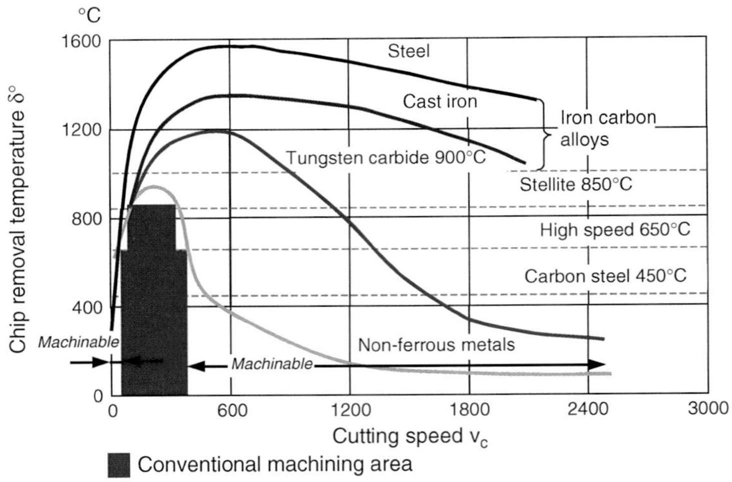

Fig. 15.4 Chip removal temperature as a result of the cutting speed [5].

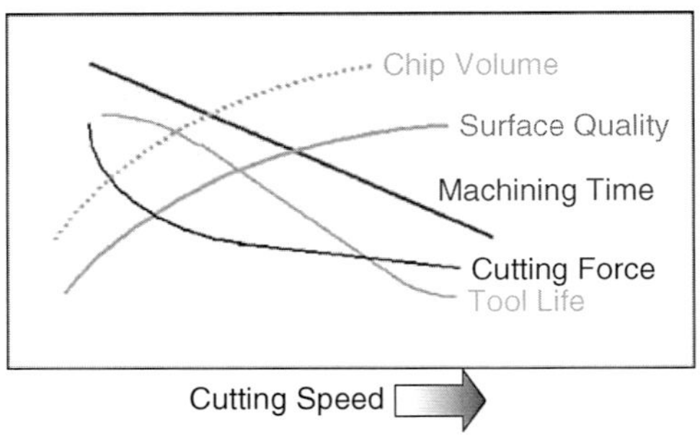

Fig. 15.5 Machining characteristics for HSM technology [6, 7].

Schematically, the overall influence of high cutting speeds on the performance of machining is shown in Fig. 15.5. In particular, HSM technology allows the material removal rate (chip volume) and surface quality to increase, and at the same time the cutting force to decrease. A smaller amount of the heat generated in the cutting zone is transferred into the workpiece and cutting tool in comparison to traditional machining. As a result, thin-walled parts can be machined with high precision. Advanced cutting tool materials are required to compensate the shortening of tool life. Low thermal and chemical stability disqualify carbides from most of high speed machining operations. Ceramics can cut up to 1000 m/min but because of extreme brittleness and poor resistance to thermal and mechanical shock their use in HSM is substantially limited. CBN for ferrous materials and PCD for non-ferrous metals are well suited for HSM due to superior wear resistance at high temperature and they can prolong tool life to acceptable levels.

The plot of the resultant shear force F_{su} vs cutting speed up to ultra-high speeds of 20 000 m/min is shown in Fig. 15.6. Shear force F_s increases up to the cutting speed v_a and the temperature within the shear zone approaches melting point but further decreases due to the thermal softening effect. On the other hand, the dynamic component of the cutting force F_m (momentum force) increases significantly. At ultra-high speed v_i force F_m reaches the value of shear force F_{sc} observed for conventional cutting. Figure 15.7 illustrates the changes of cutting force in the more practical range below v_a, i.e. between 250 m/min and 6000 m/min. It is worth noticing that for both aluminium and medium carbon steel the cutting force reaches a local minimum at $v_c = 3000$ m/min and 2000 m/min, respectively. After these speeds the cutting forces tend to increase and

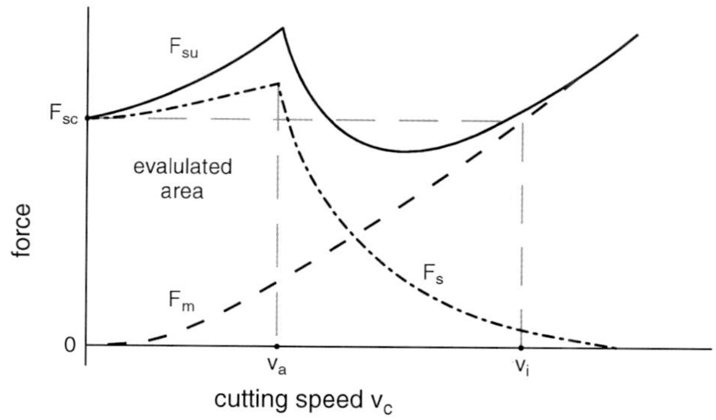

Fig. 15.6 Theoretical behaviour of the resultant cutting force and its components at ultra-high cutting speeds [3].

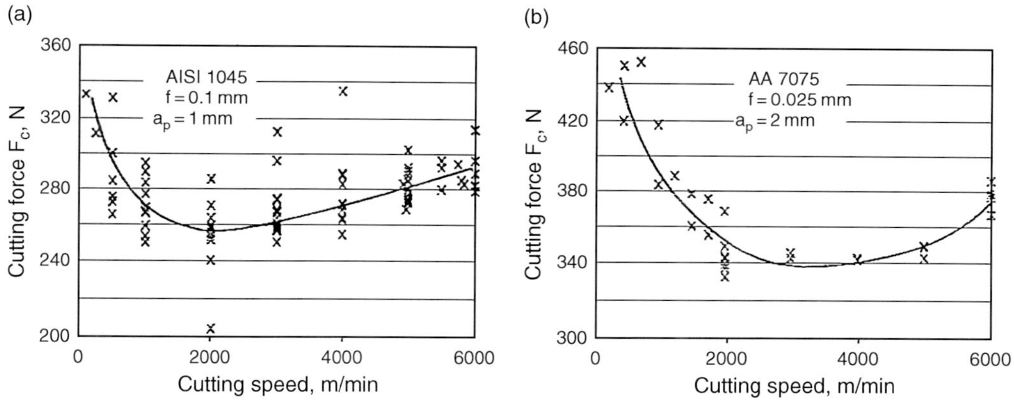

Fig. 15.7 Cutting forces in the turning of AISI 1045 steel (a) and AA 7075 aluminium (b) vs cutting speed [3].

for steel at a speed of 6000 m/min its value is the same as at the starting speed of $v_c = 250$ m/min. When milling 7050 aluminium the cutting forces diminish exponentially more visibly as depicted in Fig. 15.8b but the specific cutting force has a local maximum at a cutting speed of about 800 m/min. As a result, the milling operations on aluminium parts should be performed with cutting speeds higher than 1000 m/min. In general, the cutting speed corresponding with minimum force (Fig. 15.7a) coincides well with a minimum flank wear *VB* of 100 μm, as indicated in Fig. 15.9.

The chip formation in HSC depends strongly on the workpiece material machined. For some materials, e.g. alloy steels and nickel-based superalloys, a transition from a continuous to discontinuous (serrated, saw-tooth, shear-localized) type chip with characteristic thin, intense, shear bands between neighbouring segments was observed [1]. As reported [1], shear localization is favoured by a combination of a low value of the product of the thermal properties $\kappa\rho c$, by an hcp structure, and by high hardness, e.g. titanium alloys, nickel-based alloys and hardened alloy steels. Cutting conditions such as tool geometry and feed rate were found to influence visibly the segmentation severity and the onset of segmentation. As an example, for AISI 4340 steel (200 BHN) this transition is around 240 m/min for a 0.045 mm/rev feed. Similar transitions were also observed for Inconel 718, a nickel-based superalloy. For titanium alloys, no transition is generally observed over the entire speed range. At ultra-high speed ballistic machining (UHSM) at speeds up to 80 000 m/min, a continuous ribbon-like chip was produced when machining annealed 4340 steel, while a segmented chip was dominant when machining heat-treated AISI 4340 steel, Inconel 718 and titanium 6Al-4V [1].

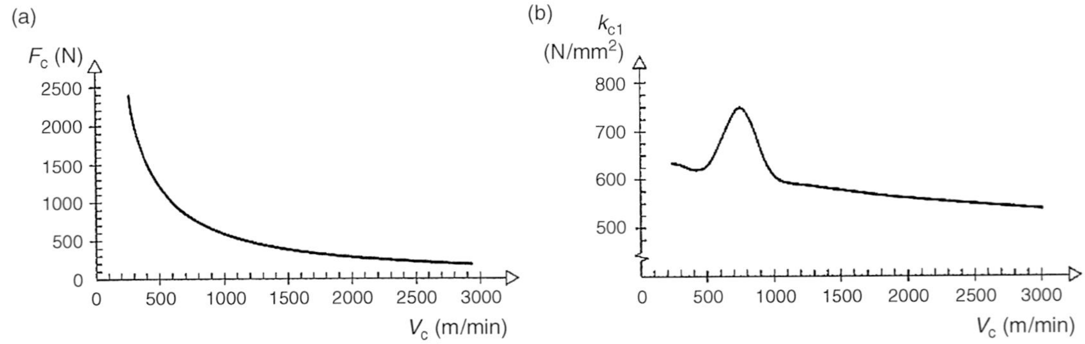

Fig. 15.8 Cutting force (a) and specific cutting force vs cutting speed for milling 7050 aluminium (b) [8].

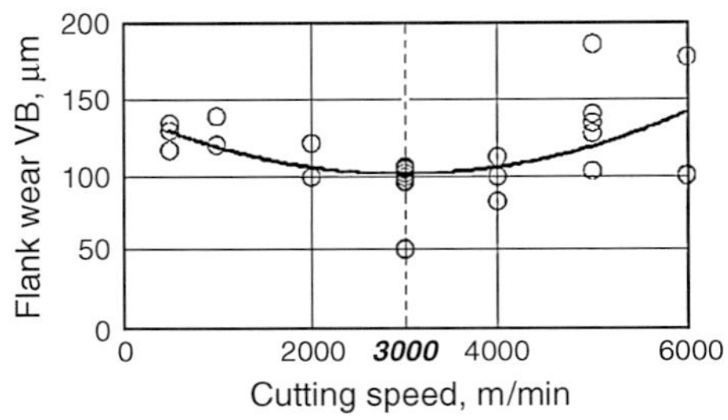

Fig. 15.9 Flank wear of the ceramic tool in the turning of AISI 1045 steel at high cutting speeds [3]. Cutting conditions: feed rate $f = 0.1$ mm/rev, depth of cut $a_p = 1$ mm.

Some exemplary microphotographs of shear-localized chips obtained for Ti-6Al-4V titanium alloy at a cutting speed of 45 m/min and AISI 304 (SS 2333) austenitic stainless steel at $v_c = 180$ m/min by means of the quick-stop technique are shown in Fig. 15.10. In the case of milling of the aluminium alloy AlZnMgCu1.5 (7075 ISO) the tendency to chip segmentation depends on the heat treatment conditions [9].

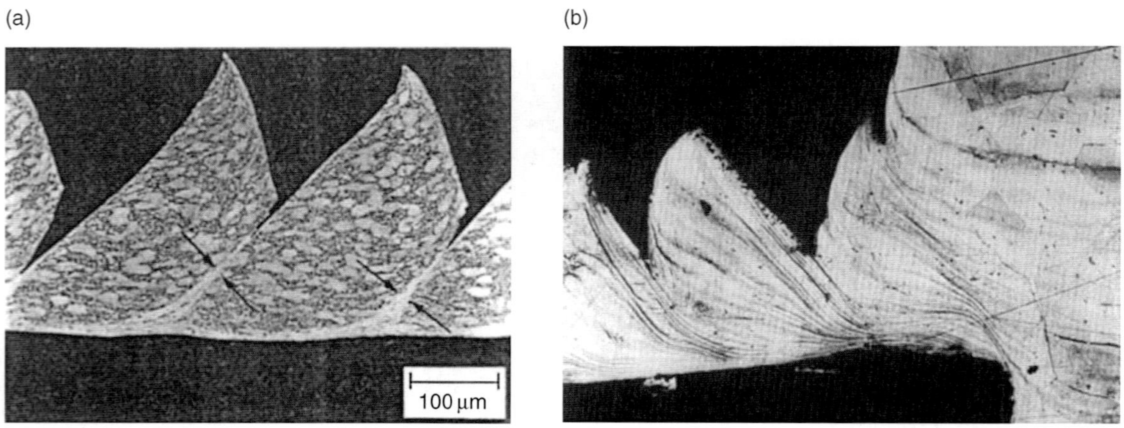

Fig. 15.10 Photomicrographs of shear-localized chips from titanium 6Al-4V at $v_c = 45$ m/min [1] (a) and austenitic stainless steel (SS 2333) at $v_c = 180$ m/min, $f = 0.3$ mm/rev, $a_p = 3$ mm, $\gamma_0 = 5°$ [8] (b).

For overaged microstructure ($Y = 225\,\text{MPa}$) continuous chips were produced even for highest cutting speeds of 7000 m/min. In contrast, for underaged material ($Y = 405\,\text{MPa}$) segmentation occurred at all cutting speeds ranging between 1000 m/min and 7000 m/min.

15.1.3 HSM technology and appliances

It is obviously known that the practical utilizing of the benefits of this evolutionary machining concept required the specific integration of machine, tooling interface and cutting tools. The most relevant criteria related to these three groups of interest are specified in Fig. 15.11.

Figure 15.11 clearly depicts that there is an array of design elements and factors which integrally contribute to the function and reliability of high speed machines, and especially high speed spindles. It should be emphasized that rotational spindle speeds, axes' accelerations and rapid traverses have to be selected coordinately and adjusted according to the specific machining tasks.

The technology that has really revolutionized HSM machining is a linear (motion) drive. Two categories of such drives are discerned, namely permanent-synchronous-brushless and synchronous-linear-induction motors. The first concept is of simple mechanical design but the second solution is better for longer, continuous, magnetic fields and long travels. Fast rotational spindle speeds of 20 000–30 000 rpm are meaningless for high speed machining unless they are coordinated with fast axes movements of up to $19.6\,\text{m/s}^2$ or 2g, and rapid traverse of about 200 m/min [7].

As shown in a plot illustrating the dependence of reduction of the machine's positioning time on the acceleration rate (Fig. 15.12) this time is cut in half when producing an acceleration of $19.6\,\text{m/s}^2$ at rapid travel of 200 m/min in comparison with a conventional machining centre. It is also important that newly

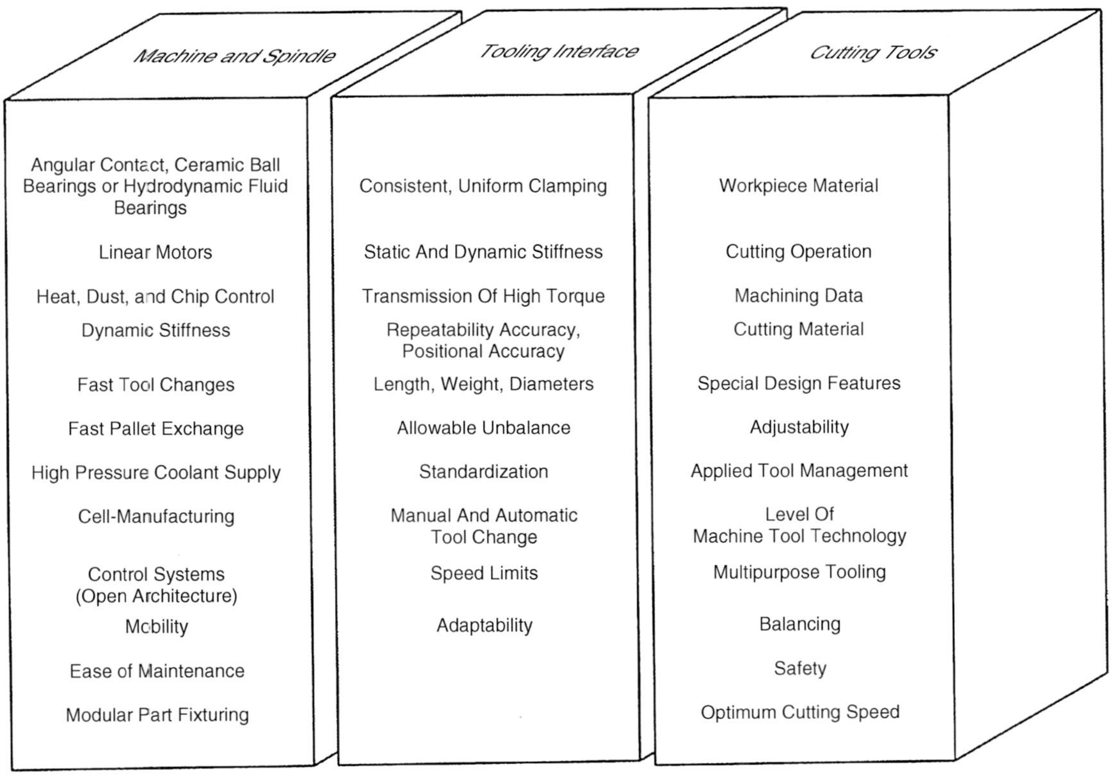

Fig. 15.11 Criteria of high speed/high velocity machining [7].

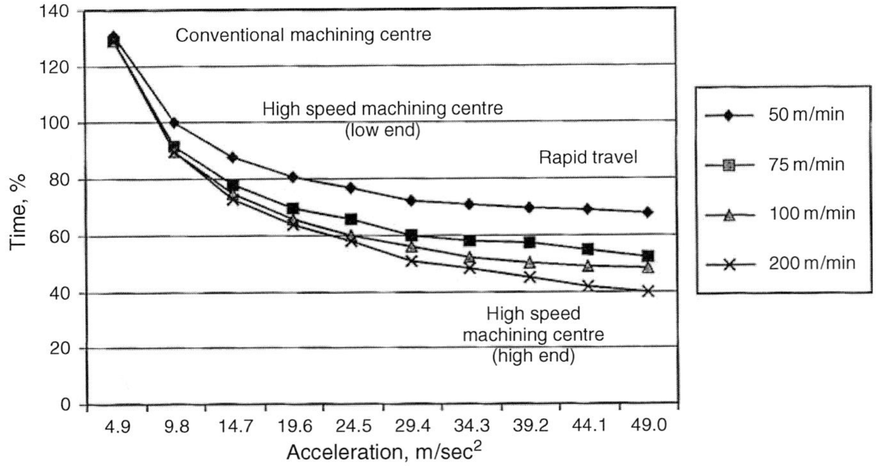

Fig. 15.12 Positioning time relative to acceleration [7].

designed ball screws with hollow shafts and double screw arrangement allow acceleration rates of up to 9.8 m/s^2 or 1g and can also be a solution for a lower range of HSM technology.

It was documented in HSM practice that the cutting tool's interface with the machine tool is most critical for stable machining [7]. Moreover, at spindle speeds above 10 000 rpm (often considered as the lower boundary of HSM) balance becomes a major issue. Cutting tool balance is also very important because of the relationship between force and velocity. For example, the centrifugal force caused by an unbalanced tool while turning at 10 000 rpm exceeds 25 times that generated by the same tool at 2000 rpm. The ISO Standard 1940 determines balance specifications through grades, and for most high speed applications the G2.5 grade is sufficient, but for spindle speeds above 20 000 rpm a G1 grade is required [10]. For example, for the G1 the rotational speed of 20 000 rpm corresponds to the maximum allowable residual unbalance of about 0.2g × mm/kg.

In the case of the commonly used HSK clamping system torque is transmitted through its taper and face as well as its inner shoulder due to pressure clamping. It can be assumed that a total of 10% of the clamping force is transmitted through the shortened taper, while 90% is transmitted through face contact. To ensure uniform and reliable gripping, as well as positional accuracy and repeatability, special tool clamping mechanisms placed into the spindle hollow are designed. They can utilize a combination of disc/helical springs or the clamping unit with gas spring as shown in Fig. 15.13. The latter consisting of a tool draw bar, spring and amplifier (Fig. 15.13b) is adapted to the tool clamping system based on a polygonal taper (so-called Capto by Sandvik Coromant [8]).

Depending on the specific requirements, various types of bearings, including high precision angular contact bearings and active magnetic and aerostatic bearings, are suitable for HSM machine spindles [13, 14]. Today, integral motor spindles (directly driven spindles) with hybrid ceramic bearings remain the most popular solution for high speed machining centres. Spindles with aerostatic bearings are preferred for ultra-precision machining due to minimal, non-repeatable runout and excellent vibration behaviour. An active magnetic bearing system is very costly but its use is justified when the diagnostic module is built into the spindle (it plays the role of a force or position sensor for monitoring and control). The data can be output from the spindle via an infrared interface.

In the group of machining centres (at least 60% of the total market) a special subgroup constitutes high speed machining centres used in small, medium and large-batch production. High speed machining centres for large-batch manufacturing with spindle speeds in the 16 000 to 24 000 rpm range, rapid traverse rates of 100 to 120 m/min and accelerations of 2 to 3g are mainly used in the automotive field [13]. For high speed positioning applications, hexapod and similar parallel kinematic or non-Cartesian machines continue

(a)

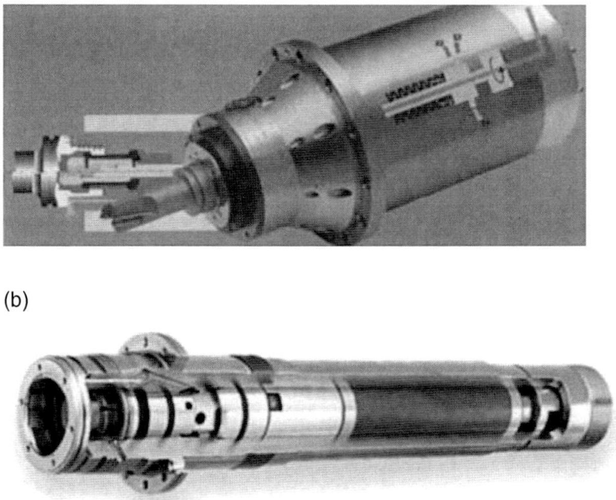

(b)

Fig. 15.13 Spindle for HSC with tool clamping system consisting of a set of disc springs (HSK63 taper, $n = 30\,000\,\text{min}^{-1}$, $P = 30\,\text{kW}$) [11] (a) and spindle mechanism with gas spring [12] (b).

to be developed. For one-off small and medium-batch manufacturing there is a diverse range of high speed machining centres. For example, aerospace machines with speeds up to 40 000 rpm are designed for machining aluminium. In the die/mould industry complex three-dimensional shapes and machined materials range from hardened steels, soft steels and aluminium to copper and graphite need to be machined with appropriate speed, accuracy and high surface finish. Practical feed rates are up to 8.0 m/min for hardened steels and up to 12 m/min for graphite.

Typical constructions of high speed machining centres are bridge-type construction, a modified bridge construction and an overhead gantry system. Figure 15.14a presents a view of a modern high speed machining centre which fulfils all the above-mentioned requirements regarding kinematical and functional aspects and guarantee reliable CNC control.

15.1.4 Basic applications of HSM technology

As indicated in Table 15.1 possible application fields of high speed machining result directly from technological advantages specified in the first column of this table. In relation to the material type, HSC cutting can be used for the entire process when machining light metals, non-ferrous metals and plastics, while a finish HSC operation is more suitable for machining steels, cast irons and difficult-to-machine alloys. Nowadays, HSC cutting is dominated by milling operations, mainly in the automotive, aircraft and die and mould industry sectors. High speed milling (HSM) applications can be divided into three categories [4, 5, 16, 17]:

1. Face milling operations performed on engine blocks or gearbox housings with rotational speeds of 1000–5000 rpm. They require moderate power of 10–20 kW.
2. Milling operations with high volume rates on monolithic parts (up to 90% of the material can be removed as chips), such as manufacturing complex, thin-walled aluminium or titanium parts in the airplane and aerospace industries with small end mills. High power of 20–40 kW and rotational speeds as high as 10 000–30 000 are necessary in these applications.
3. Manufacturing very precise parts in mechanical or electronic sectors with a low power of 1–5 kW but very or ultra-high speeds of 20 000–100 000 rpm. Tooling, copper and graphite electrodes for EDM and finishing of die and moulds in hardened state are some typical applications.

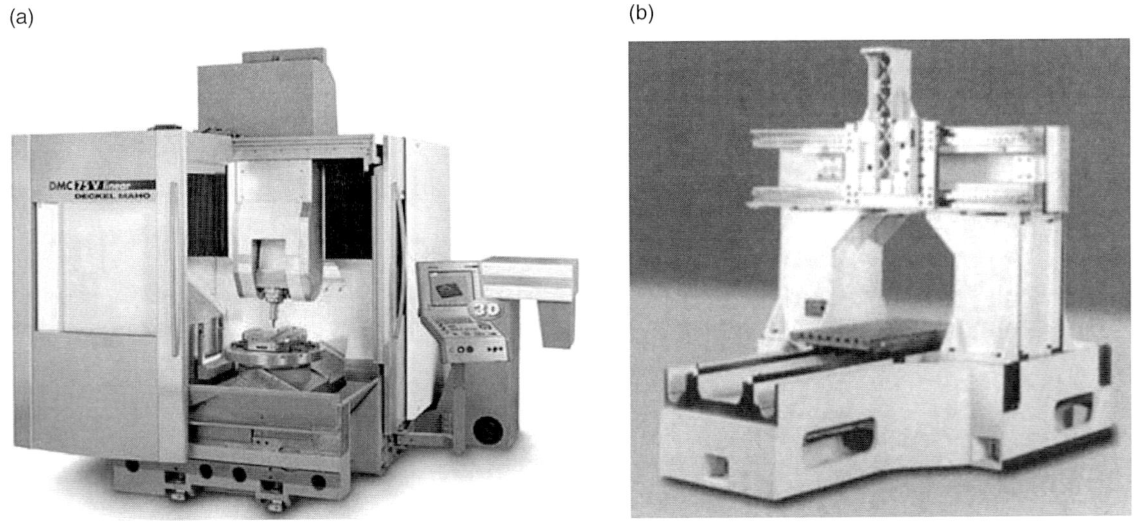

Fig. 15.14 HSM machining centre with linear drives. Technical data: rotational spindle speed = 18 000, 28 000 or 42 000 m^{-1}, rapid traverse rate along *X*/*Y*/*Z* axes 90 m/min, feed rate = 90 000 mm/min, tool magazine for 30 tools, CNC control system – Heidenhain iTNC 530 or Siemens 840D, clamping taper – HSK-A63 for automatic tool changing. (a) General view and (b) an example of monobloc bridge construction of HSC milling machine [8, 15].

Table 15.1 Application fields of high speed machining [16, 19].

Technological advantage	*Application field*	*Application examples*
Large cutting volume	light metal alloys	aircraft and aerospace production
	steel and cast iron	tool and die mould manufacturing
High surface quality	precision machining	optical industry, fine mechanical parts
	special workpieces	spiral compressors
Low cutting forces	machining of thin-walled workpieces	aircraft and aerospace products automotive industry, household equipment
High frequencies of excitation	non-vibration machining of difficult-to-machine workpieces	precision mechanics an optical industry
Heat removal by chips	non-warping machining	precision mechanics
	cold workpieces	magnesium alloys

Printed circuit drilling with micro-drills of less than 0.5 mm in diameter with spindle speeds between 300 000 and 400 000 rpm is one of a number of special operations corresponding to the third group of HSM applications. The basic technical problem emerging in such extreme machining is the relationship between the velocity of the CNC control system and the feed drive dynamics [16].

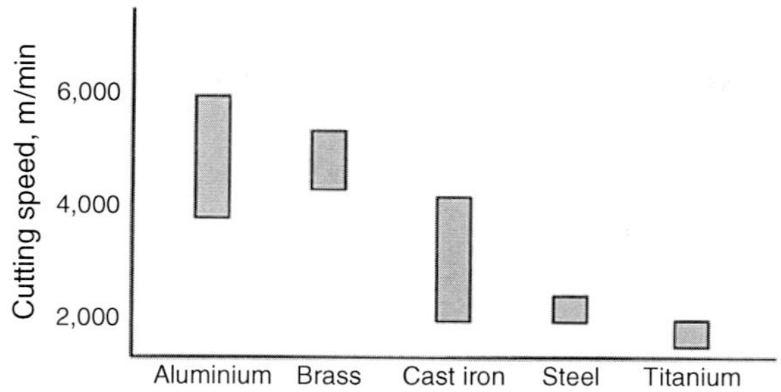

Fig. 15.15 Optimum cutting speeds for HSM machining of a variety of materials [7].

Recently, the application of HSM technology is extended to prototypes and pre-series in Al, Ti and Cu for the aerospace, electric/electronic, biomedical and defence industries. It has been reported [18] that HSM technology can compete favourably with other rapid prototyping techniques in the production of hardened steel moulds/dies for the net shape manufacture of forgings, plastic injection mouldings and die-castings.

Figure 15.15 provides recommended values of cutting speeds corresponding to the minimum machining cost (optimizing strategy described in Chapter 14) for a range of work materials of different machinability.

The benefits of HSM documented in relation to the initial investments are as follows [7]:

1. The metal removal rates increase five-fold.
2. There is up to 70% reduction in machining times.
3. There is a 25–50% decrease in machining cost.
4. There is an exponential increase in productivity.

In many cases the possibility of obtaining high quality surfaces with HSC milling results also in the shortening of some subsequent finishing abrasive processes such as grinding or polishing. This is the case especially observed in tool and die and mould making [5, 16, 20, 21]. In fact, with a rapid development of superhard cutting tool materials, the HSM technology of hardened steels creates considerable interests to die and mould manufacturing. It is expected that about 50% of traditional machining operations will be replaced by HSM ones [20]. In particular, high volume fraction (90%) CBN tools enable milling steels hardened to about 60 HRC with cutting speeds of around 1000 m/min.

Figure 15.16 illustrates the substantial reduction of the entire production time due to decreasing hand polishing operations which eliminates environmental harm from the EDM process. It is possible to obtain the dimension tolerances as low as 0.02 mm and, in turn, to reduce the total lead time to about 30–50%.

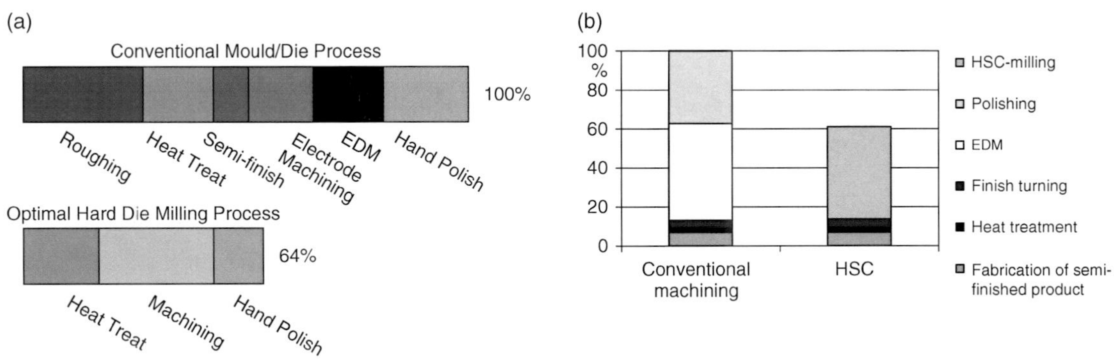

Fig. 15.16 Effects of using HSM-based processes in die and mould manufacturing [15, 21].

15.1.5 Machining of monolithic parts

As mentioned earlier, HSM reduces cutting forces and heat transfer into the workpiece, allowing machining of thin-walled parts with relatively high precision. Currently, the monolithic machining with high speeds is widely used in the aerospace industry replacing assembled sheet metal components.

This spectacular success provides a strong incentive to use this new machining technology for other applications, such as fabrication of deep-pocketed aluminium parts, pressure vessels with highly complex geometry for cold neutron reactor, the STM (scanning tunnelling microscope) body from pure molybdenum and titanium turbine blades, as shown in Fig. 15.17. For instance, aluminium parts with thin-walled structures were machined at 20 000 rpm using an end mill of 118 mm long and 11.8 mm in diameter.

Typically, in high speed milling of titanium alloys with a 13 mm end mill machine spindles operate with 4000 rpm without sophisticated controls and finishing is performed at 152 m/min, but it is possible to increase the cutting speed in super finishing, even to 300 m/min [23]. The cutting tool and the cutter path become the key factors. High levels of heat are generated. To avoid friction it is important to use end mills with high, sharp helix angles, the best coatings are TiAlN and AlTiN. Recently, HSM milling of titanium alloys is performed as wet machining with flood coolant supplying at a high pressure of 200 bar [12].

Figures 15.18 and 15.19 show examples of monolithic titanium parts used in the aircraft industry. The titanium bulkhead for the F18 fighter (Fig. 15.18) is very expensive and aerospace suppliers used machine tools with linear motors and high speed spindles to machine as quickly as possible.

Some examples of parts being machined using different machining operations are shown in Fig. 15.19.

Recently, several vital components for the Airbus A320 and the new Airbus A380 have been successfully produced by machining in solid titanium blocks using machining centres equipped with tooling by Seco Tools, Sweden [8]. For example, the first roller track beam of the A320 (Fig. 15.19c2) is machined with

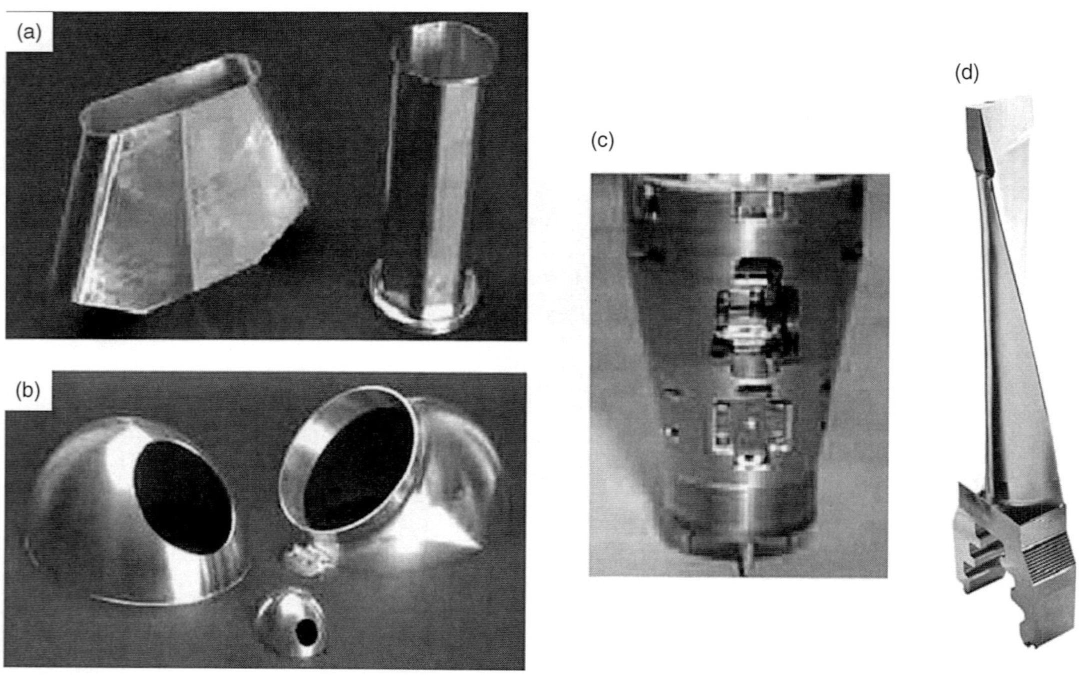

Fig. 15.17 Examples of monolithic components machined at high speeds [21, 22]: (a) deep-pocketed aluminium parts; (b) thin-walled aluminium pressure vessels for neutron reactor; (c) the STM microscope body made of pure molybdenum; (d) turbine blade made of titanium. (see Plate 14)

Fig. 15.18 Monolitic titanium bulkhead for the F18 fighter [24].

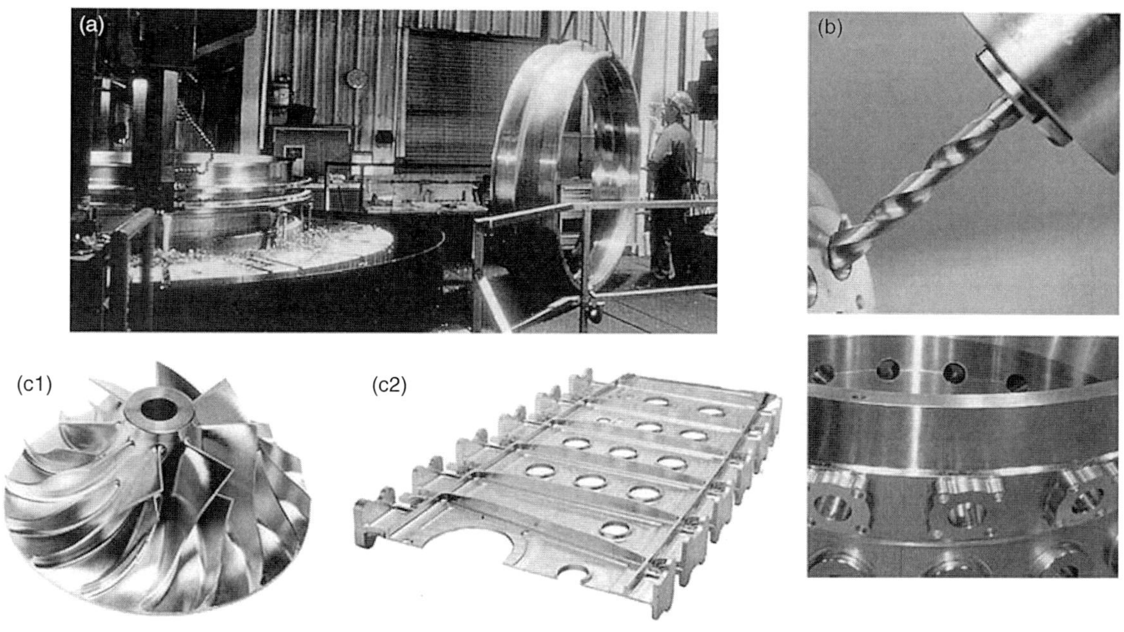

Fig. 15.19 Typical aerospace parts shaped by machining: (a) vertical turning a ring; (b) drilling holes in a case; (c1) milling a blade disc – BLISK, and (c2) milling a landing gear frame [21].

highly positive, sharp tools and inserts, which generate minimum cutting forces. Fine grain carbides are used for milling operations, such as face milling, contouring, slot milling, pocket milling, and chamfer and copy milling. All tools have internal through coolant directed on the cutting edges. The cutting speeds for titanium are between 50 and 70 m/min, providing tool lives between 30 and 60 minutes.

References

[1] R. Komanduri, D.G. Flom, M. Lee, Highlights of the DARPA Advanced Machining Research Program, *Trans. ASME J. Eng. Ind.*, Vol. 107 (1985), 325–335.
[2] G. Byrne, D. Dornfeld, B. Denkena, Advancing cutting technology, *CIRP Annals*, Vol. 52/2 (2003), 483–507.
[3] H. Schulz (ed.), *Scientific Fundamentals of HSC*, Carl Hanser Verlag, Munich, 2001.
[4] S. Smith, J. Tlusty, Current trends in high-speed machining, *Trans. ASME, J. Manuf. Science Engng*, Vol. 119 (1997), 664–666.
[5] *Die and Mould Making. Application Guide*. Sandvik Coromant, 1999 (www.coromant.sandvik.com).
[6] W. Grzesik, Developments in Metal Removal Processes, *Proc. 4th Intln Scientific Conference* 'Development of Metal Cutting', Košice, Slovakia, 2002, pp. 103–110.
[7] B.P. Erdel, *High-Speed Machining*, SME, Dearborn, 2003.
[8] www.micron-ac.com, www.gildemeister.com, www.secotools.com, www.coromant.sandvik.com, www.gildemeister.com.
[9] H. Schulz, E. Abele, A. Sahm, Material aspects of chip formation in HSC machining, *CIRP Annals*, Vol. 50/1 (2001), 45–48.
[10] B.P. Eredel, New dimensions in manufacturing, Hanser Gardner, Cincinnati, 1998.
[11] H. Sykora, J. Greif, HSC mit dem richtigen Setup, *Werksttat und Betrieb*, Vol. 133, No. 6 (2000), 84–86.
[12] *Titanium Machining – Application Guide*, Sandvik Coromant, Sweden, 2004.
[13] M. Sudhakar, New developments in high-speed machining and controls, *Machining Technology*, Vol. 15, No. 3, Third Quarter (2004).
[14] B. Möller, High-speed spindles, *EPE*, Vol. 20, No. 1–2 (1996), 24–27.
[15] T. Brauen, Wege zum FSC-Fräsen, *Werstatt und Betrieb (Scheizer Präzisions-Fertigungstechnik)*, Vol. 133, No. 6 (2000), 10–11.
[16] H. Schulz, T. Moriwaki, High-speed machining, *CIRP Annals*, Vol. 41/2 (1992), 637–642.
[17] I. Nieminen, J. Paro, V. Kauppinen, High-speed milling of advanced materials, *J. Mater. Proc. Technol.*, Vol. 56 (1996), 24–36.
[18] J.P. Urbanski, P. Koshy, R.C. Dewes, D.K. Aspinwall, High speed machining of moulds and dies for net shape manufacture, *Materials and Design*, Vol. 21 (2000), 395–402.
[19] H. Schultz, *High-Speed Machining*, Carl Hanser, Munich, 1996.
[20] M.A. Elbestawi, L. Chen, C.E. Becze, T.I. El-Wardany, High-speed milling of dies and molds in their hardened state, *CIRP Annals*, Vol. 46/1 (1997), 57–62.
[21] W. Grzesik, State-of-the-art of metal cutting processes and technologies, *Proceedings of 3rd Intln Conference on Machining and Measurement of Sculptured Surfaces*, MMSS'2003, Cracow, Poland, pp. 87–104.
[22] M.A. Davies, T.L. Schmitz, B.S. Dutterer, How to succeed at HSM, *Manufacturing Engineering*, Vol. 126, No. 3, 90–104.
[23] R.B.Aronson, HSM is not just for aluminium, *Manufacturing Engineering*, Vol. 131, No. 4 (2003).
[24] J.R. Koelsch, High-productivity machining, *Manufacturing Engineering*, Vol. 136, No. 3 (2006), 137–148.

15.2 Dry and Semi-dry Machining

15.2.1 The way to dry machining

According to the manufacturing statistics shown in Fig. 15.20 the total cost for acquiring, maintaining and disposing of coolants represents between 7% to 16(17)% of total production cost depending on the workpiece, the production structure and the production location [1, 2]. In contrast, tooling cost is within single digits (usually about 4%). Cost, as well as health and environmental issues, mandate manufacturing enterprises to drastically reduce coolant consumption and, if possible, eliminate it altogether. As a result, these trends tend to two more economically and environmentally friendly conceptions of machining, termed, respectively, near-dry and dry machining.

Table 15.2 provides complete comparison of three different cooling strategies (flood, MQL and dry) using several criteria.

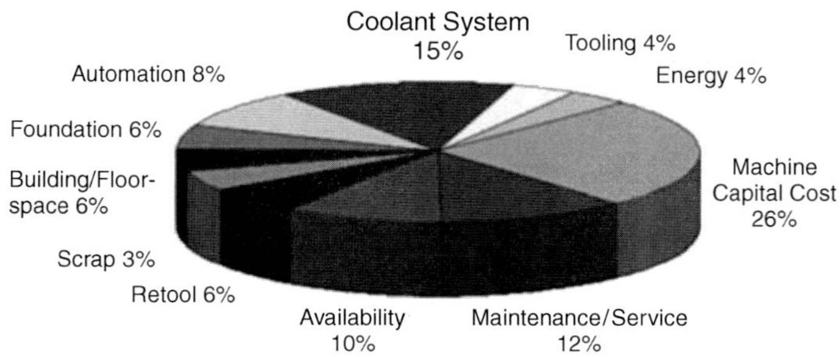

Fig. 15.20 Distribution of manufacturing costs for wet machining [3].

Table 15.2 Evaluation of different cooling strategies in machining [10].

Factor considered	*Flood cooling using circular supplying system*	*Near-dry (MQL) machining*	*Dry machining*
Cooling effect	● ● ●	□ □ ●	□ □ □
Lubricating effect	□ □ ●	● ● ●	□ □ □
Chip transport	● ● ●	□ □ ●	□ □ □
Investment costs	□ □ ●	□ □ □	● ● ●
Enterprise costs	□ □ ●	□ □ ●	● ● ●
Disposal costs	□ □ □	● ● ●	● ● ●
Coolant regeneration	□ □ □	● ● ●	● ● ●
Health aspects	□ □ ●	□ ● ●	□ ● ●
Protection against corrosion	● ● ●	□ □ ●	□ □ □
Coolant waste	□ □ □	● ● ●	● ● ●
Cleaning processes	□ □ □	□ ● ●	● ● ●
Chip recycling	□ □ □	● ● ●	● ● ●

Gradation: ● ● ● very profitable, □ ● ● profitable, □ □ ● less profitable, □ □ □ unprofitable

Picking the right fluid is not the only consideration. It is well known that getting the coolant to the cutting area is critical. A fluid can be useless if not delivered efficiently at the proper temperature and pressure, through well-positioned and appropriately shaped nozzles, and using a filtering system matched to the job. Eliminating cooling–lubricating media, due to the absence of their three essential functions, i.e. reduction of friction, absorption of the generated heat and chip evacuation, has consequences for the workpiece (lower dimensional accuracy, deterioration of surface finish, additional cleaning), the tool (decreasing tool life, severe adhesion, different chip formation, the need for chip extraction and tool cleaning) and machine tool (special designs for chip disposal, temperature control and cleaning). A substitute for these vital, positive effects cannot be found by simply turning off the flow of coolant, and special modifications in the cutting technology are necessary. As can be seen from Fig. 15.21 non-geometrically defined machining processes (grinding, honing, burnishing, EDM) are very coolant sensitive. When some of these processes can be replaced by competitive processes with geometrically defined tools, successful machining without fluids can be achieved.

It is evident from Table 15.3 that for a wide range of material–process combinations near-dry machining is essential but entire elimination of cooling–lubricating media is possible in several cases only. In general,

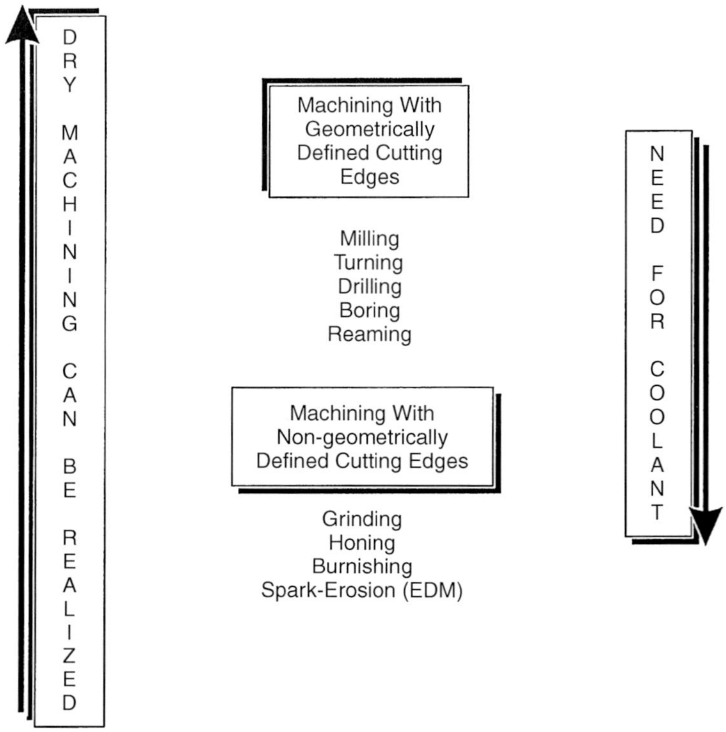

Fig. 15.21 Machining operations and their need for coolant [2].

Table 15.3 Recommendations for performing dry and MQL cutting operations [1].

material / processes	aluminium		steel		cast iron
	cast alloys	wrought alloys	High alloyed steels bearing steels	Free cutting steels quench and tem-pering steels	GG20-GGG70
drilling	MQL	MQL	MQL	dry/MQL	dry/MQL
reaming	MQL	MQL	MQL	MQL	MQL
tappening	MQL	MQL	MQL	MQL	MQL
treat forming	MQL	MQL	MQL	MQL	MQL
deep hole drilling	MQL	MQL		MQL	MQL
milling	MQL/dry	MQL	dry	dry	dry
turning	MQL/dry	MQL/dry	dry	dry	dry
gear milling			dry	dry	dry
sawing	MQL	MQL	MQL	MQL	MQL
broaching			MQL	MQL/dry	dry

Table 15.4 Most challenging machining/work material combinations for dry machining [2].

Workpiece material	*Machining operations*				
	Turning	*Milling*	*Reaming*	*Threading*	*Drilling*
Cast iron					
Steel alloys		●	●	●	
Hardened steel		●		●	
Aluminium		●	●	●	
Superalloys	●	●	●	●	●

in terms of materials, MQL is most suitable to machining operations conducted on aluminium and wrought aluminium alloys. This is due to the fact that dry machining requires a work material with a high temperature capacity and low temperature conductivity. Aluminium, for example, has a high temperature conductivity and absorbs more heat generated during machining and thereby promotes workpiece deformation resulting from thermal expansion. This makes it a difficult-to-cut material under dry conditions. As shown in Table 15.4 superalloys are not candidates for any dry machining operation.

Grey cast iron is on the opposite side of the spectrum and because of the small heat expansion coefficient, low cutting temperature and short chips, as well as the lubricating effect of the embedded graphite, it is a better choice for dry machining (especially turning and milling). MQL applies to hole-making operations (drilling, reaming, tapping and threading) seems to be irrespective of the type of work material machined. Sawing is a classical area of application for the MQL technique.

15.2.2 Dry machine tools and equipment

Removing coolants from the machining process requires special design features for the machine tools, especially to control thermal stresses and to guarantee reliable chip control and easy, fast and safe evacuation of the chips from the working space. From a practical point of view and based on previous experiences the conversion from wet to dry machining requires only a few, but decisive, modifications of existing machines, as shown in Fig. 15.22.

The machining centre shown in Fig. 15.23 can operate in three different modes, namely, wet, MQL and dry machining operations can be performed without tool spindle change. The changeover requires only a different ejector, pull rod adaptor and rotational coupling. Therefore, the high speed machining centre incorporates a two-channel system with a built-in lubrication/air generated supply close to the edges of the cutting tool in the main spindle, as illustrated in Fig. 15.24.

This concept allows the loss of mixture in the main spindle to be minimized. As a result, the above-described design feature provides a mixture volume flow within a spindle rpm ranging from 0 to 24 000 min^{-1}. The two-channel system allows, via the NC control system, the individual programming of the desired oil volume for numerous tools applied (milling cutters, drills, reamers, boring tools, treading mills and thrillers). For applications not requiring oil, as, for example, for milling of aluminium with PCD tools and grey cast iron, only supplying pressured air to the cutting tool is possible.

For dry machining another important criterion is the design of a smallest possible work area which allows the build-up of chips. On the other hand, chip disposal out of the work area and machine tool must

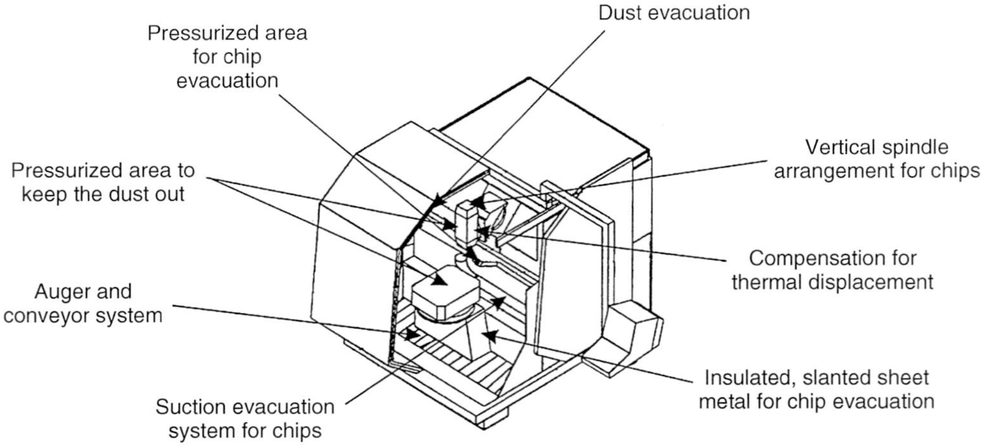

Fig. 15.22 Special functional elements of dry machining centre [2].

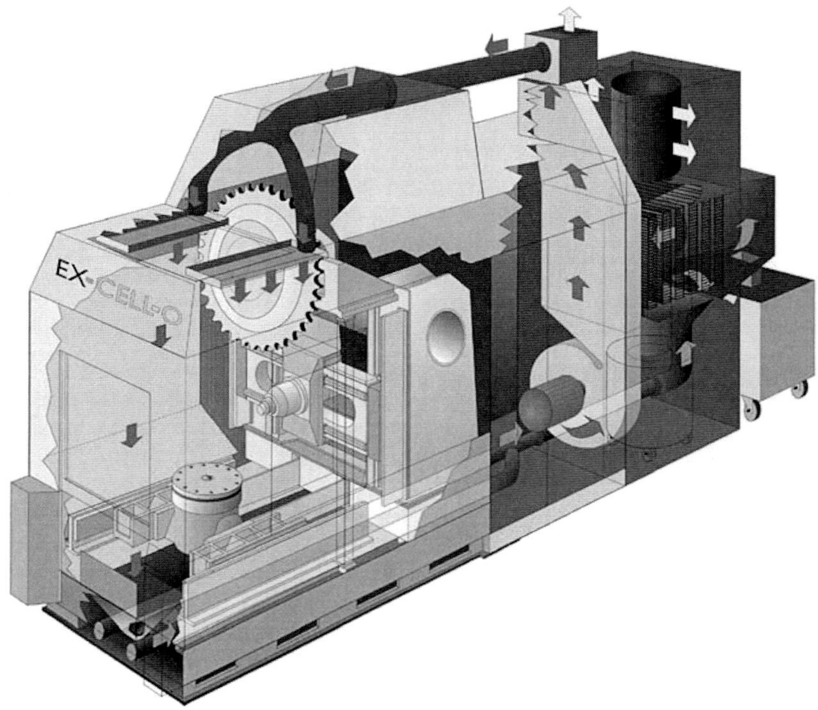

Fig. 15.23 Horizontal machining centre equipped with dry and MQL machining systems [3, 6]. (see Plate 15)

be guaranteed in the shortest possible distance. When machining grey cast iron chip disposal is done using a conventional conveyor. On the other hand, when machining aluminium alloys a safety certified exhaust system is necessary. Chips, metallic dust and oil mist are exhausted out of the work area. Exhausting causes the effect of an air curtain occurring – similar to a water curtain in the painting chambers.

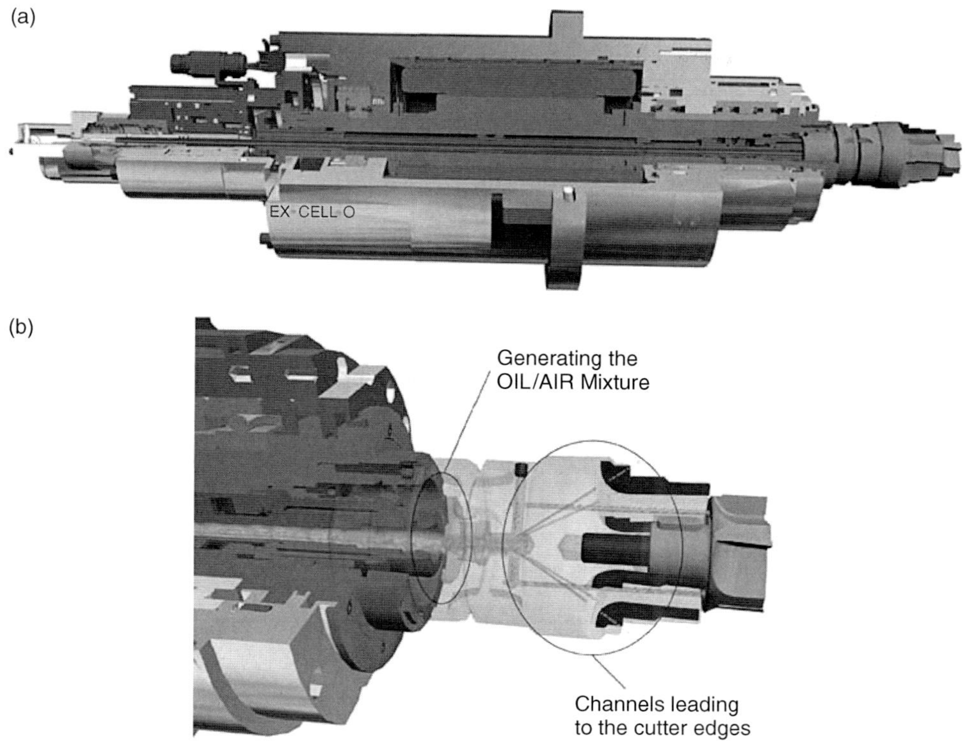

Fig. 15.24 Construction details of spindle with two-channel supplying system [3]: (a) cross-section of the spindle, (b) near-dry machining with MQL channels. (see Plate 16)

In addition, it is possible to optimize the air stream for dry machining. A high volume but steady air stream enables an optimal work area clean of dust and chips to be achieved. Improvements made within the machine layout (Fig. 15.25) and controlled guidance of the air stream result in better efficiency of the air stream and substantial reducing of its volume. The advantages observed in Fig. 15.25 are inclined sheet metal surfaces without protruding parts and bolt heads. Ideally, they should be dry.

In addition, vertical (Fig. 15.26) or oblique hanging and normal or oblique lying positions of the workpiece, depending on its weight, were tested in order to discourage chip build-up from the working zone [1, 15]. In some cases additional movement around the horizontal Z axis or workpiece tilting are necessary for an effective chip clearance from the working area. In the design shown in Fig. 15.26 chip handling and its transportation are facilitated by a totally enclosed working area and an automated chip conveyer.

A fundamental problem arising in dry drilling is removing the chips from the hole and away from the working area of the machine tool when there is no coolant to flush them out and transport. Consequently, a suction device is required or the spindle position can be changed as illustrated in Fig. 15.27. As can be seen in Fig. 15.27 two options are possible:

1. Drilling upwards by virtually turning the machine 'upside down'. In this arrangement chips drop from the hole due to the effect of force of gravity. A disadvantage is that chips are falling into the chuck and the spindle and they must be cleaned before tool changing.
2. Oblique-vertical location of the spindle, which causes the chips to fall directly onto the chip conveyor or be transported through the suction system.

Fig. 15.25 Optimal arranged work area for dry cutting.

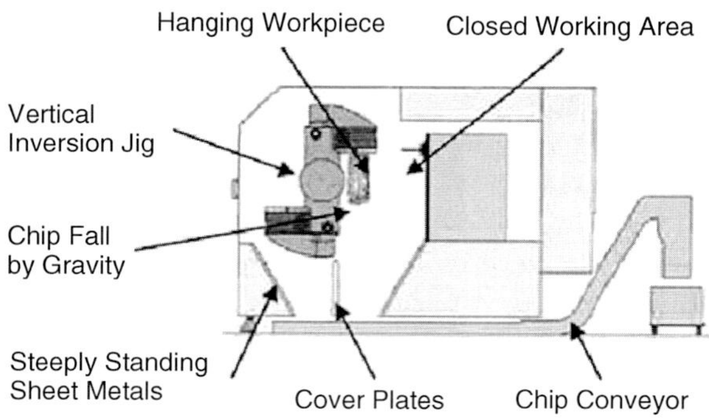

Fig. 15.26 Machining centre with vertically hanging workpiece [1].

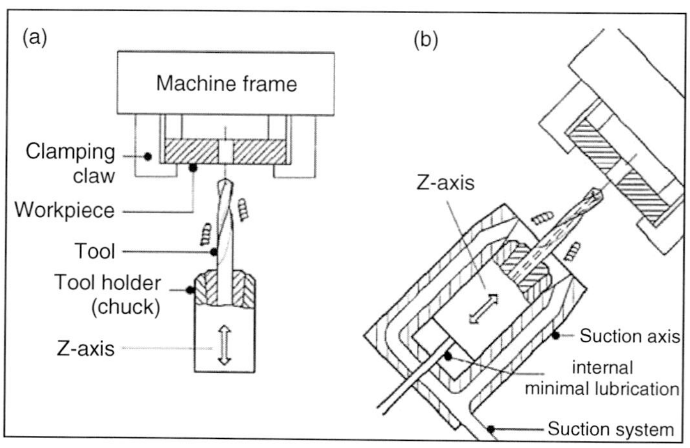

Fig. 15.27 Changes of spindle position in dry drilling [12]: (a) drilling vertically upwards; (b) oblique-vertical positioning.

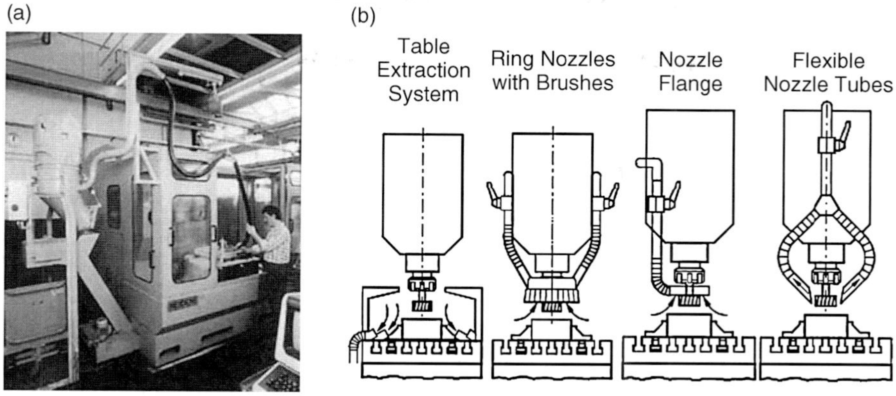

Fig. 15.28 A suction installation integrated with machine tool (a) and possible solutions of suction devices for milling applications (b) [1, 12].

In milling and drilling operations chips and particles can easily deposit on pallets, jigs or workpieces and contaminate the working area. In these cases special exhausting systems must be installed to remove the chips close to the cutting zone. For instance, Fig. 15.28a illustrates a chip suction installation consisting of swivel arm, pipe balancer and automatic waste trap on the left side. Moreover, Fig. 15.28b presents four different suction devices for chip extraction from the cutting zone. They utilize various techniques for cleaning the table and the workpiece from the chips when no cooling fluid is delivered to the machining space room. In drilling operations some exhaust hoods eject the chips in a specific direction in the working area, whereas other concepts transport the chips, via hoses, directly out of the machine tool. Recently, a chip extractor with a spring band spiral was developed which removed chips together with the MQL aerosol [1]. An interesting idea for cleaning the machine tool interior from small chips and dust is the use of a dry ice (CO_2 at −78°C) blasting device [24].

15.2.3 Dry machining operations

Dry drilling with carbide drills is the most problematic machining operation due to the serious difficulties in heat dissipation and chip transportation via helical flutes. As indicated from Fig. 15.29 dry drilling grey cast iron is manageable and yields tool life comparable to wet operation, whereas for ductile iron it decreases

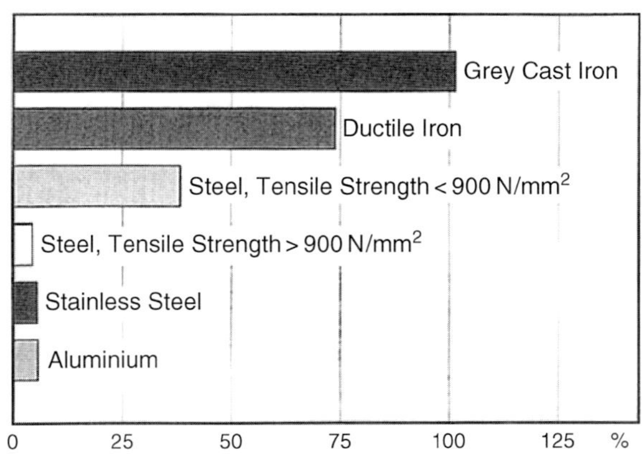

Fig. 15.29 Comparison of tool life for dry and wet drilling on different workpiece materials [2].

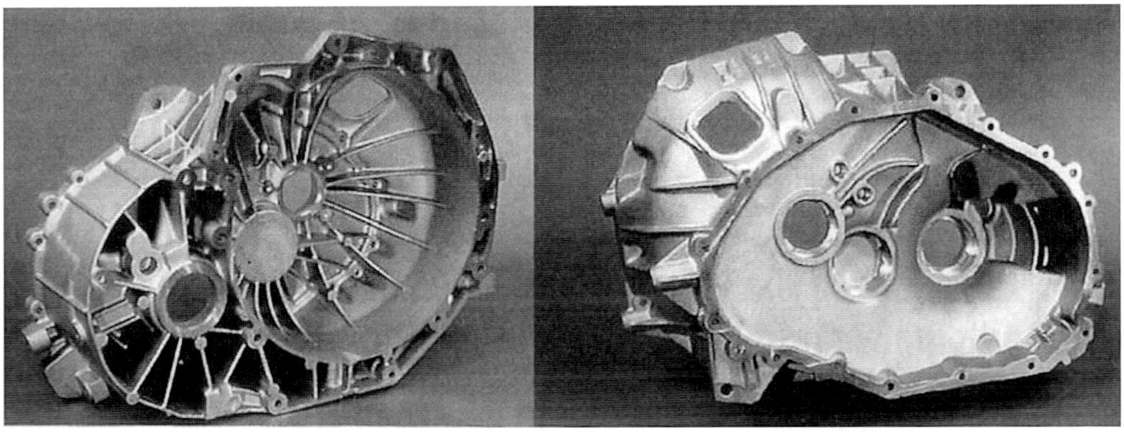

Fig. 15.30 Casings of clutch and gearbox machined on the machining centre from Fig. 15.23 [6].

by about 25%. On the other hand, quasi drilling in aluminium and its alloys is more reliable and effective with minimal coolant supplying which mitigates the tendency to adhesion. Dry and high speed drilling of end bearing caps made of a GG26Cr alloy cast iron at $v_c = 450\,\text{m/min}$ and $v_f = 6\,\text{m/min}$ (assuming $L_f = 100\,\text{m}$) is possible using spiral-fluted solid carbide drills coated with TiAlN or ceramic drills when keeping $v_c = 380\,\text{m/min}$ and $v_f = 3.13\,\text{m/min}$ [12]. TiAlN coated carbide drills were also found to be predestined for dry drilling at higher rates [13]. For example, in drilling of 8 mm holes in a 38MnVS5 steel wheel hub at 90 m/min and $f = 0.15\,\text{mm/rev}$, using such drills a cumulated total length of $L_f = 20\,\text{m}$ was achieved [12].

Figure 15.30 shows examples of automobile parts (AlSi10Mg die-cast casings) machined completely on a high speed machining centre under dry conditions. Drilling operations with PCD coated solid carbide drills of 4–13 mm diameter are performed at rotational speeds of 16 000–22 000 min^{-1} and feed rates of 0.3–0.5 mm/rev [6]. Also, multilayer and multiple coated drills can be of interest. Moreover, reaming operations are carried out at a rotational speed of 12 000 min^{-1} and a feed rate of 0.15 mm/rev. As reported in the keynotes [11] the dry reaming operations have the potential to replace grinding and honing in fine machining of bores in hardened materials. For example, dry reaming with a reamer equipped with PCD guiding pads and CBN cutting inserts results in a surface roughness Rz less than 3 μm and deviations in shape accuracy smaller than 9 μm.

The importance of the thermal problem arising in dry drilling operations of carbon and alloy steels is depicted in Fig. 15.31. During dry drilling in Ck45 steel parts the temperatures measured on the flanks of the TiN coated carbide tools near the cutting edges are higher at above 200°C in comparison with wet machining. As a result, a disturbance in chip control can occur. Ultimately, hot ribbon/snarl chips can render chip evacuation from the hole and subsequent blockading of the flutes can cause tool failure.

Comparatively, Fig. 15.32 shows how both workpiece and tool temperatures vary with the volumetric machining rating (from 100 cm^3/min at $v_c = 200\,\text{m/min}$, $a_p = 2.5\,\text{mm}$, $f = 0.20\,\text{mm/rev}$ to 350 cm^3/min at $v_c = 400\,\text{m/min}$, $a_p = 3.5\,\text{mm}$, $f = 0.25\,\text{mm/rev}$) in dry turning of medium carbon steel. It is evident that workpiece temperature decreases from 110°C to 65°C at the highest cutting speed of 400 m/min, whereas the tool temperature remains practically constant (it oscillates slightly between 65 and 70°C). This fact suggests that machining with high productivity needs the contact time between the tool and the workpiece to be as short as possible.

In dry boring operations the excessive heat generation causes the cutting tools to expand and cut bigger holes. For example, for boring an 8.5 mm blind hole of 25 mm length in SAE 1045 steel with cermet tools the diameter deviation at $v_c = 90\,\text{m/min}$ and $f = 0.2\,\text{mm/rev}$ was equal to ±5 μm. In the same operation the difference in surface roughness Rz (Fig. 15.33) measured at the top and the bottom of the hole was about 5 μm.

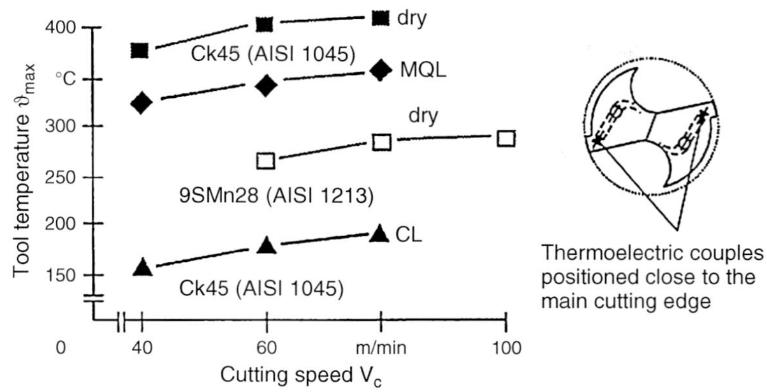

Fig. 15.31 Comparison of drill temperatures in wet (CL) and dry operations [11]. Drill diameter – 11.8 mm; feed rate – 0.2 mm/rev.

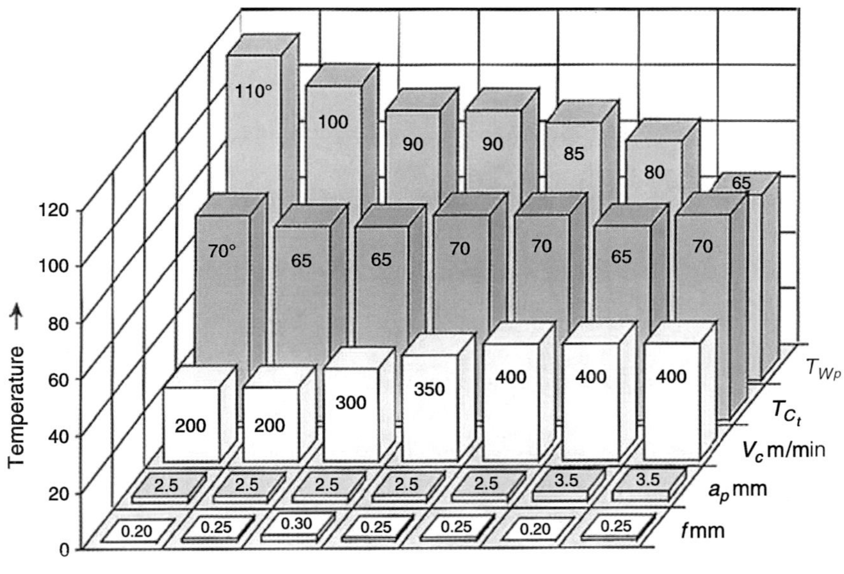

Fig. 15.32 Workpiece (T_{Wp}) and tool (T_{Ct}) temperatures vs volumetric machining rating in dry turning of C45 carbon steel [7]. Tool material – CVD-Ti(C,N)/Al_2O_3/TiN coated carbide.

Figure 15.34 illustrates how special self-lubricating coatings (soft tungsten carbide/carbon WC/C and chromium carbide/carbon CrC/C and hard/soft, e.g. titanium nitride plus soft top layer of molybdenum disulphide TiN + MoS_2) can diminish the flank wear of a face milling cutter in dry and high speed milling of an AlSi10Mg-wa alloy.

Primarily, dry milling reduces tool life of uncoated carbide cutters by 35% and this negative effect can be compensated by using inserts coated with hard/soft layers – this lessens the build-up of deposits and reduces friction. Quantitatively, hybrid coatings deposited on carbide substrates increase tool life by about 10% in dry operations compared to conventional CL technology. When the adhesion between the carbide tool and the aluminium workpiece was fully eliminated by the third-party action of a CVD-PCD coating, the tool life in the dry milling operation tested increased by about 30%.

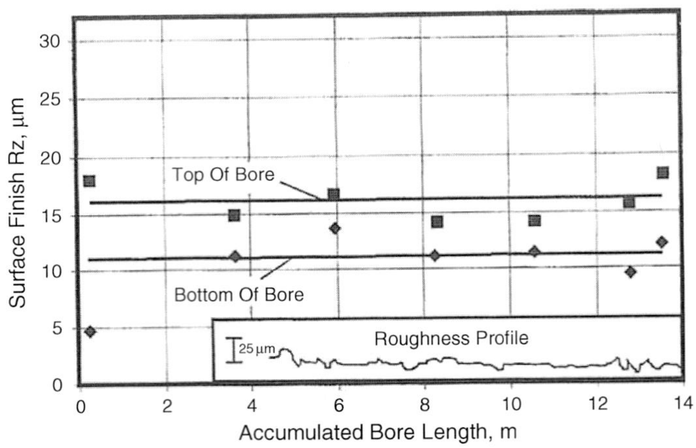

Fig. 15.33 Surface finish over bore length [2].

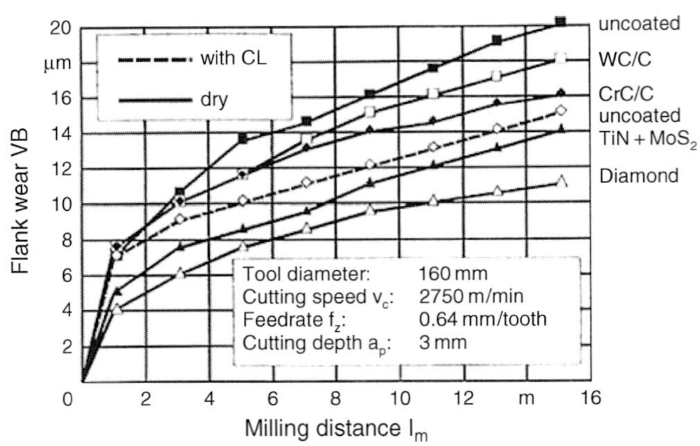

Fig. 15.34 Tool wear recorded for differently coated HC-K10 face milling cutters in dry milling of AlSi10Mg-wa aluminium casting alloys [11].

15.2.4 Introduction of near-dry machining

Wherever 100% dry machining cannot be realized for technological limitations as discussed in previous sections, minimum quantity/volume lubrication (MQL/MVL) may provide a considerable alternative to wet cutting. Typically, machining centres handle coolant volumes of 20 to 100 l/h. With MQL, only 0.03 l/h to a maximum 0.2 l/h of fluid is used [2].

However, depending on the type and main function of the fluid medium supplied to the cutting zone, a distinction can be drawn between minimum quantity lubrication (MQL) and minimum quantity cooling (MQC). Taking into consideration the two primary functions of a cooling lubricant (cutting fluid), i.e. cooling and lubricating effects in wet machining, a more general term – minimum quantity cooling lubrication (MQCL) – is appropriate for a near-dry process. It should be noted in Fig. 15.35a that MQCL is a system whose individual components, including supplying technology, MCQL media, parameter settings, cutting tools and machine tools, mutually affect the operation of all the others.

More frequently, the lubricants are vegetable oil-based, non-toxic and biodegradable fluid media which reduce friction and adhesion between the workpiece, the chip and the tool. Due to low thermal capacity

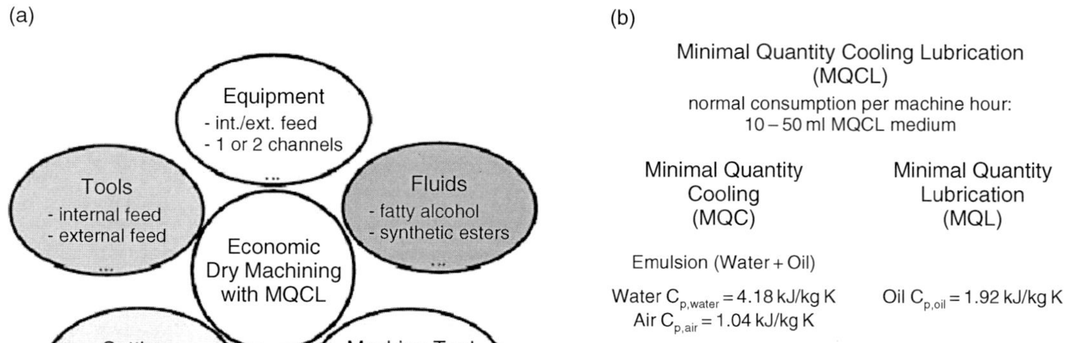

Medium	Cooling	Lubrication	Chip removal
Emulsion	excellent	good	excellent
Oil	good	excellent	good
Air pressure	little	no	little

Fig. 15.35 Structure (a) and definitions of near-dry processes [1, 11].

of both oil and air, and very small amount of the medium supplied, the direct cooling effect of the oil/air mix is very slight. As a result, the use of oil-based medium is termed an MQL strategy. In contrast, the MQC technique which utilizes emulsion, water or air (cold or liquid) provides practically no lubrication. Dry machining operations with MQC-fed systems can be essentially used to cool the tool or the machined part more efficiently. They can contribute to reducing or avoiding thermal problems which deteriorate part accuracy and surface finish.

15.2.5 MQL media and mixture supplying systems

According to up-to-date knowledge of MQL media, fatty alcohols and synthetic esters (chemically modified vegetable oil) are most commonly used as cutting lubricants in MQL applications. All important physical properties of these two groups of MQL media, their suitability to near-dry machining operations and optimal fields of applications are specified integrally in Table 15.5.

A synthetic polyol ester having a viscosity of 19 mm^2/s at 40°C was found to be the excellent cutting lubricant [17]. It has a high boiling temperature and flash point, and a low viscosity resulting in slower evaporation than a fatty alcohol. Moreover, it has the predominant adsorption ability on the freshly cut metal surfaces, which leads to the formation a thin lubricating film of oil on the workpiece and the tool [16]. It was also proven that the adsorption ability of an ester can be visibly intensified by atmospheric oxygen, as illustrated in Fig. 15.36. In this case (Fig. 15.36b) the lubricant particles, surrounded by a large amount of air containing oxygen, form tribologically effective lubricating film on the rake face of the tool and machined surface as well.

First of all, synthetic esters are biodegradable and are classified as Water Risk Category 1 due to their good toxicological characteristics. In metal cutting operations their functions are basically connected with reduction of friction and adhesion by ensuring good lubrication. Typical examples of such applications are drilling, fine boring, tapping and thread forming operations. Also in MQL turning, synthetic biodegradable esters are superior to a vegetable oil and provide the satisfactory cutting performance comparable to the effect of flood coolant supply.

At the same level of viscosity, fatty alcohols have a lower flash point and moderate lubricating effect in comparison with synthetic esters. Due to better cooling ability they are commonly used in near-dry

Table 15.5 Characteristics and application of MQL media [1].

Characteristic of MQL media		*Main applications of MQL media*	
Synthetic esters	*Fatty alcohols*	*Synthetic esters*	*Fatty alcohols*
Chemically modified vegetable oils	Long-chained alcohols made from natural Raw materials or from mineral oils	Application for machining technologies if	
• good biodegradeability • low level of hazard to water • toxicologically harmless		• primarily reduction of friction • high surface qualities are demanded • adhesive workpiece materials (build-up edge, apparent chips) • low cutting speeds and high specific area load • lubrication of supporting and/or guiding rails	• primarily heat removal • examples are: sawing, turning and milling of gray cast iron, machining of cast aluminum alloys
• high flash and boiling point with low viscosity • very good lubrication properties • good corrosion resistance • inferior cooling properties • vaporizes with residuals	• low flash and boiling point, comparatively high viscosity • poor lubrication properties • better heat removal due to evaporation heat • few residuals		

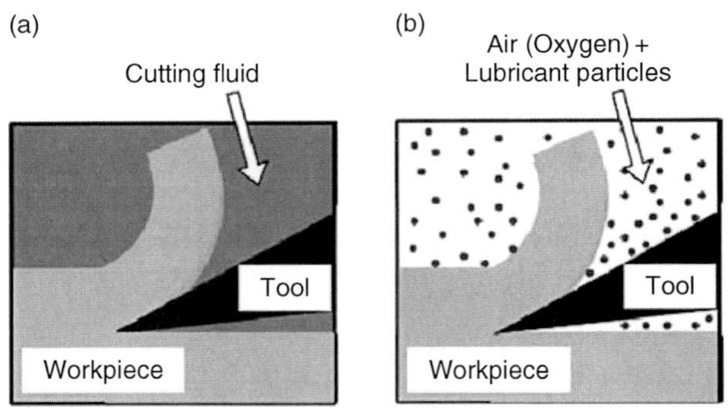

Fig. 15.36 Schematic of the action of fluid media in conventional wet machining (a) and MQL variant (b) [16].

machining of grey cast irons in which lubrication functions are served by the graphite smeared on the machined surface.

Based on specific properties of synthetic polyol esters a concept of a multifunctional fluid applicable to both machining and the machine tool parts such as the spindle bearings and hydraulic and slideway systems was developed [18]. For this purpose a fully synthetic biodegradable polyol ester of Viscosity Grade 32 was combined with some extreme pressure (EP) additives, but the satisfactory results were obtained for alkyl acid phosphate containing 8.4% P.

In near-dry machining, small lubrication particles supplying by micro-dosage pump through the capillary tube (Fig. 15.37a) are mixed with air and directed to the cutting tool from either outside the machine or through the machine spindle as shown in Fig. 15.37b.

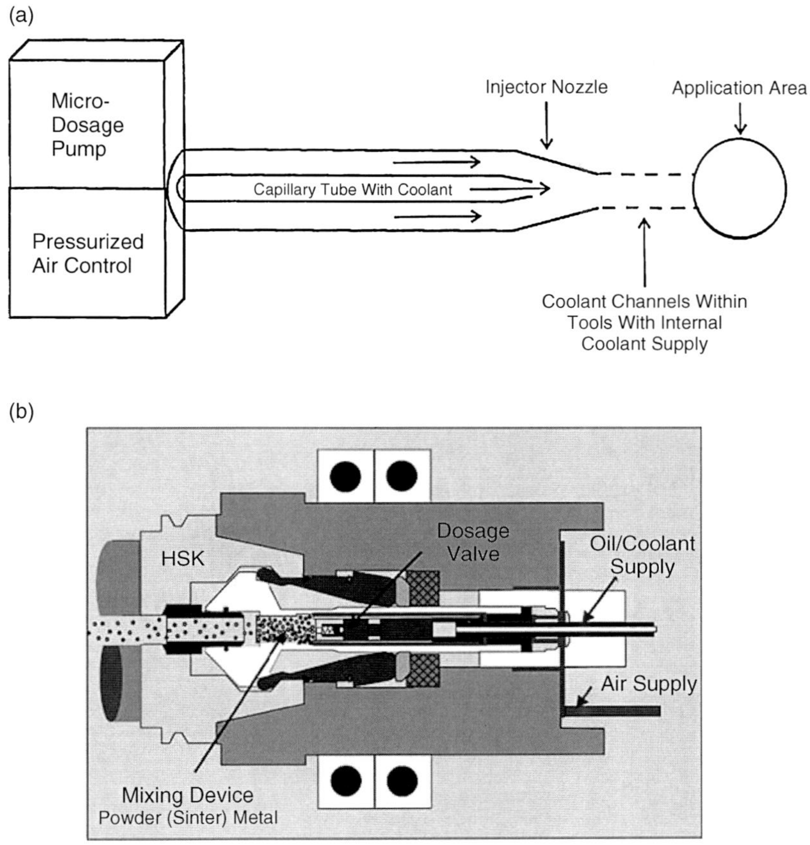

Fig. 15.37 Minimized lubrication system: a scheme (a) [2] and a practical design (b) [3].

As shown in Fig. 15.38, the oil/air mix can be induced into the tool on the cutting edge principally in four ways.

1. The external minimal lubrication when the oil/air supply is brought onto the cutting edge from the side (Fig. 15.38a). The limitation is in the length to diameter ratio of about 2 to 2.5, which practically means that no lubricant can be supplied to the cutting edges when machining long holes. This system requires the application of twin nozzles, which operate according to the Venturi principle. The inner nozzle supplies the jet of oil, while the outer nozzle surrounds the jet of oil with a shroud of air, which drags the oil along, preventing it from vaporizing. The resulting oil/air mixture at the centre of the jet achieves the desire lubricating effect at the tool surfaces. The results will only be satisfactory if the nozzles are aimed directly at the flutes of the tool.
2. Lubricating through the cutting tool using special rotary coupling between the mixed head and the mixture supplying tube, as shown in Fig. 15.38b. The disadvantage of this simple system is that for long tools, the lubricant can get stuck in the supply pipe. This obstacle can be partially overcome by forming the oil/air aerosol with nano-sized globules in the special booster [12]. As a result, the mixture can be supplied through a long spindle at rotational speeds in excess of 20 000 min^{-1}. Logically, only tools with internal coolant ducts can be applied with internal minimal lubrication.

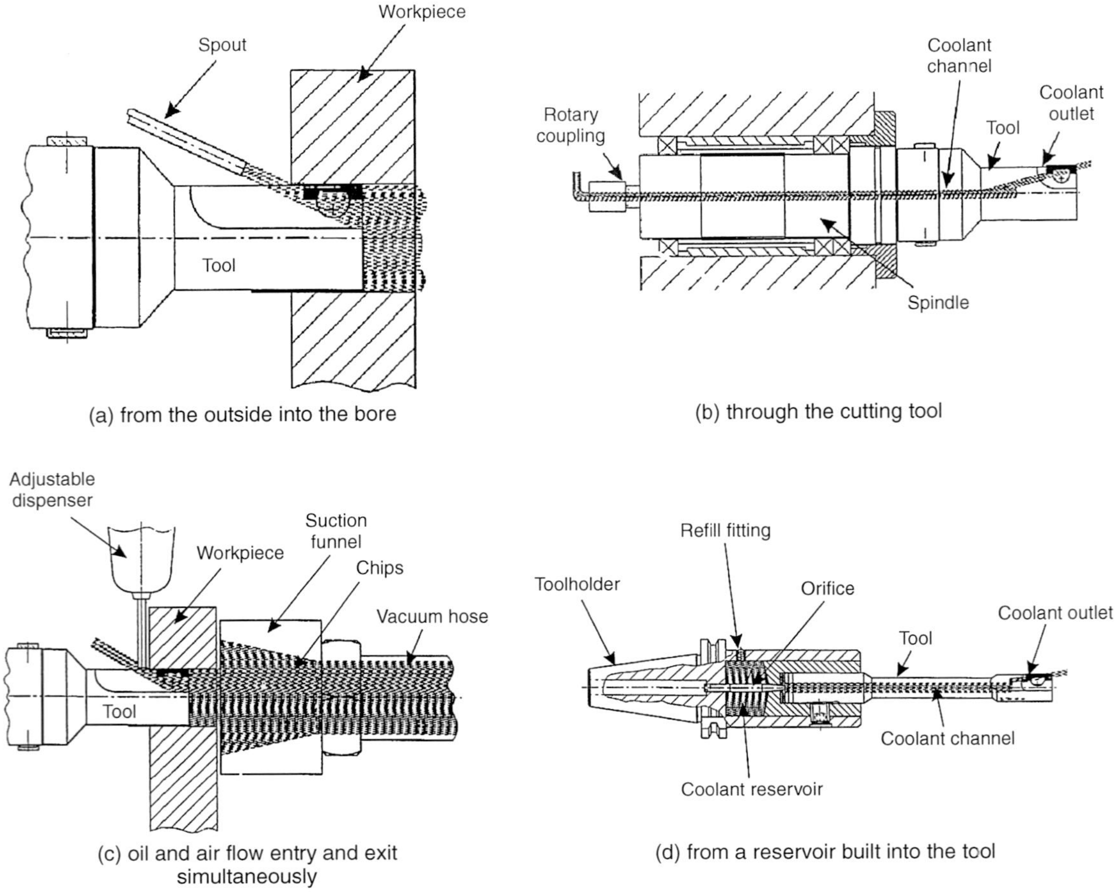

Fig. 15.38 Various methods of inducing oil/air mix [2].

3. MQL system equipped with a vacuum hose illustrated in Fig. 15.38c. Additionally, a suction hose localized at the workpiece guarantees clean and safe chip disposal.
4. The system using an oil reservoir (tank) built into the tool, as shown in Fig. 15.38d. Typically, it contains an integral pump, which is indirectly driven by the rotating spindle. When the spindle is turning, the flow of air through the nozzle draws the oil from the tank, mixes it into an aerosol and shoots down through the internal ducts of the tool to the cutting edges. In order to avoid the drip effect the chuck contains a centrifugal valve.

However, in high speed machining the centrifugal force from the high speed rotation of the conventional spindle (with rotating supplying tube) causes an oil separation effect [19]. As shown by simulation of such a mixture-supplying system, at rotational speeds of about 30 000 min^{-1} the oil drops are transported by a strong spiral flow, and a great number of oil drops stuck to the tube on the side of the mixture outlet. As a result, there is a 50% decrease in the oil supply rate at the oil/air mix outlet. This effect depends strongly on the volume and mass of oil drops, in such a way that smaller drops are able to reach a longer distance (the recommended diameter of oil drops is 2 μm). On the other hand, by using a non-rotating tube, the influence of spindle rotation on the mixture flow is substantially reduced and the rate of oil ejection from the tool ducts increases in comparison with the rotating supplying tube.

15.2.6 Near-dry machine tools and machining operations

Machine tools suitable for dry machining operations should have one internal and/or one external MQL supply, depending on the product range to be manufactured and the machining operations required. Machining centres employing tools with very small diameters should be equipped with both supply systems. In addition, the machine tool should be encapsulated, because the aerosol droplets can remain in the workspace for a long time, and equipped with a mist extraction (exhaust) unit. For the majority of automated cutting applications an internal MQL supply through the spindle is favoured, as, for example, in machining centres shown in Figs 15.23 and 15.39.

The most important issue is the integration and the control of MQL systems. Precision CNC-controlled dosing systems are integrated into the machine's spindle housing, feeding the appropriate aerosol volume to the cutting edge through ducts in the tool. Usually, when one MQL system supplies many different tools, different dosing quantities of air and fluid lubricant are required for individual tools. They have to be stored in the CNC system along with tool-related data. Because in an automated production system tools are frequently changed, it is necessary to ensure short response times of the MQL units.

In the case of a high speed machining centre from Fig. 15.23 a danger of dissipation of the oil/air mixture at high spindle rotation can be eliminated by applying a fine atomized liquid lubricating media. Since the ML media have relatively low viscosities and they can be affected by changing temperature, the lubricating effect can be reduced. Controlling the above-mentioned temperature-related viscosity change results in an optimal supply of a preset lubricating medium.

Figure 15.39 shows a typical machining centre with internal MQL supply through a hollow tool spindle for automotive applications, including such components as steering knuckles/arms (Fig. 15.38b), clutch and ABS casings and blockheads made of GK-AlSi10Mg aluminium pressure castings. In the case of machining steering knuckles the volume of lubricating fluid was reduced to 1 ml per part. Using eight machining centres in the robotized flexible manufacturing system, it was possible to produce 1400 knuckle pairs per day. Moreover one element was produced in the three setups, employing 27 machining operations per one machine tool, when the production cycle time was equal to 6 minutes and 5 seconds [14].

In order to compare the levels of thermal effects generating in dry drilling, cooling by air and lubricating by MQL the relevant temperatures were recorded on the drill flank faces by thermographic camera during its exit from the hole. It can be reasoned from the appropriate temperature fields shown in Fig. 15.40 that the maximum temperature of about 430°C recorded in dry drilling is reduced roughly twice (approximately

(a) (b)

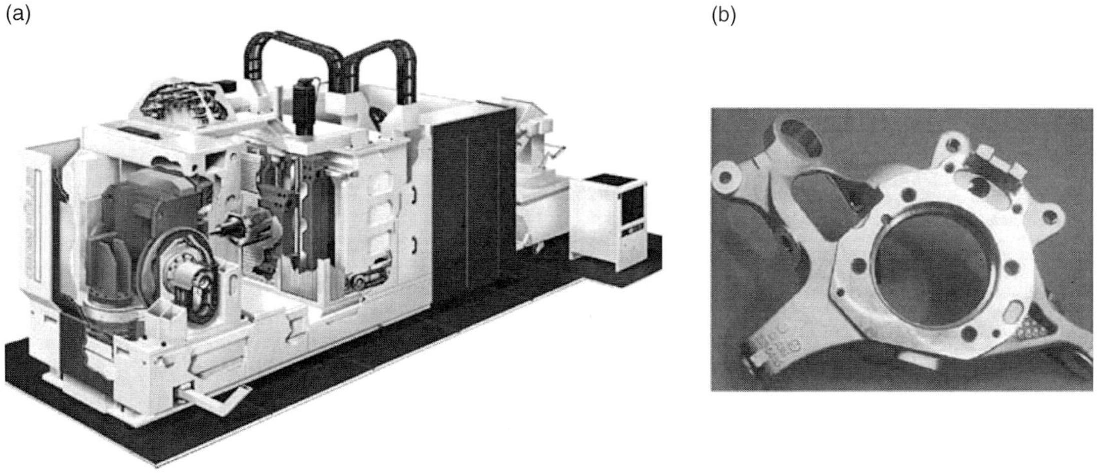

Fig. 15.39 Dry machining centre with MQL supply (a) and steering knuckle machined on it (b) [3, 14]. (see Plate 17)

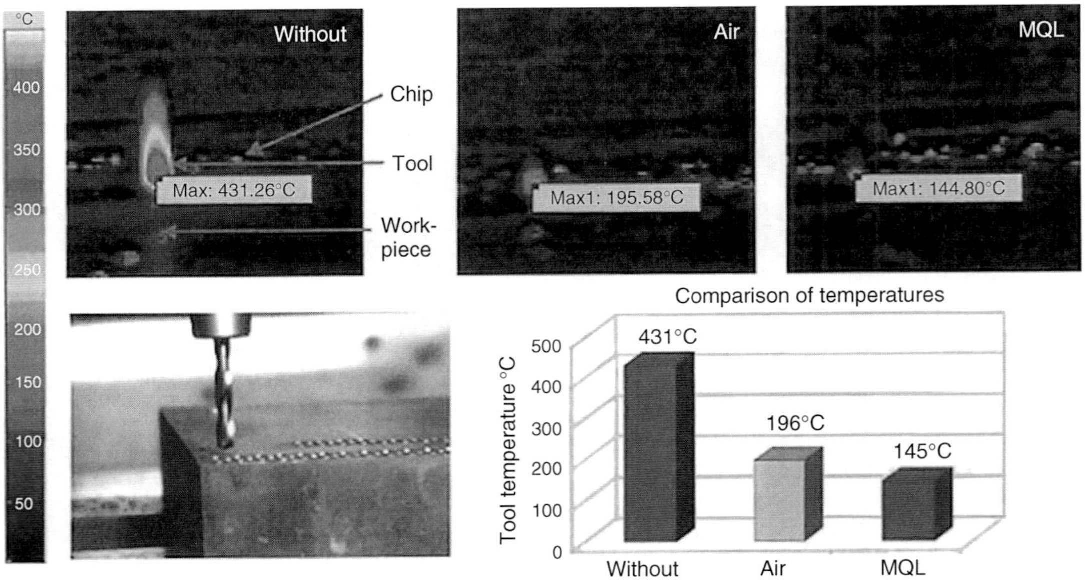

Fig. 15.40 Comparison of cutting temperatures recorded by thermographic camera in drilling at different cooling systems [9]. Workpiece material: GGG40 nodular iron, drill diameter of 8.5 mm, hole length of 42 mm, cutting conditions: $v_c = 130$ m/min, $f = 0.26$ mm/rev. (see Plate 18)

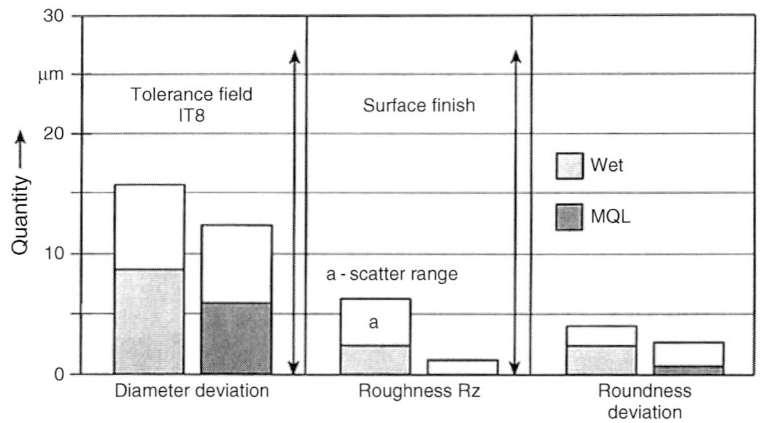

Fig. 15.41 Improving potential of fine boring (boring bar of 11.9 mm diameter and 5.5 mm long) when replacing wet operation by MQL [8]. Cutting tool material: K10 carbide when using 6% emulsion and PCD tool with MQL. Workpiece material: GD-AlSi9Cu3 aluminium alloy. Cutting conditions: $v_c = 163$ m/min, $f = 0.029$ mm/rev.

to 200°C) when applying air cooling, and down to 145°C at minimal lubrication with an oil/air flow rate of 30 ml/h [9]. It should be noted that drills with trigonal cooling ducts providing 50% higher channel cross-section, instead of conventional circular ones, were applied to intensify the cooling effect.

The improvements of the hole quality, when fine boring in a GD-AlSi9Cu3 aluminum alloy with MQL being supplied, are shown in Fig. 15.41. It can be noted that this lubricating system allows a narrower hole tolerance to be maintained, improves its roundness and, what is most visible, produces substantially better surface finish.

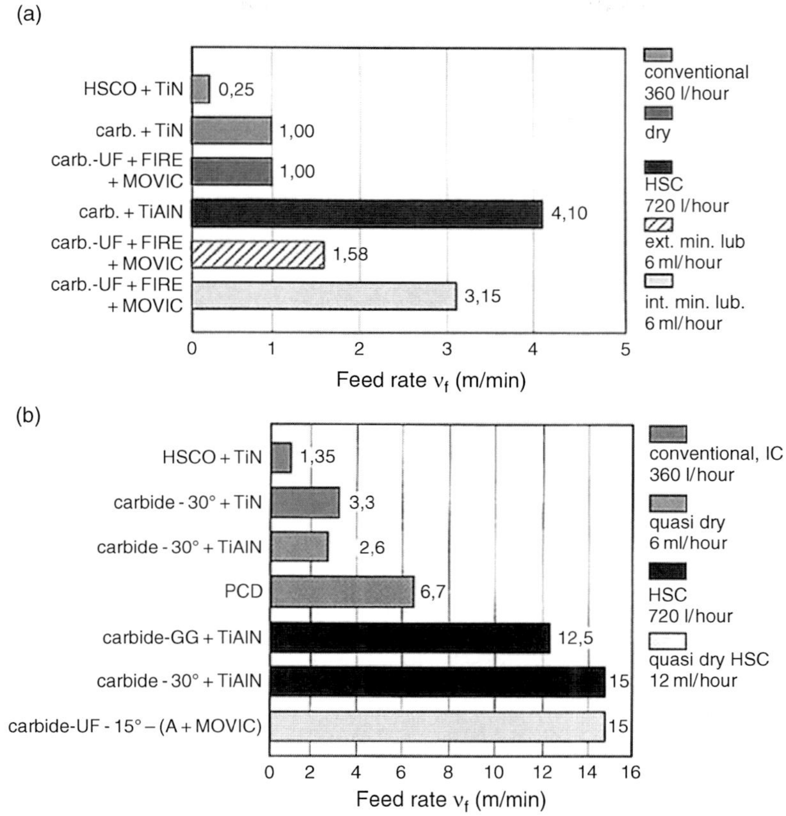

Fig. 15.42 Performance comparison between different tool materials in drilling of alloy steel (a) and Al-Si alloy (b) [12]. FIRE® and MOVIC® are trade marks by Gühring OHG, Germany and Vilab, Bern, respectively.

The diagrams in Fig. 15.42 clearly depict all previously discussed advantages of the MQL technique concerning various aspects of machining (e.g. workpiece and tool materials, tool design, coatings deposited and cooling conditions). It is evident, when comparing two MQL supplying systems at the same volume of lubricating media of 6 ml/h (Fig. 15.42a), that in drilling forged 38MnVS5 steel the most economic is an internal supplying system (it allows the double feed velocity to be used). On the other hand, the higher productivity was achieved in HSC operations when using flood cooling of 720 l/h.

The drilling of aluminium-silicon alloys is a process where dry cutting is impossible due to high ductility of the workpiece material. Without cooling and lubrication the chip sticks to the tool and breaks it shortly after the cutting begins. In the comparison (Fig. 15.42b), when drilling bores of 7.35 mm diameter and 28 mm length in an AlSi9Cu3 aluminium alloy differently coated tools with HS-Co steel, conventional and ultra-fine (UF) graded carbides and PCD were tested under flood and quasi-dry conditions. Under HSC conditions and TiAlN coated carbide tools, the best productivity was revealed for an intensive consumption of cutting fluid of 720 l/h. On the other hand, for MQL machining option and hard/soft coated UF-carbide drills the aerosol amount of 12 ml/h is enough. Simple calculations indicate that in the latter case machining requires 60 000 times less lubricant at the same productivity.

In the drilling of an aluminium-silicon alloy with 7% silicon (SAE 323) using uncoated K10 solid and PCD coated drills with an MQL supply at the flow rate of 10 ml/h, the process performance (in terms of cutting forces, tool wear and quality of holes) was similar and comparable to the results obtained with flood cooling using soluble oil (1 part oil for 25 parts of water). The cutting speed of 300 m/min and the feed rate of 0.1 mm/rev were employed [20]. The efficiency of the MQL technique in the drilling of cast aluminium alloys (ACP 5080) at a higher cutting speed of 105 m/min and higher feed rates was reported in [21]. On the other hand, it is concluded there that flood lubrication is superior at lower cutting speeds and feed velocities.

The predominant influence of heat isolating coatings (TiAlN, CrCN and TiCN) in drilling titanium alloy Ti6Al4V at 30 m/min, under near-dry conditions with external cooling ($p = 3.5$ bar, $Q = 50$ ml/h), on the maximum workpiece temperature is documented in [22]. The differences of temperatures recorded along the length of the hole for uncoated and coated drills were in the range of 150–200°C (highest discrepancies were recorded near the top surface, i.e. during tool entrance).

A very specific application of minimum quantity cooling lubrication (MQCL) is hard turning (HTMF) of a through-hardened AISI 4340 steel of 46 HRC [23]. The optimized composition of a cutting fluid was a coolant-rich (60%) one with minimal additives. During machining a thin pulsed jet of fluid with an injection rate of 2 ml/min was delivered at a high injection pressure of 20 MPa and pulsing rate of 600 pulses/min. In particular, in the cutting speed range of 40–120 m/min the cutting temperature was substantially reduced in comparison with the dry hard process (at the highest speed applied of about 400°C), even below the level recorded at wet cooling.

References

[1] K. Weinert, I. Inasaki, J.W. Sutherland, T. Wakabayashi, Dry machining and minimum quantity lubrication, *CIRP Annals*, Vol. 53/2 (2004), 511–537.
[2] B.P. Erdel, *New Dimensions in Manufacturing*, Hanser Gardner, Cincinnati, 1998.
[3] www. crosshuellerex-cell-olamb.com.
[4] B.P. Erdel, *High-Speed Machining*, SME, Dearborn, 2003.
[5] G. Byrne, D. Dornfeld, B. Denkena, Advancing cutting technology, *CIRP Annals*, Vol. 52/2 (2003), 483–507.
[6] A. Feinauer, A. Stoll, Trocken spanen in der Großserie, *Werkstatt und Betrieb*, Vol. 133, No. 1–2 (2000), 15–16.
[7] U. Rinken, K. Christoffel, Trockenbearbeitung entlastet die Umwelt, *Werkstatt und Betrieb*, Vol. 129, No. 6 (1996), 574–576.
[8] N.N., Schnell und trocken, 2. Schmalkalder Werzeugtagung am 5. und 6. November 1996, *Werkstatt und Betrieb*, Vol. 130, No. 5 (1997), 356–357.
[9] P. Hänle, D. Gsänger, Prozesssichere Werkzuege fürs Bohren mit MMS, *Werkstatt und Betrieb*, Vol. 136 (2003), No.11, pp. 58–61.
[10] R. Lamers, Der vierte Weg zur MMKS (die Minimalmengen-Kühlschmierung, Werkstatt und Betrieb, Vol. 137, No. 5 (2004), 57–59.
[11] F. Klocke, G. Eisenblätter, Dry cutting, *CIRP Annals*, Vol. 46/2 (1997), 519–526.
[12] T. Cselle, *Carbide Drills: at the Peak of Development?*, Gühring OHG, Albstadt, Germany.
[13] H.K. Tönshoff, W. Spintig, W. König, A. Neises, Machining of holes. Developments in drilling technology. *CIRP Annals*, Vol. 43/2 (1994), 551–561.
[14] J. Fleischer, J. Schmidt, M. Dyck, Es kommen trockene Zeiten, *Werkstatt und Betrieb*, Vol. 137, No. 5 (2004), 49–52.
[15] A. Feinauer, Strömungsgünstiger Arbeitsraum, *Werkstatt und Betrieb*, Vol. 133, No. 6 (2000), 37–40.
[16] T. Wakabayashi, I. Inasaki, S. Suda, H. Kokota, Tribological characteristics and cutting performance of lubricant esters for semi-dry machining, *CIRP Annals*, Vol. 52/1 (2003), 61–64.
[17] S. Suda, H. Yokota, I. Inasaki, T. Wahabayashi, A synthetic ester as an optima cutting fluid for minima quantity lubrication machining, *CIRP Annals*, Vol. 51/1 (2002), 95–98.
[18] S. Suda, T. Wahabayashi, I. Inasaki, H. Yokota, Multifunctional application of a synthetic ester to machine tool lubrication based on MQL machining lubricants, *CIRP Annals*, Vol. 53/1 (2004), 61–64.
[19] T. Aoyama, Development of a mixture supply system for machining with minimal quantity lubrication, *CIRP Annals*, Vol. 51/1 (2002), 289–299.
[20] D.U. Braga, A.E. Diniz, G.W.A. Miranda, N.L. Coppini, Using a minimum quantity of lubricant (MQL) and a diamond coated tool in the drilling of aluminium-silicon alloys, *J. Mater. Proc. Technol.*, Vol. 122 (2002), 127–138.
[21] J.F. Kelly, M.G. Cottrell, Minimal lubrication machining of aluminium alloys, *J. Mater. Proc. Technol.*, Vol. 120 (2002), 327–334.

[22] R.P. Zeilmann, W.L. Weingärtner, Analysis of temperature during drilling of Ti6Al4V with minimal quantity of lubricant, *J. Mater. Proc. Technol.*, Vol. 179 (2006), 124–127.
[23] A.S. Varadarajan, P.K. Philip, B. Ramamoorthy, Investigations on hard turning with minimal cutting fluid application (HTMF) and its comparison with dry and wet turning, *Int. J. Mach. Tools Manuf.*, Vol. 42 (2002), 193–200.
[24] W. Zielasko, Chip removal, transport and machine tool cleaning technology, *Proceedings of First International Workshop on High Performance Cutting*, Paris, 2002, pp. 53–59.

15.3 Hard Part Machining

15.3.1 Definition of basic feature of hard part machining (HM)

Traditionally, the finishing operations on the machine parts in a highly tempered or hardened state with hardness value in excess of 60 HRC are grinding processes, but recently hard cutting operations using tools with geometrically defined cutting edges have become more and more capable of replacing them and guaranteeing comparable surface finish (possible values of 0.1 μm *Ra* in CBN high precision machining) and IT (tolerance) standards (correspondingly IT3), as shown in Fig. 15.43. Their applications have spread over such leading industrial branches as automotive, roller bearing, hydraulic and die and moulds sectors. Gear wheels, geared shafts, bearing rings and other transmission parts are typically machined by turning, while high speed milling is a leading technology in the die and moulds industry.

In particular, the hard cutting process performed with ceramic or CBN tools can often cut manufacturing costs, decrease production time (lead-time), improve overall product quality, offer a higher flexibility and allow dry machining by eliminating coolants (Fig. 15.43b). The two first advantages lead to substantial shortening of the traditional technological chain with heat treatment and finish grinding after rough operation, as illustrated in Fig. 15.44.

In general, hard turning can provide a relatively high accuracy for many hard parts but sometimes important problems arise with surface integrity, especially with undesirable patterns of residual stresses and the changes of subsurface microstructure, so-called *white layer*, which reduces fatigue life of turned surfaces.

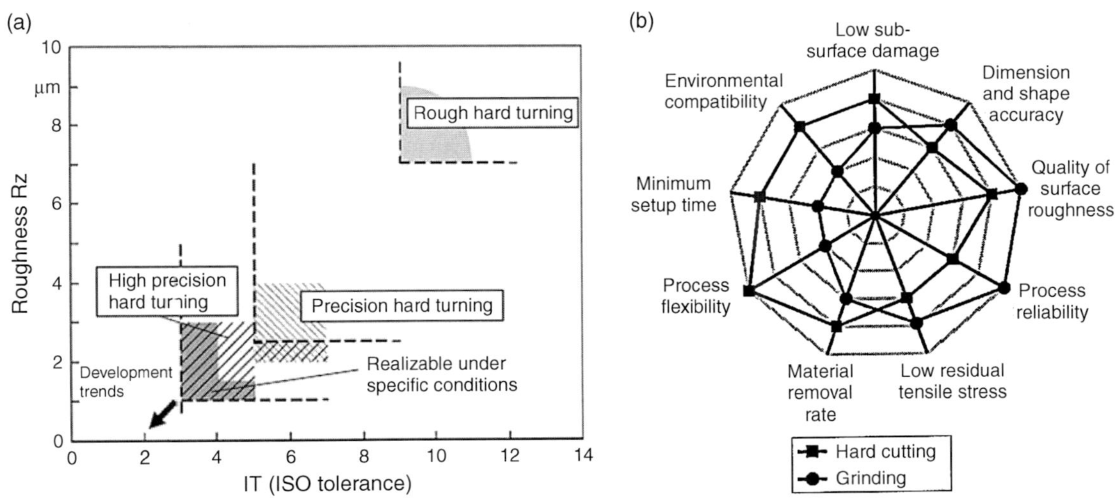

Fig. 15.43 Achievable surface roughness and ISO tolerance in hard turning (a) [1] qualitative comparison of hard turning and grinding processes [2].

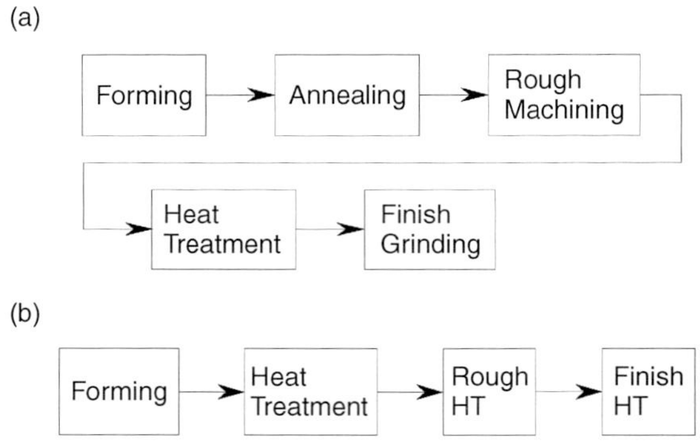

Fig. 15.44 Technological chains for conventional production process (a), and production process with hard turning (HT) operations (b).

Grinding and turning are such opposite machining operations that their full substitution is not always easy or possible. Some of the inherent differences between these machining processes are as follows [3]:

1. Hard turning is a much faster operation because it can be done in one setup and pass under dry conditions.
2. Lathes offer more production flexibility.
3. Rough and finish operations can be performed with one clamping using a CNC lathe.
4. Multiple turning operations are easier to automate through tool changes on turning centre or turning cell.
5. Since hard turning is done dry, there are no costs for coolant, its maintenance or disposal.

15.3.2 Physical aspects of hard machining

Hard machining is performed under unique technological and thermo-mechanical conditions and, as expected, the cutting process mechanisms (chip formation, heat generation, tool wear) differ substantially from those observed in machining soft materials. As shown in Chapter 3, HM is performed also as dry and HSM processes. In particular, while small depths of cut (0.05–0.3 mm) and feed rates (0.05–0.2 mm/rev) are used, both small undeformed chip thickness and the ratio of the undeformed chip thickness to the radius of the cutting edge are obtained in such processes. This geometrical relationship leads to an effective rake angle of −60° to −80° and as a result extremely high pressure is generated to remove material in the vicinity of the cutting edge. Moreover, a large corner radius causes the components of the resultant cutting force to be high in conjunction with extreme high thermal stresses, as shown in Fig. 15.45. It can be observed in Fig. 15.45c that cutting forces increase drastically with machining materials with hardness higher than about 45 HRC (this value is often referenced as a lower limit of HPM).

In particular, larger negative rake angle and tool corner radius influencing the passive force F_p increase remarkably and this effect causes an absolute stable and rigid process to be provided. This requirement has to be especially retained when using superhard tools with smoothing, multi-radii geometry, so-called wiper tools, as shown in Fig. 15.46.

The formation of saw-tooth chips (Fig. 15.47) is one of the primary characteristics in the machining of hardened steels with geometrically defined cutting tools. Catastrophic failure within the primary shear zone during saw-tooth chip formation is usually attributed to either cyclic crack initiation and propagation or to the occurrence of a thermo-plastic instability [7, 8]. For example, for the orthogonal machining of a through-hardened AISI 52100 bearing steel of 50–65 HRC with PCBN tools the onset of chip segmentation due to adiabatic shear was observed at relative cutting speeds below 1 m/s [9]. In addition, these shear bands are formed at frequencies in the range of 50 kHz to 120 kHz when cutting speed was varying from 0.35 m/s to 4.3 m/s and segment spacing becomes more periodic as cutting speed is increased. Producing saw-tooth chips in orthogonal cutting of the 100Cr6 steel of HV730 hardness at cutting speeds of 25–285 m/min and

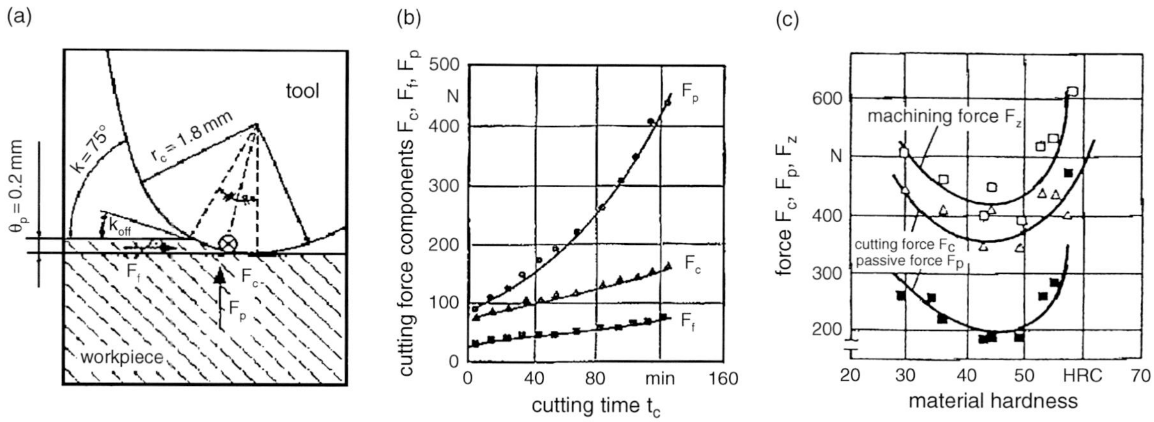

Fig. 15.45 Geometrical conditions (a) and time dependence of cutting forces (b) in HT of 16MnCr5(AISI 5115) steel of 60–62 HRC hardness with CBN tools at cutting speed of 145 m/min [4] and the influence of steel hardness on cutting forces ($v_c = 90$ m/min, $f = 0.15$ mm/rev, $a_p = 0.9$ mm) [5] (c).

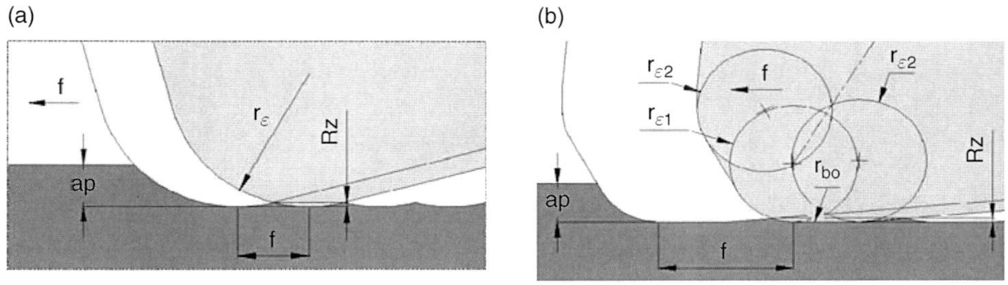

Fig. 15.46 Comparison of inserts with conventional geometry (a) and wiper shape (b) [6]. Symbols: f – feed; a_p – depth of cut; r_ε – radius of the cutting edge; $r_{\varepsilon 1}$ and $r_{\varepsilon 2}$ – radii of wiper curvature; r_{bo} – radius of smoothing part; Rz – valley-to-peak height.

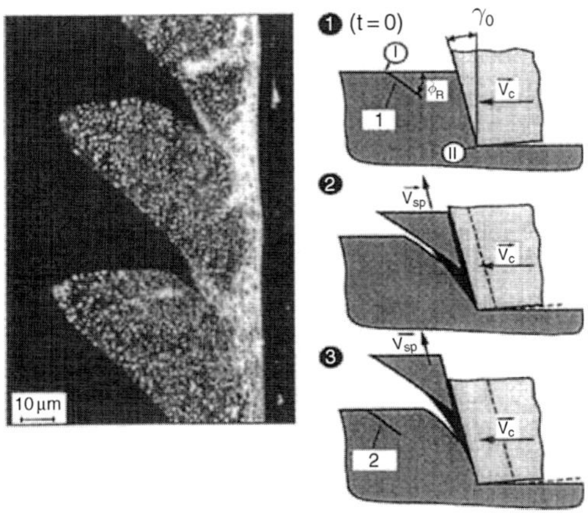

Fig. 15.47 Chip formation mechanisms for hardened 100Cr6 (60–62 HRC) steel and undeformed chip thickness of 0.05 mm (case when $h > 0.02$ mm) when using PCBN tool [12].

feed rates of 0.0125–0.2 mm/rev was confirmed by Poulachon *et al.* [10]. Moreover, Shaw *et al.* reported that in face milling of case carburized AISI 8620 steel (61 HRC) with PCBN tools at $v_c = 150$ m/min, $f = 0.13$ and 0.25 mm/rev and $a_p = 0.13$ and 0.25 mm the chip formation is of a cyclic saw-tooth type [11].

Figure 15.47 illustrates the cyclic mechanism of the formation of chip segments due to crack initiation (numbered successively by 1 and 2) when the undeformed chip thickness is higher than 0.02 mm (for very small undeformed chip thicknesses less than $h < 0.02$ mm continuous chips are formed).

When machining hardened 100Cr6 bearing steel the direct stresses σ_{VB} (Fig. 15.48a) reaching approximately 4000 MPa independent of flank wear result in high mechanical and thermal stresses extended on the machined surface of the workpiece. Thermal stresses result mainly from the friction between flank wear land and the workpiece which for the friction coefficient of 0.2–0.3 causes high tangential stress [13]. The temperature field due to friction when assuming a semi-infinite moving body with an adiabatic surface and the heat partition to the workpiece of 80% is as shown in Fig. 15.48b.

When the temperature near the machined surface exceeds the γ–α transition temperature martensite produced by friction development can form so-called white layers observed in chip micrographs.

Figure 15.49 shows how work material with increasing hardness affects the tool edge temperature in the vicinity of the flank face measured by means of a two-colour pyrometer with a fused fibre coupler.

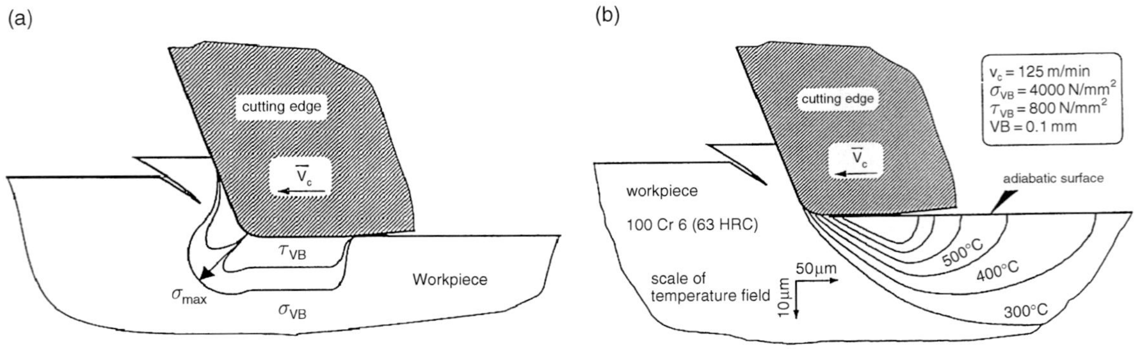

Fig. 15.48 Stress distribution before crack propagation (a) and temperature field (b) for the machining case from Fig. 15.24 (cutting speed of 125 m/min and tool wear $VB = 0.1$ mm) [13].

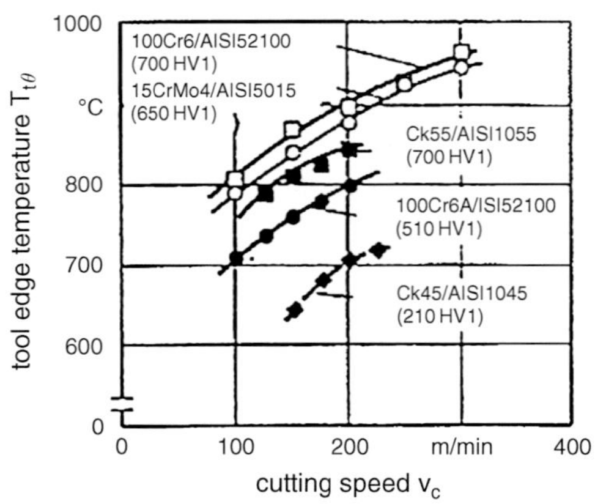

Fig. 15.49 The influence of work material with defined hardness on the tool flank temperature; depth of cut = 0.1 mm, feed rate = 0.1 mm/rev [14].

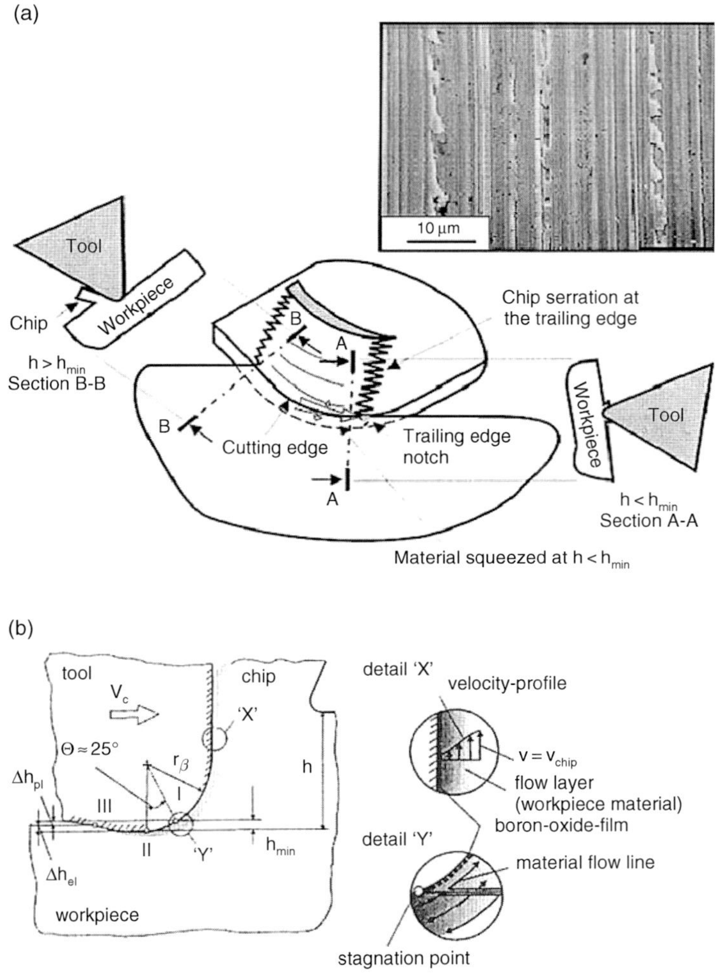

Fig. 15.50 Mechanism of material side flow during hard turning [4, 16].

There is a close relation between the tool temperature and the hardness of the work material used, and the cutting speed causes a substantial increase of the temperature [15]. For machining high carbon-chromium AISI 52100 bearing steel of highest 700V1 hardness with low content CBN (60% vol.) the temperature is about 800°C at $v_c = 100\,\text{m/min}$ and it increases to about 950°C at a cutting speed of 300 m/min.

The characteristic phenomenon of material side flow generated during hard turning operations is shown in Fig. 15.50. According to many investigations, it is attributed to the squeezing effect of the workpiece material between the tool flank and the machined surface when the chip thickness is less than a minimum value h_{min}. On the other hand, it can also be originated from the flowing of plastified material through the worn trailing edge to the side of the tool [16]. It causes substantial deterioration of surface finish because the squeezed, flake-like, hard and very abrasive material (right upper photo) is loosely attached to the generated surface along the feed marks. This negative effect is more significant for higher cutting speeds and larger tool nose radii, and is more intensive when tool wear progresses.

The material behaviour when the uncut chip thickness h is less than the minimum chip thickness h_{min} in point (I) defined by the stagnation angle $\Theta \approx 25°$ is illustrated in Fig. 15.50b. Because chip formation does not occur elastic–plastic deformation of the surface layer are observed as depicted in detail 'Y'.

In point (II) the elastic component Δh_{el} springs back after moving the tool and behind point (III) the plastic deformation component Δh_{pl} leads to the final deformation of the surface layer.

Figure 15.51 illustrates typical tool wear features observed in finish turning on CBN tools with the predominant wear effect VB_{max} and VB_C localized on the clearance face and tool corner, respectively. In many hard finishing turning operations with mixed ceramic tools the notch wear is also concentrated on the active secondary cutting (trailing) edge as shown in Fig. 15.52b. It is responsible for the profile sharpening effect [18] causing the blunt irregular initial peaks to be transformed to final sharp individual peaks from the side of the trailing edge by copying the notch grooves.

The contact area between worn PCBN tool, workpiece and chip is divided into five characteristic sections as shown in Fig. 15.52a. In the outer zones 1 and 5 non-tribological effects occur. Similarly, in zone 3, where a stable flow layer prevents tool wear they were not observed. Probably, the protective layer consisting of boron oxide films (details 'Y' and 'X' in Fig. 15.52b) is formed at high contact temperatures.

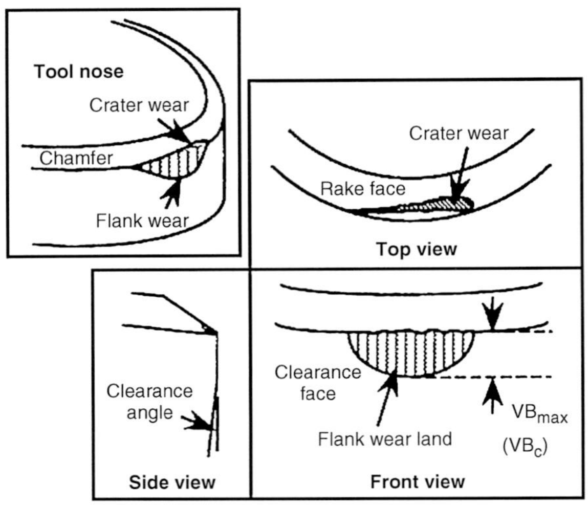

Fig. 15.51 Typical wear types in CBN finish hard turning [17].

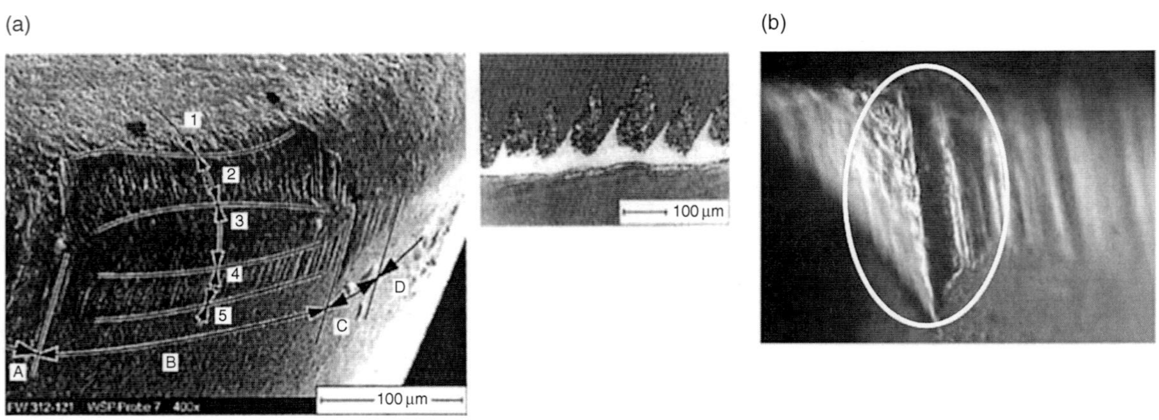

Fig. 15.52 Typical wear pattern of a PCBN tool corresponding to saw-tooth chip formation (a) [5] and developed notch wear on the secondary flank face of mixed ceramic tool (b) [18]. Cutting conditions in case (a): 16MnCr5 of 60–62 HRC, $\alpha_0 = 7°$, $\gamma_0 = -7°$, $\kappa_r = 95°$, $\varepsilon_r = 80°$, $r_\beta = 130\,\mu m$.

The flank wear concentrated in zone 4 with characteristic scouring areas is affected by severe abrasive wear. In zone 2 predominant wear mechanisms are material deformation and continuous sliding friction caused by the chip. The bottom side of the saw-tooth chips shows a developed white layer the scale of which is almost independent of tool wear progress.

15.3.3 Applications of HM technology

HM was found to be very efficient in many industrial branches such as automotive, aerospace, bearing, hydraulic and die/mould industries. It can be realized in a number of various machining operations (turning, milling, drilling, broaching, reaming and threading) performed with cermet, ceramic, PCBN and PCD tools. As a rule, CBN tools are recommended for hardnesses above 50 HRC up to about 70 HRC to generate finishes of down to $Ra = 0.3\,\mu m$. The cutting speed for HT of 60 HRC steels with CBN inserts varies from 100 to 200 m/min. Conventional values of feed rates oscillate between 0.05 and 0.20 mm/rev. Ceramics and cermets are suitable for lower hardnesses of 40–50 HRC at cutting speeds of 300–350 m/min and surface finish down to $Ra = 0.6\,\mu m$.

A range of machining operations performed on lathes on hardened transmission parts including longitudinal continuous and interrupted turning, facing and boring, and grooving and threading are shown in Fig. 15.53. Further improvement of the hard turning process can be achieved by using special single-edged tools which performed orthogonal cuts named hard plunge turning [19, 20].

Today, the machining of hard steels is often combined with high speed machining, especially when it comes to milling operations. Figure 15.54 illustrates two exemplary hard roughing milling operations on convex (Fig. 15.54a) and concave (Fig. 15.31b) surfaces using solid and indexable insert ball-nose end mills, respectively. In addition, Fig. 15.55 shows the milling of a complex structure on a five-axis CNC milling machine. In particular, CNC machine tools for hard machining have provided a very competitive alternative to grinding and EDM.

Figure 15.56 gives machining conditions for roughing, semi-roughing and finishing milling operations (in dry or finishing in HSM options) with coated solid end- and ball-nose end mills of fine-grained tool substrates.

Figure 15.57 shows a hard broaching machine (a) along with diamond broaching tool (b) for making internal splines in a hardened gear wheel (c). The broach is coated with metallically bonded diamond grains.

Fig. 15.53 Typical examples of hard turning, boring and threading operations.

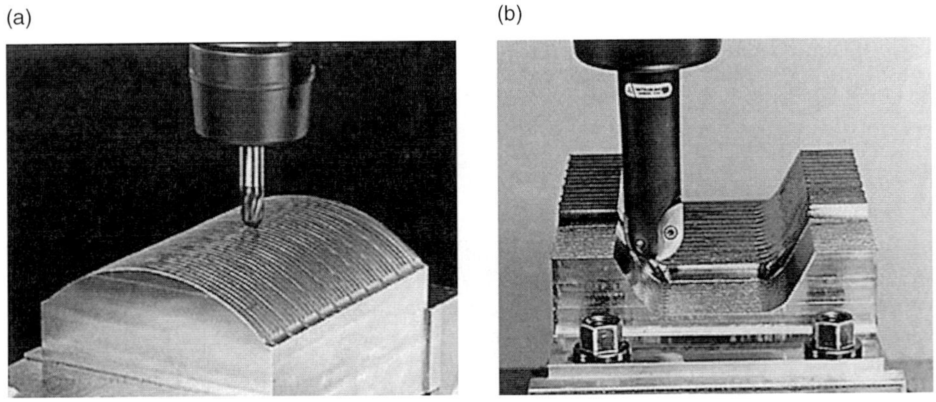

Fig. 15.54 Examples of hard milling of convex (a) and concave (b) surfaces using ball-nose end mills.

Fig. 15.55 Hard milling using five-axis machining concept.

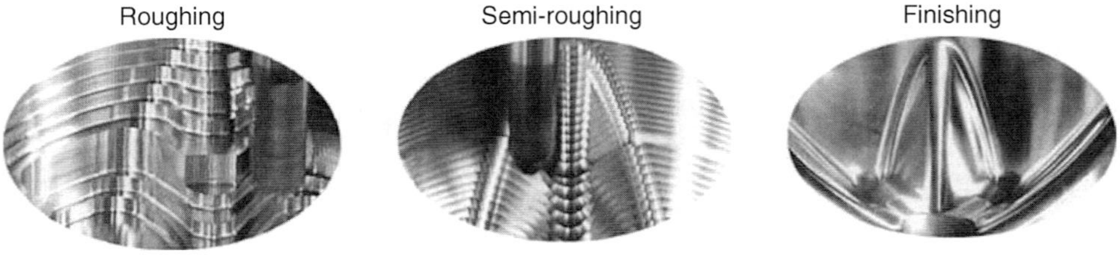

Roughing		Semi-roughing		Finishing	
Solid carbide end mill coated with TiAlN for dry machining (60° helix and six cutting edges) with drilling capability		Solid carbide two-edge, ball end mill coated with TiAlN for dry machining with drilling capability		Solid carbide two-/four-edge, ball end mill coated with PVD TiAlN-based coating for high speed milling	
a_p/a_e, mm	18/0.5	a_p/a_e, mm	0.5/0.5	a_p/a_e, mm	0.25/0.25
n, rev/min	7000	n, rev/min	6500	n, rev/min	13 000
v_c, m/min	220	v_c, m/min	200	v_c, m/min	400
v_f, mm/min	4200	v_f, mm/min	1600	v_f, mm/min	3300

Fig. 15.56 Machining conditions in milling operations on hardened steel die [21].

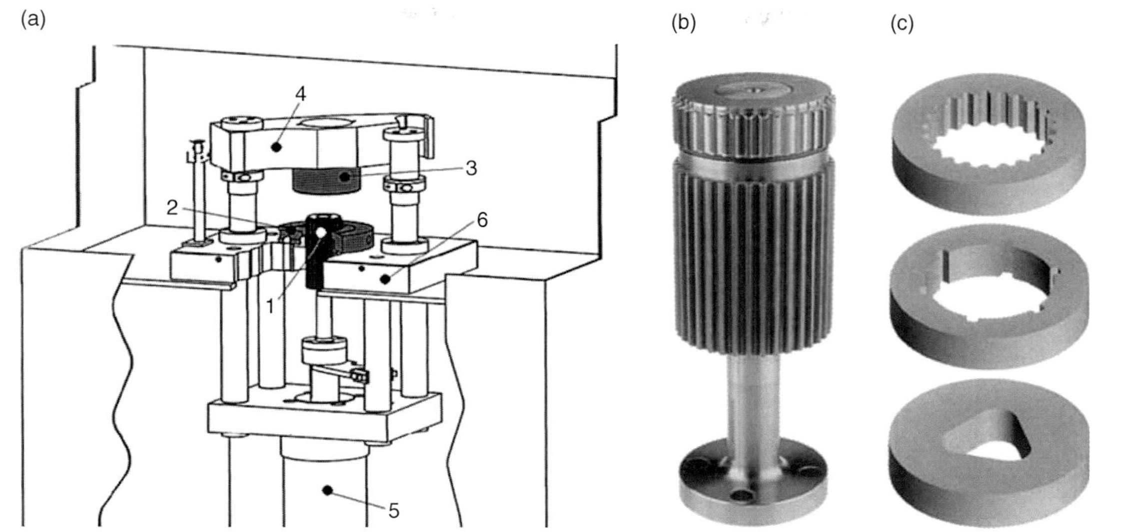

Fig. 15.57 Hard broaching of gear wheel [22]: (a) a scheme of broaching machine, 1 – broach, 2 – workpiece holder, 3 – hold-down, 4 – hold-down assembly, 5 – hydraulic piston, 6 – table, (b) PCD coated broach, (c) typical internal profiles.

It consists of the steel body that also contains the mounting flange. The centring piece (2 in Fig. 15.57a) positions the workpiece according to the profile to be shaped and checks the permissible broaching stock. It is possible to obtain a surface finish *Ra* of about 0.3 μm. Typical applications of HB include finishing a variety of internal profiles such as prismatic internal profiles, internal gear flanks, multiple key ways, polygons and spline profiles. It is reported [24] that a PCBN grade with a high PCBN content can be successfully used in dry broaching operations but from a performance point of view (cutting distance is the limiting factor) cutting speeds are substantially less than in hard turning (60 to maximum (90) 100 m/min versus 300 m/min for hard turning).

A new tendency in gear manufacturing is to use skive hobbing as a hard finishing process, possibly on a standard hobbing machine, instead of profile or generating grinding [24]. Unfortunately, it consumes high machining times which result in frequent regrinding of the tool. In this manufacturing sector combined/simultaneous machining operations, involving hard turning and cBN grinding, are performed on gear wheels and bearing components using one machine tool equipped with two machining stations [24, 25]. This specific type of complete machining will be discussed in Chapter 15.5.

Recently, fine hole-making (ISO IT4 dimensional accuracy) by hard reaming is applied as an alternative to grinding and honing. Figure 15.58 presents one- (a) and four-edge (d) reamers equipped with CBN inserts. The new design of small diameter reamer (Fig. 15.58c) with a special guide pad can stabilize itself in the hole and differ from a classical boring tool (Fig. 15.58b) which sustains a deflection due to the action of the high passive force *Fp*. For instance, it is possible to obtain the surface finish of Rz = 0.8–1.0 μm and roundness deviation less than 4 μm of the 6.4 mm hole at rotational speed of 5000 rev/min^{-1} and feed velocity of 200 m/min [23]. In the design shown in Fig. 15.58d four CBN blades are brazed in the head body, and the clamping system used reduced the radial run-out down to 3 μm. As a result, the surface roughness of *Rt* = 1.0–1.3 μm was produced.

15.3.4 Surface finish produced by HPM

This section outlines some important 2D and 3D characteristics of surface integrity including surface roughness, bearing curves and microstructural changes in the subsurface layer, and the distribution patterns of the induced residual stresses.

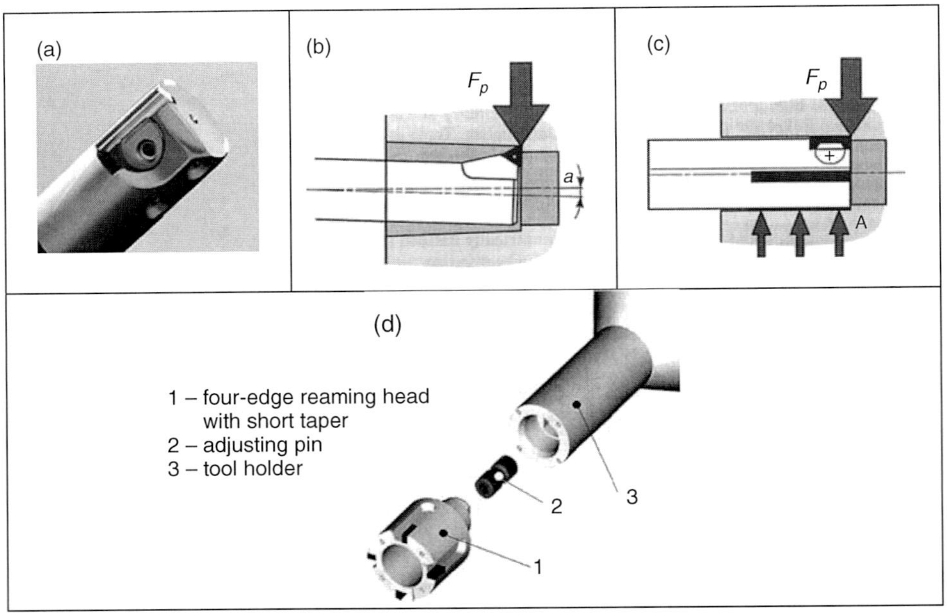

Fig. 15.58 Hard reaming with one-edge (a – c) and four-edge (d) reamers [23].

Figure 15.59 presents some characteristic surface profiles and corresponding 3D topographies for hard turning with CBN (left) and mixed ceramic (right) tools at the same cutting parameters ($v_c = 100\,\text{m/min}$, $f = 0.1\,\text{mm/rev}$ and $a_p = 0.2\,\text{mm}$). When comparing the relevant surface profiles they contain more sharp (CBN HT) or blunt (MC HT) peaks with characteristic lateral smaller flashes (more pronounced in MC HT) resulting from side flow effect shown in Fig. 15.60b.

Characteristic surface profiles generated by conventional and wiper ceramic tools are magnified in Fig. 15.60. It was observed that due to a smoothing effect wiper tools produce blunt irregularities with the RMS profile slope ranging from $R\Delta q = 1.5°$ to $5.5°$. On the other hand, more sharp profiles with $R\Delta q$ values of 5° to 10° were recorded [6]. As can be expected distinct differences in surface profile shapes will result in corresponding bearing area curves (Fig. 15.61b) which further result in their contact capabilities.

Figure 15.61 presents the deterioration of surface roughness *Ra* during a natural wear test performed at a cutting speed of 100 m/min during 30 min. It is evident that for both fresh tools the *Ra* values of 0.5–0.6 μm were obtained. Only minor changes of *Ra* parameter, near 10%, were recorded during the first 15 min of the wear tests with corresponding VB_C nose wear of about 0.1 mm. For the lowest feed rate of 0.04 mm/rev and conventional ceramic tools the values of height parameters *Ra*, *Rz* and *Rt* were equal to 0.25 μm, 1.60 μm and 1.80 μm, respectively. In general, the values of the average peak spacing *RSm* were approximately equal to the appropriate feed rates used.

In general, as shown in Fig. 15.62, the shapes of the bearing (Abbot's) curves obtained with conventional CBN and MC cutting tools are not satisfactory and the surfaces machined by hard turning need further improving by finishing belt grinding or superfinishing operations (Fig. 15.62b). Additionally, as shown in Fig. 15.62a, it is possible to generate, using a special wiper tool with varying chamfer geometry, the flatness of the BAC comparable to the honing effect.

In the contact zone, the austenite temperature of the workpiece material is reached during an extremely short time of approximately 0.1 ms and structural changes must be expected [5]. A metallurgically unetchable structure, called 'white layer', followed by a dark etching layer result from microstructural changes observed in hard machining, as, for example, on AISI 52100 steel surfaces of 60 HRC hardness (Fig. 15.63). Three lower SEM micrographs show white layer, underlying dark layer and the bulk material consisting of carbides

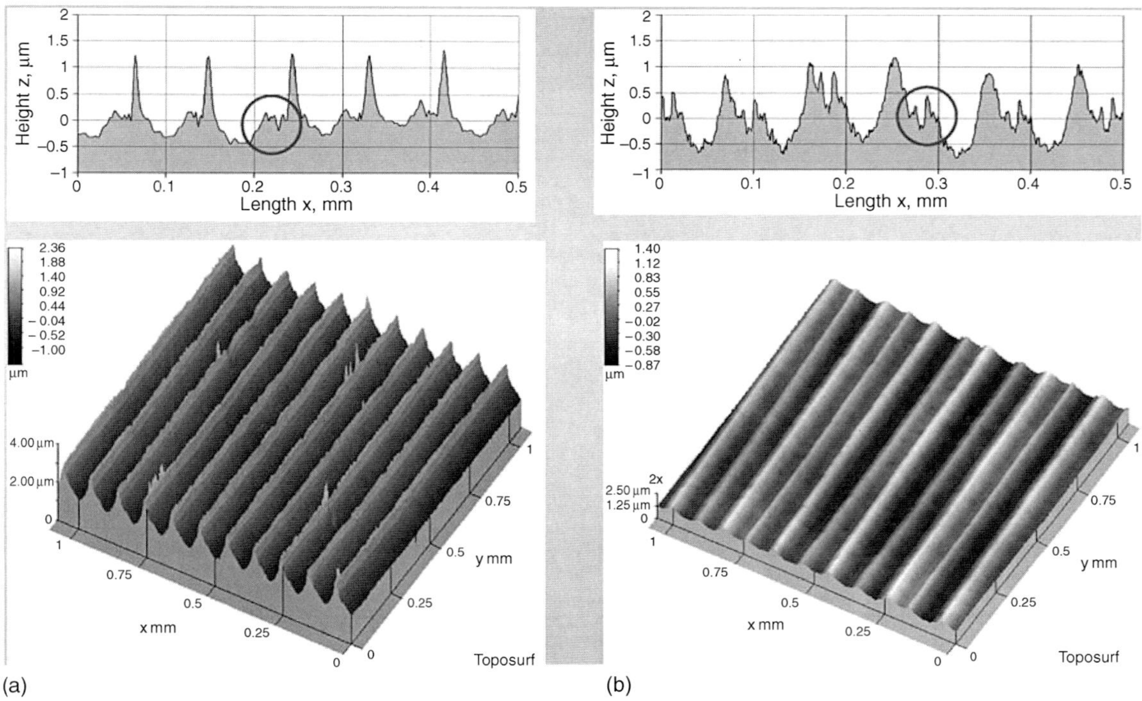

Fig. 15.59 Typical surface profiles and corresponding 3D visualizations produced in CBN (a) and MC (b) hard turning operations.

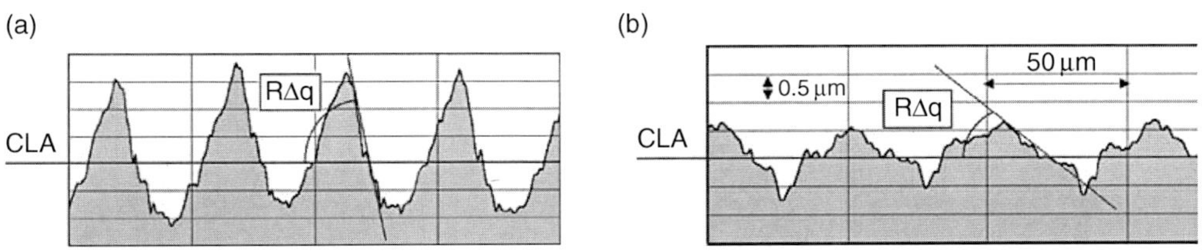

Fig. 15.60 Characteristic shapes of the profiles generated in turning with conventional (a) and wiper (b) ceramic tools for a constant feed rate of 0.1 mm/rev. Vertical magnification – 7000×. Horizontal magnification – 200×.

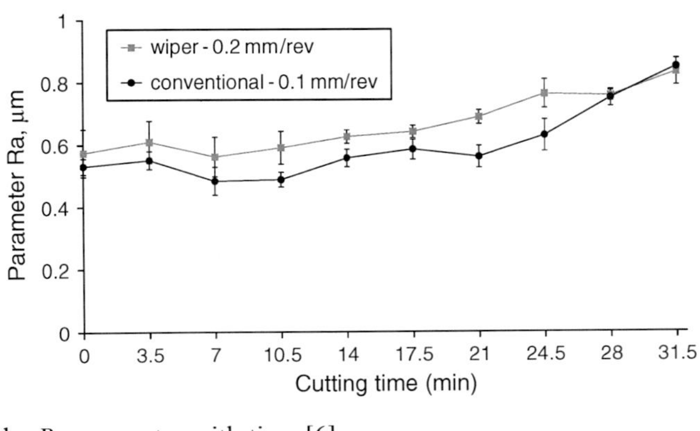

Fig. 15.61 Changes of the *Ra* parameter with time [6].

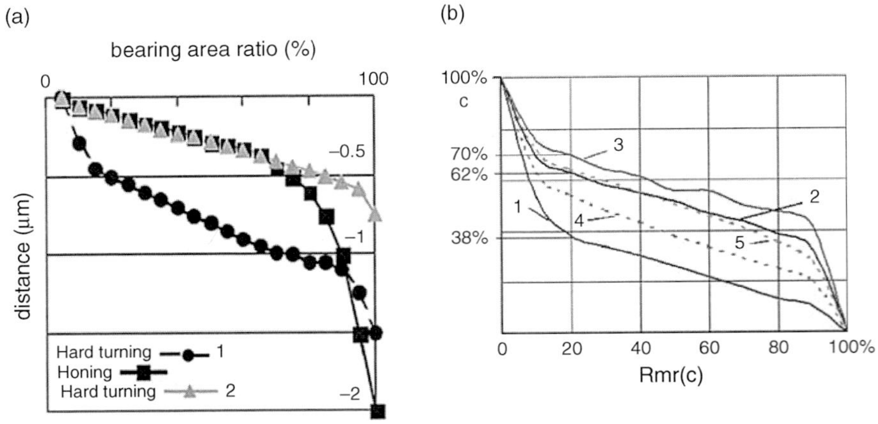

Fig. 15.62 Comparison of bearing area curves for hard turning ($1 - f = 0.1$ mm/rev; $2 - f = 0.15$ mm/rev and varying chamfer angle) and honing ($v_c = 45$/min, $h = \pm\ 0.1$ mm, tool – B15 SiBA N1 and Bubber C120) [26] (a) and differences in BAC's curves obtained in CBN (1) and MC (4) hard turning and finish abrasive processes (preliminary (2) and finish (3) belt grinding and superfinishing (5)) after [6].

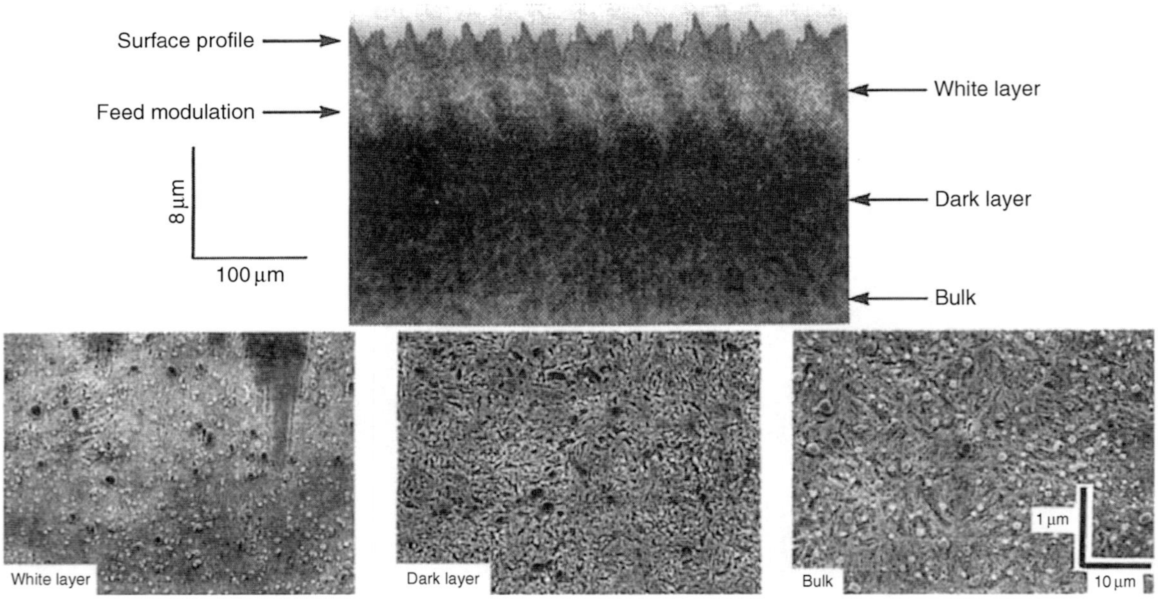

Fig. 15.63 Examples of microstructural changes in a hard turned surface of 52100 steel: (a) optical micrograph, and (b) SEM micrographs of three characteristic zones [27].

distributed in a martensitic matrix. In general, carbide size and its distribution in both white and dark layers indicate no difference from the bulk.

Corresponding microhardness distribution (using a Knoop indenter and 100 g load) in the surface layer is shown in Fig. 15.64. It should be noted that the hardness is similar in the white layer of a few microns in thickness, whereas the dark layer with an overtempered martensite is substantially softened. The total

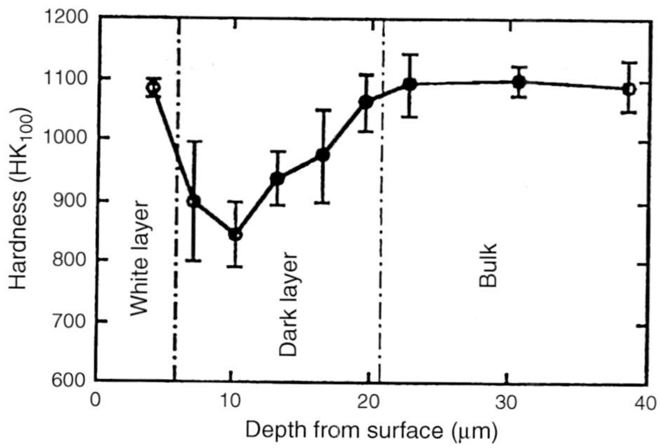

Fig. 15.64 Microhardness distribution in a hard turned surface [27].

machining–thermally affected zone is about 20 μm thickness, which is about one tenth of the depth of cut ($a_p = 200\,\mu m$). Typically, the depth of the white layer increases rapidly with tool wear and cutting speed but for the bearing steel tested a saturation is observed at $v_c = 3\,m/s$ [27].

White layers consist of over 60% austenite which is a non-etching white component in contradiction with the dark martensite scoring [5]. Due to this fact, Klocke *et al.* [24] suggested that the term 'white layer' is misleading and proposed to distinguish two groups of this structure's appearance in micrographs. When applying special etching chemicals or increased etching time, white layers of the first group remain white. On the other hand, white layers of the second group are affected by chemicals and a fine-grained martensitic structure becomes visible. Both types of white layer are distinguished by high brittleness and susceptibility to cracking and therefore they are classified as a damage of the machined components.

The changes of the distribution of residual tangential stress in the sublayer with cutting time are shown in Fig. 15.65a. It should be noted that the tensile stresses occur near the machined surface and their values increase progressively with tool wear. In addition, corresponding displacement of the point of the maximum

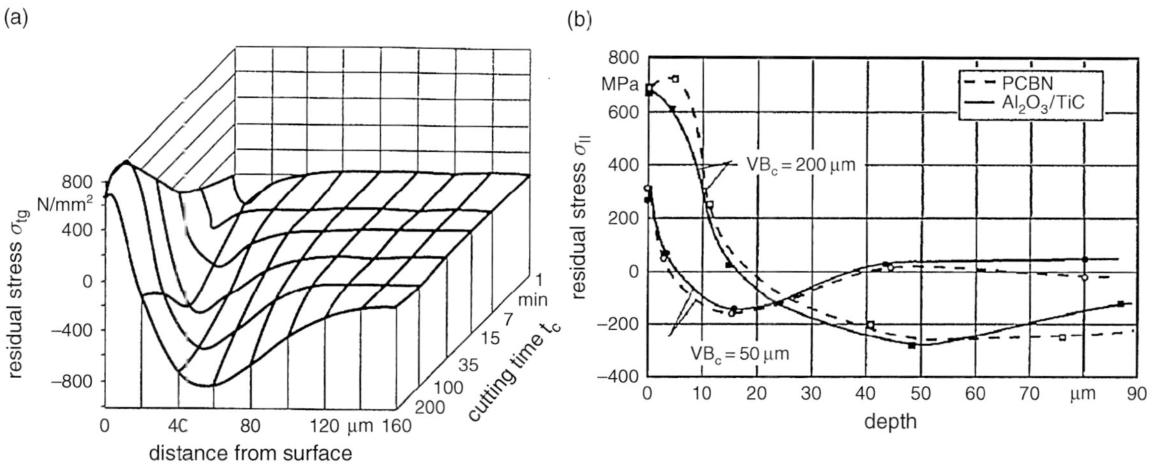

Fig. 15.65 Modification of residual stress profile in the surface layer within tool-life period [13]; work material: 16MnCr6 case-hardened steel of 62 HRC, tool material: CBN (a) and comparison of residual stress distribution when turning with PCBN and Al_2O_3-TiC ceramic tools (b) [5].

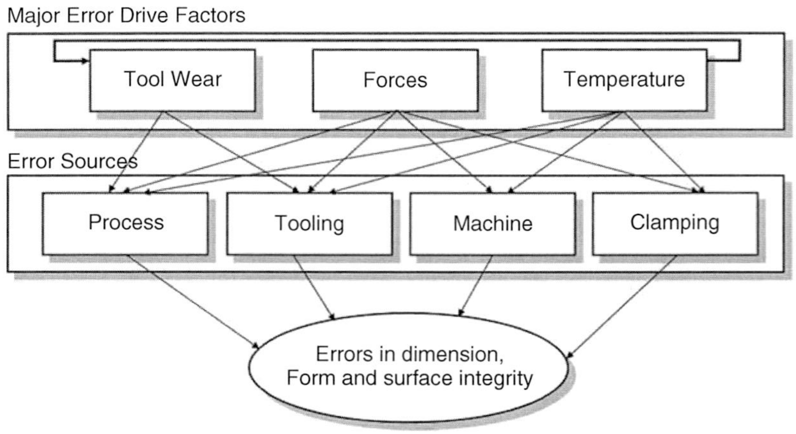

Fig. 15.66 Major error drive factor and error sources in precision hard turning [28].

compressive stresses of about −800 MPa deeper beneath the surface is observed. The effect of the influence of tool flank wear was confirmed for Al_2O_3-TiC ceramic tools and the differences in the shapes of stress distribution curves for these two cutting tool materials are marginal.

The combination of martensite and extremely fine-grained austenite structure constituents can cause different residual stresses in the white layer. In particular, the residual stresses of the dominating austenite material components are clearly shifted towards compressive stresses. However, the martensite tensile residual stresses result from lower specific volume and consequently higher density of austenite [5]. Compressive residual stresses induced by hard turning were found to improve rolling contact fatigue (RCF). Hard turned surfaces may have more than 100% longer fatigue life than ground ones with an equivalent surface finish of 0.07 μm [28] but this positive effect can be significantly reduced with the presence of a white layer.

Now, the achievable dimensional accuracy in precision hard turning using CNC machine tools with high static and dynamic stiffness, thermal stability and high precision motion control is ISO IT5[22,29]. As shown in Fig. 15.66 many factors affect the part accuracy and process stability and reliability. Three main factors, namely, tool wear, cutting forces and cutting temperature, are the driving factors providing the error sources in process, tooling, machine structure and clamping devices. For instance, when flank wear of a CBN tool reaches the value of $VB = 0.2$ mm the dimensional errors on 100Cr6 (60–62 HRC) steel parts measured in radial/axial directions increases up to 25 μm. Moreover, because the cutting temperature is at least 800°C substantial thermal expansions of both tool shaft and the workpiece are observed. It is assumed that the thermal effects can contribute to more than 50% of overall error of the machined parts. In particular, for the same machining conditions ($v_c = 160$ m/min, $f = 0.05$ mm/rev, $a_p = 0.05$ mm) thermal expansion on the CBN tool tip and workpiece can reach up to 10 and 15 μm, respectively [28].

In order to minimize possible errors in precision hard turning magnetic chucks, a solid carbide tool-holder and active error compensation and special monitoring systems are proposed.

References

[1] G. Byrne, D. Dornfeld, B. Denkena, Advancing cutting technology, *CIRP Annals*, Vol. 52/2 (2003), 483–507.
[2] F. Klocke, E. Brinksmeier, K. Weinert, Capability profile of hard cutting and grinding processes, *CIRP Annals*, Vol. 54/2 (2005), 557–580.
[3] B.P. Erdel, *High-Speed Machining*, SME, Dearborn, 2003.
[4] B.P. Erdel, *New Dimensions in Manufacturing*, Hanser Gardner, Cincinnati, 1998.

[5] H.K. Tönshoff, C. Arendt, R. Ben Amor, Cutting of hardened steel, *Annals of the CIRP*, Vol. 49/2 (2000), 547–566.
[6] W. Grzesik, T. Wanat, Surface finish generated in hard turning of quenched alloy steel parts using conventional and wiper ceramic inserts, *Int. J. Mach. Tools Manuf.*, Vol. 46 (2006), 1988–1995.
[7] J. Barry, G. Byrne, The mechanisms of chip formation in machining hardened steels, *Trans. ASME J. Manuf. Sci. Engng,* Vol. 124 (2002), 528–535.
[8] M.A. Davies, T.J. Burns, C.J. Evans, On the dynamics of chip formation in machining hard metals, *CIRP Annals*, Vol. 46/1 (1997), 25–30.
[9] M.A. Davies, C.J. Chou, C.J. Evans, On chip morphology, tool wear and cutting mechanics in finish hard turning, *CIRP Annals*, Vol. 45/1 (1996), 77–82.
[10] G. Poulachon, A. Moisan, I.S. Jawahir, On modelling the influence of thermo-mechanical behavior in chip formation during hard turning of 100Cr6 bearing steel, *CIRP Annals*, Vol. 50/1 (2001), 31–36.
[11] M.C. Shaw, A. Vyas, Chip formation in the machining of hardened steel, *CIRP Annals*, Vol. 42/1 (1993), 29–33.
[12] W. König, F. Klocke, *Fertigungsverfahren 1. Drehen, Fräsen, Bohren*, Springer, Berlin, 1997.
[13] W. König, A. Berktold, K.F. Koch, Turning versus grinding – a comparison of surface integrity aspects and attainable accuracies, *CIRP Annals*, Vol. 42/1 (1993), 39–43.
[14] T. Ueda, M. Al Huda, K. Yamada, K. Nakayama, Temperature measurement of CBN tool in turning of high hardness steel, *CIRP Annals*, Vol. 48/1 (1999), 63–66.
[15] R.C. Dewes, E. Ng, K.S. Chua, P.G. Newton, D.K. Aspinwall, Temperature measurement when high speed machining hardened mould/die steel, *J. Mater. Proc. Technol.*, Vols 92–93 (1999), 293–301.
[16] H.A. Kishawy, M.A. Elbestawi, Effects of process parameters on material side flow during hard turning, *Int. J. Mach. Tools Manuf.*, Vol. 39 (1999), 1017–1030.
[17] Y.K. Chou, Ch.J. Evans, Tool wear mechanism in continuous cutting of hardened tool steels, *Wear*, Vol. 212 (1997), 59–65.
[18] W. Grzesik, T. Wanat, Influence of tool wear on surface finish in hard turning using differently shaped ceramic tools, paper submitted to *Wear.*
[19] W. Grzesik, T. Wanat, Hard turning of quenched alloy steel parts using conventional and wiper ceramic inserts, *Trans. NAMRI/SME*, Vol. 33 (2005), 9–16.
[20] D. Huddle, Plunge turning can be a cost-effective grinding alternative, *Manufacturing Engineering*, Vol. 128, No. 4 (2002), 76–81.
[21] Manufacturing die and molds, www.coromant.sandik.com.
[22] Hard broaching, www.faessler-ag.ch.
[23] D. Kress, Erfolge furs Hartreiben und Hartfräsen, *Werkstatt und Betrieb*, Vol. 133, Nos 1–2 (2001), 64–65.
[24] F. Klocke, E. Brinksmeier, K. Weinert, Capability profile of hard cutting and grinding processes, *Annals of the CIRP*, Vol. 54/2 (2005), 557–580.
[25] H. Mushardt, P. Manger, Komplettbearbeitung mit Hartdrehen und Schleifen, *Werkstatt und Betrieb*, Vol. 134, Nos 1–2 (2001), 37–40.
[26] F. Klocke, H. Kratz, Advanced tool edge geometry for high precision hard turning, *Annals of the CIRP*, Vol. 54/1 (2005), 47–50.
[27] Y.K. Chou, Ch.J. Evans, Process effects on white layer formation in hard turning, *Trans. NAMRI/SME*, Vol. 26 (1998), 117–122.
[28] J.M. Zhou, M. Anderson, J.E. Ståhl, Identification of cutting errors in precision machining hard turning process, *J. Mater. Proc. Technol.,* Vols 153–154 (2004), 746–750.

15.4 High Performance and High Efficiency Machining

15.4.1 Basic aspects of high performance and high efficiency machining

During the last few years great emphasis has been placed on high performance cutting (HPC) as a means of achieving high levels of productivity and reduction of production costs, taking frequently into account the factors of time, precision or production volume. High performance machining can be assessed by different methods and measurement techniques, and can be perceived in different ways, qualified through different methods and achieved by various means. The implementation of HPC strategies and their ranges depends on the profiles of the manufacturing branches. For example, aircraft manufacturing tends to finish technologically advanced aircraft parts without machining residues such as coolant residues and chip wetness. During near-dry machining chips should be prevented from entering the inside of the fabricated hulls. Typically, the following aspects of HPM are identified [1–3]:

1. Reduction of non-productive times (NPT) in cutting operations.
2. Reduction of operation times by optimizing cutting parameters and tool developments (for example, by using combined and multifunction cutting tools). Specification of possible ways of total optimization of tooling systems and machining processes is shown in Fig. 15.67.
3. Consideration of dry cutting and minimum quantity lubrication (MQL).
4. Implementation of environmentally friendly production.
5. Burr reduction.
6. Rationalization of chip removal, chip transportation and machine cleaning technology.

The trend in recent years has been towards integrated processes. For example, by replacing grinding with hard turning, some process steps can be eliminated (Chapter 15.3). The requirement for integrated processes places new and demanding challenges on the design and technology of the cutting processes. The economic efficiency of production facilities is a central issue of cutting technology. In plant, the adding of value to products and workpieces only takes place during essential operation time. Figure 15.68 demonstrates how the cycle time (NPT) for machining one model workpiece can be reduced by about 50% as a result of systematic optimization of five factors: machining parameters, tools, positioning time, tool arrangement in the magazine and tool change time. On the other hand, the operation time depends on the combination of cutting parameters (medium feed rate and high cutting speed, high feed rate and medium cutting speed, high feed rate and high cutting speed) and rapid traverse rate (Chapter 15.1). It is documented that the combination of high feed rate and medium cutting speed can substantially reduce operation time. This machining strategy is known as high efficiency machining (HEM) [5]. Especially, when machining titanium and aluminium components in the aerospace sector, where volumetric rates up to 95% are common, machining at both high cutting speed and feed rate reduces production costs. The characterization of HEM in terms of potential advantages and disadvantages is shown in Fig. 15.69.

It offers many of the advantages provided by high speed machining, such as less burr formation, lower cutting forces, high material removal rates in roughing by using medium cutting speeds (up to 2000 m/min

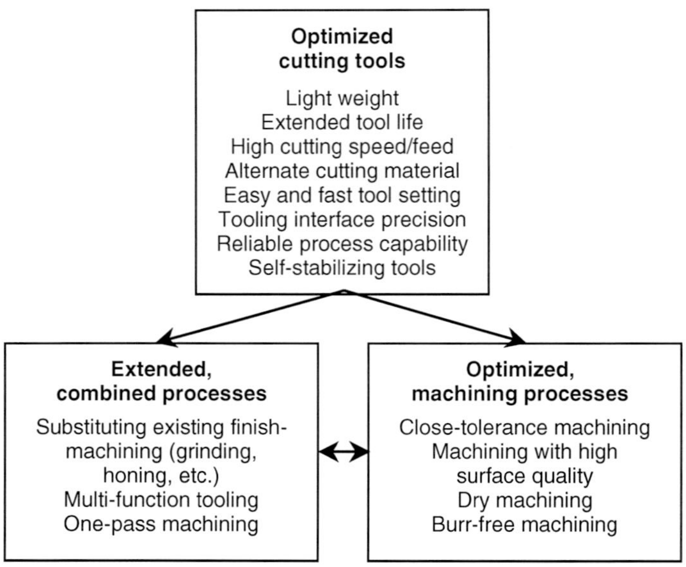

Fig. 15.67 Possibilities of optimizing cutting tools and machining processes [4].

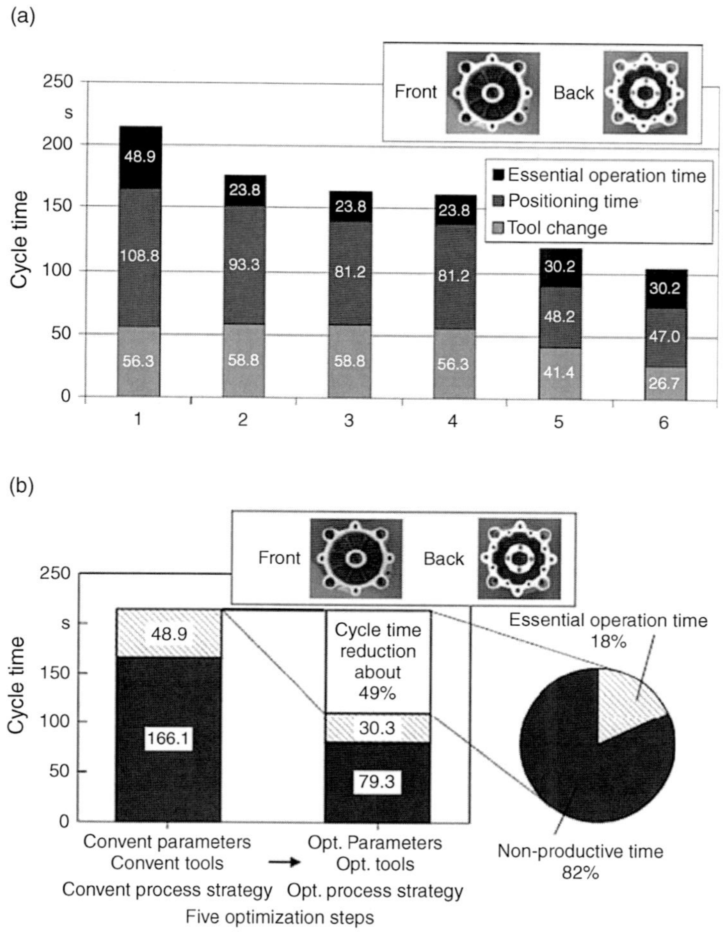

Fig. 15.68 Reduction of the cycle time for machining a model workpiece [3]. (a) five step optimizing, (b) final effect of process optimization.

when machining aluminium alloys), high feeds per tooth (up to 0.3 mm/tooth), and excellent surface quality after finishing operations. Because spindle power often imposes limits on the cutting parameters, high spindle power is necessary for high efficiency machining [5].

When spindle power is limited, as spindle speed increases, material removal rate decreases significantly. One reason is the increasing ratio of idle power to total power. Another factor is the decrease of feed per tooth, which causes higher friction during cutting. Because of these factors, roughing during HEM should be logically carried out at lower rotational speed and higher feed rate per tooth, typically of 0.4–0.5 mm/tooth without problems caused by the high section of cut. On the other hand, there is a limit to the increase in feed rate arising from machine dynamics. As an example, in machining tasks with a short travel length, only limited feed rates can be reached because of the limits on the acceleration of the axis of the machine tool.

Tool clamping systems are of critical importance for the attainment of high productivity and process quality in HEM. Both the spindle–tool-holder interface and the clamping device are important factors in obtaining good technological results. The problems connected with the centrifugal effect and its influence on reliability were discussed in Chapter 15.1. Hence, the selection of the tool-clamping device must consider a range of requirements including: high rigidity and accuracy, high transmissible torque and suitability for operations at high rpm. As a result, the most frequently used clamping devices in HEM are the

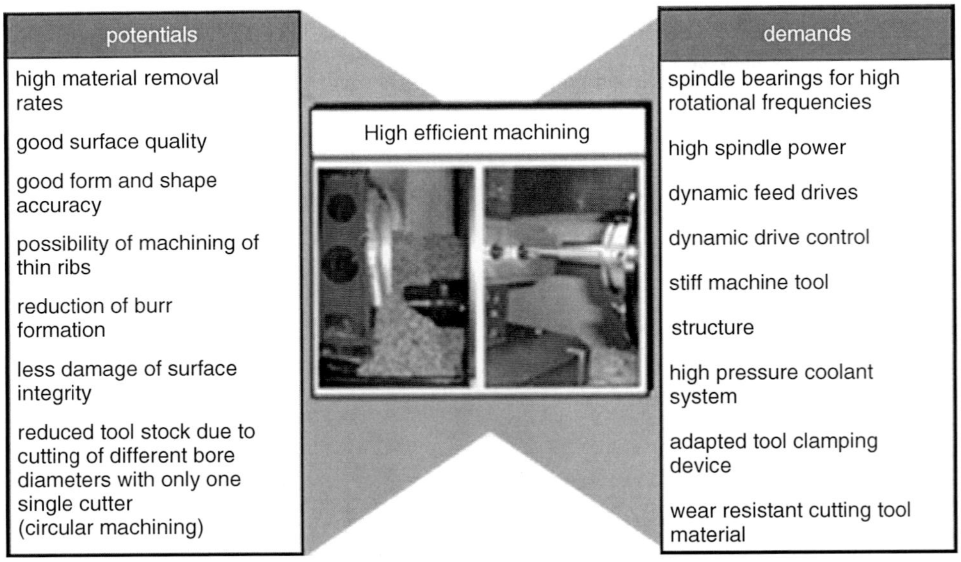

Fig. 15.69 Potential and demands of high efficiency machining [5].

hydraulic-expansion chuck and the heat-shrinking chuck but the second solution is preferable due to its lesser sensitivity to cutting vibration.

A number of tools already meet the demands of HEM, including coated and uncoated carbides, cermets, ceramics, PCD and PCBN, which are also applicable for HSM. New developments are taking place in tools with complex cutting edge geometry, like end-mill cutters and drilling tools. Further effects can arise from the application of innovative machining technologies, such as circular milling replacing drilling. It allows machining holes of different diameters by adjusting the NC program up to 1.3–2 tool diameter. This capability of CNCs reduces the time consumed by ancillary operations. In addition, improvement of chip breaking by interrupted cutting and easier chip removal should also be stated. The surface quality achieved by circular milling operations equals the result produced by reaming. Another circular milling process already used in modern manufacturing is threading (more details in section 15.4.3).

15.4.2 Machine tools and tooling

A primary trend in the machine tool world is toward higher productivity, shorter lead-time, higher accuracy, lower cost and environmentally safe operations. To satisfy these demands, machine tool builders are turning to a number of key technologies partially commented on in previous sections. Current trends for job shops are toward producing parts in small-to-medium lot sizes, often within a 615 mm or 24 inch cube, using multi-axis machine tools. In particular, implementing five-axis machining to finish the entire part, rather than applying secondary operations on a number of different machines or refixturing the workpiece, was one way to remain competitive. For instance, for milled surfaces, five-axis machines allow the CNC programmer to shape practically any complex 3D surfaces. This is especially useful for mould and die applications in which both flat-nose and ball-nose end mills can be used. In the first case more metal can be cut in one pass and scallops remaining on the machined surfaces can be reduced. This type of machining also can be of significant benefit to some aerospace applications that must generate complex mould lines or airfoil surfaces. As reported by many end users [6], the combination of high speed machining techniques, high-level CNC controls, and CAM software allows five-axis machines to be most productive and make cutting processes extremely competitive against non-traditional processes such as EDM or ECM. Five-axis machining, enabling multisided operations, makes the transition to integral and monolithic structures more flexible and economical. The potential applications of parallel kinematic machines will be outlined below.

The debate on the use of dedicated machines vs multifunctional machining centres seems to be over. Although dedicated machines are easier to maintain and cheaper, end users often require different types of single-function machines to complete all the machining processes for a part. In contrast, today's multifunctional machining centres (Chapter 15.5) can turn, drill, mill, tap, hob, and even grind the whole part on the same machine. The multitasking capability saves significant time in material handling, scheduling and job setting. Due to their flexibility and programmability machining centres can be either mixed with dedicated machines in a flexible transfer line or used exclusively in an agile machining system.

Vertical machining centres (VMC) still dominate the market due to their economic price and capacity to handle heavy, large parts. Horizontal machining centres (HMC), however, prevail in agile or flexible machining systems. They can be equipped with faster spindles and larger tool magazines. With a rotary index table, HMCs can machine as many as five faces on the same part. The structurally stiff, bridge or gantry-type machines are important for making die as well as large hard metal parts. Figures 15.70 and 15.71 show some components of such a high performance HMC, including the bed from grey cast iron and gantry column from nodular cast iron, and spindle as well as coolant supplying units, respectively. Four optional spindle speeds – 8000, 10 000, 16 000 and 24 000 min^{-1} are possible. Air/oil mist lubricates spindle ball bearings and the motor and spindle portion is cooled by water. Rapid traverses of 60(90) m/min, accelerations of 10(12) m/s^2 and 2.8(2.2) chip-to-chip time are the main dynamic features.

The machine concept of a high performance VMC includes the following design aspects [7]:

- Monoblock design in a special casting ensuring optimum dampening characteristics and high static and dynamic rigidity.
- Tilting movement of the tool excludes compensation movements during simultaneous machining.
- Unique axis configuration guarantees minimal moving masses, symmetric drives for all workpiece lengths and optimum flow of chips.
- Torque drives in all rotary axes for highest dynamics and precision.
- Cover design, machine enclosure and the simple modular axis design ensure minimum maintenance intervals, resulting in above-average machine up-time.
- Ergonomically arranged peripheral units facilitate easy maintenance.
- Workpiece handling allows loading and unloading of workpieces during the machining cycle.
- The compact machine design results in fast installation and minimum floor space requirements.
- The machine concept allows supplying both organic and synthetic coolants.

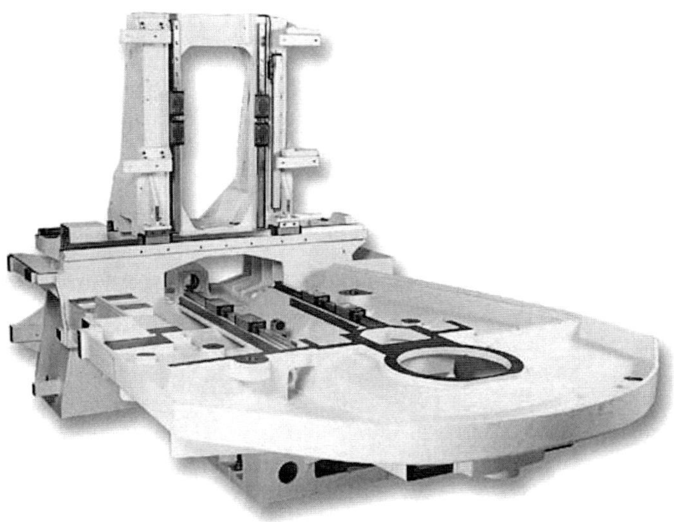

Fig. 15.70 Structure of a horizontal machining centre with gantry column [7]. Source: Heller.

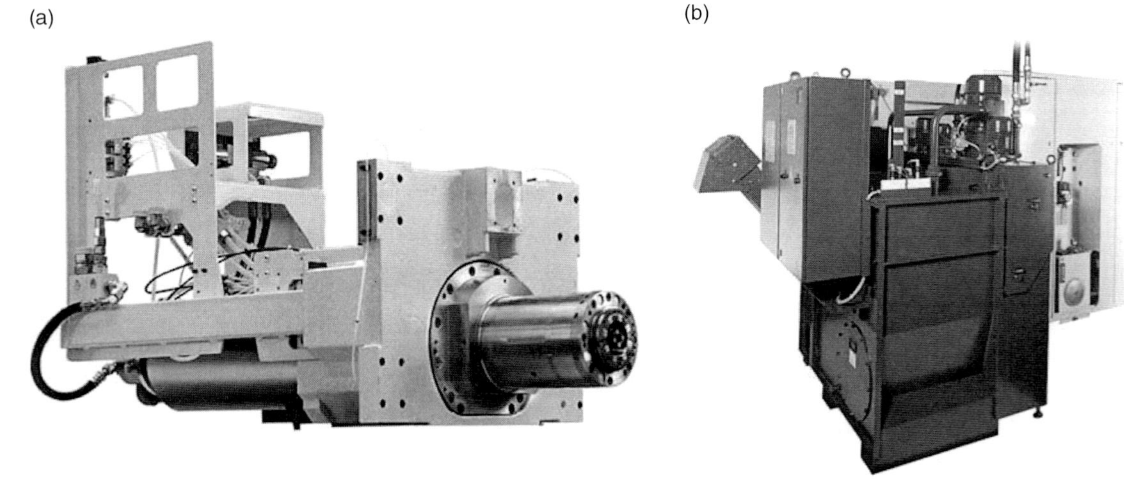

Fig. 15.71 Spindle unit of a VMC from Fig. 15.70 (a) and coolant unit (b) [7].

Typical applications of the VMC from Fig. 15.72 cover the following parts:

- Machining of turbine and compressor blades for thermal turbo-engines of different forms and dimensions.
- Efficient machining of blades in chromium-nickel steel, titanium, nickel-based alloys and titanium alloys.
- Complete machining of industrial turbine blades.

Single spindle vs twin spindle machines for turning and machining centres has been the topic of an ongoing debate in recent years. In general, European manufacturers prefer the concept of twin-spindle machining, where two parts are cut simultaneously. Large strokes for large-part machining are possible with a twin-spindle (duo) machining centre. There are inherent cost efficiencies and floor space savings when two spindles, two fixtures, and two tool changers are combined in one machine. Installation and rigging

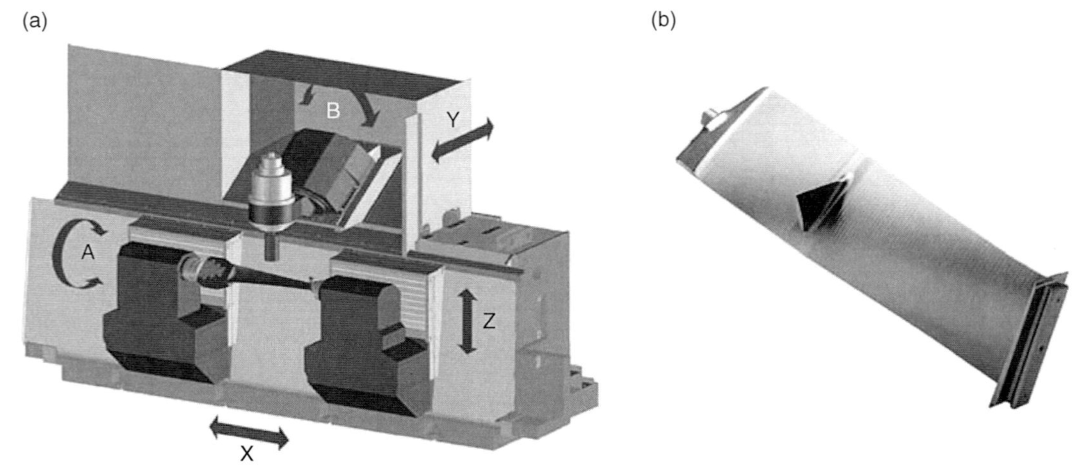

Fig. 15.72 Single spindle five-axis high performance vertical machining centre for high precision machining of freeform surfaces (a) and turbine/compressor blade (b) [7].

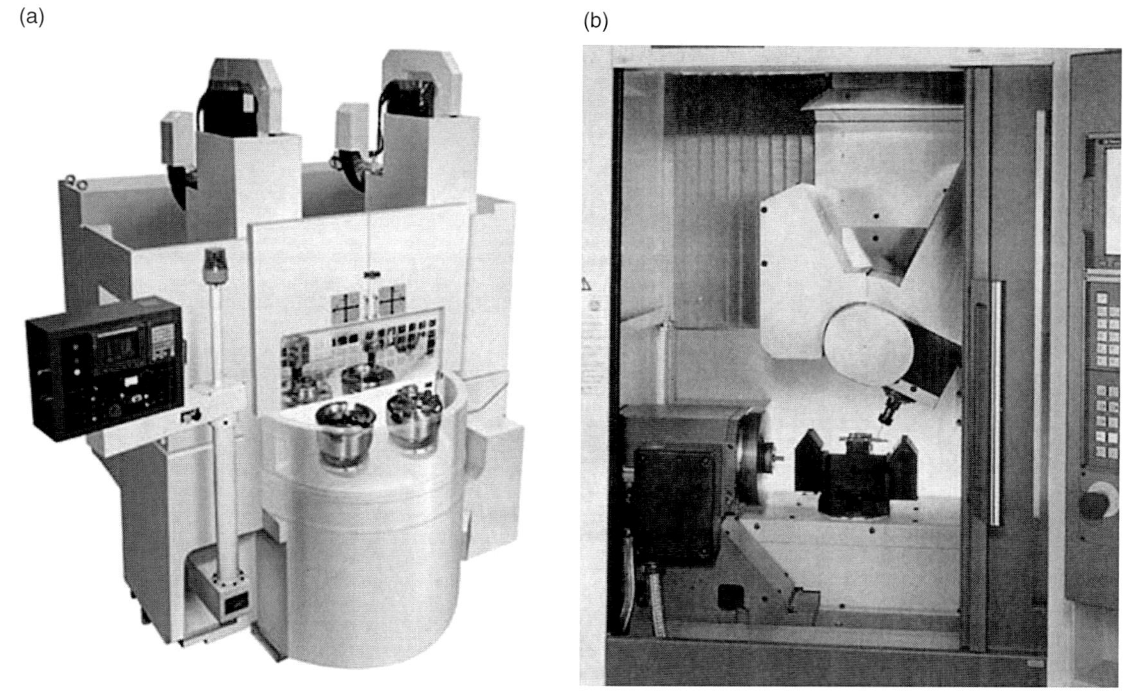

Fig. 15.73 Examples of high performance machine tools: (a) vertical CNC lathe with four-spindle chuck system (source: Kitako), (b) seven-axis vertical machining centre (source: Chiron America Inc.) [7].

costs, utilities and coolant consumption are lower also. In the CNC lathe shown in Fig. 15.73a four spindles, generally partnered as pairs, are mounted in a vertical, square pattern in a large carrier drum. When the carrier is indexed 180 degrees, two spindles move in for machining as the other two spindles move out for unloading and loading. Operation is similar to a pallet changer on a machining centre. It should be noticed that identical or different parts can be machined independently because the spindles, slides and turrets in the machine zone work independently for flexibility. As a result, this allows the machine to work AA/AA, AB/AB or AA/BB operations depending on the type of job. Unrestricted access to the spindles in the loading position makes for simple manual, robot or gantry loading. Spindle rigidity and positioning are ensured by a large diameter, precision-toothed, curved coupling between the carrier OD and bed ID. The small footprint is designed to save space on the shop floor.

A seven-axis vertical centre shown in Fig. 15.73b is a travelling-column machine with a fully integrated NC-controlled tilting spindle. It features rapid feed rate to 76 m/min (3000 ipm), acceleration/deceleration rates to 2g, and spindle rotational speeds to 40 000 rpm. Tool change time is 0.8 s tool-to-tool and 1.9 s chip-to-chip. The profitable applications include one-setup machining of orthopaedic devices and surgical instruments, dental instruments and implants, automotive turbocharger impellers and wristwatch parts. The machine can be coupled with optional bar feed capabilities.

The perpetual strive for more productivity and high precision through more economical methods and less energy expenditure has spurred a radical new machining concept termed parallel kinematic machine (PKM). It is of primary importance that the conflict between stiffness and dynamics is ideally solved by means of parallel couple kinematics. One type of PKM, also commonly called a hexapod machine (shown in Fig. 15.74), is a complete departure from the conventional basic machining-centre design. Going back to its original design for flight simulation, the hexapod, also known as the Stuart platform, has a parallel-link mechanism. Instead of the conventional bed-frame design, the dexterous Stuart-platform or strut-like robotic

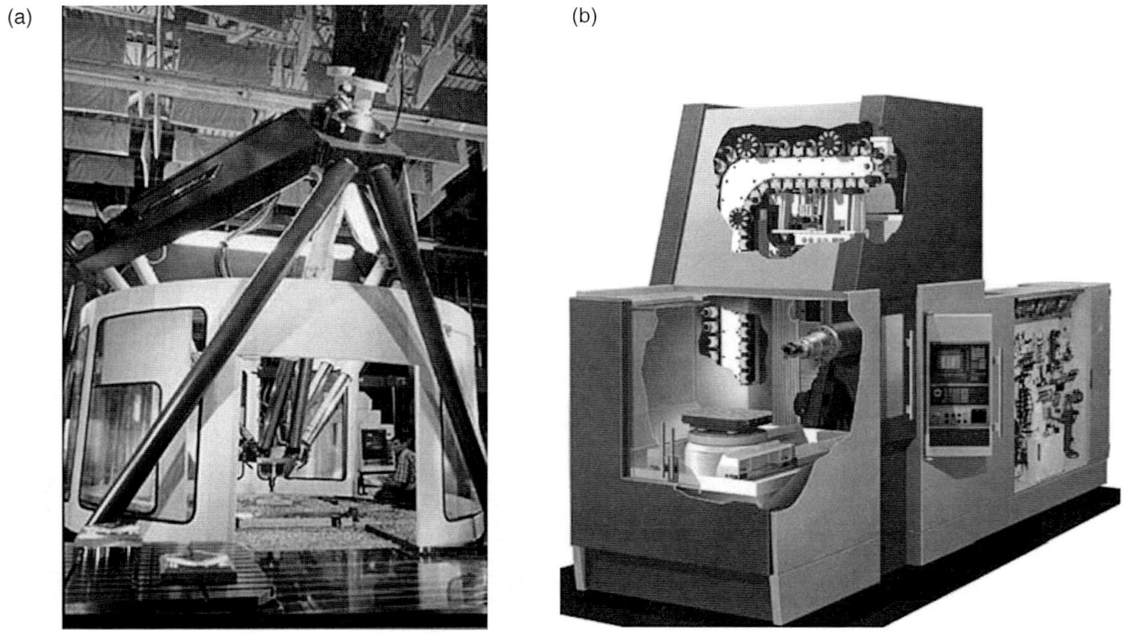

Fig. 15.74 Examples of parallel kinematic machines [7]: (a) octahedral hexapod machine installed at NIST, (b) five-axis HMC with PKM integrated with spindle motion actuation. (see Plate 19)

linkage of PKMs offers manufacturers a versatile, fast, five- or six-axis machine tool. The machines can perform various manufacturing operations, including machining, welding, plasma spraying, and coordinate measuring, as well as part and fixture manipulation [6]. The PKM machines have already been used in aerospace and the mould and die industry to perform complicated contour machining [4].

The basic features of the PKM machine shown in Fig. 15.74b, which can perform both wet and dry machining, are as follows:

- High dynamics (rapid traverse rate, acceleration, jerk), thus shorter idle times.
- High rigidity, thus shorter operating times depending on the process.
- Reduced part costs.
- Parallel kinematics with equal accuracy as Cartesian machines.
- Rotary table with torque motor/simple operations possible even without clamping.
- Short chip-to-chip time.
- Floor operated chain-type magazine.
- Optimized for high volume production.
- For application in agile manufacturing systems.

Nevertheless, hexapods are facing some difficulties because their rigidity stems from the six ball screws that share the load (Fig. 15.75). As a result, the motion of the ball screws generates a lot of heat, affecting the accuracy of the machine due to excessive thermal expansion. This can be partially solved by thermal compensation in the software and the inner cooling of hollow struts, which can lower the generated heat.

At this stage, the PKMs are still no replacement for the conventional machine tools because of their lack of stiffness (less than $50\,N/\mu m$) and machining accuracy (± 0.05 at best) [6]. Some manufacturers, however, are applying the PKM design to the spindle motion actuation in five-axis configurations as shown in Fig. 15.74b. This overall configuration has the potential to offset some PKM disadvantages while maximizing the advantages.

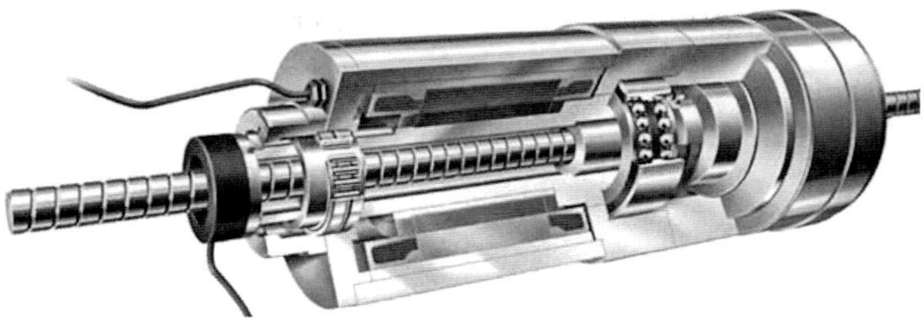

Fig. 15.75 High dynamic ball-screw feed unit for PKM machines [8].

15.4.3 Simplifying machining operations

The success of machining processes primarily depends on machine tools and cutting tools employed. The cost of tooling, as a percentage of the total production cost, has increased 25–30% during the last decade because of the use of flexible manufacturing systems. Conversely, it directly influences the other 70–75% of the production cost and it is why proper tool layouts and specifications are so important. In many cases the right tooling, and not the machine tool, often guarantees the highest accuracy and productivity on the production floor. It is postulated that modern machining processes must consider and cover the following aspects [4]:

- One-pass finish machining.
- Multi-functional tooling.
- No post-machining processes.
- Standard tools for regular production machines.
- Modular tooling systems.
- Standardization and tool management.
- High repeatability and accuracy.
- Achievable part finishes and predictable machining results.
- High cutting speeds and feeds.
- Easy and reliable tool clamping.
- Using cutting tool materials with best performance.
- Minimizing machining time.
- Minimizing tool setting.

The practical examples of modern cutting tools and tooling systems overviewed in this section, covering a number of machining processes performed in different manufacturing branches, indicate many trends in pursuing high performance machining.

Indexable drills (Fig. 5.76), starting at a diameter of 12.7 mm, are pretty economical and efficient in hole-making even to about 45 mm (1¾") because they provide high machining productivity, versatility and long, reliable tool life. Moreover, they are capable of drilling both roughing and finishing holes, as solid drilling and boring passes. The achievable hole tolerances with the design shown in Fig. 5.76 can be reduced to +0.25 mm (typically IT12–13, IT10 when presetting) and ±0.05 mm when using a stationary boring tool. The surface finish is about *Ra* 1–2 μm with a high feed and it can be diminished to *Ra* 0.5 μm with a lower feed and visible action of the wiper peripheral insert (1). As shown in Fig. 5.76, the indexable drill utilizes a synchronized cutting action of central and peripheral inserts (the corner of the peripheral insert is shifted down at a small distance from the position of the central insert) The drill makes step-by-step contact with the workpiece, avoiding high radial forces that push the drill off-centre. Initial contact made with one edge on the central insert (1) is followed by the outer edge of the peripheral insert (2) and then by the second edge of the central insert (3). This is a decisive feature in the additional ability to making cavities by plunge drilling.

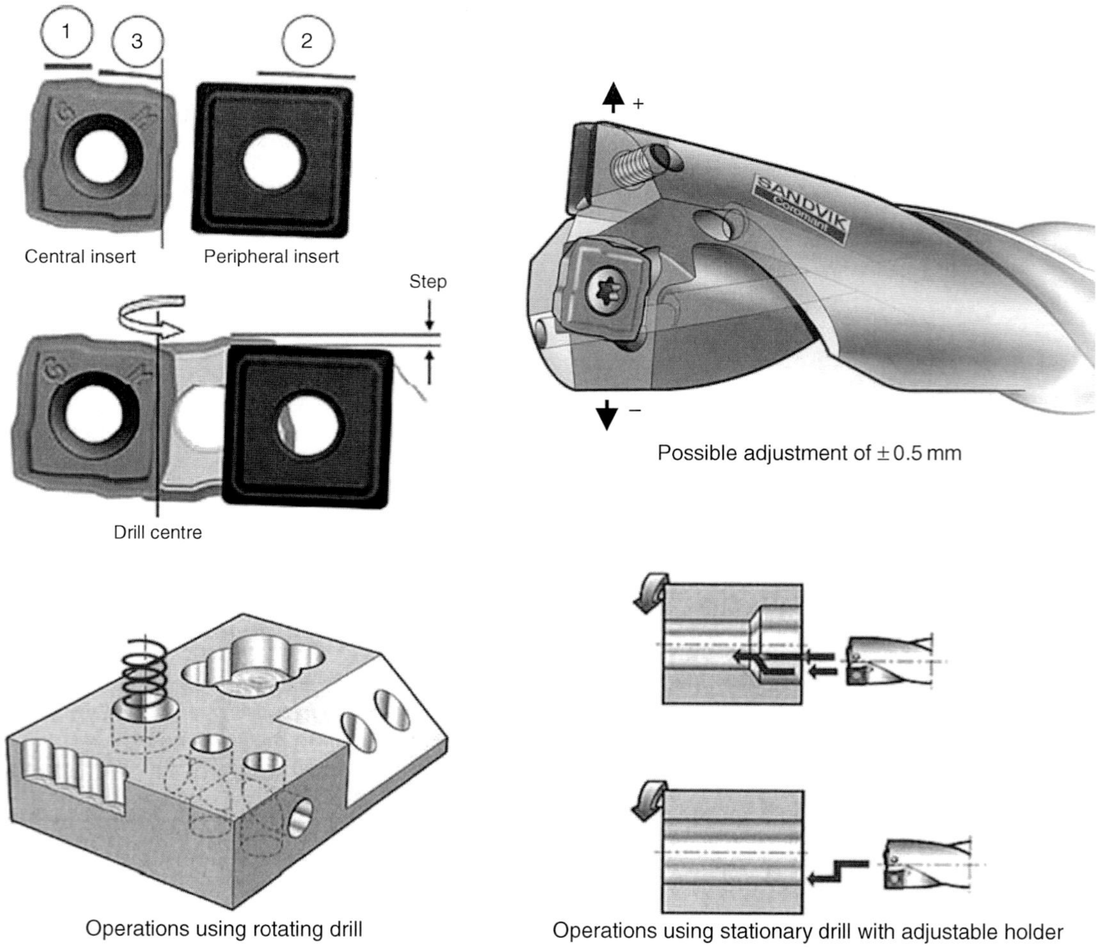

Fig. 15.76 Adjustable indexable insert drill [7] (CoroDrill 880).

Indexable drills with controlled cutting action have found many practical applications in the automotive, power plant, aerospace and die and mould industries. Some examples of applications in the automotive and power plant branches, along with cutting parameters and technological data, are shown in Figs 15.77a and b, respectively.

According to the customers' needs it is possible to assemble special boring tools consisting of a body with modular tool connection, fixed insert pockets, and fixed, adjustable and micrometric cartridges, as illustrated in Fig. 15.78. These cartridges and heads are especially suitable for using on special multiple diameter roughing and finishing boring tools.

Mostly, holes are drilled but modern CNC machine tools provide alternative methods of milling hole-making operations which can be especially profitable when the size of the hole is larger than 45 mm and both power and torque are limited. This is also an alternative when precision holes (IT7 or better) require a secondary finishing operation by a boring tool or reamer. In such cases precision solid-carbide end mills (Fig. 15.79) of smaller diameter (2–16 mm) with special geometrical features and grades can be used to perform HPM.

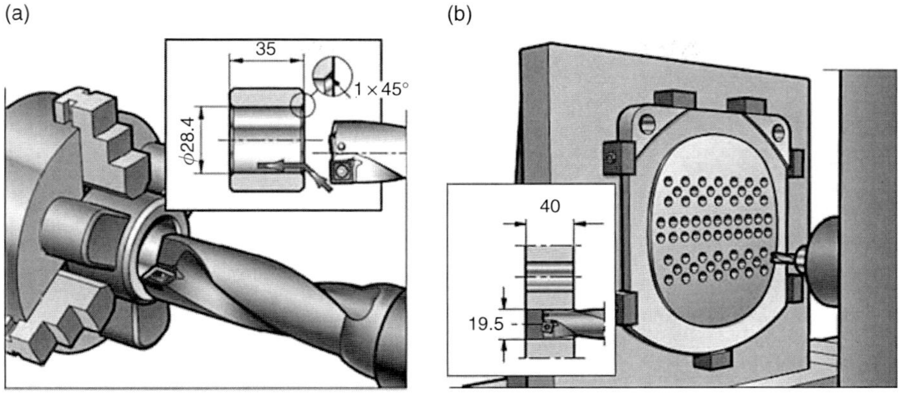

Fig. 15.77 Typical applications of adjustable indexable insert drills [7] (CoroDrill 880): (a) tri-axis (hole diameter – 28 mm, hole tolerance – 0/+0.15 mm, unalloyed steel (CMC 02.1), $v_c = 234$ m/min, $f_n = 0.14$ mm/rev, $v_f =$ 372 m/min, drilling with chamfering is possible in one stroke); (b) heat exchanger plate (hole diameter – 19.5 mm, hole tolerance – 0/+0.25 mm, ASTMA 182 unalloyed steel, $v_c = 232$ m/min, $f_n = 0.06$ mm/rev, $v_f = 135$ m/min, chip evacuation is critical).

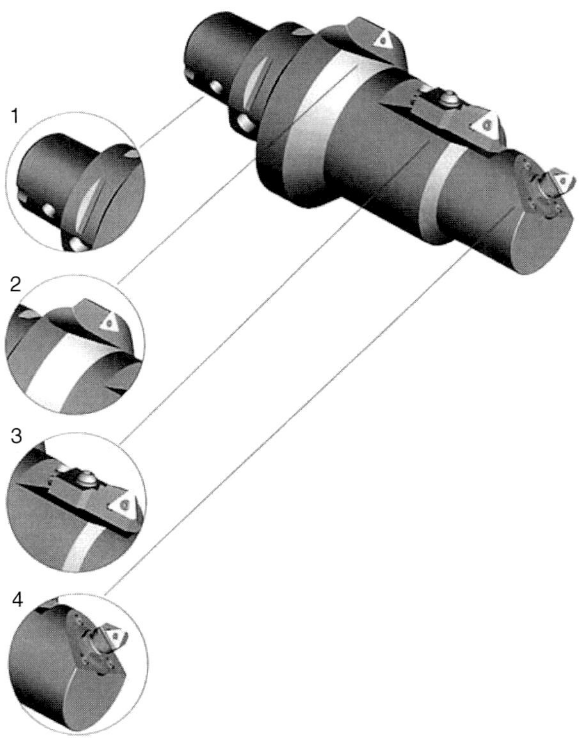

Fig. 15.78 Flexible conception of boring tools. 1 – coupling (Coromant Capto), 2 – insert pocket, 3 – radially adjustable insert cartridge, 4 – interchangeable, micro-adjustable boring head [7]. Source: Sandvik Coromant.

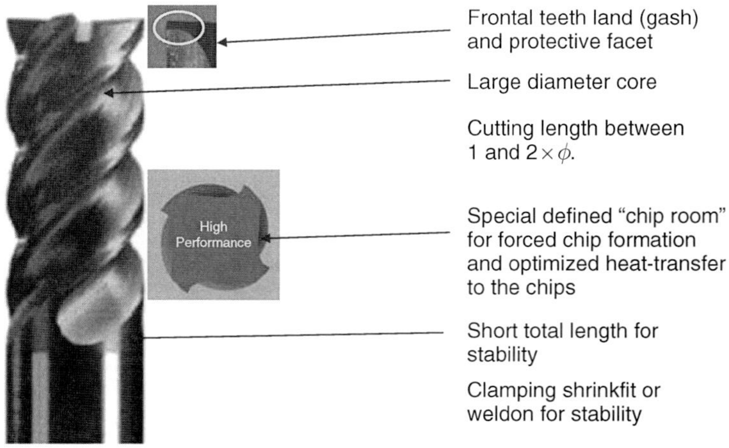

Fig. 15.79 Layout and features of a tool for HP milling and drilling operations [7]. Source: SECO Tools. Cutting parameters: $v_c = 120\,\text{m/min}$, $v_f = 127\,\text{mm/min}$, $f_z = 0.01 \times D_c = 1 \times \text{Dc}$ (HSM version $= 0.1 \times D_c$), $a_e/a_p = 6\,\text{mm}$, hole length equivalent to tool life – 35 metres.

The substrate of the tool is ultra-fine carbide grade (W/Co = 90%/10%) of 0.8 μm grain size which is coated with TiAlN monolayer film. The end mill with drilling possibility from Fig. 15.79 has special chip formers in the flutes and corner protection with a small 45° chamfer or corner radius. In particular, a gashed land or outer diameter reduction (eccentric cutting edge) prevents the cutting edges against fracturing when machining hard steel of hardness 50–65 HRC. In general, drilling feeds of up to 50% of the normal feed can be achieved and drilling is performed at feeds reduced to 5% of normal feed for $a_p > D_c$. When ramping, it is necessary to reduce feed to 50% for rake angles less than 15°, and to 25% for angles higher than 15°.

For larger diameter holes and cavities, developments in tooling, CNC machine technology and readily available canned programming routines are making helical interpolation milling (aka corkscrew milling), circular interpolation milling and plunge milling (aka *Z* axis milling) effective choices for manufacturers replacing drilling, as compared in Fig. 15.80.

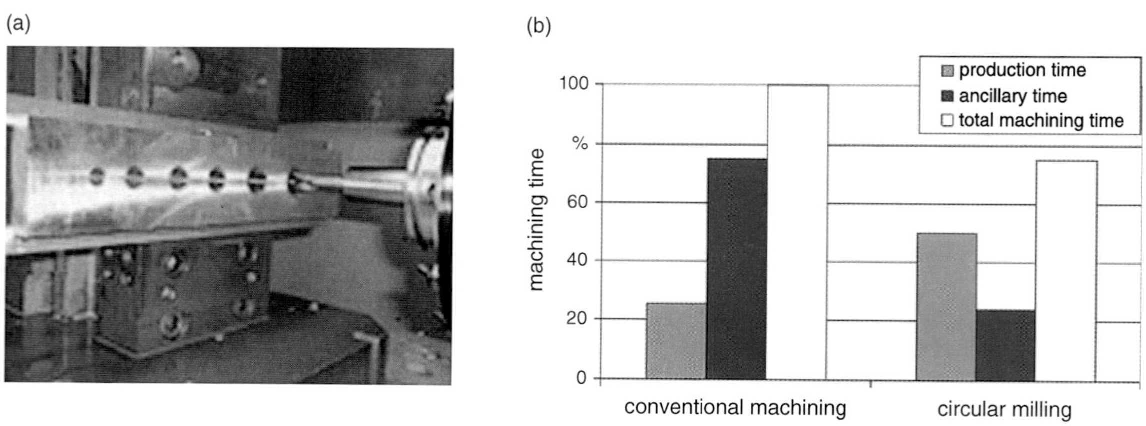

Fig. 15.80 Circular milling (CM) vs conventional machining [5]: (a) a view of CM operation, (b) time distributions in conventional and CM operations.

As indicated earlier, various methods of milling holes and closed cavities are proposed, namely [7, 9]:

1. Circular interpolation milling involving walking around OD or ID at the full depth of the cutter to open an existing hole. In this way the hole is opened up and, in some cases, finished to achieve tolerances within IT6. Suitable tools are indexable insert drills (Fig. 15.76), high feed and shoulder milling cutters (Figs 15.81, 15.82 and 15.83b), round insert milling cutter and special solid carbide end mills (Fig. 15.83c). The insert end mill shown in Fig. 15.82 is especially designed for manufacturing frame components made of non-ferrous alloys for the aerospace industry as well as machining of deep cavity faces by die and mould makers. Circular interpolation can also be performed for chamfering operations with some turning tools.
2. Helical interpolation milling (also called tornado milling) involves using a milling cutter to open the hole by ramping into a solid or an existing hole, and then spiralling down in the Z axis. The operation can be performed with a rotating tool (simultaneously with an X/Y circular motion) and as internal turn milling with a stationary tool and rotating workpiece. Suitable tools are, in addition to indexable insert drills (Fig. 15.76), indexable insert cutter (Fig. 15.83b), round insert cutter and solid carbide end mill (Fig. 15.83c), as well as any cutter with ramping ability [7]. Helical and circular interpolation milling enables the manufacturer to generate large holes with a machine of limited horsepower, say 10 hp (7.5 kW) to 15 hp (11.2 kW) [9] that is not able to push either a spade or indexable drill through to make the hole.
3. Plunge milling/drilling which can create a new hole (drilling) as well as rough out a cavity by successive plunge cuts along a shoulder. It is especially suitable for rough machining because it is power efficient, productive and spares the machine tool spindle as the cutting force is directed axially along the spindle. Suitable tools include indexable insert drill (Fig. 15.76), indexable insert cutter (Fig. 15.83b), round insert cutter and high feed milling cutter (Fig. 15.81).

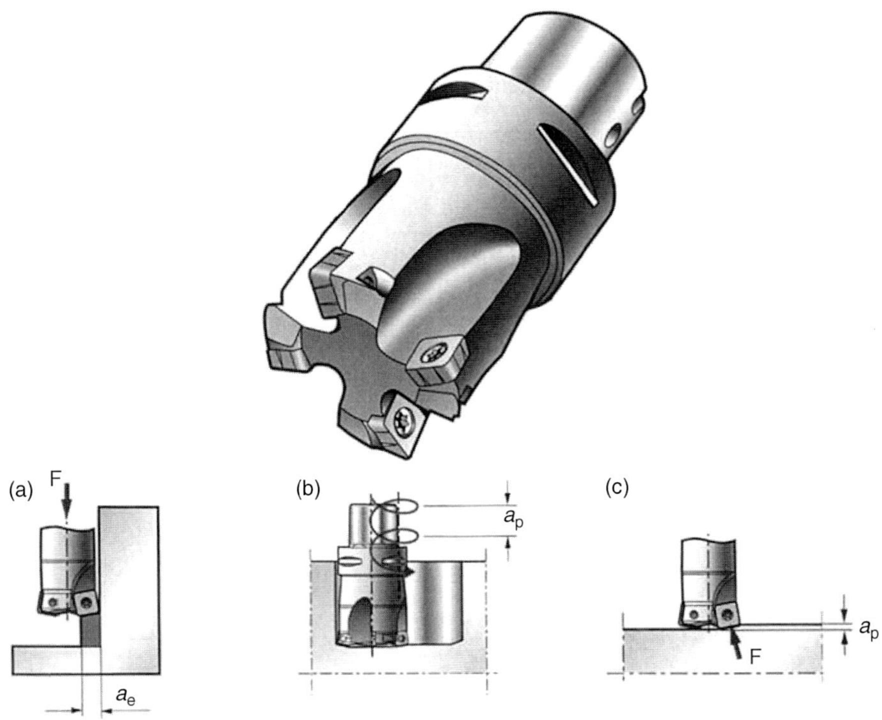

Fig. 15.81 High feed face mill and plunge milling cutter [7] (CoroMill 210): (a) plunge milling, (b) high feed helical interpolation, (c) high feed face milling.

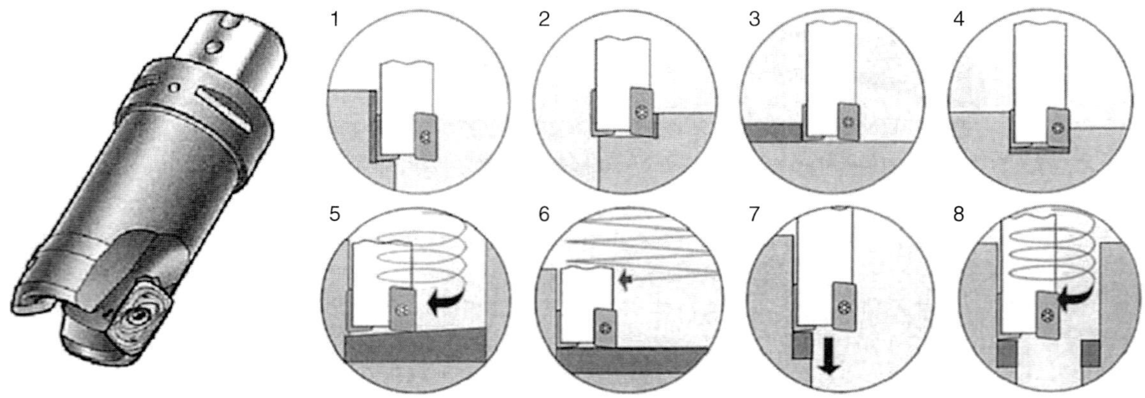

Fig. 15.82 Multifunctional high performance square end mill for non-ferrous alloys [7] (CoroMill 790): 1 – cut along the shoulder, 2 – square shoulder milling, 3 – face milling, 4 – groove milling, 5 – helical interpolation, 6 – linear ramping, 7 – plunge milling, 8 – helical feed boring.

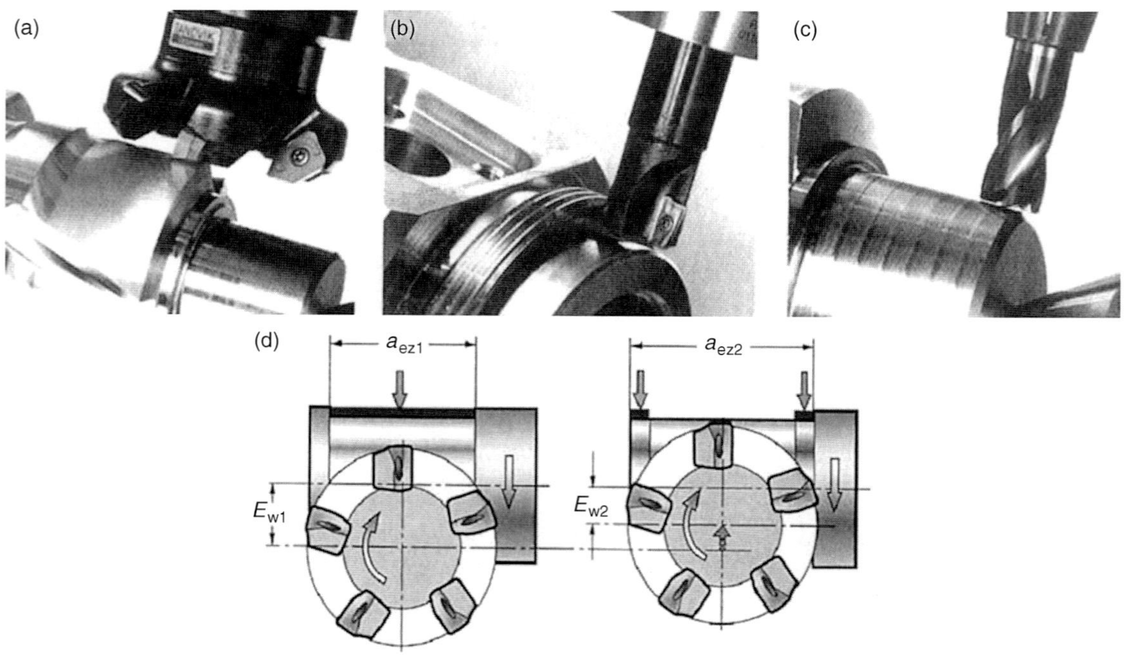

Fig. 15.83 Turn-milling operations with different cutters: (a) indexable insert milling cutter with special smoothing inserts CoroMill 245, (b) long edge milling cutter CoroMill 390, (c) solid carbide end mill CoroMill Plura, (d) cutter arrangement in two successive operations [11].

Turn milling (Fig. 15.83) is a specific machining operation performed on turning centres with rotary tool capabilities (they rightly deserve to be called turn-milling centres). This involves cutting with a rotating milling tool while the workpiece is also rotating, similar to turning [10, 11]. Specific challenges that turn milling can face are as follows:

- High metal removal rate.
- Performing interrupted cuts and obtaining short favourable chips (against the tangle chips in turning).

- Preventing the long turned parts from deflecting.
- Longer tool life because the tool has multiple inserts to divide the cutting load.
- Machining eccentric or odd shapes. In this case the radial (X axis) motion of the milling cutter can be coordinated with the rotation of the workpiece to machine profiles other than perfect circles.

In turn milling workpiece rotation provides the C axis motion that delivers the desired feed rate for the milling cutter (the workpiece speed is much slower than what is typical of turning). Y axis motion is also needed because the milling cutter cannot machine the part to size when it is on-centre. One of the limited disadvantages in turn milling is form error because regularly spaced scallops in the surface are leaved when using wiper inserts.

In general, in face turn milling one wiper insert is used (Fig. 15.83a and b) to generate the straight line contact between the cutter and the machined surface in order to create the cylindrical part of the component. To cover the full width of the cutter, the tool has to be placed with at least two offsets (Fig. 15.83d), first E_{w1} during the first revolution of the workpiece and then moved to E_{w2} for a second cut. Moreover, if the surface is wider than the cutter diameter, the cutter should be moved in axial direction to achieve the required length, but not more than 80% of the a_{ez1} per revolution. To achieve a sharp corner, the cutter must take a second pass (the corner material is cleaned away).

An increasingly popular alternative to tapping is thread milling, which traces its use back to aerospace applications in the Gemini Space Program [12]. Technological advancements in thread mills and the availability of modern CNC machine tools are enabling manufacturers to produce high quality threads. Moreover, the advantages of this method include reducing cycle time, increasing productivity and eliminating costly scrap and associated downtime that result from broken tools. Thread milling produces a superior thread because it is free cutting leading to very small, recutting chips. Tapping requires more spindle power when holes get larger and deeper. There is also a real danger of broken tapes which leads to additional costs when machining on expensive parts. A variety of the designs of thread-milling cutters is shown in Fig. 15.84. Thread mills can be of a single-point (tooth) design (Fig. 15.84c) or a multi-point design in solid (Fig. 15.84a) and indexable (Fig. 15.84b) versions. Typically, thread mills are equipped with through coolant holes, as, for example, cutters from Fig. 15.84a and c. Thread mills generate the thread profile by helical interpolation. Hence, a single-point cutter requires the same number of interpolations as there are pitches, i.e. an eight-pitch thread, 1" long needs eight circular interpolations around the workpiece. A multi-point cutter, which in fact is a series of single cutters on one body/flute, can normally complete a screw thread in one revolution of the work.

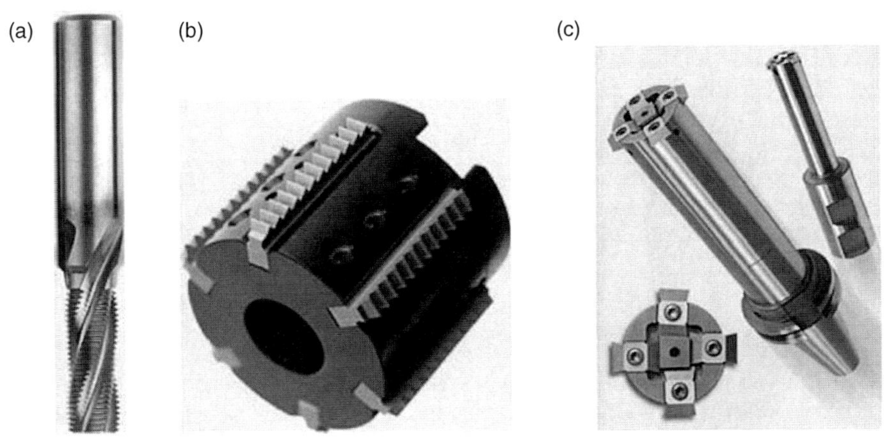

Fig. 15.84 A variety of thread-milling cutters [7]: (a) TiCN coated solid carbide thread mill, (b) indexable thread mill with six long, double-sided inserts, (c) four-tooth indexable thread mill. Source: Emuge.

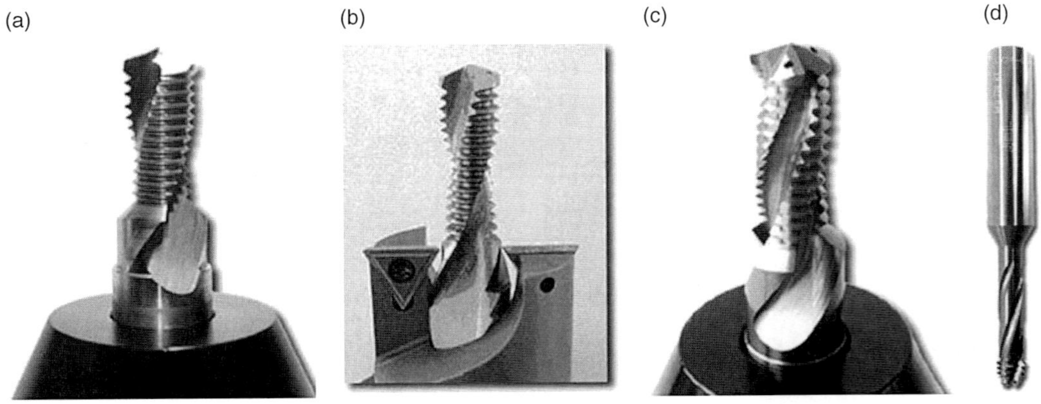

Fig. 15.85 A variety of thriller drill- and thread-milling tools [7]: (a) universal drill- and thread-milling tool, (b) drilling, chamfering and thread-milling tool, (c) drill- and thread-milling tool with through coolant holes, (d) direct circular thread-milling tool. Source: Seco Tools.

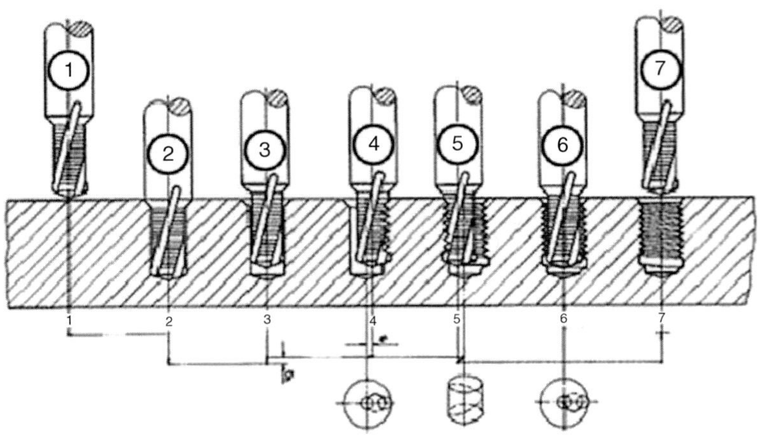

Fig. 15.86 Drilling, chamfering and thread milling with thriller drill- and thread-milling tool from Fig. 15.85a [7]. Source: Seco Tools.

A revolutionary method of producing threads by using the CNC technology is employing a combined 'thriller' thread milling tool for drilling, chamfering and thread milling in one operation. According to the manufacturer's data thread hole production time is reduced by up to 50% compared to conventional tapping due to high cutting speed and combined operations. In addition, it is possible to form blind and through holes, as well as right- and left-handed threads, with one tool. The thriller tool, made of solid carbide, presented in Fig. 15.85a, is able to form a thread during a cycle consisting of seven steps, as shown in Fig. 15.86. They include: 1 – approach, 2 – drilling and chamfering, 3 – withdraw, 4 – radial setting to nominal thread diameter – positioning loop, 5 – forward advance around the pitch with simultaneous circulation of the tool around the centre axis of the thread, 6 – radial movement back to the bore centre – retraction loop, 7 – reverse stroke from the finished thread bore.

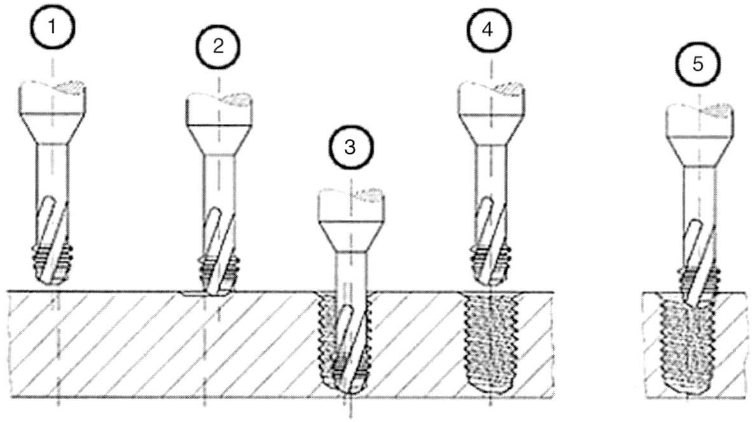

Fig. 15.87 Schematic sequence machining a thread with direct circular thread-milling tool [7]: 1 – approach, 2 – circular milling of the chamfer, 3 – circular drilling and thread milling, by the circular movements the core hole and thread is produced in one operation, 4 – reverse of the finished thread bore, 5 – circular deburring of the first thread, if necessary.

A thread milling tool designed as one central solid carbide tool, performing drilling and thread-milling operations, is equipped with two indexable inserts – a T-insert for counterboring and spot facing and a V-insert for chamfering, as shown in Fig. 15.86b. By using formed inserts, a variation of hole geometries can be produced. On modern CNC machining centres such tools are working with revolutions up to 24 000 rpm and more, and providing significant gains in time and quality improvement [6].

Newly designed thread-milling tools, shown in Fig. 15.85d, include circular (CBGF) and direct circular (DBGF) thread-milling tools performing core hole, chamfer and thread with countersinks in one operation. This tool features a higher core cross-section and narrow, geometrically optimized flutes (three to five), and a special mill-drilling tip. These construction characteristics lead to good chip formation and low cutting forces during drilling, chamfering and thread milling. In particular, when using DGGF tools, core hole and thread are milled simultaneously in a circular way, in one operation (operation 4 in Fig. 15.87). In fact, it is a one-step thread-milling method [4]. Moreover, PCD thread-milling cutters of DBGF type are suitable for the machining of lightweight materials like aluminium with cutting speeds $v_c = 100–400\,\text{m/min}$ and feeds per tooth $f_z = 0.05–0.25\,\text{mm/tooth}$ depending on the thread diameter [7], magnesium and fibrous reinforced synthetic materials. Furthermore, PCD tools are also suitable for dry machining and with MQL supply when working on aluminium.

High precision threads with high surface quality can also be produced by the thread turn-milling process, which integrates the features of single-point threading, previously described thread milling and thread whirling [13]. In practice, two different variants of this method – the traversing method with inclined (Fig. 15.88a) and parallel (Fig. 15.88b) tool axis and the plunge method (Fig. 15.88c) – can be applied.

With the thread turn-milling process the thread is produced in one pass. The rotating milling cutter has several teeth on the circumference and has a lead that is the same as the thread produced. However, the direction of lead of the tool and the workpiece is opposite. It should be noted that the drive power is shared between the tool and spindle drives. As shown in Fig. 15.88, the workpiece and the tool are revolving in opposite directions with different speeds. Thread turn milling with the traversing method can be performed with a tool slightly inclined (Fig. 15.88a) or parallel (Fig. 15.88b) to the workpiece axis. In the first arrangement the tool is inclined by twice the lead angle, which results in a high quality surface comparable to single-point threading. The parallelism of the tool and workpiece axes means that the tool must be equipped with a special form of ground inserts for profile correction. When using the plunge method the tool and workpiece are also parallel and the complete thread is produced in one pass. Therefore, the length of the

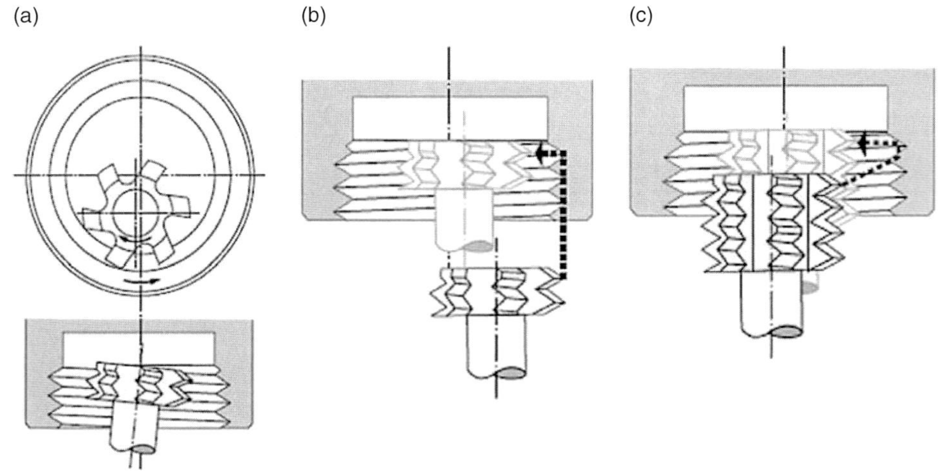

Fig. 15.88 Thread turn milling of internal threads: (a) traversing method with inclined tool axis, (b) traversing method with parallel tool axis, (c) plunge method [12].

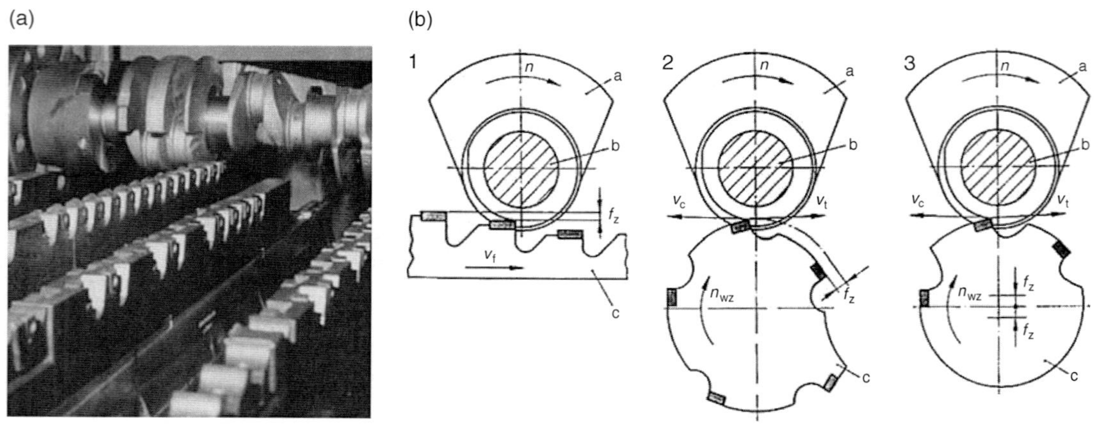

Fig. 15.89 Turn broaching of a crankshaft [14, 15]: (a) machining of bearing seats on a crankshaft with a section of linear broaches, (b) possible kinematical variants (1 – linear method, 2 and 3 – rotary method with spiral and circular tools, respectively); a – crank web balance weight, b – main crankshaft bearing journal, c – tool.

milling cutter must be at least as long as the thread. Unfortunately, the surface quality is not as good as with the traversing method but outperforms the results of conventional thread milling.

Turn broaching is an efficient method for machining steel and nodular cast iron crankshaft. Special turn-broaching machines are available for linear, circular and spiral operating methods shown in Fig. 15.89b [16, 17]. As mentioned before, there are three main methods of operating, namely:

1. Linear turn broaching, where the crankshaft axis and tool are positioned at a fixed distance. Stock removal is done by the cutting edge progression in the tool.
2. Circular turn broaching, in which the crankshaft axis is also positioned at a fixed distance but the stock removal is controlled by the vertical feed movement of the tool axis.

3. Spiral turn broaching in which, compared to the circular arrangement, the tool only rotates and the total allowance on a bearing journal is progressively removed by an appropriate layout of the inserts along the Archimedes' spiral [16].

Typically, the finishing segments in the tool are fitted with inserts embedded in adjustable cartridges that can be set to close tolerances and in the roughing part the inserts are fixed directly in pockets. The main advantages of turn broaching are short machining times, high and repeatable accuracy, close tolerances and good surface roughness by additional burnishing effect, and long tool lives resulting from short engagement times of the individual cutting edges and low thermal stresses. The further advantage of the process is its high productivity due to intermittent contact between the tool and the workpiece. The intervals are determined by the cutter spacing or necessary chip volume.

Circular turn broaching of all four main bearing journals on a V6-cylinder crankshaft made of GGG60 nodular iron was performed with a cutting speed of 180 m/min, rotational speeds ranging stepwisely from 420 to 1000 min^{-1}, feed velocity of 850–2100 mm/min, maximum undeformed chip thickness of $h_{max} = 0.48$ mm and machining time of 29 s. For comparison, machining time for linear turn broaching of a 4-cylinder crankshaft made of steel was equal to 48 s at the same cutting speed and comparable with other machining conditions [18]

It is known from machining practice that the automotive crankshaft is a unique component covering a large number of different machining processes used. Today, milling of crankshaft and camshaft blanks is a commonly used method, instead of classical turning operations, due to exceptionally high productivity and economy. As shown in Fig. 15.90 internal (a) (also called planetary milling) and external (b) milling methods can be employed for manufacturing automotive crankshaft bearings and crankpins in large-scale production. During the milling operations a slow rotation of the crankshaft by means of duplex worm drives provides the feed motion. According to the production data the number of pieces per tool life was increased by 90% and the cutting time was reduced by 43% when dry machining pin bearings on a car crankshaft of GGG50 nodular cast iron [7]. Similarly, by using circular milling cutters to machine main bearings on a diesel crankshaft of steel (UTS = 1000 MPa), the number of pieces per tool life was increased by 150% and the cutting time was cut down by 22% in comparison to conventional tools with radially arranged indexable inserts. Moreover, the grinding time can be reduced due to the high accuracy of the primarily milled bearings.

(a) (b)

Fig. 15.90 High efficiency machining of crankshafts: (a) internal milling, (b) milling of main bearings and flange using a gang of external cutters [7]. Source: Heller (a) and Ingersoll (b).

References

[1] G. Byrne, D. Dornfeld, B. Denkena, Advancing cutting technology, *CIRP Annals*, Vol. 52/2 (2003), 483–507.
[2] G. Eisenblätter, Industrial co-operation in high performance cutting, *Proceedings of First International Workshop on High Performance Cutting*, Paris, 2002, pp. 6–15.
[3] K. Weinert, H. Löbbe, High performance cutting-utilization ratio and machining strategies, *Proceedings of First International Workshop on High Performance Cutting*, Paris, 2002, pp. 23–29.
[4] B.P. Erdel, *High-Speed Machining*, SME, Dearborn, 2003.
[5] P. Andrae, High-efficiency machining, *Manufacturing Engineering*, Vol. 125, No. 4 (2000), 82–96.
[6] Machining Tech Trends. The future of machining technology and processes. Prepared by MTA of SME, 2002, www.sme.org.
[7] Internet database: www.coromant.sandvik.com, www.secotools.com, www.mitsubishicarbide.com, www.emuge.com, www.ingersoll.com, www.heller.com, www.chironamerica.com, www.starragheckert.com.
[8] F. Dürschmied, Parallel-Kinematik: Zur Serie gereift, *Werkstatt und Betrieb*, Vol. 134, No. 9 (2001), 76–81.
[9] J. Lorincz, Holemaking strategies, *Manufacturing Engineering*, Vol. 137, No. 3 (2006), 109–118.
[10] P. Zelinski, Applying turn-milling, Modern Machine Shop, www.mmsonline.com.
[11] Turn-milling with CoroMill 390, www2.coromant.sandvik.com.
[12] J. Lorinicz, Solutions come in many forms. Tapping and thread milling head the list, *Manufacturing Engineering*, Vol. 137, No. 5 (2006), 73–78.
[13] Thread-turn-milling of internal threads, www.index-werke.de.
[14] M. Müller, Neue Werkzeuge zum Fräsen und Drehräumen optimiren die Fertigung, *Werksttat und Betrieb*, Vol. 124 (1991), 551–554.
[15] M. Müller, *Drehräumen-eine neue Technologie zur Bearbeitung rotationssymmetrischer Werkstücke*, Krupp Widia GMBH.
[16] G. Augusten, K.H. Schmid, Turning/turn-broaching – a new process for crankshaft and camshaft production, *Industrial and Production Engineering*, Vol. 14, No. 2 (1991), 36–43.
[17] *Modern Metal Cutting. A Practical Handbook*, Sandvik Coromant, Fair Lawn, USA, 1996.
[18] N. Reiter, M. Müller, E. Stalwitz, Technologie des Drehräumens, *Werksttat und Betrieb*, Vol. 122 (1989), 201–206.

15.5 Multitasking and One-pass Machining

15.5.1 Background of multitasking machining

Multitasking can be defined as the ability to perform various manufacturing operations without manual intervention. In this aspect, parts are automatically handled from first to final operations and loading/unloading during machining operations. Of primary importance is the capability to perform a variety of machining operations such as turning, milling, drilling and taping and produce a complete part in one setup. The latter fact has inspired the machine tool manufacturers to introduce a new term for this type of machining, *complete machining*.

Byrne *et al.* [1] distinguished two process strategies which led to this concept, namely:

1. Integration of various machining processes into one machine tool.
2. Six side machining.

In general, three leading options of process optimization are utilized [1]:

1. Parallel processing involving two or more processes independently on a single machine, e.g. four-axis turning.
2. Hybrid processes (assisted machining) when two or more processes are coupled to achieve a specific workpiece alteration, e.g. plasma/laser aided machining (PAM/LAM) and ultrasonic assisted machining (UAM). These processes will be outlined in section 15.6.
3. Integrated processes incorporating two or more conventional processes, e.g. creep feed grinding and high speed machining.

Multitasking machining can provide three main performance benefits to manufacturing operations, namely [2, 3]:

1. Improve part accuracy.
2. Reduce cost required to manufacture each part.
3. Reduce non-value added manufacturing time.

The accuracy benefits that will be realized through multitasking machining are astounding. Multi-process tolerance stackup will be minimized due to reduced part handling. In addition, difficult datum dependent workpieces will be more accurately machined because automatic workpiece transfer and robot handling will ensure better locating repeatability.

The basic cost benefits are now achieved due to the reduction of total machines necessary to produce the same parts, include reductions of: workholding cost, in-process inventory, burden rates, necessary manpower for total number of setups, and large numbers of duplicate tools from machine to machine.

Besides the basic production benefits multitasking provides there are many commercial benefits as well. Most important of which is a substantial drop in non-value added time. In addition, big improvements in time consuming work such as programming, program verification, and manual part load and unload are observed. Moreover, other potential benefits include:

- Reduction of overall floor space required.
- Reduction of power requirements.
- Increased throughput for small to medium lot sizes.
- Increased flexibility to challenge constantly changing part features from engineering.

The first multifunctional machine tool called *Mill-Turn*, which can perform three basic machining operations, i.e. turning, drilling and milling, was produced in the mid-1990s by WZL in Linz [2]. This turning-drilling-milling centre combines the features of a four-axis turning centre and a five-axis machining centre.

The basic characteristics of a multifunctional machine tool for complete machining include three factors, namely: flexibility, short part cycle times and precision. In general, the following concepts of complete machining are used in industry:

- Classical solution based on the integration of turning, drilling, boring and milling operations with the additional possibility of performing such operations as deep hole-making, gear hobbing and turn milling [2, 4, 5].
- Combination of turning and grinding due to the fact that less time is needed for finish grinding. Moreover, the reclamping error is eliminated when two or more setups must be used [5]. In this option, complete machining can also incorporate hard turning and/or finish grinding (super finishing) with a CBN wheel on an ID grinding machine platform [6, 7]. Typically, a cutting speed higher than 60 m/s is needed to get the full benefit of CBN. Some dedicated machines equipped with CBN tools, as in the case presented in Fig. 15.96, are capable of speeds up to 200 m/s.
- Combination of high speed milling and laser technologies to machine hard and superhard (PCD, CBN) materials. Typically, HSC milling as pre-machining and then laser finishing in one setting are performed.
- Grinding with multiple wheels and spindles. Currently, this concept is extended to the ability to do OD/ID grinding and tapers, threads, and bores as well as out-of-round parts on one CNC machine with B-axis positioning [8].
- Ultrasonic-assisted machining of hard brittle (so-called advanced) materials with diamond tools. Typically, technological processes involve drilling, milling and grinding operations.

A multitude of machining processes and additional tasks (part washing before measurements, measuring on CMM's station) are listed in Fig. 15.91.

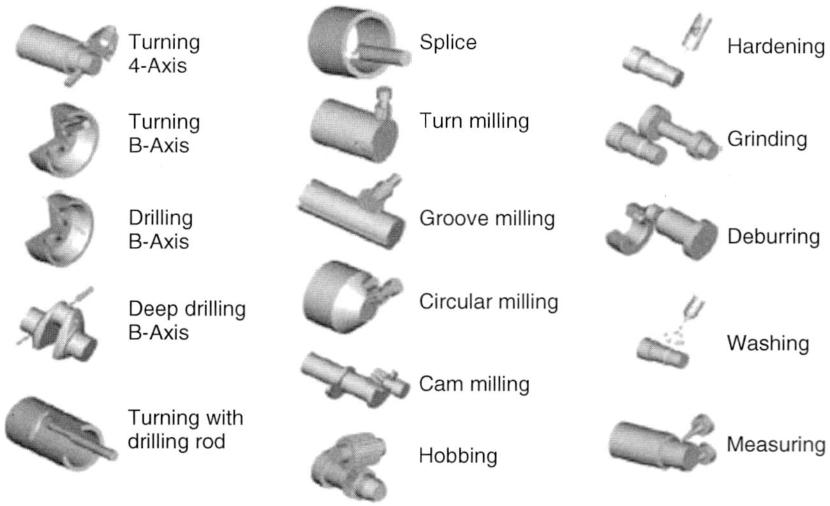

Fig. 15.91 A range of machining and auxiliary operations performed on mill turn centres [1].

15.5.2 Multitasking machines and tooling

Complete machining of some complex aircraft parts is shown in Fig. 15.92. Nowadays, the restriction in wider application of HSC technology in turning, resulting from high centrifugal forces on a part chucking system, is overcome by turn milling. An example of the application of turn-milling operation in machining heavy unbalanced parts with lateral hollow lugs is shown in Fig. 15.92a. In addition, Fig. 15.93 illustrates

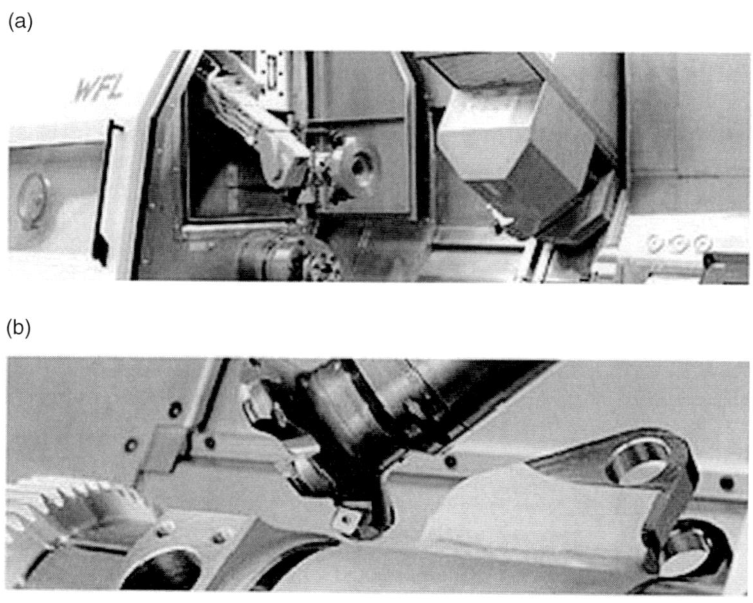

Fig. 15.92 Examples of complete machining in the aircraft sector [2]: (a) machining of aircraft component, (b) turn milling of heavy unbalanced workpiece.

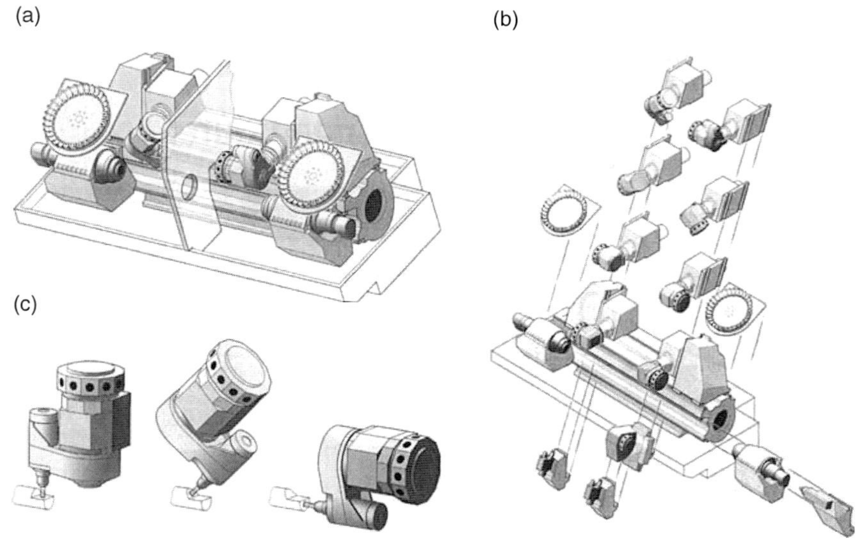

Fig. 15.93 Turn-milling multitasking modular machine tool: (a) general view, (b) a scheme of its modular structure and (c) different positions of milling cutter with angular arm [9].

a special turn-milling twin machine tool with modular structure in which angular milling heads can be positioned differently (Fig. 15.93c) depending on the machining task, i.e. for machining horizontal, oblique or face plane.

The advanced lathe from Fig. 15.93 is also suitable for milling operations in five axes. A lathe in modular construction also allows complete machining of extremely complex components at high productivity and accuracy. The milling spindles are provided with HSK taper tools from big tool magazines. The very stiff machine structure is designed for high metal removal volumes and guarantees persistently high part quality. Modern multitasking machine tools do not only provide powerful driven tools but complete high speed tool spindle heads mounted on $X/Y/Z$ axis units with 240° B axes. Figure 15.94 shows some examples of utilizing the B axis in machining non-rotational parts (below, a blade of the steam turbine is being milled when the B axis is at 30°).

The machining centre from Fig. 15.94 is the perfect five-axis solution for all requirements: aerospace, automotive, mould making, component manufacturing and medical technology. As much as five-sided machining by the universal milling head is made possible due to its automatic refitting of vertical to horizontal machining in combination with the NC rotary table. According to the requirements, the NC rotary table can be combined with a universal milling head and NC-controlled A or B axis or the option of a swivel rotary table as a five-axis solution equipped with a horizontal milling head. The swivel rotary table can also be used in combination with a palleting system to make production even more efficient.

With a counter workpiece spindle, six-side machining can be performed with rapid speeds of up to 75 m/s [1]. Using this concept, a range of components for automotive applications can be completely machined on one multitasking machining centre as shown in Fig. 15.95.

For some applications the multifunctional machine has the potential to reduce part cycle times by as much as 25%, mainly by eliminating rough grinding steps. As shown in Fig. 15.96 complete machining of a gear in the hardened state is successively performed on four machining stations – two for hard turning and two for grinding or super finishing.

After hard turning operations with CBN tools and a cutting speed of 3000 m/min (workstations 1 and 2 in Fig. 15.97) only a small machining allowance of about 20–30 μm remains for finishing grinding

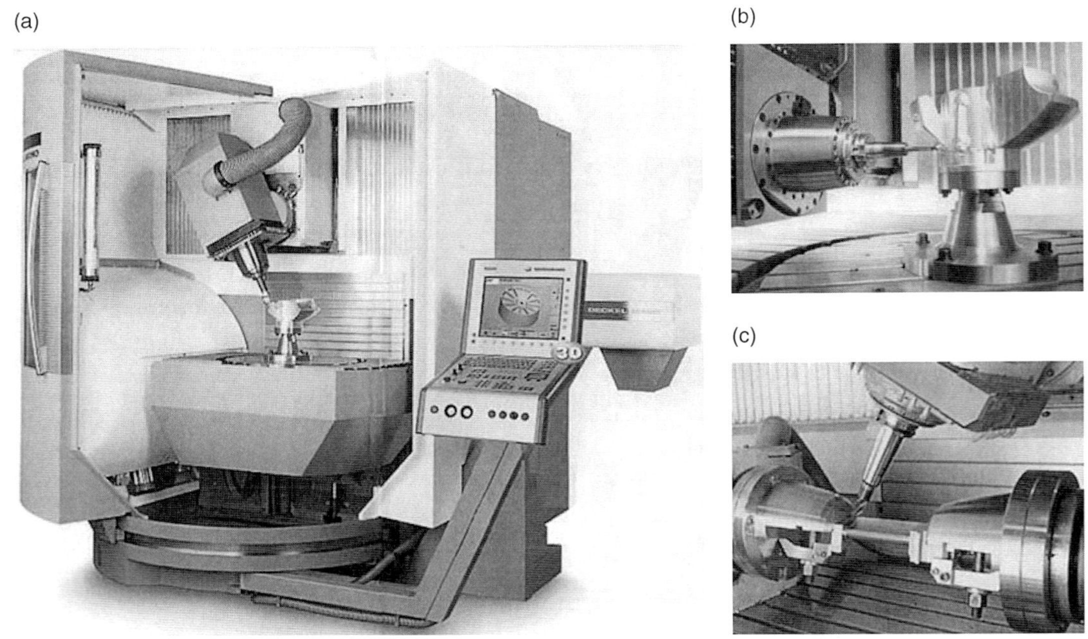

Fig. 15.94 Complete machining of non-rotational solid part (a) and (b), and steam turbine blade (c) using fast rotation of 150° around the *B* axis inclined at 30° [10].

Fig. 15.95 Exemplary applications of multitasking machining [3]: (1) flange, (2) housing of timer distribution, 3) drive train component, (4 and 5) milling operation on five-axis machining centre with 12 position turret head nd twin spindles (counter spindle), (6) milling of helical groove on a shaft.

(a)

(b)

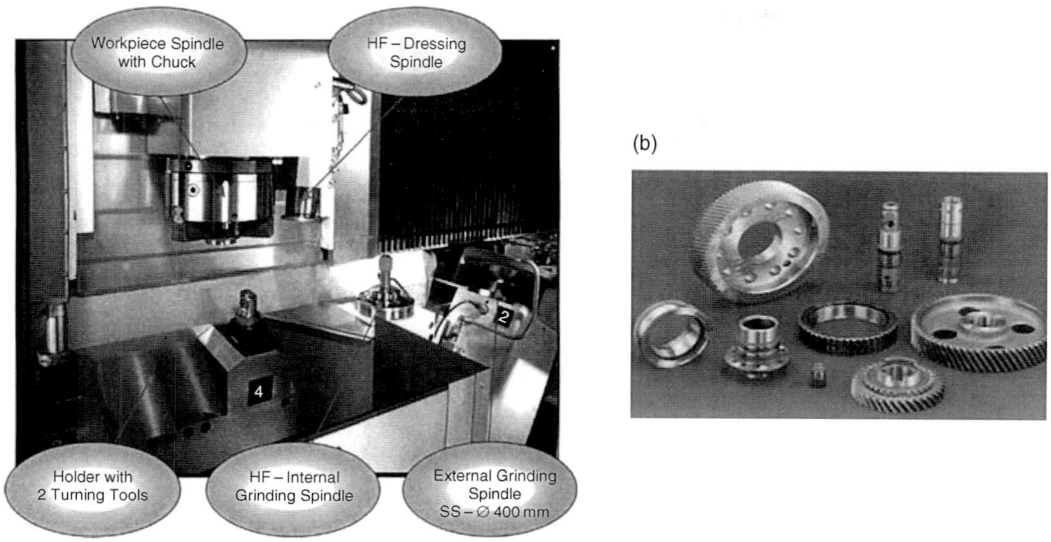

Fig. 15.96 Turning and grinding machining centre: (a) working area with two separate stations, (b) a variety of turn/grind parts [3]. (see Plate 20)

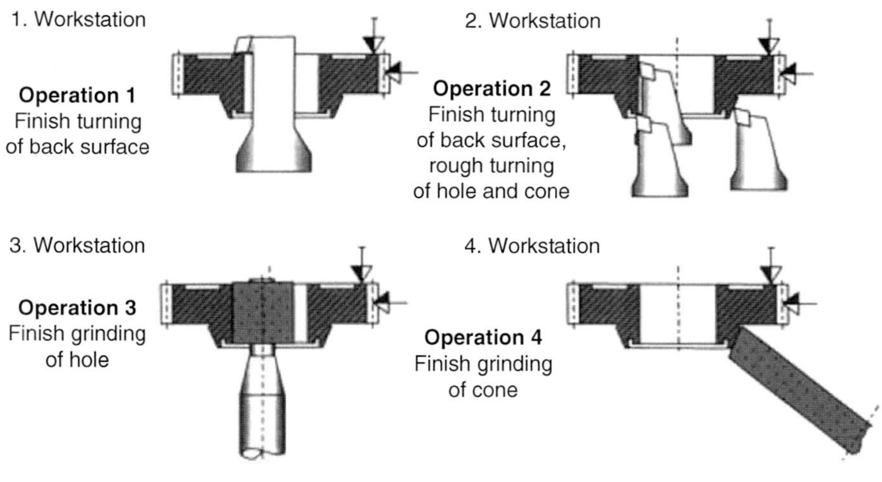

Fig. 15.97 Complete machining of a gear in the hardened state [6].

with the extremely high speed of 100 m/s, using CBN grinding wheels. Comparison of costs for three different machining strategies with or without hard turning partition is given in Fig. 15.98. It is evident from this diagram that the main benefit from hard turning is the reduction of both personal expenses and equipment cost.

A milling and laser machining centre with typical application examples is shown in Fig. 15.99. A total of six controlled axes are divided into three mechanical and three optical axes. It should be noted that laser machining is especially economic in the case of small parts, for example filigree dies and moulds and grooved tool inserts made of sintered carbide.

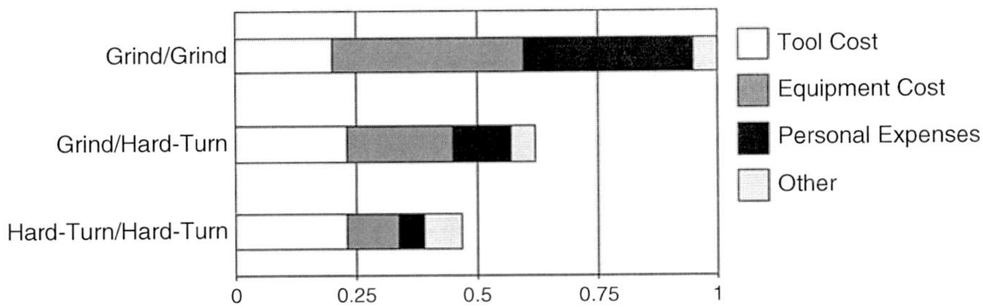

Fig. 15.98 Cost comparison of turning vs grinding [5].

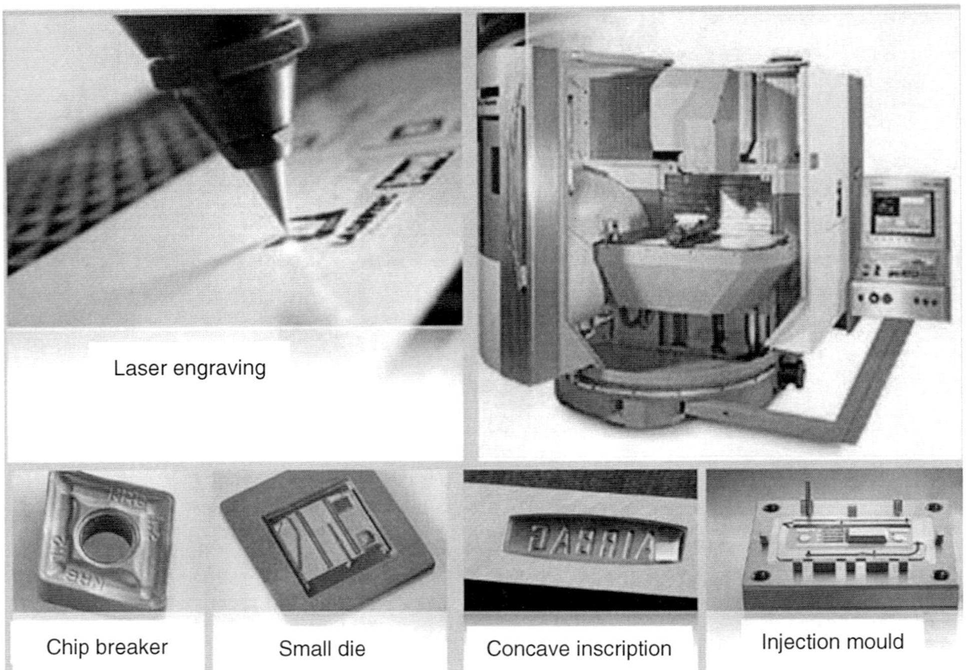

Fig. 15.99 HSC milling and laser shaping on MC with three optical and three mechanical axes [3].

In the case of the machining centre from Fig. 15.100a complete machining integrates boring, milling, grinding/honing operations. In order to perform honing operations (Fig. 15.100b) coolants should be additionally supplied through the spindle with defined and controlled pressure, as well as a PCD coated honing tool with CNC-controlled contact loads of stones, and an automatic measuring system of the hole diameter, which reliably operates in the presence of pressured coolant, should be applied. A special fine filtering system with the unit dimension of 10 μm for the coolant is installed. Similarly, the diameter of the finishing boring bar is automatically adjusted when pressured coolant flows. Figure 15.100b shows honing of the cylinder hole in the air compressor casing on this machining centre. Finally, the *Rz* roughness parameter of 3–5 μm and cylindricity error of 3–5 μm can be guaranteed.

Practical implementation of the idea of multitasking machining requires the use of a large number of turning and rotating tools, some of them with multifunctional machining tasks, as shown in Fig. 15.101.

Fig. 15.100 Five-axis machining centre (a) integrated with grinding/honing spindles (b) [11].

Special designs of these tools include: single and twin turning units on machines with the *B* axes locked at 45°, mini-turret with four tools for quicker tool change, end mills replacing drills and boring tools, internal and external threading milling cutters, short ball-nose end mills and end mills with wiper inserts for turn milling.

15.5.3 One-pass machining

One-pass machining is a special case of complete machining. It can cover any processes, whether the tool is to produce a finished part from a raw casting in one pass, or one tool is to finish several configurations in one pass (combination tooling), or if one tool is used to perform several functions (e.g. circular milling) [12]. For this purpose both multifunctional and multistage tools are used in order to perform hole-making, machining outer surfaces, chamfers and faces. The three relevant criteria of one-pass finish machining are machining time, surface finish and cost. As shown in Fig. 15.102 one-pass machining will be replaced in the future by the zero-defect concept without final part inspection.

Some classical examples of this method are complete machining of complex parts involving surface milling, boring, recessing and retract machining at high cutting speeds performed by circular milling tools, and one-pass, one-setup hard turning. It should then be noticed that HSM and hard machining frequently assist the complete machining.

A basic question in implementing the philosophy of one-pass machining is the selection of machining operations which are most applicable for the overall production flow, and can achieve the desired tolerances most productively and economically. It is evident that process substitution in surface finishing operations is classically done when the turning-hardening-grinding chain is substituted by hard turning (Fig. 15.98). Another recently developed process which replaces turning and grinding (Figs 15.103a and b, respectively) is called OD reaming, illustrated in Fig. 15.103c.

Figures 15.104–15.106 show a range of multi-step and combination tools for hole-making including boring and chamfering and facing. In the case of the engine cylinder head (Figs 15.105c, d and 15.106) valve seats/guides are generated by fine boring and the upper plane by face milling.

One-pass machining with this type of tooling reduces non-machining time (setup and tool changes), eliminates stacking up tolerances (step-by-step cutting), promotes lean manufacturing (lowering tool inventory) and complements the capabilities of high quality machine tools.

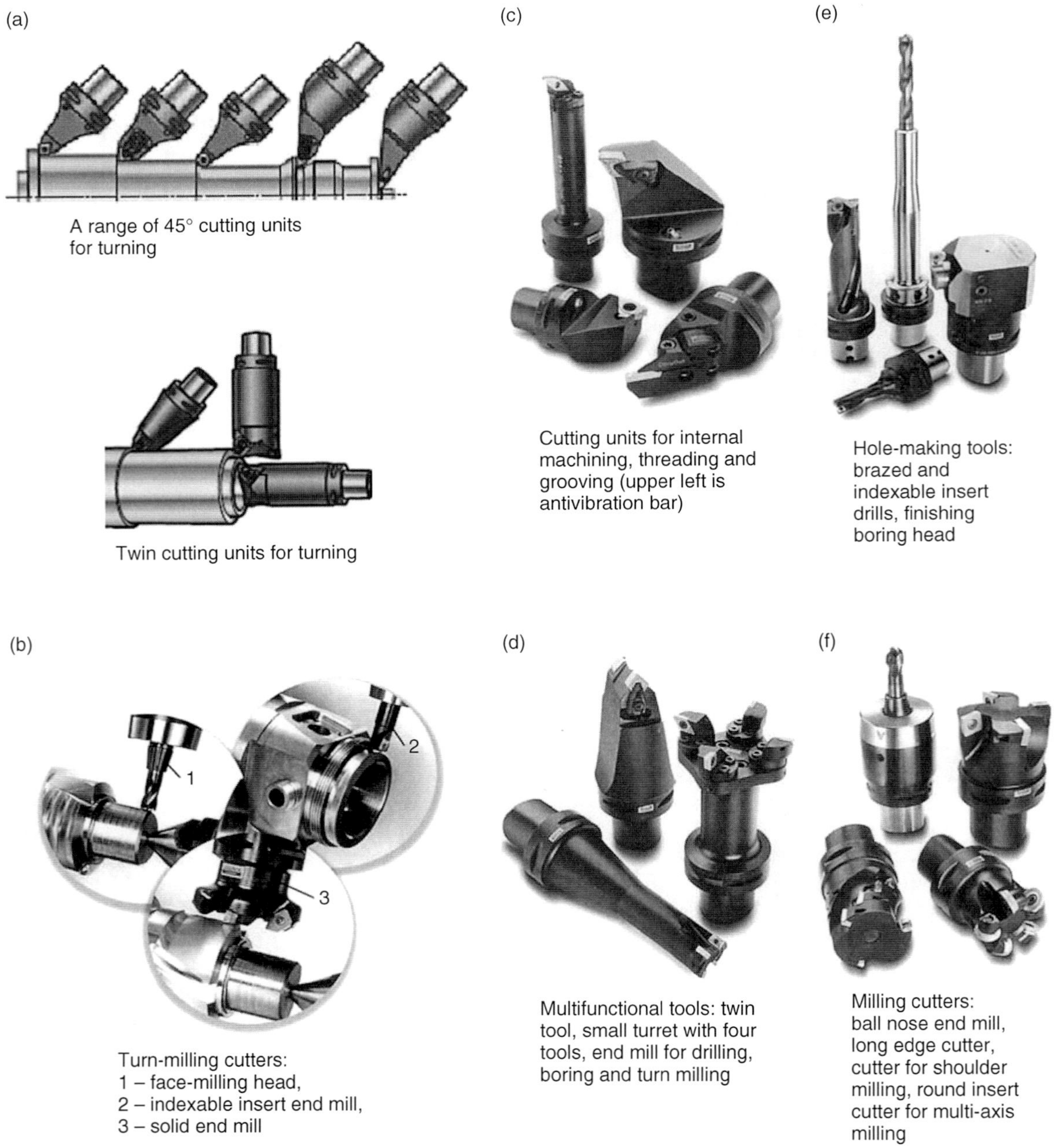

Fig. 15.101 Selected tools for multitask machining with Coromant Capto clamping system [3].

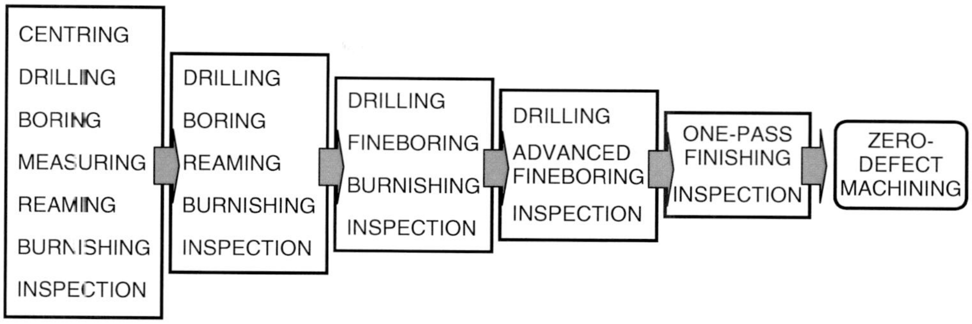

Fig. 15.102 Possible evolution of machining technologies [10].

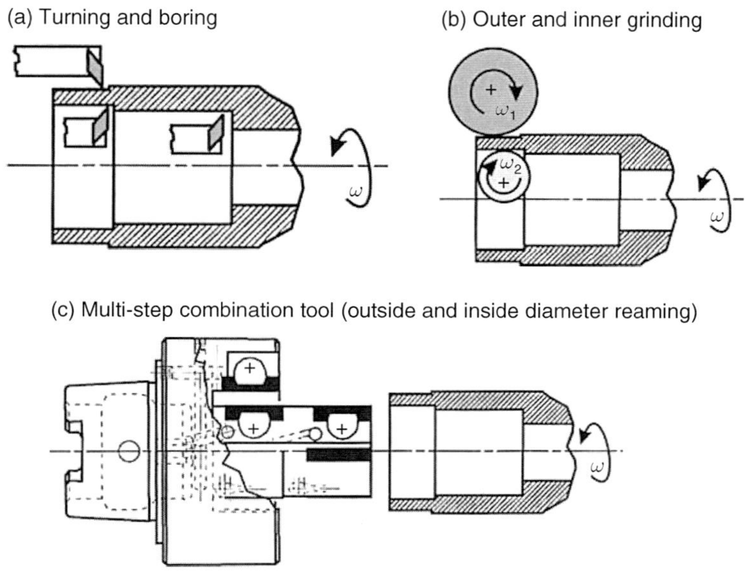

Fig. 15.103 Fine boring operation illustrating one-pass machining using a combination tool [12].

Fig. 15.104 A variety of multi-step boring tools: 1 – boring tool with various chamfering inserts; 2 – boring bars with various chamfering and facing inserts; 3 – combination boring tools with adjustable cartridges [3].

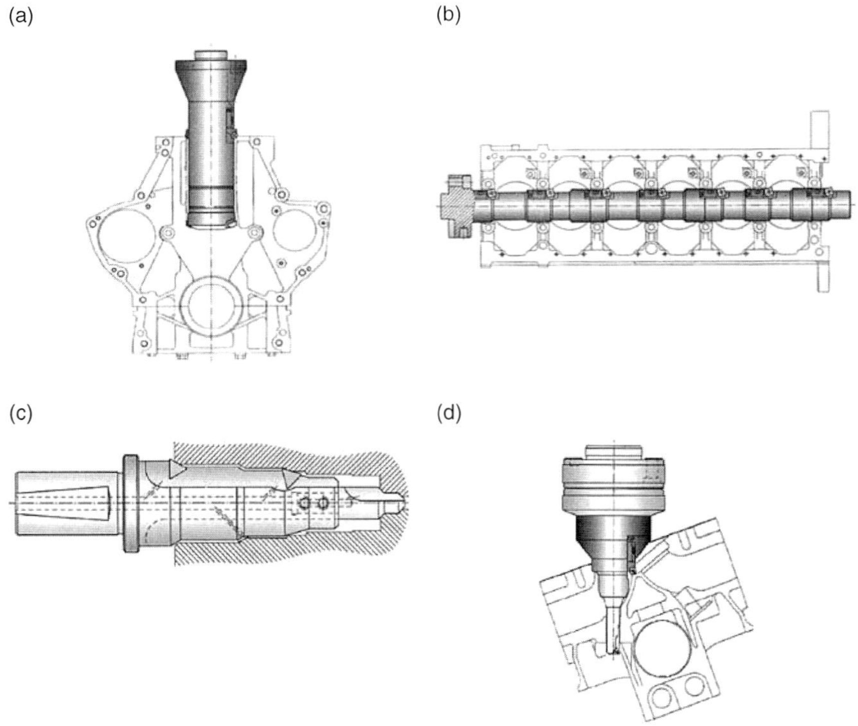

Fig. 15.105 Multi-step and multifunctional tools for machining cylinder block (a and b) and cylinder head (c and d) [3]: (a) rough boring and chamfering tool, (b) rough line boring bar, (c) chamfering and drilling tool, (d) valve seat boring tool.

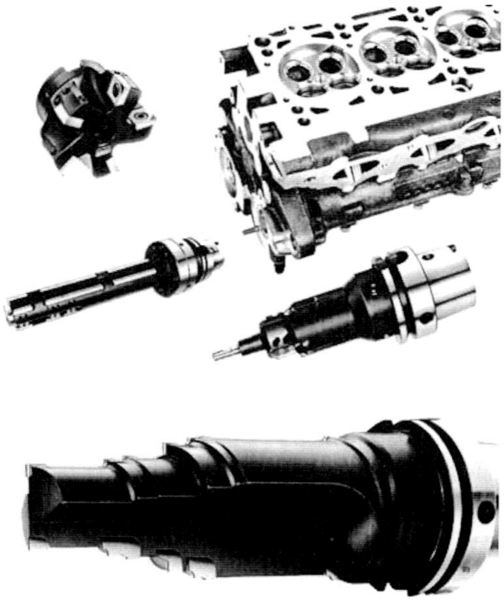

Fig. 15.106 Multitasking tools for machining of engine cylinder head [3].

References

[1] B. Byrne, D. Dornfeld, B. Denkena, Advancing cutting technology, *CIRP Annals*, Vol. 52/2 (2003), 483–507.
[2] Millturn Technologies, internet at www.wfl.at.
[3] www. coromant.sandvik.com, www.schaudtmikrosa.com, www.komet.com, www.gildemeister.com, www.heiden hain.com, www.walter-ag.com, www.makino.com, www.iscar.com.
[4] Complete machining of aircraft components at Airbus, *Machining*, Vol. 1, Issue 1, 2003, 22–30. Also available at www.wfl.at.
[5] M. Kaukal, System integration for internal machining operations, *European Production Engineering*, September 2003, 10–12.
[6] H. Mushardt and P. Manger, Komplettbearbeitung mit Hartdrehen und Schleifen, *Werkstatt und Betrieb*, Vol. 134, Nos 1–2 (2001), 37–40.
[7] Ch. Bates, Two machines, one goal, *American Machinist*, Vol. 145, No. 5 (2001), 71–75.
[8] R.B. Aronson, Grinding moves to multitasking, *Manufacturing Engineering*, Vol. 130, No. 2 (2003), 44–49.
[9] H. Link, Modular auf Grosses getrimmt, *Werkstatt und Betrieb*, Vol. 133, No. 6 (2000), 110–114.
[10] B.P. Erdel, *New Dimensions in Manufacturing*, Hanser Gardner, 1998.
[11] U. Speetzen, Multifunktionsmaschinen, *Werkstatt und Betrieb*, Vol. 133, No. 6 (2000), 50–52.
[12] B.P. Erdel, *High-Speed Machining*, SME, Dearborn, 2003.

15.6 Ultrasonically Assisted and Thermally Assisted Machining

15.6.1 Ultrasonically assisted machining and ultrasonic vibration cutting

Traditionally, ultrasonic machining (USM) is classified as one of the non-conventional machining processes which utilize a mechanical action of the vibrating tool, commonly at ultrasonic frequencies of 20 to 40 kHz [1, 2]. USM is a valuable process for the precision machining of hard and brittle materials. Material is removed by the abrasive action of a grit-loaded slurry (boron carbide, aluminium oxide and silicon carbide particles transported by water) circulating in a gap of 25–100 μm between the workpiece and the vibrating tool. The tool with relatively ductile materials, such as stainless steel, brass or mild steel, which is usually brazed to the tool-holder in order to avoid fatigue problem, is gradually fed into the workpiece to form a cavity corresponding to the tool shape. Typical arrangements for ultrasonic machining and polishing are shown in Fig. 15.107a and b, respectively.

The main elements of a USM system are: 1 – a high frequency generator, 2 – a transducer, which utilizes piezoelectric or magnetostrictive effect, 3 – a transformer (sonotrode), which is shaped to amplify the vibration output of the transducer, and 4 – a tool-holder and tool. Ultrasonic polishing can uniformly polish a precise surface layer from machined or EDMed workpieces (e.g. 0.01 mm from a 1 μm *Ra* surfaces) by using an abradable tool tip, such as graphite (Fig. 15.107b). Typical surface improvements range from 3:1 to 5:1.

USM can be used to produce cavities in most materials whether conductive, non-conductive, metallic, ceramic or composite. It is most effective on materials that are harder than 50 RC. The USM is limited to rather low material removal rates, usually less than 1.5 cm^3/min, with penetration rates between 0.2 and 6 mm/min. Machining depths of 50 mm or more are possible, but in practice depths are usually less than 5 mm. Holes as small as 0.1 mm can be drilled and the upper limit on cavity size is approximately 75 mm. The large holes, however, are preferably made by trepanning, in which the tool is shaped as a tube with wall thickness of 0.5 mm to 1.0 mm. Tolerances of ±0.25 mm and surface finishes of 0.25 μm can routinely be obtained.

As reported [3] a special ultrasonic machining centre yields a new ultrasonic technology (Fig. 15.108) enabling efficient machining of hard, brittle materials such as ceramics (Al_2O_3, ZrO_2, SiC, Si_3N_4), quartz glass, hard steels (>53/54 HRC), silicon, graphite, composites and precious stones such as sapphire and rubin. Productivity up to five times higher than that of conventional processes with a surface finish of $Ra < 0.2$ μm is possible. In order to assure maximum roundness accuracy high frequency spindles of 3000 to 60 000 rpm

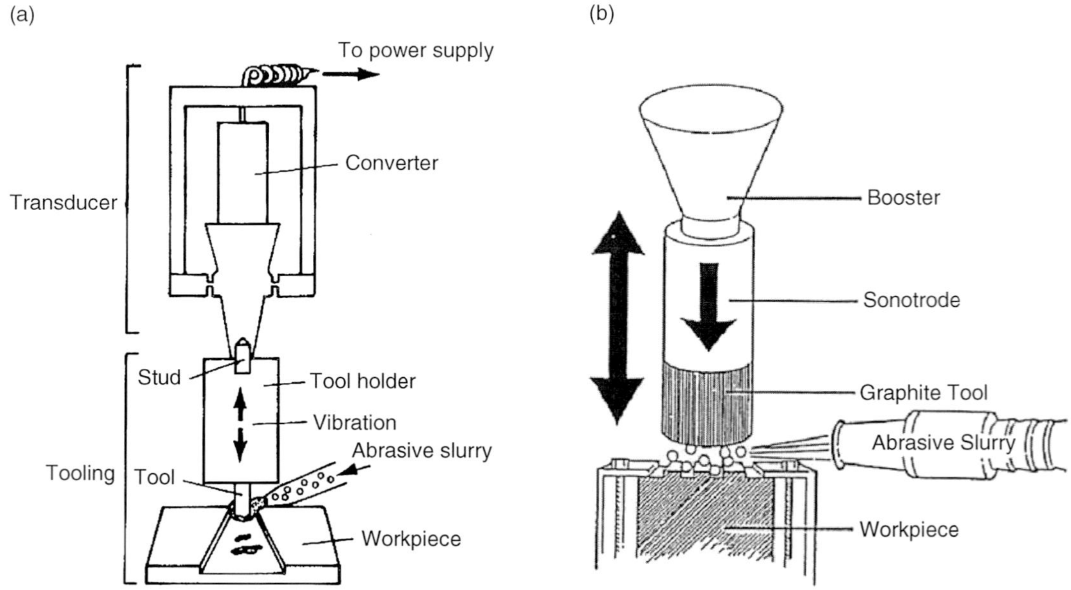

Fig. 15.107 Schematic showing components of typical ultrasonic machining (USM) and polishing installations [2].

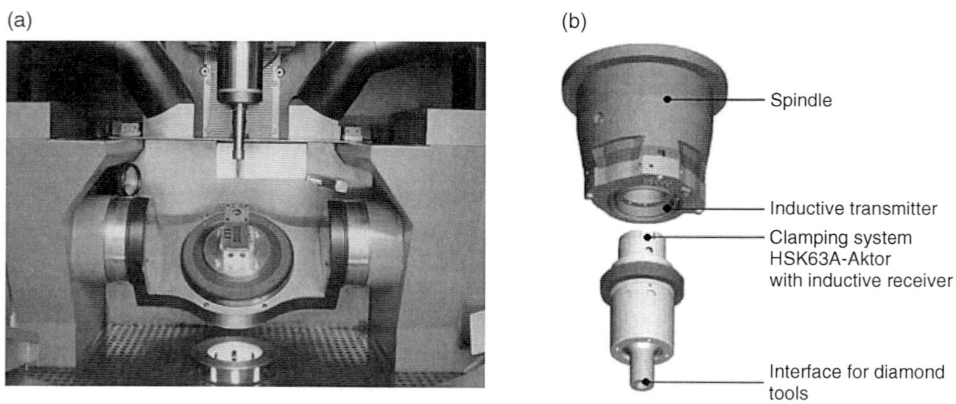

Fig. 15.108 Workpiece setup at USM centre (a) and a scheme of spindle-tool system (b) [3].

are required. This technology can be successfully applied in such industry sectors as optical, biomedical, automotive and aerospace, and in pump and housing manufacturing. Some parts shaped by USM are shown in Fig. 15.109. Diamond tools with bonded abrasive grain such as milling cutter, grinding wheel and drill (precision holes up to a diameter of 0.3 mm) are used. Automatic in-process resharpening of diamond tools is performed.

For decades, the principle of the cutting tool vibrating with small amplitudes of $a = 5$–$10\,\mu m$ and at ultrasonic frequencies of $f = 15(20)$–$50\,kHz$ is also effectively utilized to enhance traditional machining processes such as turning, drilling, threading, grinding and superfinishing, especially when machining a range of gummy materials producing intensive adhesion (built-up edge), including austenitic steels, nickel- and cobalt-based superalloys, titanium alloys and aluminium alloys [4, 5]. In order to ensure the regularity of

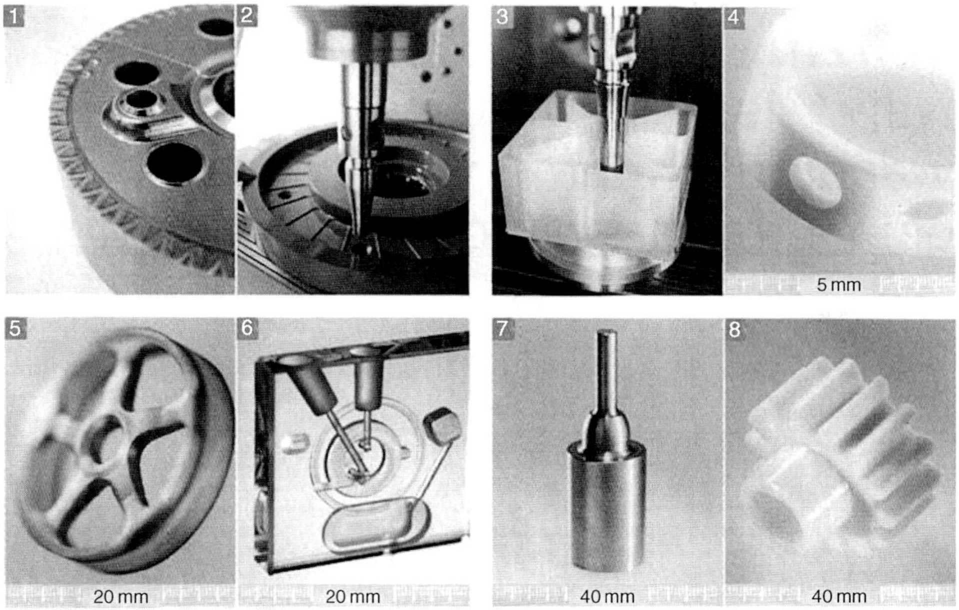

Fig. 15.109 Examples of a variety of parts machined by USM [3]: 1 – fine smoothing of a cogwheel mould made of Boehler steel of HRC63 with 70% time savings; 2 – use of Si_3N_4 ceramics in moulds for clutch discs to increase tool service life; 3 – narrow ligaments <1 mm for light-weight structures in mirror mounts made from Zerodur; 4 – ball-bearing cage made from zirconium oxide; 5 – paddle wheel with narrow ligaments <1 mm made from glass; 6 – pump insert made from silica glass, manufactured in complete machining with milling, boring and grinding with surface quality $Ra = 0.2\,\mu m$; 7 – carbide die with roughing and finishing during 20 min; 8 – cog made from aluminium oxide.

force impulses when the cutting edge of a tool vibrates in the principal or cutting direction the cutting speed should be less than the maximum vibration speed, i.e. $v_c < 2\pi\ af$ (where a is vibration amplitude and f is frequency). The main advantage of ultrasonic vibration cutting (UVC) is high part accuracy resulting from the reduction of elastic deformation in the tool/workpiece system, and lower heat generation and adhering tendency of the chip to the cutting edge [5].

The direction of ultrasonic vibrations can be parallel to one of cutting force components (F_c, F_f and F_p), i.e. tangential, axial or radial, or non-linear which is a result of superposition of vibrations in two directions – tangential and axial or tangential and radial [4]. When the cutting tool vibrates in a plane including the cutting direction and the thrust direction, an elliptical vibration mode is generated (Fig. 15.110) and the tool has a velocity component in the chip flow direction in each cutting cycle after it penetrates into the workpiece. In consequence, the frictional force between the tool rake and the chip is substantially reduced by reversing the frictional direction and thinner chips are obtained. It was documented that the surface roughness generated by the elliptical-vibration ultraprecision turning with a frequency of 20 kHz was less than $Rt = 0.02\,\mu m$ and the geometrical shape accuracy was improved. Moreover, elliptical vibration cutting suppressed burr formation.

Under real production conditions the conventional UVC processes performed on some difficult-to-cut materials, such as hardened steels, stainless steels and Ni-based alloys, cannot be used extensively due to the chipping and/or unusual wear of the cutting edge [5]. This phenomenon is caused by a collision between the tool flank and the machined surface of the workpiece when the tool moving backwards contacts the workpiece. In order to avoid this negative influence of tool vibrations a special high rigidity tool clamping system is applied and the vibrations are directed at the angle of 10–30° from the principal cutting direction

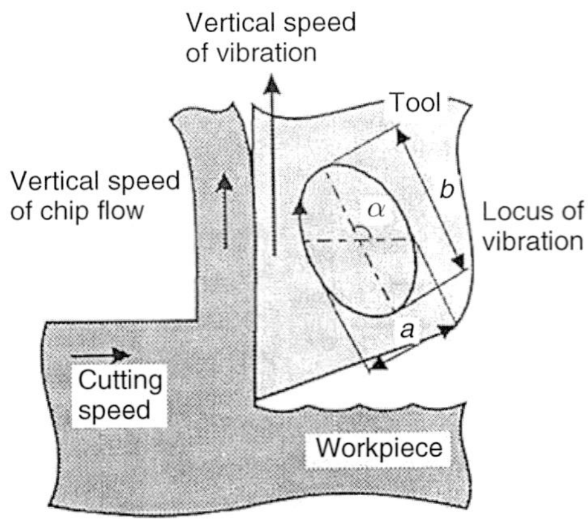

Fig. 15.110 Principle of elliptical vibration cutting [6].

towards the workpiece. In particular, for improving the rigidity of the tool the longitudinal vibration mode, in which the vibration system was stable, was applied and the tool was strongly attached to a post at four nodal points of the tool shank. As reported [5], inclined vibration cutting in which the ultrasonic vibration directions are spaced at an angle $\phi = 30°$ and $\psi = 7°$ to feed and cut directions, respectively, causes the flank wear during continuous turning of hardened stainless steel SUS440B (55 HRC), and intermittent turning of a quenched chromium-molybdenum steel SCM435 (40 HRC) is drastically reduced and surface finish is improved.

Figure 15.111a presents a drilling head in which the ultrasonic torsional mode vibration of 27 kHz frequency was superimposed on a twist drill rotating with a constant speed equal to one third of the maximum vibration speed. In particular, a newly designed torsional bolt clamped langevin-type ultrasonic transducer

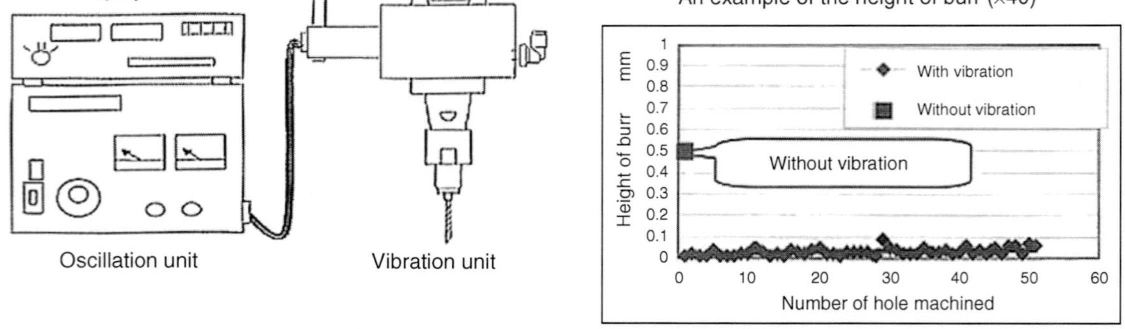

Fig. 15.111 Setup of the ultrasonic drilling head (a) and effect in reduction of burr formation (b). Work material A6063 aluminium alloy [7].

(BLT) with power of 200 W and natural frequency of 27 kHz was used. The taper shank of the drilling unit can be attached rigidly to the spindle of a CNC drilling and milling machine or CNC machining centre. A display unit, equipped with a phase locked loop control circuit, shows resonant frequency, the amplitude of the drill tip, voltage and ampere supplied, and electric power selectively during running of the drilling system. The ultrasonic torsional mode vibration cutting was successfully employed in drilling holes of 3 mm diameter in carbon fibre reinforced plastics (CFRPs) [8]. By assistance of ultrasonic torsional vibrations the occurrence of shallow craters of 50–70 μm in depth, associated with hole exit damage, was restrained and tool life was prolonged five times in comparison to conventional drilling.

Similarly, so-called ultrasonic-vibration planning equipment incorporating longitudinal mode ultrasonic vibrations was applied in finishing sharp-edged corners in deep pocket grooves of a plastic mold [9]. In this process special brazed vibration (planning) tools were used to shape right arcs in the three perpendicular grooves (in the X, Y and Z directions).

15.6.2 Laser-assisted machining

There are a number of conventional and unconventional machining processes that have been explored for machining of advance materials, such as ceramics and high temperature alloys including Ni-based and Ti-based alloys. Some examples are diamond grinding and thermally assisted machining (TAM) processes, particularly laser- and plasma-assisted machining processes, ultrasonic machining, ultrasonic abrasive machining, electrical discharge machining and abrasive water jet cutting. The concept of improving machining performance by localized heating of the workpiece has been around for at least 50 years. In particular, it has been observed that as the temperature of a material increases, its shear strength decreases which results in reducing the strain-hardening effect and the energy expended in cutting. Various hot-machining techniques have been tested, including, for example, heating in furnaces, torches, electrical resistant effect and induction heating [10]. Unfortunately, these heating techniques have not been widely applied in industry due to insufficient localization and intensity of the heating agent and predominantly because of serious problems with temperature control in a small volume of the workpiece adjacent to the cutting edge.

Effective implementation of TAM with a single-point cutting tool requires use of a controllable, highly concentrated beam of energy, capable of creating a hot spot at approximately 800–1000°C directly in front of the cutting tool. This requirement limits practically the heat source to either arc plasma jet or a restricted class of lasers. In Japan and Europe, there is significant renewed interest in arc plasma and laser-assisted machining. The principle of LAM performed with a single-point cutting tool is shown in Fig. 15.112. Recently, due to the rapid advancement of laser technology, LAM was found to be a viable option for machining of some difficult-to-machine materials including hard and severe strain-hardened metals, ceramics, glass, plastics and composites [10–12]. In LAM, the material is heated by a laser prior to the material removal action, and

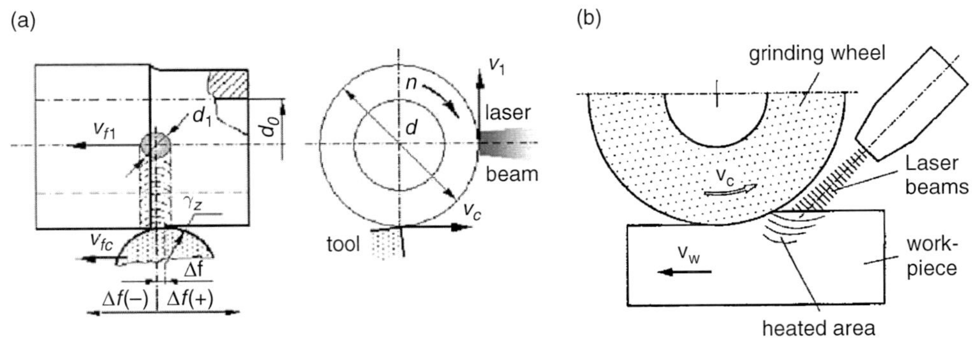

Fig. 15.112 A scheme of laser-assisted machining with single-point cutting tool (a) and laser-assisted grinding [19] (b).

subsequently the thermally plastified material is machined. As a result, this process is characterized by a number of advantages such as absence of intensive tool wear and tool breakage, chatter, machine deflection, and mechanically induced material damage typically addressed to conventional machining.

In general, two laser types, namely, CO_2 which operates at 10.6 μm and increasingly Nd:YAG laser at 1.064 μm wavelength, can be used as heat sources [12, 13]. Especially, due to better cooling, Nd:YAG laser gives higher beam quality, which allows focusing the laser to a small spot or increasing the focal length [14]. Coating the machined surface with a special paint that drastically increases absorptivity is another way to improve the energy efficiency. Moreover, a high power diode laser (HPDL) with a maximum power of 2.5 kW, which emits the shortest wavelength of 0.84–0.94 μm and the energy delivered is more readily absorbed by a target than light from the other lasers, starts to be employed. Also, an excimer laser with 0.2–0.35 μm wavelength was found to be more effective in assisting grinding and conditioning (truing, dressing and cleaning CBN wheels) than Nd:YAG lasers [19].

Typically, the LAM system is mounted on the gantry construction assembled to a CNC lathe in order to localize the laser source in the selected point on the workpiece surface and move it before the cutting tool. A scheme of supplying a CO_2 laser to a CNC lathe is shown in Fig. 15.113. Moreover, a PC-based data acquisition system is used to collect the cutting forces and the workpiece surface temperature.

The comparison of conventional and laser-assisted face-milling processes performed on an X38CrMoV53V alloy steel of 58 HRC hardness with CBN toroidal cutters ($v_c = 500$ m/min, $f_z = 0.1$ mm/tooth, $a_p = 3$ mm, $a_e = 18$ mm, CO_2 laser power of 4 kW) revealed the reduction of feed force, normal feed force and passive force of 45%, 62% and 44%, respectively. When using Al_2O_3-TiC ceramic tools this reduction was more visible and the force components were further reduced by 15% and 30% [15]. Probably this effect results from extremely low thermal conductivity (factor 5 comparing to CBN) of ceramics and different heat partitioning.

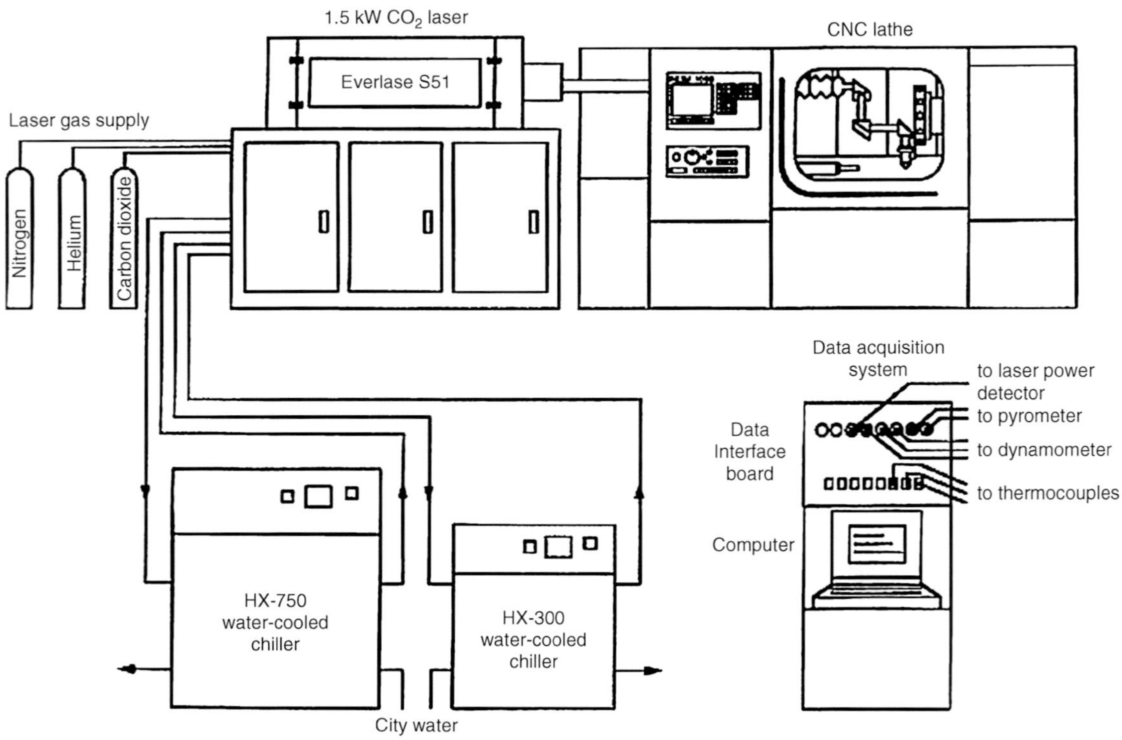

Fig. 15.113 Schematic diagram of the LAM system [10].

LAM processes were found to be very effective in machining various types of structural ceramics including silicon nitride (Si_3N_4), partially stabilized zirconia (ZrO_2) and mullite [10, 12, 15]. The selection of Si_3N_4 is based on the fact that it has the greatest potential for use in engine applications and bearings, and its mechanical properties are typical of the broad class of structural ceramics. ZrO_2 is popularly used for diesel engine parts because its thermal expansion coefficient is close to that of steel. Representative machining conditions for LAM of silicon nitride parts are 0.1–0.2 mm/rev in feed rate, 0.5–1.0 mm in depth of cut and 60–120 m/min in cutting speed [10]. While the system is capable of 1.5 kW power, usually 300–600 W in laser power was sufficient in performing a successful LAM process. For laser-assisted turning of hot-pressed silicon nitride, brittle fracture and poor surface quality were found to occur at temperatures less than 1100°C, while continuous chip formation and surface finish comparable to diamond grinding effect were recorded for temperatures exceeding 1200°C [10]. Specific cutting energy decreased with temperature and was about $6 J/mm^3$ compared with $40–100 J/mm^3$ in grinding. Therefore, LAM is less likely to induce subsurface damage than grinding. If LAM of ceramic parts could be implemented instead of grinding at typical metal cutting speeds, processing cost reductions of 60–80% could be achieved. Moreover, for silicon nitride and partially stabilized zirconia, surfaces with roughness *Ra* gradually less than 1 μm (minimum 0.3 μm Ra), comparable to those generated by grinding, and without visible subsurface cracks, were obtained [16]. Recently, the LAM method, termed 'excimer laser-assisted chemical machining' (ELACM), which uses a krypton fluoride (KrF) laser with a 248 nm wavelength, was used to intensify and improve surface integrity in machining SiC ceramic [12].

The third group of laser applications covers hybrid and sequential processes which combine laser hardening and machining passes in two different ways. They can be performed when the cutting follows directly the laser heating or it is performed after laser hardening of the workpiece, respectively [13]. In hybrid processes appropriate cutting speeds are relatively small (5–10 m/min) in order to achieve the required tempering conditions but in sequential processes the cutting speed can be increased up to the level characteristic for hard machining. In addition, in hybrid operations the centre of the hot laser spot was displaced at a distance of $d_l = 0.875–1.0$ mm in relation to the tool corner to guarantee full hardening effect of the surface layer. As reported, laser-hardened surfaces have periodically repeated irregularities (Fig. 15.114) resulting from both volumetric and structural changes characteristic of extremely high temperature gradients. In consequence, cyclic changes of microstructure and hardness along the heat source trace are observed.

Figure 15.115b shows the distribution of the HV0.2 hardness after laser hardening, LAM and CBN hard turning performed to remove the residual irregularities and dimensional inaccuracies occurring after laser hardening. During hardening the diameter of the laser beam was kept constant at 2 mm. Metallographic investigation revealed substantial thermal influences of the subsequent laser traces within semi-spherical, periodically repeated zones. It should be noted that a visible white layer is formed after LAM but its thickness is less than in sequential hard turning [13]. Finally, the anisotropic structure of bearing steel after laser hardening consists of the fine-grained martensite due to γ–α_M phase transition, and dark-etched, coarse-grained acicular tempered martensite resulting from the repeated laser heating.

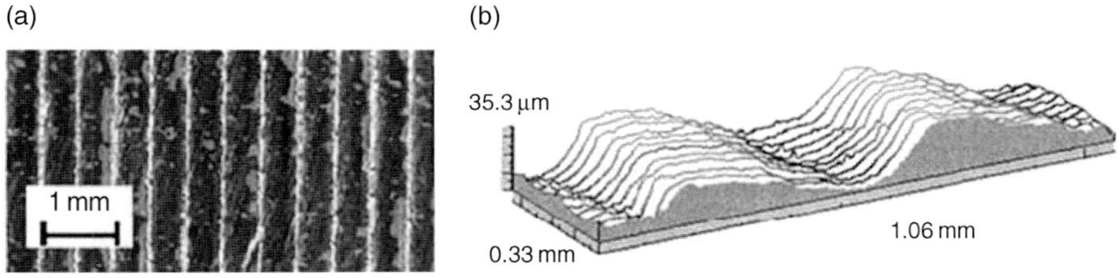

Fig. 15.114 Laser-hardened surface of a bearing steel (a) and its 3D topography (b). Workpiece material: bearing steel (1% C, 1.5% Cr) equivalent to DIN 100Cr6; tool material: CBN. LAM conditions: $P = 2250$ W, $v_l = 5.4$ m/min, $f_l = 0.50$ mm/rev, $d_l = 2.0$ mm [13].

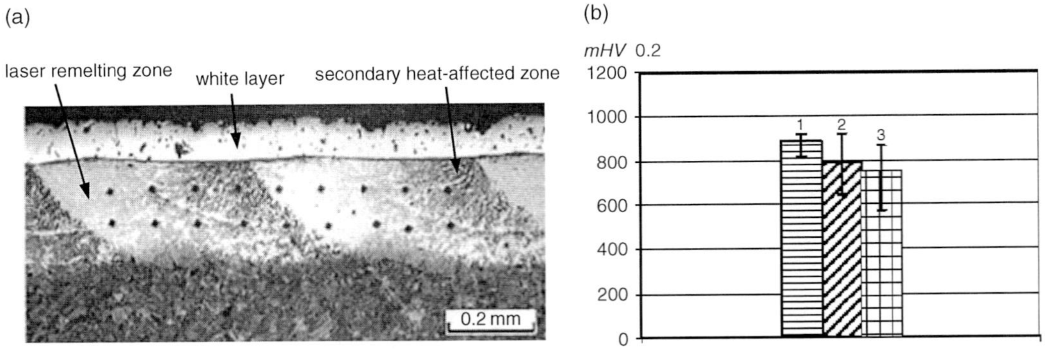

Fig. 15.115 Microstructure of the surface layer after LAM ($P = 2250$ W, $v_l = v_c = 5.4$ m/min, $f_l = f = 0.5$ mm/rev, $d_l = 2$ mm, $a_p = 0.1$ mm, $r_\varepsilon = 2.78$ mm, PCBN(K10D)) (a) and the comparison of average meso-hardnesses HV0.2 after laser hardening (1), LAM (2) and sequential machining (3) [13].

15.6.3 Plasma-assisted machining

The idea of hot machining is to generate a large difference in the relative hardness between the cutting tool and the workpiece materials, which is of crucial importance in the machining of engineering ceramics, leading to reduction of cutting forces, improved surface finish and longer tool life [9]. A hot machining technique using plasma jet heating has recently been employed to improve the machinability of glasses such as borosilicate glass and silica-alumina compound (mullite) and engineering ceramics such as alumina, zirconia and silicon nitride ceramics, sintered high speed steel and both stainless and alloy steels [10, 17, 18]. Plasma-assisted hot machining (PAM) utilizes a high temperature plasma arc to provide a controlled source of localized heat, which softens only a small portion of the workpiece. Softening of the workpiece zone just in front of the cutting tool releases very high energy densities and confines the heat. A plasma arc consists of a high velocity, high temperature stream of ionized gas capable of supporting a high current, low voltage electric arc.

A plasma torch produces this physical phenomenon by having a tungsten electrode centrally placed within a water cooled copper nozzle. Figure 15.116 illustrates schematically a plasma torch which generates a plasma jet between a tungsten electrode and nozzle by using argon and hydrogen as the working gas mixture. By adding hydrogen the surface temperature raises to 1200°C and quick temperature control is possible. A gas stream is fed down the annulus between these, the gas being ionized by a high frequency discharge between the copper nozzle and the central electrode. This is followed by a low current pilot arc and then by a high current main arc. Temperature within the arc has been reported to be in the range of 16 000°C to 30 000°C.

Figure 15.117 shows how power consumption can be reduced in plasma arc hot machining of forgings from the two steels when turning with carbide tools at different depths of cut and a constant cutting speed of 100 m/min. For maximum depth of cut equal to 5 mm the percentage reduction in the spindle power consumption reaches 25% and 21% for case (a) and (b), respectively. In addition, it was revealed that the surface temperature after each hot cut was less than 100°C, which would not induce any metallurgical damages in the surface layer. Other economic factors documented in the hot machining of stainless steel are 1.8 times gain in MRR and 1.7 times improvement in tool life.

Sintered high speed steels with a more homogeneous structure and high hardness exceeding 65 HRC are the next products for which the replacement of grinding by plasma hot machining leads to an increase in productivity. The mechanical and tribological effects occurring during PAM of sintered HSS are shown in Fig. 15.118. As the surface temperature rises above 600°C the cutting forces tend to decrease abruptly and chatter marks disappear as well. Good coincidence between the force and hardness distribution was found. At 800°C the forces are reduced by two thirds. The apparent coefficient of friction also slightly decreases

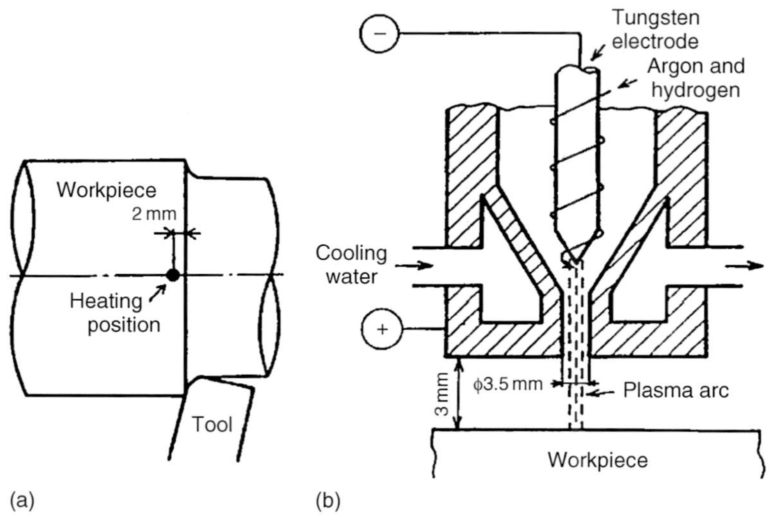

Fig. 15.116 Setup for plasma heating: (a) heating position; (b) plasma torch [17].

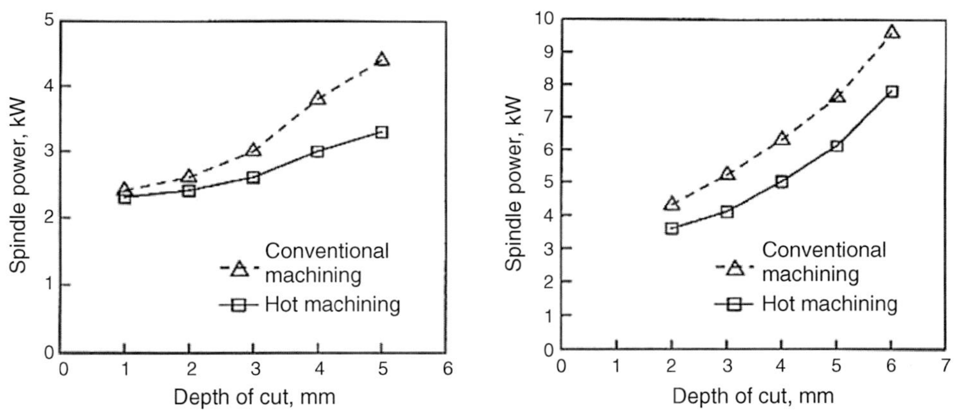

Fig. 15.117 Power consumption in conventional and hot machining of SS410 stainless steel of 200 BHN hardness (a) and alloy steel with 3.75% Cr and 2.72% Mo and 290 BHN hardness (b) [18].

with temperature at its higher levels. However, it is difficult to select an appropriate heating temperature because the occurrence of oxidation and slight segregation of the surface starts above 700°C.

High temperature alloy Inconel 718 exhibits a rapid reduction of hardness above 650°C and, therefore, it is not necessary to elevate the workpiece temperature to such a high temperature, as in LAM. Due to economic reasons, PAM of Inconel 718 provides a reasonable alternative. When the machining was performed with ceramic tools (alumina oxide reinforced with silicon carbide whiskers) under typical machining conditions, i.e. cutting speed of 2–6 m/s, feed rate of 0.1–0.3 mm/rev and plasma current of 180–300 A, the cutting forces decrease with temperature, as shown in Fig. 15.119.

It is evident in Fig. 15.119 that at a surface temperature of 500°C the resultant cutting force is reduced by about 30%. It is especially important that PAM eliminates microchipping of the cutting edge and reduces the degradation of surface roughness with time. As reported [10], the surface roughness was improved almost four-fold near the end of tool life and smoother and shiner surfaces were generated on Inconel 718 parts. Moreover, the tool life was increased by 40% although the flank wear increased slightly.

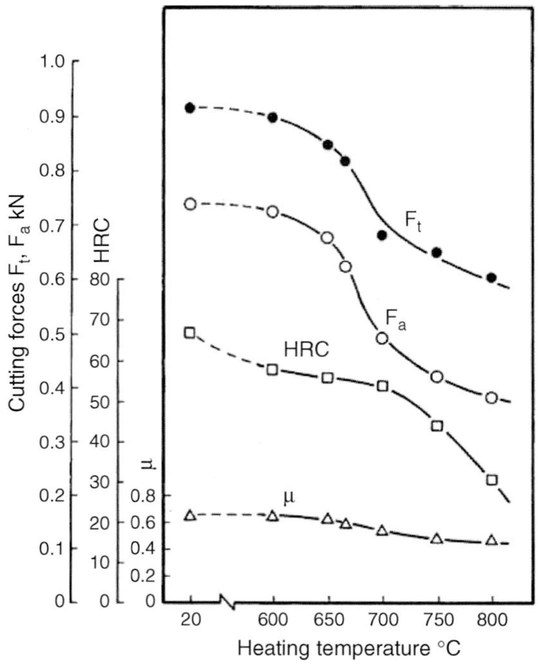

Fig. 15.118 Changes of cutting forces, apparent coefficient of friction (*m*) and Rockwell hardness (HRC) at various temperatures in machining of sintered HSS [17]. Tool: CBN (BN 300), $v_c = 20$ m/min, $f = 0.1$ mm/rev, $a_p = 1$ mm.

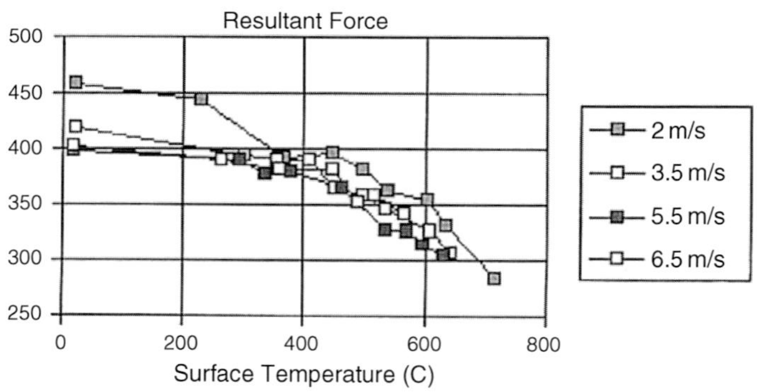

Fig. 15.119 Influence of surface temperature on the resultant cutting force in PAM of Inconel 718 [10].

References

[1] G. Boothroyd, W.A. Knight, *Fundamentals of Machining and Machine Tools*, Marcel Dekker, New York, 1989.
[2] *ASM Handbook*, Vol. 5, *Surface Engineering*, ASM Int., Materials Park, OH, 1999.
[3] Ultrasonic machining center ULTRASONIC 20 linear/HSC 20 linear (DMG 1/2006), www.gidemeister.com.
[4] W. Grzesik, *Fundamentals of Machining Metallic Materials* (in Polish), WNT, Warsaw, 1998.

[5] M. Jin, M. Murakawa, Development of a practical ultrasonic vibration cutting tool system, *J. Mater. Proc. Technol.*, Vol. 113 (2001), 342–347.
[6] T. Moriwaki, E. Shamoto, Ultrasonic elliptical vibration cutting, *CIRP Annals*, Vol. 44/1 (1995), 31–34.
[7] H. Suzuki, H. Yagashita, Burrless drilling by vibration cutting applying ultrasonic torsional mode vibration, *Transactions of MAMRI/SME*, Vol. 33 (2005), 461–468.
[8] H. Yagashita, Cutting mechanism of drilling CFRP laminates and effect of ultrasonic torsional mode vibration cutting, *Transactions of MAMRI/SME*, Vol. 34 (2006), 213–220.
[9] S. Yamada, M. Jin, M. Murakawa, Finishing of mold by ultrasonic-vibration planing method, *Transactions of MAMRI/SME*, Vol. 30 (2002), 137–144.
[10] Y.C. Shin, S. Lei, F.E. Pfefferkorn, P. Rebro, J.C. Rozzi, F.P. Incropera, Laser-assisted machining: its potential and future, *Machining Technology*, Vol. 11, No. 3 (2000), 1–6.
[11] E.O. Ezugwu, Key improvements in the machining of difficult-to-cut aerospace superalloys, *Int. J. Mach. Tools Manuf.*, Vol. 45 (2005), 1353–1367.
[12] G. Chryssolouris, N. Anifantis, S. Karagiannis, Laser assisted machining: an overview, *Trans. ASME, J. Manuf. Sci. Engng*, Vol. 119 (1997), 766–769.
[13] M. Kawalec, M. Jankowiak, A. Barbacki, Geometrical and physical properties of technological surface layer of alloy steels in conditions of hybrid laser hardening and cutting, *Problemy Eksploatacji*, No. 4 (2005), 171–180.
[14] J.R. Koelsch, J. Destefani, Lasers on the shop floor, Laser Supplement 2005, www.sme.org.
[15] W. König, F. Treppe, A.K. Zaboklicki, Laserunterstütztes Fräsen, *VDI-Z*, Vol. 134, No. 2 (1992), 43–48.
[16] P. Rebro, F.E. Pfefferkorn, Y.C. Shin, F.P. Incropera, Comparative assessment of laser-assisted machining for various ceramics, *Transactions of MAMRI/SME*, Vol. 30 (2002), 153–160.
[17] T. Kitagawa, K. Maekawa, Plasma hot machining for new engineering materials, *Wear*, Vol. 139 (1990), 251–267.
[18] G. Madhavulu, B. Ahmed, Hot machining process for improved metal removal rates in turning operations, *J. Mater. Process. Technol.*, Vol. 44 (1999), 199–206.
[19] E. Westkämper, Grinding assisted by Nd:YAG lasers, *CIRP Annals*, Vol. 44/1 (1995), 317–320.

CHAPTER SIXTEEN

MICRO-MACHINING

Contents

16.1 DEFINITION AND MINIATURIZATION

Micro- and nanotechnology are common phrases used to describe phenomena, components, products, etc. related to very small dimensions, usually based on functioning principles that integrate different scientific disciplines. Micro-machining is the most basic technology for the production of miniaturized parts and components [1]. In other words, miniaturization is a key to the manufacture of future generations of many kinds of micro-electromechanical products for a variety of industrial and medical applications, because of the continuing demand for increased functionality and reduced size. Due to this fact micro-manufacturing is sometimes used as a synonym for microprocessing [2]. From such a point of view, micro-manufacturing encompasses micro-electromechanical systems (MEMS), micro-systems technologies (MST) and, in addition, includes processes related to the production and packaging of micro-systems [3]. MEMS, which integrate mechanical elements, sensors, actuators and electronics on a single, usually silicone, device, are on the lower fringes of the micro area [4]. This area does not utilize many conventional machining operations, but a version of stereolithography and specialty plating processes to make parts. Micro-machining by precision technology such as three-dimensional micro-EDM, micro-laser machining, micro-cutting, micro-grinding, etc. can produce microscopic and mesoscopic mechanical structures of complex shapes. The demand is not only for small parts, but also for small complex features on larger parts. Currently, the following areas of interest can be observed [4–7]:

1. The medical and electronic industries are the most active in this area and in both cases the incentives are the same: making smaller parts with more capabilities. The cell phone is an obvious example and the goal is to cram more features into one unit.
2. The medical industry is under pressure to produce various repair and pain-relief devices which must be small enough to embed into the human body. Micro-sized components can lead to the progression of minimally invasive techniques for orthopaedic, cardiological and neurological applications.
3. The aerospace industry is a large end user of fasteners, fittings, sensors and various flow-control devices.

Advanced Machining Processes of Metallic Materials
ISBN: 978-0-08-044534-2

4. The automotive sector every year introduces many innovations for convenience, entertainment and safety items. As a rule, many of these features require very small motors and actuators. From a construction point of view, there is a trend to install complex fuel injection and small flow-control elements. Some of the most representative examples in this category are micro-valves and fluidic graphite channels (circuits) for fuel cell applications.

Initiatives worldwide show the increasing focus from governments and industry because of the large potential of these technologies. The exploitation of the large potential, however, depends on the ability to manufacture the products and make them accessible to a large group of customers. Therefore, product development and manufacture are essential. During the last decade a dramatic increase in the number of micro-products has been observed. The development has been initiated by the micro-electronics industry but in recent years also medical and biomedical applications have become important. It is foreseen that the total world market for micro-systems will increase from 30×10^9 USD in 2001 to 60×10^9 USD in 2005. This increase is foreseen to come mainly from IT applications as well as medical and biomedical applications. The miniaturization of products is a key element in the development of sustainable products. The Japanese are the current word leaders in micro-factory development and it is assessed that the size of the replacement market for micro-machine tools in Japan will reach $1.1 billion in 2015. Similar market growth is being projected for America, driven by the rapid expansion of the biomedical industry at 16% per year [4]. Statistics show that approximately 35% of the parts manufactured in the USA today fit in the palm of the hand [6].

Because micro-manufacturing is a fairly new industry, definitions are not yet firm. So far, three terms are commonly linked to these products: *nano* (one billionth of a unit), *micro* (one millionth of a unit), and *meso* (generally bigger than micro). Visually, a dimensional meso-/micro-scale covering the range between MEMS and miniaturized parts is shown in Fig. 16.1.

The world *micro-machining* may literally mean the machining of the dimensions between 1 μm to 999 μm but it should also consider the contemporary levels of conventional technologies. For instance, the micro-/macro-border was set around 200 μm with some variation among different manufacturing methods [1]. On the other hand, if micro likewise means 'very small', it should also indicate too small to be machined easily by normal machining. In such a sense the range of micro-machining can be placed between 1 μm and 500 μm [1].

The range of machining accuracy between 0.01 μm (lower range of precision manufacturing) and 0.001 μm (1 nm) is termed by Taniguchi as ultra-precision machining (Fig. 16.2). Figure 16.3 shows the capability of micro-machining in relation to other manufacturing processes such as laser machining, EDM, grinding and the LIGA (photo-lithography method using a synchrotron) process. It is worth noticing that micro-machining allows attaining *Ra* values in the range down to almost 5 nm for features down to 1 μm. In contrast, as shown in Fig. 16.2, for 'normal machining', e.g. CNC turning and milling machines, accuracies of 10 μm to 100 μm can be achieved.

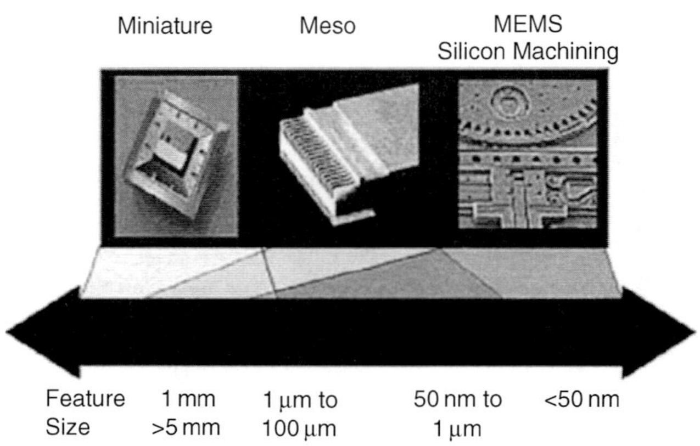

Fig. 16.1 A reference of meso-/micro- (nano-) scale [8].

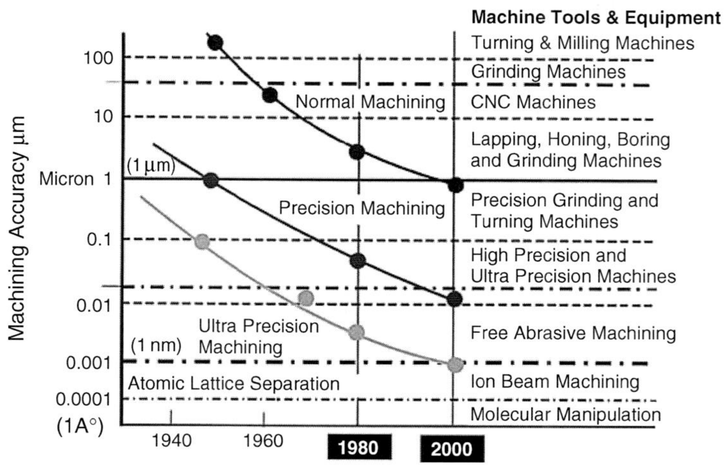

Fig. 16.2 Taniguchi's prognosis for increasing machining accuracy in the 21st century [9].

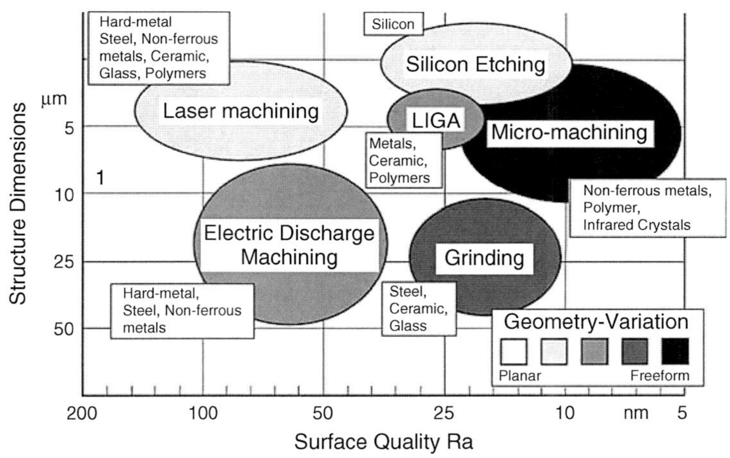

Fig. 16.3 Micro-machining relative to other machining processes [9].

Although micro-manufacturing and conventional precision machining use many of the same techniques and equipment, the goals are quite different. Precision engineering is informally defined as the production of parts with a size-to-tolerance ratio greater than 10 000. In micro-manufacturing, the goal is to produce mini- and micro-systems (typically 1 cm or less in size) with micro-components (typically 1 to 100 μm in size) with sufficient tolerances to achieve required functionality and repeatability [3]. These tolerances are usually smaller than in precision machining.

16.2 A Survey of Micro-machining Processes

Parallel to the endeavours of pushing the micro-manufacturing technique down to the order of submicron or nanometer scale, there have been numerous activities that focus on applying different micro-machining technologies to fabricate end products. This results from the fact that micro-manufacturing covers several

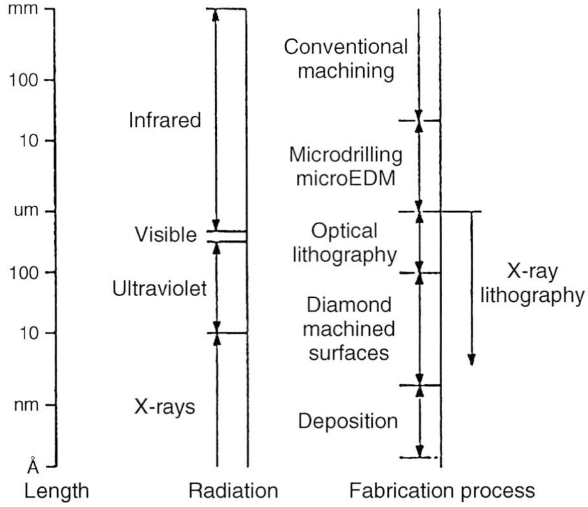

Fig. 16.4 Dimensional magnitudes covered by various micro-machining processes [3].

orders of dimensional magnitude, as depicted in Fig. 16.4, which is a comparison of this range and the electromagnetic spectrum.

As mentioned previously, many types of machining processes can be utilized to perform micro-machining. Table 16.1 selects the main methods classified according to the machining phenomena employed. Moreover, Table 16.2 provides the corresponding micro-machining techniques used in practice and depicts some advantages and disadvantages associated with them.

In processes using *mechanical force* the useless part of material is removed through plastic or brittle breakage mechanisms. To satisfy small unit removal (UR) high stresses which cause breakage of material (or equivalently highly concentrated force) should be exerted in appropriate positions on a very small area or volume of the workpiece. As a result, assuming the desired UR of about 0.1 μm (100 nm), a micro-tool with the cutting edges sharpened to a radius smaller than 1 μm is necessary. Cutting, sandblasting, grinding and ultrasonic machining belong to this category of removal mechanism, but only in cutting and grinding processes can the tool position be correlated with the machined surface. The major drawback of such micro-machining processes is the strong dependence of obtainable accuracy and machinable size on the mechanical force, particularly due to substantial elastic deformation of the micro-tool and/or the workpiece.

Table 16.1 Principal micro-manufacturing methods according to CIRP [2].

Principle	*Method*
Force	cutting, grinding, USM
Melting/vaporization	EDM, LBM, EBM
Ablation	LBM (excimer, short pulse), FIB
Dissolution	ECM, photoetching
Plastic deformation	punching, press
Solidification	moulding, casting
Lamination	stereolithography
Recomposition	electroforming

Table 16.2 Evaluation of micro-machining processes by Makino [8].

Process	*Principle*	*Minimum feature size in μm*	*Advantages*	*Disadvantages*
μ-Moulding and Casting	Solidification	500	Mass production	Springback
μ-Punching	Plastic deformation	50	Mass production	Need uniform clearance
μ-Milling and Grinding	Force	25	Good geometric accuracy and surface finish	Tool deflection/damage due to cutting forces
Stereolithography	Lamination	12	Complex 3D shape	Limited work material
Excimer laser	Ablation	10	No heat damage on surface	Limited work material
μ-[EDM and Laser]	Melting/vaporization	5	Negligible force	Possible surface damage and low MRRR
FIB	Sputtering	0.2	Stress free	Low MRR

EDM, LBM (laser beam machining) and EBM (electron beam machining) belong to the group of processes in which the energy required to remove material is consumed mainly for *melting*, and in some cases also for *vaporization*, by means of heat generated by various physical phenomena. The molten material is removed by pressurized gas and a small UR is obtained by reducing the pulse energy (mainly controlling the electrical parameters). The concentration of energy sufficient for melting is provided by the pinch effect of a short-pulse arc in high pressure gas (EDM) and by controlling the beam shape by an optical system (in both LBM and EBM processes). Because molten material can be removed with a very small force, a very thin tool can be used in EDM. As a result, the workpiece can also be thin and elastic. In all these processes the tool (shaped beam in LBM and EBM) does not contact mechanically and there is an uncertainty in controlling the workpiece dimensions. Moreover, the heat-affected layer (partially remained molten material) is formed on the machined surface. Material removal by *ablation* uses power sources (excimer and femtosecond (FS) lasers) which can emit a beam with very high quantum energy or an energy beam of which incident power density to the workpiece is extremely high. In the first case, exceeding the binding energy of atoms causes the decomposition of each molecule into atoms and their removal from the workpiece body. In general, ablation does not generate heat-affected layers and micro-shapes with high dimensional accuracy and surface layers free of defects can be machined by projecting mask patterns. Unfortunately, low MR efficiency and high equipment cost are main drawbacks. By using frequency-tripled Nd:Vanadate lasers ablation accuracies higher than 10 μm with surface qualities below 1 μm can be achieved with sufficient ablation rates for the tooling industry [10].

Removal by chemical or electrochemical *dissolution* in liquid is based on ionic reaction on the workpiece surface, providing a very small UR in the direction perpendicular to the surface. A typical example is photoetching in which the patterns are specified by a patterned mask and formed by etching. Because etching proceeds isotropically the application of this method is limited to products with low aspect ratio (mainly semiconductor devices). More suitable for 3D micro-machining is electrochemical etching performed by a pipe-type tool which specifies the required part shape. As previously, there are some reservations about part accuracy resulting from the fact that the gap filled by electrolyte reduces the precision of tool copying. On the other hand, surfaces free of any damage and residual stresses are generated because the machining force is almost zero and heat influence does not exist.

Plastic deformation is the predominant shaping mechanism in processes in which the shape of the product is specified by copying the shape(s) of a punch (a die) or a mould. Its wider applicability to mass production depends on the access to production of micro-mould/die tooling [5]. The springback effect or partial recovery after deformation limits the part accuracy. Another limit is an insufficient plasticity and usually superplasticity conditions are required in micro-/nano-machining processes.

In the case of *solidification* (injection moulding and die casting) a liquid or a paste is solidified in a mould, and the shape of the mould is replicated. Problems similar to plastic deformation emerge (for example, the springback recovery) and additionally defects such as bubbles can appear in the workpiece.

Lamination is a unique type of solidification from liquid based on layer-by-layer incremental shaping. Although various rapid prototyping processes are included in this method, today micro-machining uses the stereolithography (SL) technique. This method can be easily used for both outside and inner (cavities) profiles. A scanning beam such as a laser can generate any required 3D shape but the choice of material is very limited.

In contrast to dissolution, *recomposition* utilizes the reverse phenomenon of dissolution, i.e. metal ions in an electrolyte are deionized to become solid and form a shape specified by a mould or a substrate. A typical example is electroplating which provides very small UR and copies most problems characteristic of solidification. Mass productivity is possible due to the use of a mould but the materials that can be machined are limited.

16.3 Micro-machines and Equipment

Today, the development of micro-manufacturing equipment proceeds in two opposite ways. First, manufacturers can adapt existing machine tools and tooling systems to handle micro-manufacturing tasks. Such activities involve scaling down some units to handle smaller parts or adding options that can facilitate the micro-manufacturing applications. The other option is to design a new generation of machine tools, specifically for micro-manufacturing. In particular, part handling and fixturing is of significant importance. Since the parts are obviously fragile, special attention is necessary to avoid damage during the process (hold down and transfer by vacuum is more common). The target scale of the individual micro-machine tools can be defined as maximum cubic volume, with a major dimension of 200 mm in all axes, which is roughly the size of a typical shoebox [4]. The goal is a micro-/meso-scale machine tool (mMT) capable of producing micro-features with submicron accuracy. This section outlines some newly designed machine tools for various micro-machining operations.

The early manufacture of clocks and watches in Switzerland stimulated the development of machine tools (called Swiss-type lathes) capable of making a variety of small parts with the necessary precision. Later, these machines have evolved into Swiss-type machining centres with milling, drilling and boring capabilities. The newer trend is to design machines with the ability to finish small parts (currently 4 mm in diameter) in a single setup with lower cost. Figure 16.5 shows a high speed, high precision lathe for the machining of tiny precision parts with a maximum machining diameter of 7 mm and length of 40 mm (Fig. 16.15 in section 16.4), but a new model allows going down to subminiature parts with 0.5 mm in diameter. The machine can perform metal cutting at speeds of up to 12 000 rpm, but without the rotary guide bushing the cutting speeds go up to 16 000 rpm. Subspindles for drilling and milling deliver 10 000 rpm. The use of linear motors to drive the slide and tool post results in fast part processing and elimination of thermal distortion. A scale feedback control system for all axes offers the perfect machining configuration for small, high precision parts. The reduction of non-cutting idle time was achieved by the possibility of opening and closing the chuck without deceleration of the spindle. This effect was increased by installing two independent tool posts (Fig. 16.5b), which eliminate tool exchange. The machine can be optionally equipped with a vacuum part removal system that delivers parts segregated from the chips and carousel workpiece separator which greatly improves productivity. Rotary tools mounted into the gang tool post allow small-diameter parts requiring complex machining such as polygon turning and end face drilling, to be machined easily. Citizen sliding

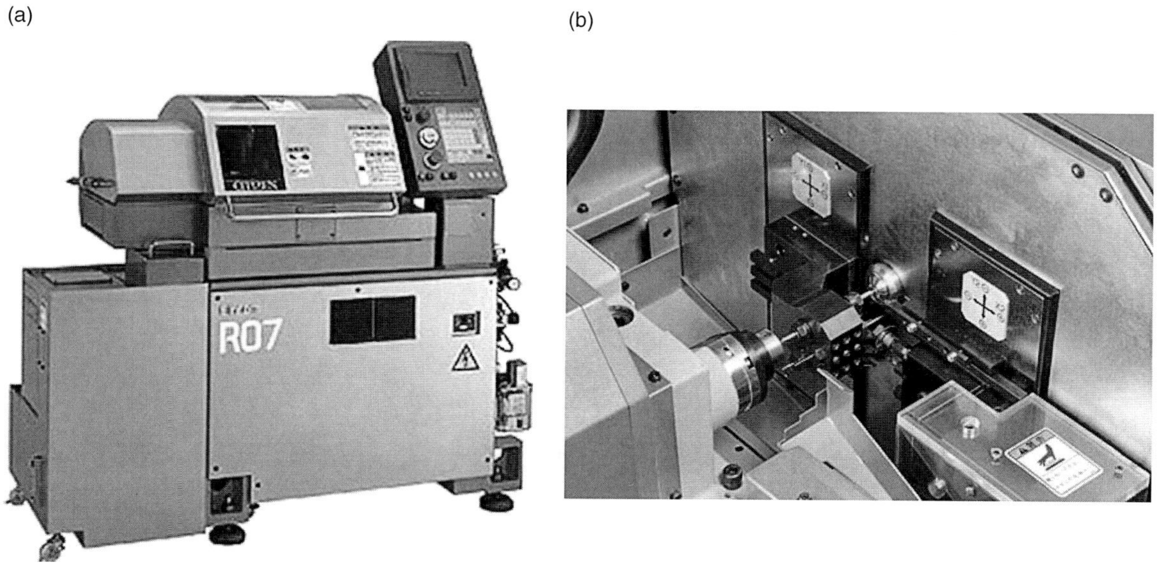

Fig. 16.5 CNC automatic micro-lathe (a) and axis arrangement (b) [11].

head machines are equipped with a Quick Start (QS™) holding system, consisting of a number of holders in the tool post with special spring-loaded clamping, developed by Sandvik for external machining [12].

Figure 16.6 illustrates the principle of the micro-EDM lathe which, from a kinematical point of view, operates similarly to a conventional lathe. This new type of micro-machine is additionally equipped with an electrode fabrication system for preparing on-the-machine complex micro-electrodes with sharp edges and corners. Both WEDG (wire electrodischarge grinding) and EDG (electrodischarge grinding) with wire and disc electrodes were applied for fabricating radii and sharp corners, respectively. For example, when the coaxial hole with a groove is machined, first the cylindrical electrode is used for boring the hole by EDM and next the electrode is reformed into a special plug shape to make the inside groove. This method can be used to machine coaxial overhung cavities, cylindrical micro-holes and even spherical surfaces. Typical parts are micro-ball bearings and joints, specially shaped nozzles and micro-screws, as depicted in Fig. 16.6b.

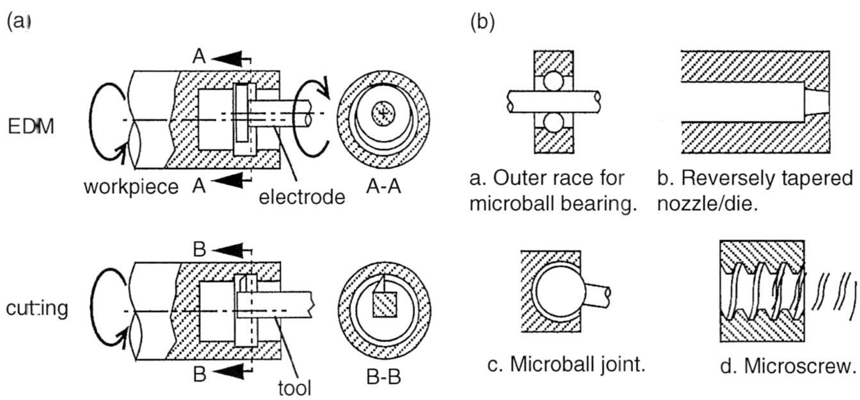

Fig. 16.6 Principle of micro-EDM lathe (a) and its possible applications (b) [13].

Micro-milling often uses a high speed machining technique with spindle rotations between 40 000 and 50 000 rpm and specialized CAM programming. But many problems arise when typical micro-machining applies to end mill sizes of less than 3.3 mm (0.13") diameter and most machined parts will fit inside a 1" (25.4 mm) cube [14]. Along with a rigid machine, the spindle must be stable to minimize tool vibration and prevent thermal expansion. This is because any vibration or runout at the tool tip deteriorate the surface finish and part accuracy and can also fracture small end mills. It is well known that small vibrations are amplified relative to tool diameter as it is reduced. For example, a vibration of 0.0001" (0.00254 mm) is a much larger fraction of a 0.010" (3 mm) end mill (1% in diameter) than a 0.5" (13 mm) end mill (0.02% in diameter) [14]. One solution is a direct change-type spindle. In fact, by eliminating a tool-holder total runout caused by tool-holder variation is substantially reduced. Usually, a spindle runout of less than 4 μm is required.

Figure 16.7 shows a CNC micro-milling machine (termed as an office mill [15]) equipped with a 50 000 rpm brushless electric micro-motor spindle with 270 W power rate. Maximum values of feed rate and cutting speed are equal to 19.2 m/min (757 ipm) and 12.7 m/min (500 ipm). It is possible to perform four-axis machining using a micro-rotary table or five-axis machining by installing a micro-trunnion table. All possible travels along X, Y and Z axes are equal to 8" (203 mm). Moreover, on this base small work cells for three-, four- or five-axis parts can be created. Positioning accuracy of $\pm 5\,\mu$m ($\pm$ 0.0002") and repeatability of $\pm 3\,\mu$m ($\pm$ 0.0001") can be achieved. An ultra-compact CNC mill was primarily designed for high production or rapid prototyping of small, precision 2D and 3D parts. The control system is equipped with a 40GB hard disk drive and Ethernet interface.

The high speed vertical machining centre shown in Fig. 16.8, called 'smart machine', was designed for automation with a special attention to minimizing vibration and thermal variations. Optionally, the

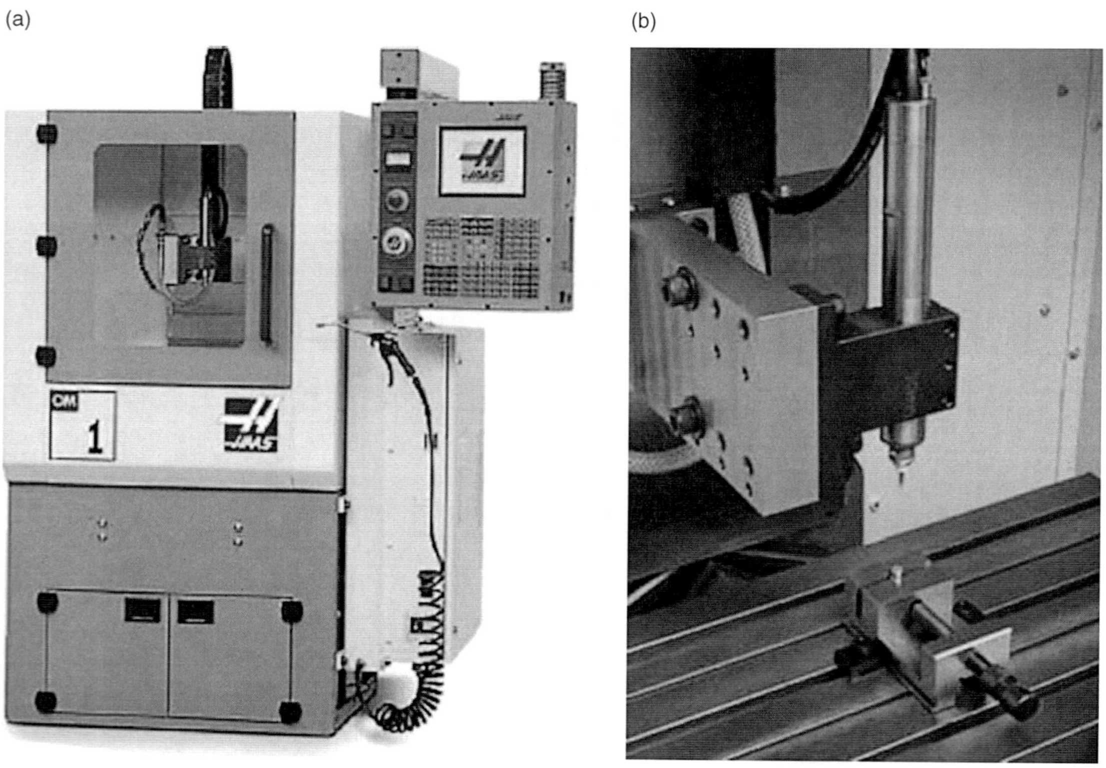

Fig. 16.7 CNC micro-milling machine (a) and view of high speed spindle (b) [15].

Fig. 16.8 High speed machining centre for micro-machining [16]. Source: Mikron.

spindle speeds are 30 000 min^{-1}, 42 000 m^{-1} or 54 000 m^{-1}. The machine shown in Fig. 16.8 has a table chuck with a palletizing system and a seven-position pallet changer as standard features. The modified bridge design of the machine allows the operator to access the machine from the front and permits the use of a pallet-changing device from the rear. The machining centre can also be served by a robot in place of the pallet changer. The VMC was equipped with several hardware and software modules, consequently called 'smart machine' modules that improve process reliability under a high speed machining regime. One of these modules, the Advanced Process System (APS), has a built-in vibration sensor in the spindle together with a spindle diagnostic module. With the help of this system, vibrations that occur during a milling process can be displayed on the monitor of the CNC control as a 'G-load' (measuring range of G-loads between 0 and 10 G). Another unique module labelled Intelligent Thermal Control (ITC) addresses the issue of thermal growth by means of a built-in intelligence on the thermal behaviour of the machine. As a result, it compensates automatically any thermal drift for all operating conditions. Moreover, a warm-up phase or pre-heat cycles are no longer necessary, and machining can be carried out with a high level of confidence in a fully unmanned mode. The third module named Remote Notification System (RNS) was installed to obtain information about the operating state of the machining centre and the process itself, regardless of the location of the user. If, for example, the operator has categorized in advance an NC message as important, the plain text information on the generation of unmanned operation is sent to the user on their mobile phone by SMS (short message service).

The micro-machining vertical centre shown in Fig. 16.9 was designed with a C-type concrete made of a special mineral cast in order to maximize the stiffness characteristics and achieve better vibration damping. The machining centre was equipped with a very stiff spindle of 160 000 rpm speed. Typical applications include the manufacturing of prototypes and small-batch parts made of such materials as steels, non-ferrous metals, plastics and ceramics with a desired surface finish of $Ra < 0.2\,\mu m$. Both three-axis and four/five-axis machining using CNC-precision indexing table with an analogue drive are possible. Achievable positioning accuracy is about $\pm 1.0\,\mu m$. The workpiece dimensions can be automatically measured by a probe and the

Fig. 16.9 Micro highest precision CNC machining centre [17]. Source: Kern-Microtechnic.

appropriate data can be transmitted using the IR transmitter. Moreover, both length and radius of the cutting tool are measured automatically using a special laser measuring system.

The micro-/meso-scale machine tool (mMT) shown in Fig. 16.10 is a high precision, three-axis CNC milling machine specifically designed to manufacture small, demanding parts and to support advanced micro-machining research. The spindle with air bearing or air turbine rotates with a very high speed of 160 000 rpm. The overall size (600 mm square footprint) and utility requirements (20 A at 120 VAC, and 6 CFM of compressed air at 100 psi) allow the mMT to be easily installed in office or laboratory settings. It provides 2 μm positioning accuracy throughout its 63 mm travel in the X, Y and Z directions by the use of high resolution (0.02 μm), high accuracy Heidenhain optical linear encoders mounted to precision-ground granite substructures under the X/Y and Z stages. This level of accuracy allows machining operations to be completed on micro-scale features that demand tight absolute tolerances In addition, the unique design of the mMT utilizes a grid encoder for the X/Y stage, improving the straightness, perpendicularity and accuracy of those stages' motion. Combination of high performance encoders with high specific-force linear motors allows high bandwidth tuning, a key ingredient in contouring performance. The ironless AC linear motors used produce zero cogging torque, reducing disturbance forces in the system and further increasing the dynamic performance of the stages. Finally, the motors provide high peak-acceleration capability, a critical feature in micro-machining applications due to unique cutting mechanics present at the micro-scale.

The machine uses a unique ball-and-vee type kinematic mounting system for both the workpiece and the spindle. This mounting system allows the workpiece and/or the spindle to be removed and replaced with submicron repeatability. This capability allows, for example, a machining operation to be interrupted, and the workpiece to be removed, inspected and replaced without the need to re-register the part. In a situation where the machine is being used for multiple projects, each project can have its own kinematic workpiece pallet. This enables multiple users to easily transition from one setup to the next without losing their part registration.

Fig. 16.10 Micro-/meso-scale machine tool with milling and RP capabilities [18]. Source: Microlution, model mMT Microlution-310S.

The integrated Delta Tau GeoBrick machine controller provides full CNC operation of the machine, supporting a standard set of G- and M-codes. In addition, the open architecture of the GeoBrick supports advanced research applications by enabling direct access to all of the controller parameters and programs. For instance, custom motion parameters, control-loop algorithms, and PLCs can be implemented. The machine also features a custom interface circuitry to enable easy user access to digital and analogue I/O, limit inputs, and flag inputs. Machine operation is achieved using included human/machine interface software. An integrated handheld pendant provides handwheel and E-stop functionality. For research use, the machine can be instrumented with a number of integrated sensors, including cutting force sensors, acoustic emissions sensors, and laser measurement heads for metrology/inspection. Typical applications of the mMT are listed in Table 16.3.

Table 16.3 Typical applications of the mMT [8].

Process	*Application example*	*Process attributes*
Silicon-based lithography	Microchips, sensors, MEMS	Batch process, high volume, 2D geometry
Laser machining	Circuit boards, stents	Flexible process, slow
Mechanical machining	Dental implants, fuel injection nozzles	Flexible process, 3D geometry
Injection moulding	Cell phone buttons, hearing aid components	High volume, plastic parts, requires mould
Stamping/forming	Microphone components, electrical connectors	High volume, sheet metal parts, requires die

Not too long ago, a hole the size of a human hair (about 0.003" in diameter) was about the smallest hole that could be made using conventional machine tools on a production basis. Today, advances in the fields of medical devices, communications, optics, electronics, computers and others have created a need for holes that are straighter, more accurate and in many cases much smaller in diameter than a human hair. Today, this need is mainly met by using a number of non-conventional hole-making techniques including laser micro-drilling, electron beam drilling and EDM micro-drilling [19].

Ultraviolet (UV) light laser is found to be predominantly a very competitive tool for micro-drilling and other micro-machining applications. It offers a combination of high peak power, high repetition rates and beam quality that makes it suitable for drilling holes as small as 10 μm (0.0004") diameter in stainless and carbon steels, titanium, ceramics, silicon and other hard materials up to about 1 mm thick.

EB drilling (Fig. 16.11) is similar to laser drilling in that energy is created and precisely focused on a workpiece to bring about highly localized melting. The focused electron beam concentrates so much power at the hole site that it creates a 'vapour capillary' surrounded by a cylinder of molten material. Typically, a CNC control, specially designed for the EB drilling process, controls the main drilling parameters and the axial movements of the workpiece and beam. The most important advantage of the EB drilling process is its ability to produce hundreds to thousands holes, up to 2000 holes per second, quickly. Modern EB drilling machines are capable of producing holes from about 0.05 mm (0.002") to 0.15 mm (0.060") in diameter in metals and other conductive materials from 0.25 mm (0.010") to 6.0 (0.250") thick [19].

The EDM micro-drilling machine (Fig. 16.12a) is essentially a sinker or ram-type EDM machine with some modifications that enable it to drill holes as small as 5 μm (0.0002") in diameter. Pure tungsten wire is the preferred electrode material for some users because of its exceptional wear resistance but also tungsten carbide can be a good choice to form small holes in the 5–10 μm range. The electrode is typically rotated while cutting in order to promote flushing of residues. An essential component of the EDM micro-drilling machine is a stereo microscope that is used to position the electrode at the workpiece surface before the start of cutting, and to monitor the EDM drilling operation. Like the EB drilling process, EDM micro-drilling is limited to conductive materials.

The micro-EDM machine, shown in Fig. 16.12, is a precision machine tool acquired from Matsushita Panasonic for micro-manufacturing applications. Micro-EDM utilizes the spark erosion phenomenon for removing the material from the workpiece through a dielectric medium. As mentioned earlier,

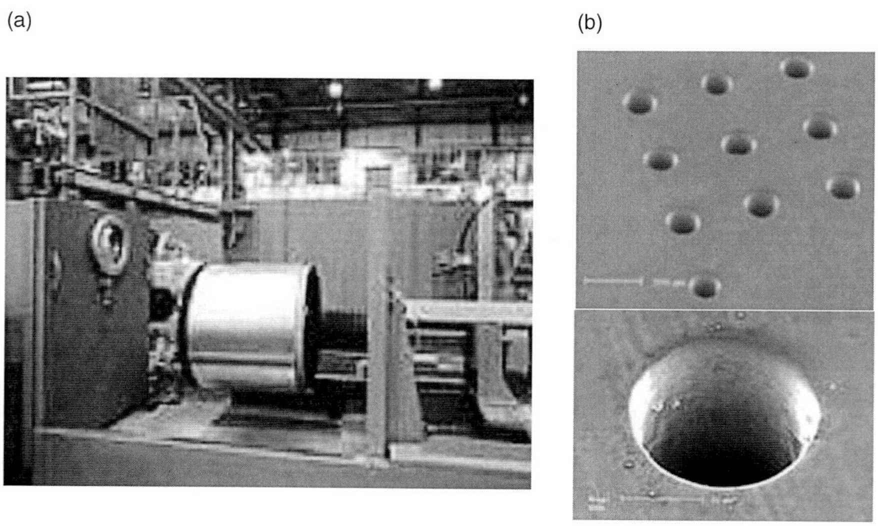

Fig. 16.11 Electron beam (EB) drilling machine [19] (a) and close-ups (b) illustrate the quality of laser-drilled 113 micrometre diameter holes. Upper: holes are trepanned in 0.55 mm thick stainless steel. Bottom: a closer view of a 113 micrometre diameter hole.

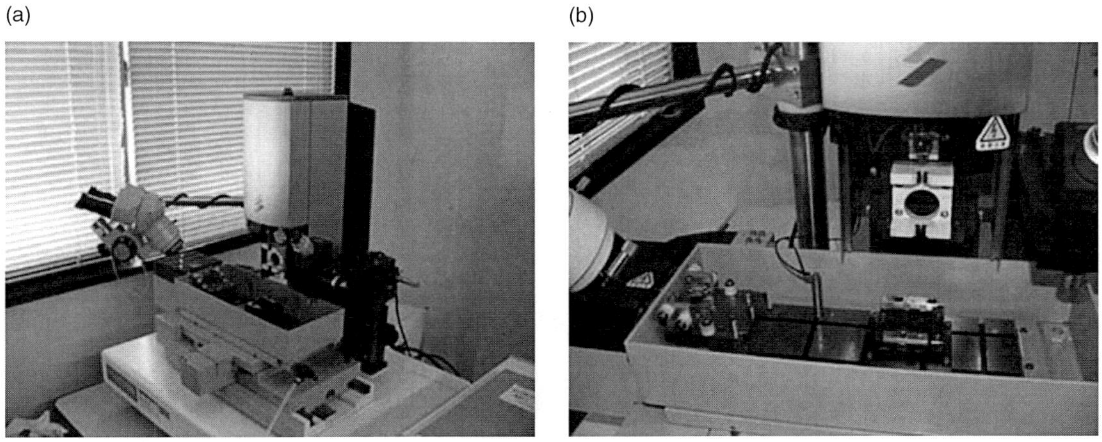

Fig. 16.12 Micro-EDM machine (a) and magnification of the working space (b) [20].

micro-electro-discharge machining can process any material that has electrical conductivity including both easy-to-machine materials such as quenched steel and difficult-to-machine ones such as carbides.

Precise positioning and machining accuracy enable the machine to produce accurate round holes with the diameter ranging from 15 to 300 μm. Along with the CAD/CAM software, complex shapes can be fabricated. This system has a built-in wire electrode-discharge grinding (WEDG) unit. With this unit, the machine can process the tool electrode to the desired diameter (or shape) or turn it into micro-shafts by itself, using the electro-discharge principle with reverse polarity.

EDM is potentially an important process for the fabrication of micro-tools, micro-components and parts with micro-features. Currently, micro-EDM technology used for manufacturing micro-features can be categorized into four different types [21]:

1. Micro-wire EDM in which a wire of diameter less than 0.02 mm cuts a conductive material.
2. Die-sinking micro-EDM which uses an electrode with micro-features to produce its mirror image in the workpiece.
3. Micro-EDM drilling employing micro-electrodes of 5–10 μm in diameter to 'drill' micro-holes in the workpiece.
4. Micro-EDM milling for producing almost any kind of 3D cavities by adopting a movement strategy for micro-electrodes of 5–10 μm in diameter similar to conventional milling.

A new application of the micro-EDM technology is the production of cores for injection moulds and carbide milling tools with a diameter smaller than 100 μm for micro-cutting cavities. In this case a CNC-controlled EDM machine, like that shown in Fig. 16.12b, is equipped with the WEDG (wire electro-discharge grinding) unit [22].

Recently, abrasive water jets generated by the suspension method (passing a pressurized suspension of abrasive particles in water through a ceramic or diamond cutting nozzle) with the use of nanometre-sized abrasive particles (a few microns down to a minimum of 50 nm) was found to be very effective in micro-drilling and profiling operations [23]. As a result, fine cutting jets with jet diameters ranging from 30 to 70 μm were produced in diamond nozzles with bore diameters between 40 and 60 μm. Several holes per second were drilled depending on the valve actuator response.

Figure 16.13 shows a high speed precision CNC milling machine for precision micro-machining and industrial engraving of a variety of materials, such as aluminium, plastics, stainless steel, graphite, ceramics and virtually any non-ferrous materials. A precision CNC milling machine or CNC router can be applied for both low volume and high mix production of such parts as front panels, nameplates, printed circuit boards (PCBs), firearm engraving, micro-drilling, 3D engraving, aluminium machining, plastics machining,

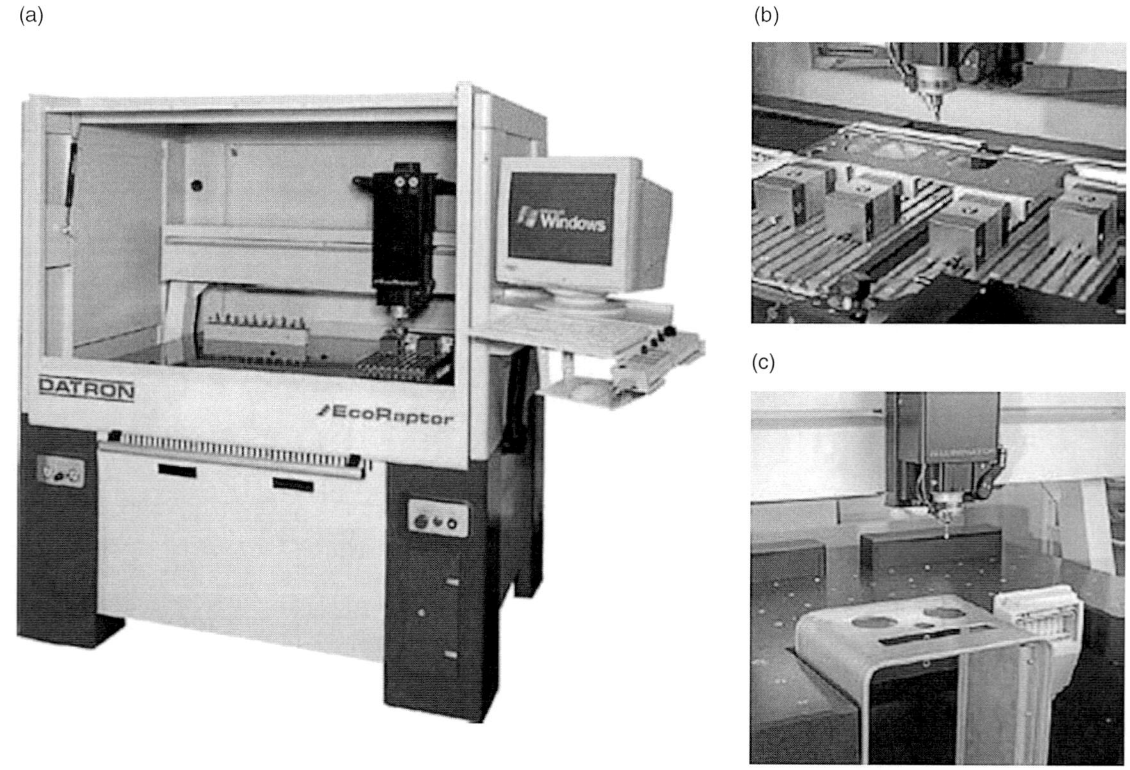

Fig. 16.13 Micro-machining CNC centre (a) and exemplary applications: electronics (b) and HSM of aluminium part (c) [24]. Source: Datron Dynamics Inc.

milling micro-moulds or micro-EDM electrodes, rapid prototyping, UID marking, as well as the production of electronics, medical and aerospace parts. The CNC machine from Fig. 16.13a with steel construction and a cast polymer concrete bed features a 60 000 rpm spindle, feeds up to 15 m/min (600 in/min) and a 10-tool changer. A Pentium PC-based control offers easy-to-use canned cycles, look-ahead buffers and extended diagnostic features. Achievable absolute accuracy is about ±0.05 mm (±0.002") and attainable resolution reaches ±4 μm (0.00016"). Special application of the micro-mill is the production of PCBs. For this purpose the machine can mill, drill and rout PCBs without part lifting from the machine table during the process. Moreover, the installation of the dust extraction control valve system leads to the production of dust-free PCBs.

The specific trend in micro-machining observed is to scale the machine tools down to micro-machines to produce micro-components [6]. This is quite a logical fact because machine tools for production of micro-parts do not necessarily have to be large to achieve high precision. Moreover, several benefits resulting from miniaturization include reduction of power, working space, materials and costs. As experienced, micro-machining centres can exhibit exceptional engineering properties when a special expensive material is used in their constructions. For example, micro-machine tools have higher natural frequencies and lower vibration amplitudes than normal-sized equivalents due to substantial reduction of mass. This leads to the possibility of fabricating parts under a wider range of spindle speeds without unstable zones. Figures 16.14a, c, d and f illustrate some miniature micro-machine tools.

Consequently, miniature machines have given an impulse to create small footprint factories, known as micro-factories [5, 6]. Suitcase-sized micro-factories were first made by the Japanese in the mid- and

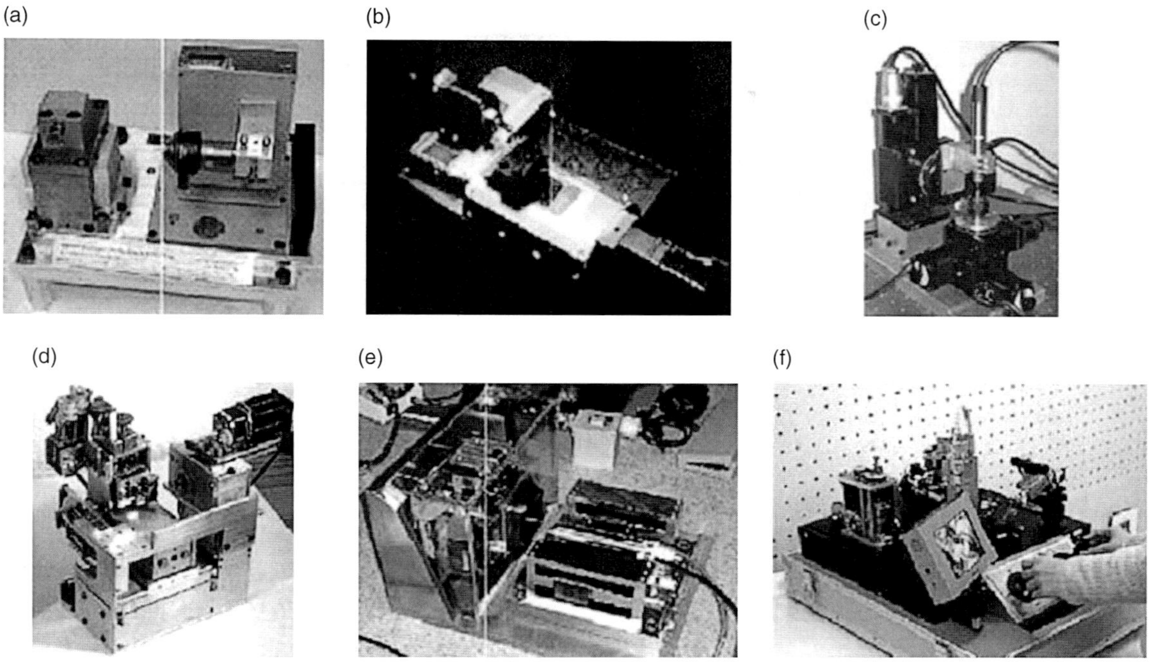

Fig. 16.14 Examples of miniature (micro-) machines (a, b, c and d) and micro-factories (e and f) [6].

late 1990s. Some examples of such micro-manufacturing complexes are shown in Figs 16.14b and e. They can have different cells with different functionalities such as micro-lathe, micro-milling and micro-press, and, for example, can be equipped with a two-fingered tweezer robot and miniature CCD cameras to manipulate micro-components. Micro-factory actuators are either piezoelectric (i.e. flexture designs) or voice coil actuators to achieve submicrometre accuracies. High speed air bearing spindles are typically used, similar to ultra-precision machines. In particular, micro-factories are expected to be suitable for the production of micro-components during military or space explorations where energy supply is very limited. The current development of micro-factories is limited by appropriate sensors and actuators, which must be small enough to be embedded within the drive and control systems. Other important problems to be solved are low structural rigidity and vibration isolation from the external disturbances to achieve high precision and desired tolerances. Two approaches to micro-factories are in development. One is a modular approach with multiply sections for assemblies and the second is a desktop-size factory for producing micro-components [5].

16.4 Examples of Micro-machining Products

In this section some representative examples of parts fabricated by means of various micro-machining techniques are selected in order to gauge the scale of their miniaturization from millimetres to micrometres and associated manufacturing limitations.

As mentioned in section 16.3 a high precision Swiss-type machining centre can produce a variety of tiny precision parts with a maximum machining diameter of 7 mm (5/32") and length of 40 mm (diameter ×2 long). Some examples of such small diameter parts with special geometrical features (nozzles, shafts, EDM copper electrodes, single-thread injector worm screws) produced by a combination of turning, drilling, milling and threading operations are presented in Fig. 16.15.

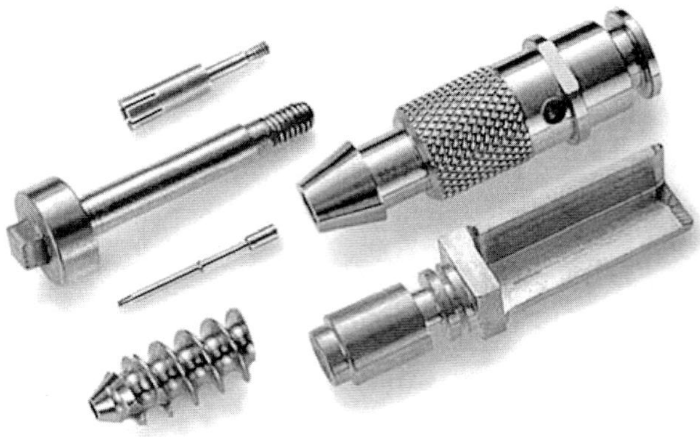

Fig. 16.15 Typical parts machined by a Swiss-type machine from Fig. 16.5 [11].

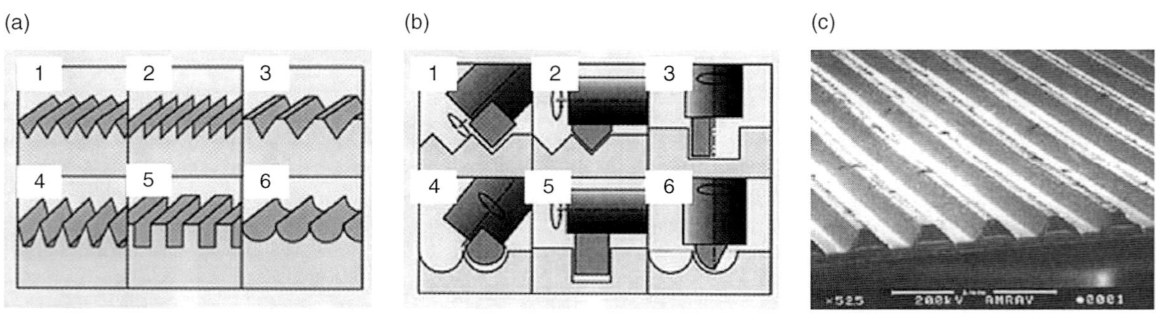

Fig. 16.16 Milling of ultra-precision micro-grooves with single crystal diamond end mills [25].

Single crystal diamond with a cutting edge radius of about 50 nm is a preferred tool material for conventional micro-cutting (100 μm size features are routinely machined) of non-ferrous materials like brass, aluminium, copper and electroless nickel. Figure 16.16a shows micro-grooves of different shapes (numbered 1–7) produced by means of an ultra-precision milling machine and single crystal diamond mills with different operating arrangements (Fig. 16.16b). This method allows the production of V-shaped grooves with high shape accuracy without burrs and surface roughness of 48 nm Rz ($P-V$) value. Other examples of the application of micro-diamond machining are flow passages for the micro-compact heat exchangers shown in Fig. 16.16c, in which surface enhancement performed is necessary to reduce convection boundary resistance. Each passage from Fig. 16.16c is 100 μm in width at the bottom and 80 μm in depth. The material is high conductivity copper foil 125 μm thick. Micro-diamond machining can also be used to complement the lithography process. As reported in [26] tungsten carbide tools with extremely small cutting edge radii can produce reproducible surface roughness lower than $Rz = 0.8\,\mu m$ when micro-milling steel miniaturized components. Recently, also special methods of ultra-precision cutting of carbon steel parts with diamond cutting tools using cryogenic cooling and ultrasonic vibrations have been proposed [27].

Figure 16.17 shows elements of a meso-scale stepper motor of 10 mm × 10 mm × 5 mm in size machined by the micro-EDM process. Micro-EDM enables high performance motors by machining difficult-to-machine materials such as neodymium iron boron and Hiperco® alloy [28].

Micro-parts such as the diesel injection nozzle, micro-electrode and miniature tuning fork, shown in Fig. 16.18a, b and c, respectively, were produced by non-conventional micro-machining methods including laser, EDG (electrodischarge grinding) and FIB methods.

Fig. 16.17 Meso-scale stepper motor machined by the micro-EDM process [28].

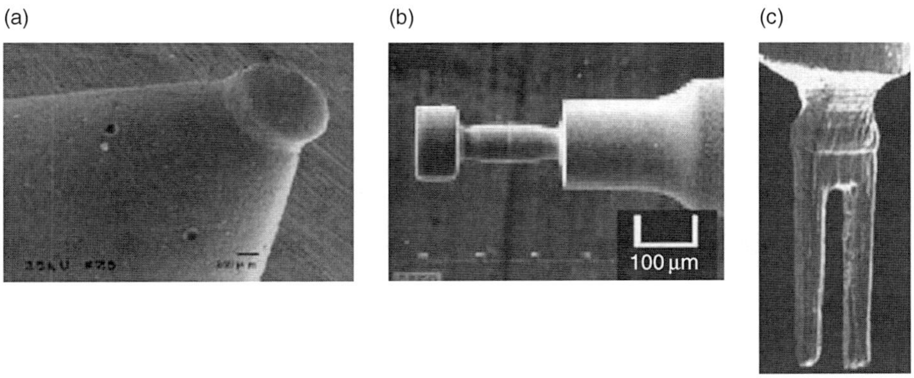

Fig. 16.18 Examples of applications for micro-machining: (a) holes in diesel injection nozzle micro-drilled by CVL (copper vapour laser), (b) micro-electrode for the machining of the groove in the hole shaped by EDG, (c) tungsten tuning fork made with FIB (focused ion beam) micro-machining [28].

Another application of micro-machining covers analysis and visualization of micro-fluidic systems and solar cells (Fig. 16.19a and b), micro-implants and micro-bevel gears produced by the DUV (deep ultraviolet lithography) method (Fig. 16.19c and d) and a range of MEMS systems and their components shown in Fig. 16.20. The production of a micro-gear from Fig. 16.19d covers several stages. First, a seed layer of the platinum or titanium is deposited on a wafer surface. Then a thick coat of the resist is deposited on the seed layer. With photosensitive resist of the appropriate viscosity, it is possible to deposit up to a 50 µm thick resist layer. An optical mask is used in conjunction with an ultraviolet light source (200 nm wavelength) to expose the resist to the desired pattern. The resist is then developed, leaving behind the exposed seed layer and resist walls. The remaining resist can be used as a mould. The mould with the metallic seed layer can be filled with metal (nickel) by electroplating.

Figure 16.21 presents some possibilities for shaping micro-parts using precision RP performed on the mMT machine from Fig. 16.10. They include: cytological nozzle with acrylic and green ceramic with

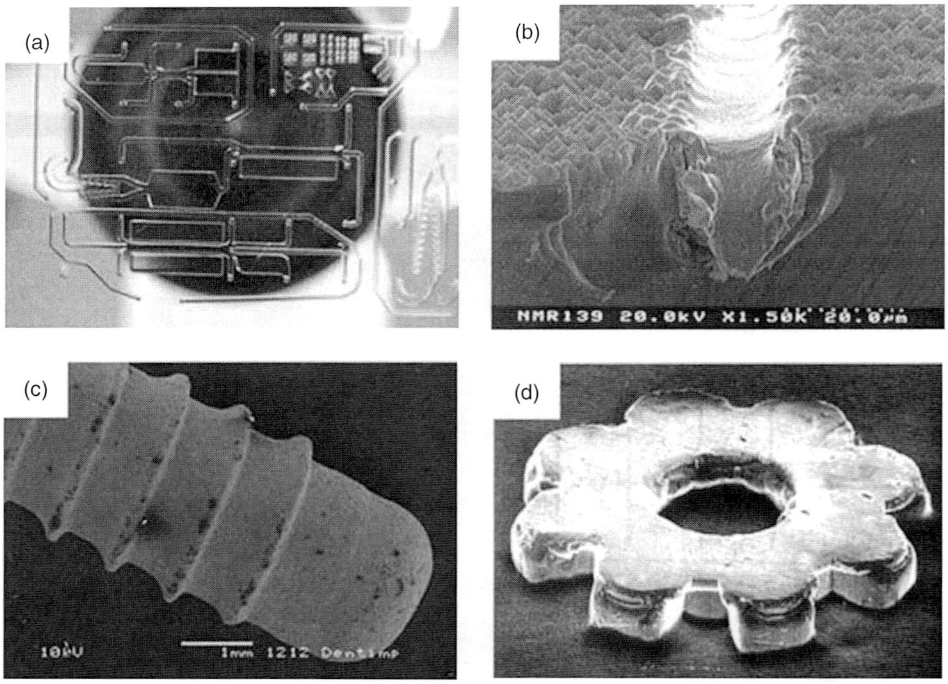

Fig. 16.19 Examples of micro-machining products [3, 28]: (a) micro-fluidic analysis system, (b) metal contact on solar cell, (c) bio-compatible surface of implant, (d) micro-gear made by DUV fabrication.

internal features below 70 μm (a), biomedical components with features smaller than 100 μm (b), aerospace nozzle with features as small as 50 μm (c) and micro-fluidic system of 50 μm in size with high precision holes and slots made of polycarbonate, calcium fluoride, Teflon and epoxy workpieces.

One of the more challenging aspects of the micro-moulding process is the problem of manufacturing the mould cavities, which often requires very high tolerances and regions of great detail. Another one is the production of core pins as small as 0.003" (0.076 mm) less than the diameter of a human hair. In many cases of thin-walled components and/or tiny feature components, the cavity-to-core error can be no more than 5 μm [5]. Current technologies capable of creating such cavities include [29]:

- Micro-EDM (electric discharge machining).
- Laser machining (laser ablation).
- Silicon reactive etching.
- LIGA (lithography and electroplating).
- Micro-STL.
- Micro-milling.

Currently, the favoured method of producing cavities for the majority of MM components is to use micro-EDM or micro-milling techniques as these are based at the cutting edge of tried-and-tested technology. LIGA techniques show much promise but the technology is currently in a state where the form for a cavity cannot be produced reliably and doubts remain over the wearing properties of cavities in nickel and other alloys. Laser machining shows potential, but very high frequency pulsed lasers are required to get optimum accuracy and surface finish.

Typical capabilities of laser ablation technology allow ablation of areas up to 100 × 100 mm with silicon reactive etching which uses a chemical reaction between a sulphur hexafluoride (SF6) plasma and an exposed

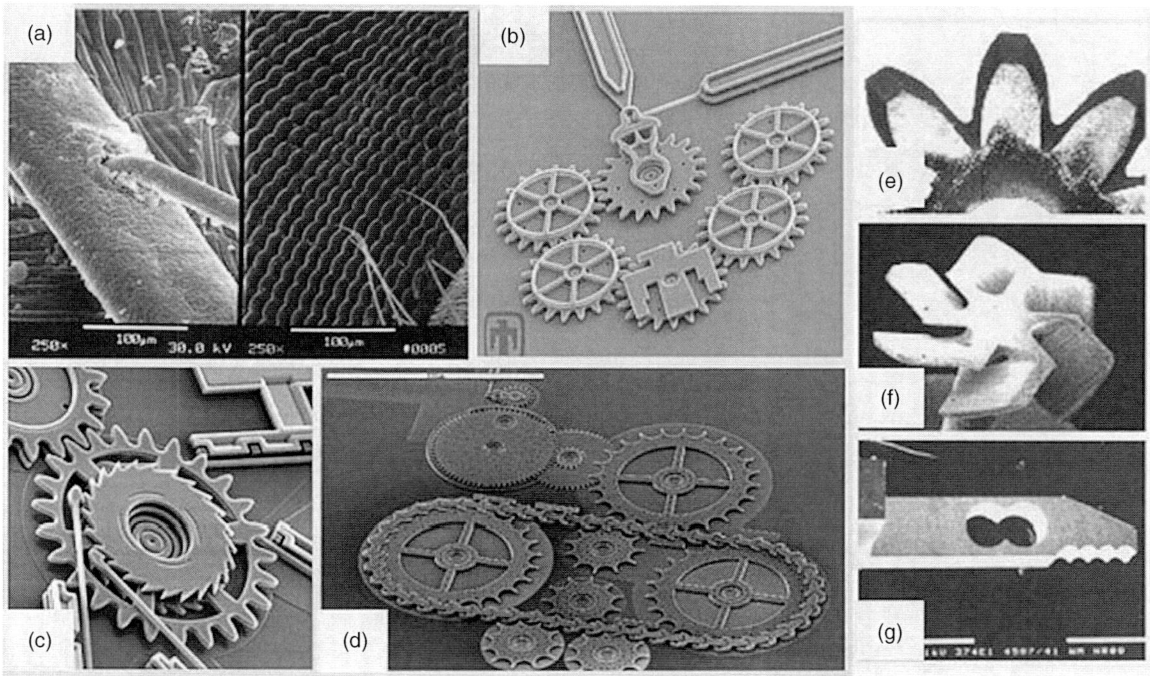

Fig. 16.20 Examples of micro-machining products [3, 28]: (a) individual lenslets of a fly's eye compared to a 25 mm diameter wire placed through a micro-drilled hole in a human hair, (b) gear train made by Sandia four-layer polysilicon surface micro-machining process, (c) ratcheting system fabricated in the five-level technology, (d) unsupported spans as long as 500 μm – Sandia National Laboratories' Summit V process, (e) Si bevel gear, (f) steel micro-propeller, (g) micro-gripper jaw with force sensor.

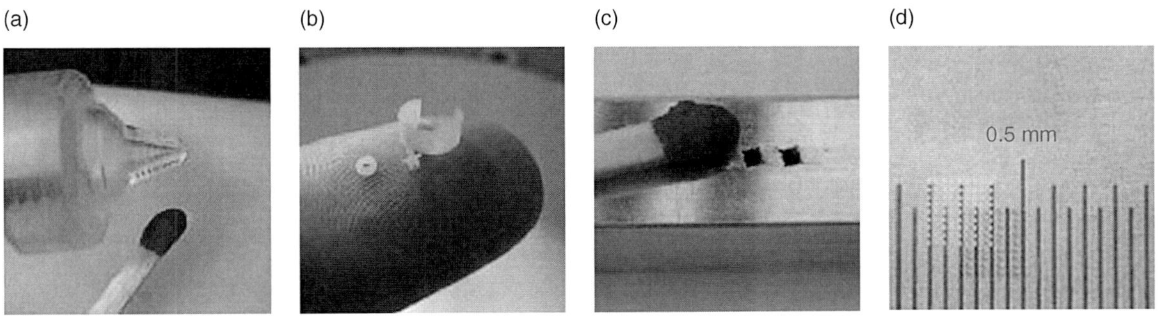

Fig. 16.21 Some parts fabricated by the RP micro-machine from Fig. 16.10 [18].

region of silicon wafer to form 3D structures. The SF6/silicon reaction forms an isotropic etch profile, unlike other irradiating techniques. Therefore, deep reactive ion etching (DRIE) incorporates alternate stages of etching and passivation (in which a fluorocarbon plasma is introduced into the process which causes a layer of fluorine-based polymer to build up on the substrate) to produce complex deep, vertical features, as shown in Fig. 16.22. An actuator for an MEMS device fabricated by the DRIE technology is shown in Fig. 16.23.

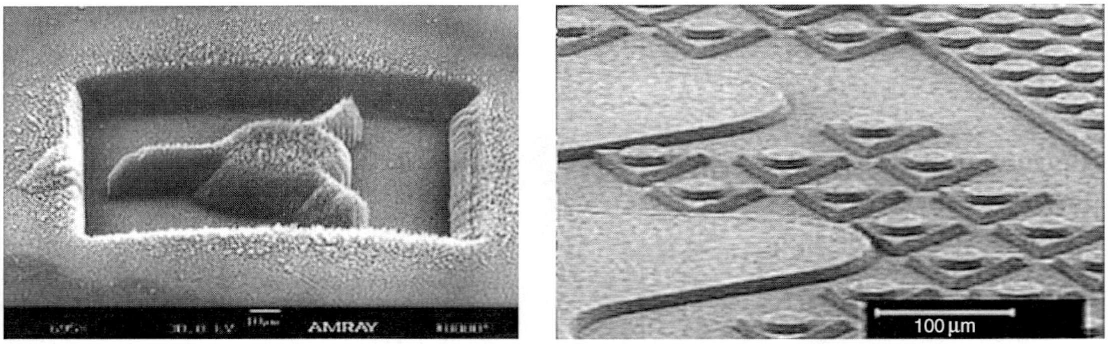

Fig. 16.22 SEM pictures of laser ablated samples [29].

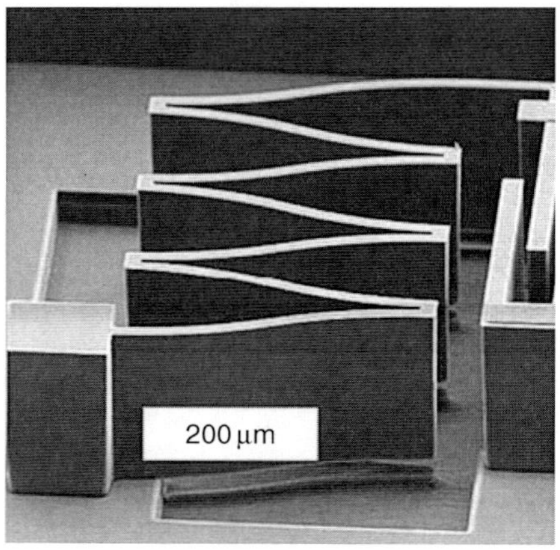

Fig. 16.23 Example of a DRIE product, an actuator for an MEMS device with tolerances of a few microns and a minimum surface roughness of 0.2–0.3 μm [29].

Sizes achievable in micro-moulding can be as low as 0.5 μm, and 50 μm length channels can be inter-connected in a flow pattern. Finish requirements are optical in nature and measured in Angstroms for optical components. The medical market is the largest for micro-moulded components for orthopaedic, cardiological and neurological applications. The automotive market employs many 'behind the dash' applications such as GPS/navigation system sensors and air bag sensors. Another market is micro-fluidics, which employs lab-on-a-chip applications for assay testing, drug discovery, and environmental testing in research laboratories and pharmaceutical and biomedical applications.

16.5 Tooling and Fixturing for Micro-machining

The size of precision micro-cutting tools determines the limit of the size and accuracy of micro-structured features. Diamond tools are often used for ultra-precision machining (see Fig. 16.16) but they are limited to machine non-ferrous materials. Therefore, micro-tools such as micro-end mills and drills are generally made from tungsten carbides with grain sizes smaller than 600 nm [22].

The tiny milling and turning tools, shown in Fig. 16.24, capable of generating features that are at least 25 μm in size, are made by sculpting carbide and HSS blanks using a focused ion beam. End mills made in

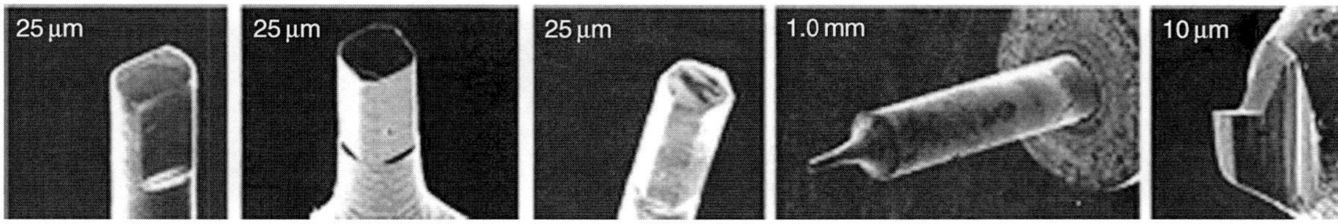

Fig. 16.24 A range of micro-end mills (first three photos) and micro-turning tools (last two photos) [30].

this way can be as small as 20 μm in diameter and turning tools can be about 10 μm wide. These tools were used in the meso-machining of miniature nuclear weapon components made of aluminium, brass and 4340 steel [30].

When shaping tools with an ion beam, there is no way of generating the fluted, complex geometry which is typical of a standard end mill. Instead, meso-machining end mills have simple geometric cross-sections as shown in Fig. 16.25. The cutting edges were obtained by intersecting two planes without a helix angle. This end mill of 25 μm in diameter with five cutting edges was used to make a 25 μm wide × 25 μm deep channel (upper insert) in aluminium. A typical depth of cut was 1 μm when milling slots between 20 and 30 μm wide in the three materials were selected.

Comparably, the smallest drill diameter today is 10–20 μm. Such miniature drills with the ultra-fine carbide grades are used by toolmakers in the printed circuit board (PCB) industry. To facilitate the use of extremely tiny drills a rise in productivity spindles with ultra-high speeds of 300 000 rpm is necessary [32].

Carbide micro-milling tools of 100 μm diameter (like the smallest commercially available tools) fabricated on the EDM machine from Fig. 16.12, equipped with the WEDG unit working with a brass, are shown in Fig. 16.26. The cutting part of micro-tool is positioned eccentrically in relation to the cylindrical shank and the backside is inclined at the clearance angle of 10°. The resulting surface roughness was $Ra = 0.06$ μm, which is appropriate to obtain sufficient sharpness of the cutting edges. Because the WEDG allows the

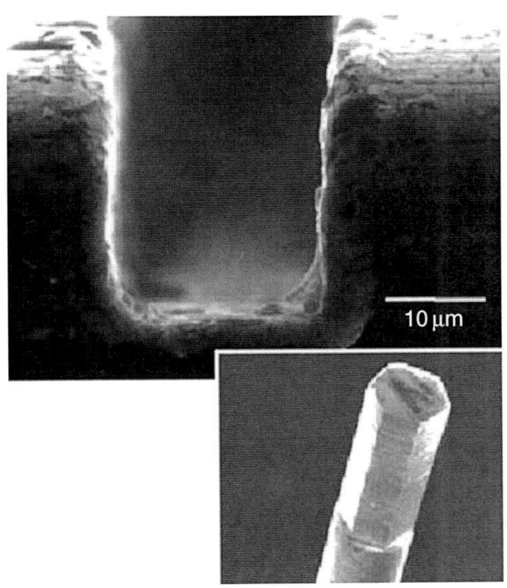

Fig. 16.25 Magnification of a 25 μm end mill tool from Fig. 16.26 and 25 μm wide × 25 μm deep channel (above) in aluminium [29].

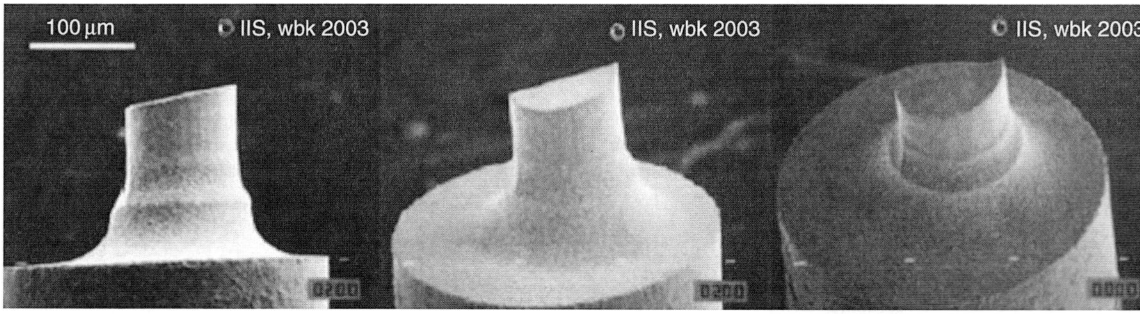

Fig. 16.26 EDM-machined tungsten carbide milling tools of 100 µm diameter [22].

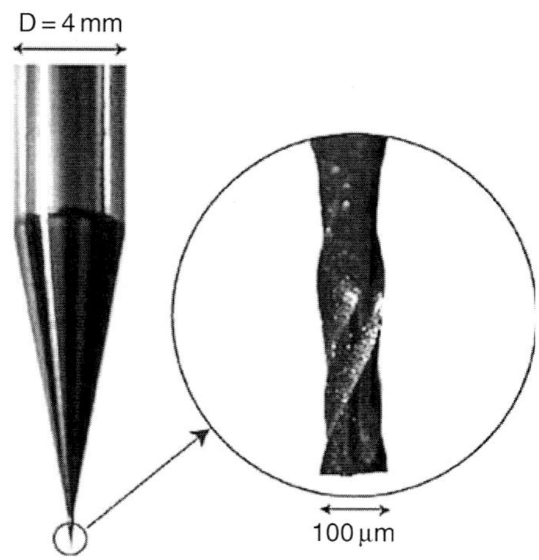

Fig. 16.27 Tungsten carbide micro-end mill with two flutes [22].

production of electrodes with a diameter of about 20 µm, it is a real opportunity to go down with tool miniaturization and produce tools with a diameter of 50 µm or even 20 µm, similar to FIB technology [22].

Commercially available tungsten carbide micro-end mills with helix flutes produced by grinding can be as small as 50 µm in diameter. Figure 16.27 illustrates a typical two-fluted micro-end mill of 100 µm in diameter at the end of conical-tool tip.

Nowadays, several of the major cutting tool manufacturers offer tool systems specifically dedicated to the micro-manufacturing performing on Swiss-type or small CNC machines [12]. Iscar offers tools with multidirectional grooving inserts as narrow as 0.5 mm (0.0197"), a multifunctional carbide tool of 4 mm in diameter (PICCO-MFT) for drilling, facing, internal and external turning and threading, and miniature carbide coolant-fed drills and ISO turning tools with shanks as small as 10 mm (0.375"). Mitsubishi Carbide produced smallest solid micro-grain carbide, coolant-through drills of 1–2.8 mm (0.040–0.110") in diameter for drilling in aircraft aluminium, 6Al4V titanium, 304L stainless steel and 4130 steel. The drills are coated with a PVD-Al,TiN layer with high resistance to adhesive welding. Also solid-carbide end mills starting at 0.1 mm in diameter are available. Sandvik Coromant introduced a CoroTurn XS programme of boring inserts and holders designed for turning, copying, grooving, pre-parting, profiling and threading of components with bores down to 1 mm. In addition, a number of external tools are capable of doing the appropriate operations on extremely small parts. The production programme includes solid end mills as small as 0.4 mm (0.016") and solid drills of 1.5 mm (0.059") diameter. Moreover, MA Ford offers off-the-shelf drills down to 0.1 (0.05) mm (0.0039"(0.002")) diameter, die-mould ball end mills of 0.4 mm (0.0156") and traditional four-flute end mills of 0.12 mm (0.0050") in diameter.

In order to use micro-tools 1/4" or smaller, high frequency spindles of 60 000 rpm and feed rates up to 1000 inches are required, because fragile micro-tools are too easily broken by conventional vertical machining centres (VMC) with big hulking motors designed with the torque to run 3" fly cutters. For high speed machining of aluminium, for instance, or precision industrial engraving with micro-tools, a high speed CNC router, such as that shown in Fig. 16.13, is needed. The effect is that high speed machining, micro-hole drilling or micro-drilling and industrial engraving with small high speed end mills, drills, thread mills and engraving tools can be performed as intended at super high speeds.

Some types of micro-tools for various machining operations are selected in Fig. 16.28 and their dimensions and typical applications are listed in Table 16.4.

Seco Jabro Tools [12] developed the MINI range of miniature solid carbide end and ball-nose mills of 0.1–2.0 mm in diameter for use in the manufacture of small components. They cover all groups of workpiece materials, i.e. soft and hardened steels, stainless steels, aluminium, titanium and copper. Additionally, special MINI DIAMOND end mills of 0.2–2.0 in diameter, coated with CVD PCD films, were specially designed for machining high abrasive materials like graphite.

Kennametal's concept, shown in Fig. 16.29, was named KM Micro [12]. It allows types of machining operations such as turning, boring, cut-off and drilling to be performed. The system consists of a clamping unit and a cutting unit. The quick-change cutting unit can be simply replaced by a cutting unit that contains a new insert. The clamping unit can be a square shank (Fig. 16.29b) or a round shank. Bolt-on flanges

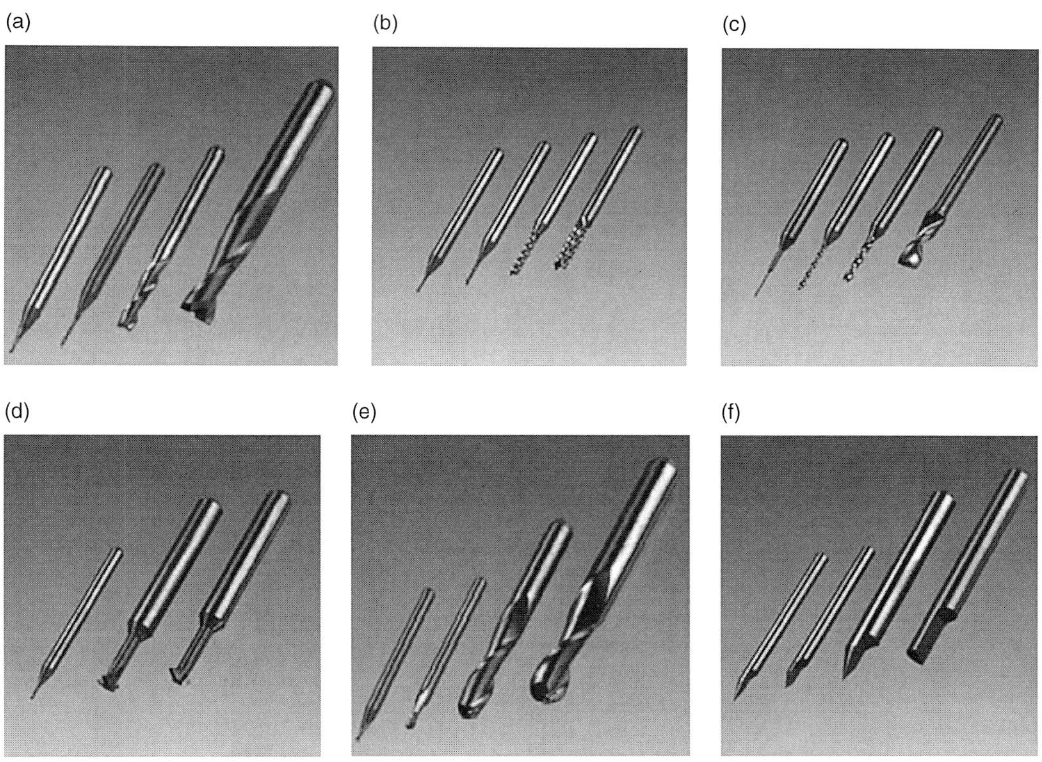

Fig. 16.28 Micro-tools for HSM [24]: (a) double-fluted end mills, (b) fish tail end mills for PCB milling, (c) micro-grain carbide drills with 1/8" shank, (d) thread mills, (e) ball nose end mills, (f) standard engraving tools.

Table 16.4 Characteristics of micro-tooling from Fig. 16.28 [24].

Type of micro-tool	*Diameter in mm*	*General use*
Double-fluted end mills	0.3–1.5	Aluminium, brass, duro-plastics, copper
Fish tail end mills for PCB milling	0.6–3.0	Printed-circuit material, e.g. FR4
Micro-grain carbide drills	0.4–3.0	Aluminium, brass, plastics, copper, etc.
Thread mills	3 and 6	Non-ferrous materials
Ball nose end mills	3, 6 and 8	3D and model-making applications
Standard engraving tools with conical tool tip	3 and 6	Aluminium and plastics

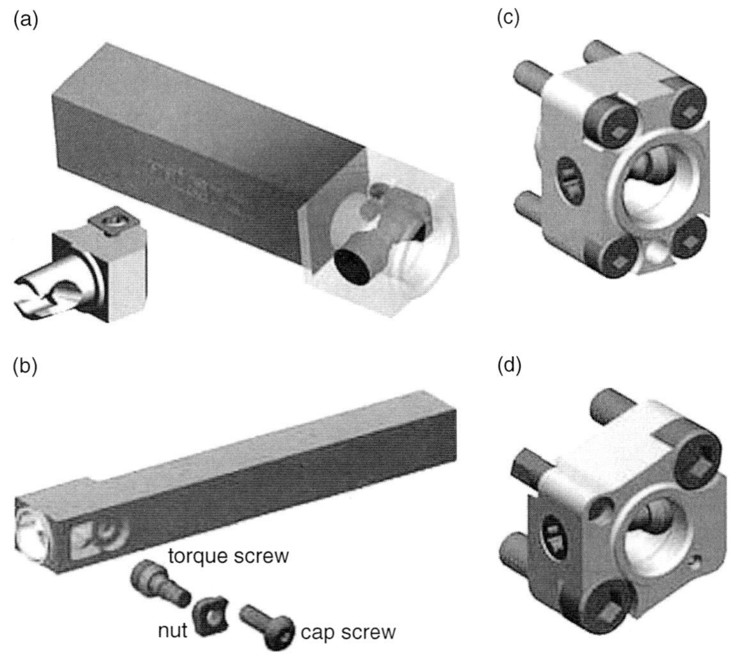

Fig. 16.29 KM Micro Quick-Change Tooling [12]: (a) clamping system and micro-quick-change cutting unit, (b) square shank adaptor, (c) universal flange clamping unit, (d) flange attachment unit.

(Fig. 16.29c and d) are also available to be used with special mounting blocks. Their use on Swiss-type machines often means that one or more additional positions can be added in the existing tooling envelope and, as a result, machine uptime decreases at a rate of 25–40%. It should be noted that both square and round shanks can be used in any tool slot, turret or block with a dimension matching the system size.

Part rigidity, which often causes problems in micro-machining and fixturing, is one of the most important aspects of micro-machining. The traditional methods of holding parts often do not apply to small-scale parts and unconventional methods are often necessary. During micro-machining, the cutting pressure is much lower and the parts are much more likely to fly off the machine and a substantially lower holding force can be used. This allows fixturing parts with glue, double-sticky tape or wax. Removing the fixture is as important as holding the part. If the parts have thin walls, they can easily be broken when the fixture is removed. Although a vice is not used, referencing can still be used in multiple setups. One method of referencing is machining locating holes in the part as it is being machined. Furthermore, to avoid part movement, components are often filled in with wax as they are made.

One of the simplest methods of fixturing involves using double-sticky tape which is mounted to the back of the part and fastens it on a plate on the table of the machine. It is applied and removed easily and allows a sufficient holding force for micro-machining and rapid prototyping. Disadvantages include excessive vibration when larger cuts are made and the tendency to lift the part when milling and drilling with high helix tools. Both coolant and heat work against double-sticky tape and it is predominantly used in dry machining or with air cooling.

Wax gives additional rigidity when used for either stabilizing or mounting components. Typically, the piece is put into place and then hot wax is dripped onto the workpiece. Similar to double-sticky tape, hot wax is poured on the mounting surface and the workpiece is clamped into place. Waxes can provide an extremely strong holding force when cooled. When they need to be removed, the workpiece is either pulled free or heated to allow the wax to run off. Wax is also used to support weak parts as they are machined, as shown in Fig. 16.30.

Glues like cyanoacrylates (superglues) and Loctite can also hold parts in place very well. They are easy to mount on a clean surface and prevent vibration of the part but are also somewhat difficult to remove. Typically, acetone is used to remove the superglue after machining is completed.

Sandwiching is a traditional method of fixturing that often works well with thin plates. The part is placed between a base component mounted to the table or vice and a top component that is bolted or clamped into place. The top component maintains pressure on the thin workpiece and prevents it from buckling as it is cut. It allows holes to be drilled through the plate or milling to occur effectively.

Soft jaws are often mounted on the vice and the desired profile is machined into the jaws, as, for example, a round or hex profile.

The example from Fig. 16.30 illustrates fixturing a complex part with a two-sided setup. The two parts made were an aluminium butterfly and a Plexiglas butterfly. The overall dimensions of the butterfly are 0.575 × 0.375 × 0.125" (1.461 × 0.953 × 0.318 cm). The wing thickness of the butterfly is 0.004" (0.0102 cm). Because of the thin wings and the features on both sides of the butterfly, support was required while machining. The butterfly needed to be held in the workpiece until the last operation because it would be very difficult to hold onto once removed. Accurate positioning of the workpiece was necessary to prevent breaking through the wings, which could happen if the piece was displaced in the X, Y or Z direction during the flipping process. A block of aluminium (for the aluminium butterfly) or Plexiglas (for the Plexiglas butterfly) was cut to the nominal thickness. Then, it was mounted to an aluminium block in the vice. Two dowel pin holes were milled into the piece as well as the fixture in the vice to locate the block. The top of the butterfly was profiled using Mastercam X with a Haas Office Mill and an NSK spindle. First, a roughing pass was taken with a 0.062" (0.158 cm) diameter ball end mill. Two dowel pin holes were put into the piece as well as the fixture in the vice. Finishing passes were taken with a 0.010" (0.0254 cm) diameter ball end mill. Next, the cavity formed was filled in with a softer Mitee-Grip wax. The wax was heated with a soldering iron and then allowed to fill the top surface of the butterfly. This wax was used to support the

Fig. 16.30 From left to right: top being cut, one side filled in, second side cut with wax removed [14]. (see Plate 21)

butterfly when the bottom side was cut. The workpiece was flipped over and dowel pins were used to locate it. Machining then began from the other side. Another roughing pass was taken with parameters identical to the first roughing pass. The finishing pass was done with a 0.010" (0.0254 cm) diameter ball end mill. Throughout this finishing pass, the bottom was profiled. During the last contour, the end mill was taken around the perimeter of the part, cutting it free from the workpiece. At this stage the butterfly was solely held into position by wax. The wax was then removed with hot water and mineral spirits.

16.6 Metrology for Micro-machining Processes and Products

In the micro-machining world appropriate meso- and micro-measuring systems are necessary because part features are often too small to use a conventional contact profilometer or laser. Recently, a special laser system that can simultaneously measure six degrees of freedom using a meso-scale laser head was developed to perform error measurement on an mMT platform [4]. Similarly, conventional machine tool positioning devices are not the best alternative at this scale and a new solution employs a device based on voice-coil technology, typically used for high speed scanning applications that include tissue scanning, where fast acceleration and high positioning resolution are required. For such applications, a piezoelectric-based actuation system can be suitable as well. As reported, both an optical microscope for the observation of the micro-cutting process when very tiny tools are cut, i.e. end mills of about 20 μm in diameter and turning tools of about 10 μm wide (machining under the microscope), and an electron microscope inspection of machined parts can be an integrated part of the micro-machine tools [30]. Inspection techniques in measuring very small micro-parts require customized vices, tweezers and fixturing. Measuring of micro-components is very important for the assembly of functional devices. Due to the miniature size of micro-components, the human eye alone is not sufficient to observe micro-component defects or micro-system performance. Consequently, sophisticated vision and optical systems are required to examine the accuracy and surface finish of such final components. Generally, the smallest touch-trigger probe tip for tactile probing is 0.3 mm in diameter with a stylus length in the 2–3 mm range. However, touch-trigger probes in combination with optical image processing sensors, analogue scanning probes, and lasers have created a whole class of versatile multi-sensor machines that can be configured for both shop-floor and quality-control laboratory measurement and inspection [31].

As reported in previous sections, recovery of the Swiss Machining technology centres results in the manufacture of a wide range of components, from 0.5 mm diameter through to 85 mm diameter. Materials machined include aluminium alloys, stainless steels and plastics such as PTFE. To measure a vide variety of Swiss machined precision components with tight tolerances two non-contact measuring systems from Vision Engineering can be used as shown in Fig. 16.31. Kestrel, a two-axis manual system (Fig. 16.33a), is preferred for rapid, shop-floor gauging applications, whereas Hawk, an automatic three-axis platform (Fig. 16.31b), was designed for more challenging measurement tasks where pre-programmed inspection routines can be run. Inspectors use the Kestrel as a shop-floor measuring tool because it is simple to use, robust and provides fast results. The Kestrel is located alongside the CNC plant, allowing first off components to be taken directly from the machine for immediate inspection and measurement. Vision Engineering's Hawk 5000 VED is the first non-contact measuring microscope which offers the option of both automated video measurement and manual optical measurement, all within the same measuring routine, providing users with the ability to perform higher volume automated batch video measurement routines, or high accuracy optical measurements for one-off or critical parts. This is a motorized three-axis measuring system with automatic video edge detection (VED) integrated with a QC5000 PC package, which allows for measurement routines to be programmed and stored for future use. Typically, the Hawk is located in the QC laboratory where it is used by specially trained quality personnel.

Recently, multi-sensor technology allows consolidating on large-scale, multiple, unique applications on the same measuring platform and combining measurement and inspection capabilities that satisfy the requirements of diverse departments such as R&D, engineering and manufacturing on one machine.

(a) (b)

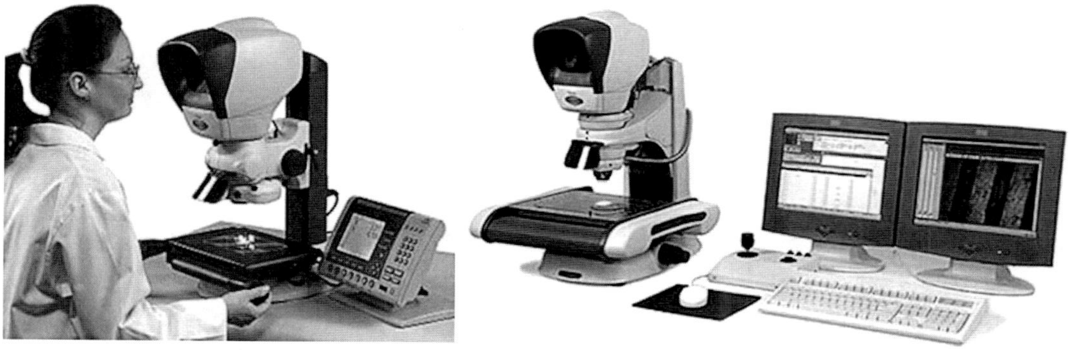

Fig. 16.31 Vision Engineering equipment [32]: (a) Kestrel 200 manual two-axis toolmaker's measuring microscope, (b) Hawk 5000 VED automated video and optical measurement microscope.

Fig. 16.32 Werth Check® CNC shop-floor multi-sensor coordinate measuring system [32].

Figure 16.32 shows a shop-floor multi-sensor coordinate measuring system with automatic image analysis and CNC positioning. Specific design features are: fixed bridge, moving table configuration, multi-sensor CMM; rigid granite construction; high accuracy ball bearing guideways, and linear measuring scales with low thermal coefficient of expansion. Measuring range: $X = 400$ to 1500 mm; $Y = 400$ to 1000 mm; $Z = 150$ to 300 mm. Available measuring sensors include: non-contact image processing (IP) video sensor with fixed telecentric objective; zoom sensor; laser sensor; 3D-Patch, contact-Renishaw point-to-point touch trigger and dynamic scanning sensors; fibre optic (micro-) sensor and contour sensor. Applications cover such parts as plastic connectors and housings; stamped/formed parts; printed circuit boards (PCBs); plastic injection moulding components; small dies; medical component, tools and fixtures. In particular, the fibre probe sensor features a tip with a diameter as small as 25 μm, long stylus length (often in excess of 50 mm) and extremely low gauging force (typically in the μN regime) for non-destructive measurements of optical-quality surfaces, deep bores, and extraordinary small holes and elements. In addition, the fibre probe tip can be illuminated to measure blind or closed features, counterbores, and other geometries/profiles that cannot be measured by transmitted light. Typical applications of this sensor include, for example, tiny turbine blade cooling holes, microscopic-diameter diesel fuel injector spray holes, and infinitesimal cranio-facial surgical screws often of 0.3 mm in diameter or less.

(a) (b) (c)

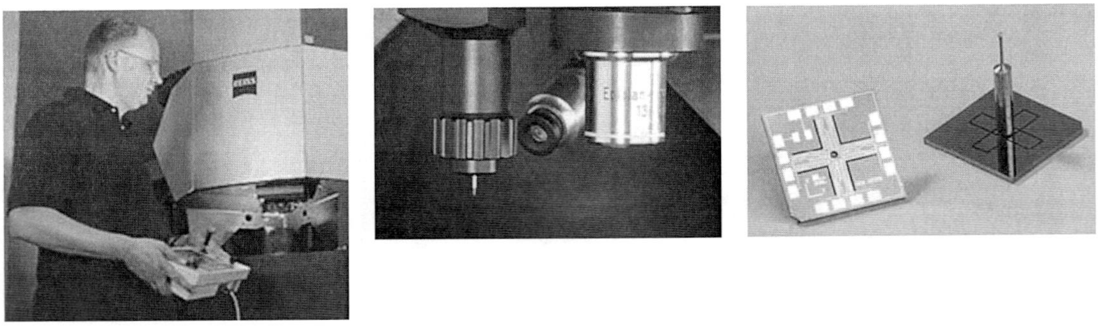

Fig. 16.33 Zeiss F25 CMM machine with 3D touch scanning sensor (a), multi-sensor system (b), and optical sensor for 2D measurements (c) [32].

Micro-systems and MEMS require new measuring systems, in which both the sensors (scanning and touch sensors) and the workholding forces can be tolerated by extremely delicate parts. According to Carl Zeiss [32] the vertical walls of micro-parts can be captured optically employing tactile coordinate metrology. As an example of such measuring systems, Fig. 16.33 shows a Zeiss F25 CMM measuring machine equipped with 3D micro-probe which exerts measuring forces that are about 1/400th of the probing force of a large industrial machine. This is because the probe forces were reduced to less than 0.5 mN/μm. The measuring machine is supported on air bearings with a measuring volume of one cubic decimetre. The ultra-precise kinematics combined with the highly accurate measuring system enable measuring uncertainty of 250 nm at a resolution of 7.8 nm. Using minimal probe forces, this resolution, along with optimum control of the linear drives, enables touch measurements even in bores less than one millimetre in diameter. A touch, passive measuring stylus was developed based on a 6.5 × 6.5 mm silicon chip membrane and integrated piezo-resistive elements. The stylus permits measurements in both single-point and scanning mode. The 3D micro-stylus is designed for stylus diameters of 50–500 μm and stylus tip diameters of 100–700 μm. With a free shaft length of up to 4 mm, it is possible to measure in small, deep-lying structures with probe forces of less than 0.5 mN/μm.

The combination of touch and optical sensors (Fig. 16.33b) enables measurements of 2D and 3D structures in the same coordinate system. An additional camera aids visualization when probing very small features, thus simplifying part programming. The time-tested CALYPSO package from Carl Zeiss enables the evaluation of captured data using standard measuring software with an intelligent user interface that can also be used on all other coordinate measuring machines.

The ViScan camera sensor (Fig. 16.33c) from the standard programme, combined with an objective lens based on those used in Zeiss microscopy, is used as an optical sensor for 2D measurements. These optics, which are optimized for depth of field and have minimized distortion, ensure accurate measuring results with maximum resolution in reflected and transmitted light.

The production of 3D shapes by micro-machining makes their testing and handling quite difficult. It should be noted that flat surfaces on steel parts can be more easily measured with non-contact or rarely with contact devices than, for instance, complex 3D surfaces generated on the moulded components. Figure 16.34 shows an inspection station consisting of an Olympus BX-41 optical microscope with a Prior H-101 motorized X, Y, Z stage, a data-processing PC, and a Lumenera LU125 CMOS mono-chromatic camera to capture high resolution images.

Component manufacturers and micro-moulders require similar inspection machines with identical fixtures to validate near-micron tolerance in micro-components. In addition, extremely clean HEPA-filtered air-controlled rooms are necessary for repeatability or measurements. It is an obvious standard for micro-sized parts to insist on the process capability of 1.33 C_{pk} or better with respect to the performance of appropriate devices.

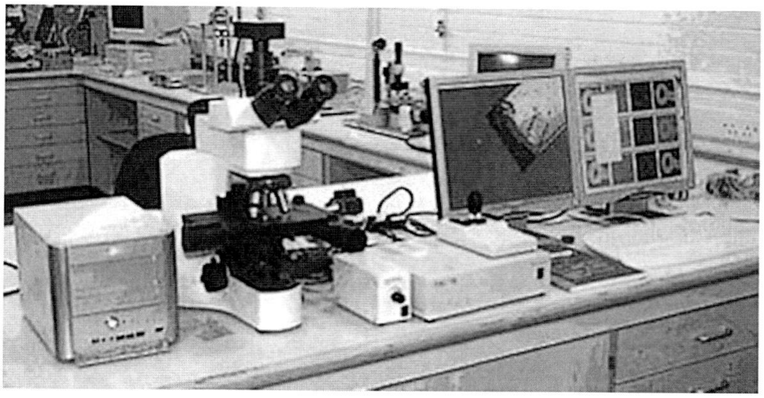

Fig. 16.34 Inspection station for micro-moulded components [5].

References

[1] T. Masuzawa, H.K. Tönshoff, Three-dimensional micromachining by machine tools, *Annals of the CIRP*, Vol. 46/2 (1997), 621–628.
[2] T. Masuzawa, State of the art of micromachining, *Annals of the CIRP*, Vol. 49/2 (2000), 473–488.
[3] Edited by R.C. Dorf and A. Kusiak, *Handbook of Design, Manufacturing and Automation*, John Wiley & Sons, New York, 1994.
[4] R.B. Aronson, Micromanufacturing is growing, *Micromanufacturing*, 2004, pp. M1–M18, www.sme.org.
[5] D. Bibber, Is going micro worth the effort?, *Manufacturing Engineering*, Vol. 137, No. 6 (2006), 115–123.
[6] J. Chae, S.S. Park, T. Freiheit, Investigation of micro-cutting operations, *Int. J. Mach. Tools Manuf.*, Vol. 46 (2006), 313–332.
[7] *ASM Handbook, Surface Engineering*, Vol. 5, ASM International, Materials Park, 1994.
[8] J. Shanahan, Trends in micro machining technologies, Makino, www.makino.com.
[9] G. Byrne, D. Dornfeld, B. Denkena, Advancing cutting technology, *Annals of the CIRP*, Vol. 52/2 (2003), 483–507.
[10] A. Gillner, J. Holtkamp, C. Hartmann *et al.*, Laser applications in microtechnology, *J. Mat. Proc. Technol.*, Vol. 167 (2005), 494–498.
[11] High-speed, high-precision machining of tiny precision parts, www.marucit.com/products/R07.html.
[12] www.iscar.com, www.mitsubishicarbide.com, www.maford.com, www.kennametal.com, www.secotools.com, www2.coromant.sandvik.com.
[13] T. Masuzawa, K. Okajima, T. Taguchi, M. Fujino, EDM-lathe for micromachining, *Annals of the CIRP*, Vol. 51/1 (2002), 355–358.
[14] M. McCormick, Ch. DeBoer, Micromachining techniques, fixturing and end mill selection in high-precision VMC parts, *Machining Technology*, No. 3, 2006.
[15] D. Hayes, Office machine overview, Haas Intl. Sales Training.
[16] 'Smart machine' – a new dimension in milling technology, www.mikron-ac.com.
[17] Kern Micro Höchstpräzisions-CNC-Bearbeitungszentrum, www.kern-microtechnic.com.
[18] What is micro-manufacturing?, Microlution 310-S. Precision 3-Axis CNC Micro-Milling Machine, www.microlution-inc.com.
[19] L. Rakowski, Non-traditional methods for making small holes, www.mmsonline.com/articles/060202.html.
[20] Micro-EDM machine, http://www.ntu.edu.sg/mae/research/programmes/precision/Micro-EDM.htm.
[21] D.T. Pham, S.S. Dimov, S. Bigot, A. Ivanov, K. Popov, Micro-EDM – recent developments and research issues, *J. Mater. Proc. Technol.*, Vol. 149 (2004), 50–57.
[22] J. Fleischer, T. Masuzawa, J. Schmidt, M. Knoll, New applications for micro-EDM, *J. Mater. Proc. Technol.*, Vol. 149 (2004), 246–249.
[23] D.S. Miller, Micromachining with abrasive waterjets, *J. Mater. Proc. Technol.*, Vol. 149 (2004), 37–42.
[24] High-speed precision CNC milling machines, www.datrondynamics.com.

[25] K. Sawada, T. Kawai, Y. Takeuchi, T. Sata, Development of ultraprecision micro grooving (manufacturing of V-shaped groove), *JSME Int. J.,Series C*, Vol. 43(1) (2000), 170–176.
[26] H. Wuele, V. Hüntrup, H. Tritscher, Micro-cutting of steel to meet new requirements in miniaturization, *CIRP Annals*, Vol. 50/1 (2001), 61–64.
[27] E. Brinksmeier, R. Gläbe, Advances in precision machining of steel, *CIRP Annals*, Vol. 50/1 (2001), 385–388.
[28] Meso-machining, Sandia National Laboratories, www.sandia.gov.
[29] Micromoulding tooling, www.ukmig.com/ukmig/about.html.
[30] P. Zelinski, Machining under the microscope, www.mmsonline.com/articles/010504/html.
[31] J. Lorincz, Measuring micro parts, *Manufacturing Engineering*, Vol. 135, No. 5 (2005), www.sme.org.
[32] www.visioning.com (Rowan Precision, New Milford, UK), www.werth.de (Werth Messtechnik GmbH, Giessen, Germany), www.zeiss.de/imt (Carl Zeiss IMT, Germany).

CHAPTER SEVENTEEN

NANOMANUFACTURING/NANOTECHNOLOGY

Contents

17.1 DEFINITION AND STATE OF THE ART OF NANOMANUFACTURING

The term nanotechnology was coined by a Japanese researcher Norio Taniguchi in 1974 to term precision machining with a tolerance of a micron or less. Today, the term has come to apply to the technology of controlling the structure of materials down to a few atoms or molecules to combine nano-scale with ultra-precision accuracy (Fig. 17.1). As depicted in Fig. 17.1, nanomanufacturing requires the integration of nano-scale material science with manufacturing methodologies to solve multi-domain (mechanical, thermo-fluid, electrical, optical, chemical, biological, etc.) and multi-scale (nano-, micro-, meso-, and macro-scale) problems.

The nano-scale is about a thousand times smaller than the micro-scale, that is, about 1/80 000 of the diameter of a human hair. Approximately three to six atoms can fit inside a nanometre, depending on the atom. The prefix *nano* means ten to the minus ninth power, or one billionth. Some popular definitions of nanotechnology, more or less adequate, with various aspects of individual science and applications, are provided below in order to show the transformation of its meaning during the last 20 years [2, 3].

Merriam-Webster's Collegiate Dictionary definition: nanotechnology is the art of manipulating materials on an atomic or molecular scale especially to build microscopic devices (as robots).

The About.com definition at the physics portal: nanotechnology means the development and use of devices that have a size of only a few nanometres. Research has been carried out into very small components, which depend on electronic effects and may involve movement of a countable number of electrons in their action. Such devices would act faster than larger components. Considerable interest has been shown in the production of structures on a molecular level by suitable sequences of chemical reactions. It is also possible to manipulate individual atoms on surfaces using a variant of the atomic force microscope.

Webopedia's definition of nanotechnology: a field of science whose goal is to control individual atoms and molecules to create computer chips and other devices that are thousands of times smaller than current technologies permit. Current manufacturing processes use lithography to imprint circuits on semiconductor materials. While lithography has improved dramatically over the last two decades to the point where some manufacturing plants can produce circuits smaller than one micron (1000 nanometers), it still deals with aggregates of millions of atoms. It is widely believed that lithography is quickly approaching its physical limits. To continue reducing the size of semiconductors, new technologies that juggle individual atoms will

Advanced Machining Processes of Metallic Materials
ISBN: 978-0-08-044534-2

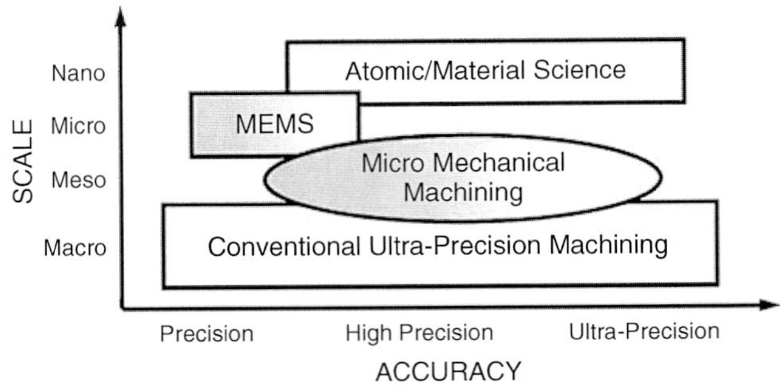

Fig. 17.1 Dimensional size versus accuracy in machining [1].

be necessary. This is the realm of nanotechnology. Although research in this field dates back to Richard P. Feynman's classic talk in 1959, the term nanotechnology was first coined by K. Eric Drexler in 1986 in the book *Engines of Creation*. In the popular press, the term nanotechnology is sometimes used to refer to any sub-micron process, including lithography. Because of this, many scientists are beginning to use the term molecular nanotechnology when talking about true nanotechnology at the molecular level.

Whatisit.com definition: Nanotechnology, or, as it is sometimes called, molecular manufacturing, is a branch of engineering that deals with the design and manufacture of extremely small electronic circuits and mechanical devices built at the molecular level of matter. The Institute of Nanotechnology in the U.K. expresses it as 'science and technology where dimensions and tolerances in the range of 0.1 nanometre (nm) to 100 nm play a critical role'. Nanotechnology is often discussed together with micro-electromechanical systems (MEMS), a subject that usually includes nanotechnology but may also include technologies higher than the molecular level.

Since 2000, the US Government, through the National Nanotechnology Initiative (NNI) has defined nanotechnology to mean working at the atomic, molecular and supermolecular level, in the length scale of approximately 1–100 nm range, in order to understand, create and use materials, devices and systems with fundamentally new properties and functions because of their small structure. This is because in the 10–20 nm range, the physical, chemical and biological characteristics of matter can be significantly altered which lead to the unique properties of nanomaterials and improving product functionality [4].

Nanomanufacturing aims at building material structures, components, devices/machines and systems with nano-scale features in one, two or three dimensions by making use of the following approaches (Fig. 17.2):

- Bottom-up directed self-assembly of nanostructure building blocks from atomic, molecular and supramolecular levels.
- Top-down high resolution processing, ultra-precision (nano-) machining, fragmentation methods (mechanical milling, spark erosion, etc.), lithography (optical, UV, e-beam, scanning probe microscope based with 1–10 nm), positioning assembly. Top-down refers to processes that start with bulk material and removing what is not needed.
- Engineering of molecules and supramolecular systems, molecules as devices 'by design', nano-scale machines.
- Hierarchical integration with larger-scale systems.

It is generally agreed that four generations of nanoproducts will be developed in the period 2000–2020, as shown in the right scheme in Fig. 17.2. These include passive nanostructures, active nanostructures, systems of nanosytems and heterogeneous molecular nanosystems [3, 4]. The first and simplest group utilizes such manufacturing techniques as fragmentation, patterning, restructuring of bulk materials and lithography. Many passive nanostuctures like *nanostructured coatings*, dispersion of nanoparticles, surface nanopattering, *ultra-precision engineering* and bulk materials incorporating nanometer-sized particles such as polymers, metals

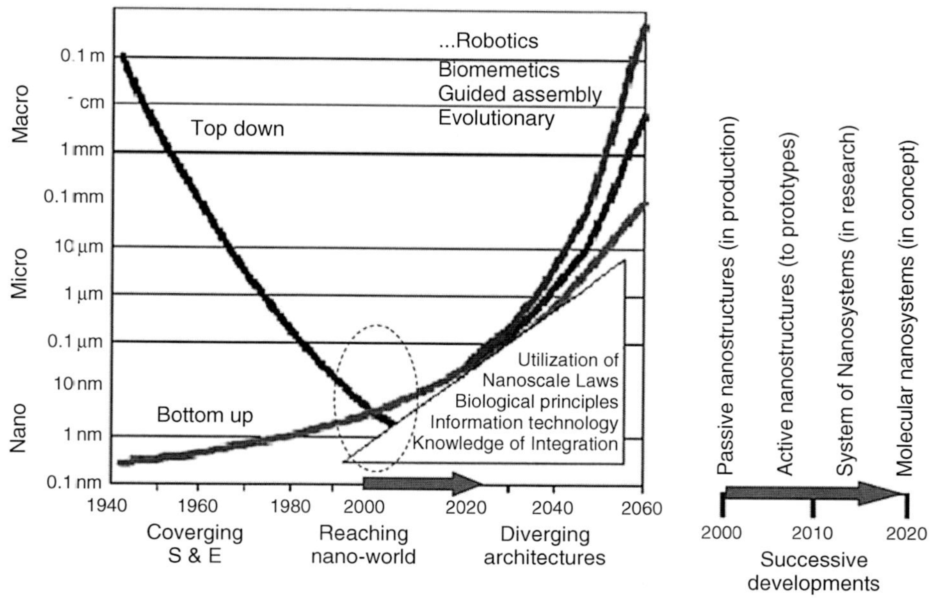

Fig. 17.2 Reaching nano-world and system creation [3].

and ceramics are already manufactured. In this book only some types of passive nanostructures (indicated above by italic) and top-down processes are discussed.

Active nanostructures which change their state in time during operation are expected to emerge over next five years. As an example, this group involves new transistors, targeted drugs, chemicals, actuators, molecular machines and light driven molecular motors. Manufacturing requires understanding of interfaces, field and boundary control, positioning assembly, integration, directed self-assembly, templating and designing new molecules.

The third stage, called systems of nanosystems, involves combining passive nanostructures into simple devices or machines like nanorobots. Its development will be based on new techniques in system engineering, device architecture, integration, multi-scale self-assembly and in-situ processing.

The most advanced generation of products will incorporate heterogeneous molecular nanosystems in which each molecule will have a specific structure and function. Molecules will be used as devices, and fundamentally new functions will emerge from their engineering structures and architectures. Nano-scale machines, subcellular interventions, directed and multi-scale self-assembling and human/machine interface (HMI) can be included in this category.

As an example, Fig. 17.3 compares gears fabricated in different scales, i.e. micro- versus nano-scale, and using passive and active nanosystems. It is well known that gears as a macro-scale mechanism have been used for millennia to transmit mechanical power and information. In the nano-scale the design of gear transmissions is challenged by high surface to volume ratio, many specific physical influences (friction, stiction, surface tension, electrostatic forces, etc.) and numerous hurdles in their manufacturing (repeatability, costs, mass production). At the same time, nature offers robustly producible, effective molecular mechanisms, such as a biomotor for rotational motion shown in Fig. 17.3c.

Figure 17.4a shows a second hand drive mechanism that runs on a watch. It uses the world's smallest plastic wheel made from a carbon nanofibre and plastic resin composite. The wheel of 0.2 mm diameter (Fig. 17.4b) was fabricated with a plastic injection moulding technique and special microparts design and production technology, as well as nanotech-applied technology used in watch manufacturing. The practical application of this technology opens up the possibilities of expansion and enhancement of computers, precision devices, medical devices and vehicle parts.

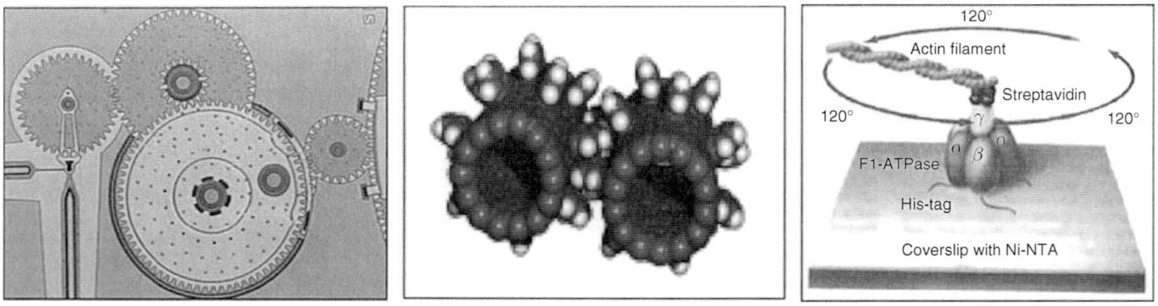

Fig. 17.3 Comparison of micro- and nanoproducts of different generations: (a) micro-scale gear, (b) nano-scale gear and (c) biological nanomotor [5]. (see Plate 22)

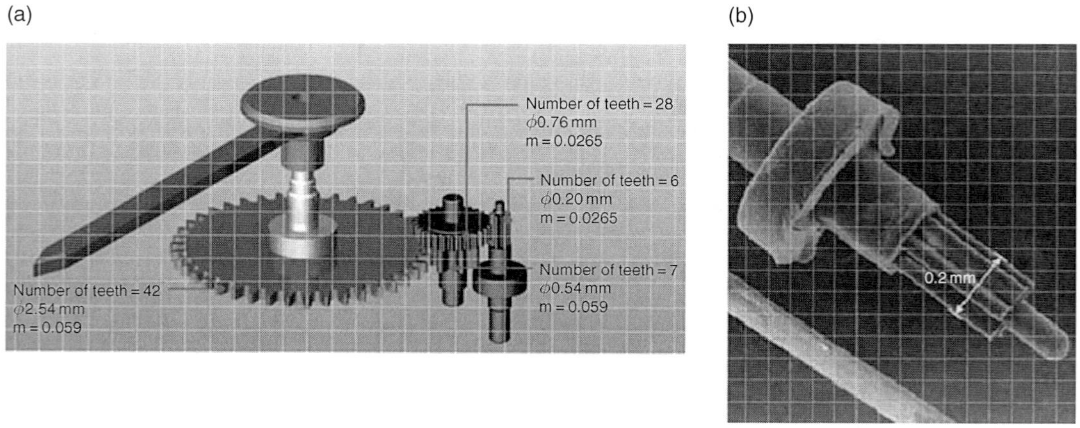

Fig. 17.4 Exemplary nanomanufacturing products: (a) nanogear, (b) world's smallest carbon nanofibre composite-made wheel [6]. (see Plate 23)

Work on nano-scale is currently in progress at universities, major corporations, private starts-up and military and NASA labs (these include self-assembled nanoparticles, nanotubes, nanorings, nanowires, nanocomposites, and nanorobotics). The goals of the research range from biotech applications to nanoassemblers, nanocomputers, and nanochips along with molecular electronics. There are also more advanced scenarios for the potential applications of nanotechnology. The transition of nanotechnology research into manufactured products is limited today to the passive nanostructures as depicted in Fig. 17.2; however, some products entered the market relatively quickly and already have significant impact.

17.2 Ultra-precision Machines and Nano-scale Machining Operations

Almost all of the major machine tool builders in Japan, the USA and Europe offer at least one machine with the capability of using nanotechnology. This was often reflected in the model name: Nano Mycenter, Pyramid Nano, Nanotech, RoboNano and so on. Micro, super, hyper and mega are also used to designate machines in the nanotechnology category. Although vertical machining centres and milling machines are the common machine category described by nanotechnology, various models of lathes, grinders and electrical discharge machines are also given this functional label. A milling machine with nanotechnology capability is likely to have one, or all, of the following features [7]:

- Linear motor drives in the major axes.
- Hydrostatic guideways.
- Some means to compensate for or eliminate sources of heat.
- Very high speed spindles (40 000 rpm or higher).
- Granite base and/or column.
- Laser-based position feedback.
- CNC unit capable of processing and displaying nanometer units.
- Non-contact cutting tool deflection or measurement system.

Machining to submicron accuracy strongly relies on advanced control unit technology. For example, a popular choice for the nanotechnology Fanuc Series 31i CNC features high speed processing to support nanometre interpolation. Nanometre interpolation means that axis moves programmed in micron units are calculated in nanometre units for smoother, more accurate contouring moves. Interpolation cycle time is faster than 1000 blocks of look-ahead for high accuracy contouring with 0.4 millisecond block processing time. High resolution encoders enabling 16 million pulses per revolution also contribute to obtaining ultra-high accuracy and a so-called nano-smoothing feature for five-axis machining. As a result, surfaces produced on the workpiece are visibly smoother, and largely free of cusps and ridges produced with conventional machining. Nanotechnology allows milling, turning and grinding operations to produce mirror-like finishes without polishing.

In turning, the highest spindle speeds ranging from 3000 to 5000 rpm are required for precision finishing cuts to obtain cutting speeds of 120 to 915 m/min (400 to 3000 ft/min) [8]. Moreover, fine stepless feeds are desired to provide gradual entry of the tool into the workpiece, especially at high cutting speeds. For ultra-precision turning such lathes are unsuitable. To achieve desired stiffness new designs of ultra-precision lathes utilize aerostatic bearings for the spindle and hydrostatic bearings for the bedway. The lathe shown in Fig. 17.5 is capable of performing feeds at nanometre levels and very fine depth of cut, which are necessary to

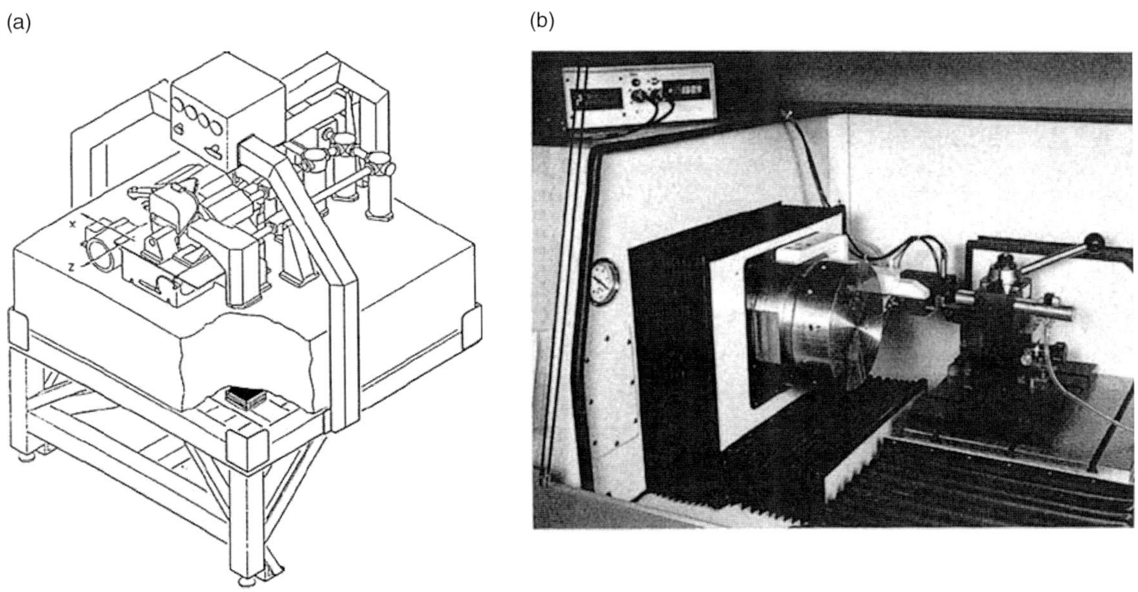

Fig. 17.5 Ultra-precision lathes for ductile-regime machining of germanium and silicon (a) and non-axisymmetrical optics (b) [8].

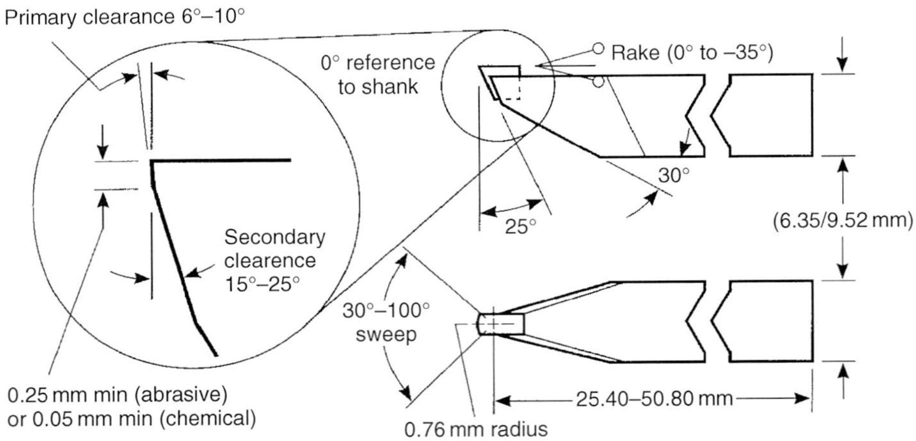

Fig. 17.6 Single-crystal diamond tool used in ultra-precision turning of brittle materials [8].

machine brittle materials such as germanium and silicon in ductile mode (known as ductile machining). The cutting tools used are single-crystal diamonds shown in Fig. 17.6. They are prepared by chemical machining to obtain the cutting edge radius of about 50 nm. After finishing on germanium aspheric lenses, mirror-like surfaces with a 1 nm *Ra* were produced. Precision tolerances for silicon, germanium, aluminium and copper are bilateral of the order of +20 nm for a linear dimension of 20 mm. Ultra-precision tolerances are unilateral from +5 nm (lower) to +10 nm (upper) for a linear dimension of 20 mm [8].

The miniature ultra-precision lathe shown in Fig. 17.7 is capable of producing nanosurface finishes required in ultra-high precision components. The lathe features a small workspace of 20 mm by 20 mm, making it extremely cost effective and suitable for the fabrication of miniature ultra-precision components, e.g. inserts for micro-lenses. Some important technical data concerning drive system and spindle are specified in Table 17.1.

Fig. 17.7 Miniature ultra-precision lathe Microtool Series UPL-2020 [9].

Table 17.1 Characteristic features of ultra-precision lathe from Fig. 17.7.

	Ultra-precision XZ stage unit
Guide way	sliding contact journal with DLC coating
Driving motor (*X*, *Z*)	AC servo motor with gear of 1/33, 30 W, 0–5000 rpm, with 16 bits incremental encoder
Stroke	20 mm (*X*), 20 mm (*Z*)
Resolution	5 nm resolution
	Ultra-precision spindle unit
Bearing	air journal, radial rigidity 24.5 N/μm, axial rigidity 35.3 N/μm, max radial road 61.7 N, max axial road 88.2 N, radial run-out 0.05 μm, axial run-out 0.03 μm, air supply pressure 490 kPa, air consumption 6 l/min
Driving motor	DC coreless motor, 1000–12000 rpm, 300 gf-cm (at 10 000 rpm)
Base	0.5 μm flatness, 500 mm (W) 450 mm (D) 100 mm (H)
Light	special order, white colour LED, 2500 lux

As with turning, ultra-precision milling of mirror-like surfaces is performed by means of single-crystal diamond milling cutters. The structure of an ultra-precision milling machine with the spindle rotating at 23 000 rpm is shown in Fig. 17.8a. When using diamond pseudo-end mills, it is possible to manufacture sculptured surfaces such as toroids and paraboloids (Fig. 17.8b) with a typical feed and depth of cut of 2 μm/rev and 75 μm, respectively.

The Makino Hyper2J ultra-high accuracy micromachining centre (Fig. 17.9a) was created, designed and built to meet the extremely rigorous machining needs in the manufacturing of precise and detailed parts, including: medical instruments, biotechnology products, semiconductor devices, fibre optical

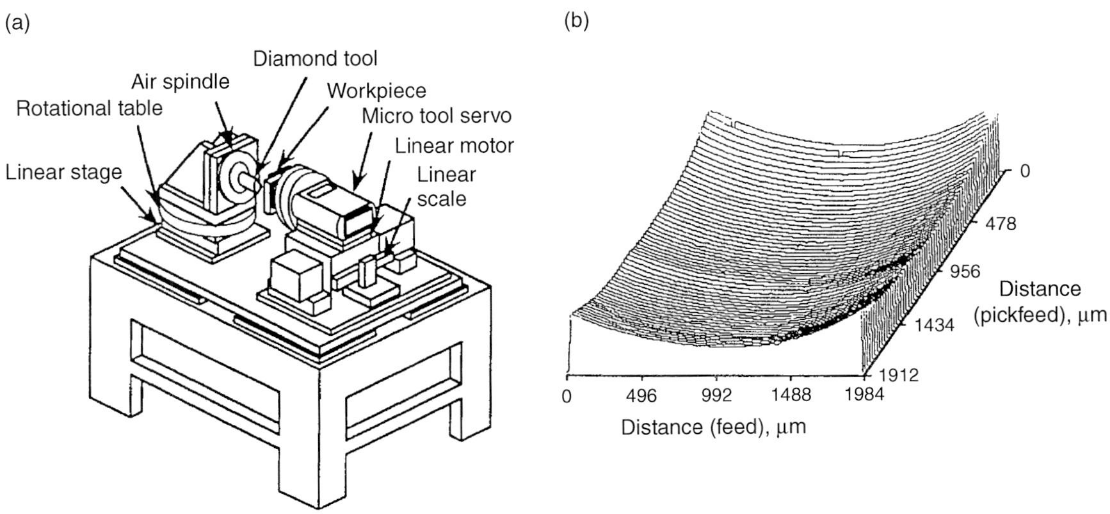

Fig. 17.8 Ultra-precision milling machine (a) for sculptured (paraboloidal (b)) surfaces [8].

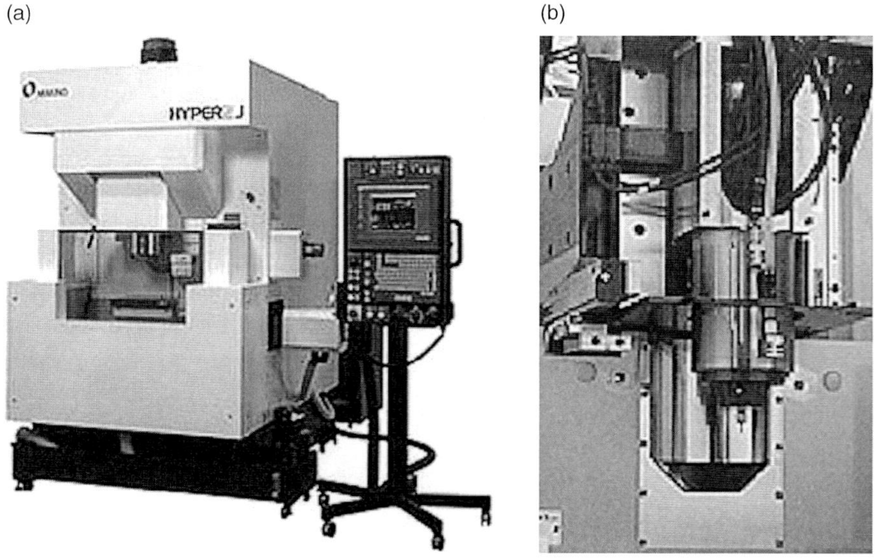

Fig. 17.9 Ultra-high accuracy micromachining centre (a) equipped with 40 000 rpm spindle (b) [10]. Source: Makino.

communications and optical products, such as liquid crystal displays and spherical lenses. The vertically oriented machine achieves accuracies and surface finishes that have been previously unattainable. Each axis is capable of submicrometre movement. Equipped with a 0.5 nm (0.000019") scale feedback (a 2 nm scale feedback system is optionally available) and built to exacting alignments, the micromachining centre provides a positioning accuracy of ±0.3 μm (±0.000012"), with a repeatability accuracy of ±0.2 μm (±0.0000078"). Moreover, a Super Geometric Intelligence (SGI.4) servo control was conceived specifically to produce exceptionally smooth machined surfaces. This is especially important in high feed rate machining of complex 3D shapes involving continuous tiny blocks of NC data. The micromachining centre is by default equipped with a very low vibration, direct-chucking 40 000 rpm spindle, with an optional 170 000 rpm spindle available for ultra-fine machining applications. The direct chucking eliminates a tool-holder or collet, resulting in minimizing rotational vibration. Additionally, the granite base below the machine reduces the influence of floor vibration and temperature changes. Hereby, an exceptional surface finish and extended tool life can be obtained. Both the contact tool length measurement system and a non-contact spindle position measuring device are installed in order to correct the tool tip position and minimize microscopic level differences in machined surfaces. Moreover, the entire machine can be capsulated as a 'thermal chamber' for exacting control of both environmental and machine influences on its submicron accuracy.

The KERN Pyramid Nano five-axis machining centre shown in Fig. 17.10a has a symmetric portal frame allowing maximum rigidity and thermo-symmetric stability. A special material, KERN ARMORITH® with low thermal conductivity, was used to build a main frame and table base as illustrated in Fig. 17.10b. The machining centre was developed for applications that require maximum accuracy and surface finish on larger workpieces from hardened steels (up to 64 HRC), graphite and ceramics. Optionally, 36 000 and 50 000 rpm vector-controlled milling spindles are available. Acceleration is $10\,m/s^2$ (1 g) with a feed rate of 30 m/min. In particular, hydrostatic drives and guideways of the X, Y and Z axes allow maximum surface finishes in nano-precision. Positioning scatter of ±0.3 μm, positioning tolerance of ±0.4 μm, precision on the workpiece of ±1.0 μm, resolution of 0.1 μm and surface finish $Ra = 0.05$ μm are the representative indexes of achievable precision. Automatic measuring by laser of tool length, radius and concentric accuracy, automatic workpiece measuring with a touch probe system and infrared data transmission are basic metrological procedures. As shown in Fig. 17.10b an intelligent temperature management system within ±0.25 K involves a separate

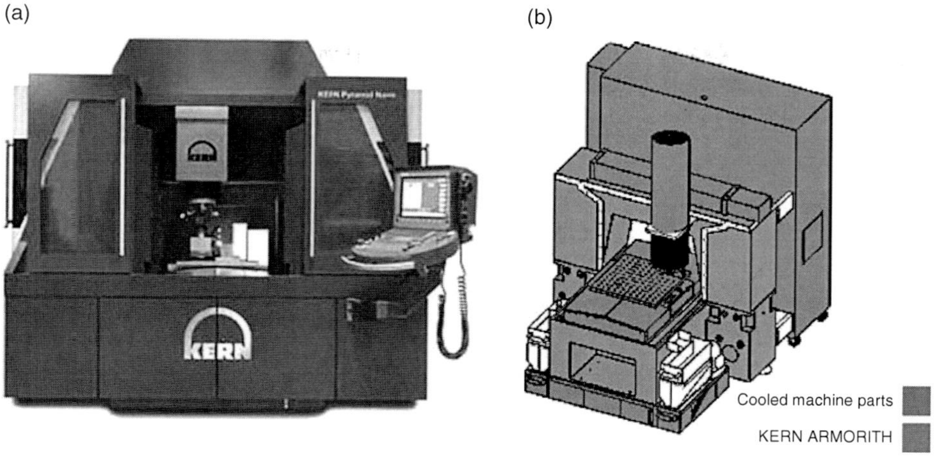

Fig. 17.10 Nanoprecision CNC machining centre (a) and sketch showing cooled zones and parts made of patented KERN ARMORITH® material (b) [11]. Source: KERN.

water cooling system for the milling spindle, hydraulic unit, and the electrical cabinet. This cooling effect was achieved by circulating hydraulic oil flows through machine base, guideways and drives of the axes.

CNC Heidenhain controls are fully equipped with all software packages, such as up to five axes interpolation, pre-programmed cycles for rigid tapping, various drilling and milling cycles, subprogramming, teach-in programming, cylinder barrel programming, graphical simulation, tool administration, tool radius correction, etc.

Figure 17.11a and b show superprecision two-/three-axis lathes with 10 000 rpm spindle and five-axis machining centre with tilting spindle of 8000 rpm, respectively. The chuck lathe is dedicated to the watch industry and high precision engineering, whereas the main application of the machining centre includes aviation, automotive and high precision industries. For the five-axis machining centre relevant resolutions are equal to 0.1 μm for three linear X, Y and Z axes and 0.0001° for B and C axes. As in many similar machine tools, a tool breakage detection system, tool measurement system and inspection probe are installed.

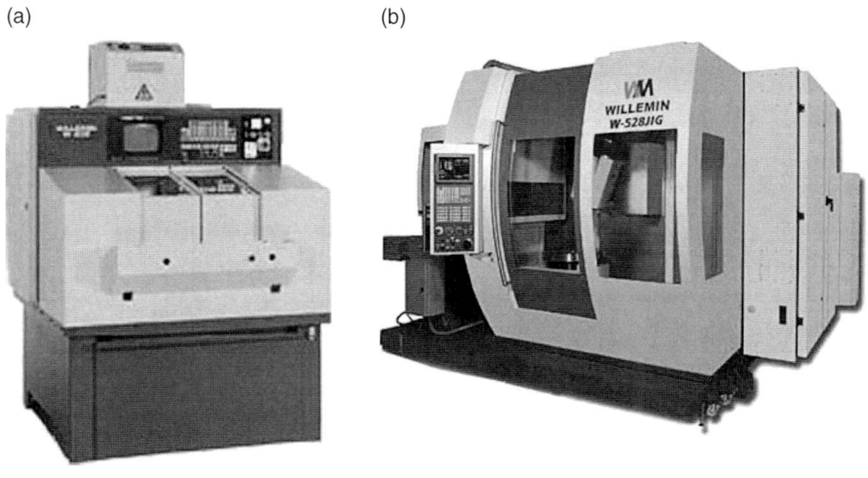

Fig. 17.11 Superprecision two–three-axis chuck lathe (a) and five-axis machining centre (b) [12]. Source: Willemin-Macodel.

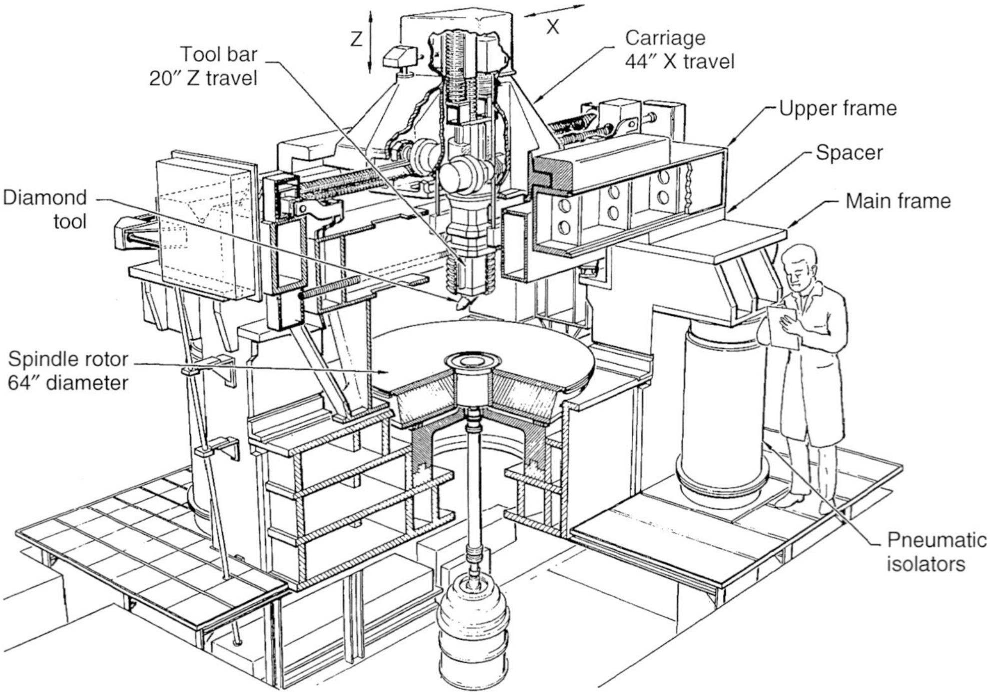

Fig. 17.12 A concept of large optics vertical diamond turning machine [13].

Some specific types of ultra-precision machine tools include the Large Optics Diamond Turning Machine (LODTM) developed at the Lawrence Livermore National Laboratory in the USA, shown in Fig. 17.12, and the Cranfield Nanocentre diamond turning machine (Fig. 17.13) developed in the UK. It is an obvious fact that for such ultra-precision machines thermal effects are the most important factor contributing to dimensional errors and apparent non-repeatability.

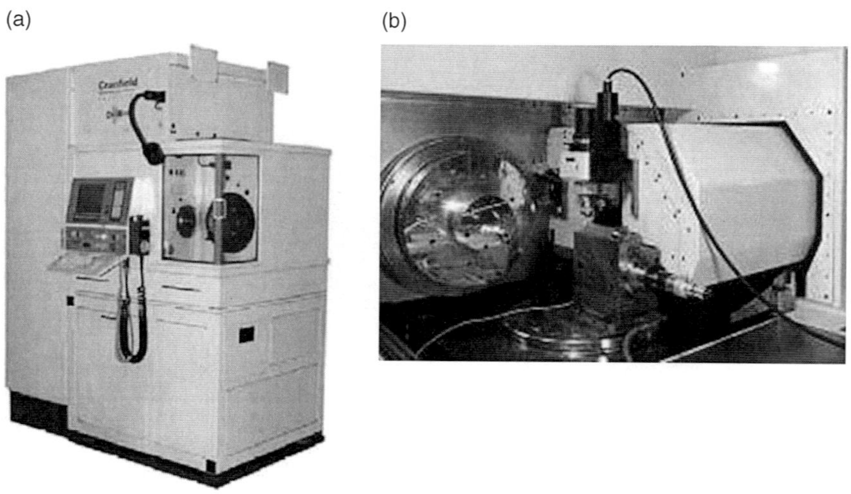

Fig. 17.13 Ultra-high precision diamond turning DeltaTurn 40 (a) and spindle and tool post (b) [15].

Therefore, in the LODTM, which was developed to diamond machine large optical components with a surface finish of the order of 2 to 4 nm *Ra*, temperature was controlled in a specially environmentally controlled room which included a high velocity air shower controlled to 0.005°C [14].

The DeltaTurn machine incorporates novel design features to ensure:

- High static and dynamic stiffness coupled with minimized thermal sensitivity and precise motion control.
- Control of tool wear and thermal drifts.
- Precise control of the relative tool and components positions.

Moreover, a single X–Y grating scale is mounted on one axis, on the plane of the cutting tool and workspindle axis, and a two-axis scanning head mounted on the other axis. To ensure that the workpiece dimensional tolerance of 1 μm on a 70 mm diameter is achieved, an automated part handling device is used to locate each finished component into a dedicated measuring system, which is integral with the machine, allowing 100% inspection. Data for length and diameter measurements is fed to the CNC controller, which can make appropriate tool position offsets. Tool wear and failure is identified by continually measuring cutting forces.

Figure 17.14 shows an ultra-precision grinding machine with a 6000 rpm air spindle and two linear and one rotary axes, labelled Tetraform 'C'. This is a new and radical concept in the design of machine tools and measurement machines which are required to work with ultra-precision accuracy at very high speed [17]. This is achieved through the use of an internally damped space frame with all the loads carried in closed loops. The resulting design gives very high static stiffness coupled with exceptional dynamic stiffness. The Tetraform concept allows the machining of grinding brittle materials to nanometre surface finish with very low levels of subsurface damage, at realistic removal rates. When machining 'optical' surfaces in a 65 HRC tool steel, surface finishes of 4 nm *Ra* were obtained with a 25 μm depth of cut and using a 75 μm grit CBN wheel [16].

An ultra-precision free-form three-axis grinding machine shown in Fig. 17.15 has a built-in surface profilometer. It is a new machine concept where advanced mass production and ultra-precision technologies are combined. Simple, highly symmetrical structures and low mass, high power density and temperature-controlled subsystems are employed using positioning control with 1 nm resolution. A dual head rotary encoder, mounted close to the table face, guarantees rotary movement with a 0.01 arc sec resolution. Moreover, in order to measure the quality of ground surfaces, an in-situ profilometer with 10 nm resolution was built into the machine frame. The accuracy of this in-situ profilometer is better than any comparatively sized coordinate measuring machine. In particular, the BoX® (Big OptiX) machine has been developed in

Fig. 17.14 Ultra-precision grinding machine Tetraform 'C' [14, 16].

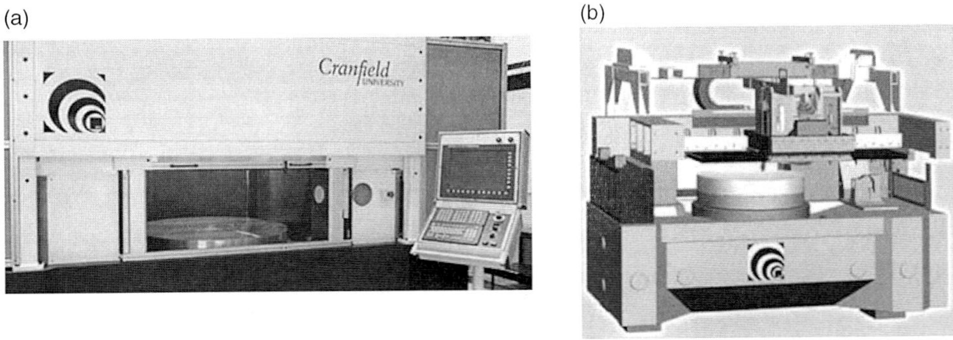

Fig. 17.15 BoX® ultra-precision free-form grinding and measuring machine (a) and relevant FEM model (b) [15].

response to the significant increase in the number of large mirror segments required for the next generation of extra large telescopes (ELTs), such as the James Webb Space Telescope, which will replace the Hubble Telescope at the end of the decade. In general, NURBS interpolation permits the grinding of flats, spheres, aspheres, off-axis forms and truly free-form surfaces.

Advances in the technology for ultra-precision processing require developing the nanostructuring technology which utilizes scanning probe microscopy (SPM), particularly scanning tunnelling microscopy (STM) and atomic force microscopy (AFM) [18]. SPM has been used not only for measurement of surface topography with atomic resolution but also for nanostructuring in which the probing tip can be utilized as a microtool with nano-scale movement. This technology enables modification of a solid surface topography by various methods, such as manipulation of single atoms, transfer of molecules, dip nanolithography and nanorubbing techniques.

In SPM-based direct mechanical nanomachining methods, indication and cutting of solid materials, such as silicon, gallium, arsenide and some other metal surfaces are possible. Unfortunately, in the case of hard materials using a conventional probing tip as a cutting tool is limited to only several nanometres in depth of cut. Figure 17.16b presents a novel technique of the nanomachining of single-crystal silicon using a pyramidal diamond tip bonded to a nanomachining cantilever (Fig. 17.16a). The AMF cantilever was fabricated based on the combination of photolithography and hot-filament chemical vapour deposition (HF-CVD) technique. As shown in Fig. 17.16b a silicon wafer was mounted with its <110> orientation along the X and Y axes of the AFM piezo-scanner. The machining direction was perpendicular to the silicon lever-length direction,

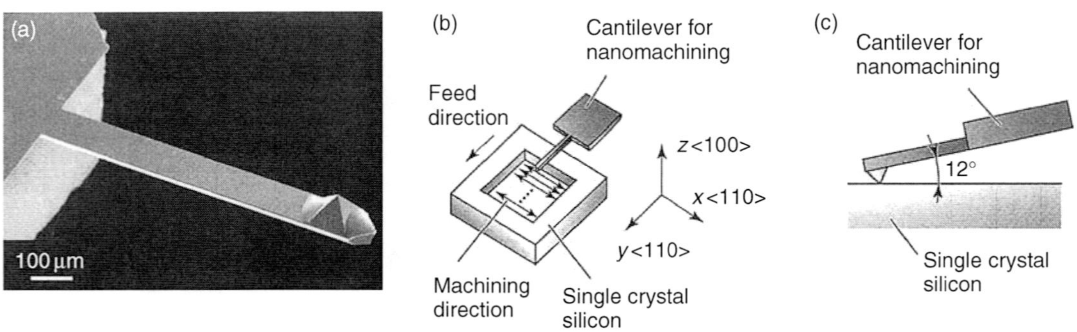

Fig. 17.16 Nanomachining of single-crystal silicon using AFM: (a) SEM image of diamond-tip cantilever, (b) scheme of nanomachining [18].

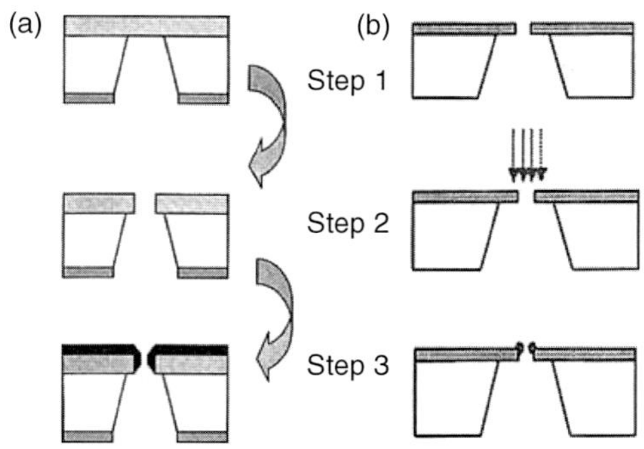

Fig. 17.17 The procedure to make nanoholes (a) and reduce nanohole size (b) [19].

and parallel to the <110> direction of the silicon surface. In addition, the cantilever was tilted at 12° (Fig. 17.16c) in order to avoid contact of its base with the sample. As reported [18], optically smooth surfaces with *Ra* parameters about 0.5 nm were generated, but after a longer machining distance of 500 mm.

The next bottleneck in nanomachining is making single-digit nanometer holes in a controllable way in different materials, with different size distribution and hole shapes. Basically, the top-down methods, such as electro-beam (EB), focused ion beam (FIB), X-ray, deep UV, atomic force microscope, etc., and various lithography and etching methods can be applicable to fabricate nanometer size holes [16, 19, 20]. The general method of fabrication of the single-digit nanoholes is illustrated in Fig. 17.17a. As indicated in Fig. 17.17a it includes three steps, as follows:

1. Preparation of free-standing nano-scale thin film using the method of epitaxial deposition and electro-chemical etching.
2. Making submicro holes of 0.2–0.02 μm by using a Fib or AFM probe with a nano pin as a puncher.
3. Reducing the hole diameter to less than 10 nm by epitaxial deposition, FIB or e-beam induced deposition, micro-coating or thermo-assisted shrinkage.

On the other hand, reducing the hole diameter (Fig. 17.17b) requires the following operations:

1. Adjustment of an electro-beam to a diameter of 100 nm and then pointing the beam to make a pre-hole of about 50 nm.
2. Focusing EB in an SEM to deposit carbonaceous materials on the edge of the hole.
3. Reduction of the hole diameter to the desired size of about 2 nm.

17.3 Examples of Nano-products

The ultra-precision and nanotechnology processes include [16, 20]:

- Single-point diamond and CBN turning.
- (Multi-point) fixed abrasive processes, e.g. diamond and CBN grinding, honing, belt polishing.
- Free abrasive (erosion) processes, e.g. lapping, polishing, float polishing, CMP (mechano-chemical, chemo-mechanical processes).
- Chemical (corrosion) processes, e.g. etch machining (perhaps as part of photo- or electro-lithography).
- Biological processes, e.g. chemolithotrophic bacteria etch processing of inorganic materials.
- Energy beam processes (removal, accretion and surface transformation processes) including photolithography, X-ray lithography, micro-EDM, electron beam machining (EBM), laser beam machining (LBM), excimer and femtosecond lasers, scanning tunnelling microscopy (STM) and atomic force microscopy (AFM).

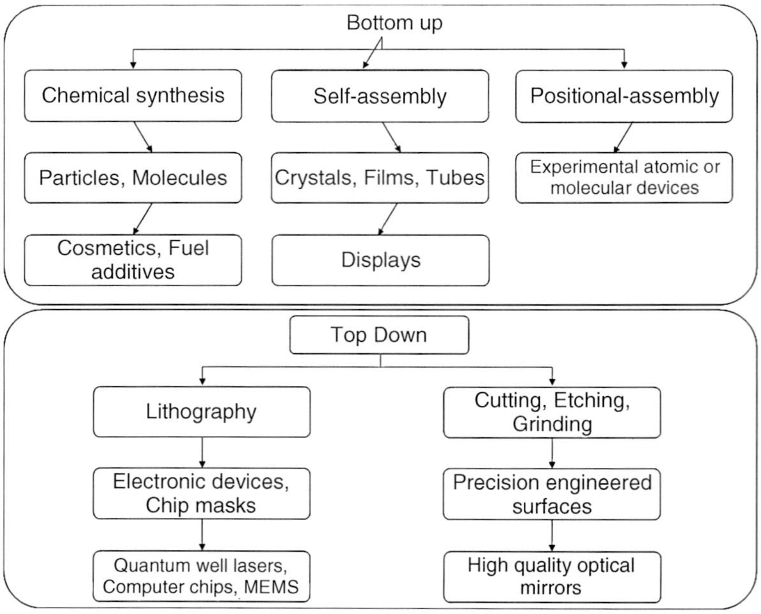

Fig. 17.18 A range of nano-manufacturing techniques applied in bottom-up and top-down approaches [21].

Figure 17.18 specifies some important nano-manufacturing techniques grouped for both bottom-up and top-down strategies.

Some examples of ultra-precision diamond turning process on aluminum and silicon components are shown in Fig. 17.19. By applying ultrasonic vibration to a single-crystal diamond tool, a ductile regime in the machining of brittle glasses (soda-lime glass, quartz glass, fused silica) can be obtained even for 2 μm depth of cut which results in a surface roughness *Ra* of 100 nm [16, 22, 23].

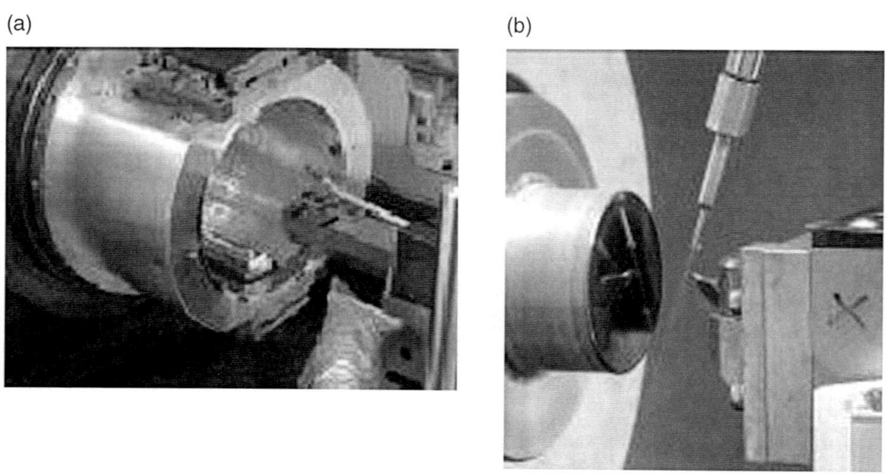

Fig. 17.19 Examples of diamond turning on the machine from Fig. 17.13 [15]: (a) diamond turned aluminium re-imager components of James Webb Space Telescope, (b) diamond turned silicon optical component on the Moore 'Nion' machine.

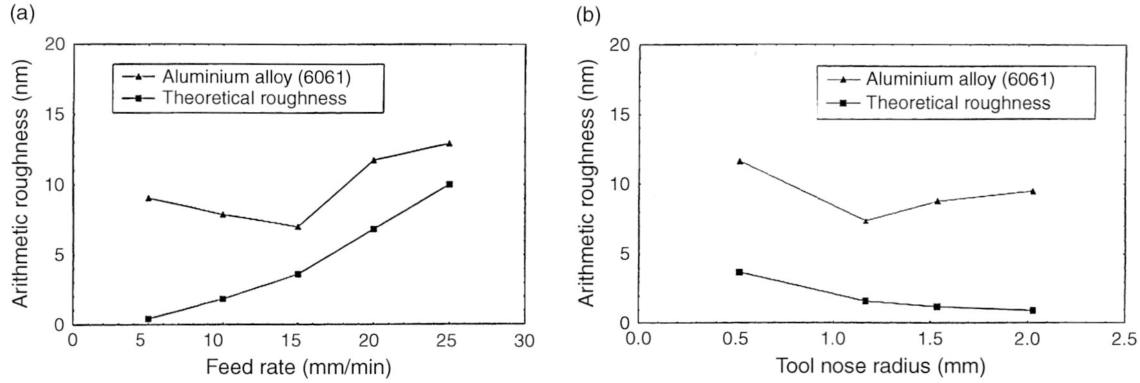

Fig. 17.20 Nanosurface roughness in single-point diamond turning of 6061 aluminium alloy [23]: (a) influence of feed rate at spindle speed of 2000 rpm, 2 μm depth of cut and tool nose radius of 0.5 mm, (b) influence of tool nose radius at spindle speed of 2000 rpm, feed rate of 75 μm/rev and 2 μm depth of cut.

It was documented [23] that in single-point diamond turning (SPDT) the predominant factors influencing surface finish are: spindle speed, feed rate (Fig. 17.20a), tool nose radius (Fig. 17.20b), relative tool/work vibration, tool interface and material swelling. In particular, SPDT performed under a finer feed rate, a larger tool nose radius, a higher spindle speed and a greater amplitude of vibration and is more susceptible to the effect of tool interference.

Optical quality surfaces, comparable to those generated in ultra-precison diamond turning, produced on high alloy refractory and stainless steels can be obtained by a combination of ultra-precision CBN turning, ultra-precision burnishing, performing micro-deformation of the turned surfaces, and oriented polishing [24]. For example, after ultra-precision turning of DIN 1.4301 austenitic stainless steel with 0.5 μm grain-sized single-crystal CBN tools, the *Ra* parameter of 50–60 nm was achieved, and further it was reduced to 20–25 nm and 10–20 nm after burnishing and polishing with the paste containing 0.25 μm size grains, respectively. Moreover, sufficient reflectiveness and imaging ability of the surfaces were provided.

Diamond turning has been shown to be an appropriate technique for the production of germanium IR optics, which have aspheric or 'hybrid' diffraction surfaces [14]. Figure 17.21a shows a diamond turned germanium test piece of <111> orientation produced using an in-feed of 4 μm/rev, on which both ductile and brittle regions can be identified as shown in Fig. 17.21b. In this case, surface finish of around 2 nm *Ra* can

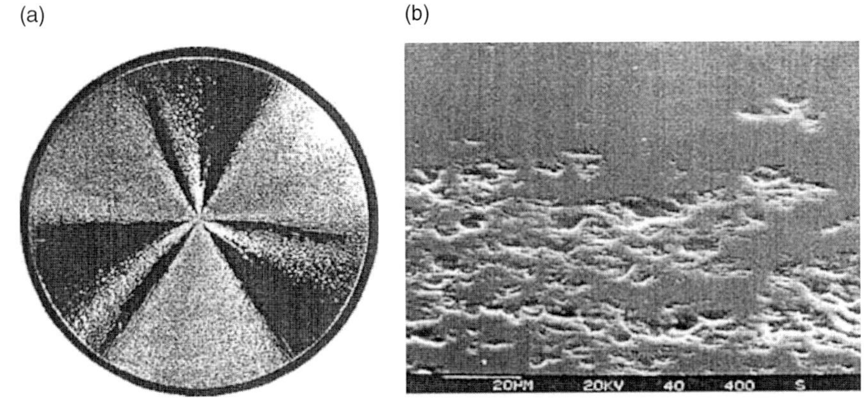

Fig. 17.21 Diamond turned germanium (a) and detail showing the boundary between ductile and brittle regions [14].

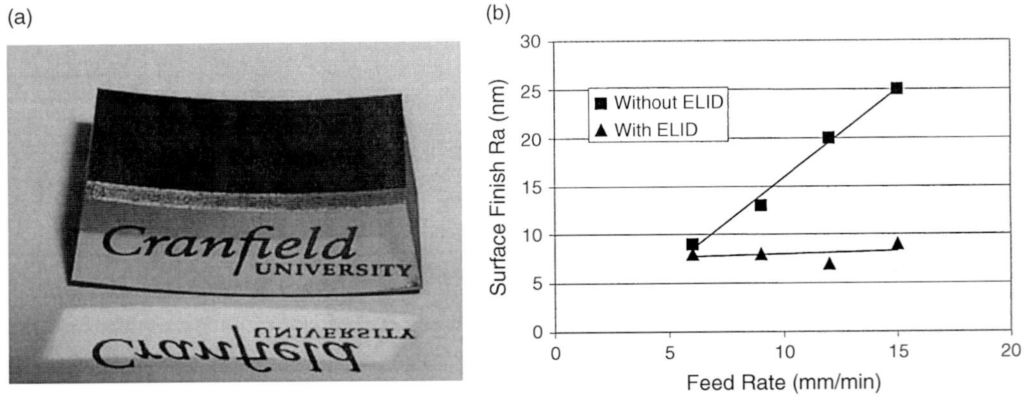

Fig. 17.22 Example of an optical quality surface produced in a single grinding cycle on Tetraform 'C' (a) and variation in surface finish produced in a CBN grinding with and without ELID [17].

be achieved with little evidence of surface fracture. In order to avoid the brittle fracture occurrence, resulting mainly from tool wear, diamond grinding with electrolytic in-process dressing (ELID) of a metal bond CBN wheel was applied. Such improvement allows obtaining optical quality, defect-free surfaces with a surface finish of 1 nm *Ra*. As depicted in Fig. 17.22 this effect can also be visibly observed when grinding with ELID M50 hardened bearing steel using a 75 μm CBN wheel, 25 μm depth of cut and 39 m/s wheel speed.

For ultra-precision finishing of optics and brittle materials such as optical glass BK7, Zerodur and single crystal silicon a nano-abrasive machining process, which is capable of obtaining a few nanometres form accuracy by corrective figuring, was proposed as shown in Fig. 17.23a.

This process is similar to water jet abrasive machining or liquid honing but by controlling the collision (erosion) energy and the collision angle it was possible to decrease the abrasion rate to a few nanometres per minute and, as a result, to create ductile mode abrasion with a few nanometres surface roughness. By applying CNC-controlled circular motion machining, in which the scanning motion is correlated with local form accuracy by a special NC programme, with parameters specified in Fig. 17.22, the surfaces on optical glass with a flatness of $PV = 30$ nm and surface roughness of 1.5 nm *Ra* were produced.

A number of nano-products including single nano-components and both top-down and bottom-up types of NEMS (Nano-Electro-Mechanical-Systems) produced by employing different ultra-precision techniques are shown in Figs 17.24 and 17.25.

The rotor (Fig. 17.24a) is 100 microns in diameter and 7 microns in thickness. It rotates up to 10 000 rpm. A bimorph thermal actuator (Fig. 17.24b) was fabricated using two types of polyimides with different thermal expansion coefficients and a metallic micro-heater in between them. The key concept is to coordinate simple motions of many micro-actuators in order to perform a task. The dimensions of the cantilever were 500 micrometres in length, 100 micrometres in width, and 6 micrometres in thickness. DNA molecules forming DNA tweezers (Fig. 17.24d) were attracted between a pair of sharp tips having radius around 10 nm. The molecules attached to tips were coated with aluminium. Therefore, it was possible to retrieve a molecular bundle from water and to observe it by electron microscope. Bio-molecular motors (Fig. 17.24e), such as actin-myosin, kinesin-tubulin, and FoF1-ATPase, can work as a nano-actuator. This figure shows transportation of a kinesin (motor molecule) coated micro- or nanostructure on the patterned rail molecule (actin or tubulin) on a glass substrate. A micro-tooth (Fig. 17.24f) with 20 micron height and 10 micron width was formed by ICP-RIE process.

The image (Fig. 17.25a) is a colourized scanning electron micrograph (SEM) of a high aspect ratio MEMS device fabricated out of single-crystal silicon by the MEMS and Nanotechnology Exchange fabrication network. The device itself is made from a silicon-on-insulator (SOI) wafer where a relatively thick single-crystal silicon device layer (shown in yellow and blue in the image) has been etched into a complex micromechanical element that can move under force. The underlying area, shown in red in the image,

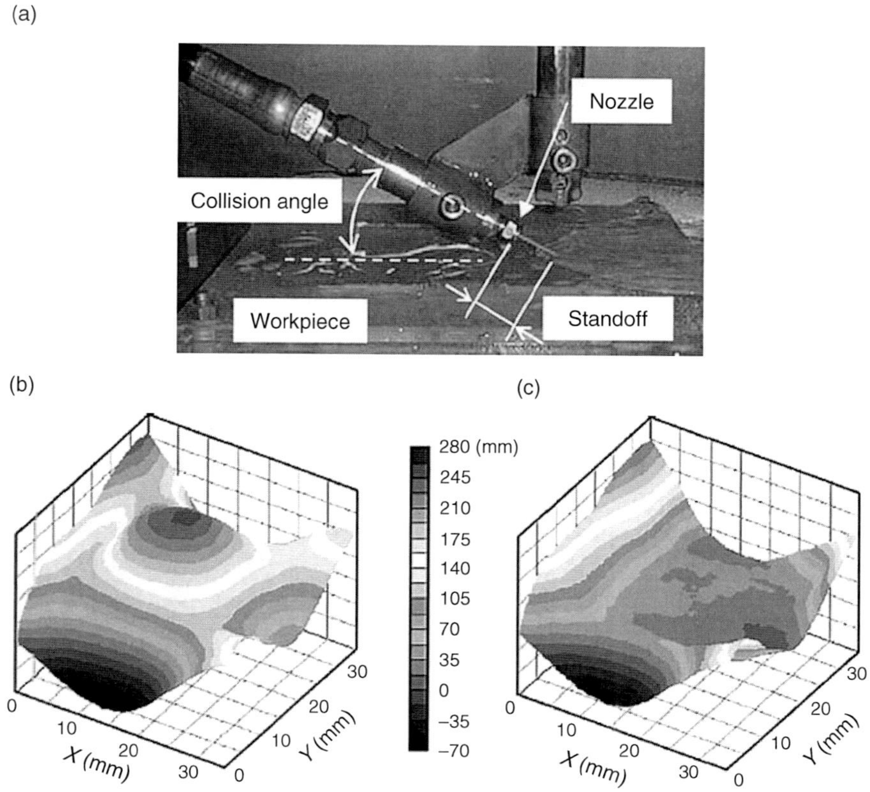

Fig. 17.23 Setup of nozzle and workpiece in nano-abrasion machining (a) and experimental result of local material removal in corrective figuring (b) and (c) using abrasive grains of 0.6 μm size, nozzle of 1 mm diameter, 4 MPa ejecting pressure and 20 mm stand-off, collision angle of 90° and 1 mm radius of circular motion [25].

is the SOI handle wafer which also has been etched to form a circular hole completely through the substrate. This image demonstrates the complexity of micromechanical devices that can be realized using the MEMS and Nanotechnology Exchange.

This eccentric image (Fig. 17.25b) is a colourized scanning electron micrograph (SEM) of mechanical gear which is smaller than a human hair. The gear was made of polycrystalline silicon using surface micromachining techniques and is no longer attached to the substrate. However, the gear did eventually lodge onto an electrical bond pad (shown in green) due to surface forces. This image illustrates the extraordinary capability of MEMS technology to enable the realization of extremely small mechanical elements. This device was made through the MEMS and Nanotechnology Exchange fabrication network.

The image (Fig. 17.25c) is a colourized scanning electron micrograph (SEM) of a salient-pole electrostatically actuated micromotor made from polycrystalline silicon using surface micromachining techniques. The central rotating element of the motor (e.g. the rotor) is the circular structure in the middle (shown in blue) held to the substrate by the central bearing (which is shown in red). Properly phased voltage potentials are placed on the motor stators (typically 120 degrees advanced in phase sequentially around the stators), which are equally spaced around the perimeter of the rotor (also shown in blue) and these applied voltages on the stators cause the central rotor to turn around the bearing at extremely high angular velocities. This device was made via the MEMS and Nanotechnology Exchange fabrication network.

The image (Fig. 17.25d) is a colorized scanning electron micrograph (SEM) of microstructure made using extremely short wavelength exposure techniques of a thick radiation sensitive polymer layer which was spun onto the substrate and developed after exposure to reveal an extraordinarily high aspect ratio microstructure.

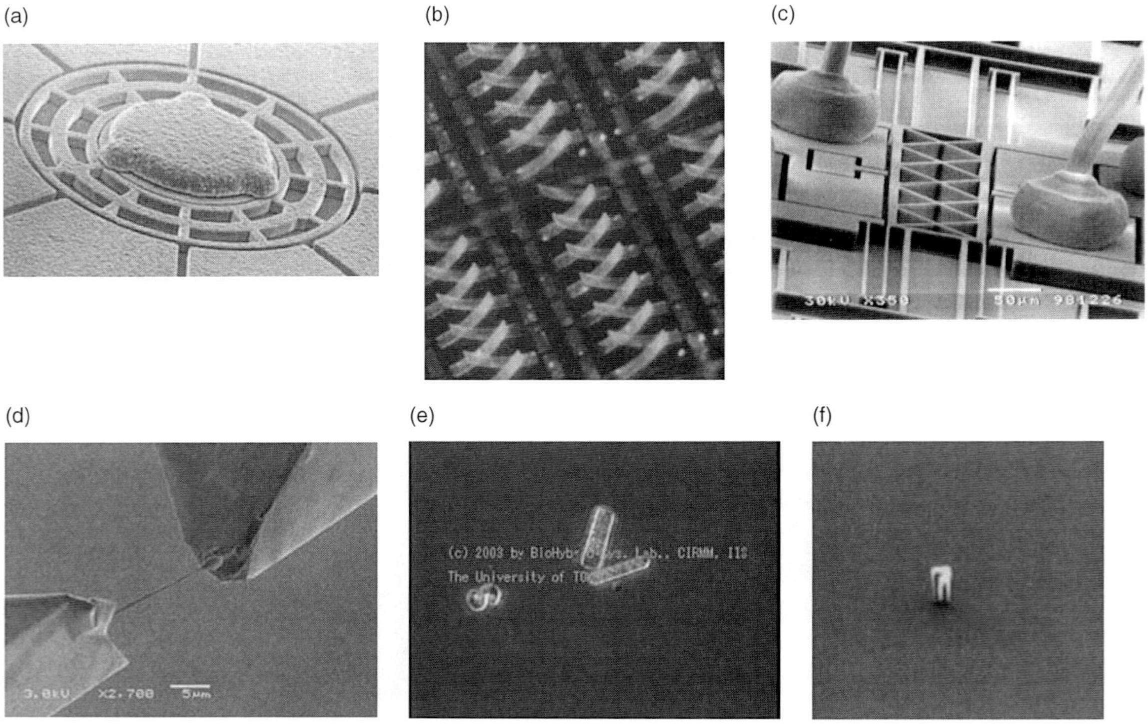

Fig. 17.24 Examples of nanoproducts: (a) micro-motor, (b) ciliary motion system, (c) micro-electromechanical digital–analogue converter (MEMDAC), (d) DNA tweezers, (e) biomolecular motors, (f) micro-tooth [26].

The central post (shown in orange) is hundreds of microns high and only a few microns in diameter. The surrounding angular-shaped structures are also a few microns wide, separated from one another by a few microns, and are also hundreds of microns high. This structure was also made by employing the MEMS and Nanotechnology Exchange fabrication network.

Nanotechnologies can be regarded as interdisciplinary and are primarily enabling technologies. They will pervade all sectors of industry and have a significant impact on the global economy. Nearly all products will have some nanocomponents embedded in them. They already exist in IT and communications products and are rapidly extending into the consumer and healthcare products. Nanotechnologies will produce new manufacturing paradigms that will require forward thinking of industry and government into the building of appropriate infrastructures, which can maximize the benefits and opportunities offered by the technology. It is evident that the commercialization on micro- and nanotechnologies is necessary to bridge the gap between technology providers, users and customers. The realization of nanoproducts produced using top-down processes follows a similar commercialization path as microproducts. On the other hand, this traditional path will have to be significantly modified for nanoproducts produced using bottom-up processes.

There are no comprehensive market surveys dedicated to nanotechnology products although many of them exist for micro-systems and micro-products. The fastest growth will be for nanomaterials. Estimates predict an annual growth rate for nanoporous materials, nanocomposites, thin films and coatings to be about 15% [21].

Table 17.2 gives an indication of the timescale for the evolution of new nanoproducts and systems over next 25 years based on current developments. It is estimated that there are over 500 major companies worldwide now involved in producing nanoproducts either as materials, components or embedded systems. Some prognoses have shown that the disruptive nature of nanotechnology can radically alter the incremental path to commercialization and irreversibly change the existing manufacturing methodology.

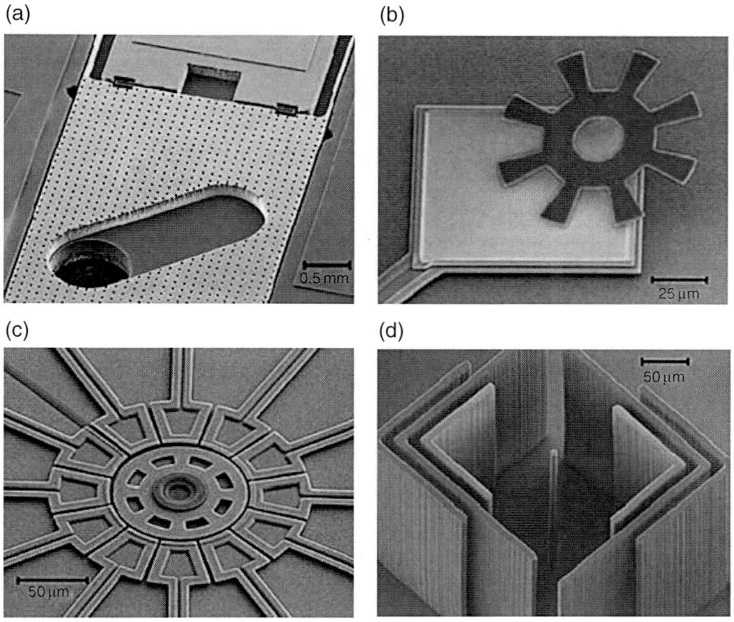

Fig. 17.25 Examples of nanoproducts: (a) a high-aspect ratio MEMS device, (b) SEM of mechanical gear, (c) SEM of a salient-pole electrostatically actuated micromotor, (d) SEM of high-aspect ratio microstructure with central post [27]. (see Plate 24)

Table 17.2 Evolution of commercially available nanoproducts [21].

Segment	*Now*	*1–2 years*	*1–5 years*	*20–25 years*
Electronics	High density microchips Fast high density computers Communication	HD displays Minute disc drives Integrated systems Smart hand-held devices	Nanochips Organic Semiconductors HD memory chips	Molecular high performance computers Self-assembly devices
Materials	Catalysts Sunscreens Cosmetics Coatings	Carbon nanotubes Structures universally used	Fuel cells Anticorrosion coatings	Precoded protein Catalysts Designer catalysts
Life sciences	Biosensors Chemical sensors	Biochips Lab-on-chip	Biochips for proteomics Drug delivery systems	Customized design for diagnostics
Sensor medicine	Self-monitoring devices	Advanced devices integrated with RF systems	Designer drugs Device implants Patient monitoring	Robotic surgery Self-diagnostics and repair
Instrumentation	Smart systems Personal devices Cell phones, etc.	Environmental monitoring systems Integrated systems	Intelligent monitoring Advanced smart systems	Robotic surgery Global monitoring Human/machine interfaces

17.4 NANOMETROLOGY

Many manufacturing processes in nanotechnology operate at the atomic level. In order to achieve adequate control of the motion of systems that are used in such processes, for example for atomic manipulation, dimensions and displacements need to be known at the atomic level or even below. In general, nanotechnology processes need to be controlled for surface integrity and dimensional accuracy, which constitutes the new field of *nanometrology*. Atomic force microscopy (Fig. 17.26) is the latest device developed for that reason but laser and X-ray interferometers are also used in nanometrology [20].

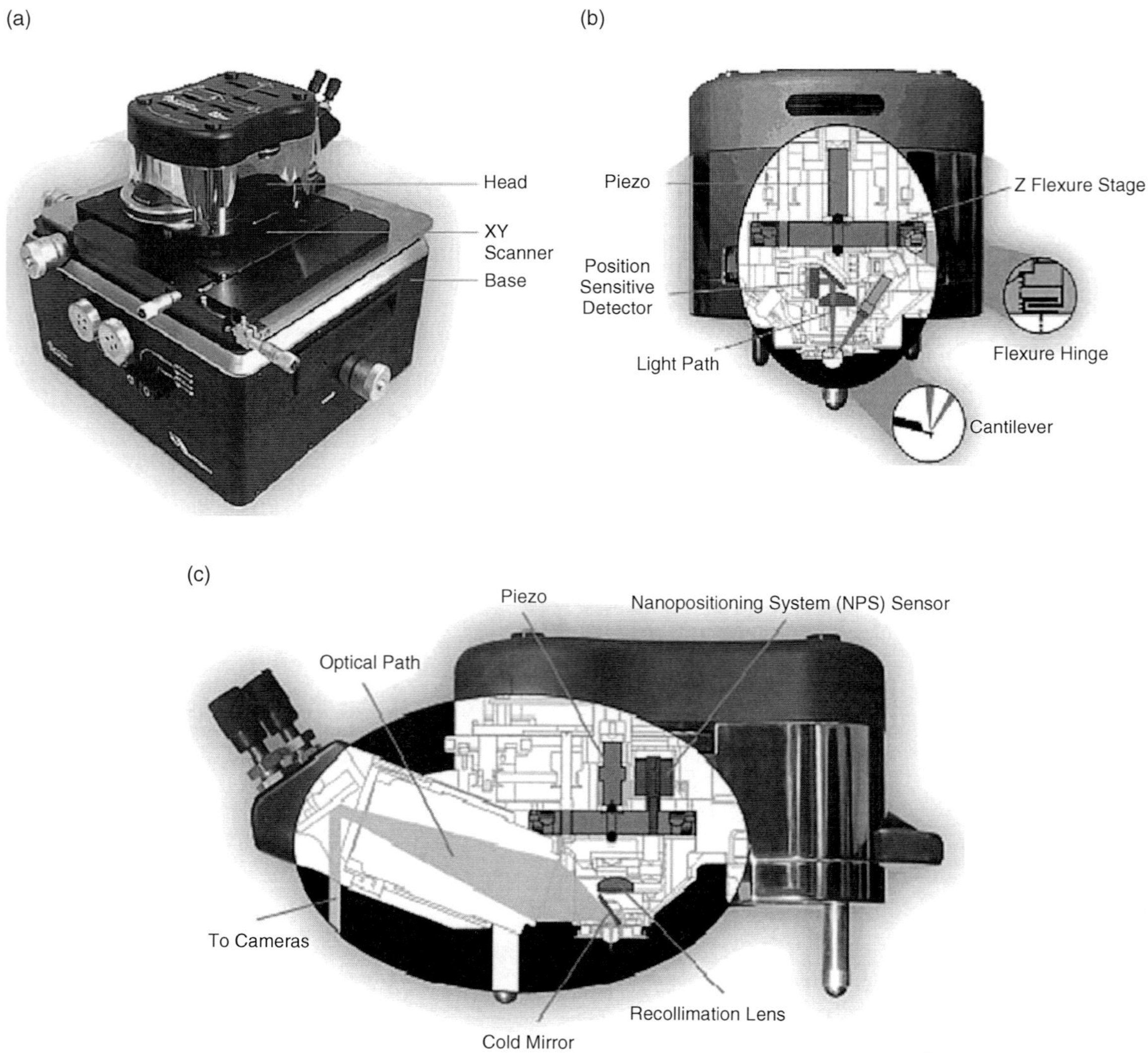

Fig. 17.26 Atomic force microscope (AFM): (a) general view, (b) head, (c) XY scanner. Source: Asylum Research [28]. (see Plate 25)

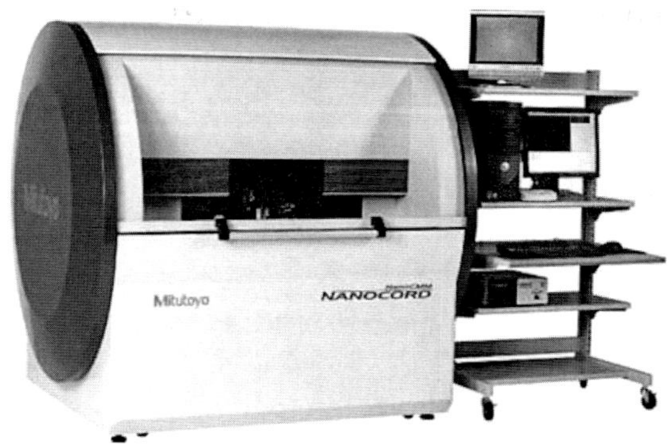

Fig. 17.27 M-NanoCoord minute form measuring system [29]. Source: Mitutoyo.

There are basically two methods of achieving the accuracy of position and displacement at the atomic level [16]. First, by using a laser interferometer as an ultra-sensitive displacement transducer and subdividing the fringes to an extremely fine level. The problem in the uncertainties of subdivision of optical fringes can be overcome by using an X-ray interferometer. Using the diffraction properties of X-rays in a silicon crystal an interferometer can be constructed with a fringe spacing of about 0.2 nm, but displacement measurements are very slow. A specific area of nanometrology is the form measurement, e.g. the roundness or the flatness of the machined workpiece. This kind of metrology is very demanding and difficult to achieve, because there are many sources of uncertainty involved, where the measuring of roughness, dimensions and shape is essential for the functionality of the structure.

The measurements of surface texture on supersmooth surfaces can be performed using a wide range of scanning probe microscopes (SPMs), scanning tunnelling microscopes (STMs) and atomic force microscopes (AFMs). These types of probe are essential to enable the coordinate system of the component under test to be linked to the coordinate system of the measuring instrument. Unfortunately, such probing systems are essential for qualitative measurements. A recent trend has been for special instrumentation to be developed for extending the measuring range of one axis of an AFM which normally has a small, but cubical, working volume. With form measurements at the nanometre level of accuracy special techniques that enable the error of the component to be separate from the error of the measuring system can be used to great effect. A newly developed form measuring system in nano-scale is shown in Fig. 17.27.

The AFM shown in Fig. 17.26 was designed with sensored, 90 μm, closed-loop positioning of scan axes, with sensor noise less than 0.6 nm average deviation in a 0.1 Hz–1 kHz bandwidth (less than 0.3 nm and 0.06 nm for Z axis and Z height) and sensor non-linearity less than 0.5% [29]. It has built-in advanced features such as nanolithography and 3D rendering to display and visualize 3D images in real time while both scanning and off-line processing. The head (Fig. 17.26b) has a sensored optical lever with diffraction limited optics and a low coherence light source which virtually eliminates interference artifacts. The nanopositioning system (NPS^{TM}) for the Z axis provides precise measurements of the cantilever position for accurate force and topography measurements. The AFP uses a flexured scanner (Fig. 17.26c) and NPS sensors which measure the exact position of X and Y axes. They correct for hysteresis and creep, providing flat scans and the ability to accurately zoom and offset with one mouse click.

The measuring system shown in Fig. 17.27 is able to accurately measure minute forms such as MEMS products, electro-devices, precision dies and moulds for subminiature electronic components, and aspherical lenses and microscopic holes. The following features can be distinguished:

- A newly developed ultra-high accuracy main unit.
- A vision probe as a standard accessory.
- The LI-HU (laser probe) is especially suitable for non-contact, high speed measurement of minute forms.

- The CHR Probe (confocal probe) is especially suitable for measuring transparent workpieces and highly inclined workpieces.
- The Minute Form Probe (ultra-low measuring force, contact-type probe) is especially suitable for non-destructive measurement of soft workpieces.
- The UMAP 100 touch signal probe with a micro-stylus as small as 15 nm, especially suitable for measuring micro-holes.

It should be noticed that the resolution of these various accessories is given in fractions of a nanometre. As reported in section 16.5 the F25 coordinate measuring machine from Carl Zeiss IMT (Fig. 16.35) is also capable of measuring submicron/nanometre features. This ability can be based on the facts that the touch scanning sensor with the resolution of 0.001 μm (1 nm) and ultra-precise, incremental length measurement systems with glass-ceramic scales with a resolution of 0.0078 μm (7.8 nm) were installed.

References

[1] J. Chae *et al.*, Investigation of micro-machining operations, *Int. J. Mach. Tools Manuf.*, Vol. 46 (2006), 313–332.
[2] Nano-definition, www.Nanoindustries.com/define.html.
[3] D. Keenan, New frontiers in nanomanufacturing, http://thenanotechnologygroup.org.
[4] W. Ahmed, M. Jackson, Nanomanufacturing – the future?, *Int. J. Nanomanuf.*, Vol. 1, No. 1, www.inderscience.com/ijnm.
[5] Ch.C. Doumandis, Nano-manufacturing: technical advances, research challenges and future directions, www.isnm1006.org/forms/NanoManufacturing_review.pdf.
[6] Second Hand Drive Mechanism, Kitagawa Industries Co., Ltd.
[7] M. Albert, Japan International Machine Tool Fair (JIMTOF) puts 'nanotechnology' under magnification, www.mmsonline.com/articles/010506.html.
[8] *ASM Handbook, Surface Engineering*, Vol. 5, ASM International, Materials Park, 1994.
[9] Miniature ultra-precision lathe Microtool Series UPL-2020, www.microlution-inc.com.
[10] The Makino Hyper2J ultra-high accuracy micromachining center, www.makino.com.
[11] Kern Pyramid Nanorázisions-CNC-Bearbeitungszentrum, www.kern-microtechnic.com.
[12] Super precision machine tools, www.willemin-macodel.com.
[13] Edited by R.L. Shell and E.L. Hall, *Handbook of Industrial Automation*, Marcel Dekker, New York, 2000.
[14] J. Corbett, D.J. Stephenson, The control of surface integrity by precision machining and machine design, *Proceedings of 1st Int. Congress of Precision Machining*, Usti nad Labem, Czech Republik, 2001, pp. 31–43.
[15] http://www.cranfield.ac.uk/sas/materials/precisionengineering/facilities.htm.
[16] J. Corbett, P.A. McKeown, G.N. Peggs, R. Whatmore, Nanotechnology: international developments and emerging products, *CIRP Annals*, Vol. 49/2 (2000), 523–545.
[17] D.J. Stephenson, D. Veselovac, S. Manley, J. Corbett, Ultra-precision grinding of hard steels, *Prec. Eng.*, Vol. 25 (2001), 336–345.
[18] N. Kawasegi, N. Takano, D. Oka *et al.*, Nanomachining of silicon surface using atomic force microscope with diamond tip, *J. Manuf. Sci. Engng*, Vol. 128 (2006), 723–729.
[19] J. Zhou, G. Yang, Nanohole fabrication using FIB, EB and AFM for biomedical applications, *Int. J. Prec. Eng. Manuf.*, Vol. 7, No. 4 (2006), 18–22.
[20] A.G. Mamalis, Recent advances in nanotechnology, *J. Mater. Proc. Technol.*, Vol. 181 (2007), 52–58.
[21] D. Tolfree, Commercialising nanotechnology concepts-products-markets, *Int. J. Nanomanuf.*, Vol. 1, No. 1 (2006), 117–133.
[22] J. Gan, X. Wang, M. Zhou *et al.*, Ultraprecision diamond turning of glass with ultrasonic vibration, *Int. J. Adv. Manuf. Technol.*, Vol. 21 (2003), 952–955.
[23] C.F. Cheung, W.B. Lee, Characterization of nanosurface generation in single-point diamond turning, *Int. J. Mach. Tools Manuf.*, Vol. 41 (2001), 851–875.
[24] A.G. Mamalis, A.I. Grabchenko, M. Horvath *et al.*, Ultraprecision metal removal processing of mirror-surfaces, *J. Mater. Proc. Technol.*, Vol. 108 (2001), 269–277.

[25] O. Horiuchi, J. Ikeno, H. Shibutani, H. Suzuki, Y. Mizukami, Nano-abrasion machining of brittle materials and its application to corrective figuring, *Prec. Eng.*, Vol. 31 (2007), 47–54.
[26] http://www.fujita3.iis.u-tokyo.ac.jp/index.php?Gallery.
[27] MEMS and Nanotechnology Exchange, https://www.mems-exchange.org/.
[28] F25 measuring nanometers, www.Zeiss.com/imt.
[29] MFP-30 Atomic Force Microscopes, www.AsylumResearch.com.

CHAPTER EIGHTEEN

Sensor-assisted Machining

Contents

18.1 Sensors and System Architecture

It is well known that steadily increasing demands for higher accuracy and productivity lead to the implementation of large-scale process automation and consequently towards fully automated and unattended flexible manufacturing systems. This process automation is attempted in the face of variations in the manufacturing environment such as work material, tooling, machinery performance, rates of production, order of process elements, etc. In classical manufacturing a large number of these processes are machining based, involving various cutting operations. To ensure the optimum performance of such complex systems, sensors are needed for in-process monitoring and feedback for controllers. As shown in Fig. 18.1, the focus of monitoring is on the machine (diagnostics and performance monitoring), the tools or tooling (state of wear, lubrication, alignment), the workpiece (geometry and dimensions, surface features and roughness, tolerances, metallurgical damage), or the process itself (chip formation, temperatures, energy consumption).

All the above-mentioned four focus areas are subject to monitoring needs, often with competing requirements for time response or location of sensors. However, in order to establish the states of the individual elements of machining systems a multiple-sensor system is necessary for process monitoring as illustrated in Fig. 18.2. This approach is of special importance for monitoring processes that are categorized as precision manufacturing, i.e. for which the tolerances on form, dimension or surface features are very tight [2]. In the case of machining, these processes are characterized by very small material removal rates, very low process power consumption and the requirement that the sensor does not intrude in the process or require substantial modification of the machine. Because the most important element of the sensor system is the signal processing methodology for feature extraction and decision-making, an intelligent (knowledge-based) monitoring system can be implemented [1, 2, 3]. This monitoring system is capable of extracting multifaceted information about the machining environment and perform simultaneous multiphenomena sensing. The currently available sensors for machining environment monitoring are mostly designed for observing a single phenomenon, such as static or dynamic components of the resultant cutting force or acoustic emission (AE) signals. As a result, they do not integrate the advantages such as compactness, multifunctionality, ease of use, etc. Advanced sensor design, coupled with signal processing technique, permits improved information

Advanced Machining Processes of Metallic Materials
ISBN: 978-0-08-044534-2

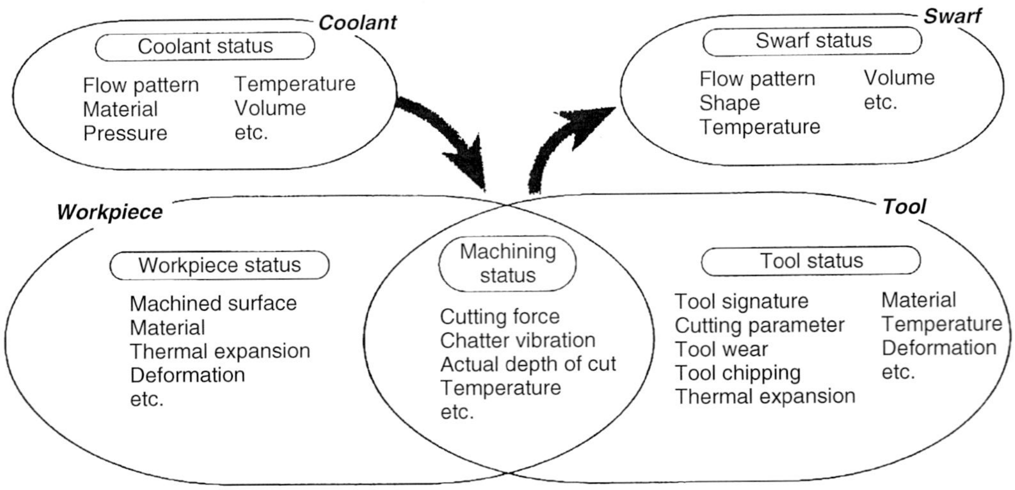

Fig. 18.1 Sensing objective during cutting [1].

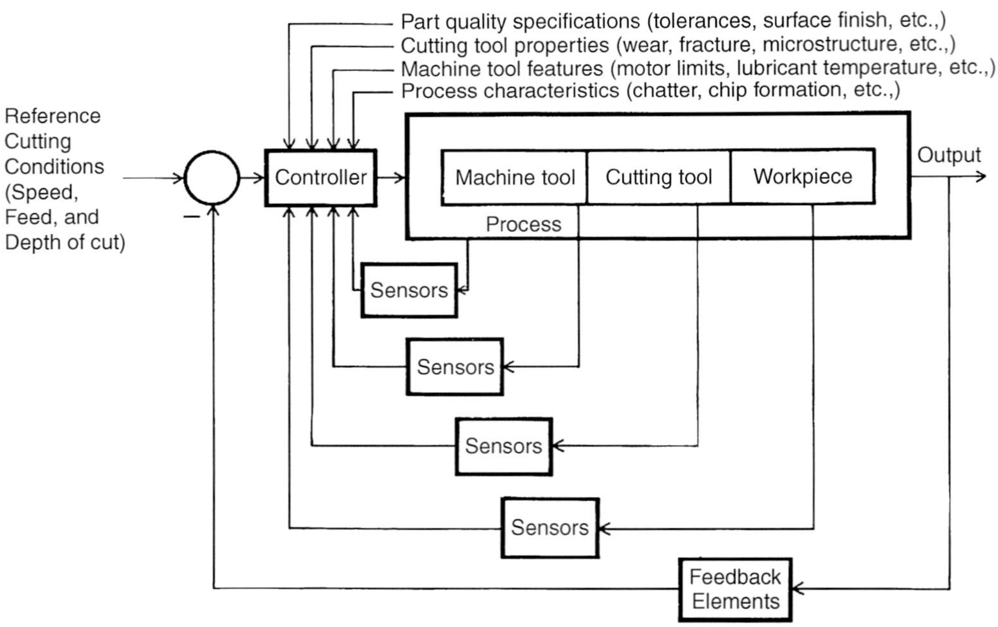

Fig. 18.2 A scheme of a multiple-sensor drive machining process control system [4].

about the process condition, enabling process optimization and control. The desirable performance features of in-process sensors can be listed as follows [1]:

- Simultaneous multiphenomena sensing functions to extract multifaceted information enabling a total understanding of the manufacturing environment.
- High response and sensitivity to extract small signals from the status of the monitored object under the influence of various disturbances arising during machining.

- Compact sensor structure to mount the sensor device near the machining point and extract the necessary information for precise recognition of the machining status.
- Resistance and stability against outer disturbances.
- Easy to use and maintenance free.

Figure 18.3, after Hoshi [2, 5], shows the results of a survey of different monitoring methods for a variety of machining operations. This survey covered the following measurements or sensors for edge chipping, fracture and wear related to cutting tools, and poor hole quality obtained by drilling: touch sensors, load amperage, vibration, torque/force limiter and acoustic emission. The bulk of the processes monitored are drilling and the most frequently used approaches of monitoring are tool touch (with 100% of success) and load amperage (80% of success). Vibration sensing and acoustic emission (AE) are the next most often applied, to be 75% and 33% successful, respectively.

New demands were placed on sensors and monitoring systems for advanced machining technologies such as high speed machining or hard machining which allow the production of precision components. They include the following challenges [6]: low cutting energy when very small layer is removed from the workpiece, response time in milli- or submillisecond ranges for extremely high spindle speeds approaching 100 000 rpm, difficulties in the installation of sensors in complicated tooling systems, protection against high energy pieces in case of tooling failure, producing additional noise to sensor signal due to dynamic effects of tool entry and exit, additional noise produced by MQL system and difficulty with failure sensing for coated tools in dry machining. In particular, some in-process sensors for precision or ultra-precision machining shown in Fig. 18.4a can be capable of improving the process control over critical process variables which leads to tightening control limits and increasing productivity. As a rule, if the percentage change of the feature of the tool or workpiece being monitored decreases, the reduction of the effective signal for monitoring

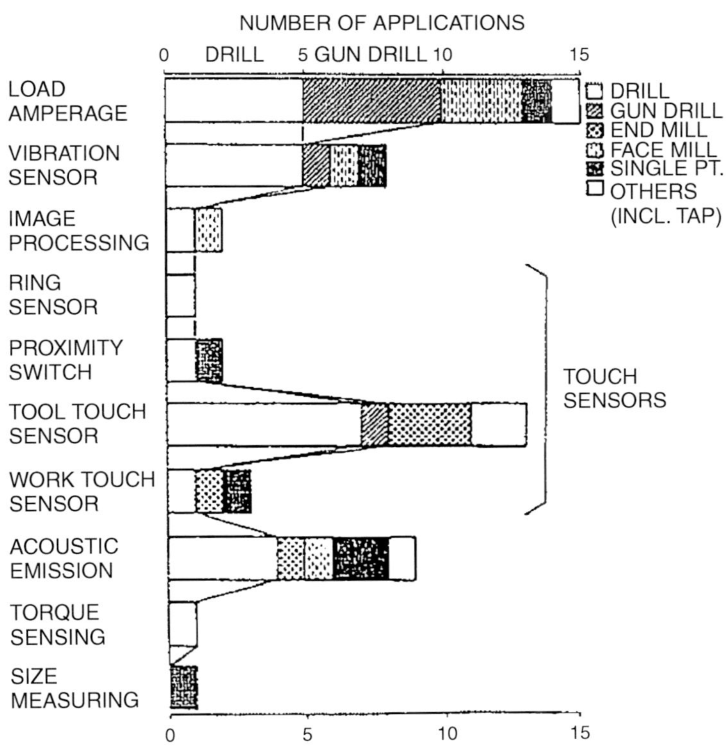

Fig. 18.3 Monitoring methods in manufacturing [5].

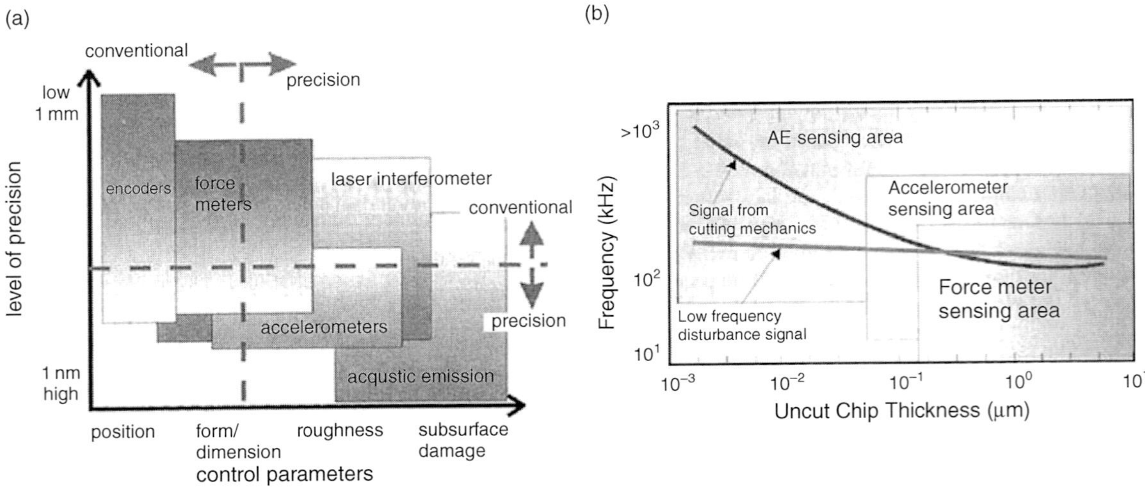

Fig. 18.4 Sensor application vs level of precision and error control parameters (a) and the influence of noise on the effective signal in monitoring HSM [6].

follows. For instance, the signal to noise ratio is reduced due to machining with very small uncut chip thickness, as shown in Fig. 18.4b. Moreover, Fig. 18.4b illustrates the suitability of three sensors (force, acceleration and AE) for monitoring of high speed machining with decreasing chip thickness value.

Examples of some most frequently used sensors (AE, vibration, power and displacement) for machining and grinding operations are shown in Figs 18.5–18.7. In particular, piezoelectric sensors (shown in Fig. 18.5) can be used for numerous applications including force, strain, pressure, vibration and acoustic emission measurements. They are highly sensitive, inherently stable, reliable, rugged, and have wide frequency response, uniquely wide dynamic range, small size and practically unlimited life. Typically, sensors are mounted on the turret head housing and the tool post in turning, on the spindle head in milling, on the spindle quill in drilling and fitted on the headstock or dresser unit in grinding [3, 7]. The sound sensor from Fig. 18.6c can also be mounted directly on the end of the spindle of a grinding machine with wireless signal transmission. It should be noted that the location of the sensing transducer of an AE sensor can influence the magnitude of the signal detected. In general, the amplitude decreases as distance from the cutting point increases and is sharply attenuated by passing through a contact surface.

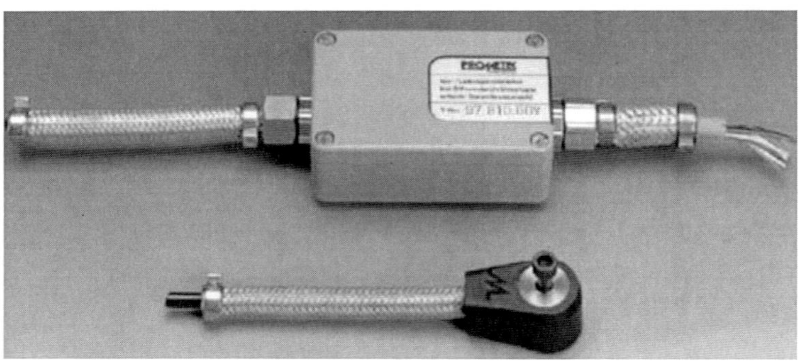

Fig. 18.5 Ultra-sound and vibration piezoceramic AE sensor AE 30, Type S; preamplifier KSV [8].

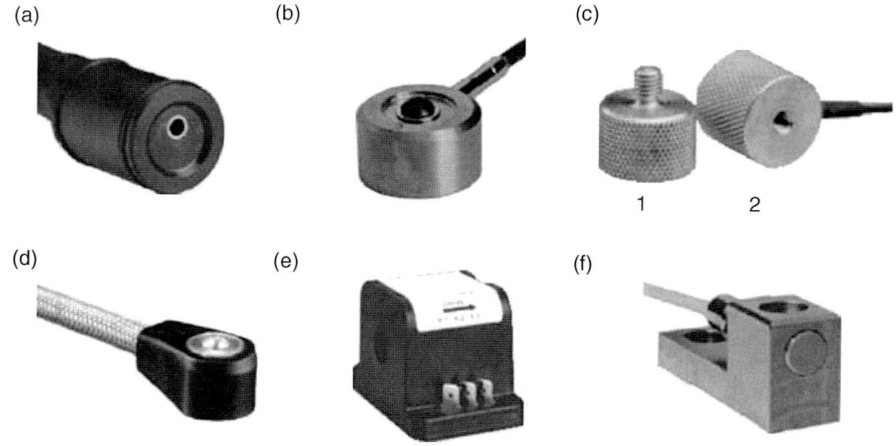

Fig. 18.6 A range of sensors used for monitoring metal cutting and grinding operations: (a) fluid sound sensor for structure-borne sound and acoustic emission via coolant flow from workpiece, (b) wideband sensor for structure-borne sound and acoustic emission, (c) wireless sensor for structure-borne sound and acoustic emission (1 – sensor and transmitter, 2 – receiver), (d) acoustic emission and vibration sensor, (e) hall current sensor for effective power, (e) inductive displacement sensor for force and elongation [8].

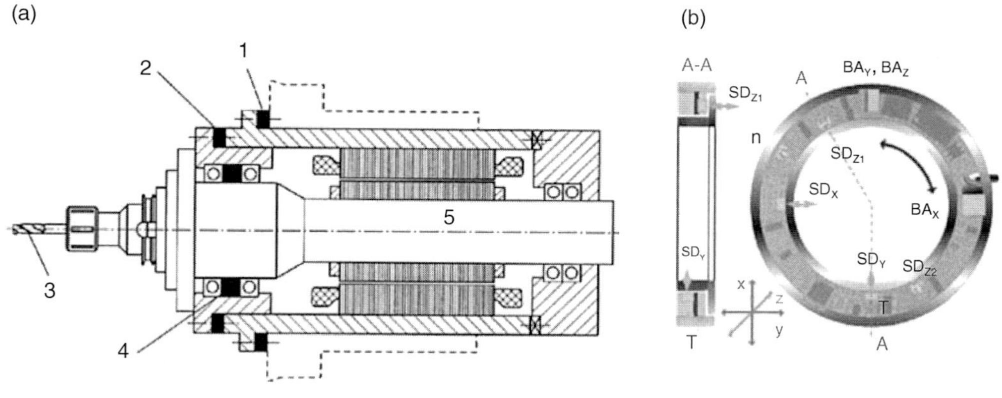

Fig. 18.7 Construction of direct spindle (a) integrated with measuring ring (b) and force sensors: 1 – three-axis measuring ring for forces acting on spindle flange, 2 – three-axis measuring ring for forces acting on axial flange, 3 – cutting tool, 4 – force sensor on spindle bearing, 5 – spindle motor [9].

The measuring ring shown in Fig. 18.7b is capable of measuring ten spindle features with nine sensors, as follows:

- 3D shaft displacement (SD) at 2 μm resolution with four inductive displacement sensors.
- 3D bearing acceleration (BA), one radial up to 10 kHz with ICP output signal, one radial, one axial up to 1 kHz.
- Shaft speed sensor (n), one impulse/revolution.
- One temperature sensor (T).
- Counts time when spindle-on-power.
- Counts time when spindle-on-speed.

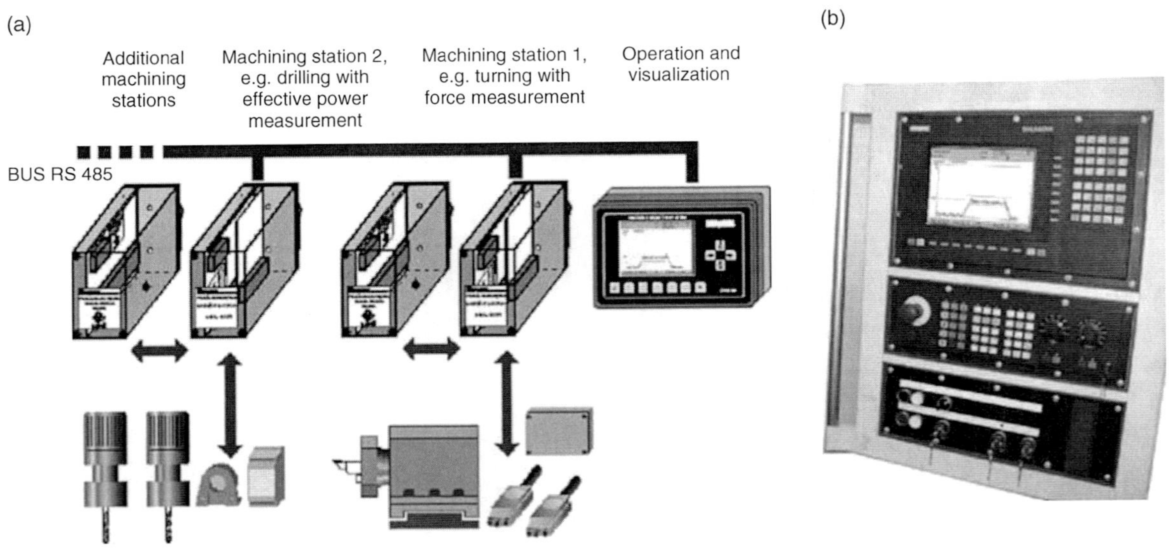

Fig. 18.8 Structure of a PROMOS tool monitoring system (a) and operator panel (b) [8]. Source: PROMETEC GmbH.

Vibration, sound, ultrasonic vibration and acoustic emission are actually all vibration measurements although the frequency range differs in each of these monitoring methods and, in addition, sound is airborne vibration when all others are mechanical vibrations of the machine structure. The frequency range in vibration measurements is typically from about 1 Hz to about 10 kHz (16 kHz or even 20 kHz is used as a limit). In sound measurements the frequency range is from 20 Hz to 20 kHz, and in ultrasonic vibration the frequency ranges from 20 kHz to about 80 kHz. On the other hand, acoustic emission starts where ultrasonic vibration ends up and ranges to about 1 MHz [10]. The vibration signals (often together with thrust force) were found to be very effective for monitoring drill wear and breakage for a wide range of drill sizes. However, a higher frequency range from 0.5 to 40 kHz is more suitable for very thin drills (for drills of 1 mm diameter the natural frequency is about 25 kHz). Acoustic emission is a phenomenon which occurs when, for different reasons, a small surface displacement of a material surface is produced. AE sensors are used for the tool wear and failure detection occurring in various metal cutting processes but the appropriate frequency ranges are different. For example, the 200 kHz sensor is suitable for tool wear monitoring, whereas the 800 kHz sensor can be used for tool breakage detection [10].

The tool monitoring system shown in Fig. 18.8a reliably monitors machining operations such as turning, drilling, milling, sinking, reaming, thread cutting, broaching etc. and also offers, depending on the configuration, the following functions:

- Rapid detection of machine collisions.
- Detection of random tool breakage.
- Prevention of tool breakage from overload.
- Wear detection.
- Tool contact/workpiece contact.
- Process visualization and optimization.

The monitoring system from Fig. 18.8 consists of four modules, namely:

- Sensors that record the loads subjected to the tool during processing and send the appropriate signals to the monitor module.
- Monitor modules which process the sensor signals, compare the current signal with the set limits or with the learned limit values and generate information for the machine operator or the controller.

- Machine interface modules which are the interface between monitor modules and the machine controllers.
- The operator panel module (Fig. 18.8b) or the operating and visualization software in an open NC control giving fast and direct access to all of the parameters and settings in the monitoring system.

In multiple-sensor monitoring systems the fundamental problem is to determine what the significant information for feature recognition is. The raw signals obtained from the AE, force and current sensors must be processed to provide useful information on the cutting state because their interpretation is very difficult due to the randomness of these signals. In the time domain the signal feature can be represented by a number of statistical parameters such as the root mean square (RMS) value, arithmetic mean, standard deviation, and kurtosis as well as other statistical characteristics of the distribution of the signal. On the other hand, in the frequency domain fast Fourier transform (FFT) is the most frequently used signal analysis technique. A number of FFT-based functions such as power spectrum, autocorrelation, cepstrum and also two-channel functions including cross-correlation, cross-spectrum, frequency response as well as some multi-signal frequency response function with more than two channels [2, 3, 10]. Wavelet transform (also together with artificial neural networks (ANNs)) is described as a good solution in the time–frequency domain and both continuous and discrete wavelet transforms are used for tool breakage detection using spindle and feed current signals. In particular, it may be a perfect tool for many applications requiring automated monitoring of manufacturing operations, i.e. micro-drilling operations due to obtaining signals without high frequency components.

There are many approaches to sensor data fusion and decision-making but besides the linear discriminant function (LDF) one effective technique is the use of neural networks [10, 11–13]. When using multiple-sensor systems the information provided by the process models (for example, force-, accoustic emission- and temperature-based models) should be synthesized in order to determine the best estimates for the process variables. It was documented in [12] that a complex network with two hidden layers provides the lowest mean square error (MSE) for the wear estimate in comparison with estimates based on the outputs from different sensors and a simple network with one hidden layer. Figure 18.9 shows the schematic of the operation of a tool condition monitoring based on sensor fusion with a complex three-layered neural network. The system employs multiple sensors, i.e. acoustic emission (AE), piezo-quartz force and electro-magnetic current sensors. Inputs to the sensing system include features of the acoustic emission signal (autoregressive (AR) parameters and amplitude of energy within certain predefined frequency bands), cutting force (amplitude of energy within certain predefined frequency bands) and average motor current. The data, time based and frequency based, is sampled and analysed in real time. The signals from the sensors used are processed using a multichannel autoregressive (AR) series model or fast Fourier transform (FFT) technique. The resulting AR coefficient matrices or power spectral density are passed to the feature extraction block, in which the features which are most sensitive to tool wear are selected. Then, the six most significant features are fed into a previously trained ANN to generate final decision on the state of the cutting tool. As a result of this signal processing procedure up to 97% reliability was achieved in correctly identifying the wear state of the tool during single-point turning of carbon steel [2].

In case of CNC machine tools the measured sensor signals are processed by real-time monitoring and control algorithms, and the corrective actions are taken accordingly by the CNC control system. The corrective actions may consider manipulation of spindle speed, feed, tool offsets, compensation of machine tool positions, feed stop and tool change depending on the process monitoring and control application. Such a sensor-assisted cutting is called intelligent machining and the appropriate monitoring system is called intelligent monitoring system (IMS) [15]. Its basic feature is that the CNC control system must open to allow integration of a user-developed real-time application program. From the definition, the intelligent system should be able to perform automatically the following tasks [14]:

- Test and select the best configuration of sensors and signal processing methods, and, as a result, select measured signal feature which indicates the best correlation with the monitored phenomenon.
- Automatically model a relationship between the extracted feature and the relevant phenomenon without human interaction due to self-learning abilities.
- Represent the acquired knowledge in a human comprehensive form.

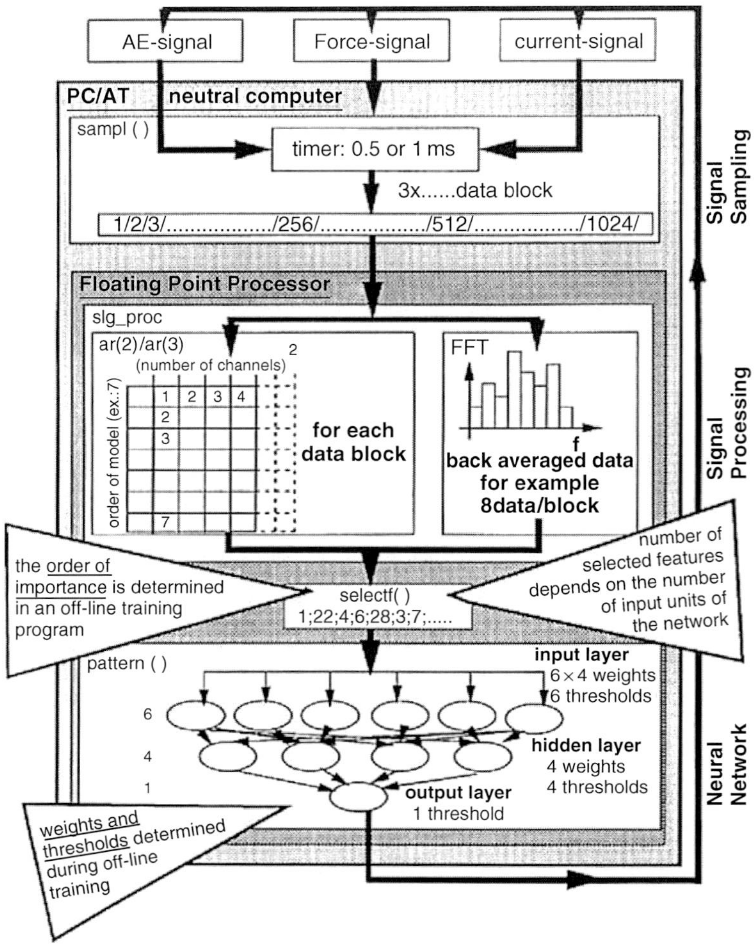

Fig. 18.9 Schematic of operation of the monitoring system [11].

A concept of the structure of an intelligent monitoring system is shown in Fig. 18.10, which shows that the software module includes three steps in assessing the final state of the monitored phenomenon. In the first, feature-extraction step, several features of the measured signals are estimated and further the machine tool operator selects the most important feature based on personal experience (the feature-selection step). Finally, the previously selected features are integrated within a decision-making procedure (feature-integration step).

Another way to implement the IMS is developing new hardware/software systems in order to shift the human/machine tool operator position to a higher level of supervision. The first approach postulates the implementation of an intelligent monitoring system designer (IMSD) which consists of several modules with the same functions as IMS from Fig. 18.10 but employing advanced processing methods and algorithms (for example, FFBP neural networks, genetic algorithms, neuro-fuzzy logic systems). In particular, such a system reduces the time required for final decision regarding the state of the monitored object or phenomenon. The second approach is based on the implementation of the intelligent tool (IT) which is a software/hardware system acting in a defined environment (machine tool, workpiece) and automatically enabling assessment of its current condition, prediction of its state in the near future and assessment of the state of other phenomena based on the information available in the considered environment [14].

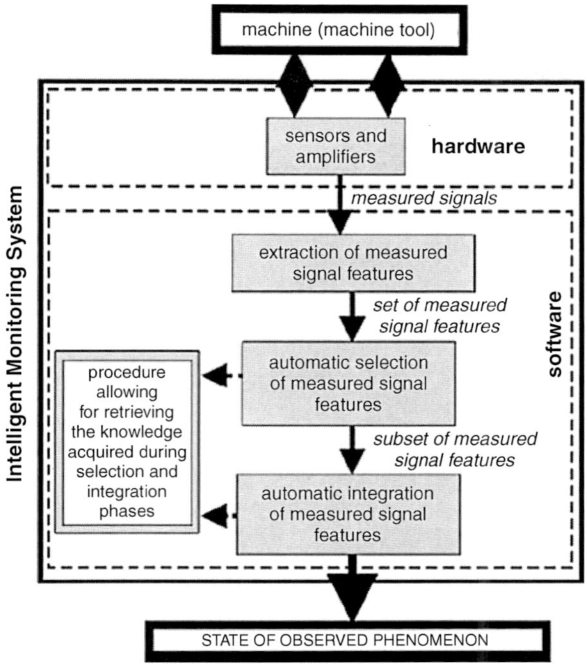

Fig. 18.10 Block diagram of an intelligent monitoring system (IMS) [14].

18.2 Practical Examples of Monitoring Systems for Metal Cutting Applications

The machining process has always required supervision, and in the past this function has usually been the responsibility of the machine tool operator. However, with the increasing use of automated manufacturing systems, ending with fully unmanned systems, there is a visibly growing need for systems which can automatically monitor the machining or other manufacturing processes. Continuous in-process monitoring is the key to zero-defect production in the automated factory of the future. For machine tools these systems are termed *tool monitoring systems* (TMS) or *tool condition monitoring* (TCM) systems. The main goal in the use of sensor systems for tool condition monitoring in machining and grinding is to enhance productivity, and hence increase competitiveness by maximizing tool life, minimizing machine downtime, reducing scrappage and preventing damage [16]. Commercially available TCM systems are used to detect tool breakage, collision, wear and missing tool and states for machine tools. Figure 18.11 shows the configuration of a complete monitoring system coupled to the turning centre which could be either a highly sophisticated stand-alone machine, or just a part of a large manufacturing cell or flexible manufacturing system.

Figure 18.12a illustrates the structure of a tool monitoring system which can be used to monitor turning processes on CNC lathes, and drilling and milling processes on machining centres. This system utilizes a strain gauge sensor mounted in the tool-holder (a) to measure torque, axial and radial forces. For effective process monitoring it is important that the signal used should vary progressively as the tool is wearing, and not just at the point when it breaks. It is well known that during a drilling operation the axial force component provides a better indication of the cutting edge's condition as a function of wear rather than the torque value. This differential allows more accurate setting of realistic alarm levels. Electronics embedded into the tool-holder include a pre-processing unit and A/D converter. Sensor signals (electrical signals) are

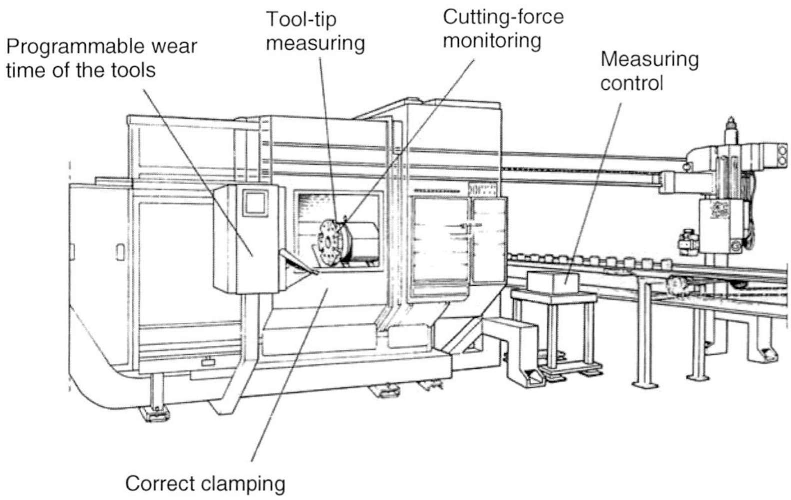

Fig. 18.11 Some of the monitoring systems on highly sophisticated turning centres [17].

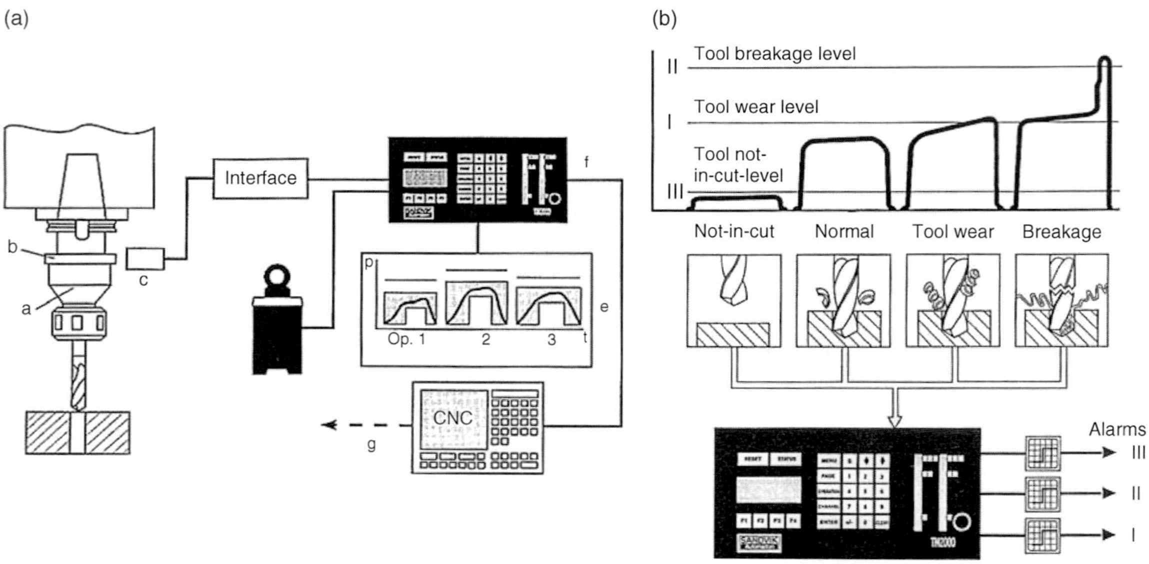

Fig. 18.12 Drilling process monitoring using sensors for torque and force measurements within the tool holder and current sensor [17, 18]: (a) structure of monitoring system, (b) monitoring strategy.

transmitted to the signal-processing device (single- or multi-channel monitoring unit) by means of a wireless signal transmission unit consisting of the ring (b) and the head (c). Once the signal is received, the processing device can immediately initiate action by the machine controller if the tool is worn, broken or not-in-cut. Figure 18.12b shows how continuous monitoring of the axial force can be used to trigger four alarm levels:

- Level I can be used to detect progressive tool wear. The alarm signal can be used to initiate a tool change after completion of the cutting operation.
- Level II can be utilized for tool breakage. This signal should be used to stop the machine immediately when the breakage occurs.

- Level III is planned for detection whether the tool is in-cut or not-in-cut. This may indicate that either the tool or the component is missing.
- Level IV can be used for crash protection. Like level II, this alarm signal should stop the machine immediately and, in so doing, protect the machine tool.

Figure 18.13 schematically depicts the methods by which the signals from piezoelectric sensors are evaluated by a monitoring system for tool breakage, and tool wear and unpredictable collisions. In the case of turning shown in Fig. 18.13a, piezoelectric strain sensors are attached to specific points on the turret housing of the turning centre in such a way that they produce strong signals on the cutting action. At the moment of tool breakage, there is a significant rise in the cutting force, followed by a sharp drop, as illustrated in Fig. 18.13a. This pattern is a tell-tale sign of tool breakage and forms the basis on which the system works.

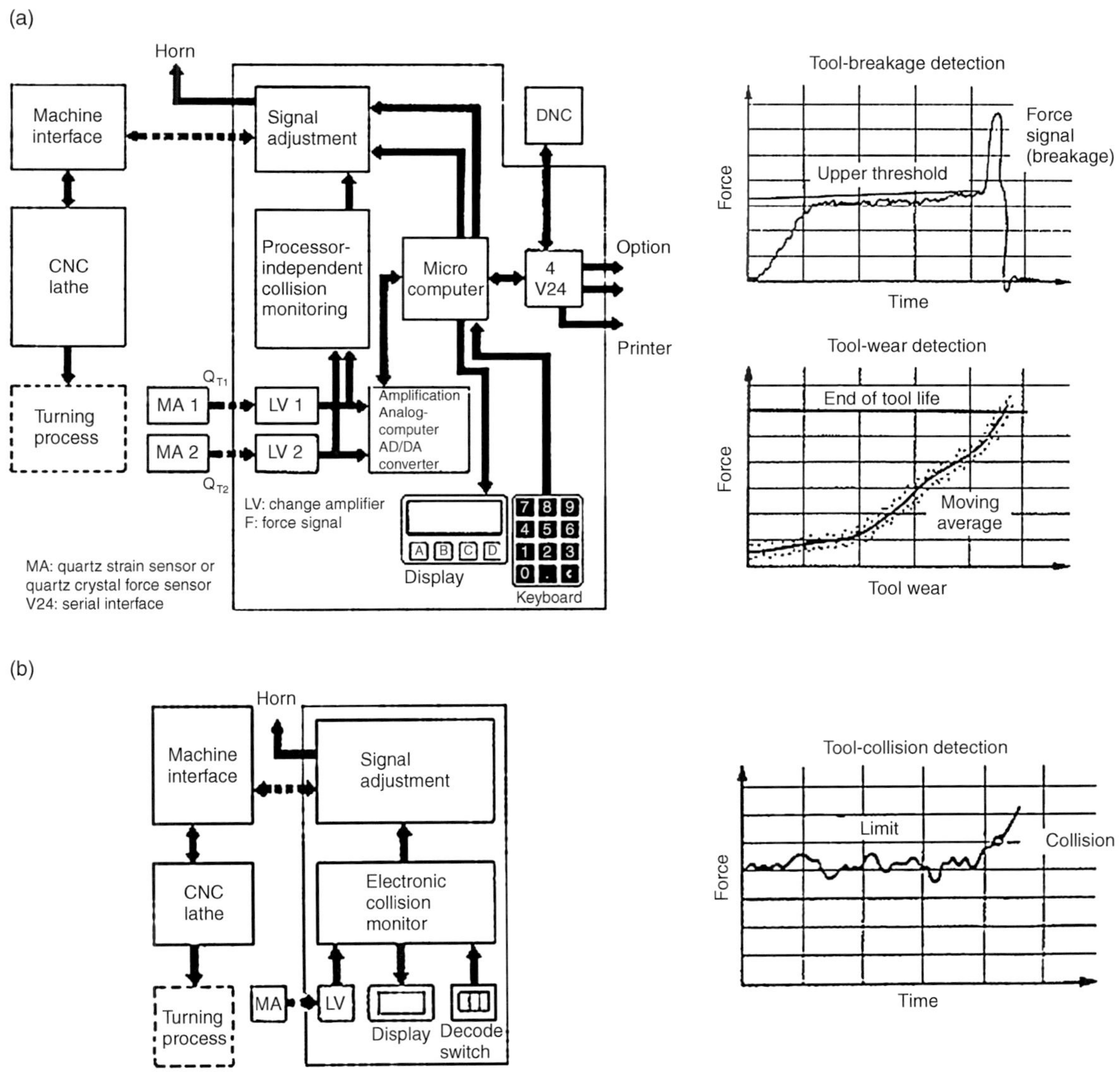

Fig. 18.13 Tool-condition monitoring systems [17]: (a) the tool-breakage and tool-wear detection system, (b) the collision-monitoring system.

The digital breakage-monitoring system continuously monitors the upper and lower force values and if these values deviate from the norm because of the breakage, a command is issued to stop the feed motor. As a result, this action prevents any continued operation with a damaged tool.

The tool wear monitor is also a digital system based on the rise in the force level that occurs over the lifetime of the cutting edge and operates by checking the rise against the force differential measured on a reference cutting edge operated until blunt. The wear threshold (end of tool life in Fig. 18.13a) is obtained by adding the initial force value of a new cutting edge to the force differential of the reference tool. Once this limit is exceeded, the system indicates that the tool is worn and should be replaced.

A schematic diagram of typical circuitry and a graph of the expected force change during a collision are shown in Fig. 18.13b. The operating principle applying to a turning centre is that the strain transducer measures the deformation in the turret during machining and then compares it with a preset threshold value. If the measurement exceeds the threshold, a command is given within milliseconds to stop the feed motor in order to avoid damage to the machine or workpiece.

Two typical applications of the acoustic emission (AE) sensors in tool-condition monitoring performed on turning and machining centres are shown in the diagrams in Fig. 18.14. It should be noted there that the

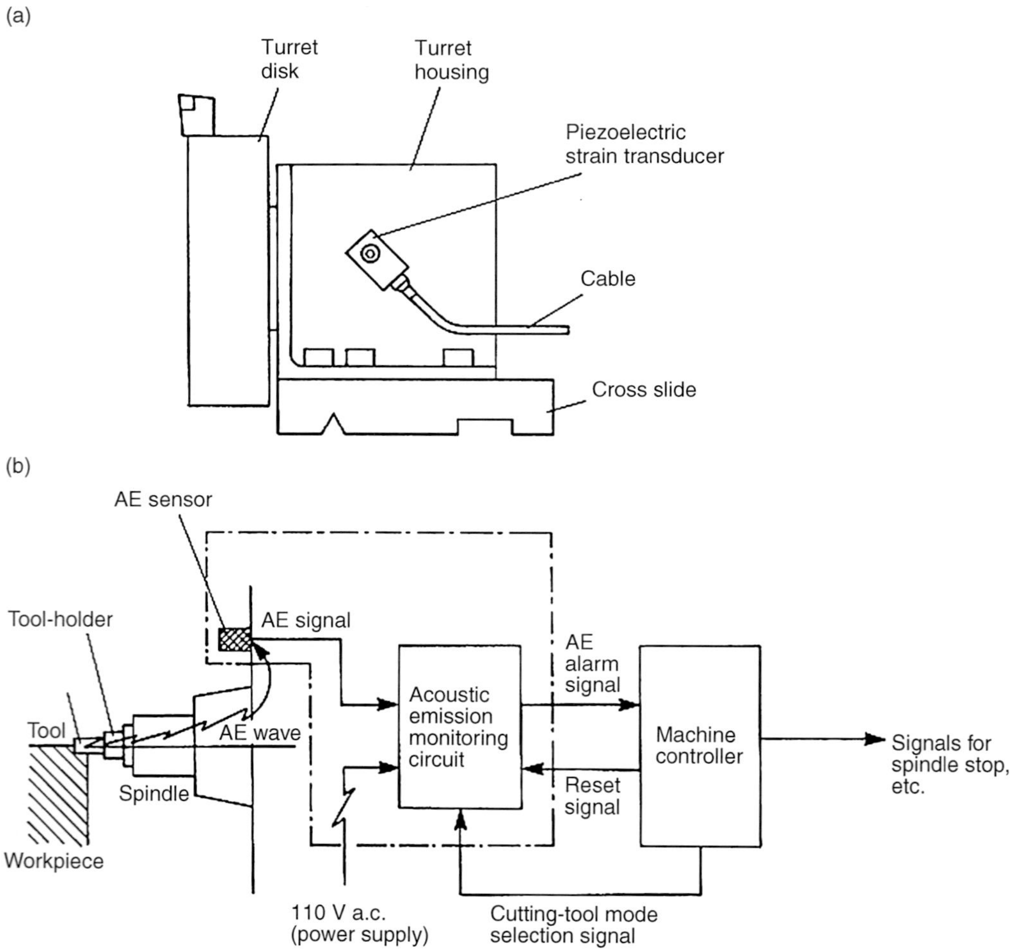

Fig. 18.14 Typical applications of the acoustic emission principle for tool-condition monitoring [17]: (a) a turning-centre application of AE, (b) a machining-centre application of AE.

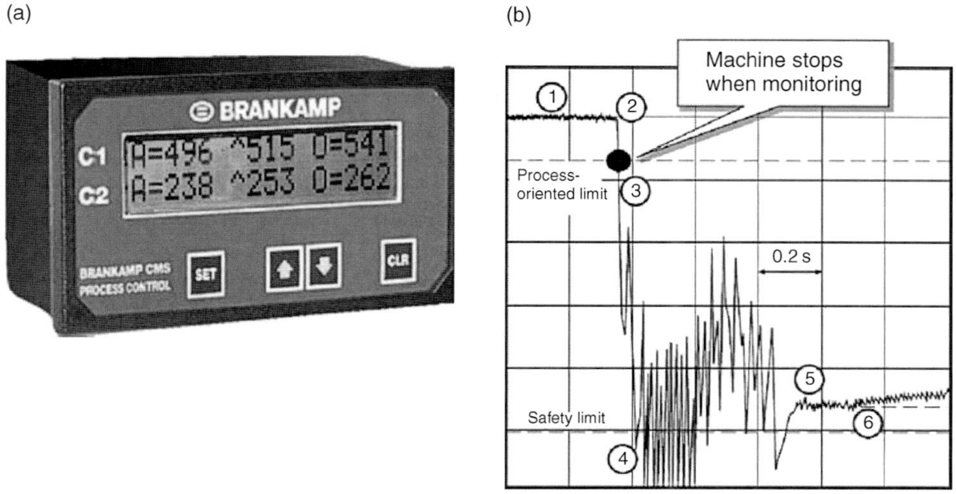

Fig. 18.15 Collision monitoring for machine tools [19]: (a) control panel, (b) collision detection. 1 – rapid traverse motion, 2 – first contact of tool and workpiece, 3 – abrupt increase in processing force, 4 – NC programme starts, five-spindle jammed when main spindle motor is overloaded, 6 – motor current is switched off by fuse.

AE waves travel through the cutter, spindle and concrete, rather than through the workpiece, etc. In the first case (Fig. 18.14a) the AE signal is picked up by a piezoelectric strain transducer fitted at some convenient position on the turning-centre turret housing. By using a pulsor, an artificial signal can be generated which simulates the vibrations associated with a tool breakage. The pulsor enables the operator to localize the AE sensor accurately in the optimum position and the amplifier gain to be adjusted for maximum sensitivity of the monitoring system. This procedure means that tool breakage is reliably detected, since the signal will be at the same level as the simulated breakage signal transmitted from the pulsor. The tool monitor continuously recognizes the current state of the tool during the relevant part of the CNC programme. If the tool fails, the alarm is instantly activated, and if any tool disturbance is detected then cutting action stops immediately.

The monitoring system for collision detection (CMS) using a one- or two-channel control panel during the event of CNC programme is overviewed in Fig. 18.15a. The CMS can be suitable for a range of CNC machine tools including CNC lathes, machining centres, and drilling, broaching, turn broaching and grinding machines using several different sensors, such as magnetic inductive sensors, piezoelectric force and strain sensors and AE sensors with an amplifier. The monitor display shows the actual and maximum force values and the switch-off limit. The rapid traverse is stopped just after the first contact of the tool and the workpiece (point 2 in Fig. 18.15b) resulting in a sudden increase in the measured strains (force) as in the previously discussed case shown in Fig. 18.13b. This resulted in spindle jamming and overloading the main spindle motor. After several tens of seconds the CNC programme reacts by switching off motor current via a fuse due to crossing the safety limit adjusted.

Hard turning, which is often used in the manufacturing industry as an economic alternative to grinding, belongs to the class of processes with unpredictable reliability [20]. As a result, conventional wear-monitoring systems are not suitable for finishing hard turning operations due to a conglomeration of phenomena, such as chip formation, tool wear and surface finish, which exhibit unique behaviour. It is proposed to monitor flan and crater wear during hard turning by means of a force-based monitoring system supported by a dynamic artificial intelligence (AI) technique (dynamic neural networks). In addition, a self-organizing map (SOP) analysis, also known as Kohonen maps, was applied to determine possible correlations between process variables and remove occurring disturbances from the force signals [20].

A real problem arising in precision hard turning is the compensation of dimensional errors induced by tool wear and thermal expansion of both tool and workpiece. Typically, thermally induced errors are

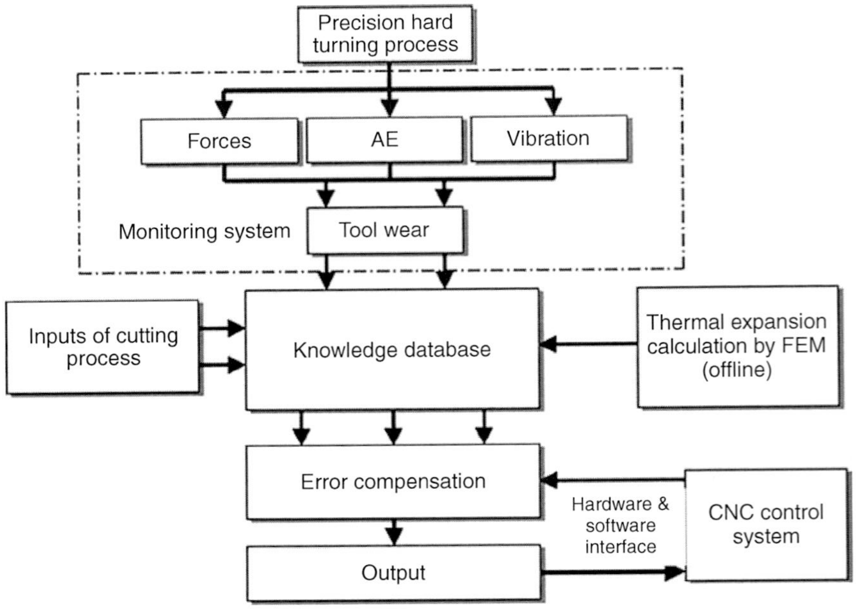

Fig. 18.16 Method of dimensional error compensation in precision hard turning [5].

compensated by a pre-calibrated error compensation method, whereas tool wear induced error is suitable to be compensated with active error compensation. An integration of these two compensation methods in one system, linking with a CNC control system, is shown in Fig. 18.16. A tool wear monitoring system was installed to identify the relationship between tool wear and sensor signals, and corresponding thermal expansions of the tool and workpiece were predicted by FEM computation. As a result, a new NC code was generated and transferred to the CNC control system through the interface between the monitoring and CNC control systems.

The highest hierarchical level of manufacturing systems is intelligent manufacturing also termed sensor-based intelligent manufacturing [15, 22]. It is capable of performing not only the requirements of ordinary manufacturing systems but also makes manufacturing in alien worlds, such as in the atomic-scale world (nanotechnology) or even in space. In addition, intelligent manufacturing enables creation of self-evolving manufacturing systems that use automatic data collection and self-learning, and supports the development of creativity by allowing unskilled designers to manufacture complex parts easily. The required knowledge is basically composed of knowledge about manufacturing phenomena (physical process model) and knowledge about available manufacturing processes. The fundamental structure of the sensor-based intelligent manufacturing system is shown in Fig. 18.17. The main components of the system are knowledge, sensors and actuators. From the phenomena occurring in manufacturing, the dominant physical quantities are extracted by sensors and inserted as an input to the physical model. As a result, the current state of the process is identified and its future state can be predicted. The present and predicted values are compared with the desired values provided from the design system. Any discrepancies between these values cause appropriate modifications of the parameters and the required motions of the actuators are calculated and sent as command values to controllers. Moreover, all activities described are memorized, judged and evaluated by a separate monitoring function. The control algorithms for each function are modified enabling the self-evolution of the manufacturing system.

A structure of an intelligent machining centre, which was designed and constructed using the conceptual methodology described above, is illustrated in Fig. 18.18. In the intelligent machining centre four main functions are essential, namely: detection of deformation, calculation of compensation values, implementation of

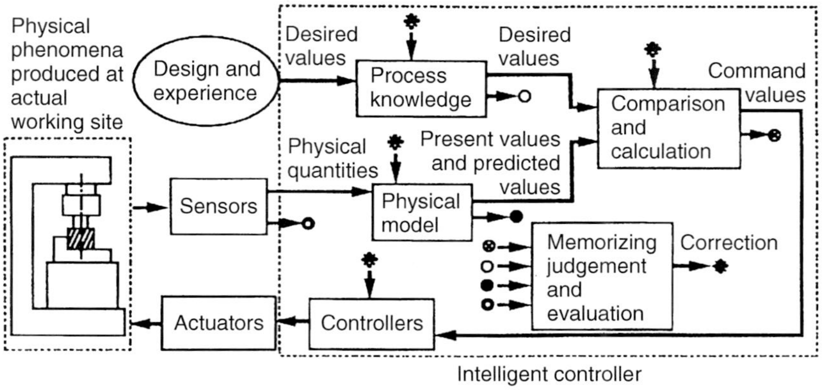

Fig. 18.17 Fundamental structure of the sensor-based intelligent manufacturing system [22].

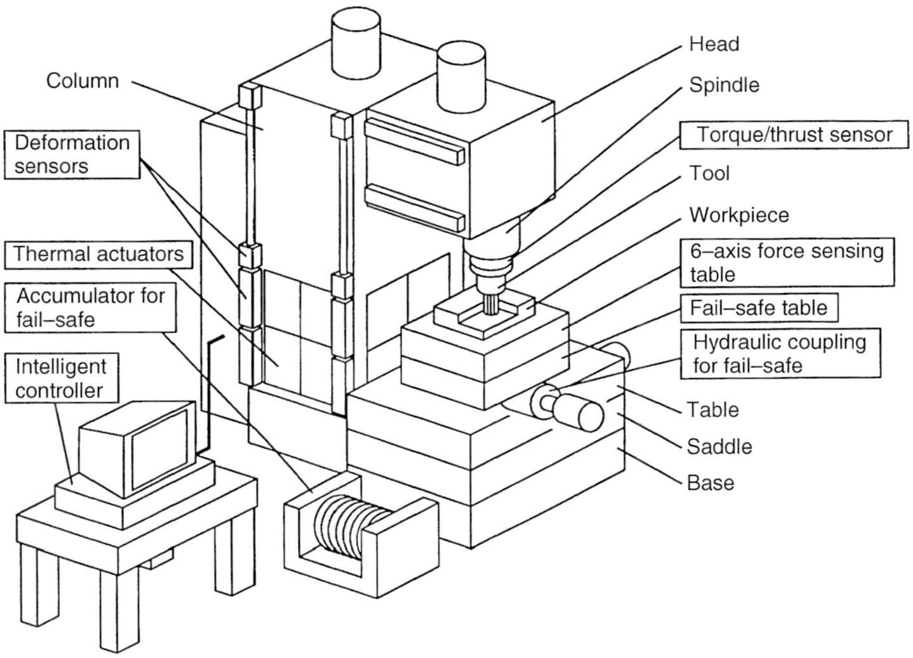

Fig. 18.18 Hardware structure of the intelligent machining centre [22].

compensation and counter-measurements. This machining centre implements the functions of compensating thermal distortion by thermal actuation, intelligent machining by force/torque control, fail-safe operation by using hydraulic coupling and external control not only by a conventional NC controller but also by workstations, connected through the internet or artificial satellite. The spindle itself has the ability to measure the force and torque within the selected rates and fail-safe performance to protect the system against excessive torque. The protection of the head, the table and the saddle against the effects of abnormal loads was achieved by fitting a one-directional force sensor and a fail-safe element to them. Deformation sensors were coupled to the surfaces of the column and the head to measure the mechanically and thermally induced deformations

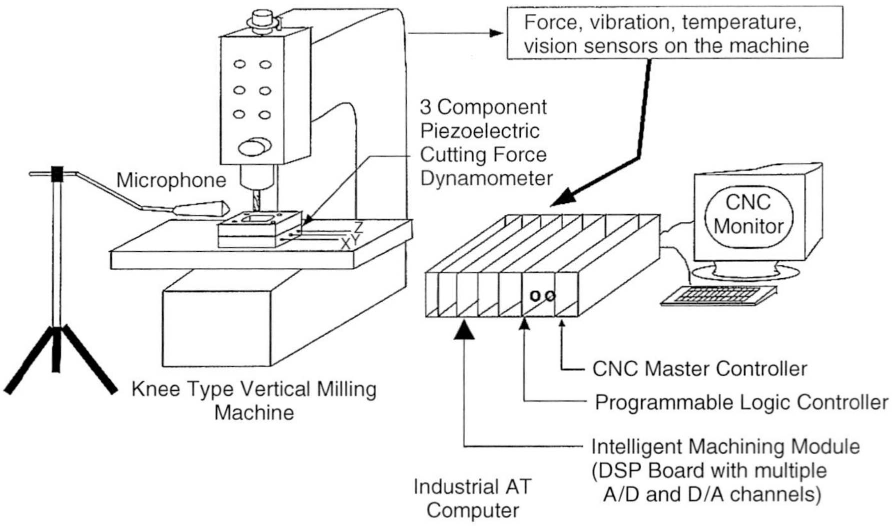

Fig. 18.19 Sensor-assisted intelligent machining system [15].

of the machine structure. Moreover, several temperature sensors are put onto the surfaces to detect their temperatures and heating by the environment. The column itself is constructed as an active structure that means the correct geometry is guaranteed by commands of the intelligent controllers. Vision and sound sensors are installed outside the machining centre to monitor the overall process performance.

Basically, in intelligent manufacturing (machining) the intelligent controller is used together with the standard NC controller as shown in Fig. 18.19.

In particular, this intelligent machining system incorporates an intelligent machining module (IMM) which runs on existing commercial CNC systems, in this case on a digital signal processing (DSP) board with multiple signal processing A/D and D/A channels. Various intelligent machining tasks, such as adaptive control, tool condition monitoring, and process control, can be simultaneously performed in this system. The user can reconfigure the system by using script commands from the supplied signal processing and data collection library. The IMM is configured to communicate with commercial semi-open CNCs through the PC-CNC communication links and software. Some five-axis machining centres controlled by a FANUC™ CNC with a PC interface perform adaptive control (AC), chatter detection and tool-breakage detection [15]. The IMM sends feed and spindle speed change, machining stop, tool change, tool offset and other commands accepted by the CNC controllers.

Adaptive control (AC) systems for metal cutting processes are a logical extension of CNC systems. The term AC denotes a control system that measures certain output process variables, such as spindle deflection or force, torque, cutting temperature, vibration amplitude and power. AC systems use a variety of sensors and control strategies and can be classified as adaptive control with constraints (ACC), adaptive control with optimization (ACO) and geometric adaptive control (GAC). It should be noted that the ACO system is the most general but it is difficult to implement, thus the suboptimal ACC and ACG strategies are often used [23]. The ACC-type systems are well fitted for rough cutting, and the GAC-type systems are typically used in finishing operations, more successfully in grinding. An ACC system is basically a feedback loop where the feed adapts itself to the actual cutting force and varies according to changes in work conditions as cutting proceeds. In GACs part quality is maintained in real time by compensating for the machine tool temperature as well as for the deflection and wear of the cutting tool. The purpose of the compensation may be to improve the part dimension accuracy or the surface quality. In general, commercial AC systems used in production today for rough milling and turning are of the ACC type. Typically, ACC systems are based on

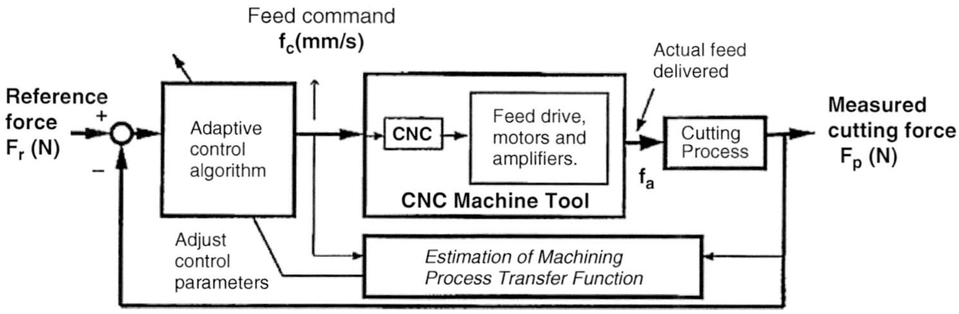

Fig. 18.20 Block diagram of a general adaptive control system in machining [15].

the measurements of a single process variable, such as force, torque, or motor current and try to maintain that variable at some predetermined value. A block diagram of an ACC system with the maximum cutting force as a constraint is shown in Fig. 18.20.

The adaptive control algorithms determine new feed command to minimize the force errors between the reference and measured values. Most ACC systems attempt to maximize the metal removal rate (MMR) by operating at a maximum feed rate compatible with maintaining a reference constant load on the cutter. An example of such a system shown in Fig. 18.21 operates also on the principle of maintaining a constant cutter force during the machining operation. The feed rate is reduced to compensate the force increase due to appropriate decreases in the process variables (for example, due to increased workpiece hardness or depth of cut) or air gaps detected. In the latter case, when the actual cutter force is below a low threshold value, the CNC controller generates a new command to double or triple the feed rate.

New trends in AC systems involve variable-gain AC systems which can adapt controller gains to changing process parameters, ACO with on-line tool wear sensing using AE sensors and their complete integration into CNC systems, as well as into CAD/CAM/CIM systems in the future with broader applications of expert and process monitoring and diagnostics capabilities.

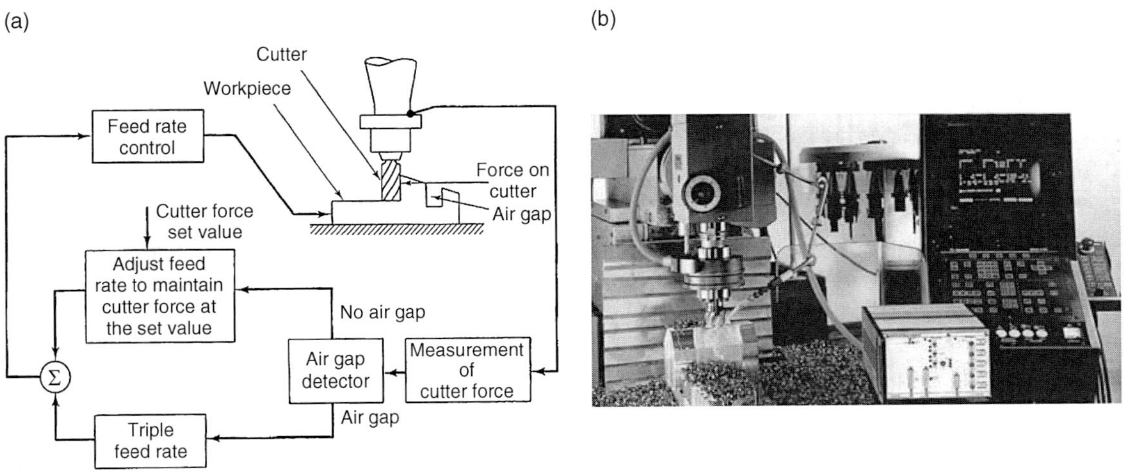

Fig. 18.21 Configuration of typical ACC machining system [24] (a) and its implementation on CNC milling machine with piezoelectric force and torque sensors [25] (b).

18.3 Touch-trigger Probing and Laser Measuring Systems

Generally, the measurement process can be divided into four sequential tasks: positioning, probing, measuring and evaluating [26]. The basic task of probing is when the target point is within the probing's measuring range and a physical linkage between the touching element and the surface is established. It should also be distinguished between hard probes, touch-trigger probes and measuring probing systems. In hard probing the sensor for detecting the contact between measured workpiece and probe is the operator or a current measuring instrument. For a touch-trigger probing system, the final output is Boolean which indicates contact or no contact. A trigger signal is generated when a surface is touched by the moving tip ball or when the tip ball leaves the surface again. In contrast, measuring systems provide quantitative information about the position of the tip ball with respect to the reference point, which is needed for scanning. In general, four main modes of signal transmission are applied, namely: optical (infrared), radio frequency (RF), inductive and laser ones.

A huge variety of probing systems for performing different measurement tasks on the shop floor, as well as in the metrological environment, have been developed. In the case of coordinate measuring machines (CMM) installed in special metrological labs, probing systems must ensure reproducibility of the sensing operation even in the submicrometre range. Recently, two additional trends in the application of probing systems in the manufacturing environment, including on-machine inspection probing performing on CNC machine tools, and incorporating high speed CNC coordinate measuring machines into manufacturing plants in conjunction with CNC machine tools, are observed [25, 27]. It is possible to couple directly CNC measuring machines (for example, those from the MACH CMM series by Mitutoyo, which operate at feeds of 18 000 mm/s and 1.8g acceleration [25]) with some CNC machine tools for concurrent inspection of machining operations. They can provide pre-/post-machining feedback to machine tools for machining adjustments. In addition, if an automatic part handler/feeder is integrated into the CMM machines, it operates between the CNC machine tool and CMM and completes the manufacturing system. Typical applications of on-machine probing systems include setup, measuring and inspection tasks for both workpiece and tool, usually by programmed probing routines during workpiece machining, as shown in Fig. 18.22.

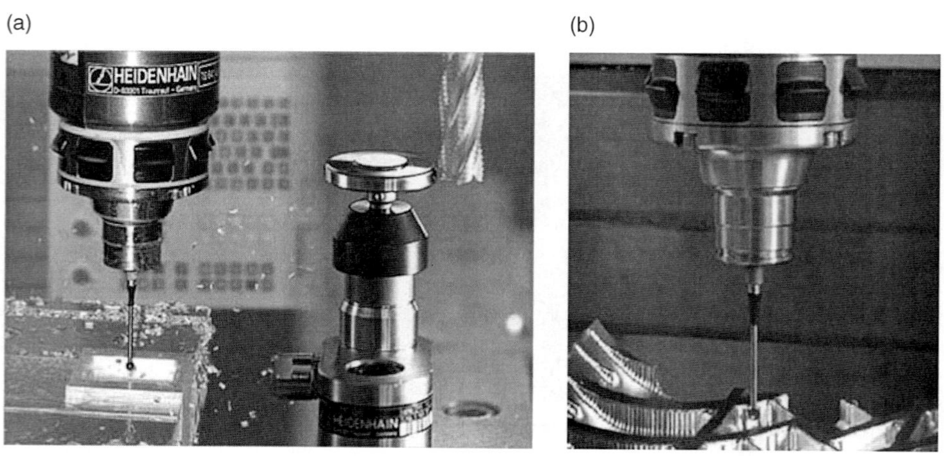

Fig. 18.22 Examples of on-machine probing for tool and part measurements: (a) measuring workpiece between machining cycles and tool presetting, (b) part inspection [25].

According to the feedback from modern manufacturing plants, there are some standard applications of touch-trigger, on-machine probing measuring systems, which globally provide multiple benefits, as follows:

- Fast setups and automated machining of complex geometry parts.
- Identification of the type and size of parts, and their alignment in the fixture.
- Simplification of traditional fixtures by coupling a pallet fixturing system with machine/tool probes because the probe locates everything and compensates before machining.
- Tool presetting and blunt or broken tool detection.
- Checking tools, verification of setups and adjusting offsets.
- Substantial reduction of scrap rate almost to zero.
- Detection of a pallet without clamping by providing error message to the control system.
- Routine machine calibration.

The 3D touch probes shown in Fig. 18.23 are specially suited for installation particularly on milling machines and machining centres. Probes from Figs 18.23a and b are mounted in the machine spindle and can be used for CNC-controlled machine tools with manual (a) or automatic (b) tool changer. They can perform workpiece alignment, datum setting, workpiece measurement and digitizing 3D surfaces.

The probe from Fig. 18.23b features an integrated blasting unit: the probing point can be cleaned of loose particles with the aid of compressed air or cooling liquids through three jets at the bottom of the probe. Chip accumulation in pockets allows automatic measuring cycles during unattended operation. The blasting unit can only work on machines with a compressed-air or cooling fluid duct through the spindle. Figure 18.23c shows a 3D touch-trigger probe with a revolvable touching element and an optical switch as a sensor used to measure length and diameter of the tool mounted in the spindle during rotation or while it is stationary. The tool touch probe is mounted in a T-slot on the table. The optical switch consists of an LED, a lens system mounted onto the stylus and a different photo-diode detecting the movement of the light spot caused by the movement of the lens system.

The NC control automatically saves the results of measurement in the tool tables. This probe can be used to measure tool wear during machining between any two machining steps and automatically compensate it during subsequent machining operation, and also to check a tool due to tool-breakage monitoring. Probe accuracy of $\pm 15\,\mu m$ and probe repeatability from one direction of $2\sigma \pm 1\,\mu m$ at a probing velocity of 0.4 m/min are achievable.

As shown in Fig. 18.24, 3D touch probes can differently scan the surface model and transmit the data to create the CNC programme for the relevant part. Depending on the CNC control or the digitizing software available, the scanning process can generate a meander, unidirectional linear or contour line path.

Fig. 18.23 3D touch probes for machine tools [25]: (a) probe with cable (hard-wired) signal transmission, (b) probe with infrared signal transmission, (c) probe for tool measurement. Source: Heidenhain.

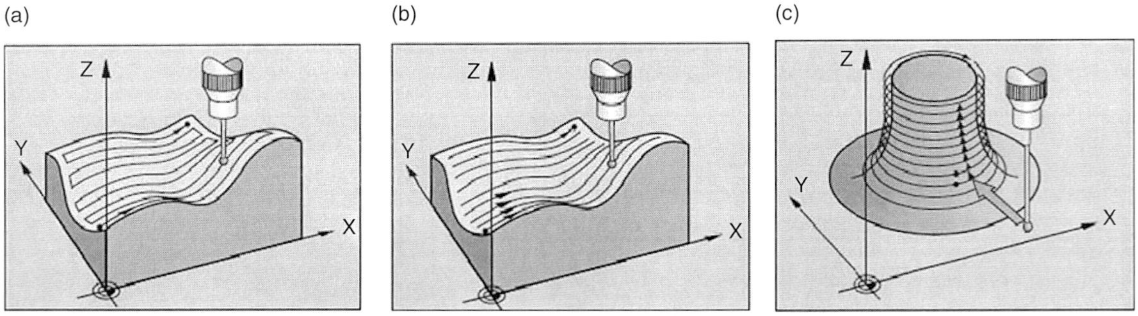

Fig. 18.24 Types of digitizing: (a) meander digitizing, (b) line-by-line digitizing, (c) digitizing along contour lines [25]. Source: Heidenhain.

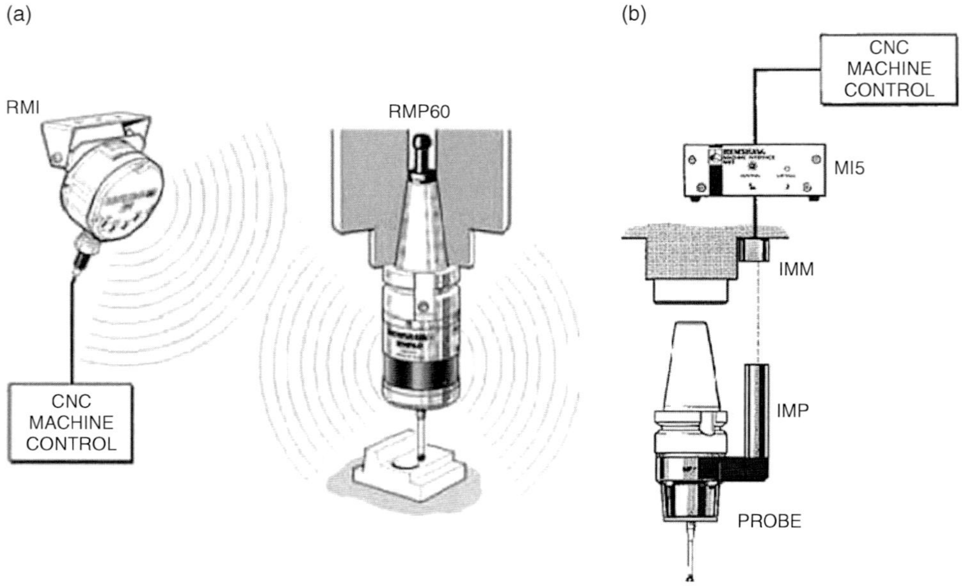

Fig. 18.25 Radio transmission (a) and inductive transmission systems (b) [25]. Source: Renishaw.

A radio transmission system (Fig. 18.25a) provides communication between the probe and the machine's controller for a long distance up to 15 m path length. The radio transmission probe receives signals from machine control and transmits probe status signals to the CNC controller in 'operating mode' using the radio machine interface (RMI). The RMI converts it into a form compatible with the CNC programme. Operating modes include workpiece measurement and setup on medium to large horizontal, vertical and gantry machining centres, five-axis machines, twin spindle machines and vertical turret lathes. An inductive transmission system (Fig. 18.25b) works by passing power and probe signals across a small air gap between two induction modules – inductive probe module (IPM) and machine tool module (IMM). Probe signal is converted into a form accepted by the CNC controller by means of machine interface unit (M15). Inspection systems with inductive transmission of signals can be installed on machining centres and lathes (not recommended for retrofit installations).

Three-dimensional touch probes with infrared (optical) transmission signal (Fig. 18.26) transmit the trigger signal over an infrared light beam using a transmitter/receiver unit. This feature makes it ideal for use on machines with automatic tool changers. It can be mounted in both vertical and horizontal spindles as

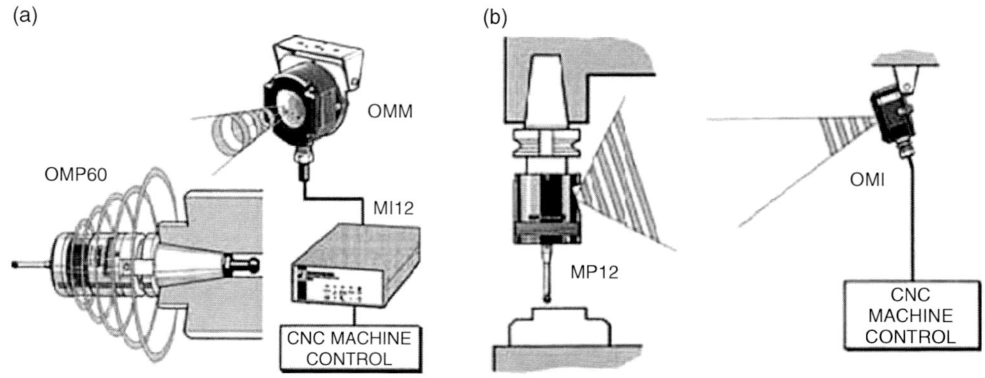

Fig. 18.26 Trigger signal transmission for touch probes using optical transmission: (a) horizontal machining centre, (b) vertical machining centre [25]. Source: Renishaw.

well as in swivel heads. The transmitter/receiver features two multicolour LED indicators that continuously display the condition of the infrared transmission and the touch probe (deflection and battery capacity). In the optical transmission system shown in Fig. 18.26a signals are transmitted to the optical machine module (OMM) which is hard-wired to a machine interface unit (MI12). The functions of the OMM and MI unit can be integrated into a single module – the optical machine interface (OMI), as shown in Fig. 18.26b. Probes specially designed for high speed machine tools can be used in this signal transmission system.

Tool setting probing systems for CNC lathes, termed high precision pull-down arm (HPPA) and high precision removable arm (HPRA), are shown in Figs 18.27 and 18.28, respectively. The HPPA is a manually operated 'pull-down, push-up' system which is permanently located within the horizontal turning centre and used for tool setting operations. Automatic locking of the arm by special rotary device enables the probe's stylus to be relocated to within 5 μm (2σ). On the other hand, the HPRA is a 'plug-in' arm which is

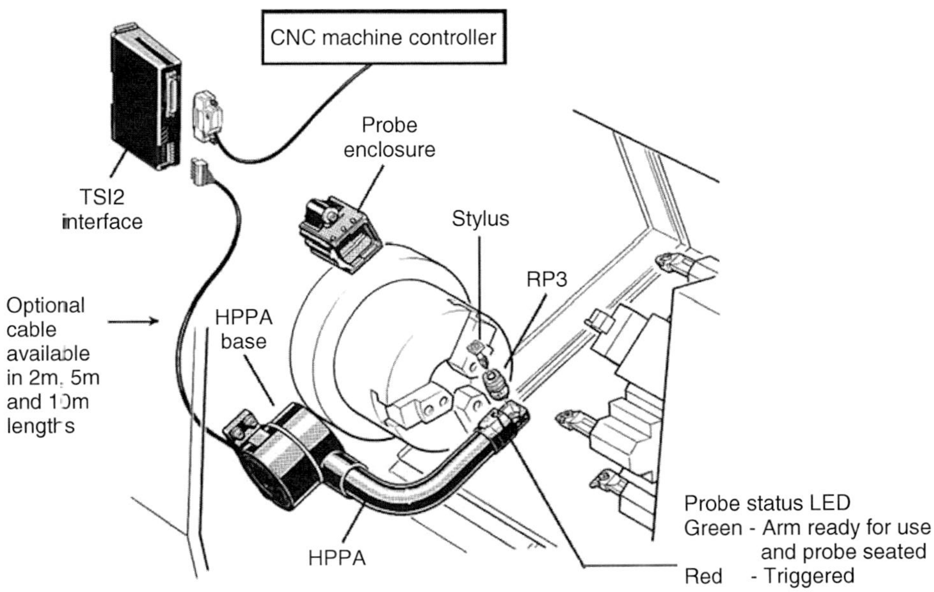

Fig. 18.27 HPPA pull-down arm [25]. Source: Renishaw.

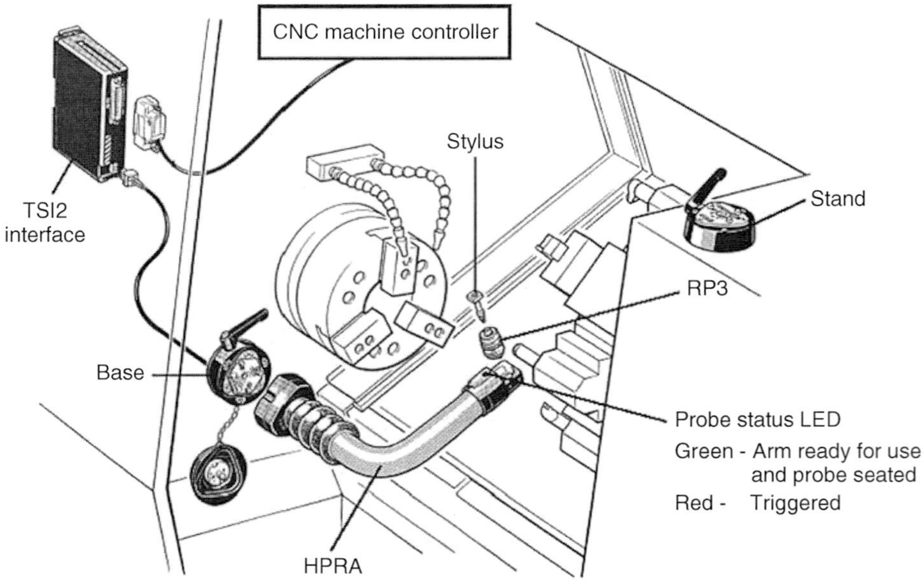

Fig. 18.28 HPRA removable arm [25]. Source: Renishaw.

manually located inside the machine for tool setting, and removed once it is completed. The variation of arm location is about 5 μm (2σ). In addition, the high precision motorized (electrically powered) arm (HPMA) is available for in-process automatic tool setting and broken tool detection. The HPMA tool setting system is fitted to a horizontal machining centre with a multi-pallet changer.

The Laser Control, in-process tool management system shown in Fig. 18.29a monitors tools in real time within the machining environment and certifies that the correct tool is intact within the specified tool-holder.

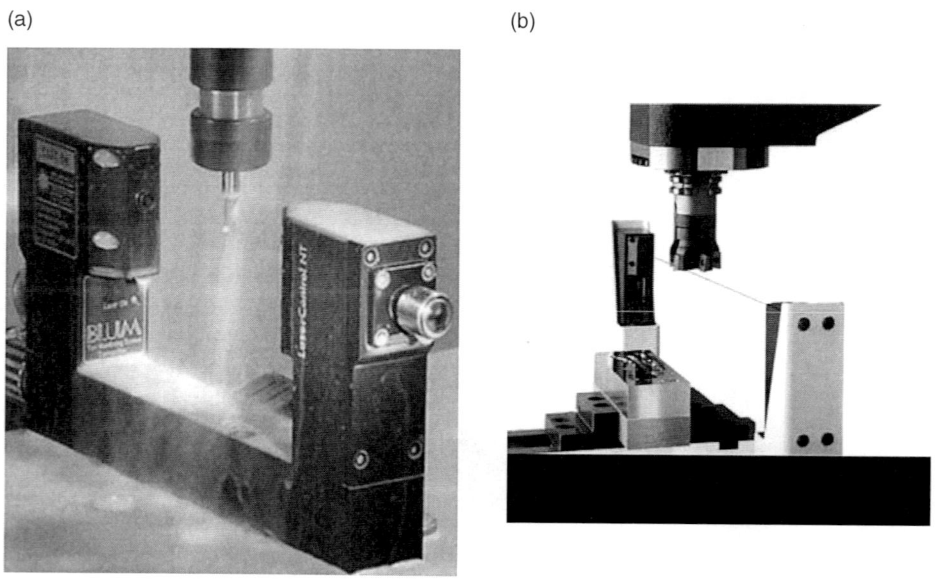

Fig. 18.29 Laser system for monitoring of tool breakage (a) and optical presetting of cutting tool (b) [25, 28]. Source: Blum LMT and Renishaw.

It also verifies length, diameter, roundness, and run-out with a repeatability of better than 1 μm. Moreover, this non-contact tool measurement system allows real tool length and flight circle deviation, as well as thermal drift of machine CNC axis and tool, to be determined and compensation automatically programmed into CNC control. Tool wear can be detected with an extremely accurate form control cycle, which compensates for a worn tool or change in a spare tool from the tool magazine. The system operates as part of the machining cycle and does not involve setup time, downtime or require the spindle to stop for tool measurement. Delicate or small tools can be measured without reducing their life through contact. For example, the TL Micro and TL Nano laser systems by Renishaw (Fig. 18.29b) permit the measurements of tools whose diameters are at least 0.03 mm with an accuracy repetition of down to ±0.2 μm. The laser systems can be installed in a CNC machining centre, and a precision of a few microns in measuring and tool setting is commonly achievable. They can be suited to the standard spindle speeds and HSC spindles with over 30 000 rpm. It should be noticed that the deviations of the centreline (flight circle deviation) and length positions of tool arising at high spindle speeds can be determined and compensated automatically into the CNC control programme.

Measuring cycle routines include the length of concentric tools, such as twist drills, starting from 0.05 mm diameter, tap drills, engraving needles and reamers; eccentric tools, such as end mills, face mills, milling cutters of unknown length and diameter; and single cutting edge monitoring at the rounded edges of ball-nose cutters with cylindrical or conical shafts. Software measuring routines have been developed for Siemens, Fanuc, Heidenhain, Num, Fadal, Boston Digital, Okunta, Haas and other CNC controls. The laser system can be easily retrofitted to vertical and horizontal machining centres, including today's ultra-high speed models.

Advances in software for programming probing routines are one of the main drivers simplifying probe use. In particular, new CAD/CAM software can produce a probing plug-in for process control and inspection at the same time as the machining tool paths are generated [27]. The plug-in supports probing routines to set tool length and diameter, perform workpiece identification and setup, in-process tool offset control and part inspection. The entire process can be fully simulated before machine prove-out. Consequently, on-machining probing becomes a natural part of technological process and the probe is used as any other tool in the magazine.

Scanning probes provide faster data acquisition needed for form measurement, and due to this fact they are particularly suited to determining functional fits including cylinders, bores, valve seats, gears, etc., where variation and deviation are critical. As reported, conventional scanning systems generally take measurements at less than 20 mm/sec but high scanning systems allow the same task to be performed at a scanning speed of about 100 mm/s, i.e. faster by a factor of 10 or more [29]. An example of a probe suitable for very fast scanning is shown in Fig. 18.30a. In connection with the CMM controller it enables so-called Renscan DC, which means high speed scanning with compensation of dynamic influences. The SP25M probe uses the parallel kinematic system which is assembled out of one linear and two rotational axes resulting in a nearly hemispherical measuring range of 0.5 mm radius. In addition, it employs an X–Y pivot mechanism featuring a compact membrane (planar) spring. By optical measurements via two mirrors on the scan module all movements of the stylus can be measured without any inter-axis error and with resolution of about 0.1 μm. The probe uses three scanning modules optimized for different stylus lengths, and the longest version permits a total reach of nearly 400 mm.

A new technology known as *machine vision* was implemented in manufacturing to replace human vision because the requirements of many manufacturing processes have surpassed the limits of human eyesight. Manufactured items are often produced too quickly or with tolerances too small to be analysed by human eye. In a typical machine vision application, a video camera positioned on the production line captures an image of the part to be inspected and sends it to the machine vision computer. The computer then uses sophisticated image analysis software to extract information from images and generate decisions about those images, such as [25]: locating objects accurately even within complex or confusing scenes, inspecting objects to ensure quality and consistency, identifying objects by analysing their shapes or by reading serial numbers on their surfaces, and automatically making measurements. Once the vision system has processed the image and made every necessary analysis, the inspection result is then communicated to other equipment on the factory floor, such as an industrial controller, a robotic arm, a deflector which removes the part from the assembly line or a positioning table that moves the part. Machine vision sensors offer manufacturers highly accurate and cost-effective methods to help improve quality on the factory floor with the latest vision sensors for in-process inspection systems. Advances in machine-vision gauging systems include more powerful sensors and

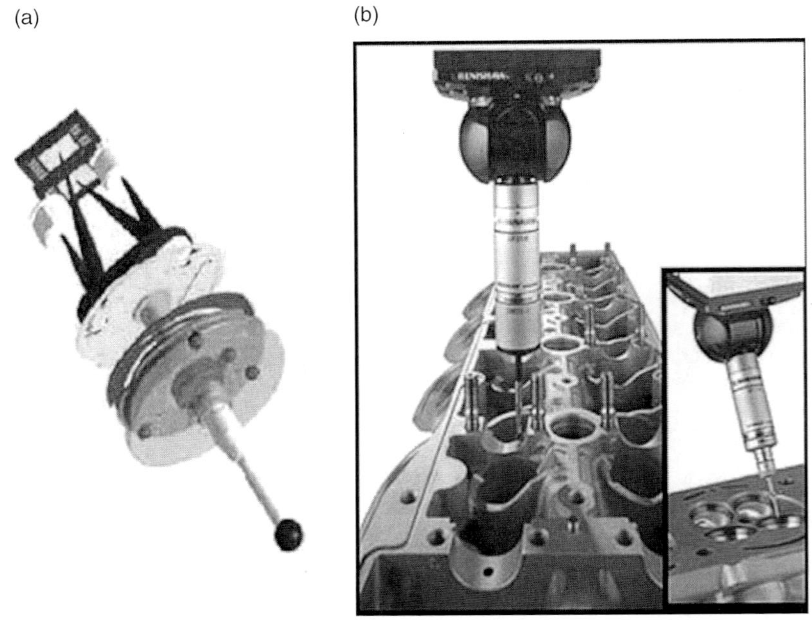

Fig. 18.30 High speed scanning probe; Renishaw's SP25M (a) and its use in measurements of cylinder bores [25, 29]. Source: Renishaw.

laser-based gauging systems. With the latest techniques, manufacturers can improve shop-floor measurement with systems featuring lower cost, greater durability, increased precision, and ease-of-use that can make machine-vision sensors for gauging a more viable alternative to contact gauging. High precision applications such as the automotive and aerospace industries demand in-process gauging systems capable of delivering the goods when manufacturers employ vision systems on the factory floor. Laser-based inspection/measurement systems shown in Fig. 18.31 use CCD technology similar to those of vision systems but employ a laser light source with very high precision for gauging applications instead.

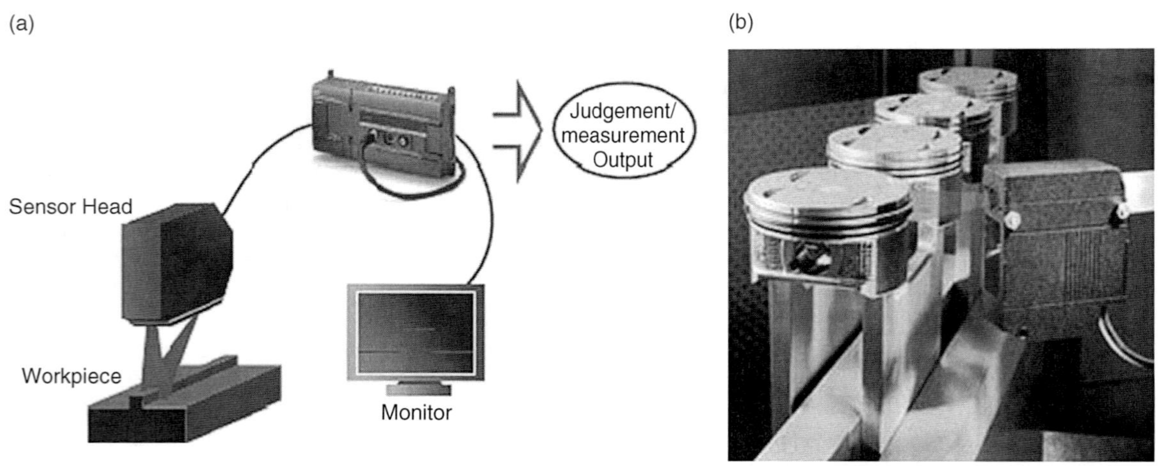

Fig. 18.31 In-line inspection using vision sensors [25, 30]: (a) Omron Z500 profile measurement system, (b) measurements of pistons. Source: Omron Electronics LLC.

Vision sensors for gauging usually hold cost advantages compared to PC-based vision systems. In some industrial factory applications resolution requirements for sensor gauging are also not as high as those provided by higher-end PC-based vision systems that measure submicron-sized components for industries including semiconductors or nanotechnology.

The Z500 laser-based sensors shown in Fig. 18.31a enable higher precision measurements because for laser-based sensors their own laser light source behaves like a highly controlled light element, a pinpoint beam with approximately a 10 μm spot diameter. Using vision sensors, users primarily perform pattern matching for surface defects, but the laser light source sensor creates almost a cross between a sensor and the 2D vision CCD element with high precision. When inspecting the parts, the system can perform gauging or measuring with the ability to measure the height of the object, by moving the object or the sensor, within several microns. The CCD element monitors the profile of the reflected light, so light comes down like a fan-type, wide beam, and then it can measure diversely shaped objects (also with round shapes) in a stable manner.

18.4 Sensor-guided and Intelligent/Smart Tools

Recently, cutting tools that can monitor the machining process and communicate the actual status are of a great interest to automated manufacturing. In general, many types of sensors described previously can be installed into tooling, depending on the accuracy and difficulty of the monitoring process. Using sensor(s), transmitter and appropriate electronics with subprogrammes residing in the machine control systems, creates intelligent, high precision and highly productive cutting tools. A scheme of the closed-loop electronics for such an intelligent tool is shown in Fig. 18.32.

The built-in servomotor, shown in Fig. 18.33, actuates the motion of the tool by harmonic drive, proving a four-fold oversampling at 800 kHz for fast and trouble-free operation. Moreover, this tool is designed with additional machine axis control which can perform rather complex contours. New features of intelligent tools include communication with the machine control system and the possibility for self-adjusting in automatic mode to variations resulting from critical events during machining processes.

The sensor-guided boring bar, termed *smart or intelligent tool*, with the ability to correct the cutting-insert location, is shown in Fig. 18.35. This design was the result of implementing 'agile line boring' (called BOA,

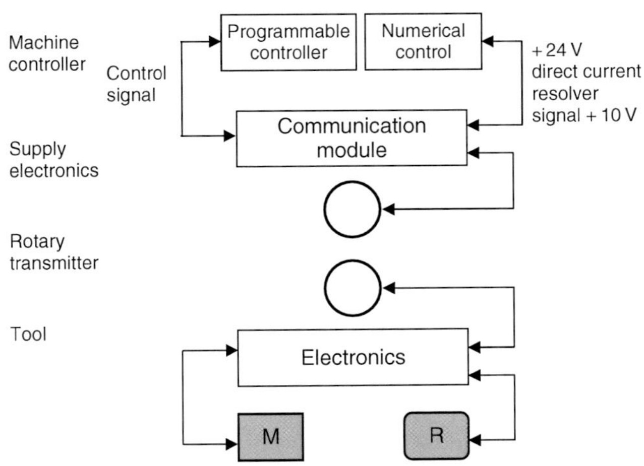

Fig. 18.32 Closed-loop control electronics used for intelligent tooling [31].

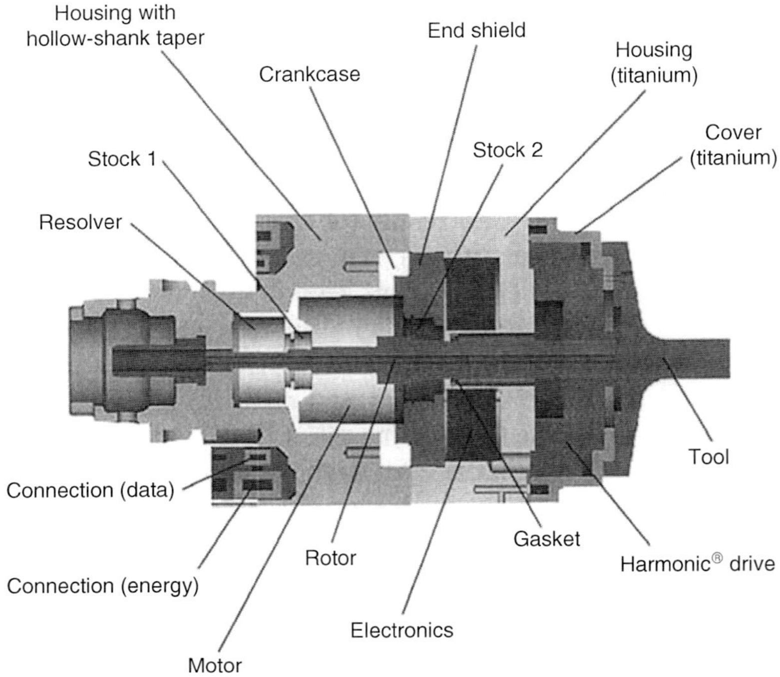

Fig. 18.33 Electronically activated feed-out system for actuating tool motion [31].

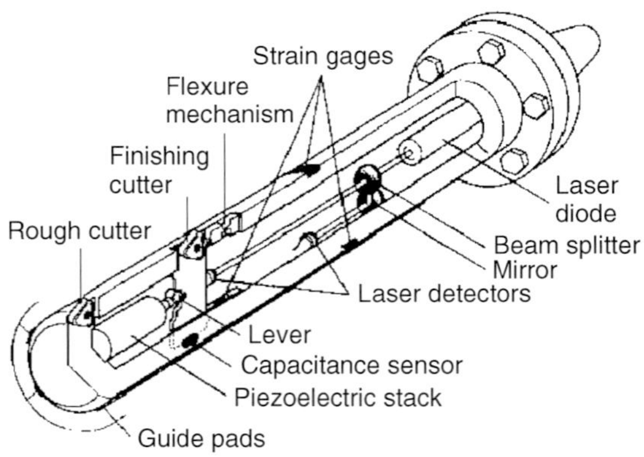

Fig. 18.34 Sensor-guided boring bar [32].

for Boring with Optimal Accuracy [32]) for cam and crank bores which are especially difficult to machine due to very high length, and both dimensional and geometric precision. Protection against error-causing vibrations and elimination of tool deflections were two construction challenges. In addition, even a 3–6 m distance between an off-board controller and a tool in conventional machine causes unacceptable signal delays. So, active machining control requires a new tool sensor and controller to continuously sense and provide feedback. The tool shown in Fig. 18.34, rotating at 5000 rpm, has a built-in computer, a laser guidance system and piezoelectric sensors housed within the tool body. This was particularly important

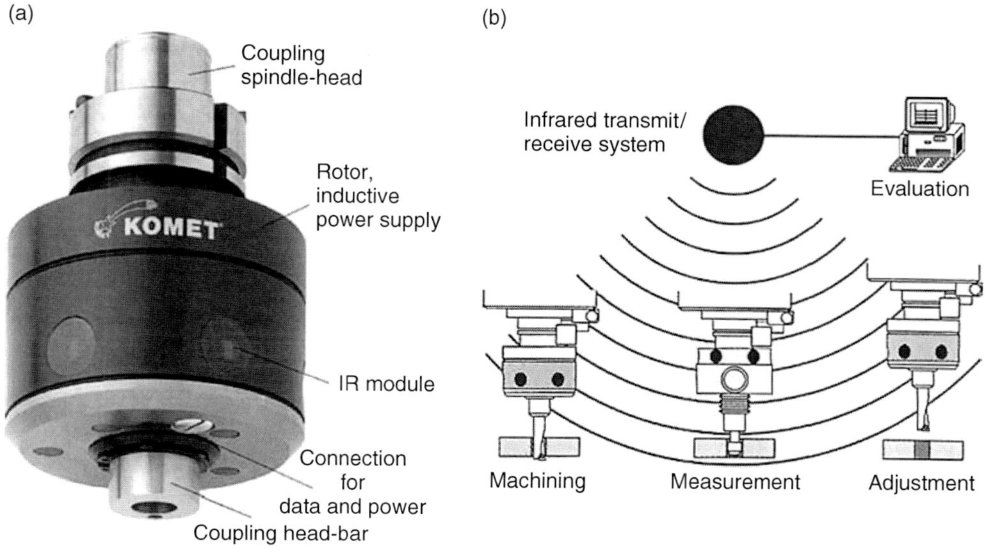

Fig. 18.35 Micro-adjustable boring head with IR module (KOMET-M042) and closed-loop micro-adjustable system [25]. Source: Komet.

because the smart (intelligent) tool – the boring bar – had to be hollow to contain actuators and sensors for real-time control. The tool has two inserts, one for roughing the bore and one for the finishing operation. The finishing insert, mounted on a flexure member inside the hollow tool, connects to a piezoelectric actuator with a very high frequency response. Tool controller outputs feed into the piezoelectric actuator, which turns them into linear impulses, and transfers to the flexure member through a lever. The lever moves the finishing insert radially just enough to cancel out the effect of vibration in the boring bar, workpiece, or machine tool. Strain gauges on the tool OD provide more loading and displacement data. Inside is an ultra-precision laser system that measures vibratory displacement and an actuator that corrects the cutting-insert location.

Figure 18.35 shows a micro-adjustable boring head with a direct absolute measuring system fitted to the adjustment slide, a servomotor and an infrared electronic transmission and receiving system (Fig. 18.35b). The micro-adjustable system termed KomTronic®-Electronic Compensating System by Komet [25] is incorporated into the closed loop. The external inductive power supply is provided. This enables the system to perform fully automatic adjustments of micron range in the diameter without manual intervention. A resolution of 1 μm in the diameter along with high repeatability and reliability is achievable when used on machining centres, flexible lines, and special and transfer lines.

Figure 18.36 shows a new generation of micro-adjustment boring systems featuring dynamic balancing and the *U* axis drive. Apart from dynamic balancing, the drive torque is minimized due to diverting the centrifugal forces onto reactive forces exerting on the housing and building up the balancing element and the slide tool system into the drive line. The KomTronic® *U* axis modular system mainly consists of a compact single slide that is driven by means of a servomotor and treaded spindle. A measuring system is incorporated into the feed unit. Data and power transmission are non-contact. This allows the cutting edge length to be adjusted radially to the axis of rotation (*Z*). Hence, if connected to the NC machine control system, the *U* axis can be interpolated with the *Z* axis. Programming is performed using standard NC programming language. The boring system can be classified as an intelligent tooling system because it can be used for multiple machining operations on machining centres with intelligent placement of the cutting edges.

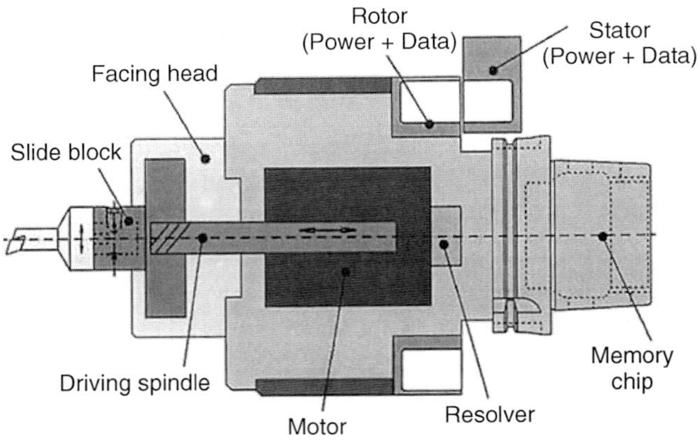

Fig. 18.36 Actuated micro-boring system – KomTronic® *U* axis system with dynamic balancing feature [25]. Source: Komet.

REFERENCES

[1] H. Shinno, H. Hashizume, In-process monitoring method for machining environment based on simultaneous multiphenomena sensing, *CIRP Annals*, Vol. 46/1 (1997), 53–55.
[2] D.A. Dornfeld, In process recognition of cutting states, *JSME Int. Journal, Series C*, Vol. 37, No. 4 (1994), 638–649.
[3] H.K. Tönshoff, J.P. Wulfsberg, H.J.J. Kals, W. König, C.A. van Luttervelt, Developments and trends in monitoring and control of machining processes, *CIRP Annals*, Vol. 37/2 (1988), 611–621.
[4] A. Thangaraj, P.K. Wright, Drill wear sensing and failure prediction for untended machining, *Robotics & Computer-Integrated Manufacturing*, Vol. 4, Nos 3/4 (1988), 429–435.
[5] T. Hoshi, Automatic tool failure monitoring in drilling and thread tapping. *Proc. III Intln. Conf. on Automatic Supervision, Monitoring and Adaptive Control in Manufacturing*, CIRP, Rydzyna, Poland, 1990, p. 41.
[6] J. Leopold, *Werkzeuge für die Hochgeschwindigkeitsbearbeitung*, Carl Hanser Verlag, München, 1999.
[7] J. Tlusty, G.C. Andrews, A critical review of sensors for unmanned machining, *CIRP Annals*, Vol. 32/2 (1983), 563–572.
[8] Modular process monitor system PROMOS, www.prometec.com.
[9] A. Kirchheim, G. Schaffner, P. Wolfer, Monitoringspindeln für Bearbeitungszentren, *Werkstatt und Betrieb*, Vol. 6 (2000), 50–53 (Schweizer Präzisions-Fertigungstechnik).
[10] E. Jantunen, A summary of methods applied to tool condition monitoring in drilling, *Int. J. Mach. Tools Manuf.*, Vol. 42 (2002), 997–1010.
[11] D.A. Dornfeld, Neutral network sensor fusion for tool condition monitoring, *CIRP Annals*, Vol. 39/1 (1990), 101–105.
[12] G. Chryssolouris, M. Domrose, P. Beaulieu, Sensor synthesis for control of manufacturing processes, *Trans. ASME J. Eng. Ind.*, Vol. 114 (1992), 158–174.
[13] R. Teti, S.R.T. Kumara, Intelligent computing methods for manufacturing systems, *CIRP Annals*, Vol. 46/2 (1997), 629–652.
[14] A. Sokołowski, J. Kosmol, Selected examples of cutting process monitoring and diagnostics, *J. Mater. Proc. Technol.*, Vol. 113 (2001), 322–330.
[15] Y. Altintas, *Manufacturing Automation*, Cambridge University Press, 2000, p. 234.
[16] G. O'Donnell, P. Young, K. Kelly, G. Byrne, Towards the improvement of tool condition monitoring systems in the manufacturing environment, *J. Mater. Proc. Technol.*, Vol. 119 (2001), 133–139.
[17] G.T. Smith, *Advanced Machining. The Handbook of Cutting Technology*, IFS Publications/Springer-Verlag, UK, 1989.
[18] E. Vogelbacher, Werkzeuge in Bearbeitungszentren mit modularen Sensoren überwachen, *Werkstatt und Betrieb*, Vol. 127, No. 5 (1994), 390–394.
[19] BRANKAMP CMS, Collision monitoring for machine tools, www.brankkamp.com.

[20] C. Schaffer, H. Kratz, P.S. Heyns, F. Klocke, Development of a tool wear-monitoring system for hard turning, *Int. J. Mach. Tools Manuf.*, Vol. 43 (2003), 973–985.
[21] J.M. Zhou, M. Andersson, J.E. Ståhl, Identification of cutting errors in precision hard turning process, *J. Mater. Proc. Technol.*, Vols 153–154 (2004), 746–750.
[22] Y. Hatamura, T. Nagao, M. Mituishi, M. Nakao, Actual conceptual design process for an intelligent machining center, *CIRP Annals*, Vol. 44/1 (1995), 123–128.
[23] *Tool and Manufacturing Engineers Handbook*, Vol. 1, *Machining*, SME, Dearborn, 1983.
[24] M.P. Groover, *Automation, Production Systems, and Computer Integrated Manufacturing*, Prentice-Hall, London, 1987.
[25] www.kiestler.com, www.mitutoyo.com, www.zeiss.itm.com, www.renishaw.com, www.heidenhain.com, www.omron.com, www.cognex.com, www.komet.de.
[26] A. Weckenmann, T. Estler, G. Peggs, D. McMurtry, Probing systems in dimensional metrology, *CIRP Annals*, Vol. 53/2 (2004), 657–684.
[27] J. Destefani, On-machine probing, *Manufacturing Engineering*, Vol. 133, No. 5 (2004).
[28] www.blum-novotest.com
[29] D. Zayia, Probing technology moves ahead. Scanning offers opportunities for speeding part inspection, *Manufacturing Engineering*, Vol. 132, No. 5 (2004).
[30] P. Waurzyniak, Vision sensors for gaging, *Manufacturing Engineering*, Vol. 134, No. 1 (2005).
[31] B.P. Erdel, *High-Speed Machining*, SME, Dearborn, 2003.
[32] M.C. Tomlinson, Machining systems: revolution, not evolution, *Manufacturing Engineering*, Vol. 123, No. 2 (2000).

CHAPTER NINETEEN

Virtual/Digital and Internet-based Machining

Contents

19.1 Overview of the Manufacturing Evolution

Major paradigms created for the purpose of manufacturing technology during the last three decades are shown in the time scale in Fig. 19.1. The beginning of this challenge is associated with the idea of computer integrated manufacturing (CIM), as the way to automate, optimize and integrate the manufacturing operations of the total system of manufacturing. During the 1980s, CIM naturally expanded due to a wider adoption of robotics and artificial intelligence (AI) techniques, as well as unattended flexible manufacturing systems (FMS) in order to reduce factory floor and labour costs. Because the investments in robotics and FMS did not provide considerable promise, the development in manufacturing was made in respect of in-process quality control, also called total quality management (TQM). In consequence engineering science was supported by organizational science [1]. By the late 1980s the organizational sciences of TQM, just-in-time (JIT), concurrent engineering (CE) and lean manufacturing were successfully combined with the engineering sciences of CIM. The fundamental distinction between concurrent engineering and concurrent manufacturing is that the former is for products while the latter is concerned with processes [2]. In particular, many concurrent manufacturing strategies (popularly meaning operating at the same time), such as outsourcing, full-service supply, offshore manufacturing, joint ventures, and globalization were implemented in different settings. Thus, the best companies extended concurrent manufacturing and TQM to a level that allowed factory floor manufacturing to be 'seamlessly' connected to the desires of the consumer. Open-architecture manufacturing and agile manufacturing consequently emerged as new manufacturing paradigms. These terms emphasized quickly reconfigurable enterprises that could respond to the new customer demands, such as delivery, quality and variety [1].

By the mid-1990s, internet-based manufacturing was the natural extension of these paradigms, emphasizing the sharing of design and manufacturing services on the internet. The availability of the internet, videoconferences and relatively convenient air travel seem to pave the way for increased global commerce. Nevertheless, improving the communication between the customer and the designer as well as between the designer and the manufacturer remain vital for obtaining short time to market. Enterprise integration, which means the integration of all divisions in a manufacturing-in-the-large enterprise, appears to be a leading issue. When this idea becomes extended beyond such intercorporation, the agreements with outside corporations will lead to the 21st century virtual corporation, which seems to be more ephemeral version of monolithic business. It should be noticed that the virtual corporation must rely on the core competency skills of each participant but all participants must become more experienced in partnering skills.

Advanced Machining Processes of Metallic Materials
ISBN: 978-0-08-044534-2

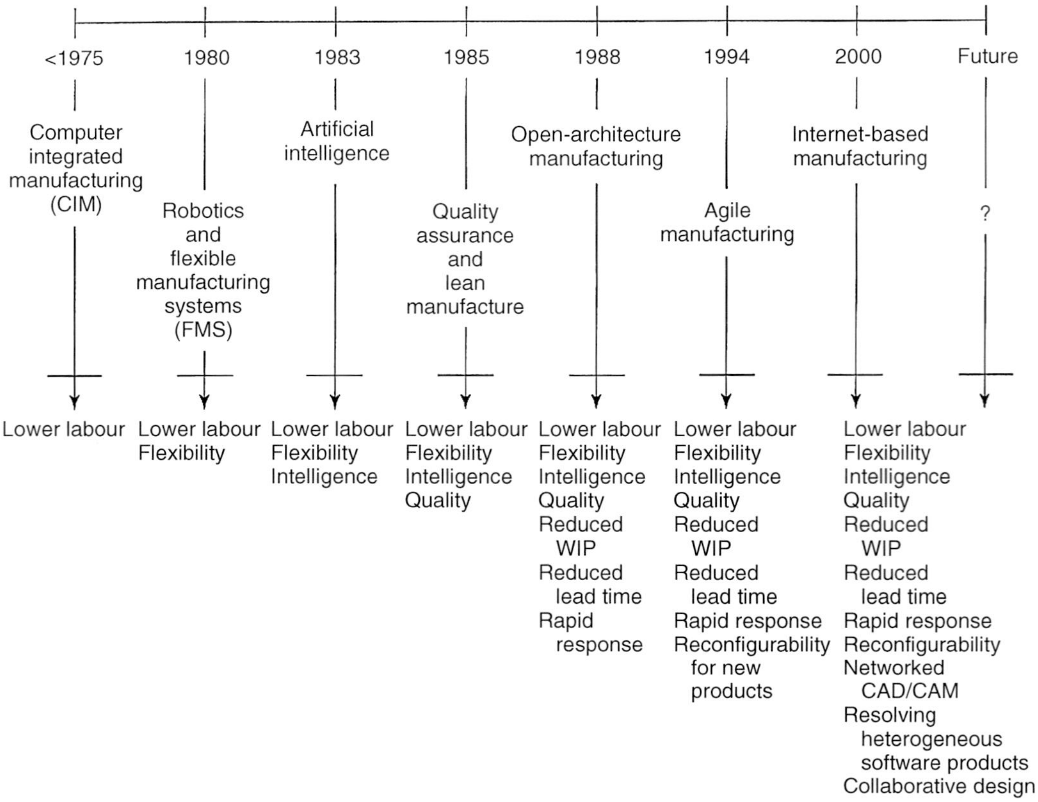

Fig. 19.1 Major paradigms in manufacturing developed in the US [1].

Manufacturing enterprises are rapidly learning how to achieve good integration of equipment, people and operations via digital computer technology or broader use of information technologies (IT) [3]. Computer technology is now developing in at least three areas vital to future manufacturing: holonic systems, virtual reality and intelligent systems. These integration activities include integration of the process planning and production planning, simulation of manufacturing systems, agile manufacturing, fast redesign of new products, modelling of manufacturing equipment performance by means of, for example, the MHI (machine/human interface), functional product analysis, virtual/digital machining and inspection algorithms, etc. One emerging concept of the future manufacturing enterprise is the digital factory, aiming for digital verification of product design, production and assembly before the foundations for the actual factory or even before investing in prototypes. It should be emphasized that the *digital manufacturing* environment is capable of combining the power of the digital product with the power of digital processes and resources. For example, the digital factory implemented in automotive manufacturing can play the role of linking the car engine development with production through the intermediate step of production planning [3]. As an example, the roadmap for implementation of the digital factory concept including also social impacts in the planning is shown in Fig. 19.2. As we can see in Fig. 19.2 an essential element of the roadmap for future manufacturing is the integration of manufacturing processes by reconfigurable machine tools with open-architecture control systems (OACS) to minimize non-productive times. Moreover, disparate sensor systems, as part of OACS, will contribute to the development of intelligent machining systems with learning ability. An additional effort should be made to assess part quality with operating machining systems.

Figure 19.3 illustrates a conception of manufacturing integration by the virtual manufacturing (VM) system composed of a virtual factory (virtual world), an integrated database and functions for engineering

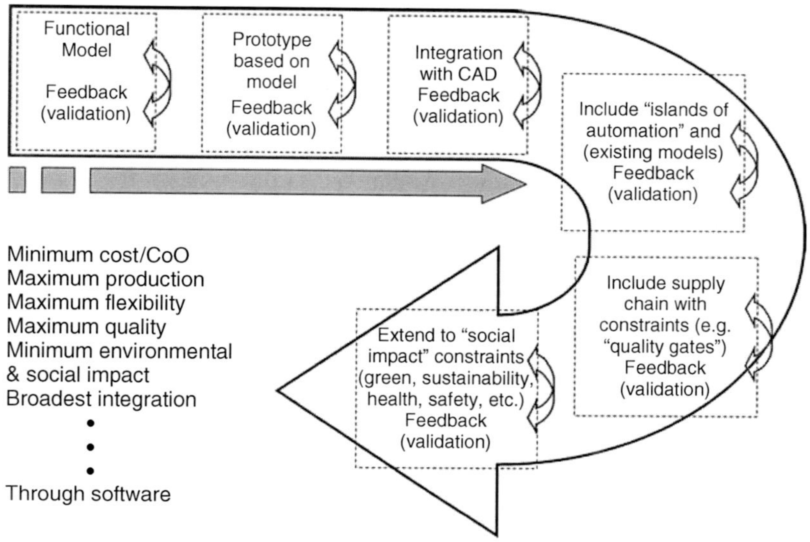

Fig. 19.2 Roadmap for implementation of digital factory concept to create the entire manufacturing enterprise [3].

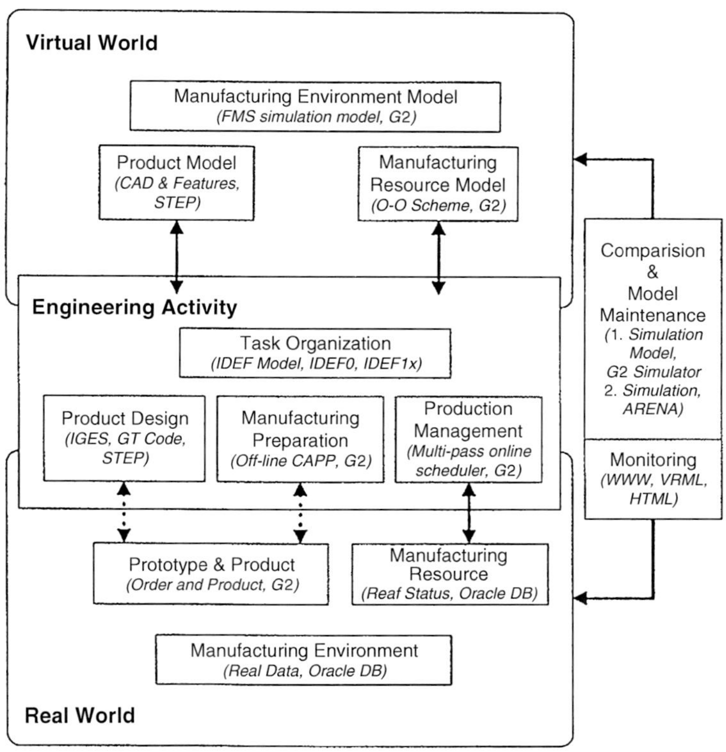

Fig. 19.3 Total manufacturing integration by virtual manufacturing concept [4, 5].

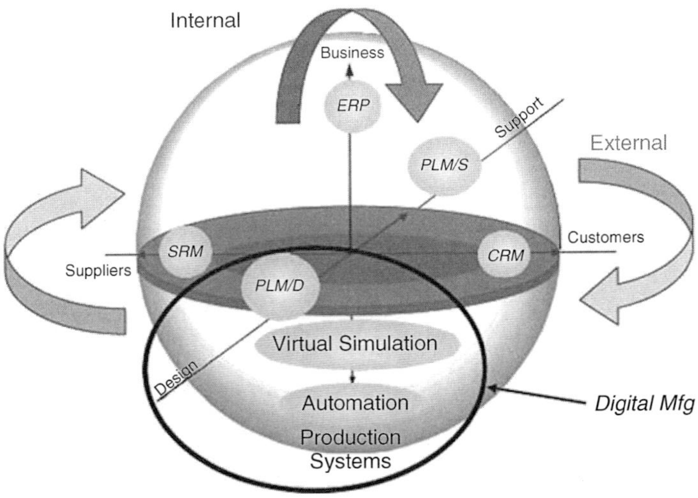

Fig. 19.4 Position of digital manufacturing in the PLM environment [12].

activities. As discussed above, activities necessary for product realization are first performed in the virtual world, where all the manufacturing objects specified and relevant processes are modelled. These models allow the product behaviour (functionality) to be predicted and product manufacturability to be evaluated. Sometimes, simultaneously and immediately after virtual activities, real manufacturing operations are performed in the real world. Anyway, comparison of models with reality and various model maintenance operations are needed. It is possible, using the concurrent engineering method to organize an entire chain of engineering activities from product design to production management, because such an environment provides required information or knowledge whenever and wherever it is requested.

It was proven that digital manufacturing can be used most effectively when it is incorporated into a broad enterprise product lifecycle management (PLM) environment as shown in Fig. 19.4. Definitions of the PLM vary but currently it can be defined as a software solution that encompasses in one environment all of the design, engineering and collaboration, and business-process management solutions required to bring a product from concept through manufacturing to end of life [29]. As mentioned above digital manufacturing is the newest latest component in the PLM universe and it is believed that it bridges the gap between design and production technology and will creatively act on agility and product quality. According to the newest statistics, aerospace companies such as Boeing (777 and 787 airliners) and Dassault Aviation (Falcon 7X striker) achieve production efficiency improvements of as much as 20% over conventional production planning processes due to the virtual 3D environment [29].

In earlier versions of PLM tools, CAD digital designs were integrated with CAM applications to enable the fabrication of machine parts. CAM tools enabled virtual cutters to be driven over digitally rendered part surfaces that would eventually be converted to machine code that CNC controllers used to drive physical (real) cutters to machine parts. The numerical control (NC) function is now a standard module within most PLM solution sets today. Along with the CAM functionality of NC came the first simulation tools to virtually model the machine parts in fabrication process. These were applications that a manufacturing engineer and NC programmer could use to build virtual models of the machine tool, workpiece and cutter to simulate the motion of the machine, removal of material, and check for interference and collisions. These CAM simulation tools were the precursors of the powerful simulation tools used today in digital manufacturing (DM). At the heart of DM are powerful next-generation 3D simulation applications that enable manufacturing engineers to virtually simulate exact models of machines, robots, conveyor lines, workcells, and specific production equipment required to fabricate, assemble, and install parts, subassemblies, and components of the product [12, 13]. Additionally, the actual factory environment of the production process can be modelled including buildings, production lines, transportation, workflow, and other facilities that represent the complete physical production environment.

The growing importance of environment and social aspects has resulted in the life cycle concept and sustainable industrial production (SIP) which denotes that products/processes complexity requires a combination of virtual factory and a physical (real) factory. The integration of these networks was termed the self-innovating factory (SIF) and the relevant model containing 'manufacturing' of current products and innovations leading to 'future' products/processes generations was named 'Manu*f*uturing' [7]. So far this concept has been implemented in the field of machine tools capable of producing more usable and valuable goods, e.g. compressors and electric motors.

19.2 Digital/Virtual Manufacturing

It is worth reminding that virtual manufacturing (VM) is a concept involving computerized manufacturing activities with appropriate models and simulations instead of objects and their operations in the real environment. As a result, VM enables estimating manufacturing processes in many applications without using real facilities. The basic assumption in the development of VM was that every manufacturing system can be decomposed into two subsystems: a real and physical system (RPS) and a real and informational system (RIS). Figure 19.5 shows the architecture of a virtual manufacturing system operating by analogy on the basis of a virtual and informational system (VIS) and a virtual and physical system (VPS). These systems communicate

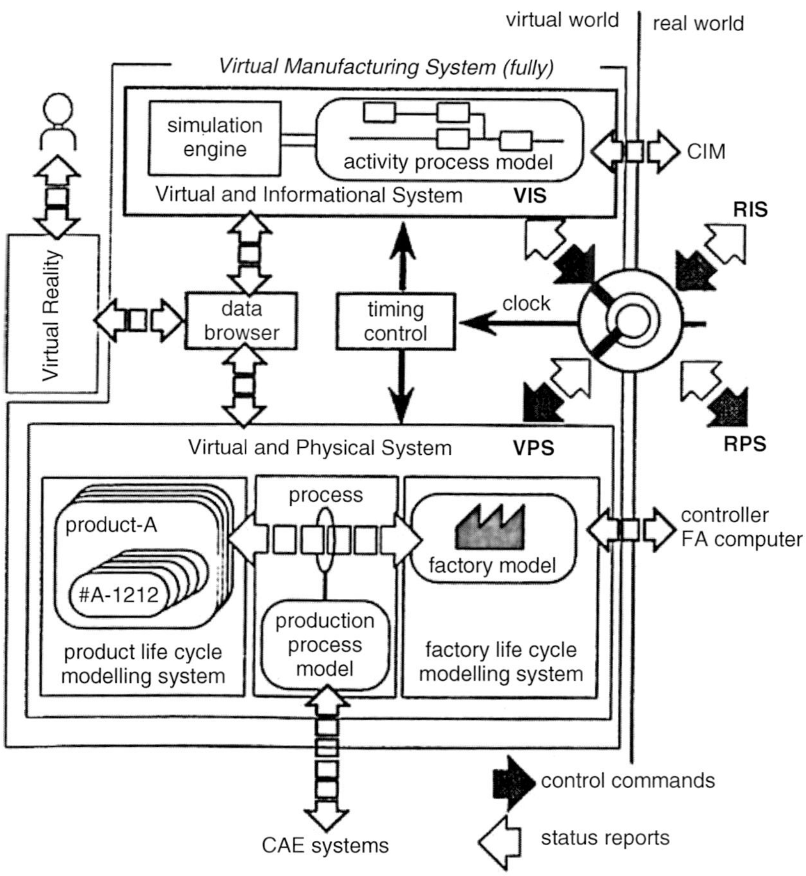

Fig. 19.5 Architecture of virtual manufacturing systems [6].

through the communication junction capable of changing the direction of information flow from and to a real and information system (RIS) and a real and physical system (RPS). The VM system has a timekeeper that synchronizes the VPS and the VIS. In addition, the VM system has another communication interface for its monitoring by human operator. The VPS (lower part of a block scheme) consists of a factory model, product models and production process models. The production process models are used to determine the possible interactions between the factory model and each variant of product models, as well as appropriate CAE system. A factory model and available product models should consider the factory life cycle and the product life cycles, respectively. On the other hand, the VIS (upper part of a block scheme) includes an activity process model which controls decision-making processes and the information flow in this system, and the simulation engine which interprets the process description and executes decision-making processes. As can be seen in Fig. 19.4 the VIS is a gateway system connecting the VPS and the computer integrated manufacturing (CIM) in the real word.

The word *virtual* has now many different technical meanings, such as virtual reality, virtual prototyping, virtual processes (including machining processes) and virtual machines (including machine tools).

In order to design a machine tool with better performance under different conditions through a comprehensive simulation of a machine tool's operation, the concept of a 'virtual machine tool' (VMT) was undertaken using existing and emerging fundamental knowledge and technologies in the areas of mechanical and manufacturing engineering and computer science [8]. The concept is aimed at the following specific tasks:

1. To develop a comprehensive software environment for simulating and evaluating a range of processes executed on alternative topological machine configurations, and
2. To explore advanced new technologies and active control methods for error compensation and vibration suppression in machining operations to significantly improve the performance of machine tools.

Basically, the VMT is a multi-area software test bed on the World Wide Web (WWW). Users can interact with the software to evaluate possible machine tool designs. Later, the more advanced versions allow the user to download CAD graphics files of the geometry and add 'off-the-shelf', as well as custom components to build a functional machine. To simulate the machine tool's performance technical issues of both fundamental (the physical models of the machine and of the process) and pragmatic (development of the modular software structure, allowing the composition of arbitrary machine/process models based on component libraries). In other words, the models developed become appropriate parts of a comprehensive library whose entries will ultimately be used to compile the simulation model. The interrelationship between the models in the context of the closed-loop model of the machining process is shown in Fig. 19.5.

In Fig. 19.6 the machine tool is viewed as consisting of two general functional entities. The first describes the behaviour of the machine tool's structure under the cutting process imposed temperature and force field, while the other characterizes the motions of the machine's moving elements with the associated errors. Both entities are influenced by the NC command codes issued by the kinematic model. The motions, errors, force and temperature fields yield a relative motion between the tool and the workpiece, which is the input to the cutting process model, in addition to auxiliary input variables. The cutting process model evaluates the resulting force and temperature variables and the characteristics of the machined part.

The virtual machine (VM) module in the form of a graphical machine tool simulator is often an integrated part of a CAD/CAM package working in one single integrated platform [9]. First, the VM enables NC programmers to graphically simulate and automatically test programs against collisions during post-processing. Second, it avoids and corrects possible programming errors that may have resulted in costly machine tool collisions and defective parts. Moreover, the integration of the VM and graphic NC post-processing (it simulates the block-to-block machine tool motion instructions generated by the NC post-processor in a real-time mode) causes that the graphic engine provides real-time feedback to the post-processor and some critical manufacturing issues such as machine collisions and over-travel are automatically detected and repaired. As a result, this type of software can enable NC programmers and machine tool operators to explore off-line advanced features of the new machine tools and test the applicability of appropriate NC programs in a virtual environment. Moreover, the same database (for example, operating within the CATIA V5/DELMIA environment) can be utilized to store a control emulator (CE) module, seamlessly integrated with virtual

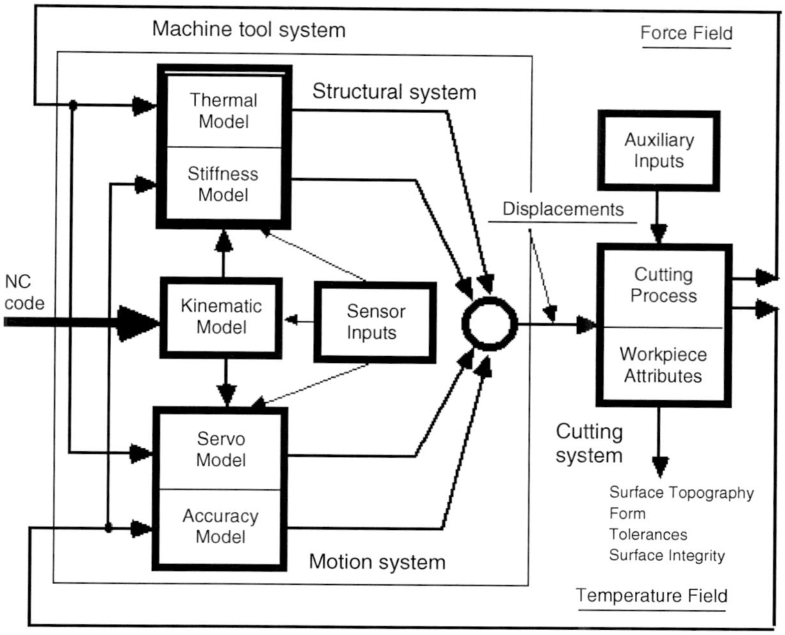

Fig. 19.6 A scheme of virtual machine tool (VMT) structure [8].

Fig. 19.7 Simulation of five-axis milling of complex parts using hyper MILL® 5AXIS CAM package [9]: (a) milling of curvilinear channels; (b) envelope (3D contouring) milling of die for a tyre; (c) rough milling of a blisc.

machine and NC post-processing. The CE package allows the users the association between machine code data (MCD) and specific machining operation within the NC program to determine, and by reading MCD, to emulate the actions of CNC machines and controllers. Figure 19.7 shows some examples of simulations of five-axis milling performed on different parts including deep curvilinear channels, the sculptured surface of a tyre-die and a blade disc (blisc) using a CAD/CAM package.

Programming software for machine tools enables manufacturers to add new features easily for improving metal cutting efficiency. With the latest updates to CAD/CAM and NC simulation/verification software, manufacturers can take advantage of an array of improvements including expanded support of multitasking machine tools, full machine 3D simulation capabilities, plus new tool paths for multi-axis and high speed machining [15]. For instance, for five-axis machining, new capabilities include a wider range of five-axis strategies for roughing and finishing, faster calculation times, improved point distribution, and easier data management. The five-axis roughing strategies now include machining to or from a point, orientation through a line or curve, and programming using a reference surface.

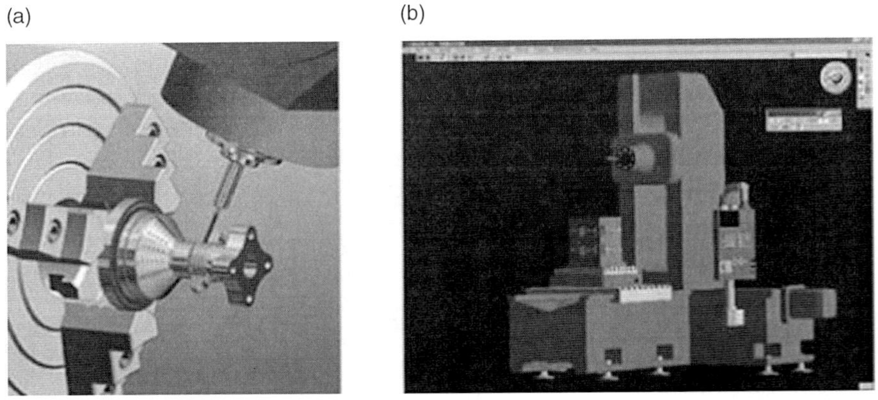

Fig. 19.8 Advanced 3D simulation capabilities: (a) machining hole on mill-turn centre using EdgeCAM 11, (b) 3D simulation of HMC using GibbsCAM [15].

Multi-axis improvements included in the latest EdgeCAM 11 add new functionality to support four- and five-axis simultaneous milling on the latest-generation mill-turn machines, as well as improved drilling routines for automatically finding and machining holes through cylindrical (Fig. 19.8a) and conical faces. Multi-axis additions from CNC Software Inc. include new multi-axis functionality and tool paths for HSM and hard milling to its latest Mastercam X Maintenance Release 2 software.

The updated Mastercam software adds a new engine to the advanced multi-axis tool paths, which provides a customized, streamlined interface that is fine-tuned to specific applications, such as impellers and turbine blades. Depending on which tool path is being used, Mastercam's parameter pages display only parameters applicable to the specific tool path type. Among advanced 3D simulation capabilities, PartMaker 8 and GibbsCAM provide a new full-machine simulation module which enables photorealistic 3D models of machine tools in action (Fig. 19.8b). The simulations, based on actual 3D solid models, can help users pinpoint potential errors and collisions on the complex multitasking turn-mill and Swiss-type turning machines. With the new system, the machine model being simulated incorporates machine specific tool-holders and attachments to ensure that any possible collisions that might occur on the machine will be detected off-line on the user's PC. As a result, CNC programmers and machinists spend less time setting up new jobs to ensure there are no collisions on the machine. In addition, Gibbs introduced a new GibbsCAM Machine Simulation option for multitasking (MTM), mill/turn, and turning machines, allowing an entire machine tool motion of a CNC program to be validated in an accurate simulation. New FX technology by Esprit CAM for CAD-to-CAM feature exchange allows users to automatically capture design intent, defining what is being machined and allowing programming parts much more quickly and easily. The FX technology provides portions of the original CAD Feature Tree directly inside the Esprit CAM system's interface, including original design elements with features, tolerances, material properties, surface finishes, and administrative data. The system enables mapping the CAD features and their associated properties into machinable features providing a complete definition of what is being machined [15]. The Virtual NC package by DELMIA [10] enables manufacturers to validate the post-processed NC program off-line, in a digital environment, leaving the actual machine tool in production. Once the tool path is verified and optimized, the programmer can validate the entire NC process using a model of the machining centre including tool changer and material handling devices. The program cycle time, tool life and surface finish are optimized based on the analysis data collected during the machining cycle. Virtual NC's built-in, 'Mimic' configurable controller emulator provides continuous monitoring and feedback of the machining cycle and automatically creates user-defined reports. A multi-axis gauge detection system is implemented to compare the machined part with the pre-defined design part and record the gouges onto a table. By selecting a gouge from the table its COI (centre of interest) is zoomed in automatically while dynamical rotating, translating or snapping the slicing plane allows better visualization of the gauge. Furthermore, by integrating the Virtual NC environment with a data model called the Manufacturing Hub, manufacturers can store, manage and reuse all product, process and resource information required throughout the product life cycle.

In order to meet the demands of the new manufacturing paradigm, manufacturers adopt the next-generation of digital manufacturing tools like the Dassault Systems DELMIA V5 platform [12, 13]. This manufacturing solution set represents an integrated suite of tools that provide manufacturing process design, tool and fixture design, factory and production system design, visualization, and automation through powerful 3D virtual simulation tools. This allows the manufacturing engineer to design, synchronize and validate production lines, robotic workcells, machining centres, production equipment, and control systems completely prior to the purchase, installation, and commissioning of a single piece of physical equipment [12]. The immediate effect and benefit for the manufacture is a substantial reduction in manufacturing life cycle time to launch a product and significant cost savings realized by the virtual validation and commission of production systems. Figures 19.9 and 19.10 show virtual simulations of a robotized VMC and two robotized workcells simulated by means of DELMIA V5 and Tecnomatix simulation platforms.

The new vertical machining centre DMC 635 V with its 8000 rpm spindle, tool magazine with 20 pockets, high rapid traverses and high tech controls offers customers benefits without compromises. Large travels with

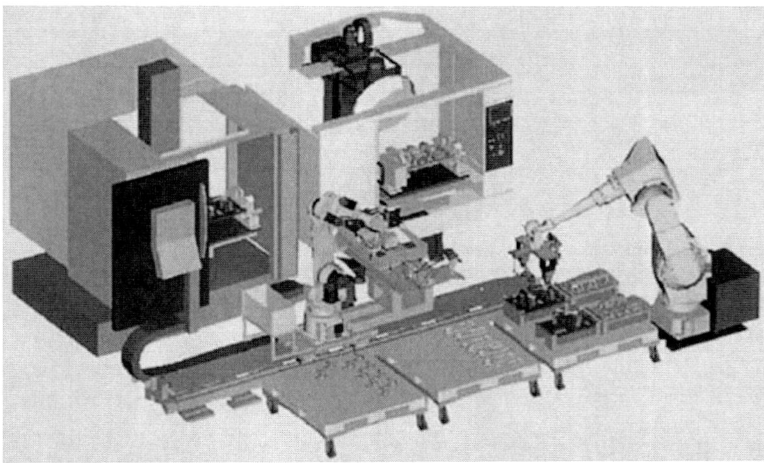

Fig. 19.9 Virtual simulation of robotized vertical machining centre [9].

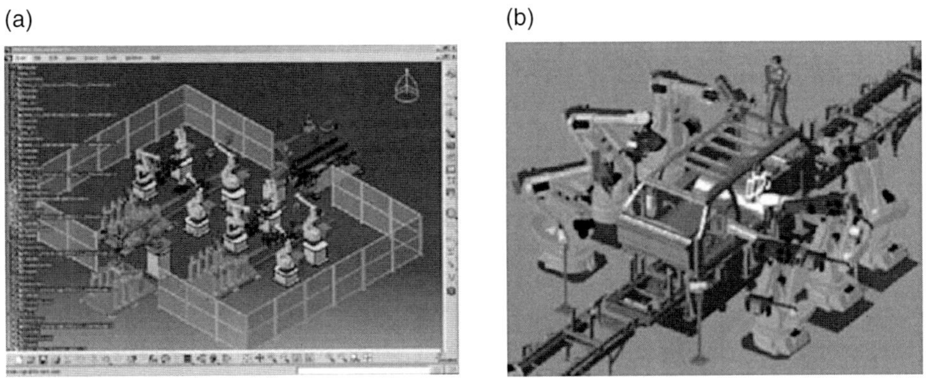

Fig. 19.10 Examples of virtual robotized workcells: (a) DELMIA V5 virtual workcell [12], (b) automotive workcell's 3D simulation from Tecnomatix [14]. (see Plate 26)

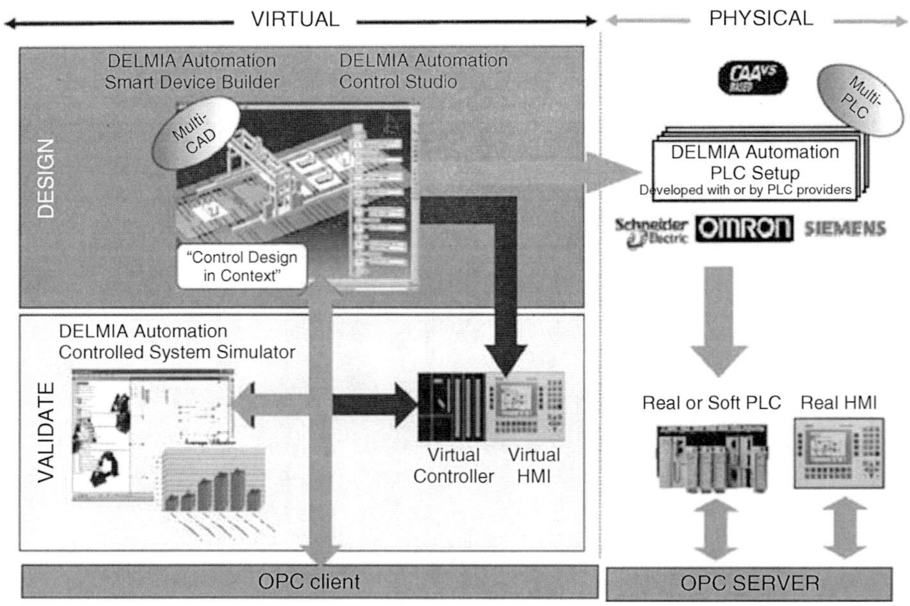

Fig. 19.11 DELMIA V5 Automation Platform [12].

low space requirement and top performances at low costs are achieved. Rapid traverses of 30 m/min and the fast tool changer with double gripper guarantee excellent performance. The CNC controls of Siemens, Heidenhain or Fanuc offer high user friendliness.

In the virtual 3D world created by the PLM technologies involving solid model product building, digital mock-up and manufacturing process simulation, the final link to the factory floor is created by the connection to machine control systems. For digital manufacturing the problem arises of how accurate information capable of driving control systems for all production equipment can be generated, i.e. automation systems and controllers can be merged. When using the DELMIA V5 Automation platform shown in Fig. 19.11, it is possible to develop and validate an appropriate programmable logic controller (PLC) program along with the logic and motion control of the automated system. The 3D digital environment provides a collaborative workspace for mechanical design, manufacturing, tooling and control to share knowledge, exchange system features and attributes, integrate domain process information, and react to engineering change and version updates.

As shown in Fig. 19.11 the DELMIA V5 Automation Platform is comprised of several modules, namely:

- Smart Device Builder to create actuators, sensors, motors and drives that build the virtual automated system.
- The LCM studio being a PLC programming environment to produce a generic PLC program independently of the PLC hardware platform.
- The controlled System Simulator which allows the user to simulate, debug and validate a complete PLC program against virtual equipment.
- The LCM Control Setup providing the user the capability to translate (post-process) control logic created with the LCM.

Another optional module is the HMI Control Panel Design which is used with the first three modules to create a virtual human/machine interface (HMI) panel.

The virtual factory still remains rather a futuristic object but due to recent advances in collaborative product development using the 3D visualization software, virtual reality and product life cycle management (PLM) systems, larger automakers (General Motors Corp., Ford Motor Co., Daimler-Chrysler AG and Toyota Motor Corp.) and aerospace (Boeing, Airbus) manufacturers are seriously evolving in digital manufacturing to streamline product development. In these manufacturing sectors, such visualization tools as 3D mock-up, virtual reality and virtual prototyping systems are widely deployed. Virtual reality (VR) and simulation software tools are expected to reduce drastically product development costs through the elimination of expensive physical prototypes. Especially, automotive, aerospace and medical device manufacturers are seeking to deploy sophisticated, collaborative systems for product development planning and virtual simulations for designing plant layouts and within manufacturing cells (Fig. 19.10). The CAVE (Computer Automated Visualization Environment), developed in the early 1990s by the Electronic Visualization Laboratory at the University of Illinois at Chicago, first found commercial applications by automakers and aircraft manufacturers [17]. For example, one of large VR system configurations is capable of projecting a massive immersive VR image on walls up to 2.4×7.3 m. Over the next few years software capable of providing a true 1:1, or human-scale, immersive visualization environment appeared. In addition, projection and display technologies have improved to the point where it is possible to create easily high resolution stereoscopic images – seeing an image in each eye with depth and volume just the same as in the real world. In particular, large-scale displays enable manufacturing teams to collaborate in much the same way they used to work around drafting tables, but with realistic, full-scale 3D models. Recently, an interface was developed by Tecnomatix [9] to integrate the eMPower VR software, which allows on-screen stereoscopic representations of objects and large-scale projections with its planning platform to enable users to review and verify the status of layout planning, logistics and cells within manufacturing facilities. To follow this trend, Airbus uses virtual factory simulation software by Delmia Corp. [9, 16] for managing tasks, including material flow, ergonomics simulations and assembly studies, on its next-generation A380 super jumbo jet. Moreover, practically all world-known automakers implement extensively VR visualization in automotive design and collaboration of manufacturing teams.

Currently, virtual reality environments are increasingly used in industry to realistically simulate new manufacturing processes before they are tested on the factory floor. As an example, Figs 19.12a and b show an

(a)

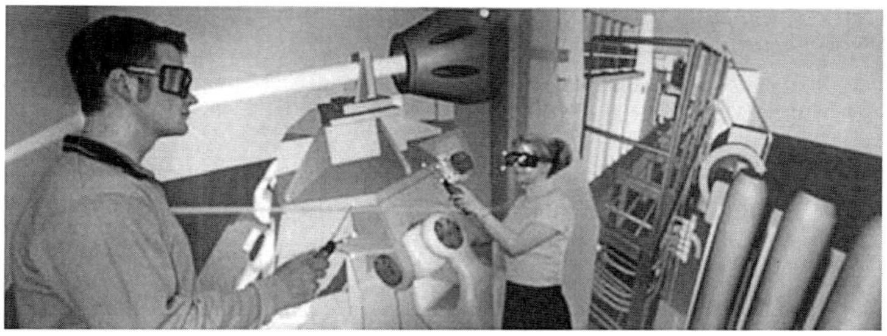

(b)

Fig. 19.12 Application of VR technique to simulate a machine tool [17] (a) hexapod machine tool (b) [9].

analysis of a machining centre from five different sides using the CAVE VR system and using a virtual reality 'immersive environment' to simulate the action of a hexapod machine tool, respectively. As mentioned in Chapter 15.4, octahedral hexapod machine tools represent the first major impact in machine tool design with an innovative support system technology.

19.3 Internet-based Manufacturing

The internet is a global information system that provides an efficient communication medium for connecting computers and computing resources around the world. The major internet services and applications include [20]: electronic mail, file transfer protocols, electronic discussion groups, search tools, and the World Wide Web (WWW), a hypertext-based information medium. Commercial entities employ the internet widely for various forms of e-commerce and/or e-business, and recently also for the future manufacturing environment called electronic-manufacturing (e-manufacturing). Linking the plant floor machines to the company business or even supply chains is still a complex challenge for both the machine tool manufacturers and customers. Using the internet for control with a series of services such as File Transfer Protocol (FTP), Telnet, WWW, electronic mail to transmit the information via networks allows exploiting interfaces with many multimedia resources. It is expected that future wireless plant-floor networks will involve remotely controlled manipulators and robots on a larger scale, as well as CNC machines and other similar industrial equipment. According to a current expertise, e-manufacturing is especially suited to the activities developed during the product development cycle in integrated and collaborative CAD/CAPP/CAM environments [21]. In general, the creation of e-manufacturing required a number of IT-based collaborative virtual and electronic-activities (e-activities) involving the more general term electronic-work (e-work). The range of e-work covers such applications as telerobotics, telemanufacturing, teleoperation and remote services.

The University of California-Berkeley has developed an integrated CAD/CAPP/CAM software environment termed CyberCut [1, 23], which is defined as 'an experimental fabrication test bed' for an internet-accessible, computerized machining service. A software test bed is simply defined as a web-based interface to an application program [22]. First, the client/designer can create mechanical components, beginning with a Java-based, CAD system that can be downloaded to their web browser, as illustrated in Fig. 19.13. The design files are then submitted to the server for process planning. CyberCut has an integral *macro-planner* that comprises a set of machining rules specific to the fabricating machine and when the design is completed, the resulting part geometry can be sent to the *micro-planner*, which converts the geometry into a machining script. A remote designer can use an open-architecture CNC machining system and stereolithography (SLA) or selective laser sintering (SLS) fabricating systems for the rapid fabrication of prismatic mechanical parts with tolerances in the range ± 0.05 mm. The CyberCut uses extended Solid Interchange Format (SIF) based on the Destructive Solid Geometry (DSG), termed DCG-IF, which directly mimics the successive material removal from the initial stock volume. The DSG language allows the necessary Boolean operations on subvolumes to be specified.

There are a number of ways to connect a machine's tool to the network but generally three different types of boxes are used to plug CNCs into traditional RS232 ports [22]. The differences result from the connection with computer and the placement of the box. They are as follows (Fig. 19.14):

1. *A USB-to-serial adapter.* This box plugs directly into the back of a computer and must be located within 15′ (4.6 m) of the PC. The box may have anywhere from one to 16, or even more, ports to plug CNCs into. Generally, these boxes are chosen because they are the least expensive and easiest to set up.
2. *An Ethernet-to-serial adapter.* These adapters are similar to the USB-to-serial adapters, except that they plug into the network instead of the USB port on the back of a PC. This type of box is ideal if the computer running the DNC software is not located centrally to the CNCs it will control. Multiple hubs can be located throughout the shop to control different groups of CNCs. Installing the devices will take at least a basic knowledge of networking, and the boxes cost slightly more than the USB-to-serial adapters.

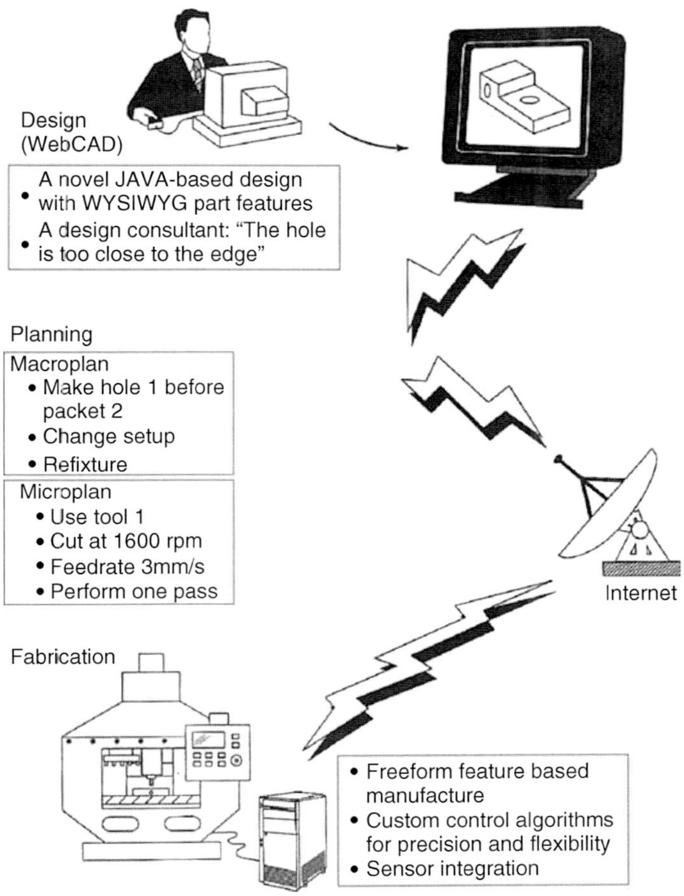

Fig. 19.13 The CyberCut project: integrated design, planning and fabrication [1].

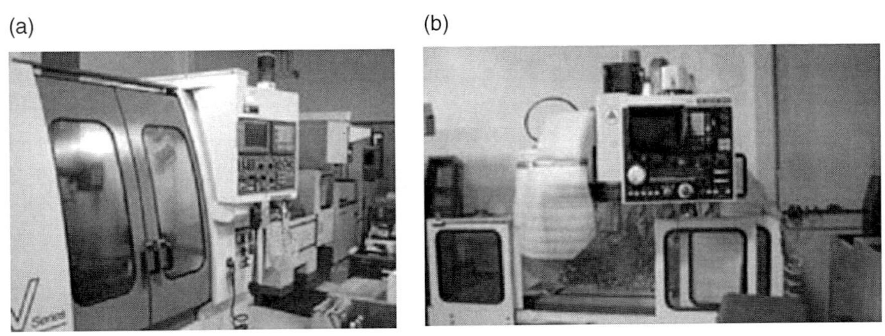

Fig. 19.14 Wireless communication systems connect CNC machines with the network [22].

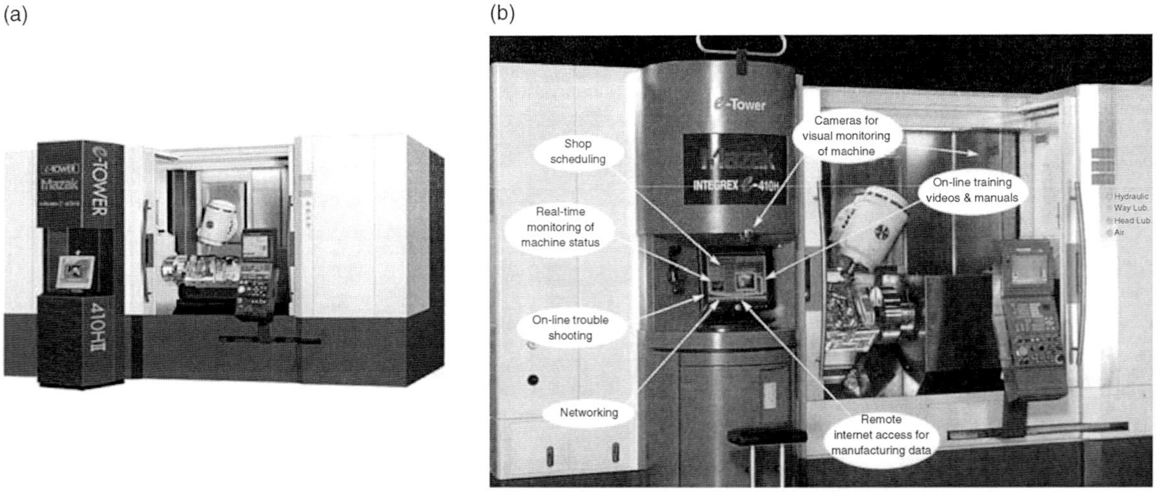

Fig. 19.15 Multitasking machining centre Integrex e-410H II (Mazak Corp.) equipped with an internet information terminal (e-Tower) [19]: (a) general view, (b) functions of information terminal. (see Plate 27)

3. *A wireless-to-serial adapter.* This box, mounted directly onto a CNC, talks to the computer system through a standard 802.11b/g network (often called WiFi). Wireless access points should be mounted in the ceiling to provide a wireless signal to the shop, and if the shop is large enough it may need to mount multiple access points in different areas. Each of these access points will have an Ethernet line which runs back to the network. There is still a serial connection between the wireless adapter and the CNC. The main drawback is that the high level of complexity required a qualified person in both wired and wireless networks to help sort through IP addresses and encryption settings.

Figure 19.15 presents the multitasking machining centre by Mazak Corp. equipped with the Mazatrol Matrix CNC and an internet information terminal named e-Tower (this series of innovative multitasking machining centres was indicted by Integrex). All Integrex® *e*-Series machines are equipped with the unique e-Tower, which plays the role of information centre for setup support, maintenance support, manuals and work scheduling, either in-office or off-site. The information terminal features a comprehensive range of information technology (IT) to make the machining centre a cyber production module, which can be used independently or integrated into an advanced cyber factory developed recently by Mazak Corp.

They can be listed as follows:

1. *Management functions.* All information, such as current machine operation status, job status and others can be monitored in an office or off-site. Moreover, the operator has convenient machine shop access to an extremely large amount of information.
2. *Work scheduling.* Jobs to be performed can be transferred from a PC in the office to inform the machine operator.
3. *Mobile messenger.* Machine operational status can be remotely monitored by using a mobile phone. It can quickly and easily check the completion schedule of a job over a mobile phone.
4. *Setup support.* e-Tower reports the current machine status and job status, scheduling of new jobs to be performed accurately and easily. The monitor screen can display test cutting of a workpiece feature that is difficult to see and optimize machine setup. Digital machine manuals are conveniently accessible to answer any kind of question. Maintenance Support and Alarm Recovery resolve small problems without the necessity of calling a service engineer, which minimizes machine idle time.
5. *Verbal messages.* Various instructions and reports are made verbally to the operator. Available languages are English, German, French, Italian, and Japanese.

6. *Image transmission.* Pictures showing machining that the operator cannot clearly see or the status of machine units, such as the tool changer, can be sent to the e-Tower monitor or an off-site location.
7. *Fingerprint identification system.* This prevents unauthorized operators from operating the machine.

In recent years, the operating reliability of machine tools and controls has improved continuously but often machine downtime leads to long production breaks due to simple programming or parameter problems. Several days' standstill before the technician service arrives, and analyses and solves the problem, is unacceptable. On the other hand, when remote diagnostics arises, the service communicates with the control over modem or ISDN, analyses the control system and repairs it immediately. For example, Heidenhain [18] offers the TeleService and TNCdiag PC-based software to perform quick and simple remote diagnosis including the CNC control system, the inverter system and even the driving motors. As shown in Fig. 19.16 the TeleService software installed on a network enables remote operation and remote monitoring of a CNC control system connected to the network. Moreover, the TNCdiag software enables the service technician to find errors in the drives (even ambient conditions leading to the errors can be investigated using dynamic display of status signals).

The Center for Intelligent Maintenance Systems (IMS) has developed the test bed, which demonstrates three fundamental features of an e-manufacturing system: transforming data from devices or machines to networked user systems using XML (eXtensible Markup Language), HTML (HyperText Markup Language), and SQL (Structured Query Language) through the IMS platform; predicting machine performance using the Watchdog Agent for prognostics; and synchronization of performance information with e-business, supply chain (including CRM, SCM, ERP) software, and manufacturing execution systems (MES) system tools. This activity started with machine-level integration, followed by installation of the workcell. Finally, the test bed consists of two machining centres, robots, and a network connected in a complete e-manufacturing cell. The IMS Center's e-maintenance tools provide predictive intelligence to monitor machine and process performance through the internet and wireless communication systems to prevent them from unexpected breakdown. In addition, the systems can compare products' performance through globally networked monitoring systems that allow companies to focus on degradation monitoring and prognostics rather than fault detection and diagnostics.

The system includes a Watchdog Agent that provides continuous monitoring and prognostics of asset degradation, and enables companies to rapidly evaluate assets' performance. Its output represents the quantitative measures of the performance of the product or equipment at a given state. By comparing performance value at different states, it can predict and assess the condition and performance of complex and sophisticated

Fig. 19.16 Remote diagnosis with TeleService [18]. Source: Heidenhain.

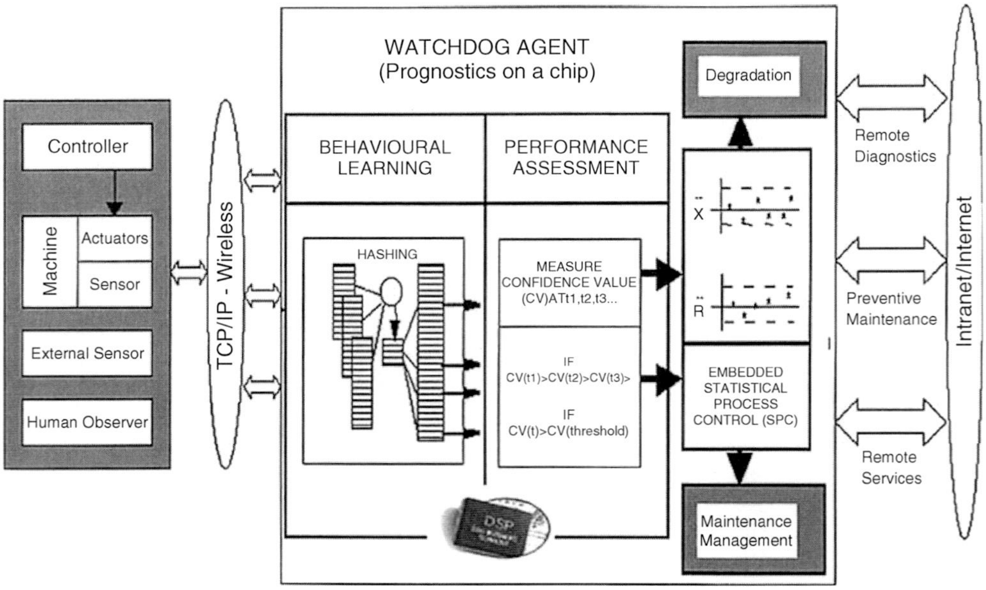

Fig. 19.17 The IMS 'Smart Watchdog Agent' for predictive e-maintenance [24, 25].

machine systems. To effectively apply this methodology in various kinds of products and machines, its integration with the working environment, such as sensors, actuators, controllers, and human interfaces, needs to be further developed. The IMS 'Smart Watchdog Agent' for predictive e-maintenance and its working principle is shown in Fig. 19.17.

Since the watchdog algorithm correlates multiple inputs of relevant data, the system can compute performance in mathematical form. The system also uses hashing to give a better view of multiple parameters on spindle performance, such as current, temperature, and vibration. Finally, the synchronization element of the IMS test bed provides the required integration with e-business systems, including applications such as customer relations management (CRM), supply chain management (SCM), and other e-commerce systems. By knowing the degradation of machines in the production floor, the operation supervisor can estimate its impacts on the materials flow and volume and synchronize it with ERP systems. For example, when tools are worn out, the information can be directly channelled to the tool providers and update the tool needs for tool performance management.

In order to support the manufacturing globalization activities on a large scale, the remote manufacturing system technology needs to be formalized to define the underlying issues for remotely integrated manufacturing. Figure 19.18 shows the remote performance assessment system for machine tool concept and its key functions. A remote intelligent reasoning agent serves a central intelligence unit for data acquisition, reasoning, remote diagnostics, telemaintenance and remote training system. The system could support remote facilities to ensure the performance of manufacturing equipment, quality of operations, and productivity of the entire plant, in particular, for those facilities located in developing countries. In a global environment, it is necessary for companies to have tools for remote access of data and an ability to analyse and troubleshoot manufacturing remotely. This need is addressed by developing an internet-based surface texture analysis and information system, and a peer-to-peer data-sharing tool for sharing, analysing and storing surface texture data, analysis results, documents and reports [32]. A block scheme of an internet-based surface texture analysis is shown in Fig. 19.19. The users interact with the system using a web browser as the interface. It has the capability of remote access for data and analysis and the users can store data files into a database system and then load them to the analysis system to perform analysis remotely. Finally, the users can search, view and share information regarding surface and form metrology (roundness, cylindricity, straightness and flatness),

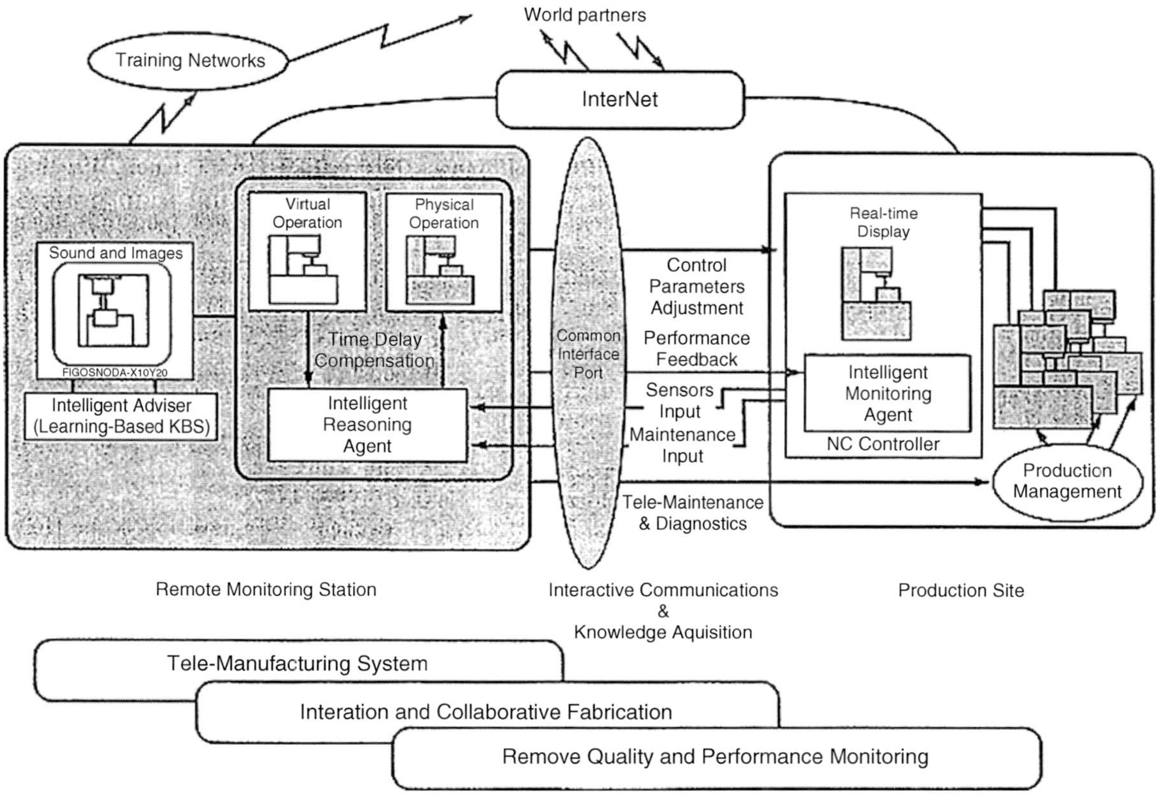

Fig. 19.18 Remote performance assessment system for machine tools [26].

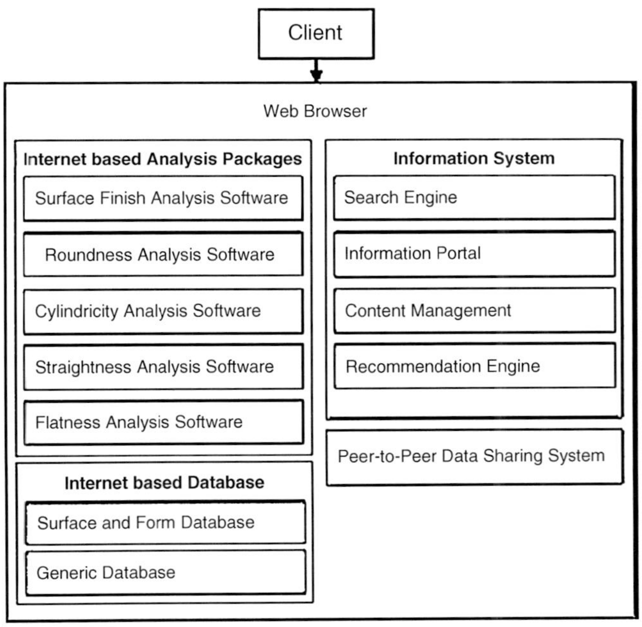

Fig. 19.19 A framework for internet-based surface texture analysis [32].

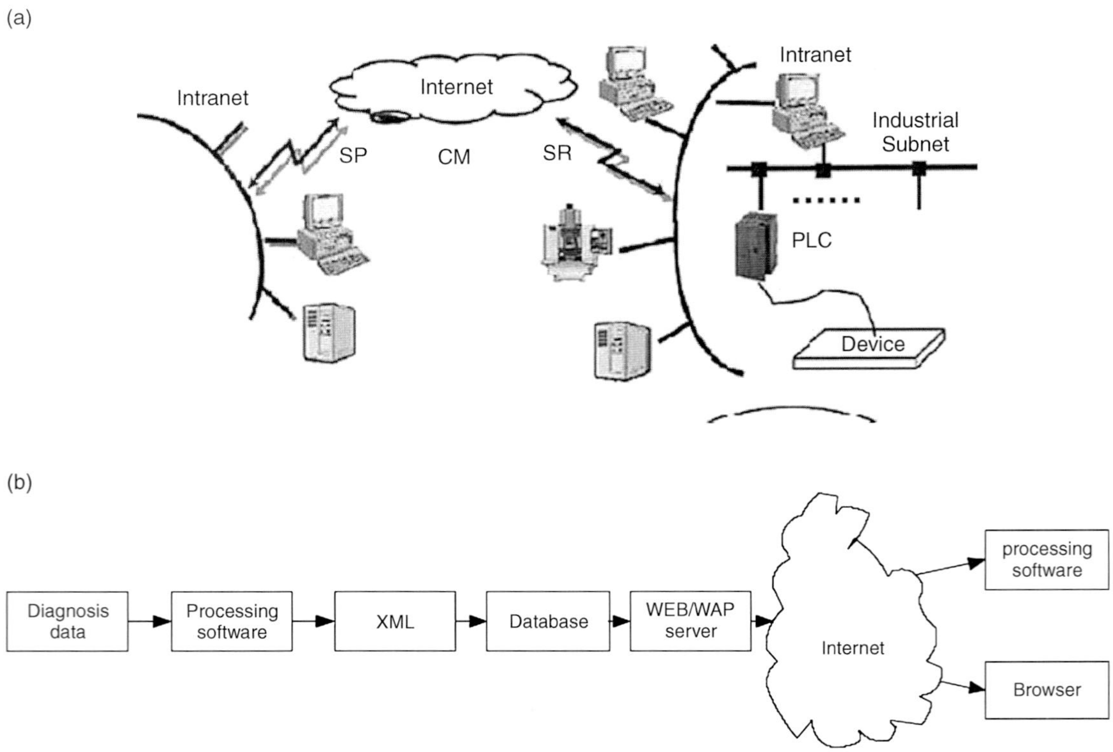

Fig. 19.20 Teleservice via internet-based remote access [30]: (a) trial system, (b) XML-based system.

and retrieve data from different peers on the network. It is postulated that the internet-based surface texture analysis system can serve as a medium for quick deployment of solutions.

A manufacturing-oriented teleservice is one of the most promising strategies for next-generation manufacturing (NGM). In general, remote access to industrial equipment via the internet requires the basic unit of a teleservice consisting of three kinds of components: a service provider (SP), a service receiver (SR) and a communication medium (CM) between the SP and the SR [30], as shown in Fig. 19.20a.

In the system architecture from Fig. 19.20a, the worldwide internet functions as the basic CM. To avoid unreliable security and speed a hierarchical communication network can be built in order to involve the more popular and mature enterprise intranet and various industrial subnets into the CM system in the local SP and SR sides. As a result, a trial communication network system integrating the global internet, existing intranet and the to-be-built industrial subnet was designed as shown in Fig. 19.20a. To ensure security, the industrial device as well as its PLC controller is instigated first via an industrial subnet and then a local intranet to access the worldwide internet.

Currently, teleservice for remote machine monitoring and troubleshooting is not widely used in machine tools and the machining sector, although many Japanese and American companies have offered such capabilities as an option for the high performance machine tools since 2000. However, in the future this type of service and maintenance will become extremely important due to overseas activities under a globally integrated manufacturing enterprise. As a result new terms – teleservice engineering and telemanufacturing – are proposed to formalize the functions, facilities and structures of these emerging production concepts [25, 27]. Teleservice engineering is addressed to support a globally integrated digital service enterprise including manufacturers and customers. A typical teleservice engineering system is shown in Fig. 19.16. In particular, multimedia-based tools such as watchdog neural chip can improve the effectiveness of digital maintenance and diagnostics.

On the other hand, telemanufacturing is referred to as an activity via communication networks and especially across information superhighways resulting in performing operations in real time and processes necessary to the design and production of goods. The main goal of telemanufacturing is to enhance the flexibility of manufacturing enterprises, especially those of medium and small sizes [27].

Telemanufacturing differs from other manufacturing philosophies such as computer-integrated, holonic or virtual manufacturing because it requires an in-house control system to coordinate different service providers, possibly via the internet, and various floor functions of manufacturing systems. A pilot remote manufacturing system (telemanufacturing) termed WebTurning was implemented for producing rotational parts in a CNC turning centre [21]. The turning centre teleoperation system, based on a client-server architecture, is composed of a graphic user interface for teleoperation, WebCam monitoring system for image/video capture which monitors CNC screens, and WebCNC teleoperation/video server of the CNC machine. Moreover, the WebTurning system is composed of two modules: servers represented by the programs located on the Linux platform and connected with the CNC machine tool through an Ethernet network interface, and the client represented by a Java applet.

Today, the development of communication technologies such as GSM (global system for mobile communication), CDMA (code division multiple access) and both WEB (WWW – World Wide Web) and WAP (wireless application protocol) makes the remote machine maintenance possible from anywhere at anytime. In the system with a WEB/WAP server and XML (eXtensible Markup Language) (Fig. 19.20b) users can not only get the diagnosis data by computer but also data, image and video of the machine by mobile phone [31].

Figure 19.21 illustrates a futuristic image of an intelligent manufacturing system (IMS) in which an artificial satellite-based global communication network is used to connect the head office with a CAD/CAM design centre, intelligent production management system and intelligent production cell represented by

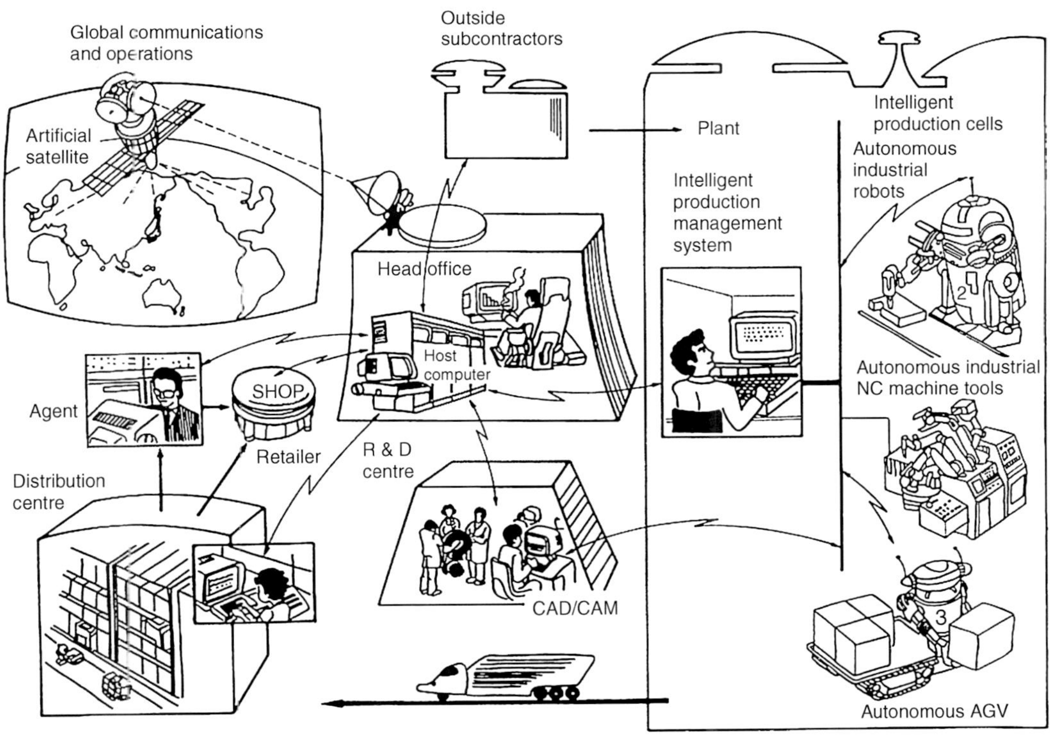

Fig. 19.21 Image of future IMS [28].

the autonomous distributed manufacturing system via host computer. In addition, the IMS involves other activities such as product distribution and interactive communication with outside subcontractors.

References

[1] P.K. Wright, *21st Century Manufacturing*, Prentice Hall, 2001.
[2] B.P. Erdel, *High-Speed Machining*, SME Publish, Dearborn, 2003.
[3] M.G. Merchant, D.A. Dornfeld, P.K. Wright, Manufacturing – its evolution and future, *Transactions of NAMRI/SME*, Vol. 33 (2005), 211–218.
[4] F. Kimura, Product and process modelling as a kernel for virtual manufacturing environment, *CIRP Annals*, Vol. 42/1 (1993), 147–150.
[5] K.I. Lee, S.D. Noh, Virtual manufacturing system – a test-bed of engineering activities, *CIRP Annals*, Vol. 46/1 (1997), 347–350.
[6] M. Onosato, K. Iwata, Development of a virtual manufacturing system by integration product models and factory models, *CIRP Annals*, Vol. 42/1 (1993), 475–478.
[7] C.R. Boër, F. Jovane, Towards a new model of sustainable production: ManuFuturing, *CIRP Annals*, Vol. 45/1 (1996), 415–420.
[8] The Virtual Machine-Tool, www.mech.northwestern.edu/MFG/AML/VMT/index.html.
[9] www.icam.com, www.gildemeister.com, www.ugs.com/products/tecnomatix.
[10] DELMIA VIRTUAL NC, www.delmia.com.
[11] A virtual reality machine tool, www.nist.gov/public_affairs/gallery/vrmanu.html.
[12] DELMIA V5 automation platform: merging digital manufacturing with automation, ARC white paper, www.arcweb.com.
[13] Digital Manufacturing in PLM environments, A CIMdata white paper, www.delmia.com.
[14] P. Waurzyniak, Digital Manufacturing Taking Hold. Collaborative 3-D visualization tools help manufacturers sharpen focus, *Manufacturing Engineering*, Vol. 130, No. 1 (2003).
[15] P. Waurzyniak, CAD/CAM software advances, *Manufacturing Engineering*, Vol. 137, No. 5 (2006).
[16] P. Waurzyniak, Going virtual. Lower-cost systems expanding audience for new virtual reality systems, *Manufacturing Engineering*, Vol. 128, No. 5 (2002).
[17] Productionstechnik Sachsen (Virtual Reality Center Production Engineering, TU Chemnitz/IWP), 2005, p. 28.
[18] Diagnostics for HEIDENHAIN controls, www.heidenhain.com.
[19] Mazak
[20] T. Bailey, P. Renton, M.A. Elbestawi *et al.*, Internet based implementation of multi-axis milling process simulation and optimization. Available at: http://mmri.eng.mcmaster.ca/como/docs/cirp2000/Cirp2000_final.html.
[21] A.J. Alvares, J.C.E. Ferreira, WebTurning; teleoperation of a CNC turning center through the Internet, *J. Mater. Proc. Technol.*, Vol. 179 (2006), 251–259.
[22] J. Lorinicz, Wired and wireless systems, *Manufacturing Engineering*, Vol. 138, No. 1 (2007), www.sme.org.
[23] P. Wright, D. Dornfeld, Cybercut: a networked machining service, *Trans. NAMRI/SME*, Vol. 26 (1998), 281–286.
[24] P. Waurzyniak, Moving toward the e-factory, manufacturing industry takes first steps toward implementing collaborative e-manufacturing systems, *Manufacturing Engineering*, Vol. 127, No. 5 (2001).
[25] J. Lee, Teleservice engineering in manufacturing: challenges and opportunities, *Int. J. Mach. Tools Manuf.*, Vol. 38 (1998), 901–910.
[26] J. Lee, Overview and perspectives on Japanese manufacturing strategies and production practices in machinery industry, *Int. J. Mach. Tools Manuf.*, Vol. 37 (1997), 1449–1463.
[27] L.L. Abdel-Malek, C. Wolf, P.D. Guyot, Telemanufacturing: a flexible manufacturing solution, *Int. J. Product. Economics*, Vols 56–57 (1998), 1–12.
[28] Y. Fukuda, Trends on manufacturing systems, *Int. J. Japan Soc. Prec. Eng.*, Vol. 27, No. 4 (1993), 291–294.
[29] P. Schmitt, PLM in aerospace, *Manufacturing Engineering*, Vol. 138, No. 3 (2007).
[30] F. Qiao, H. Schlange, H. Meier, W. Massberg, Internet-based remote access for a manufacturing-oriented teleservice, *Int. J. Adv. Manuf. Technol.*, Vol. 31 (2007), 825–832.
[31] W. Wang, P.W. Tse, J. Lee, Remote machine maintenance through Internet and mobile communication, *Int. J. Adv. Manuf. Technol.*, Vol. 31 (2007), 783–789.
[32] S.H. Bui, B. Muralikrishnan, J. Raja, A framework for Internet-based surface texture analysis and information system, *Prec. Eng.*, Vol. 29 (2005), 298–306.

CHAPTER TWENTY

SURFACE INTEGRITY

Contents

20.1 SUPERFICIAL LAYER AND SURFACE INTEGRITY

The solid surface of a tool, a part of a machine, an element or even a finished product are subjected to special physical and chemical processes which cause that the service properties of the superficial layer obtained are substantially better than those of the core or substrate. The difference between the surface of a solid and the core of the material (bulk material), in terms of structure and properties, can be considered in the following aspects [1]:

- A distinct energy condition causing the state of elevated energy and enhanced adsorption activity.
- Combination of mechanical, thermal, electrical, physical and chemical effects on the surface during manufacturing processes.
- Cyclic or continuous mechanical, thermal, chemical or physical action of the environment in which the part is exploited during its service.

The technology of surface improvement involves generally various manufacturing and processing technologies such as casting, metal forming, machining and heat treatment processes but predominant effects can be achieved using special surface engineering techniques including finishing, plating and coating technologies [2]. As a result, technological surface layers with changes in the microstructure, mechanical properties, residual stresses, chemical compositions, etc. are produced by application of various manufacturing methods, either independently or jointly. However, it should be noted that the real surfaces contain a set of interruptions of macro- and microscopic continuity, consisting of crevices, porosity and irregularities in structure, laminar or mosaic in character, or caused by the inclusion of foreign matter [1].

The layer of metal or alloy deformed by the manufacturing or fabricating process – physically (by heat, force, diffusion or foreign atoms), chemically (e.g. by oxidation) and structurally situated below the physically pure surface is the *subsurface layer* or *superficial layer* if it is relative to the core. From a standardized metrological point of view, a superficial layer is defined as a layer of material being processed, limited by the real surface of the part from the top, which includes this surface and the portion of the material beneath the real surface with different physical, and sometimes chemical, properties in comparison with the initial properties of the bulk material. The models of the superficial layer after machining differ among themselves chiefly by the degree of detail or by comprising a different number of phenomena occurring in the process. This leads to

Advanced Machining Processes of Metallic Materials
ISBN: 978-0-08-044534-2

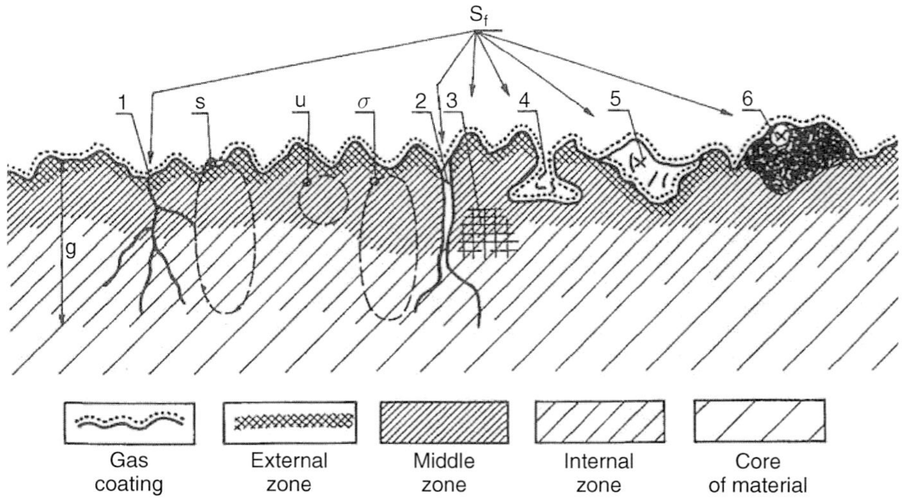

Fig. 20.1 Diagram showing a set of factors characterizing the superficial layer of a solid, against the background of a three-zone model of the superficial layer [1]: *g* – thickness of superficial layer; *s* – structure of superficial layer; *u* – strengthening of the middle zone; σ – residual stresses in the superficial layer; S_f – flaws of the superficial layer; 1 – microfracture; 2 – fissure; 3 – microshrinkage; 4 – porosity; 5 – gap; 6 – inclusion.

creating three-zone up to eight-zone models (through four-, five- and six-zone models) of the superficial layer [1]. As shown in Fig. 20.1 the simplest three-zone model consists of the following layers:

1. The external zone with a thickness of 0.001 to 0.02 μm formed by a layer of foreign particles (for example, by particles of a tool material, coolant or lubricant residues, perspiration, dirt, dust, etc.) mixed with spalled particles of the core material.
2. The middle zone with thicknesses ranging from 0.5 to 500 μm and consisting of strongly deformed grains of the bulk material, which in many cases may be significantly textured. Its structure has predominant influence on the service properties of the machined part.
3. The internal zone consisting of grains which were not permanently deformed but have a visibly different structure than the core, e.g. resulting from transformations induced by heat. The sharp transition between the internal zone and the core is practically not observable but it can be localized by the penetration of residual stresses or strain-hardening effect.

Under normal environmental conditions (Fig. 20.2), the oxide layer usually lies on top of the work-hardened layer and surface oxide layers are generally covered with adsorbed layers of gas and moisture. Finally, the outermost surface of the metal may be covered with contaminants, such as dirt, dust, grease,

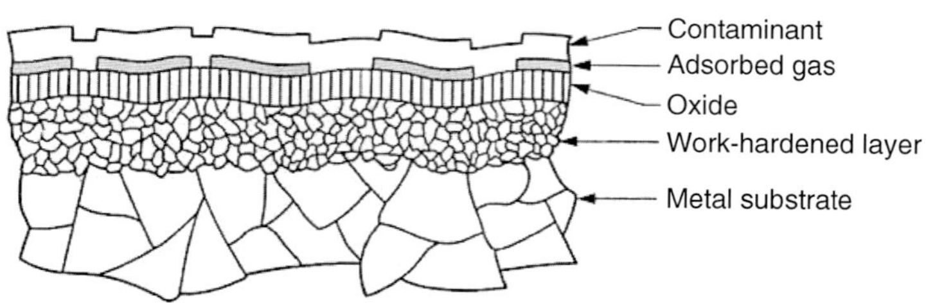

Fig. 20.2 Schematic illustration of the cross-section of the surface structure of metals [3].

lubricant residues, cleaning-compound residues, and pollutants from the environment. The thickness of the individual layers depends on processing conditions and the environment. Thus surfaces generally have properties that are very different from those of the substrate. For example, the oxide on a metal's surface is generally much harder than the bulk metal. On the other hand, non-uniform surface deformation or severe temperature gradients during manufacturing operations usually cause residual stresses in the work-hardened layer.

The state and attributes of a manufactured surface that influence performance are described by the term *surface integrity*. Surface integrity can be defined as:

1. The inherent or enhanced condition of a surface produced in a machining or other surface generation operation [4], or
2. The topological, mechanical, chemical and metallurgical 'worth' of a manufactured surface and its relationship to functional performance [5].

Surface integrity can be simplistically divided into two parts: first, the external aspects of topography, texture and surface finish and second, the internal subsurface aspects of metallurgy, hardness and residual stress [5]. The concept of surface integrity, originally defined for grinding operations [4], can be extended to any finishing operations to encompass six different groups of key factors: visual, dimensional, residual stress, tribological, metallurgical and other factors as illustrated in Fig. 20.3. It should be noted that all the parameters involved in the finishing process have a direct influence on the surface integrity of the part. On the other hand, the six groups of key factors presented in Fig. 20.3 are not random, but are rather the deterministic outcome of the manufacturing process performed. In consequence, the determination of basic relationships between the mechanical, thermal and chemical aspects of a finishing process and the six groups of key factors characterizing surface integrity is crucial to successful improvement of a finishing process. Some important components of surface integrity will be discussed in the next sections.

Apart from surface integrity the term *surface finish*, which describes the geometric feature (topography) of surfaces, is commonly used in the manufacturing practice. It is evident, when considering surface topography,

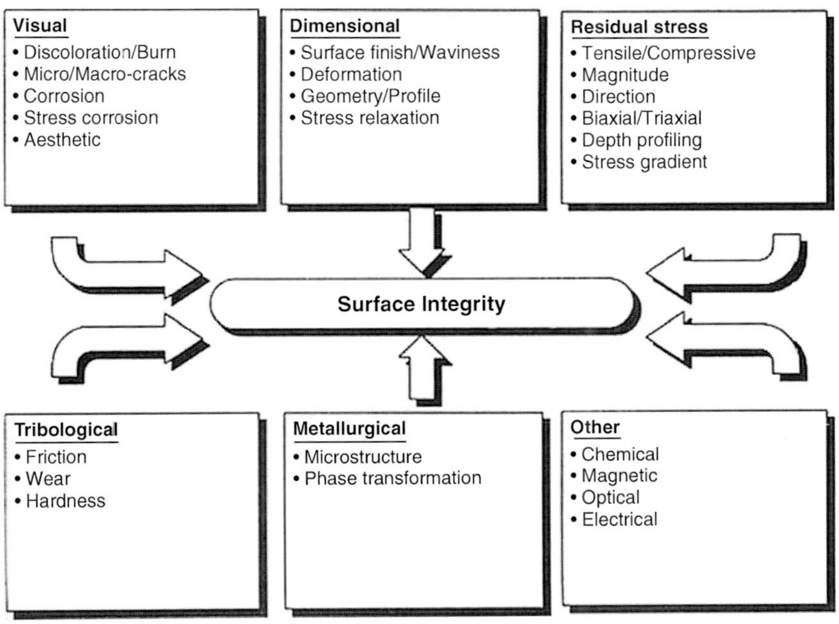

Fig. 20.3 The six groups of key factors that define the surface integrity of a finished material [6].

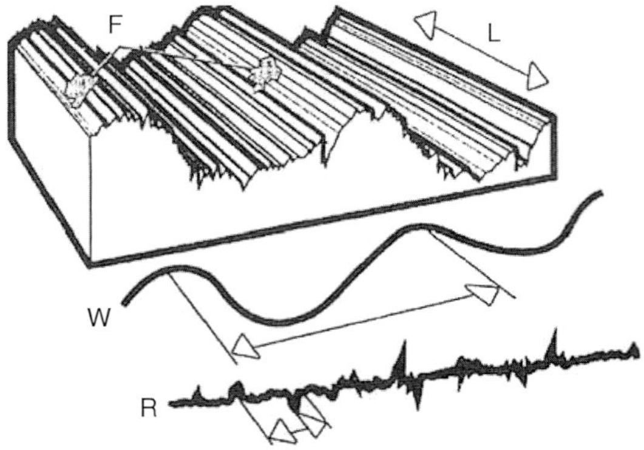

Fig. 20.4 Enlarged view of the topography of a machined surface. *F* – flaws, *L* – lay, *R* – roughness, *W* – waviness.

that most engineered surfaces have regular and irregular spacings that tend to form a characteristic pattern or *texture* on the surface, as shown in Fig. 20.4. This surface texture is generated by a combination of various factors that are due either to the manufacturing process or the material being machined. Of the factors relative to the material, the most fundamental is its microstructure. The direction of the dominating pattern, generated

(a)

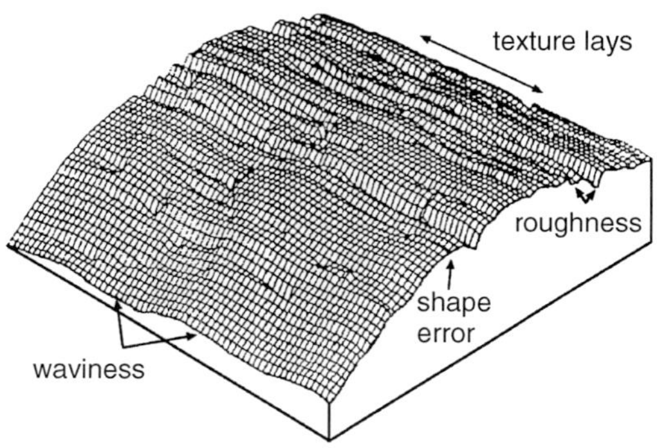

(b)

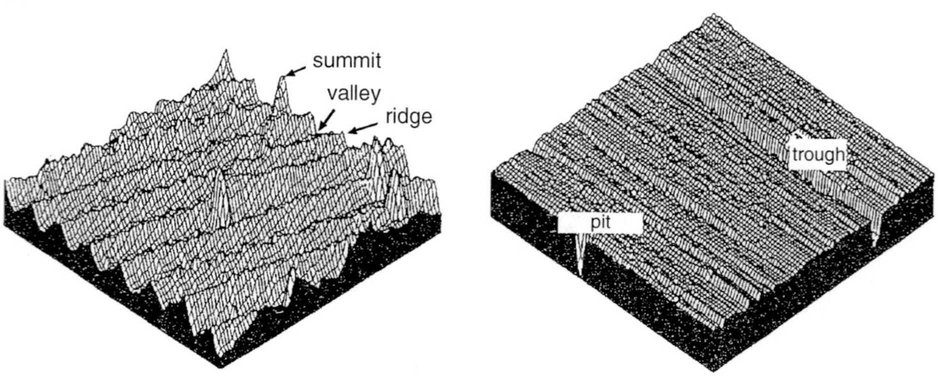

Fig. 20.5 Illustration of the basic components of a 3D surface.

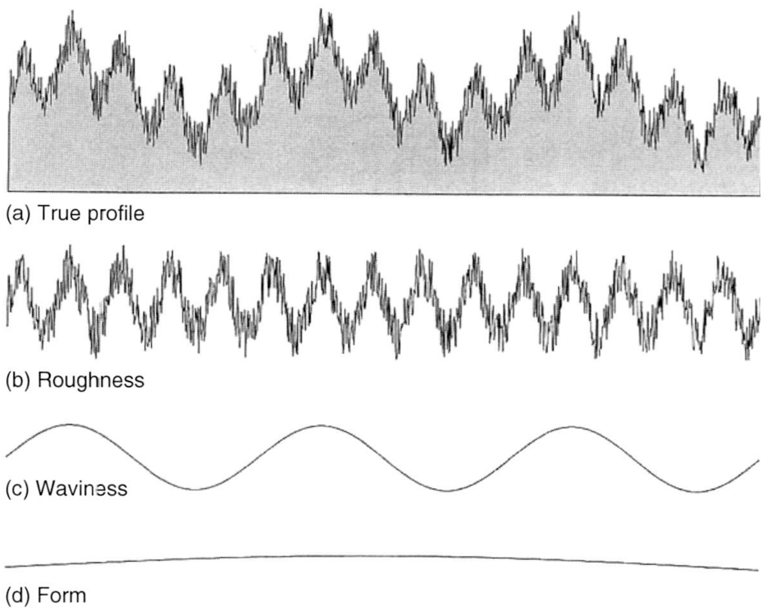

Fig. 20.6 A 2D surface profile represents the combined effects of roughness, waviness and form [7].

by the machining method, is characterized by *surface (texture) lay* (symbol *L* in Fig. 20.4). International Standards ISO 1302 (2001) provide six texture lay classes with appropriate symbols (parallel, perpendicular, circular, radial, crossed and multi-directional). Moreover, individual *flaws* (symbol *F* in Fig. 20.4) can be distinguished by a special surface inspection technique. These flaws include foreign material inclusions, scratches, cracks, micro-holes and other unintentional defects of the real machined surface.

As shown in Figs 20.4–20.6 the topography of a surface is made up of a combination of three features (orders), namely [5–7]:

1. *Surface roughness* refers to the high frequency irregularities on the surface caused by the interaction of the material microstructure and the cutting tool action. This relates directly to the manufacturing unit event (the inherent generating mechanisms) and describes the irregularities caused by each feed rate, abrasive grit, particle, or spark.
2. *Surface waviness* refers to the medium frequency irregularities (the texture) on the surface on which the surface roughness is superimposed. Waviness is caused by such things as the instability of the cutting tool, workpiece deflection and drive screw errors.
3. *Surface form* is the general shape of the surface, neglecting both roughness and waviness, which are caused principally by thermal distortion, slideway error, wear or flexure of the machine/workpiece interface.

As shown in Fig. 20.5a, in the micron or even nanometre range, the surface of all metals has a series of summits (peaks) created during the manufacturing process. In addition (Fig. 20.5b), valleys and ridges or pits and troughs can be distinguished among individual patterns depending on the machining process used.

In practice, roughness, waviness and form are never separate but only superimposed on top of each other. So, these distinctions are therefore qualitative not quantitative. Most surfaces are the result of a combination of these three characteristics (Fig. 20.6). However, roughness results from the manufacturing process rather than machine tool, waviness is attributed to the individual machine (usually imbalance in the grinding wheel, lead screw inaccuracy and lack of rigidity) and form errors are caused by mutual effects such as insufficient fixturing of the part, straightness errors of the guideways or thermal distortion of both the tool and the workpiece.

20.2 Surface Roughness Evaluation

The characterization and measurement of surface texture is very complex because the character of a machined surface involves the three dimensions of space. The real surface geometry is so complicated that a finite number of parameters cannot provide a full description. In addition, any numerical assessment of a surface finish will be influenced by the direction in which measurements are taken in relation to the lay, and by an arbitrary distinction between roughness and waviness. Figure 20.7 provides a basic overview of the typical terminology used for surface texture evaluation relating to the most common standard methods based on the centre or mean line (M) system. Some methods evaluate the profile-height dimensions, some evaluate the longitudinal dimensions and some the form of irregularities.

Roughness parameters which are numerical values assigned to a surface can be calculated in either 2D or 3D forms. 2D surface profile analysis has been widely used in the labs and shop floors for more than half of a century. In recent years, an increased need for 3D surface topography analysis occurred due to the need for evaluating a product's actual performance. There is also an increased interest in using surface metrology as a failure analysis tool instead of classical material evaluation or to analyse wear mechanisms in greater detail. This is an important reason for introducing new parameters for surface evaluation. For example, the standard ISO 4287:1997 gives details of 14 roughness parameters [5], whereas Taylor Hobson Ltd (THL) lists 24 parameters that are commonly used by industry. It should be noted that ISO 4287:1997 defines three series of 14 parameters each, i.e. P-parameters for the unfiltered profile, R-parameters for the roughness (filtered) profile and W-parameters for the waviness profile [9]. Filtering to obtain the R and W profiles is introduced in the same standard.

Surface roughness parameters are normally categorized into three groups according to their functionality, i.e. amplitude parameters (peak or valley, P–V, heights), spacing parameters (spacing of the irregularities on the surface) and hybrid parameters (determined by a combination of amplitude and spacing). Amplitude parameters reveal variations in the profile height of a surface. Spacing parameters are sensitive to variations in the wavelength of the measured profile and hybrid parameters measure both features. In contrast, Griffiths [5] proposes the 2D roughness parameters to be divided into six classes that describe various attributes of the surface profile, including amplitude parameters, amplitude distribution parameters, slope parameters, spatial parameters, other parameters and motif parameters.

Some important surface roughness parameters, being conceptually well known and clearly described in appropriate literature, selected from the three functional groups are listed in Table 20.1.

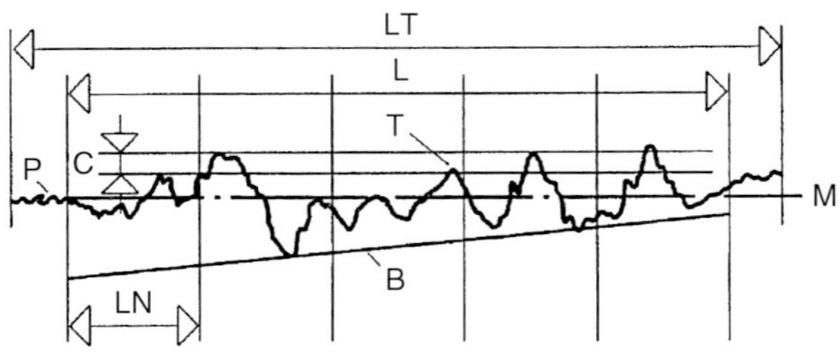

Fig. 20.7 Typical terminology used for evaluation of surface texture [8]. L – evaluation length, P – profile, T – top line within sampling length, B – bottom line within sampling length, M – mean line, LN – sampling length (cut-off), LT – traverse length, C – profile section level.

Table 20.1 List of the 2D surface roughness parameters according ISO 4287 and THL [5, 7, 9].

	Profile height parameters	
Ra	Arithmetic average height	SL
Rq	Root mean square (RMS) average	SL
Rv	Maximum depth of valleys	SL
Rp	Maximum height of peaks	SL
Rt	Total height of the profile (*Rv* + *Rp*)	EL
Rz	Maximum height of the profile	SL
Rsk	Skew (skewness)	SL
Rku	Kurtosis	SL
Rpk	Reduced peak height (Fig. 20.8)	EL
Rk	Core roughness depth	EL
Rvk	Reduced valley depth	EL
Rmr(*c*)	Material ratio at depth '*c*'	EL
*Mr*1	Upper material ratio (%)	EL
*Mr*2	Lower material ratio (%)	EL
	Spacing parameters	
Rsm	Average peak spacing	SL
HSC	High spot count (count(s))	EL
Pc	Local peak count (count/cm)	EL
	Hybrid parameters	
$R\Delta a$	Centre line average (CLA) slope (deg)	SL
$R\Delta q$	RMS slope (deg)	SL
$R\lambda a$	CLA wavelength	SL
$R\lambda q$	RMS wavelength	SL

SL – sampling length, EL – evaluating length

A set of 2D amplitude distribution parameters is determined based on the amplitude distribution function (ADF) and the bearing area curve (BAC) shown in Fig. 20.8a. The ADF is a histogram where the value $p(y)$ represents the fraction of heights lying in the stratum between y and $(y + dy)$. If the ADF is integrated the BAC, named also Abbott–Firestone curve or material ratio curve, is obtained. The third and fourth momentums of the ADF are defined as the skew (*Rsk*) and kurtosis (*Rku*), respectively.

Kurtosis and skew define surface shape and form in such a way that different manufacturing processes produce different *Rsk*/*Rku* envelopes [5]. Figure 20.8b shows five bearing area curve (BAC) parameters developed by the German car industry for surfaces generated by the plateau-honing process and defined in ISO 13565-2:1996. The method of deriving these parameters is based on a best fit line over 40% of the BAC central portion. The parameter *Rk* characterizes the long-term running surface that will influence the performance and life of the engine cylinders. Moreover, the parameters *Rpk* and *Rvk* represent the top portion of the surface that will be worn quickly when an engine starts for the first time and the oil-retaining capacity of the surface provided by the deep valleys, respectively.

In many instances, the parameters determined from a single 2D trace fail to characterize a surface satisfactorily because real surfaces interact in three dimensions, not in two as mostly assumed in engineering problems. Such simplification can lead to incorrect process design regarding, for example, the influence of

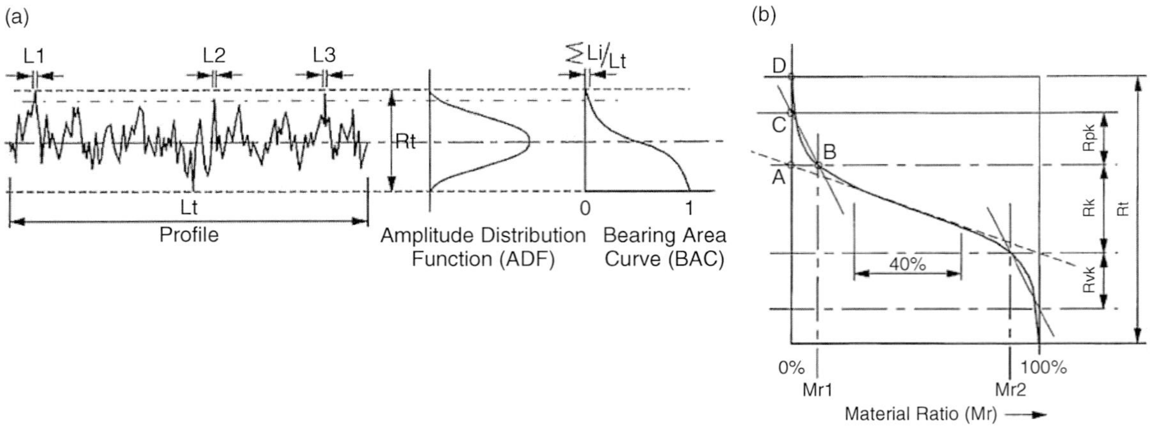

Fig. 20.8 Specification of ADF and BAC (a) and the bearing area parameters (b) [5].

surface texture direction on deep drawing friction [5]. Surface texture parameters calculated over an area are referred to as 3D, with a further subdivision into areal parameters, derived by 2D parameters, and only three uniquely defined 3D parameters. These parameters are denoted by 'S' instead of 'R' to indicate that they are calculated over a surface. At present, 3D parameters are not still standardized but it is most likely that a set of 18 3D parameters will be adopted as currently defined in 'EUR 15178EN Provisional' [5, 7, 9]. The three unique 3D parameters are *Str*, *Sal* and *Std* spatial parameters based on the evaluation of the normalized areal autocorrelation function (AACF) and the areal power spectral density (APSD). The *Str* (texture aspect ratio of the surface) measures the isotropy of a surface, the *Sal* (fastest decay autocorrelation length) describes the most significant length component of texture, and the *Std* (texture direction of the surface) indicates the direction of the significant lay of the surface.

The requirement of identifying and characterizing surface features according to function created new texture characterization methods such as various random-process functions (principally the autocovariance function (ACVF) and its Fourier transform, the power spectral density function (PSDF)), fractals, wavelets, change trees and neural networks [9, 10]. In particular, machined surfaces are likely to be multifractal (with characteristic corner frequencies) because they are usually produced by several rough and finishing processes [10, 11]. As reported in [10] the fractal dimension *D* vs the CLA average wavelength and corresponding CLA average profile slope allow differentiating both cutting and abrasive operations performed during a multi-stage technological process.

A typical engineering surface consists of a range of spatial frequencies corresponding to different wavelengths and the problem is to select appropriate robust filters to separate roughness from waviness and form [13]. One recent development in this regard is the use of wavelets for analysing the multi-scale features in engineering surfaces. By applying discrete wavelet transform (DWT) multi-scale approximation of original surface texture can be extracted at different resolution levels. Similarly, the lifting wavelet algorithm was found to be a powerful toolbox in appropriate separation of the roughness component of high precision surfaces which can include some other components that do not belong to the fundamental roughness [14].

The final surface roughness obtained during a practical machining operation may be considered a sum of the theoretical (ideal) and natural components [15]. The first one is the result of the geometry of the tool and the feed or cutting speed, whereas the natural component is affected by a number of negative effects such as built-up phenomenon, tool wear, chatter, inaccuracies in machine-tool movement, and so on. Moreover, the resulting dynamic and static stability of the total process system is a vital consequence of the quality of the surface texture achieved. The theoretical surface roughness can be calculated for turning and milling operations. In case of longitudinal turning operations with sharp-cornered and round-cornered tools, the

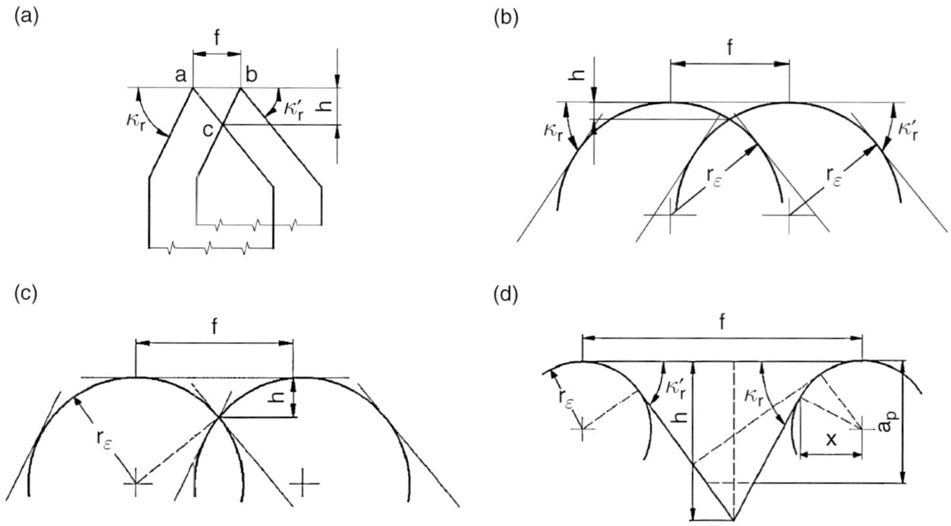

Fig. 20.9 Types of feed marks produced by cutting tools with different kinematical–geometrical features [12].

theoretical maximum achievable value of surface roughness can be calculated for four characteristic models illustrated in Fig. 20.9. They are the following:

1. Linear reproduction (Fig. 20.9a)

$$R_{zt} = f\frac{\sin\kappa_r \sin'\kappa'_r}{\sin(\kappa_r + \kappa'_r)} \tag{20.1}$$

Theoretical values of maximum and arithmetic average roughness are equal to $R_{maxt} = R_{zt}$ and $R_{at} = R_{maxt}/4$.

2. Arc reproduction for feed $f \leq 2r_\varepsilon \sin\kappa'_r$ (Fig. 20.9b)

$$R_{zt} = r_\varepsilon - \sqrt{r_\varepsilon^2 - \frac{f^2}{4}} \approx \frac{f^2}{8\,r_\varepsilon} \quad \text{or} \tag{20.2}$$

$$R_{at} = \frac{f^2}{18\sqrt{3}r_\varepsilon} \approx \frac{0.0321 f^2}{r_\varepsilon} \tag{20.3}$$

3. Arc-linear reproduction of the 1st type for $f \geq 2r_\varepsilon \sin\kappa'_r$ (Fig. 20.9c)

$$R_{zt} = r_\varepsilon(1 - \cos\kappa'_r) + f\sin\kappa'_r\cos\kappa'_r - \sin\kappa'_r\sqrt{f\sin\kappa'_r(2r_\varepsilon - f\sin\kappa'_r)} \tag{20.4}$$

4. Arc-linear reproduction of the 2nd type for $x > r_\varepsilon \sin\kappa_r$ (Fig. 20.9d)

$$R_{zt} = r_\varepsilon(1 - \cos\kappa'_r) + \frac{f - r_\varepsilon(\sin\kappa_r + \sin\kappa'_r) + r_\varepsilon ctg\kappa_r(ctg\kappa_r - ctg\kappa'_r)}{ctg\,\kappa_r + ctg\,\kappa'_r} \tag{20.5}$$

Under finishing conditions with small feeds Brammertz [12, 16] implicitly provides a material partition that remains on the surface due to the effect of the minimum undeformed chip thickness (h_{min}). Thus, the surface roughness value is given by

$$R_t = \frac{f^2}{8r_\varepsilon} + \frac{h_{\min}}{2}\left[1 + \frac{r_\varepsilon h_{\min}}{f^2}\right] \tag{20.6}$$

For peripheral milling the peak-to-valley height and the arithmetic mean value of the feed marks are given by the following expressions [14]:

$$R_{zt} = \frac{f_t^2}{8R_c} \tag{20.7}$$

$$R_{at} = \frac{8f_t^2}{18\sqrt{3}R_c} \tag{20.8}$$

More accurate expression for the peak-to-valley roughness is as follows:

$$R_{zt} = \frac{f_t^2}{8\left(R_c \pm \frac{f_t z_c}{\pi}\right)} \tag{20.9}$$

Where R_c is the cutter radius, f_t is the feed per tooth and z_t denotes the number of teeth on the cutter.

In Eqn (20.9) sign (+) is valid for up-milling and (−) for down-milling cutters.

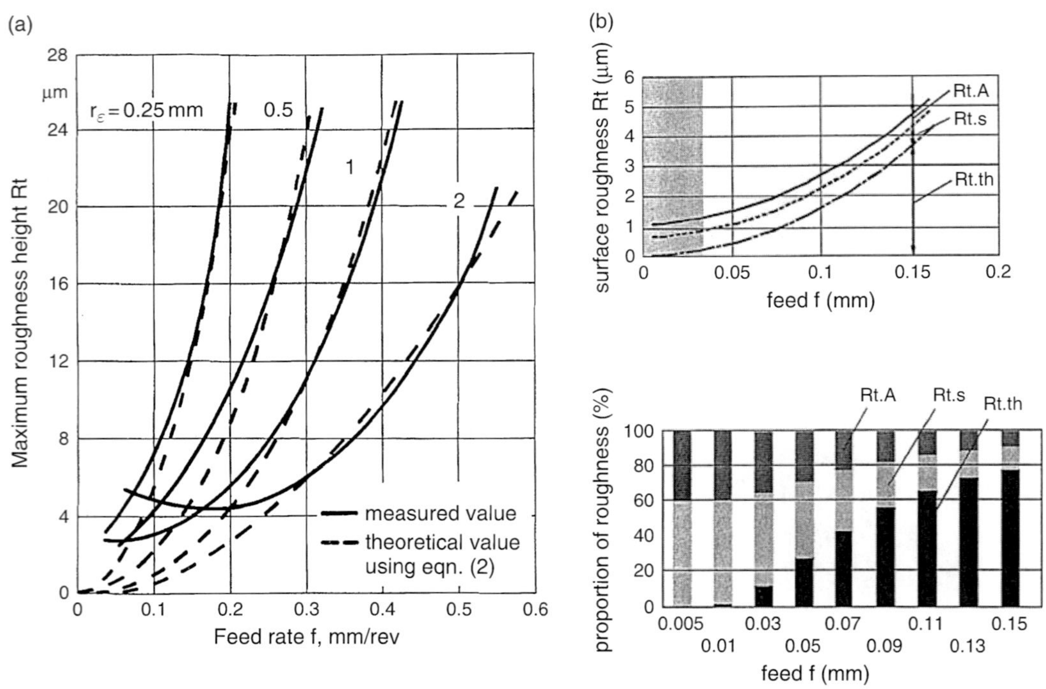

Fig. 20.10 Influence of feed rate and tool nose radius (a) and other components (b) on the maximum roughness height Rt [17, 18].

The comparison between predicted and measured (natural) values of the surface roughness is illustrated in Fig. 20.10a. This figure clearly shows that the actual roughness of the steel specimens is close to the ideal calculated from Eqn (2) only for higher feed rates, whose value depends on the radius of tool corner. Better fitting for small feed rates can be achieved by using Brammertz's model expressed by Eqn (6) with a ratio of $h_{min} \approx 0.1 \times r_n$ (r_n is the cutting edge radius). As shown in Fig. 20.10b it is possible to attribute magnitudes of the different factors influencing the surface roughness *Rt* related to feed, tool geometry (*Rt.th* – roughness component related to the tool profile), tool defects (*Rt.s* – roughness component resulting from tool defect size) and machine vibration (*Rt.A* – asynchronous machine tool error). It was revealed that in hard turning the achievable surface roughness is mainly influenced by the size and shape of defects on the cutting edge for feeds lower than 50–80 μm and by the macroscopic tool geometry for feeds higher than 100 μm [18]. In addition, at feeds lower than 50 μm machine vibration becomes a significant contributor to roughness height and the effect of the tool profile (ideal component of the surface roughness predicted by Eqn (6)) can be neglected [19].

The difference between surface forms obtained in different machining processes can be visibly reflected in skew/kurtosis space as illustrated in Fig. 20.11. Finishing abrasive processes such as grinding, honing and lapping produces negative skewed surfaces with high kurtosis that have good bearing properties. In contrast typical cutting processes such as turning and milling and EDM producing a cusp-shaped topography tend towards low kurtosis and a positive skew value which indicate good gripping or locking. This tendency is confirmed in Fig. 20.12 in which some 3D parameters, including *Ssk*, are provided for topographies of differently machined surfaces.

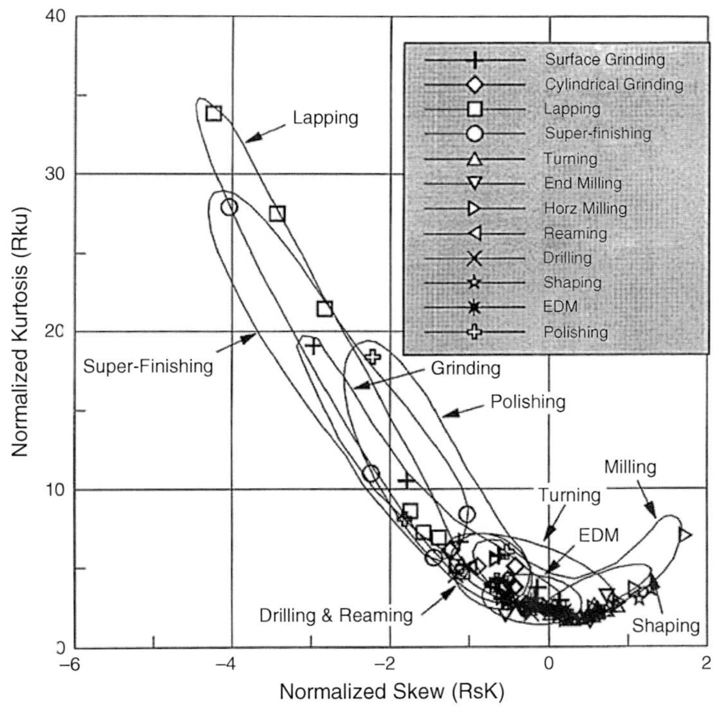

Fig. 20.11 Graph of 2D skew (*Rsk*) against kurtosis (*Rku*) for a range of manufactured surfaces [5].

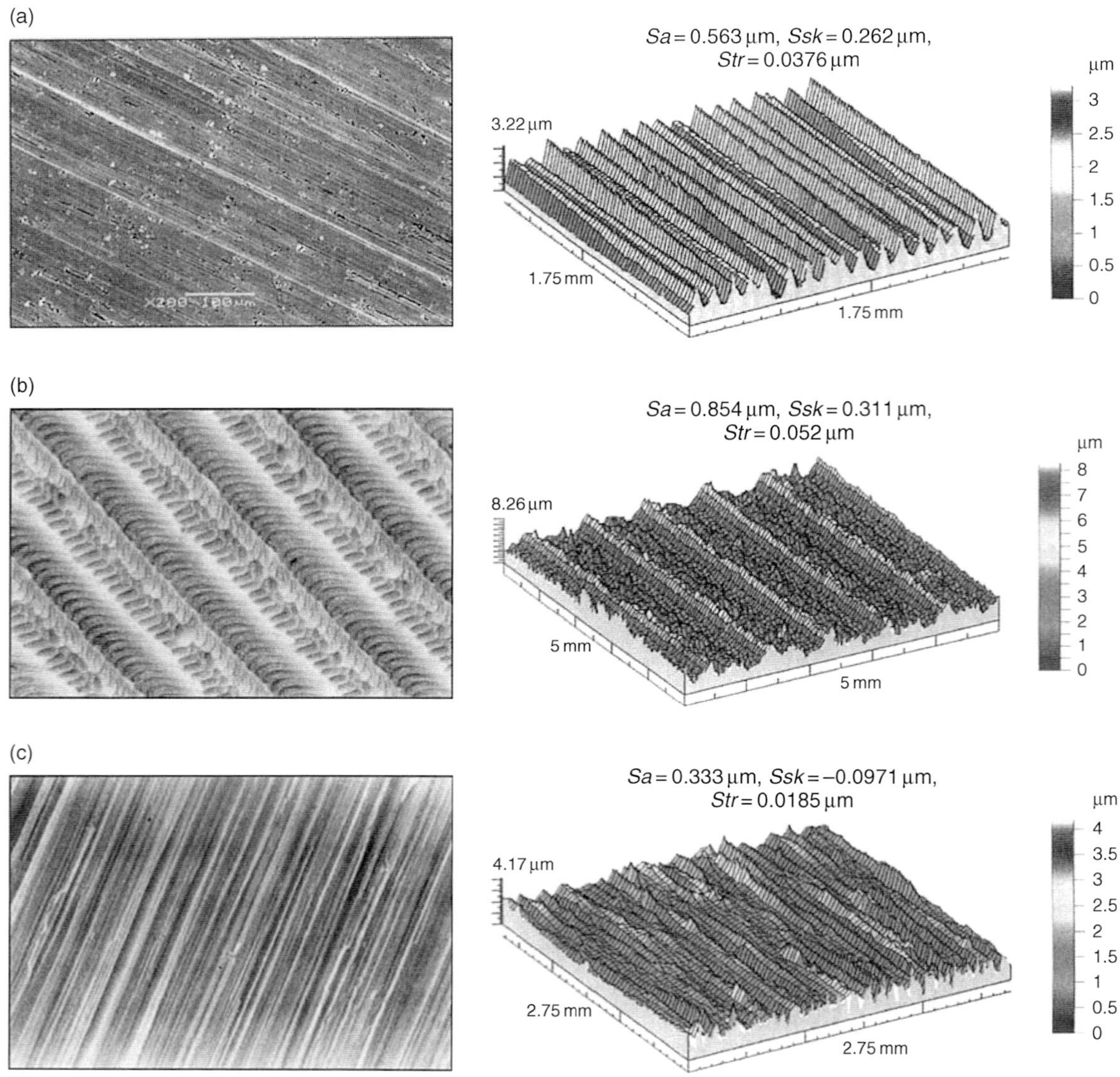

Fig. 20.12 Examples of 3D surfaces after: precision diamond turning of aluminium alloy, $v_c = 6.6\,\text{m/s}$, $f = 0.1\,\text{mm/rev}$ (a), ball-nose end milling of tool steel, $v_c = 1.8\,\text{m/s}$, $a_e = 1\,\text{mm}$ (b), plain peripheral grinding of alloy steel, $v_c = 22.1\,\text{m/s}$, $v_{ft} = 0.4\,\text{m/s}$ (c) [20]. (see Plate 28)

20.3 Surface Roughness Measurements

Surface finish and texture can be measured in a large number of ways ranging from simply tracing a fingernail across a surface to the use of a scanning electron microscope (SEM). In general, the range of surface topography measuring equipment can be divided into two classes, those used on- or in-line and those used in the laboratory. The first are based on typical 2D stylus profilometry and light scattering, whereas the second group utilizes 3D profilometry and both scanning electron (SEM) and atomic force (AFM) microscopes

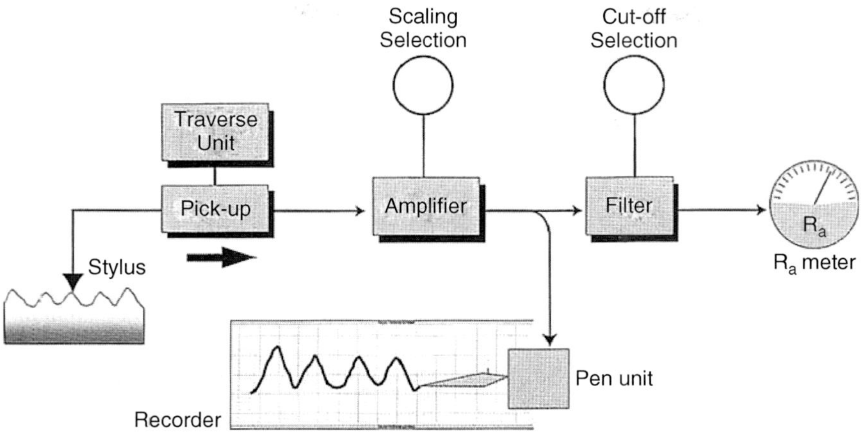

Fig. 20.13 Diagram showing the chief constituents of a stylus-type surface texture measuring instrument [7].

[5, 7, 11]. Areal techniques are rapidly growing in use but, apart from light scattering, are generally used on-line. The most common method of assessing texture involves mechanically drawing a stylus across a surface, i.e. using a contact stylus profilometer. The basic components of this type of instrument are depicted schematically in Fig. 20.13.

The pick-up supports the skid that rides over the surface. The stylus moves over the surface, driven by the traversing unit, and the gauge converts its vertical movements into proportional variations of an electrical signal. The transducer signal is amplified and processed prior to the output and is used to operate a recorder. From this filtered signal is derived, for example, an *Ra* value, which is displayed on a screen or printout. The sensitivity required of the gauge is such that it should respond to a stylus movement of less than 0.01 μm [7]. Those applicable for surface metrology are gauges with analogue (position or motion sensitive) transducers and gauges with digital (laser or phase grating interferometric (PGI)) transducers, as shown in Fig. 20.14a and b, respectively. The gauge with a variable inductance transducer, used in many models of surface measuring

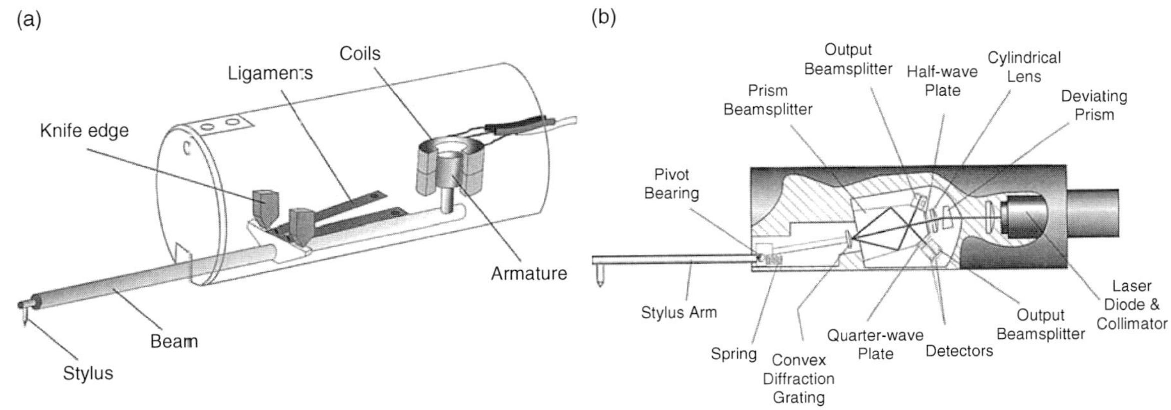

Fig. 20.14 Examples of analogue and digital transducers: (a) variable inductance gauge, (b) schematic diagram of the phase grating interferometric (PGI) gauge [7].

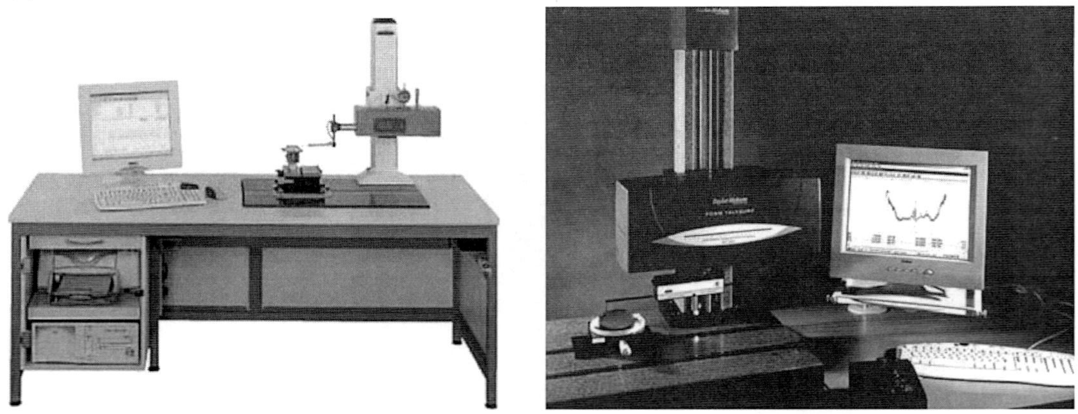

Fig. 20.15 Hommel Tester T8000 (a) and Form Talysurf PGI 1240 (b) instruments [21, 22].

instruments, offers the range to resolution ratio from 1000:1 up to more than 64 000:1, whereas the PGI gauge enables the output signal to be interpolated to give a basic resolution of 0.8 nm and a range of 12.5 mm.

A wide range of profilometer instruments are available including a small, compact and portable 2D unit, through the stationary roughness measuring stations (Fig. 20.15) up to 3D precision instruments. As shown in Fig. 20.15 typically newer instruments have extra facilities including computer analysis and programming for automatic measurements, but in principle the operation remains the same. The stationary, PC-based roughness measuring station from Fig. 20.15a can be localized whether in the measuring room or on the production line. It allows calculation and representation of the roughness parameters according to international standards and due to the modular concept can be extended to topography or contour measurements depending on the workpiece size and specific measuring task. The Form Talysurf PGI 1240 applying the PGI gauge from Fig. 20.14b is the benchmark for the form measurement industry, providing simultaneous assessment of form, dimension and texture with a single traverse. Key features include system noise reduced to less than 1 nm (Rq), 0.1 μm straightness over 200 mm traverse and X axis horizontal data spacing to 0.125 μm. The PGI Aspherics system also includes an environmental cabinet to reduce the effects of airborne vibration.

3D surface measurements can also be performed with stylus and non-contact instruments. Typically, in order to be able to measure in 3D, a traverse stepper motor driven table is added to a 2D profilometer (sometimes termed $2^{1/2}$D). Thus, a 3D scan contains 256 × 256 data (65k) points obtained by moving the stylus in the X and Y directions, whereas a 2D trace contains 16k data points. Further, a typical software package for the analysis of 3D surface data including visual (for example, 3D plots with heights coded by colour) and analysis functions (for instance, 3D parameter calculations, Gaussian and 2CR filtering, zoom, rotation and inversion of heights) is necessary. Some of the limiting characteristics of this method are resolution and distortion of data due to suppressing the natural surface contours in all planes of measurement. An alternative is non-contact measurement using different non-contact gauges such as capacitance gauge, laser triangulation (Fig. 20.16a), microscopy techniques such as SEM, atomic force (AFM) and confocal microscopy and scanning interferometer (Fig. 20.16b).

Laser triangulation is one of the simplest non-contact measurement systems. A focused laser beam is projected onto a surface and the point that is illuminated on the surface is imaged onto a position sensitive device that is calibrated in terms of Z height at the surface. The vertical resolution is limited to about 1 μm. The most successful new surface topography instrument is the so-called 'white' light interferometer [23]. A piezo drive system is used to 'scan' the objective lens about a focal point. The sample to be measured is usually positioned on a precision X/Y stage. As the image system is 'traversed' through its range by the piezo

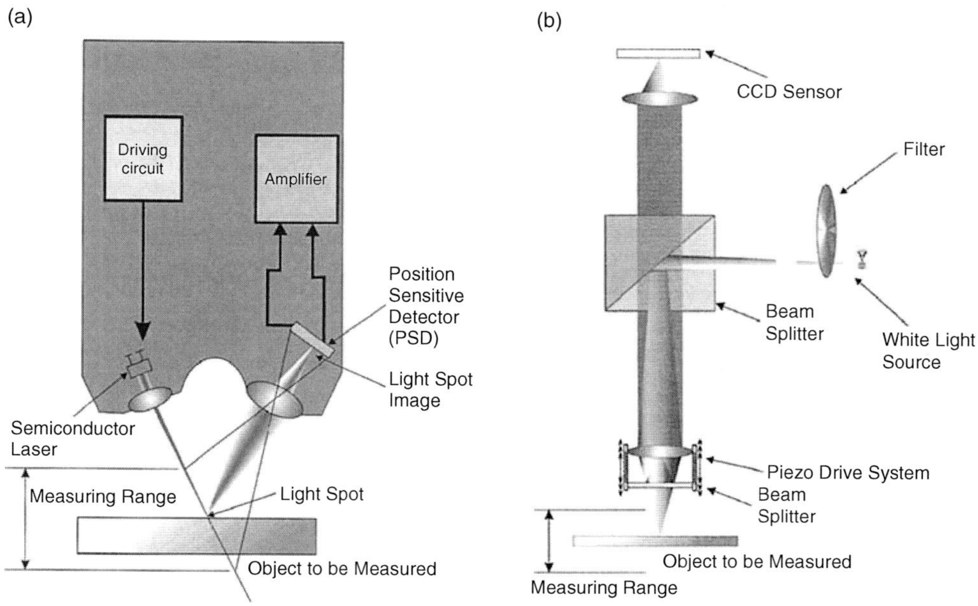

Fig. 20.16 Non-contact measurement systems [7]: (a) laser triangulation (reflectance method), (b) scanning interferometer.

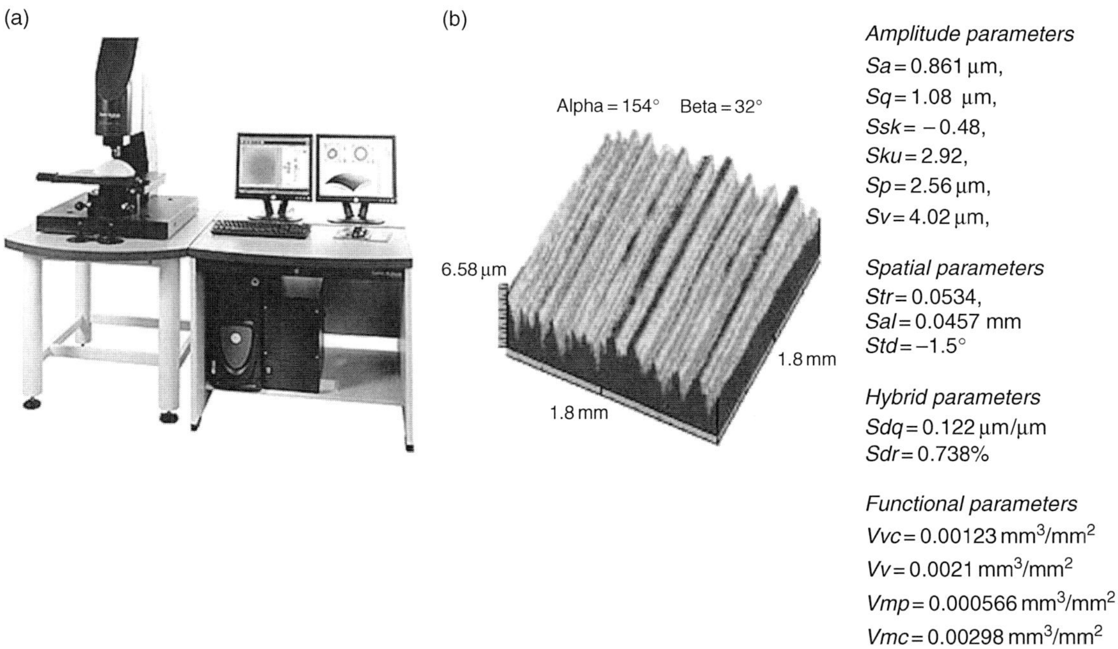

Fig. 20.17 Non-contact 3D surface profiler (scanning white light interferometer) TalySurf CCI 6000 (a) and an example of 3D measurement of horizontally milled surface [21, 24].

drive system the focal point is noted for each pixel in the CCD array. As a result a large number of points (typically 1024×1024) can be measured with very high lateral resolution of ca. 0.3 μm and vertical resolution typically less than 0.1 nm in a few seconds. An example of this type instrument is the CCI interferometer shown in Fig. 20.16b.

The TalySurf CCI (Fig. 20.17) is an advanced 3D non-contact optical metrology tool used for advanced surface characterization. This instrument guarantees a true topographical representation of a surface with 0.01 nm Z resolution over a full scan range plus a 0.4 μm lateral resolution. A measurement with over 1 000 000 data points can be completed in less than 10 seconds. This family of TalySurf CCI products utilizes patented coherence correlation interferometry (CCI) technology to provide ultra-high resolution interferometric measurements for non-contact surface roughness, step-heights, form, shape, angular and critical dimension results. A special algorithm is used to find the coherence peak position of an interference pattern produced by a selectable bandwidth light source. It can be adapted to measure and characterize a wide range of surfaces with reflectivity between 0.35 and 100%, including polished, ground, etched, transparent, rough, shiny and multi-dimensional surface types. Automatic analysis routines can be carried out using Talymap 3D software with manipulation possibilities such as high resolution zooming, profile levelling, defect removal and profile extraction. Parameter sets include area and volume, counting and sorting, roughness and waviness in both 2D and 3D and automatic step-height calculation. Data presentation tools include photo-realistic images in full colour, meshed axonometric projections and contour diagrams.

20.4 Properties of Subsurface Layer

Conventional and non-traditional machining operations produce a variety of alterations of the subsurface layer illustrated schematically in Fig. 20.18. It should be noted that the bulk material/altered material zone (AMZ) is gradual, and often there is no clear demarcation line between them. In reality, the AMZ consists of a set of subzones corresponding to various subsurface effects and features. The primary important changes, strongly influenced by machining operations and their parameters, are: microstructure and hardness profile changes, as well as the induced residual stresses. The principal causes of surface alterations are as follows:

1. The high temperature or high temperature gradients developed during the machining process.
2. Plastic deformation and plastically deformed debris.
3. Chemical reactions and subsequent absorption into the machined surface.

If the basic alteration mechanism involves intensive plastic deformation, the AMZ will be work hardened. With ductile materials, this deformation can result in plastic flow to produce cracks as well as debris, laps,

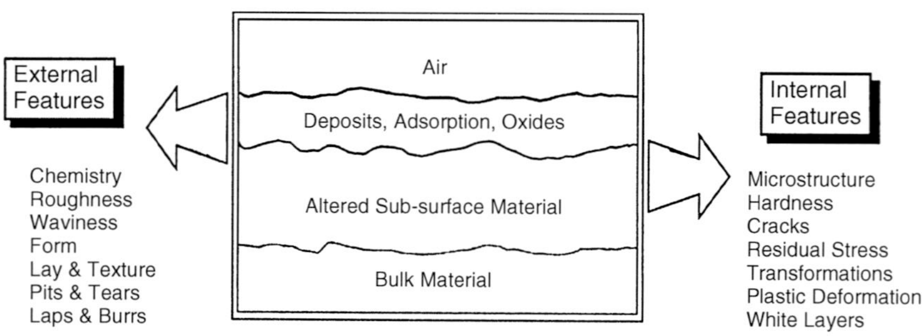

Fig. 20.18 Schematic section through a machined surface [5].

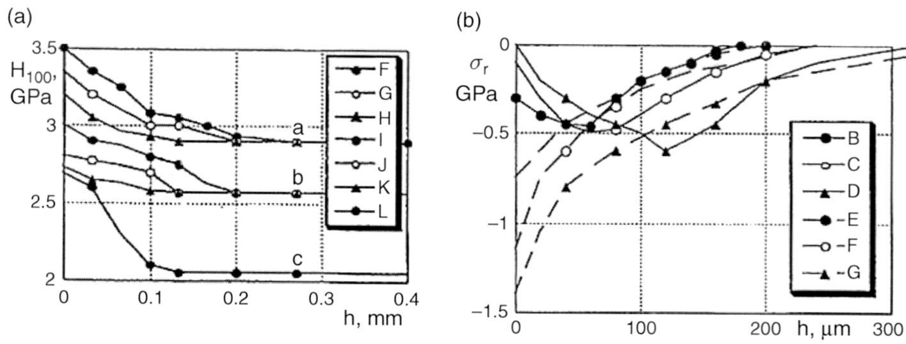

Fig. 20.19 Alterations of subsurface layer after burnishing [26]: (a) surface layer microhardness H_{100} versus steel structural condition and burnishing force. Structural condition: a – sorbite, b – ferrite + pearlite, c – ferrite. The burnishing force $P = 100$ N (H, K); $P = 200$ N (G, J); $P = 300$ N (F, I, L). Workpiece material is carbon steel AISI 1045, (b) tangential (B, C, D) and axial (E, F, G) residual stresses in surface layer. $P = 50$ N (B, E); 100 N (C, F); 200 N (D, G). Alloy steel AISI 5140, HRC 52.

built-up edge and burrs. Typical plastic deformation observed on the photomicrographs include the distorted grain structure near the surface, where the elongated grains are virtually parallel to the surface (this layer is termed the fragmented layer because the high strain is in excess of 10), and the deformed layer subjected to lower plastic deformation in which a structure is still evident. Both these layers together make the plastically deformed AMZ or the mechanically affected zone (MAZ) [5]. The depth of the MAZ is reasonably constant across the surface in constant unit event processes like machining and burnishing (Fig. 20.19). For typical cutting operations it ranges from 0.02 to 0.05 mm for finishing and from 0.075 to 0.12 mm for roughing operations [25].

The variation in microhardness with the thickness of the surface layer after burnishing of the carbon steel AISI 1045 shown in Fig. 20.19a revealed that the maximum strain-hardening rate occurs for ferrite, austenite and martensite structures, and thicknesses of the fragmented zone with high dislocation density and entire MAZ zone are usually 0.02–0.04 mm and 0.2–0.5 mm (0.1–0.3 mm for hardened steels), respectively.

Residual stress-generating mechanisms can be simplistically represented by three separate models:

1. Thermal phase transformation called thermal or hot model (Fig. 20.20a) in which the residual stress is caused by a volume change. If the change in phase causes a decrease in volume, the surface layer is under tension and a tensile residual stress is induced. On the other hand, an increase in volume results in the compressive residual stress. The model is adequate for the conventional heat treatment of steel.
2. Thermal/plastic deformation called mixed model (Fig. 20.20b) in which heat causes expansion of the surface layer and this expansion is relieved by plastic flow, which is restricted to the surface layer. When cooling, the surface layer contracts resulting in a tensile residual stress.
3. With predominant mechanical plastic deformation, called mechanical or cold model (Fig. 20.20c), for which the residual stress is compressive because the surface layer is compacted by mechanical action. This model applies to burnishing (Fig. 20.19b) in which significant tangential compressive stresses with maximum value at a certain depth below the surface are generated.

As a result of the action of the tool edge upon the work material, a field of stress is created if the yield point is exceeded, as shown in Fig. 20.21a. When the action of the stresses created ceases at point A, a contraction of the extended layers appears. Because the plastically extended layer will not be able to return to the preliminary state, only the elastically extended lower layers will affect the plastically deformed layer. As a result, a compressive stress state will be generated in the upper part of this zone.

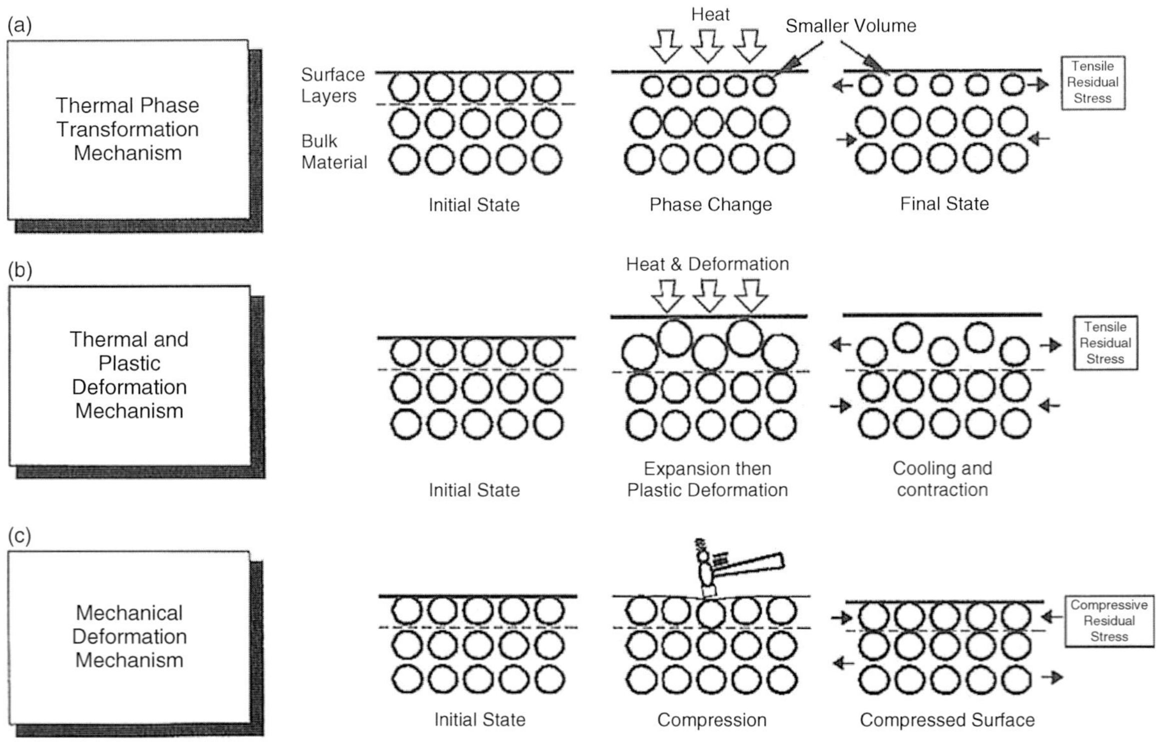

Fig. 20.20 Three residual stress models [5].

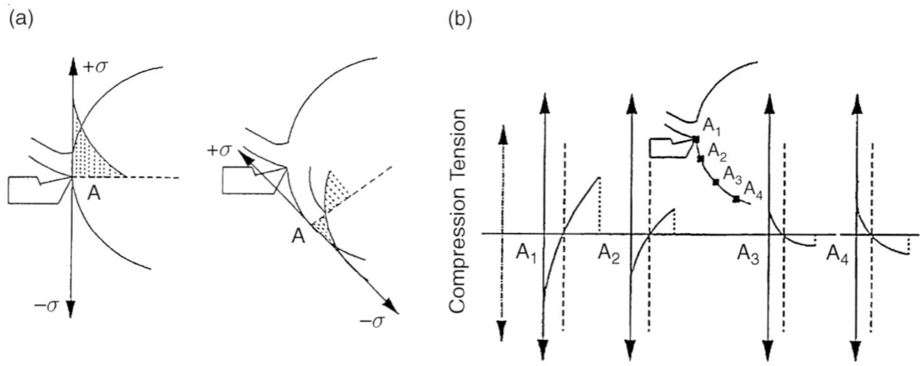

Fig. 20.21 Schematic view of appearance of residual stress: (a) mechanical and (b) thermal stress [27].

As a consequence of the heat generated during the machining process a temperature field in the workpiece materials appears. If the temperature reaches values at which compressive stresses will exceed the yield point, a permanent deformation will occur in the subsurface zone, as shown in Fig. 20.21b. As the temperature fields moves, for example from point A_1 to A_2, a reduction of stress differences is observed and after some time (point A_3) stresses in the outer layer change the sign. After complete cooling (point A_4), the state of

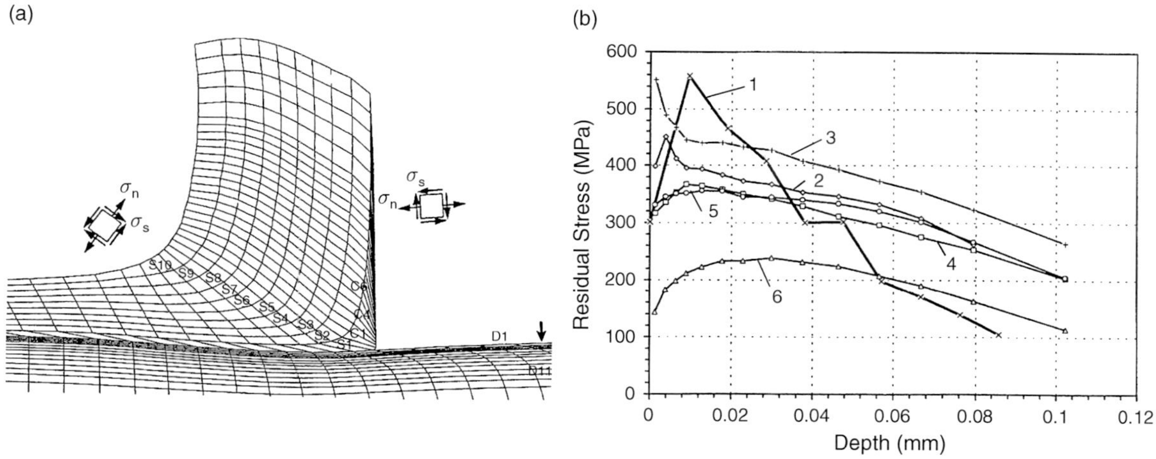

Fig. 20.22 FEM model (a) and distributions of simulated residual stresses in the cutting direction (b) for $\gamma_0 = -2°$, $0°$, $5°$ and $15°$ and perfectly sharp tools (3–6), $\gamma_0 = 5°$ and worn tool (2), and the comparison with X-ray diffraction measurements (1) after [28].

tension will appear in the upper part of the deformed zone while the deeper zones adjacent to the bulk material will be in a state of compression. It is clear from the foregoing analysis that a complete calculation of residual stresses due to machining is very difficult, since a combination of mechanical, thermal and tribological (frictional) loadings, as well as possible phase transformation effects. In general, experimental methods used can be divided into destructive methods based on removing the stressed layers by chemical etching (with or without current supply) and non-destructive methods – X-ray or neutron diffraction techniques – and the ultrasonic method. Recently, the FEM simulation (Fig. 20.22) has been developed along with typical mechanical and thermal analyses.

In Fig. 20.22a, the deformed finite element meshes include the shear zone (elements S1–S10), the tool/chip contact zone (elements C1–C15) and the cut surface near the cutting tool (elements D1–D11). These elements are used to show the distributions of normal and shear stresses and other deformation characteristics along these three sections. The X-ray diffraction measurements and finite element predictions of the distributions of residual stresses in the AISI 1020 carbon steel workpiece using fresh tools with variable rake angle ($\gamma_0 = -2°$, $0°$, $5°$ and $15°$) and the worn cutting tool with 5° rake angle are illustrated in Fig. 20.22b. It can be concluded that for perfectly sharp tools the peak tensile residual stresses are localized on the cut surface (course 3 for negative rake angle) or under the cut surface, especially when tools with high positive rake angle were used (course 6). The worn tool generates higher residual stresses of about 80 MPa on the cut surface. Typical residual stress patterns produced in different cutting and abrasive processes on steel specimens are shown in Fig. 20.23. Figure 20.23 shows the influence of the cutting speed on the residual stress component σx parallel to the cutting direction when planing AISI 304 stainless steel under orthogonal machining conditions. High tensile stresses of about 700 MPa which decrease visibly in the depth direction are present at the surface. At 400 to 600 μm depth the region with compressive residual stresses is reached. In addition, at lower cutting speed the tensile stresses penetrate deeper into the workpiece. A typical residual stress pattern produced by carbide face milling for a sharp tool and for cutters with varying degrees of cutter wear (0.2 and 0.4 mm) is shown in Fig. 20.23b. The sharp cutter has a stress of zero at the surface and it becomes compressive below the surface. In contrast, for the worn cutters, tensile stress is evidently induced at the surface and it changes into compressive stress at greater depths. During hard machining white layers and residual stress formation are influenced considerably by the sharpness of the cutting tool. With sharp tools the surface microstructure remains much the same as the bulk material and the surface residual stresses tend to be compressive as indicated by curve 1 in Fig. 20.23c.

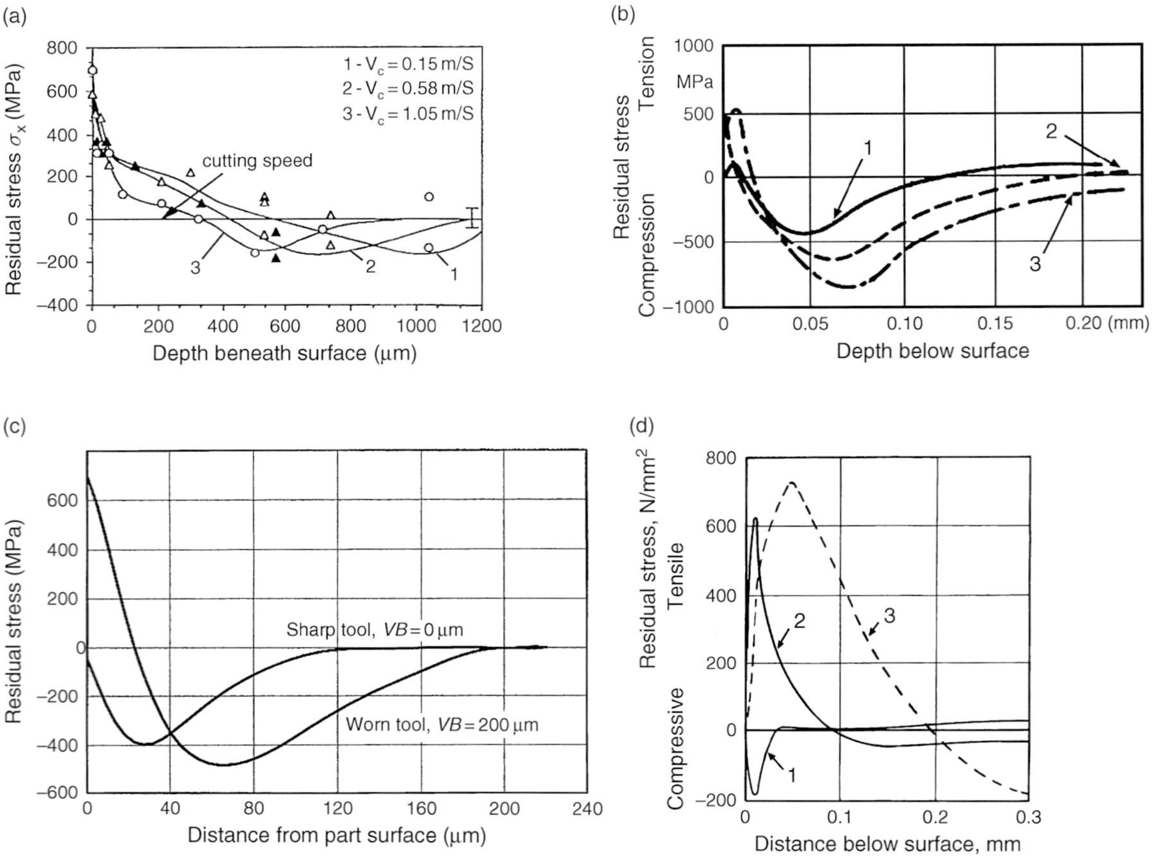

Fig. 20.23 Residual stress profiles produced in different machining processes: (a) after planning of AISI 304 steel ($a_p = 0.08$ mm, $\gamma_0 = 0°$) [29], (b) after face milling of AISI 4340 steel (R_c52), ($v_c = 55$ m/min, $f_t = 0.13$ mm/tooth, $a_p = 1$ mm) [25]. 1 – sharp tool, 2 and 3 – 0.20 and 0.40 mm wear land, (c) after turning of hardened steel with sharp and worn tools [15], (d) induced in the workpiece by various grinding conditions [15].

As the tool flank increases, friction increases and the thermal load on the surface layer grows. This leads to the development of white layers and a greater tendency to induce tensile residual stresses in the surface layer of the hard machined part as depicted by curve 2 in Fig. 20.23c. In general, grinding processes generate extremely high tensile residual stresses (Fig. 20.23d) but by careful selection of grinding conditions, the occurrence of tensile residual stress near the surface of the workpiece can be largely eliminated. Figure 20.23d shows residual stress patterns for low stress (1), conventional (2) and high stress (3) grinding conditions (the latter is observed in dry grinding with a relatively hard wheel). In particular, for low stress grinding, soft and open wheels are used at lower cutting speeds than for conventional grinding, typically 18 m/s rather than 30 m/s [14].

Table 20.2 lists the service, functional and performance criteria that have been influenced by manufacturing surface technology which generates required external and internal features specified in Fig. 20.18. The ticks indicate those surface features that substantially influenced functional performance of differently engineered surfaces. According to Stout [30] *engineered surfaces* are produced in specific ways that alter the surface and subsurface layers to give required performance.

Table 20.2 The influence of subsurface features on performance [5].

		External features	Internal features
Functional performance feature influenced	Seals	✓	
	Accuracy	✓	✓
	Cleanliness	✓	
	Reflectivity	✓	
	Tool life	✓	✓
	Load carrying	✓	✓
	Creep	✓	✓
	Magnetism		✓
	Electrical resistance	✓	
	Assembly	✓	✓
	Fluid flow	✓	
	Joints	✓	✓
	Friction	✓	✓
	Wear	✓	✓
	Lubrication	✓	
	Fatigue	✓	✓
	Corrosion	✓	✓
	Stress corrosion	✓	✓
	Strength	✓	✓
KEY	✓ = proven link between features and performance		

References

[1] T. Burakowski, T. Wierzchoń, *Surface Engineering of Metals. Principles, Equipment, Technologies*, CRC Press, London, 1999.

[2] W. Grzesik, *Advanced Protective Coatings for Manufacturing and Engineering*, Hanser Gerdner, Cincinnati, 2003.

[3] S. Kalpakjian, *Manufacturing Engineering and Technology*, Addison-Wesley Publishing Company, 1998.

[4] M. Field, J.F. Kahles, Review of surface integrity of machined components, *CIRP Annals*, Vol. 20/2 (1971), 153–163.

[5] B. Griffiths, *Manufacturing Surface Technology. Surface Integrity and Functional Performance*, Penton Press, London, 2001.

[6] *ASM Handbook*, Vol. 5, *Surface Engineering*, ASM Int., Materials Park, 1994.

[7] *Exploring Surface Texture, A Fundamental Guide to the Measurements of Surface Finish*, Taylor Hobson Ltd, 2003.

[8] *Modern Metal Cutting. A Practical Handbook*. Sandvik Coromant, Fair Lawn, 1996.

[9] L. de Chiffre, P. Leonardo, H. Trumpold, D.A. Lucca *et al.*, Quantitative characterization of surface texture, *CIRP Annals*, Vol. 49/2 (2000), 635–652.

[10] W. Grzesik, S. Brol, Hybrid approach to surface roughness evaluation in multistage machining process, *J. Mater. Proc. Technol.*, Vol. 134 (2003), 265–272.

[11] T.R. Thomas, *Rough Surfaces*, Imperial College Press, London, 1999.

[12] W. Grzesik, *Fundamentals of Machining of Metallic Materials* (in Polish), WNT, Warsaw, 1998.

[13] J. Raja, B. Muralikrishnan, S. Fu, Recent advances in separation of roughness, waviness and form, *Prec. Eng.*, Vol. 26 (2002), 222–235.

[14] X.Q. Jiang, L. Blunt, K.J. Stout, Application of the lifting wavelet to rough surfaces, *Prec. Eng.*, Vol. 25 (2001), 83–89.

[15] G. Boothroyd, W.A. Knight, *Fundamentals of Machining and Machine Tools*, CRC Press, Boca Raton, 2006.

[16] P.H. Brammertz, Die Entstehung der Oberflächenrauheit beim Feindrehen, *Industrie Anzeiger*, Vol. 2 (1961), 25–32.
[17] W. König, F. Klocke, *Fertigungsverfahren 1. Drehen, Fräsen, Bohren*, Springer, Berlin, 1997.
[18] F. Klocke, E. Brinksmeier, K. Weinert, Capability profile of hard cutting and grinding processes, *CIRP Annals*, Vol. 54/2 (2005), 557–580.
[19] M.M.W. Knuefermann, A model for surface roughness in ultraprecision hard turning, *CIRP Annals*, Vol. 54/1 (2005).
[20] K.E. Oczoś, V. Liubimov, *Geometrical Structure of Surface* (in Polish), Technical University of Rzeszów, 2003.
[21] http://www.taylor-hobson.com/talysurfcci_request.htm.
[22] http://www.hommel-etamic.com/cps/rde/xchg/SID-26EE34DB-AE2B0117/hommelwerke/hs.xsl 4192.htm).
[23] R.J. Hocken, N. Chakraborty, C. Brown, Optical metrology of surfaces, *CIRP Annals*, Vol. 54/2 (2005).
[24] Fully revised and extended Form Talysurf PGI range, *Precision News*, Fall/Autumn, 2003, www.taylor-hobson.com.
[25] Edited by T.J. Drozda, C. Wick, *Tool Manufacturing Engineers Handbook*, Vol. 1 *Machining*, SME, 1983.
[26] J.M. Walker, *Handbook of Manufacturing Engineering*, Marcel Dekker, Inc., New York, 1996.
[27] J. Kaczmarek, *Principles of Machining by Cutting, Abrasion and Erosion*, Peter Peregrinus Ltd, Stevenage, 1976.
[28] A.J. Shih, Finite element analysis of the rake angle effects in orthogonal metal cutting, *Int. J. Mech. Sci.*, Vol. 38 (1996), 1–17.
[29] Ch. Wiesner, Residual stresses after orthogonal machining of AISI 304: numerical calculation of the thermal component and comparison with experimental results, *Metal. Trans. A*, Vol. 23A (1992), 989–995.
[30] K.J. Stout, L. Blunt, A contribution to the debate on surface classifications – random, systematic, unstructured, structured and engineered, *Int. J. Mach. Tools Manuf.*, Vol. 41 (2001), 2039–2044.

CHAPTER TWENTYONE

Troubleshooting in Machining

Contents

21.1 Problems in Machining Operations

As discussed in individual chapters, many types of cutting problems can occur, but the following seem to be of general importance, namely:

- Causes of abnormal wear on the cutting edge(s) of the cutting tool;
- Tool life shorter than expected;
- Occurrence of chatter and excessive vibration;
- Cutting accuracy beyond the tolerances demanded;
- Unsatisfactory chip control;
- Defective production equipment; and
- Others.

If one of the above-selected problems eventuates and the cutting accuracy decreases, it will cause an increase of the rate of inferior production, resulting in higher percentage of scrap products and wastes. In particular, production stops lead to increasing the idle time and expensive equipment can break down and the repair costs will escalate. Finally, the workplace will not be sufficiently safe for workers.

Machining has various characteristics that define troubleshooting patterns. These characteristics originate from historical background, equipment background and some concepts of experienced personnel [1]. The following are the distinctive patterns relating to productivity:

- Subsequent troubleshooting;
- Preventive troubleshooting;
- Improvement of existing production systems;
- Advancement to high-level productivity; and
- Pursuit of the ultimate system and productivity.

A historical troubleshooting approach is to fix and repair the problem in post-mode, i.e. after it occurs. Currently, this type of troubleshooting does not permit improvement of productivity and technological advancement because of unforeseeable and irreparable damages in modern automated or unmanned manufacturing systems. Preventive troubleshooting is especially important for automated machining systems where complications will arise at any time (see Chapters 18 and 19), and if it is possible to predict and then prevent problems from eventuating. However, a small fault can be multiplied into a sizeable problem which can cause the damage of expensive production equipment or tragic repercussions to workers. The third type of

Advanced Machining Processes of Metallic Materials
ISBN: 978-0-08-044534-2

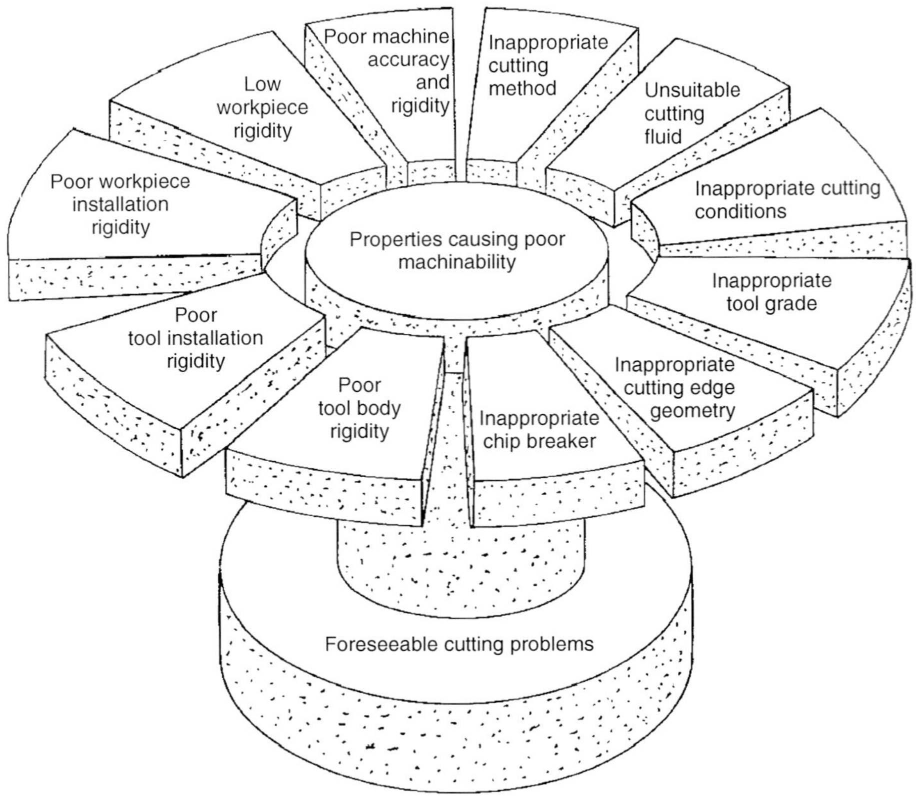

Fig. 21.1 Foreseeable cutting problems [1].

troubleshooting can lead to higher productivity by improving each element of the machining systems, as, for example, to obtain higher rigidity of the cutting tool, machine tool and the workpiece. This can be achieved by installation of appropriate hardware and software. The idea of the fourth type of troubleshooting is that basically offensive action can create the best defensive position. To accomplish a high level of productivity, it is necessary to implement both subsequent and preventive troubleshooting patterns. Roughly, it needs a broader environment to be implemented, with such areas as cutting tool selection, selection of cutting conditions, maintenance of machine tools, education of technicians and engineers, well-organized (safe) and clean working environment, and production technology focused on highly efficient equipment. The last pattern is intended for the ultimate solution leading to zero troubles in equipment, quality, work damage, capacity loss and cutting performance. In general, it requires the realization of previous troubleshooting patterns and introduction of efficient management of technical data (see Chapter 19). Figure 21.1 shows foreseeable cutting problems gathered from production experience. This data can be utilized as a learning tool to prevent the same or similar problems from materializing in the future.

21.2 Practical Examples of Troubleshooting

As an example, some frequently observed problems and their solutions in turning, drilling and milling operations are listed in Table 21.1.

Table 21.1 Troubleshooting for turning [2].

Typical tool failure	*Countermeasures*		
	Tool grade	*Cutting conditions*	*Tool geometry*
Flank wear 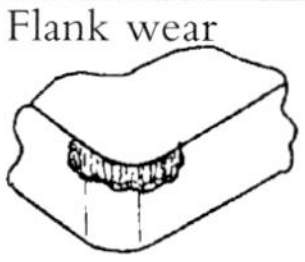	Change to more wear resistant grades P.M.K30 → 20 → 10	Reduce cutting speed Change to appropriate feed	Increase relief angle Increase end cutting edge angle. Increase corner radius Select free-cutting chip breaker
Crater wear 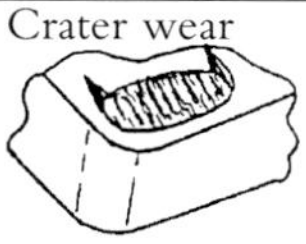	Change to more wear resistant grades P.M.K30 → 20 → 10	Reduce cutting speed Reduce feed	Increase rake angle Select an appropriate chip breaker
Notch wear 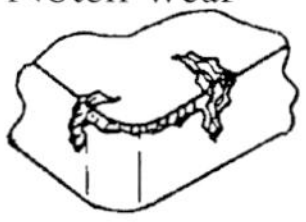	Change to more wear resistant grades P.M.K30 → 20 → 10	Reduce cutting speed Reduce feed	Increase rake angle Increase side cutting edge angle
Fracture 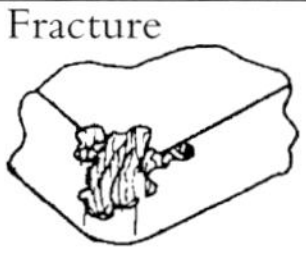	Change to tougher grades P.M.K30 → 20 → 30	Reduce feed Reduce depth of cut	Reduce rake angle Select a chip breaker with high edge strength Increase side cutting edge angle Select larger shank size
Chipping 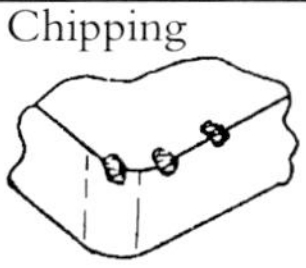	Change to tougher grades. P.M.K30 → 20 → 30	Reduce feed Increase cutting speed	Reduce rake angle Select a chip breaker with high edge strength Increase side cutting edge angle Select larger shank size
Flaking 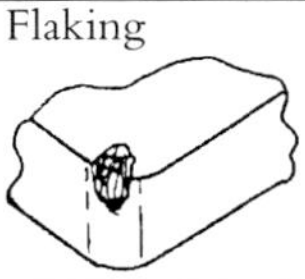	Change to tougher grades P.M.K30 → 20 → 30	Reduce feed Reduce cutting speed	Reduce rake angle Increase corner radius Increase honing width
Plastic deformation 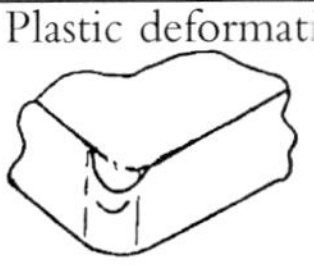	Change to more wear resistant grades P.M.K30 → 20 → 10	Reduce cutting speed Reduce feed Reduce depth of cut Supply cutting fluid in adequate volume	Increase relief angle Increase rake angle Reduce corner radius Reduce side cutting edge angle Select a free-cutting chip breaker
Chip welding 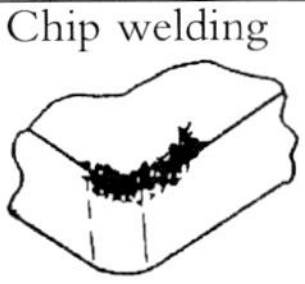	Use a grade which has a low tendency to adhere to work material Cemented carbide Coated carbide or cermet	Increase cutting speed Increase feed	Increase rake angle Select a free-cutting chip breaker
Thermal cracking 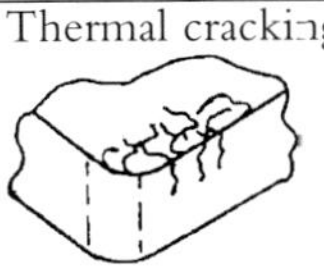	Change to more impact resistant grades P.M.K10 → 20 → 30	Reduce cutting speed Reduce feed Use a cutting fluid Supply cutting fluid in adequate volume	Increase rake angle Select a free-cutting chip breaker

Table 21.2 Troubleshooting for the solid drill and soldering drill [2].

	Problem	Cause	Countermeasure
Abnormal wear	Relief surface	Inappropriate cutting speed	Increase the cutting speed if abnormal wear is around the centre Lower the cutting speed if abnormal wear is on the periphery
		Inappropriate cutting fluid	Check the filter Use the cutting fluid of superior lubricity (increase the dilution rate)
	Margin	Inappropriate cutting speed	Lower the cutting speed
		Regrinding timing, insufficient reground amount	Shorten the regrinding timing
		Insufficient rigidity of machine and workpiece	Change the clamp method to one with rigidity
		Insufficient drill rigidity	Use smallest possible overhang
		Inappropriate cutting fluid	Check the filter Use the cutting fluid superior in lubricity (increase the dilution rate)
		Intermittent cutting when entering cut and breaking through	Avoid interruption at entry and exit Lower the feed during entering and leaving the workpiece
Chipping and fracture	Chisel section (centre of drill cutting edge)	Insufficient rigidity of the drill	Reduce the drill overhang as much as possible Increase the feed at entry when the low speed feed is selected in standard cutting condition range Use a bushing or a centre drill
		Insufficient rigidity of machine and workpiece	Change the clamp method to one with rigidity
		Inappropriate entry into the workpiece	Avoid interruption at entry into the workpiece Lower the feed at entry
		High workpiece hardness	Lower the feed rate
		Inappropriate honing	Check if honing has been made to the centre of cutting edge
	Peripheral cutting edge	Insufficient drill rigidity	Lower the cutting speed Increase the feed at entry when the low speed feed is selected in standard cutting condition range
		Inappropriate drill mounting accuracy	Check the run-out accuracy after drill installation (0.03 mm or less)
		Insufficient machinery and workpiece rigidity	Change the clamp method to one with rigidity. Lower the feed during entering and leaving the workpiece
		Inappropriate honing	Check if honing has been made to the cutting edge periphery

Table 21.2 (Cont.)

Problem		Cause	Countermeasure
Chipping and fracture	Margin	Insufficient machine and workpiece rigidity	Change the clamp method to one with rigidity
		Insufficient drill rigidity	Use smallest possible overhang Use a bushing or centre drill
		Regrinding timing and insufficient amount of reground stock.	Shorten the regrinding timing
		Intermittent cutting when entering or exiting the cut.	Avoid interruption at entry and exit Lower the feed when entering and leaving the workpiece
Breakage		Tendency to cause chipping or develop abnormal wear	Check the failure mode condition before breakage and find out the wear and chip countermeasures
		Chip packing in the drill flutes	Review the cutting conditions For internal coolant supply, raise the supply pressure of cutting fluid Use drill pecking for deep holes
		Insufficient machine output	Review the cutting conditions Use the machine with high power
Insufficient hole accuracy		Insufficient rigidity of the machinery and workpiece	Change to the clamp method with rigidity
		Inappropriate drill installation accuracy	Check the run-out accuracy of drill mounting (0.03 mm or less)
		Chip packing in the flutes	Review the cutting conditions Raise the cutting oil supply pressure Use peck-feed for deep holes
		Inappropriate edge sharpening accuracy	Check the edge shape accuracy
Long and tangle chips		Inappropriate cutting conditions	Increase the feed rate
		Inappropriate honing	Provide the appropriate honing
		Cutting edge with chipping or breakage	Review the chipping measures

Table 21.3 Troubleshooting in milling [2].

Trouble	*Possible causes*	*Countermeasures*
Rapid wear of cutting edge	Improper insert grade selection (insufficient wear resistance)	P30 → P20 → cermet, coated grade (steel cutting) K20 → K10 → coated grade (cast iron cutting)
	Excessive cutting speed	Select cutting speed suited for work material and insert grade
	Inadequate feed	Use standard cutting condition in catalogue as guide
Rapid chipping of cutting edge	Improper insert grade selection (insufficient toughness)	Cermet → P20, P30 (steel cutting), K10 → K20 (cast iron cutting)
	Cutting hard material and unfavourable surface condition	Decrease cutting speed Use cutter with strong cutting edge
	Excessive feed	Proper selection of conditions, using recommended cutting conditions in catalogue as guide
	Excessive pressure applied on cutting edge	Proper selection of engaging angle
	Machining difficult-to-cut material	Use a negative/positive-type cutter with large corner angle
Fracturing	Cracking due to thermal shock	Select insert grade of stronger thermal shock resistance Decrease cutting speed
	Too thin insert against cutting force	Change the TAC mill type
	Continuous use of excessively worn insert	Shorten replacement standard time of insert
	Cutting hard material	Use cutter of stronger tip Use cutter of larger corner angle
	Obstruction to chip flow	Use cutter with good chip discharge properties Select insert grades difficult for chips to adhere Cemented carbides → cermets, coated grades Use cutting fluid or air
	Excessively slow cutting, too fine feed	Select cutting speed and feed optimized for insert grade and work material
Excessive chip welding or build-up on cutting edge	Cutting soft material such as aluminium, copper, mild steel	Use cutter with large rake angle
	Cutting stainless steel	P30 → coated grade
	Use of cutter with negative rake or too small rake angle	Use cutter with large rake angle

Table 21.3 (Cont.)

Trouble	*Possible causes*	*Countermeasures*
Rough finish	Effect of built-up edge	Increase cutting speed Appropriate cutting depth (finish allowance) Change insert grade, K grade → M grade → P grade → cermet For cast irons: P grade → coated grade → cermet For steels: K grade → coated grade
	Effect of face cutting edge run-out	Proper installing of inserts Use insert of high dimensional accuracy Class K → class E
	Continuous use of excessively worn insert	Shorten replacement standard time of insert
	Remarkable feed marks	Feed per revolution to be set within flatland width Use wiper insert-type cutter Use cutter exclusively for finishing
Chattering	Unstable clamping of workpiece	Check clamping method of workpiece
	Cutting of welded construction of thin steel plate	Adopt cutter of large rake angle and small corner
	Excessive cutting condition	Re-examine allowable chip removal rate according to motor HP
	Face milling of narrow width workpiece	Use cutter of small cutter diameter and with many teeth
	Too many simultaneous cutting teeth engagement	Reduce number of teeth or adopt irregular pitch cutter

REFERENCES

[1] K. Karino, *Troubleshooting for Cutting*, Mitsubishi Materials, Ibaraki, Japan, 1998.
[2] www.tungaloy.de (Toshiba Tungaloy European Platform).

Index

Edwards Brothers Malloy
Ann Arbor MI. USA
March 20, 2013